Student's Solutions Manual

Fred Safier
City College of San Francisco

---

to accompany

Raymond A. Barnett, Michael R. Ziegler, and Karl E. Byleen's

Analytic Trigonometry with Applications

Eleventh Edition

PUBLISHER	Laurie Rosatone
ACQUISITIONS EDITOR	Joanna Dingle
DEVELOPMENT EDITOR	Anne Scanlan-Rohrer
ASSISTANT CONTENT EDITOR	Beth Pearson
EDITORIAL ASSISTANT	Elizabeth Baird
SENIOR CONTENT MANAGER	Karoline Luciano
SENIOR PRODUCTION EDITOR	Kerry Weinstein
COVER DESIGNER	Madelyn Lesure
COVER PHOTO	© Brazil2/Vetta/Getty Images, Inc.

Founded in 1807, John Wiley & Sons, Inc. has been a valued source of knowledge and understanding for more than 200 years, helping people around the world meet their needs and fulfill their aspirations. Our company is built on a foundation of principles that include responsibility to the communities we serve and where we live and work. In 2008, we launched a Corporate Citizenship Initiative, a global effort to address the environmental, social, economic, and ethical challenges we face in our business. Among the issues we are addressing are carbon impact, paper specifications and procurement, ethical conduct within our business and among our vendors, and community and charitable support. For more information, please visit our website: www.wiley.com/go/citizenship.

ISBN 978-1-118-11583-1

10 9 8 7 6 5 4 3 2

CONTENTS

Chapter 1 Right Triangle Ratios

EXERCISE 1.1 Angles, Degrees, and Arcs

1. Since one complete revolution has measure 360°, $\frac{1}{2}$ revolution has measure $\frac{1}{2}(360°) = 180°$;
$\frac{1}{6}$ revolution has measure $\frac{1}{6}(360°) = 60°$; $\frac{3}{8}$ revolution has measure $\frac{3}{8}(360°) = 135°$;
$\frac{7}{12}$ revolution has measure $\frac{7}{12}(360°) = 210°$.

3. 45° corresponds to $\frac{45}{360}$, or $\frac{1}{8}$ of a complete counterclockwise revolution.
150° corresponds to $\frac{150}{360}$, or $\frac{5}{12}$ of a complete counterclockwise revolution.
270° corresponds to $\frac{270}{360}$, or $\frac{3}{4}$ of a complete counterclockwise revolution.

5. Since 92° is between 90° and 180°, this is an obtuse angle.

7. Since 5° is between 0° and 90°, this is an acute angle.

9. A 180° angle is called a straight angle.

11. None of these since 210° is larger than 180°.

13. An angle of one degree measure is formed by rotating the terminal side of the angle $\frac{1}{360}$ of a complete revolution in a counterclockwise direction from the initial side.

15. Since $21' = \frac{21°}{60}$ and $54'' = \frac{54°}{3{,}600}$, then $25°21'54'' = \left(25 + \frac{21}{60} + \frac{54}{3{,}600}\right)° \approx 25.365°$ to three decimal places.

17. Since $8' = \frac{8°}{60}$ and $5'' = \frac{5°}{3{,}600}$, then $11°8'5'' = \left(11 + \frac{8}{60} + \frac{5}{3{,}600}\right)° \approx 11.135°$ to three decimal places.

19. Since $28' = \frac{28°}{60}$ and $10'' = \frac{10°}{3{,}600}$, then $195°28'10'' = \left(195 + \frac{28}{60} + \frac{10}{3{,}600}\right)° \approx 195.469°$ to three decimal places.

21. $15.125° = 15°\,(0.125 \times 60)' = 15°7.5' = 15°7'(0.5 \times 60)'' = 15°7'30''$

23. $79.201° = 79°(0.201 \times 60)' = 79°12.06' = 79°12'(0.06 \times 60)'' = 79°12'4''$

25. $159.639° = 159°(0.639 \times 60)' = 159°38.34' = 159°38'(0.34 \times 60)" = 159°38'20"$

27. There are two methods.

a. Convert the first to decimal degree form and compare with the second.

$$47°33'41" = \left(47 + \frac{33}{60} + \frac{41}{3{,}600}\right)° \approx 47.561°$$

Since $47.561° > 47.556°$, the $47.561°$ angle, or $47°33'41"$, is larger.

b. Convert the second to DMS form and compare with the first.

$$47.556° = 47°(0.556 \times 60)' = 47°33.36' = 47°33'(0.36 \times 60)" = 47°33'22"$$

Since $47°33'41" > 47°33'22"$, $47°33'41"$ is larger.

29. To compare α and β, we convert α to decimal form. Since $23' = \frac{23°}{60}$ and $31" = \frac{31°}{3{,}600}$, then

$47°23'31" = \left(47 + \frac{23}{60} + \frac{31}{3{,}600}\right)° \approx 47.392°$. Since $47.392° > 47.386°$, we conclude that $\alpha > \beta$.

31. To compare α and β, we convert α to decimal form. Since $27' = \frac{27°}{60}$ and $18" = \frac{18°}{3{,}600}$, then

$125°27'18" = \left(125 + \frac{27}{60} + \frac{18}{3{,}600}\right)° = 125.455°$. Since $125.455° = 125.455°$, we conclude that $\alpha = \beta$.

33. To compare α and β, we convert β to decimal form. Since $30' = \frac{30°}{60}$ and $50" = \frac{50°}{3{,}600}$, then

$20°30'50" = \left(20 + \frac{30}{60} + \frac{50}{3{,}600}\right)° \approx 20.514°$. Since $20.512° < 20.514°$, we conclude that $\alpha < \beta$.

35. $47°37'49" + 62°40'15" \rightarrow$ DMS

$110°18'4"$

37. $90° - 67°37'29" \rightarrow$ DMS

$22°22'31"$

39. Since $C = 2\pi r$ and 2π is somewhat more than 6, we may estimate the circumference as a bit more than 6 times 10, or 63. To two decimal places:

$$C = 2\pi(10) \approx 2(3.1416)10 \approx 62.83$$

41. Since $C = \pi d$ and π is close to 3, we may estimate the diameter as $\frac{C}{3}$ or $\frac{6}{3}$, that is, 2. To two decimal places:

$$d = \frac{C}{\pi} \approx \frac{6}{3.1416} \approx 1.91$$

43. Since $\frac{s}{C} = \frac{\theta}{360°}$, then

$$\frac{s}{1000 \text{ cm}} = \frac{36°}{360°}$$

$$s = \frac{36}{360}(1000 \text{ cm}) = 100 \text{ cm}$$

45. Since $\frac{s}{C} = \frac{\theta}{360°}$, then

$$\frac{25 \text{ km}}{C} = \frac{20°}{360°}$$

$$C = \frac{360}{20}(25 \text{ km}) = 450 \text{ km}$$

47. Since $\frac{s}{C} = \frac{\theta}{360°}$ and $C = 2\pi r$, then $\frac{s}{2\pi r} = \frac{\theta}{360°}$

$$\frac{s}{2(\pi)(5{,}400{,}000 \text{ mi})} \approx \frac{2.6°}{360°}$$

$$s \approx \frac{2(\pi)(5{,}400{,}000 \text{ mi})(2.6)}{360}$$

$$\approx 250{,}000 \text{ mi (to nearest 10,000 mi)}$$

49. Since $\frac{s}{C} = \frac{\theta}{360°}$ and $\theta = 12°31'4'' = \left(12 + \frac{31}{60} + \frac{4}{3{,}600}\right)^{\circ} = 12.518°$, then

$$\frac{50.2 \text{ cm}}{C} \approx \frac{12.518°}{360°}$$

$$C \approx \frac{360}{12.518}(50.2 \text{ cm}) \approx 1{,}440 \text{ cm (to nearest 10 cm)}$$

51. Since $\frac{A}{\pi r^2} = \frac{\theta}{360°}$, then $\frac{A}{\pi(25.2 \text{ cm})^2} = \frac{47.3°}{360°}$

$$A \approx \frac{47.3}{360}(\pi)(25.2 \text{ cm})^2 \approx 262 \text{ cm}^2$$

53. Since $\frac{A}{\pi r^2} = \frac{\theta}{360°}$, then $\frac{98.4 \text{ m}^2}{\pi(12.6 \text{ m})^2} = \frac{\theta}{360°}$

$$\theta = \frac{98.4}{(\pi)(12.6)^2} \cdot 360° = 71.0°$$

55. Make a sketch.

Following the hint, note:
$\angle BAX$ has measure one-half of the measure of $\angle BOC$. $\angle BOC$ has the same measure as arc BC, one-fifth of 360°.
Thus, $\angle BAX = \frac{1}{2} \cdot \frac{1}{5}(360°) = 36°$.
$\angle ABX$ has measure one-half of the measure of $\angle AOD$. $\angle AOD$ has the same measure as arc AD, two-fifths of 360°. Thus, $\angle ABX = \frac{1}{2} \cdot \frac{2}{5}(360°) = 72°$.
Finally, since $\angle ABX + \angle BAX + \angle AXB = 180°$,
$\angle AXB = 180° - \angle ABX - \angle BAX = 180° - 72° - 36° = 72°$.

57. Since $\frac{s}{C} = \frac{\theta}{360°}$ and $C = 2\pi r$, then $\frac{s}{2\pi r} = \frac{\theta}{360°}$

$$r = \frac{s}{2\pi} \cdot \frac{360°}{\theta}$$

$$\approx \frac{11.5 \text{ mm}}{2(\pi)} \cdot \frac{360}{118.2} \approx 5.57 \text{ mm}$$

59. Since $\frac{s}{C} = \frac{\theta}{360°}$ and $C = 2\pi r$, then $\frac{s}{2\pi r} = \frac{\theta}{360°}$

$$s = 2\pi r \cdot \frac{\theta}{360°}$$

$$\approx 2(\pi)(5.49 \text{ mm}) \cdot \frac{119.7}{360} \approx 11.5 \text{ mm}$$

In Problems 61–67 we use the diagram and reason as follows: Since the cities have the same longitude, θ is given by their difference in latitude.

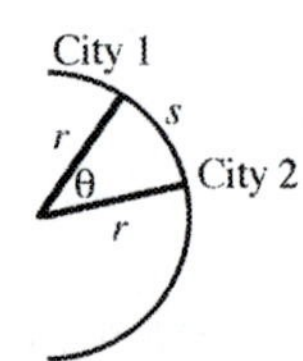

61. Since $\frac{s}{C} = \frac{\theta}{360°}$ and $C = 2\pi r$, then $\frac{s}{2\pi r} = \frac{\theta}{360°}$; $\theta = 47°40' - 37°50' = 9°50' = \left(9 + \frac{50}{60}\right)°$

$$s = 2\pi r \cdot \frac{\theta}{360°} \approx 2(\pi)(3960 \text{ mi}) \cdot \frac{9 + \frac{50}{60}}{360} \approx 680 \text{ mi}$$

63. Since $\frac{s}{C} = \frac{\theta}{360°}$ and $C = 2\pi r$, then $\frac{s}{2\pi r} = \frac{\theta}{360°}$; $\theta = 40°50' - 32°50' = 8°$

$$s = 2\pi r \cdot \frac{\theta}{360°} \approx 2(\pi)(3960) \cdot \frac{8}{360} \approx 553 \text{ mi}$$

65. To find the length of s in nautical miles, since 1 nautical mile is the length of 1' on the circle shown in the diagram, we need only find how many minutes are in the angle θ. Since

$$\theta = 47°40' - 37°50' = 9°50' = (9 \times 60 + 50)', \ \theta = 590'$$

Therefore, $s = 590$ nautical miles.

67. To find the length of s in nautical miles, since 1 nautical mile is the length of 1' on the circle shown in the diagram, we need only find how many minutes are in the angle θ. Since

$$\theta = 40°50' - 32°50' = 8° = (8 \times 60)', \ \theta = 480'.$$

Therefore, $s = 480$ nautical miles.

69. (A) Since $\frac{s}{C} = \frac{\theta}{360°}$ and $C = 2\pi r$, then $\frac{s}{2\pi r} = \frac{\theta}{360°}$; $\theta = 8°$ and $r = 500$ ft.

Hence

$$s = 2\pi r \cdot \frac{\theta}{360°} = 2\pi \cdot 500 \frac{8°}{360°} \approx 70 \text{ ft.}$$

(B) The arc length of a circular sector is very close to the chord length if the central angle of the sector is small and the radius of the sector is large, which is the case in this problem.

71. Since $\theta = \frac{360° s}{2\pi r}$, then $\theta = \frac{360°(864{,}000)}{2\pi(886{,}000{,}000)} = 0.056°$.

73. Since $\theta = \frac{360° s}{2\pi r}$, then $0.242° = \frac{360°(3{,}200)}{2\pi r}$.

Hence $r = \frac{360°(3{,}200)}{2\pi \cdot 0.242°} = 758{,}000$ mi.

EXERCISE 1.2 Similar Triangles

1. The measures of the third angle in each triangle are the same, since the sum of the measures of the three angles in any triangle is 180°. Thus, if $A + B + C = 180°$ and $A' + B' + C' = 180°$ and $A = A'$ and $B = B'$, then $C = 180° - (A + B) = 180° - (A' + B') = C'$.

3. Since there are two significant digits in each measurement, there can be only two significant digits in the calculated result. Round 4.76 to 4.8.

5. Since $\frac{a}{a'} = \frac{b}{b'}$ by Euclid's Theorem, then $\frac{5}{7} = \frac{15}{b'}$, $b' = \frac{7(15)}{5} = 21$

7. Since $\frac{a}{a'} = \frac{c}{c'}$ by Euclid's Theorem, then $\frac{12}{2.4} = \frac{c}{18}$, $c = \frac{(18)(12)}{2.4} = 90$

9. Since $\frac{b}{b'} = \frac{c}{c'}$ by Euclid's Theorem, then $\frac{52{,}000}{8.5} = \frac{18{,}000}{c'}$, $c' = \frac{(8.5)(18{,}000)}{52{,}000} = 2.9$

11. Two similar triangles can have equal sides if, and only if, they are congruent, that is, if the two triangles would coincide when one is moved on top of the other.

13. Since the triangles are similar, the sides are proportional and we can write

$$\frac{a}{0.47} = \frac{51\text{ in}}{1.0} \qquad \frac{c}{1.1} = \frac{51\text{ in}}{1.0}$$

$$a = \frac{(0.47)(51\text{ in})}{1.0} = 24\text{ in} \qquad c = \frac{(1.1)(51\text{ in})}{1.0} = 56\text{ in}$$

15. Since the triangles are similar, the sides are proportional and we can write

$$\frac{b}{1.0} = \frac{23.4\text{ m}}{0.47} \qquad \frac{c}{1.1} = \frac{23.4\text{ m}}{0.47}$$

$$b = \frac{23.4\text{ m}}{0.47} = 50\text{ m} \qquad c = \frac{(1.1)(23.4\text{ m})}{0.47} = 55\text{ m}$$

17. Since the triangles are similar, the sides are proportional and we can write

$$\frac{a}{0.47} = \frac{2.489 \times 10^9\text{ yd}}{1.0} \qquad \frac{c}{1.1} = \frac{2.489 \times 10^9\text{ yd}}{1.0}$$

$$a = (0.47)(2.489 \times 10^9\text{ yd}) = 1.2 \times 10^9\text{ yd} \qquad c = (1.1)(2.489 \times 10^9\text{ yd}) = 2.7 \times 10^9\text{ yd}$$

19. Since the triangles are similar, the sides are proportional and we can write

$$\frac{a}{0.47} = \frac{8.39 \times 10^5\text{ mm}}{1.1} \qquad \frac{b}{1.0} = \frac{8.39 \times 10^5\text{ mm}}{1.1}$$

$$a = \frac{(0.47)(8.39 \times 10^5\text{ mm})}{1.1} = 3.6 \times 10^{-5}\text{ mm} \qquad b = \frac{8.39 \times 10^5\text{ mm}}{1.1} = 7.6 \times 10^{-5}\text{ mm}$$

21. We make a scale drawing of the triangle, choosing a' to be 2.00 in, $\angle A' = 70°$, $\angle C' = 90°$. Now measure c' (approximately 2.13 in) and set up a proportion. Thus,

$$\frac{c}{2.13\,\text{in}} = \frac{101\,\text{ft}}{2.00\,\text{in}}$$

$$c \approx \frac{2.13}{2.00}(101\text{ ft}) \approx 108\text{ ft}$$

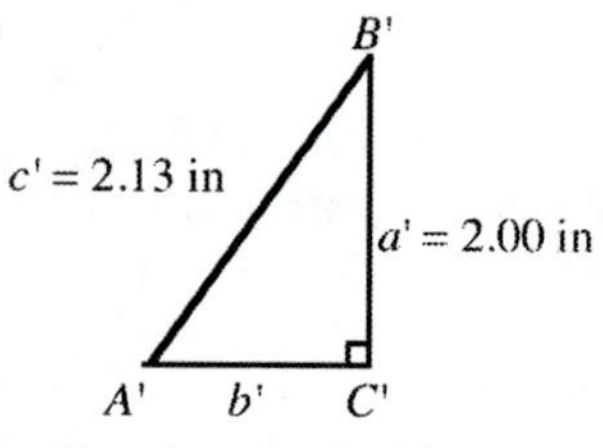

Drawing not to scale

23. Make a sketch.

The three angles of triangle *ACP* are $\angle ACP$, $\angle PAC$, $\angle APC$. The three corresponding angles of triangle *BCP* are $\angle PBC$, $\angle BCP$, $\angle BPC$. Both triangles are right triangles. $\angle PAC = \angle BCP$ since both angles are complementary to $\angle B$ of the original triangle.

As discussed in the text, these observations are sufficient to show that the triangles *ACP* and *BCP* are similar.

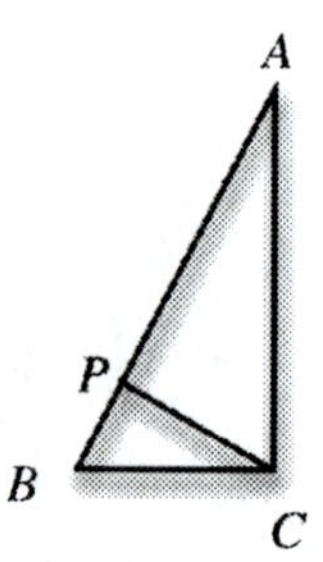

25. Make a sketch.
The three angles of triangle *PAD* are $\angle P$, $\angle D$, $\angle PAD$. The three corresponding angles of triangle *PCB* are $\angle P$, $\angle B$, $\angle PCB$. $\angle D = \angle B$ since both angles are inscribed in the circle and subtend the same arc. Clearly $\angle P$ is the same in both triangles.

As discussed in the text, these observations are sufficient to show that the triangles *PAD* and *PCB* are similar.

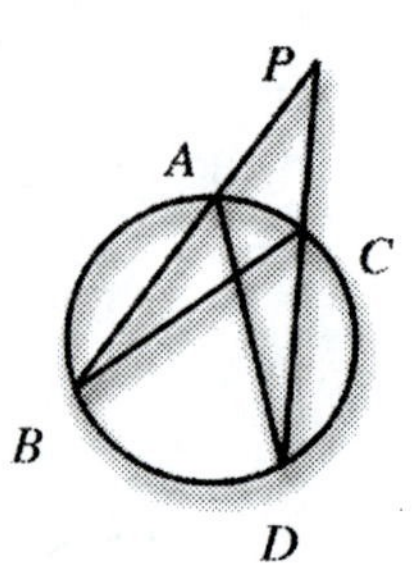

27. In the drawing, we note that triangles *LBT* and *LNM* are similar.
$MN = 9$ ft.

$NB = \frac{1}{2}$(length of court) $= \frac{1}{2}(78\text{ ft}) = 39$ ft. $TB = 3$ ft. BL is to be found. Let

$BL = x$. Then, $\frac{BL}{NL} = \frac{TB}{MN}$.

$NL = NB + BL$. Thus, $\frac{x}{39+x} = \frac{3}{9}$.

$$9(39+x)\frac{x}{39+x} = 9(39+x)\frac{3}{9} \text{ (clear of fractions)}$$

$$9x = (39+x)3 = 117 + 3x$$

$$6x = 117$$

$$x = 19.5\text{ ft}$$

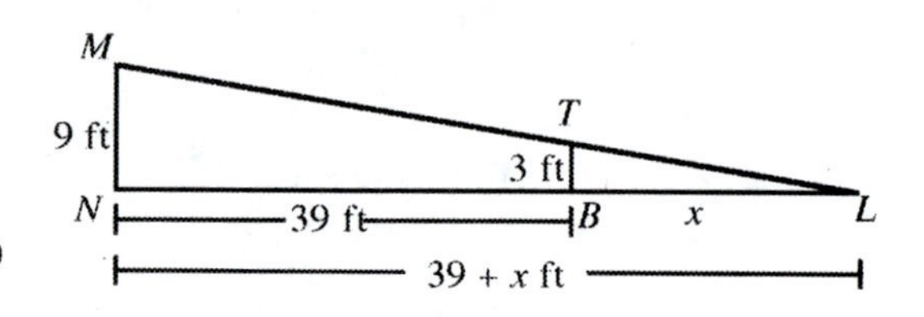

Drawing not to scale

29. Since the triangles *ABC* and *DEC* in the figure are similar, we can write $\frac{AB}{DE} = \frac{AC}{CD}$. Then,

$$\frac{AB}{5.5\text{ ft}} = \frac{24\text{ ft}}{2.1\text{ ft}}$$

$$AB = \frac{5.5}{2.1}(24\text{ ft}) = 63\text{ ft}$$

31. We reason that

$$\frac{\text{Length of flagpole 2}}{\text{Length of shadow 2}} = \frac{\text{Length of flagpole 1}}{\text{Length of shadow 1}}$$

$$\frac{x}{44\text{ ft}} = \frac{20\text{ ft}}{32\text{ ft}}$$

$$x = \frac{20}{32}(44\text{ ft}) = 28\text{ ft}$$

33. In this figure, S represents the surveyor's eye level, P the top of the pole, and T the top of the tree. $SC = 5.7$ feet, $PD = 8$ feet, $AU = BD = SC$. The height of the tree is given by $TA + AU$. $SB = 16$ feet, $SA = 200$ feet.

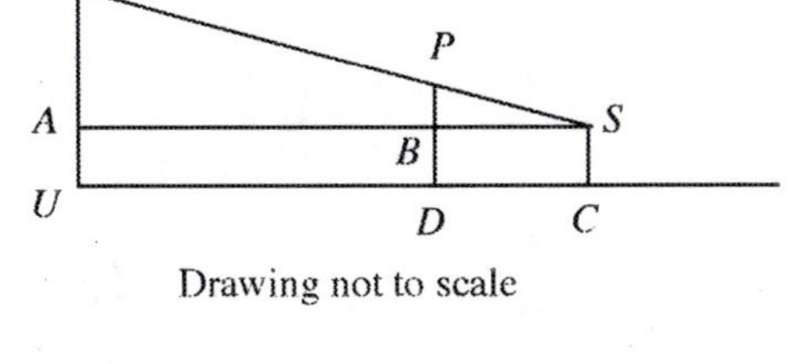

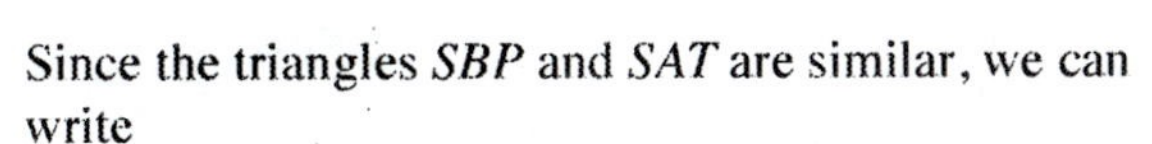

Since the triangles SBP and SAT are similar, we can write

$$\frac{TA}{SA} = \frac{PB}{SB}$$

$$\frac{TA}{200\text{ ft}} = \frac{8-5.7}{16}$$

$$TA = \frac{2.3}{16}(200\text{ ft}) = 28.75\text{ ft}$$

Thus the height of the tree is $TA + AU = 28.75 + 5.7 = 34$ feet.

35. Let us make a scale drawing of the figure in the text as follows: pick any convenient length, say 2 in, for $A'C'$. Copy the 15° angle CAB and 90° angle ACB using a protractor. Now, measure $B'C'$ (approximately 0.55 in) and set up a proportion. Thus,

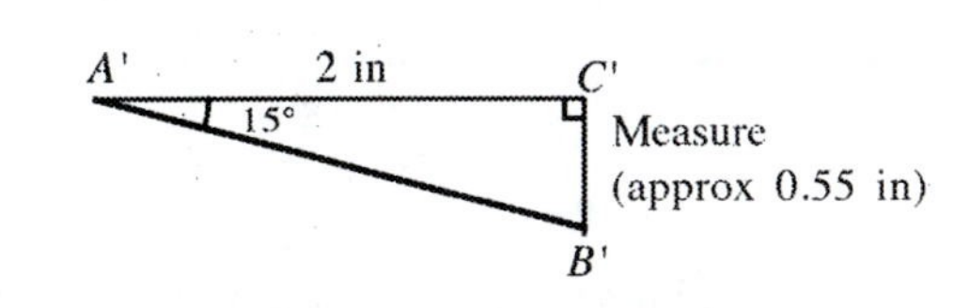

$$\frac{x}{0.55\text{ in}} = \frac{4.0\text{ km}}{2\text{ in}}$$

$$x \approx \frac{0.55}{2}(4.0\text{ km}) \approx 1.1\text{ km}$$

37. (A) Triangles PAC, FBC, ACP', and ABF' are all right triangles. Angles APC and BFC are equal, and angles $CP'A$ and $BF'A$ are equal--corresponding angles of parallel lines cut by a transversal are equal (see Text, Appendix C.1). Thus, triangles PAC and FBC are similar, and triangles ACP' and ABF' are similar, since, in each case, the angles of each triangle are equal.

(B) Since the triangles PAC and FBC are similar, corresponding sides are proportional. Hence,

$$\frac{AC}{PA} = \frac{BC}{FB} \text{ or } \frac{AB+BC}{PA} = \frac{BC}{FB} \text{ or } \frac{h+h'}{u} = \frac{h'}{f}$$

Since the triangles ACP' and ABF' are also similar, corresponding sides are also proportional. Hence,

$$\frac{AC}{CP'} = \frac{AB}{BF'} \text{ or } \frac{AB+BC}{CP'} = \frac{AB}{FB} \text{ or } \frac{h+h'}{v} = \frac{h}{f}$$

(C) Adding the results in part (B) yields

$$\frac{h+h'}{u} + \frac{h+h'}{v} = \frac{h'}{f} + \frac{h}{f}$$

or

$$\frac{h+h'}{u} + \frac{h+h'}{v} = \frac{h+h'}{f}$$

Dividing both sides by $h + h'$, we obtain

$$\frac{1}{u} + \frac{1}{v} = \frac{1}{f}$$

(D) Use the formula just derived with $u = 3$ m $= 3000$ mm and $f = 50$ mm. Then

$$\frac{1}{3000} + \frac{1}{v} = \frac{1}{50}$$

$$\frac{1}{v} = \frac{1}{50} - \frac{1}{3000}$$

$$\frac{1}{v} = \frac{59}{3000}$$

$$v = \frac{3000}{59} \approx 50.847 \text{ mm}$$

39. Label the figure in the text as shown at the right. Triangles ADG and FEB are similar, since their angles are equal.

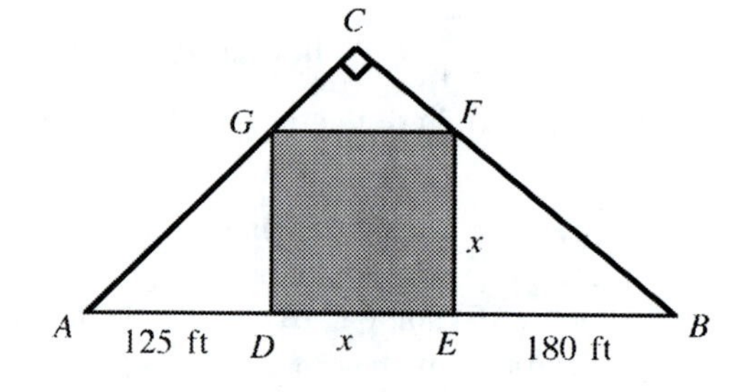

Hence $\frac{AD}{DG} = \frac{FE}{EB}$.

$$\frac{125}{x} = \frac{x}{180}$$

$$x^2 = 125 \cdot 180$$

$$x = 150 \text{ ft}$$

41. Label the figure in the text as shown at the right. Triangles ABC and CBD are similar. For, if angle ACD has measure α, the angles of ABC have measures θ, $\alpha + \theta$, $180° - (\alpha + 2\theta)$. But the angles of CBD have measures θ, $180° - (\alpha + 2\theta)$, and $180° - [\theta + 180° - (\alpha + 2\theta)] = \alpha + \theta$. Since the triangles are similar, corresponding sides are proportional. Thus

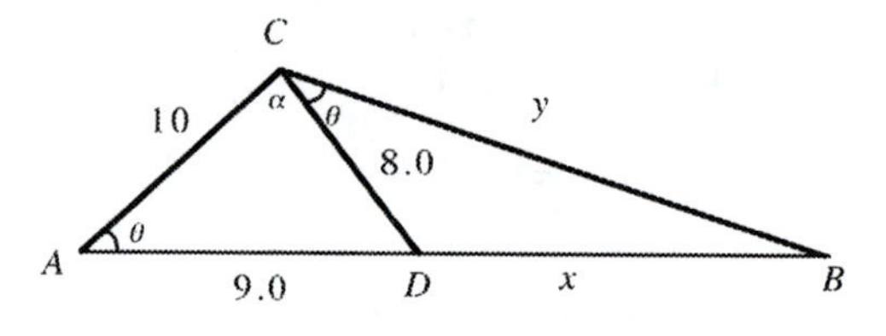

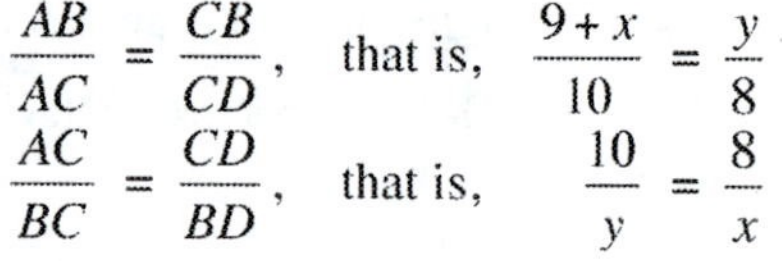

$$\frac{AB}{AC} = \frac{CB}{CD}, \quad \text{that is,} \quad \frac{9+x}{10} = \frac{y}{8}$$

$$\frac{AC}{BC} = \frac{CD}{BD}, \quad \text{that is,} \quad \frac{10}{y} = \frac{8}{x}$$

Solving yields, in turn:

$$72 + 8x = 10y$$

$$10x = 8y$$

$$x = 0.8y$$

$$72 + 8(0.8y) = 10y$$

$$y = 20$$

$$x = 16$$

EXERCISE 1.3 Trigonometric Ratios and Right Triangles

1. Make a careful drawing.

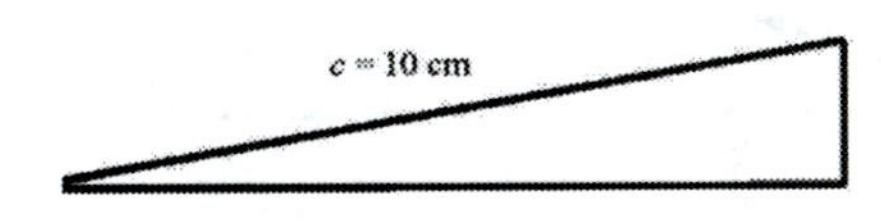

Measure $b \approx 1.7$ cm, $a \approx 9.8$ cm.

Therefore, $\sin \theta = \frac{b}{c} \approx \frac{1.7}{10} \approx 0.17$,

$\cos \theta = \frac{a}{c} \approx \frac{9.8}{10} \approx 0.98$.

3. Make a careful drawing.

Measure $b = 5$ cm, $a \approx 8.7$ cm.

Therefore, $\sin \theta = \frac{b}{c} = \frac{5}{10} = 0.50$,

$\cos \theta = \frac{a}{c} = \frac{8.7}{10} \approx 0.87$.

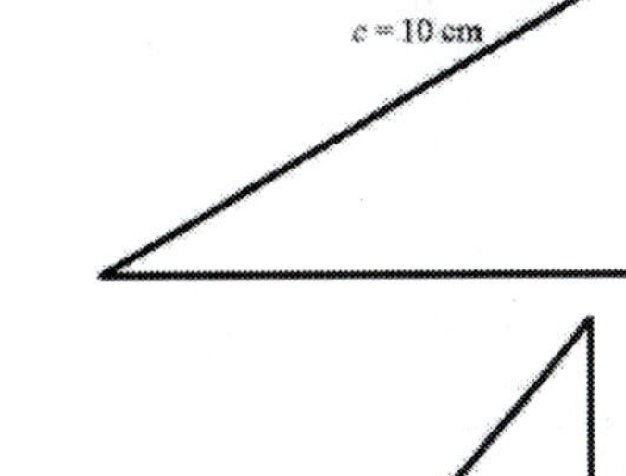

5. Make a careful drawing.

Measure $b \approx 7.7$ cm, $a \approx 6.4$ cm.

Therefore, $\sin \theta = \frac{b}{c} = \frac{7.7}{10} \approx 0.77$,

$\cos \theta = \frac{a}{c} = \frac{6.4}{10} \approx 0.64$.

7. Make a careful drawing.

Measure $b \approx 9.4$ cm, $a \approx 3.4$ cm.

Therefore, $\sin \theta = \frac{b}{c} = \frac{9.4}{10} \approx 0.94$,

$\cos \theta = \frac{a}{c} = \frac{3.4}{10} \approx 0.34$.

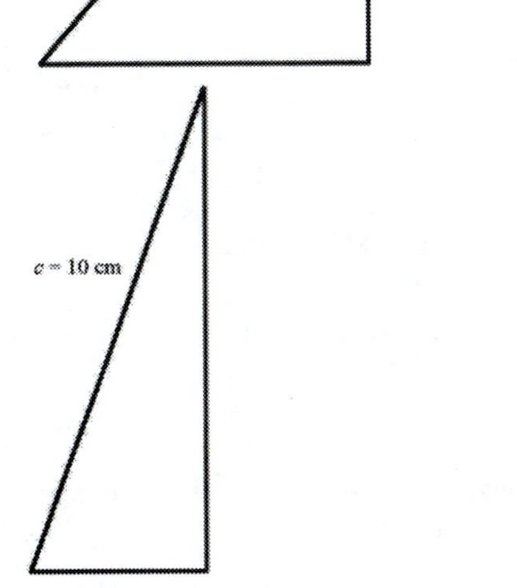

9. Make a careful drawing.

Measure $b \approx 2.7$ cm.

Therefore, $\tan \theta = \frac{b}{a} = \frac{2.7}{10} \approx 0.27$.

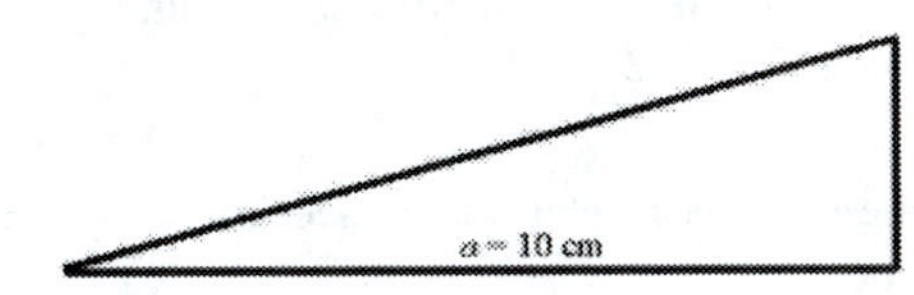

11. Make a careful drawing.

    Measure $b \approx 3.6$ cm.

    Therefore, $\tan\,\theta = \frac{b}{a} = \frac{3.6}{10} \approx 0.36$.

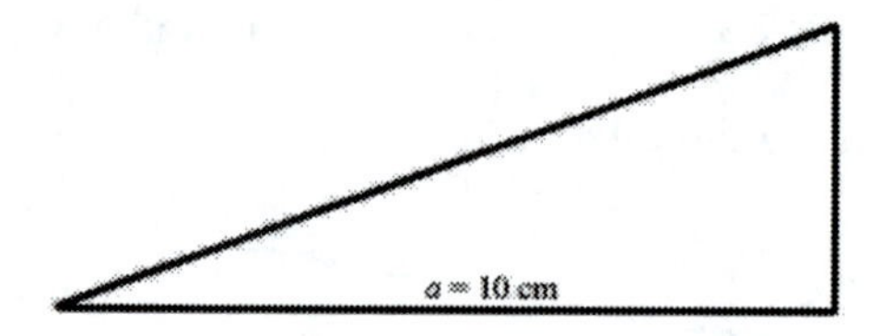

13. Make a careful drawing.

    Measure $b \approx 4.7$ cm.

    Therefore, $\tan\,\theta = \frac{b}{a} = \frac{4.7}{10} \approx 0.47$.

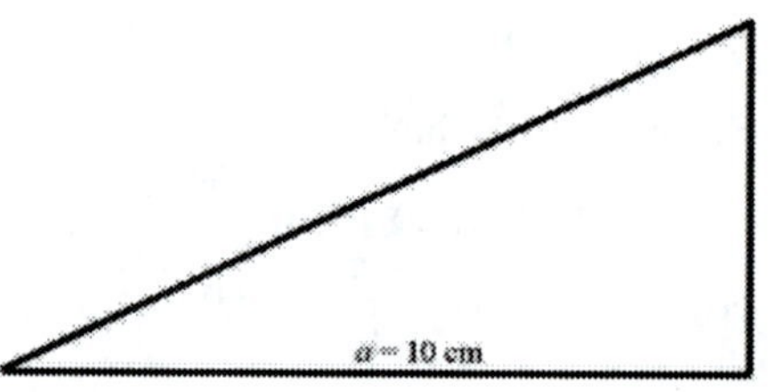

15. Make a careful drawing.

    Measure $b \approx 5.8$ cm.

    Therefore, $\tan\,\theta = \frac{b}{a} = \frac{5.8}{10} \approx 0.58$.

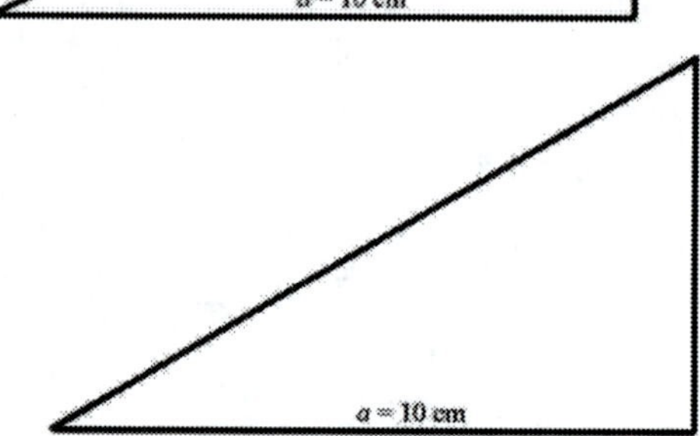

17. Set calculator in degree mode and use cos key.
    $\cos 54.3° = 0.584$.

19. Use the reciprocal relationship $\cot\,\theta = \frac{1}{\tan\theta}$. Set calculator in degree mode, use tan key, then take reciprocal.

    $$3°45' = \left(3+\frac{45}{60}\right)° = 3.75°$$
    $$\cot 3°45' = \cot 3.75° = 15.3$$

21. Use the reciprocal relationship $\csc\,\theta = \frac{1}{\sin\theta}$. Set calculator in degree mode, use sin key, then take reciprocal.

    $\csc 26.9° = 2.21$

23. Set calculator in degree mode and use sin key.
    $\sin 30.5° = 0.508$

25. Set calculator in degree mode and use tan key.
    $\tan 69.8° = 2.72$

27. Use the reciprocal relationship $\sec\,\theta = \frac{1}{\cos\theta}$. Set calculator in degree mode, use cos key, then take reciprocal.

    $$78°55' = \left(78+\frac{55}{60}\right)° = (78.91666\ldots)°$$
    $$\sec 78°55' = \sec(78.91666\ldots)° = 5.20$$

29. If $\cos\theta = 0.5$, then
$\theta = \cos^{-1} 0.5 = 60°$

31. If $\sin\theta = 0.8125$, then
$\theta = \sin^{-1} 0.8125 = 54.34°$

33. $\theta = \arcsin 0.4517$
$= 26.85° = 26°50'$

35. $\theta = \tan^{-1}(2.753)$
$= 70.037° = 70°2'$

37. The triangle is uniquely determined. Angle α can be found using $\tan\alpha = \frac{a}{b}$; angle $\beta = 90° - \alpha$; The hypotenuse c can be found using the Pythagorean Theorem or by using $\sin\alpha = \frac{b}{c}$.

39. The triangle is not uniquely determined. In fact, there are infinitely many different size triangles with the same acute angles--all are similar to each other.

41. *Solve for the complementary angle:* $90° - \theta = 90° - 58°40' = 31°20'$
Solve for b: Since $\theta = 58°40' = \left(58 + \frac{40}{60}\right)° = (58.666...)°$ and $c = 15.0$ mm, we look for a trigonometric ratio that involves θ and c (the known quantities) and b (the unknown quantity). We choose the sine.

$$\sin\theta = \frac{b}{c}$$
$$b = c\sin\theta = (15.0 \text{ mm})(\sin 58.666...°) = 12.8 \text{ mm}$$

Solve for a: We choose the cosine to find a. Thus,

$$\cos\theta = \frac{a}{c}$$
$$a = c\cos\theta = (15.0)(\cos 58.666...°) = 7.80 \text{ mm}$$

43. *Solve for the complementary angle:* $90° - \theta = 90° - 83.7° = 6.3°$
Solve for a: Since $\theta = 83.7°$ and $b = 3.21$ km, we look for a trigonometric ratio that involves θ and b (the known quantities) and a (the unknown quantity). We choose the tangent.

$$\tan\theta = \frac{b}{a}$$
$$a = \frac{b}{\tan\theta} = \frac{3.21 \text{ km}}{\tan 83.7°} = 0.354 \text{ km}$$

Solve for c: We choose the sine to find c. Thus,

$$\sin\theta = \frac{b}{c}$$
$$c = \frac{b}{\sin\theta} = \frac{3.21 \text{ km}}{\sin 83.7°} = 3.23 \text{ km}$$

45. *Solve for the complementary angle:* $90° - \theta = 90° - 71.5° = 18.5°$
Solve for a: Since $\theta = 71.5°$ and $b = 12.8$ in, we look for a trigonometric ratio that involves θ and b (the known quantities) and a (the unknown quantity). We choose the tangent.

$$\tan\theta = \frac{b}{a}$$
$$a = \frac{b}{\tan\theta} = \frac{12.8 \text{ in}}{\tan 71.5°} = 4.28 \text{ in}$$

Solve for c: We choose the sine to find c. Thus,

$$\sin\theta = \frac{b}{c}$$
$$c = \frac{b}{\sin\theta} = \frac{12.8 \text{ in}}{\sin 71.5°} = 13.5 \text{ in}$$

47. *Solve for θ:* $\sin\theta = \frac{b}{c} = \frac{63.8\text{ ft}}{134\text{ ft}} = 0.476$

$\theta = \sin^{-1} 0.476 = 28.4° = 28°30'$ to nearest 10'

Solve for the complementary angle: $90° - \theta = 90° - 28°30' = 61°30'$

Solve for a: We will use the tangent. Thus, $\tan\theta = \frac{b}{a}$

$$a = \frac{b}{\tan\theta} = \frac{63.8\text{ ft}}{\tan 28°30'} = 118\text{ ft}$$

49. *Solve for θ:* $\tan\theta = \frac{b}{a} = \frac{132\text{ mi}}{108\text{ mi}} = 1.22$

$\theta = \tan^{-1} 1.22 = 50.7°$ to nearest 0.1°

Solve for the complementary angle: $90° - \theta = 90° - 50.7° = 39.3°$

Solve for c: We will use the sine. Thus, $\sin\theta = \frac{b}{c}$

$$c = \frac{b}{\sin\theta} = \frac{132\text{ mi}}{\sin 50.7°} = 171\text{ mi}$$

51. The calculator was accidentally set in radian mode. Changing the mode to degree, $a = 235\sin(14.1) = 57.2$ m and $b = 235\cos(14.1) = 228$ m.

53. (A) If $\theta = 11°$, $(\sin\theta)^2 + (\cos\theta)^2 = (\sin 11°)^2 + (\cos 11°)^2 = (0.1908\ldots)^2 + (0.9816\ldots)^2 = 1$

(B) If $\theta = 6.09°$, $(\sin\theta)^2 + (\cos\theta)^2 = (\sin 6.09°)^2 + (\cos 6.09°)^2 = (0.106\ldots)^2 + (0.994\ldots)^2 = 1$

(C) If $\theta = 43°24'47''$, $(\sin\theta)^2 + (\cos\theta)^2 = (\sin 43°24'47'')^2 + (\cos 43°24'47'')^2$
$= (0.687\ldots)^2 + (0.726\ldots)^2 = 1$

55. (A) If $\theta = 19°$, $\sin\theta - \cos(90° - \theta) = \sin 19° - \cos(90° - 19°)$
$= \sin 19° - \cos 71° = 0.3256 - 0.3256 = 0$

(B) If $\theta = 49.06°$, $\sin\theta - \cos(90° - \theta) = \sin 49.06° - \cos(90° - 49.06°)$
$= \sin 49.06° - \cos 40.94° = 0.7554 - 0.7554 = 0$

(C) If $\theta = 72°51'12''$, $\sin\theta - \cos(90° - \theta) = \sin 72°51'12'' - \cos(90° - 72°51'12'')$
$= \sin 72°51'12'' - \cos(17°8'48'')$
$= 0.9556 - 0.9556 = 0$

57. *Solve for the complementary angle:* $90° - \theta = 90° - 83°12' = 6°48'$

Solve for b: We choose the tangent to find b. Thus, $\tan\theta = \frac{b}{a}$

$b = a\tan\theta$
$= (23.82\text{ mi})(\tan 83°12')$
$= 199.8\text{ mi}$

Solve for c: We choose the cosine to find c. Thus, $\cos\theta = \frac{a}{c}$

$$c = \frac{a}{\cos\theta} = \frac{23.82\text{ mi}}{(\cos 83°12')} = 201.2\text{ mi}$$

59. *Solve for θ:* $\tan\theta = \dfrac{b}{a} = \dfrac{42.39\text{ cm}}{56.04\text{ cm}}$; $\tan\theta = 0.7564$; $\theta = \tan^{-1} 0.7564 = 37.105° = 37°6'$

Solve for the complementary angle: $90° - \theta = 90° - 37°6' = 52°54'$

Solve for c: We will use the sine. Thus, $\sin\theta = \dfrac{b}{c}$

$$c = \frac{b}{\sin\theta} = \frac{42.39\text{ cm}}{\sin 37°6'} = 70.27\text{ cm}$$

61. *Solve for θ:* $\sin\theta = \dfrac{b}{c} = \dfrac{35.06\text{ cm}}{50.37\text{ cm}} = 0.6960$

$\theta = \sin^{-1}(0.6960) = 44.11°$

Solve for the complementary angle: $90° - \theta = 90° - 44.11° = 45.89°$

Solve for a: We choose the cosine to find a. Thus, $\cos\theta = \dfrac{a}{c}$

$$\begin{aligned} a &= c\cos\theta \\ &= (50.37\text{ cm})(\cos 44.11°) = 36.17\text{ cm} \end{aligned}$$

63. (A) In right triangle OAD, $\sin\theta = \dfrac{\text{Opp}}{\text{Hyp}} = \dfrac{AD}{OD} = \dfrac{AD}{1} = AD$

(B) In right triangle OCD, $\tan\theta = \dfrac{\text{Opp}}{\text{Adj}} = \dfrac{DC}{OD} = \dfrac{DC}{1} = DC$

(C) In right triangle ODE, $\csc\theta = \csc OED = \dfrac{\text{Hyp}}{\text{Opp}} = \dfrac{OE}{OD} = \dfrac{OE}{1} = OE$

65. (A) As θ approaches 90°, AD approaches OD, which has measure 1. Thus, $\sin\theta\ (= AD)$ approaches 1.

(B) As θ approaches 90°, EC approaches being parallel to the x axis. Thus, DC increases without bound, so $\tan\theta\ (= DC)$ increases without bound.

(C) As θ approaches 90°, OE approaches OF, which has measure 1. Thus, $\csc\theta\ (= OE)$ approaches 1.

67. (A) As θ approaches 0°, OA approaches OB, which has measure 1. Thus, $\cos\theta\ (= OA)$ approaches 1.

(B) As θ approaches 0°, EC approaches being parallel to the y axis. Thus, ED increases without bound, so $\cot\theta\ (= ED)$ increases without bound.

(C) As θ approaches 0°, OC approaches OB, which has measure 1. Thus, $\sec\theta\ (= OC)$ approaches 1.

69. Label the figure in the text as shown at the right. From the Pythagorean theorem, in the right-side right triangle

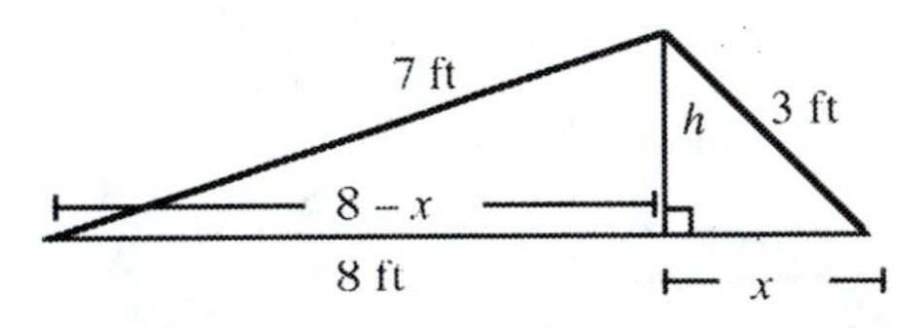

$h^2 + x^2 = 3^2$

In the left-side right triangle

$h^2 + (8 - x)^2 = 7^2$

Thus, in turn,

$$\begin{aligned} h^2 + x^2 &= 9 \\ h^2 + 64 - 16x + x^2 &= 49 \\ 64 - 16x + 9 &= 49 \\ x &= 1.5 \end{aligned}$$

$h = \sqrt{9 - x^2} = \sqrt{9 - 1.5^2} = \sqrt{6.75}$

The area of the triangle is then given by

$$A = \frac{1}{2}bh$$
$$A = \frac{1}{2} \cdot 8\sqrt{6.75}$$
$$A = 4\sqrt{\frac{27}{4}} = 6\sqrt{3}\text{ ft}^2$$
$$A = 10.4\text{ ft}^2$$

71. Label the figure in the text as shown at the right. Note that the triangle is isosceles, hence the right-hand angle also has measure θ.

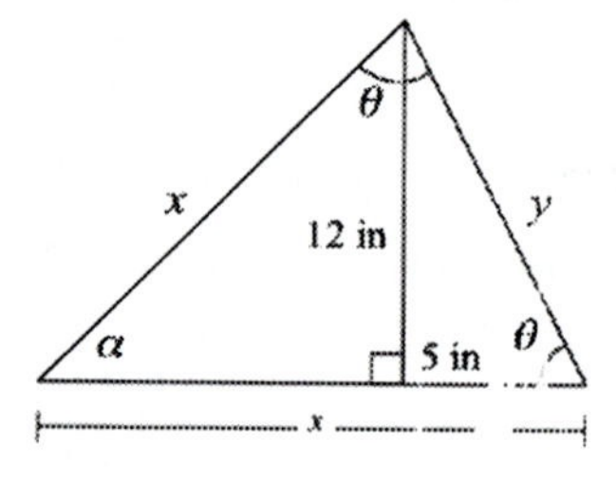

From the Pythagorean theorem, in the right-side right triangle

$$12^2 + 5^2 = y^2$$
$$y^2 = 169$$
$$y = 13$$

Then, in this triangle

$$\sin\theta = \frac{12}{13}$$
$$\theta = \sin^{-1}\frac{12}{13}$$
$$\theta = 67.38^\circ$$

Since $\alpha + \theta + \theta = 180^\circ$, $\alpha = 180^\circ - 2\theta = 180^\circ - 2(67.38^\circ) = 45.24^\circ$. In the left-side right triangle

$$\sin\alpha = \frac{12}{x}$$
$$x = \frac{12}{\sin\alpha}$$
$$x = \frac{12}{\sin 45.24^\circ}$$
$$x = 16.9$$

Finally, the area of the triangle is given by

$$A = \frac{1}{2}bh$$
$$A = \frac{1}{2}(16.9)12$$
$$A = 101.4\text{ in}^2$$

73. Make a sketch.

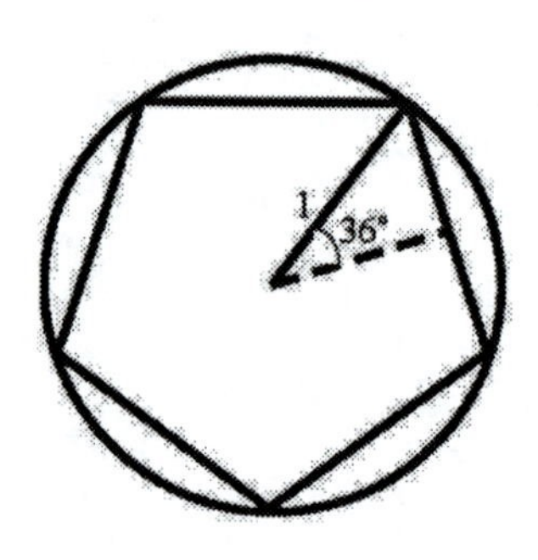

Since the pentagon is regular, it may be viewed as consisting of 10 right triangles congruent to the one shown. Thus, it has area equal to 10 times the area of one triangle. The legs of this triangle have lengths given by $b = 1\sin 36^\circ$ and $a = 1\cos 36^\circ$. Thus, the area of the triangle is $\frac{1}{2}(1\sin 36^\circ)(1\cos 36^\circ)$.

Then the area of the pentagon is $10\left[\frac{1}{2}(1\sin 36^\circ)(1\cos 36^\circ)\right] =$

$5\sin 36^\circ\cos 36^\circ$. The area of the circle is given by $\pi(1)^2 = \pi$.

Thus, the ratio of the area of the pentagon to the area of the circle is given by.

$$\frac{\text{area of pentagon}}{\text{area of circle}} = \frac{5\sin 36^\circ \cos 36^\circ}{\pi} \approx 75.7\%$$

EXERCISE 1.4 Right Triangle Applications

1. $\sin 61^\circ = \frac{\text{Opp}}{\text{Hyp}} = \frac{x}{8.0\,\text{m}}$

   $x = (8.0\text{ m})(\sin 61^\circ) = 7.0\text{ m}$

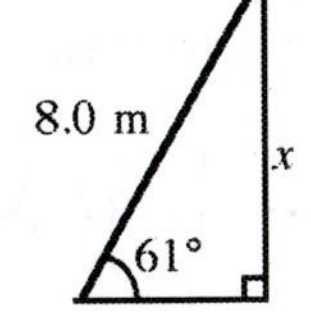

3. We first sketch a figure and label the known parts.

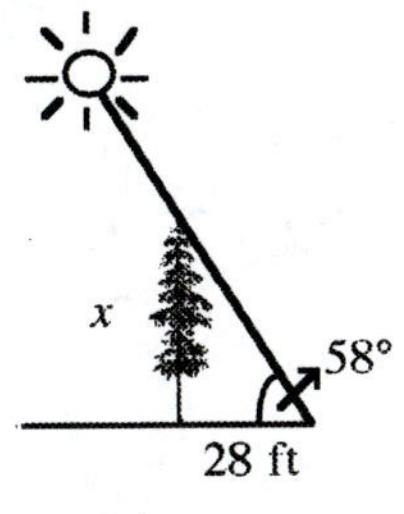

$$\tan 58^\circ = \frac{\text{Opp}}{\text{Adj}} = \frac{x}{28\text{ ft}}$$

$$x = (28\text{ ft})(\tan 58^\circ) = 45\text{ ft}$$

5. We first sketch a figure and label the known parts.

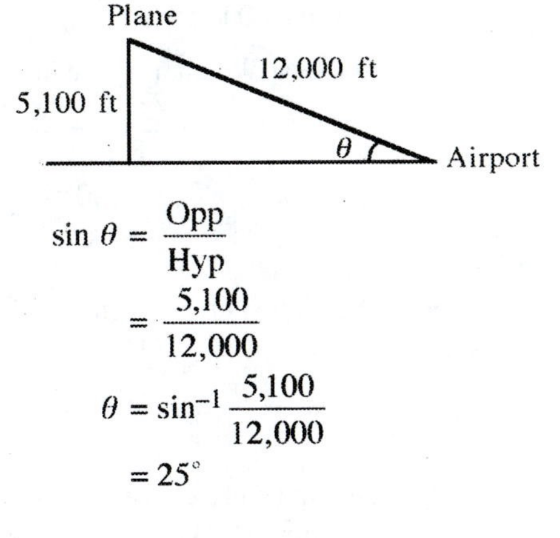

$$\sin\theta = \frac{\text{Opp}}{\text{Hyp}} = \frac{5{,}100}{12{,}000}$$

$$\theta = \sin^{-1}\frac{5{,}100}{12{,}000} = 25^\circ$$

7. We first sketch a figure and label the known parts.

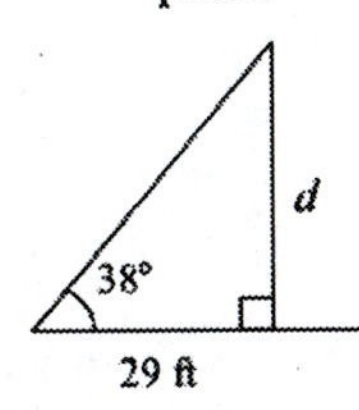

$$\tan 38^\circ = \frac{\text{Opp}}{\text{Adj}} = \frac{d}{29\text{ ft}}$$

$$d = (29\text{ ft})\tan 38^\circ = 23\text{ ft}$$

9. We first sketch a figure and label the known parts.

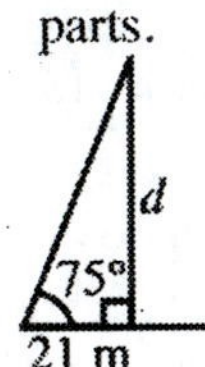

$$\tan 75^\circ = \frac{\text{Opp}}{\text{Adj}} = \frac{d}{21\text{ m}}$$

$$d = (21\text{ m})\tan 75^\circ = 78\text{ m}$$

11. We first sketch a figure and label the known parts.

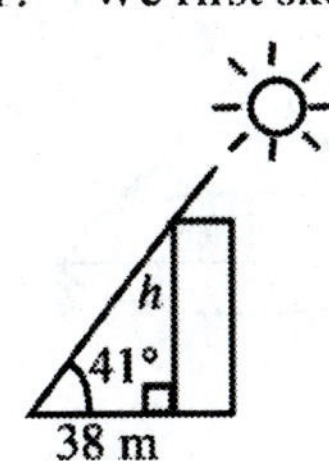

$$\tan 41^\circ = \frac{\text{Opp}}{\text{Adj}} = \frac{h}{38\text{ m}}$$

$$h = (38\text{ m})\tan 41^\circ = 33\text{ m}$$

13. $\cot \theta = \dfrac{\text{Adj}}{\text{Opp}}$

$\cot 18°20' = \dfrac{x}{70.0\text{ m}}$

$x = (70.0\text{ m})(\cot 18°20') = 211\text{ m}$

(If line p crosses parallel lines m and n then angles α and β have the same measure. Thus, $\theta = 18°20'$.)

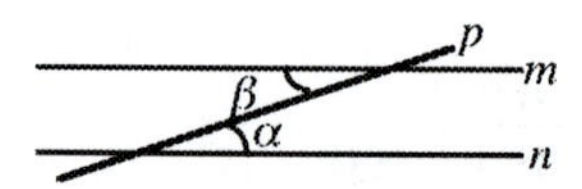

15. We first sketch a figure and label the known parts.

$$\tan 15° = \frac{\text{Opp}}{\text{Adj}} = \frac{x}{4.0\text{ km}}$$

$x = (4.0\text{ km})(\tan 15°) = 1.1\text{ km}$

17. We first sketch a figure and label the known parts.

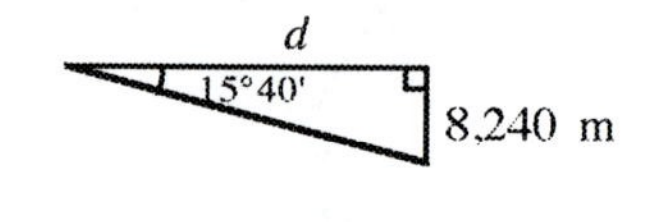

$$\tan 15°40' = \frac{\text{Opp}}{\text{Adj}} = \frac{8{,}240\text{ m}}{d}$$

$$d = \frac{8{,}240\text{ m}}{\tan 15°40'} = 29{,}400\text{ m or } 29.4\text{ km}$$

19. We first sketch a figure and label the known parts.

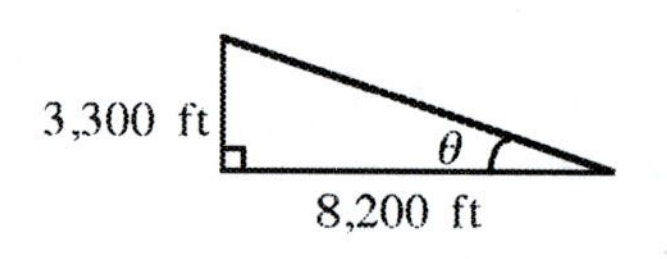

$$\tan \theta = \frac{\text{Opp}}{\text{Adj}} = \frac{3{,}300\text{ ft}}{8{,}200\text{ ft}} = 0.40\ldots$$

$\theta = \tan^{-1}(0.40\ldots) = 22°$

21. (A) In triangle ABC, $\angle\theta$ is complementary to 75°, thus $\theta = 15°$.

$BC = x =$ roof overhang

$AC = 19$ ft

$$\tan \theta = \frac{\text{Opp}}{\text{Adj}} = \frac{x}{19\text{ ft}}$$

$x = (19\text{ ft})(\tan 15°) = 5.1\text{ ft}$

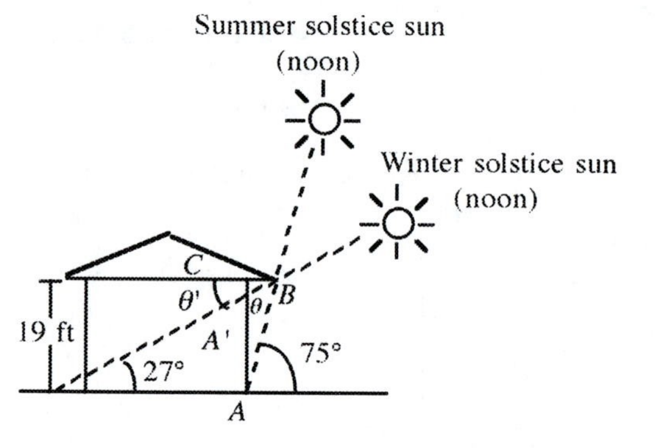

(B) In triangle $A'BC$, $\theta' = 27°$

$A'C = y =$ how far down shadow will reach

$$\tan 27° = \frac{\text{Opp}}{\text{Adj}} = \frac{y}{5.1\text{ ft}}$$

$y = (5.1\text{ ft})(\tan 27°) = 2.6\text{ ft}$

23. We note: by the symmetry of the cone, ATC is an isosceles right triangle, hence

$\angle TAC = \angle TCA = 45°$

Since the mast is perpendicular to the deck, TBA and TBC are also right triangles, and since each has a 45° angle, these are also isosceles right triangles. Then it follows that $TB = AB$ and $TB = AC$. Since the length TB is given as 67.0 feet, the diameter

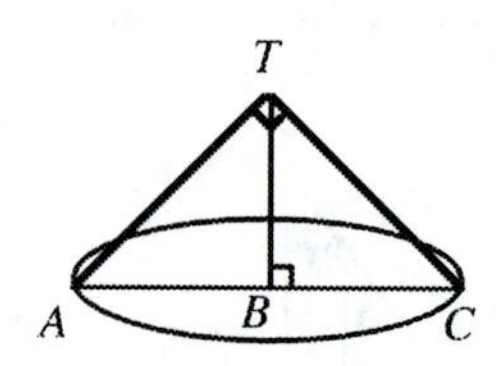

$AC = AB + BC = 67.0\text{ feet} + 67.0\text{ feet}$
$= 134.0\text{ ft.}$

25. Label the required sides AD and AB. Then

$AD = AC + CD = AC + 18.$

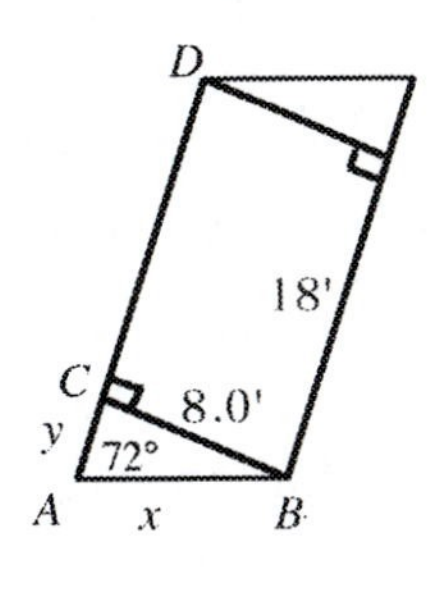

In right triangle ABC,

$$\sin A = \frac{BC}{AB} \qquad \tan A = \frac{BC}{AC}$$

$$\sin 72° = \frac{8.0\text{ feet}}{x} \qquad \tan 72° = \frac{8.0\text{ feet}}{y}$$

$$x = \frac{8.0\text{ feet}}{\sin 72°} = 8.4\text{ feet} \qquad y = \frac{8.0\text{ feet}}{\tan 72°} = 2.6\text{ feet}$$

Thus the sides of the parallelogram are $AB = 8.4$ feet and $AD = 18$ feet + 2.6 feet ≈ 21 feet.

27. (A) We note that $\angle TSC = 90° - \alpha$, hence $\angle C = \alpha$.

Thus, in triangle CST,

$$\cos \alpha = \frac{\text{Adj}}{\text{Hyp}} = \frac{r}{r+h}$$

(B) $(r+h)\cos \alpha = (r+h) \cdot \dfrac{r}{r+h}$

$$r\cos \alpha + h\cos \alpha = r$$

$$h\cos \alpha = r - r\cos \alpha$$

$$h\cos \alpha = r(1 - \cos \alpha)$$

$$r = \frac{h\cos\alpha}{1 - \cos\alpha}$$

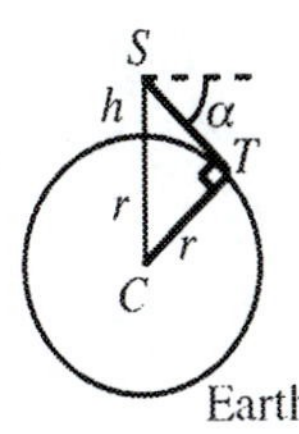

(C) $r = \dfrac{(335\text{ miles})\cos 22°47'}{1 \cos 22°47'} = 3960$ miles

29. We note that since ABC is an isosceles triangle,

$$\angle FCA = \frac{1}{2}\angle BCA = \frac{1}{2}(2\theta) = \theta,$$

and also,

$$AF = \frac{1}{2}AB = \frac{1}{2}(6.0\text{ mi}) = 3.0\text{ mi}.$$

We are to find AC, the radius of the circle.

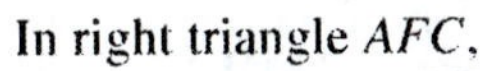

In right triangle AFC,

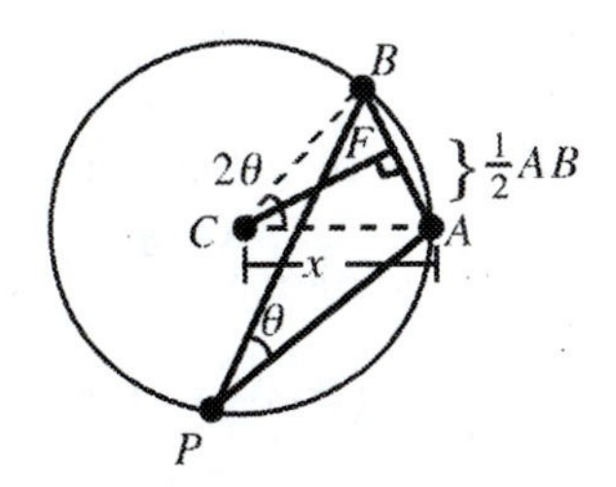

$$\sin \angle FCA = \frac{AF}{AC} \qquad \sin 21° = \frac{3.0\text{ mi}}{x}$$

$$\sin \theta = \frac{3.0\text{ mi}}{x} \qquad x = \frac{3.0\text{ mi}}{\sin 21°} = 8.4\text{ mi}$$

31. In the figure, r denotes the radius of the parallel of latitude, R the radius of the earth, i.e., the radius of the equator. Clearly, $\cos \theta = \dfrac{\text{Adj}}{\text{Hyp}} = \dfrac{r}{R}$.

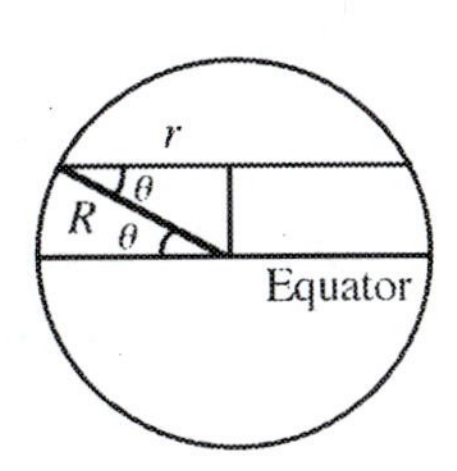

Since $L = 2\pi r$ and $E = 2\pi R$, we have

$$r = R\cos \theta$$

$$2\pi r = 2\pi R\cos \theta$$

$$L = E\cos \theta$$

E is given as 24,900 miles, In the particular case of San Francisco, $\theta = 38°$. Hence,

$$L = (24{,}900\text{ mi})\cos 38° = 19{,}600\text{ mi}$$

33. (A) A lifeguard can run faster than he can swim, so it would seem that he should run along the beach first before entering the water. [Parts (D) and (E) suggest how far the lifeguard should run before swimming to get to the swimmer in the least time].

(B) Let $PB = y$ and $PS = x$. In the right triangle SPB,

$$\csc\theta = \frac{x}{c} \qquad \cot\theta = \frac{y}{c}$$

$$x = c\csc\theta \qquad y = c\cot\theta$$

Thus the lifeguard runs a distance $d - y = d - c\cot\theta$ at speed p.

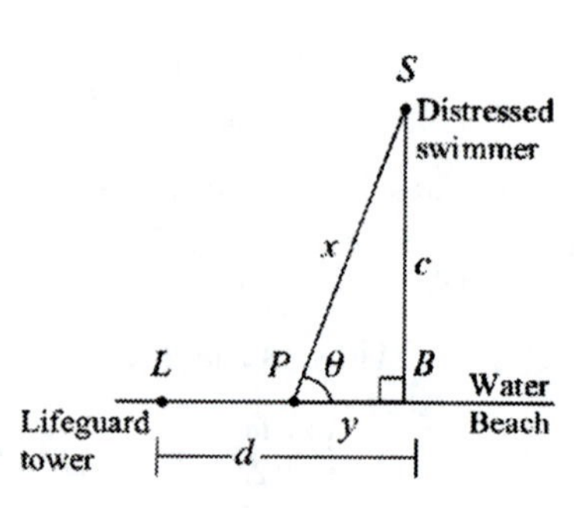

This requires a time $t_1 = \dfrac{\text{distance run}}{\text{rate run}} = \dfrac{d - c\cot\theta}{p}$

The lifeguard then swims a distance $x = c\csc\theta$ at speed q.

This requires a time $t_2 = \dfrac{\text{distance swum}}{\text{rate swum}} = \dfrac{c\csc\theta}{q}$

Hence total time $T = t_1 + t_2 = \dfrac{d - c\cdot\cot\theta}{p} + \dfrac{c\csc\theta}{q}$

(C) $T = \dfrac{(380\text{ m}) - (76\text{ m})\cot 51°}{5.1\text{ m/sec}} + \dfrac{(76\text{ m})\csc 51°}{1.7\text{ m/sec}} = 119.97$ sec

(D) T decreases, then increases as θ goes from 55° to 85°. T has a minimum value of 116.66 sec when $\theta = 70°$.

(E) We have already found that the distance run, $LP = d - c\cot\theta$.
Hence, $LP = 380\text{ m} - (76\text{ m})\cot 70° = 352$ m.

35. (A) Since the pipeline costs twice as much in the water as on the land it appears that the total cost of the pipeline would depend on θ.

(B) We note that the pipeline consists of ocean section TP, and shore section $PW = 10\text{ mi} - SP$.

Let $x = TP$ and $y = SP$. In right triangle SPT, $\cos\theta = \dfrac{4\text{ mi}}{x}$, $\tan\theta = \dfrac{y}{4\text{ mi}}$

$x = (4\text{ mi})\sec\theta \qquad y = (4\text{ mi})\tan\theta$

Thus,

C = the cost of the pipeline

$$= \begin{pmatrix}\text{cost of ocean}\\ \text{section per mile}\end{pmatrix}\begin{pmatrix}\text{number of}\\ \text{ocean miles} = x\end{pmatrix} + \begin{pmatrix}\text{cost of shore}\\ \text{section per mile}\end{pmatrix}\begin{pmatrix}\text{number of shore}\\ \text{miles} = 10 - y\end{pmatrix}$$

$$= \left(40{,}000\frac{\$}{\text{mi}}\right)(4\text{ mi})\sec\theta + \left(20{,}000\frac{\$}{\text{mi}}\right)[10 - (4\text{ mi})\tan\theta]$$

$$C = 160{,}000\sec\theta + 20{,}000(10 - 4\tan\theta)$$

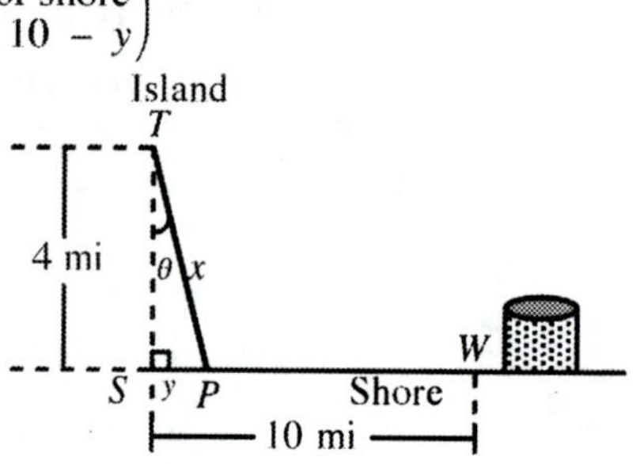

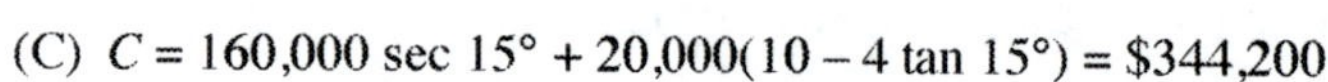

(C) $C = 160{,}000\sec 15° + 20{,}000(10 - 4\tan 15°) = \$344{,}200$

(D) As θ increases from 15° to 45°, C decreases and then increases.
C has the minimum value of \$338,600 when $\theta = 30°$.

(E) From part (A), the shore section $= 10 - y = 10 - (4\text{ mi})\tan\theta = 10 - 4\tan 30° = 7.69$ miles.
The ocean section $= x = (4\text{ mi})\sec\theta = 4\sec 30° = 4.62$ miles

37. A simple way to solve the system of equations

$\tan 42° = \dfrac{y}{x}$ $\qquad \tan 25° = \dfrac{y}{1.0+x}$ for y is to clear of fractions,

then eliminate x from the resulting equivalent system of equations.

$x \tan 42° = y \qquad (1.0 + x)(\tan 25°) = y$

$$x = \frac{y}{\tan 42°}$$

Therefore, $\left(1.0 + \dfrac{y}{\tan 42°}\right)(\tan 25°) = y$

$$\tan 25° + \frac{\tan 25°}{\tan 42°}y = y \quad \text{(Distributive property)}$$

$$\tan 25° = y - \frac{\tan 25°}{\tan 42°}y = \left(1 - \frac{\tan 25°}{\tan 42°}\right)y$$

$$y = \frac{\tan 25°}{1 - \dfrac{\tan 25°}{\tan 42°}} = 0.97 \text{ km}$$

39.

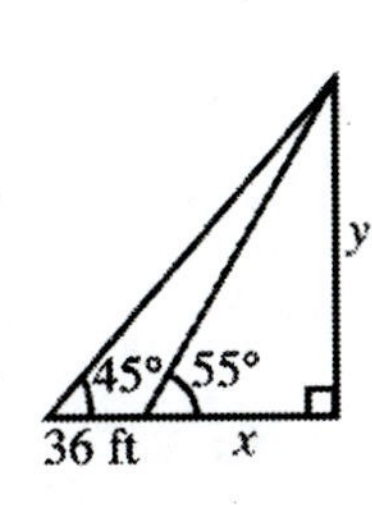

Using the figure, we can write

$$\frac{x}{y} = \cot 55° \qquad \frac{x + 36\text{ ft}}{y} = \cot 45°$$

Then $\dfrac{x + 36\text{ ft}}{y} - \dfrac{x}{y} = \cot 45° - \cot 55°$

$$\frac{36\text{ ft}}{y} = \cot 45° - \cot 55°$$

$$36 \text{ ft} = y(\cot 45° - \cot 55°)$$

$$y = \frac{36\text{ ft}}{\cot 45° - \cot 55°}$$

$$y = 120 \text{ ft}$$

41. Labeling the diagram with the information given in the problem we note: We are asked to find d = how far apart the two buildings are, and h = the height of the apartment building. Note also that $x + h = 847$, so that $x = 847 - h$.

In right triangle SLT, $\tan 43.2° = \dfrac{x}{d} = \dfrac{847 - h}{d}$.

In right triangle SLB, $\tan 51.4° = \dfrac{h}{d}$.

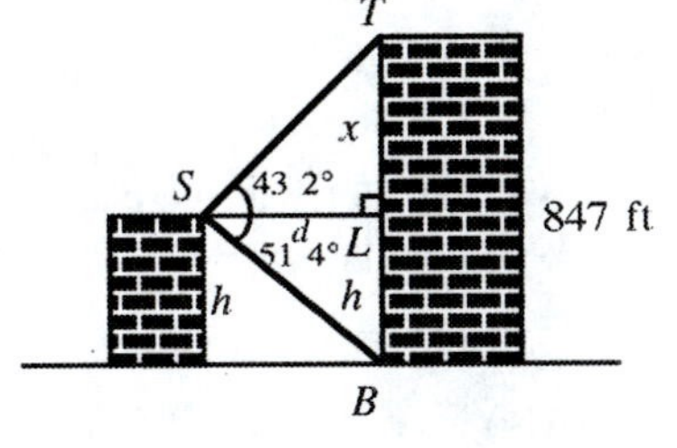

We solve the system of equations $\tan 51.4° = \dfrac{h}{d}$ and $\tan 43.2° = \dfrac{847 - h}{d}$ by clearing of fractions, then eliminating h.

(1) $d \tan 51.4° = h$ and $d \tan 43.2° = 847 - h$

Adding, $\quad d \tan 51.4° + d \tan 43.2° = 847$

$$d(\tan 51.4° + \tan 43.2°) = 847$$

$$d = \frac{847}{\tan 51.4° + \tan 43.2°} = 386 \text{ ft apart}$$

Substituting in (1), $h = d \tan 51.4° = (386 \text{ ft}) \tan 51.4° = 484$ ft high.

43. In the figure, LC represents the lighthouse, FB represents the flagpole, $LC = AB = 175$ ft. Angle ALF is given with measure $40°20'$; Angle ALB with measure $44°40'$.

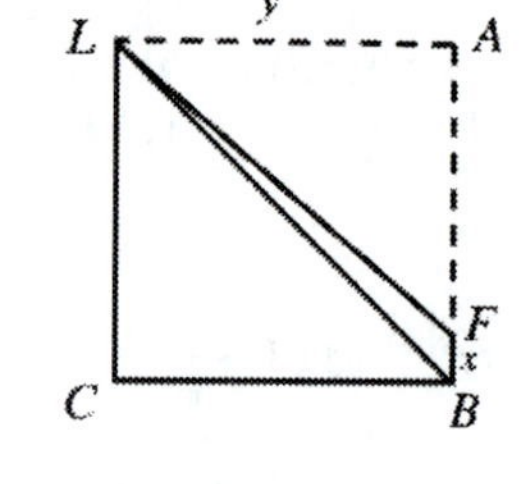

In right triangle LAB, $\tan 44°40' = \dfrac{AB}{LA} = \dfrac{175}{y}$.

Since $AB = 175$, if $FB = x$, then $AF = 175 - x$.

In right triangle LAF, $\tan 40°20' = \dfrac{AF}{LA} = \dfrac{175-x}{y}$.

We solve the system of equations $\tan 40°20' = \dfrac{175-x}{y}$ and $\tan 44°40' = \dfrac{175}{y}$ by clearing of fractions, then eliminating y.

$$y \tan 44°40' = 175 \text{ and } y \tan 40°20' = 175 - x$$

Then $y = \dfrac{175}{\tan 44°40'}$ and $\dfrac{175}{\tan 44°40'} \tan 40°20' = 175 - x$.

Thus, $x = 175 - \dfrac{175}{\tan 44°40'} \tan 40°20' = 25$ ft.

45. We are given $t = 2$ sec, $v = 11.1$ ft/sec, $\theta = 10.0°$. Thus,

$$g = \frac{v}{(\sin\theta)t} = \frac{11.1 \text{ ft/sec}}{(\sin 10.0°)(2 \text{ sec})} = 32.0 \text{ ft/sec}^2$$

47. We first sketch a figure and label the known parts. From geometry we know that each angle of an equilateral triangle has measure 60°.

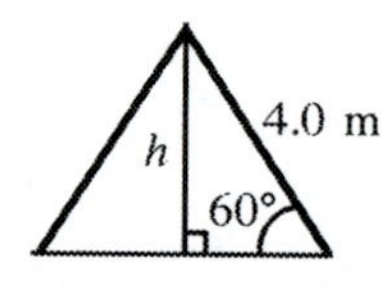

$$\sin 60° = \frac{h}{4.0 \text{ m}}$$

$$h = (4.0 \text{ m})(\sin 60°)$$

$$= 3.5 \text{ m}$$

49. We first sketch a figure and label the known parts. We note that since ABC is an isosceles triangle

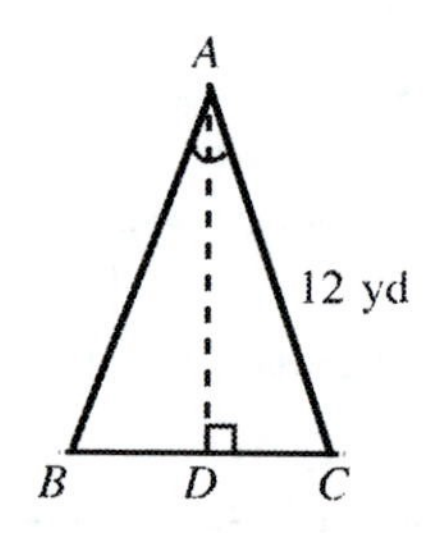

$$\angle DAC = \frac{1}{2}\angle BAC = \frac{1}{2}(48°) = 24°$$

Also base $BC = 2 \cdot DC$

In triangle DAC,

$$\sin DAC = \frac{DC}{12 \text{ yd}}$$

$$DC = (12 \text{ yd})(\sin 24°)$$

Hence the base

$$BC = 2 \cdot DC = 2(12 \text{ yd})(\sin 24°)$$

$$= 9.8 \text{ yd}$$

51. We note that since AB is a side of a nine-sided regular polygon,

$$\angle BCA = \frac{1}{9}(1 \text{ revolution}) = \frac{1}{9}(360°) = 40°.$$

Since ABC is an isosceles triangle,

$$\angle FCA = \frac{1}{2}\angle BCA = \frac{1}{2}(40°) = 20°,$$

and also,

$$AF = \frac{1}{2}AB = \frac{1}{2}x.$$

Therefore, in right triangle AFC, $\sin 20° = \dfrac{(1/2)x}{8.32 \text{ cm}}$

$$x = 2(8.32 \text{ cm})(\sin 20°) = 5.69 \text{ cm}$$

53. From the Pythagorean theorem:

Since $AB = s$ and $MB = \dfrac{s}{2}$

$$AM^2 = AB^2 + MB^2$$

$$= s^2 + \left(\frac{s}{2}\right)^2 = \frac{5s^2}{4}$$

$$AM = s\frac{\sqrt{5}}{2}$$

Since $CN = CM = \dfrac{s}{2}$

$$NM^2 = CN^2 + CM^2$$

$$= \left(\frac{s}{2}\right)^2 + \left(\frac{s}{2}\right)^2 = \frac{2s^2}{4}$$

$$NM = s\frac{\sqrt{2}}{2}$$

From the fact that $NA = MA$, thus triangle AMN is isosceles: AE bisects NM, hence

$$ME = \frac{1}{2}NM = \frac{1}{2} \cdot s\frac{\sqrt{2}}{2} = s\frac{\sqrt{2}}{4}$$

From the Pythagorean theorem, once again:

$$AE^2 + EM^2 = MA^2$$

$$AE^2 + \left(s\frac{\sqrt{2}}{4}\right)^2 = \left(s\frac{\sqrt{5}}{2}\right)^2$$

$$AE^2 + \frac{2s^2}{16} = \frac{5s^2}{4}$$

$$AE^2 = \frac{18s^2}{16}$$

$$AE = s\frac{3\sqrt{2}}{4}$$

From the fact that triangle AMN is isosceles, once again:

$$\angle NMA = \angle MNA = \frac{1}{2}(180° - \theta)$$

Since $\angle NFM = \angle AEN = 90°$ and $\angle NMF = \angle ENA = \dfrac{1}{2}(180° - \theta)$, triangles NMF and ANE are similar. Hence,

$$\frac{NF}{NM} = \frac{AE}{AN} = \frac{AE}{MA} = s\frac{3\sqrt{2}}{4} \div s\frac{\sqrt{5}}{2} = \frac{3\sqrt{2}}{2\sqrt{5}}.$$

Finally, $\sin\theta = \dfrac{NF}{NA} = \dfrac{NF}{NM} \cdot NM \div NA = \dfrac{3\sqrt{2}}{2\sqrt{5}} \cdot s\dfrac{\sqrt{2}}{2} \div s\dfrac{\sqrt{5}}{2} = \dfrac{3\sqrt{2}\sqrt{2}}{4\sqrt{5}} \cdot \dfrac{2}{\sqrt{5}} = \dfrac{3}{5}$

55. Redraw and label the text figure as follows:

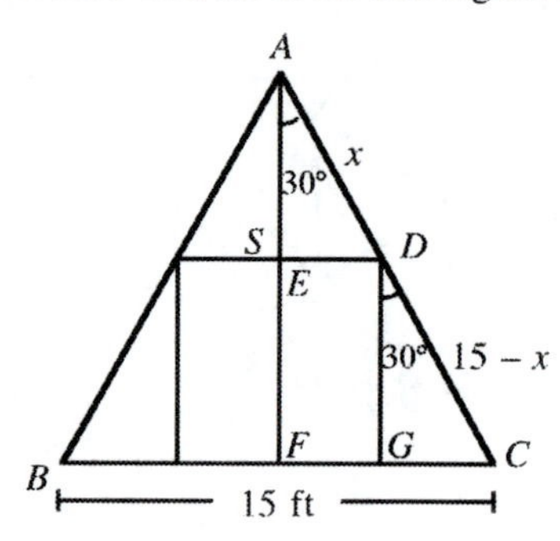

Note: AF bisects angle BAC as well as the square.

$$DE = \frac{1}{2}s \qquad DG = s$$

Let $AD = x$, then $DC = 15 - x$.
In triangle ADE,

$$\sin 30° = \frac{1}{2} = \frac{DE}{x} = \frac{s/2}{x}$$

Thus $x = s$.
In triangle DGC,

$$\cos 30° = \frac{\sqrt{3}}{2} = \frac{DG}{DC} = \frac{s}{15-x} = \frac{s}{15-s}$$

Then $\sqrt{3}(15 - s) = 2s$

$$15\sqrt{3} - \sqrt{3}s = 2s$$

$$15\sqrt{3} = (2 + \sqrt{3})s$$

$$s = \frac{15\sqrt{3}}{2+\sqrt{3}} = 7.0 \text{ ft}$$

57. We reason as in the text, Section 1.1, Example 4.

$$\frac{S}{2\pi r} = \frac{\theta}{360°}$$

$$\frac{2\pi r}{S} = \frac{360°}{\theta}$$

$$r = \frac{360°}{2\pi\theta}S$$

S = approximate diameter of moon = $\frac{1}{3.5}$ (diameter of earth)

$$= \frac{1}{3.5}\,\frac{\text{circumference of earth}}{\pi}$$

$$= \frac{1}{3.5}\,\frac{24{,}000}{\pi}$$

$\theta = 0.5°$

Thus the distance from earth to moon is approximated by

$$r = \frac{360°}{2\pi(0.5)}\,\frac{1}{3.5}\,\frac{24{,}000}{\pi}$$

$$r = 250{,}000 \text{ mi}$$

CHAPTER 1 REVIEW EXERCISE

1. $2°9'54'' = (2 \cdot 60 \cdot 60 + 9 \cdot 60 + 54)'' = 7{,}794''$

2. Since a circumference has degree measure 360, $\frac{1}{8}$ of a circumference has degree measure $\frac{1}{8}$ of 360; that is, $\frac{1}{8}(360) = 45$ degrees, written 45°.

3. Since $\frac{a}{a'} = \frac{c}{c'}$ by Euclid's Theorem, then $\frac{a}{4} = \frac{20{,}000}{5}$; $a = \frac{(4)(20{,}000)}{5} = 16{,}000$.

4. Since $20' = \frac{20°}{60}$, then $36°20' = \left(36 + \frac{20}{60}\right)^{\circ} \approx 36.33°$ to two decimal places.

5. An angle of degree measure 1 is an angle formed by rotating the terminal side of the angle 1/360th of a complete revolution in a counterclockwise direction from the initial side.

6. First find the third angle in each triangle. In the first, $A + B + C = 180°$, thus $A = 180° - B - C = 180° - 120° - 39° = 21°$. In the second, $A + B + C = 180°$, thus $B = 180° - A - C = 180° - 21° - 120° = 39°$. In the third, $A + B + C = 180°$, thus $C = 180° - A - B = 180° - 21° - 39° = 120°$. Hence all three angles are equal in all three triangles. Hence, all three triangles are similar.

7. No. Similar triangles have equal angles.

8. The sum of all of the angles in a triangle is 180°. Two obtuse angles would add up to more than 180°.

9. We draw a figure (the one shown is not drawn to scale) and label the known parts. Since triangles BTA and LMY are similar, we have

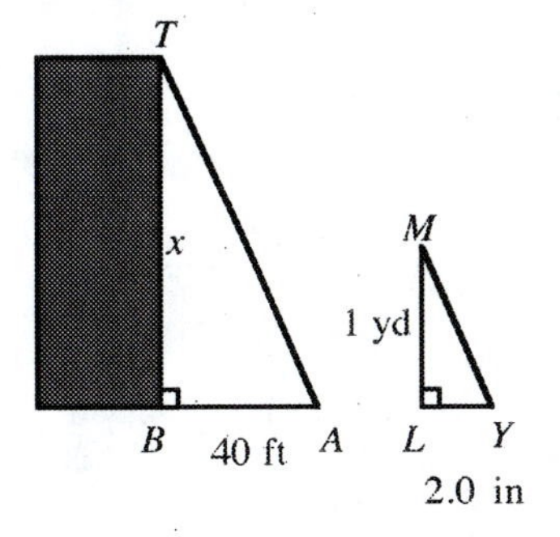

$$\frac{BT}{LM} = \frac{BA}{LY}$$

$$\frac{x}{1 \text{ yd}} = \frac{40 \text{ ft}}{2.0 \text{ in}}$$

$$\frac{x}{36 \text{ in}} = \frac{40 \text{ ft}}{2.0 \text{ in}}$$

$$x = \frac{(40 \text{ ft})(36 \text{ in})}{2.0 \text{ in}}$$

$$= 720 \text{ ft}$$

10. (A) $\frac{b}{c}$ (B) $\frac{c}{a}$ (C) $\frac{b}{a}$ (D) $\frac{c}{b}$ (E) $\frac{a}{c}$ (F) $\frac{a}{b}$

11. *Solve for the complementary angle:* $90° - \theta = 90° - 35.2° = 54.8°$

Solve for a: We choose the cosine to find a. Thus,

$$\cos \theta = \frac{a}{c}$$
$$a = c \cos \theta$$
$$= (20.2 \text{ cm})(\cos 35.2°) = 16.5 \text{ cm}$$

Solve for b: We choose the sine to solve for b. Thus,

$$\sin \theta = \frac{b}{c}$$
$$b = c \sin \theta$$
$$= (20.2 \text{ cm})(\sin 35.2°) = 11.6 \text{ cm}$$

12. Since $\frac{s}{C} = \frac{\theta}{360°}$, then $\frac{8.00 \text{ cm}}{20.0 \text{ cm}} = \frac{\theta}{360°}$ and $\theta = \frac{8.00}{20.0}(360°) = 144°$

13. In 20 minutes the tip of the hand travels through 1/3 of the circumference of a circle of radius 2 in (since it travels through an entire circumference in 60 minutes). Thus, $s = \frac{1}{3}C$ and $C = 2\pi r$.

That is, $s = \frac{1}{3}(2\pi r) = \frac{1}{3}(2\pi)(2 \text{ in}) = \frac{4}{3}\pi \text{ in} \approx 4.19 \text{ in}$.

14. There are two methods.
   a. Convert the first to decimal degree form and compare with the second.
   $27°14' = \left(27 + \frac{14}{60}\right)^\circ \approx 27.23°$
   Since $27.25° > 27.23°$, $27.25°$ is larger.
   b. Convert the second to DMS form and compare with the first.
   $27.25° = 27°(0.25 \times 60°) = 27°15'$
   Since $27°15' > 27°14'$, $27.25°$ is larger.

15. (A) 67°42'31"
   67.709°

   (B) 129.317° → DMS
   129°19'1"

16. (A) 82°14'37" – 16°32'45" → DMS
   65°41'52"

   (B) 3(13°47'18" + 95°28'51") → DMS
   327°48'27"

17. Since $\frac{a}{a'} = \frac{b}{b'}$ by Euclid's Theorem, then

$$\frac{4.1 \times 10^{-6} \text{ mm}}{1.5 \times 10^{-4} \text{ mm}} = \frac{b}{2.6 \times 10^{-4} \text{ mm}}$$

$$b = (2.6 \times 10^{-4} \text{ mm}) \frac{4.1 \times 10^{-6}}{1.5 \times 10^{-4}} = 7.1 \times 10^{-6} \text{ mm}$$

18. (A) $\cos\theta$ (B) $\tan\theta$ (C) $\sin\theta$ (D) $\sec\theta$ (E) $\csc\theta$ (F) $\cot\theta$

19. Two right triangles having an acute angle of one equal to an acute angle of the other are similar, and corresponding sides of similar triangles are proportional. Thus, in the similar triangles,
   $\sin\theta = \frac{\text{Opp}}{\text{Hyp}} = \frac{\text{Opp}'}{\text{Hyp}'}$ are the same quantity.

20. *Solve for the complementary angle:* $90° - \theta = 90° - 62°20' = 27°40'$

   Solve for b: We choose the tangent to find b. Thus, $\tan\theta = \frac{b}{a}$

$$b = a\tan\theta$$
$$= (4.00 \times 10^{-8} \text{ m})(\tan 62°20')$$
$$= 7.63 \times 10^{-8} \text{ m}$$

   Solve for c: We choose the cosine to find c. Thus, $\cos\theta = \frac{a}{c}$

$$c = \frac{a}{\cos\theta}$$
$$= \frac{4.00 \times 10^{-8} \text{ m}}{\cos 62°20'} = 8.61 \times 10^{-8} \text{ m}$$

21. (A) If $\tan\theta = 2.497$, then $\theta = \tan^{-1} 2.497 = 68.17°$.
   (B) $\theta = \arccos 0.3721 = 68.155° = 68°10'$
   (C) $\theta = \sin^{-1} 0.0559 = 3.2045° = 3°12'16"$

22. If a calculator is set in degree mode, then it yields cos(5.47) = 0.9954... Therefore, display (b) is in degree mode and display (a) must be in radian mode.

23. *Solve for θ:* We choose the tangent to solve for θ. Thus, $\tan\theta = \dfrac{b}{a} = \dfrac{13.3\text{ mm}}{15.7\text{ mm}} = 0.8471$

$$\theta = \tan^{-1} 0.8471 = 40.3°$$

Solve for the complementary angle: $90° - \theta = 90° - 40.3° = 49.7°$

Solve for c: We choose the sine to solve for c. Thus, $\sin\theta = \dfrac{b}{c}$

$$c = \frac{b}{\sin\theta} = \frac{13.3\text{ mm}}{\sin 40.3} = 20.6\text{ mm}$$

24. $40.3° = 40°(0.3 \times 60)' = 40°20'$ (to nearest 10'); $90° - \theta = 90° - 40°20' = 49°40'$

25. We first sketch a figure and label the known parts. From geometry we know that each angle of an equilateral triangle has measure 60°.

$$\sin 60° = \frac{h}{10\text{ ft}}$$

$$h = (10\text{ ft})(\sin 60°) = 8.7\text{ ft}$$

26. Since $\dfrac{s}{C} = \dfrac{\theta}{360°}$ and $C = 2\pi r$, then

$$\frac{s}{2\pi r} = \frac{\theta}{360°}$$

$$\frac{s}{2(\pi)(1500\text{ ft})} \approx \frac{36°}{360°}$$

$$s \approx \frac{2(\pi)(1500\text{ ft})(36)}{360} \approx 940\text{ ft}$$

27. Since $\dfrac{A}{\pi r^2} = \dfrac{\theta}{360°}$, then $\dfrac{A}{\pi(18.3\text{ ft})^2} = \dfrac{36.5°}{360°}$

$$A = \frac{36.5}{360}\pi(18.3\text{ ft})^2 = 107\text{ ft}^2$$

28. *Solve for θ:* $\theta = 90° - (90° - \theta) = 90° - 23°43' = 66°17'$

Solve for a: We choose the cosine to find a. Thus, $\cos\theta = \dfrac{a}{c}$

$$a = c\cos\theta$$
$$= (232.6\text{ km})(\cos 66°17') = 93.56\text{ km}$$

Solve for b: We choose the sine to find b. Thus, $\sin\theta = \dfrac{b}{c}$

$$b = c\sin\theta$$
$$= (232.6\text{ km})(\sin 66°17') = 213.0\text{ km}$$

29. *Solve for θ:* We choose the cosine to find θ. Thus, $\cos\theta = \dfrac{a}{c} = \dfrac{2{,}421\text{ m}}{4{,}883\text{ m}} = 0.4958$, so $\theta = \cos^{-1} 0.4958 = 60.28°$.

Solve for the complementary angle: $90° - \theta = 90° - 60.28° = 29.72°$

Solve for b: We choose the sine to find b. Thus, $\sin\theta = \dfrac{b}{c}$

$$b = c\sin\theta$$
$$= (4{,}883\text{ m})(\sin 60.28°) = 4{,}241\text{ m}$$

30. Use the reciprocal relationship $\csc\theta = 1/\sin\theta$. Set calculator in degree mode, use sin key, then take reciprocal. $\csc 67.1357° \approx 1.0853$

31. We note: triangles PFH and PBT are similar, hence

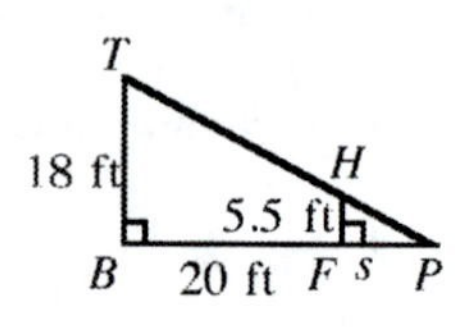

$$\frac{PF}{FH} = \frac{PB}{BT}$$

$$\frac{s}{5.5\text{ ft}} = \frac{20\text{ ft} + s}{18\text{ ft}}$$

$$5.5(18\text{ ft}) \cdot \frac{s}{5.5\text{ ft}} = 5.5(18\text{ ft}) \cdot \frac{20\text{ ft} + s}{18\text{ ft}}$$

$$18\,s = 5.5(20\text{ ft} + s) = 110\text{ ft} + 5.5\,s$$

$$12.5\,s = 110\text{ ft}$$

$$s = \frac{110\text{ ft}}{12.5} = 8.8\text{ ft}$$

32. We first sketch a figure and label the known parts. Since triangles ACB and DCE are similar, corresponding sides are proportional and we can write

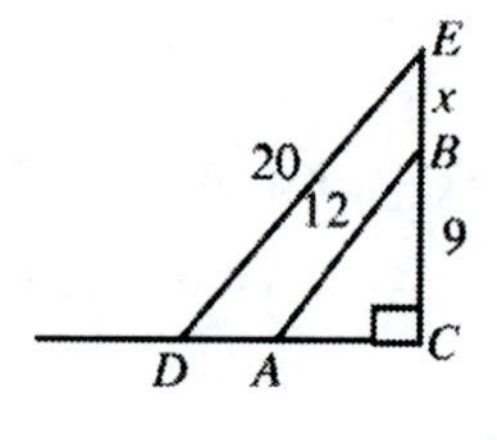

$$\frac{BC}{AB} = \frac{EC}{DE}$$

$$\frac{9}{12} = \frac{x+9}{20}$$

$$x + 9 = \frac{9}{12}(20)$$

$$x + 9 = 15$$

$$x = 6 \text{ feet higher}$$

33. We first sketch a figure and label the known parts.

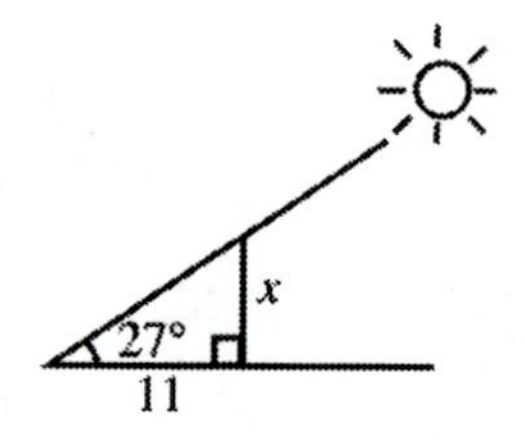

Drawing not to scale

$$\tan 27° = \frac{\text{Opp}}{\text{Adj}} = \frac{x}{11\text{ ft}}$$

$$x = (11\text{ ft})\tan 27° = 5.6\text{ ft}$$

34. Using the figure, we can write

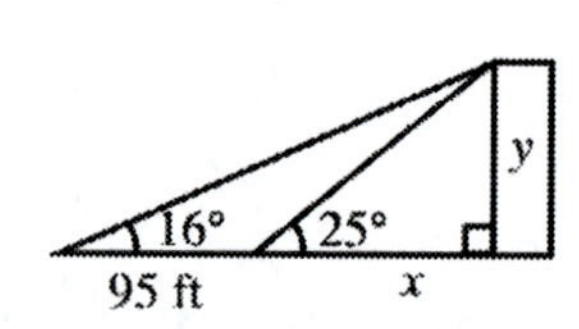

$$\frac{x}{y} = \cot 25° \qquad \frac{x + 95\text{ft}}{y} = \cot 16°$$

Then

$$\frac{x + 95\text{ft}}{y} - \frac{x}{y} = \cot 16° - \cot 25°$$

$$\frac{95\text{ft}}{y} = \cot 16° - \cot 25°$$

$$95\text{ ft} = y(\cot 16° - \cot 25°)$$

$$y = \frac{95\text{ft}}{\cot 16° - \cot 25°}$$

$$y = 71\text{ ft}$$

35. In right triangle ABC, the length of the ramp = AB.

$$\sin A = \frac{BC}{AB}$$

$$AB = \frac{BC}{\sin A}$$

$$= \frac{4.25 \text{ ft}}{\sin 10.0°} = 24.5 \text{ ft}$$

The distance of the end of the ramp from the porch = AC.

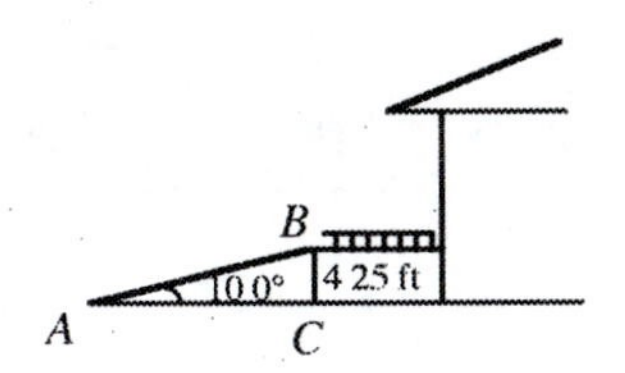

$$\tan A = \frac{BC}{AC}$$

$$AC = \frac{BC}{\tan A}$$

$$= \frac{4.25 \text{ ft}}{\tan 10.0°} = 24.1 \text{ ft}$$

36. From the figure it is clear that $\tan \theta = \frac{a}{b}$. Given: percentage of inclination $\frac{a}{b} = 4\% = 0.04$, then $\tan \theta = 0.04$; $\theta = 2.3°$

Given: angle of inclination $\theta = 4°$, then $\frac{a}{b} = \tan 4° = 0.07$ or 7%

37. Since $\frac{s}{c} = \frac{\theta}{360°}$ and $C = 2\pi r$, then $\frac{s}{2\pi r} = \frac{\theta}{360°}$; $s = 2\pi r \cdot \frac{\theta}{360°}$. Since the cities have the same longitude, θ is given by their difference in latitude.

$$\theta = 44°31' - 30°42' = 13°49' = \left(13 + \frac{49}{60}\right)°.$$

Thus, we have

$$s \approx 2(\pi)(3960)\frac{13 + \frac{49}{60}}{360} \approx 955 \text{ miles}$$

38. We first sketch a figure and label the known parts. From the figure we note:

$$h + 1400 \text{ ft} = 2800 \text{ ft}$$

$$h = 1400 \text{ ft}$$

$$\tan 64° = \frac{2800 \text{ ft}}{g}$$

$$\tan \theta = \frac{h}{g} = \frac{1400 \text{ ft}}{g}$$

Thus, $$g = \frac{2800 \text{ ft}}{\tan 64°}$$

and $$\tan \theta = \frac{1400 \text{ ft}}{\left(\frac{2800 \text{ ft}}{\tan 64}\right)}.$$

Then, $$\tan \theta = \frac{1400 \tan 64°}{2800}$$

$$= \frac{1}{2}\tan 64° = 1.025$$

$$\theta = 46°$$

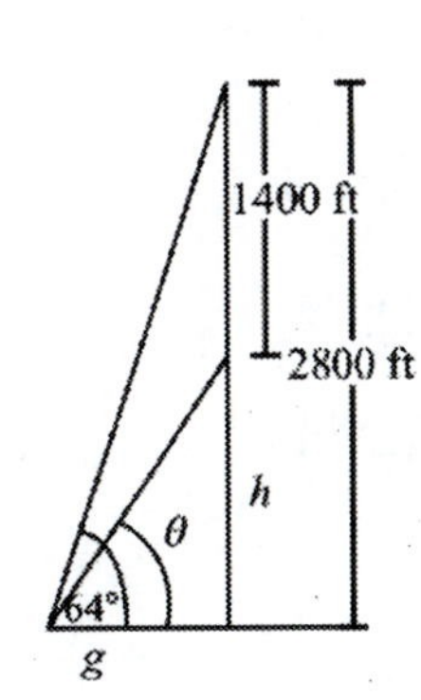

39. 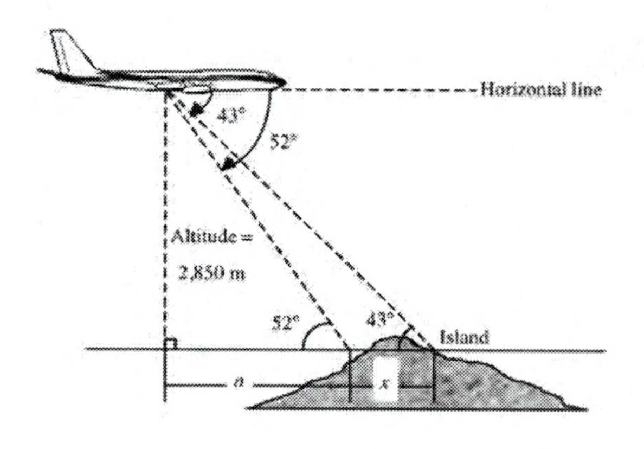

We note: $\cot 52° = \dfrac{a}{2{,}850\text{ m}}$, $\cot 43° = \dfrac{a+x}{2{,}850\text{ m}}$.

Then $\cot 43° - \cot 52° = \dfrac{a+x}{2{,}850\text{ m}} - \dfrac{a}{2{,}850\text{ m}}$

$= \dfrac{a+x-a}{2{,}850\text{ m}} = \dfrac{x}{2{,}850\text{ m}}$

$x = 2{,}850\text{ m}(\cot 43° - \cot 52°) = 830\text{ m}$

40. We note:

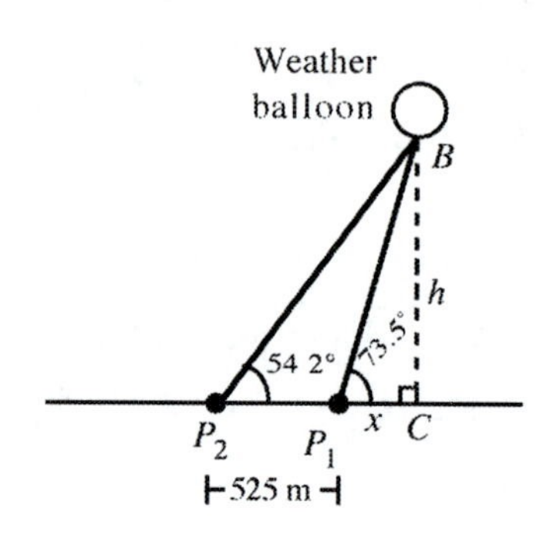

In right triangle BCP_1, $\cot 73.5° = \dfrac{x}{h}$

In right triangle BCP_2, $\cot 54.2° = \dfrac{x+525\text{ m}}{h}$

Then, $\cot 54.2° - \cot 73.5° = \dfrac{x+525\text{ m}}{h} - \dfrac{x}{h}$

$= \dfrac{x+525\text{ m}-x}{h}$

$= \dfrac{525\text{ m}}{h}$

$$h = \frac{525\text{ m}}{\cot 54.2° - \cot 73.5°} = 1{,}240\text{ m}$$

41. We use the figure and the notation of Problem 31, Ex. 1-4.

r = radius of the parallel of latitude
R = radius of the earth
θ = latitude

$\cos\theta = \dfrac{r}{R}$, $r = R\cos\theta$

L = length of the parallel of latitude
$L = 2\pi r$
$L = 2\pi R\cos\theta$

To keep the sun in the same position, the plane must fly at a rate v sufficient to fly a distance L in 24 hours, thus

$$v = \frac{L}{24} = \frac{2\pi R\cos\theta}{24}$$

$$v \approx \frac{2\pi(3960\text{ mi})\cos 42°50'}{24\text{ hr}} \approx 760\text{ mi/hr}$$

42. (A) Since α and β are complementary (see diagram) $\beta = 90° - \alpha$.

(B) In the right triangle containing r, h, and α, $\tan\alpha = \dfrac{r}{h}$. Thus, $r = h\tan\alpha$.

(C) Using similar triangles, we can write $\dfrac{H}{h} = \dfrac{R}{r}$. Then

$$H = \frac{R}{r}h,$$

$$H - h = \frac{R}{r}h - h = h\left(\frac{R}{r} - 1\right) = h\left(\frac{R-r}{r}\right) = \frac{h}{r}(R-r)$$

Since $\dfrac{r}{h} = \tan\alpha$, $\dfrac{h}{r} = 1 \div \dfrac{r}{h} = 1 \div \tan\alpha = \cot\alpha$. Hence, we can write $H - h = (R - r)\cot\alpha$.

43.

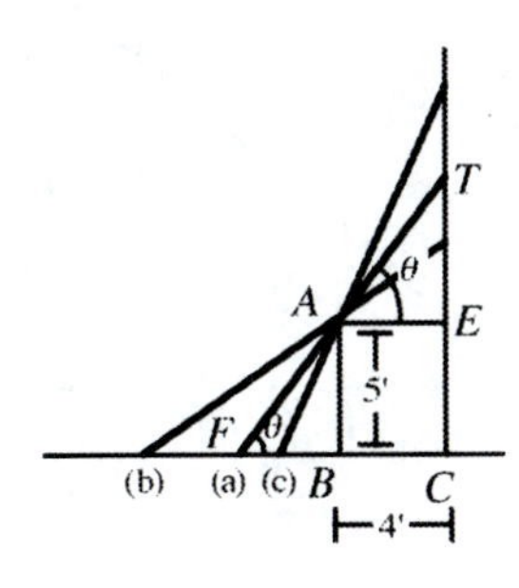

(A) See the figure. If the shortest ladder is represented in position (a), then as its foot moves away from the fence (position (b)) the ladder must get longer. And, as its foot moves toward the fence (position (c)) the ladder must also get longer.

(B) The length of the ladder is given by $FA + AT$. In triangle FBA, $\sin\theta = \dfrac{5}{FA}$,

thus $FA = \dfrac{5}{\sin\theta} = 5\csc\theta$. In triangle AET, AE is drawn parallel to BC, thus, $AE = BC = 4$.

Then, since triangle AET is similar to triangle FBA, $\cos\theta = \dfrac{4}{AT}$, thus $AT = \dfrac{4}{\cos\theta} = 4\sec\theta$.

Thus $\ell = FA + AT = 5\csc\theta + 4\sec\theta$.

(C)

θ	25	35	45	55	65	75	85
L	16.24	13.60	12.73	13.08	14.98	20.63	50.91

(D) L decreases and then increases. L in Table 1 has a minimum value of 12.73 when $\theta = 45°$.

(E) Make up another table for values of θ close to 45° and on either side of 45°.

44. Label the figure in the text as shown at the right.
Let $CD = h$, $AD = x$, then $DB = 200 - x$.

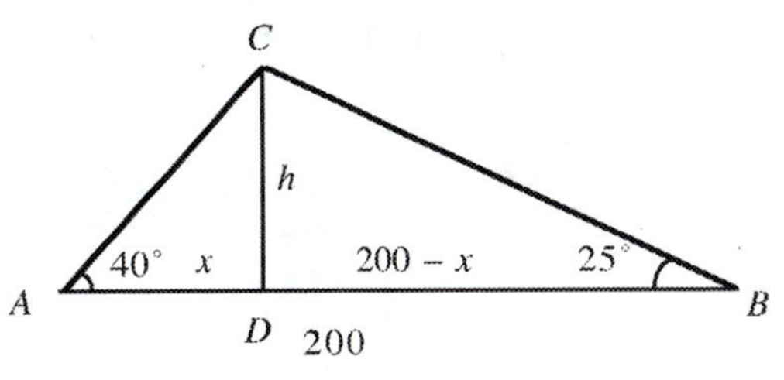

In triangle ADC,

$$\frac{x}{h} = \cot 40°$$

In triangle BDC,

$$\frac{200 - x}{h} = \cot 25°$$

Adding these, we obtain:

$$\frac{x}{h} + \frac{200 - x}{h} = \cot 40° + \cot 25°$$

$$\frac{200}{h} = \cot 40° + \cot 25°$$

$$h = \frac{200}{\cot 40° + \cot 25°}$$

Then $\dfrac{h}{AC} = \sin 40°$, so $AC = \dfrac{h}{\sin 40°}$

$\dfrac{h}{BC} = \sin 25°$, so $BC = \dfrac{h}{\sin 25°}$

Finally, $AC = \dfrac{h}{\sin 40°} = \dfrac{200}{\sin 40°(\cot 40° + \cot 25°)} = 93 \text{ ft}$

$BC = \dfrac{h}{\sin 25°} = \dfrac{200}{\sin 25°(\cot 40° + \cot 25°)} = 142 \text{ ft}$

45. Subdivide and label the figure in the text as shown at the right. Since the angles of a quadrilateral $BCDE$ have a sum of 360°,

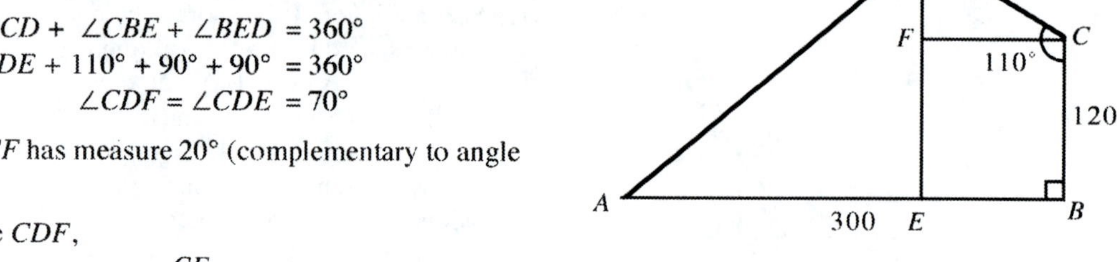

$$\angle CDE + \angle BCD + \angle CBE + \angle BED = 360°$$
$$\angle CDE + 110° + 90° + 90° = 360°$$
$$\angle CDF = \angle CDE = 70°$$

So the angle DCF has measure 20° (complementary to angle CDF).

Then, in triangle CDF,

$\dfrac{DF}{100} = \cos 70°$, $\quad \dfrac{CF}{100} = \sin 70°$

$DF = 100 \cos 70°$ $\quad CF = 100 \sin 70° = BE$

The original quadrilateral can be viewed as consisting of the triangle ADE plus the trapezoid $BCDE$. The area of the triangle is given by

$$A_1 = \frac{1}{2}bh = \frac{1}{2}AE \cdot DE = \frac{1}{2}(300 - 100 \sin 70°)(120 + 100 \cos 70°)$$
$$= 15{,}885 \text{ sq. ft.}$$

The area of the trapezoid is given by

$$A_2 = \frac{1}{2}h(b_1 + b_2) = \frac{1}{2}CF(BC + DE)$$
$$= \frac{1}{2}(100 \sin 70°)(120 + 120 + 100 \cos 70°)$$
$$= 12{,}883 \text{ sq. ft.}$$

Adding, we obtain

$$A_1 + A_2 = 15{,}885 + 12{,}883 = 28{,}768 \text{ sq. ft.}$$

Chapter 2 Trigonometric Functions

EXERCISE 2.1 Degrees and Radians

1. An angle has radian measure 1 if, when placed with vertex at the center of any circle, it subtends an arc of length equal to the radius of the circle.

3. An angle is in standard position if its vertex is at the origin of a rectangular coordinate system and its initial side is along the positive x axis.

5. $180°/57.3° \approx 180°/60° = 3$

7. $120°/57.3° \approx 120°/60° = 2$

9. $350°/57.3° \approx 350°/60° \approx 6$

11. $(-60°)/57.3° \approx (-60°)/60° = -1$

13. $1 \text{ rad} \approx 57.3° \approx 60°$

15. $-2 \text{ rad} \approx -2(57.3°) \approx -110°$

17. $3 \text{ rad} \approx 3(57.3°) \approx 170°$

19. $\pi/2 \text{ rad} = 180°/2 = 90°$

21. $30°$ is $\frac{1}{6}$ of $180°$, so corresponds to $\frac{1}{6}\pi$ radians.
$60°$ is $\frac{1}{3}$ of $180°$, so corresponds to $\frac{1}{3}\pi$ radians.
$90°$ is $\frac{1}{2}$ of $180°$, so corresponds to $\frac{1}{2}\pi$ radians.

23. $225°$ is $\frac{5}{4}$ of $180°$, so corresponds to $\frac{5}{4}\pi$ radians.
$270°$ is $\frac{3}{2}$ of $180°$, so corresponds to $\frac{3}{2}\pi$ radians.
$315°$ is $\frac{7}{4}$ of $180°$, so corresponds to $\frac{7}{4}\pi$ radians.

25. π radians corresponds to $180°$, so $\frac{\pi}{2}$ radians corresponds to $180°/2$ or $90°$, and $\frac{3\pi}{2}$ radians corresponds to $\frac{3}{2}(180°)$ or $270°$.

27. π radians corresponds to $180°$, so $\frac{2\pi}{3}$ radians corresponds to $\frac{2}{3}(180°)$ or $120°$, $\frac{4\pi}{3}$ radians corresponds to $\frac{4}{3}(180°) = 240°$, and 2π radians corresponds to $2(180°)$ or $360°$.

29. Angles coterminal with $90°$ have measures $90° + 360°$ or $450°$ and $90° - 360°$ or $-270°$.

31. Angles coterminal with $120°$ have measures $120° + 360°$ or $480°$ and $120° - 360°$ or $-240°$.

33. Angles coterminal with $\frac{\pi}{4}$ rad have measures $\frac{\pi}{4} + 2\pi$ or $\frac{9\pi}{4}$ rad and $\frac{\pi}{4} - 2\pi$ or $-\frac{7\pi}{4}$ rad.

35. Angles coterminal with $-\frac{5\pi}{6}$ rad have measures $-\frac{5\pi}{6} + 2\pi$ or $\frac{7\pi}{6}$ rad and $-\frac{5\pi}{6} - 2\pi$ or $-\frac{17\pi}{6}$ rad.

37. $\theta_r = \frac{\pi}{180°}\theta_d$
$= \frac{\pi}{180°}3°$
$= \frac{\pi}{60}$ Exact form
≈ 0.05236 rad to four significant digits

39. $\theta_r = \frac{\pi}{180°}\theta_d$
$= \frac{\pi}{180°}110°$
$= \frac{11\pi}{18}$ Exact form
≈ 1.920 rad to four significant digits

41. $\theta_r = \frac{\pi}{180°}\theta_d$
$= \frac{\pi}{180°}36°$
$= \frac{\pi}{5}$ Exact form
≈ 0.6283 rad to four significant digits

43. $\theta_d = \frac{180°}{\pi \text{ rad}}\theta_r$
$= \frac{180°}{\pi}2$
$= \frac{360°}{\pi}$ Exact form
$\approx 114.6°$ to four significant digits

45. $\theta_d = \frac{180°}{\pi \text{ rad}}\theta_r$
$= \frac{180°}{\pi}(0.6)$
$= \frac{108°}{\pi}$ Exact form
$\approx 34.38°$ to four significant digits

47. $\theta_d = \frac{180°}{\pi \text{ rad}}\theta_r$
$= \frac{180°}{\pi} \cdot \frac{2\pi}{7}$
$= \frac{360°}{7}$ Exact form
$\approx 51.43°$ to four significant digits

49. Since $\theta = \frac{s}{r}$ in radian measure, we have

(A) $\theta_r = \frac{2\text{ m}}{5.0\text{ m}} = 0.4$ rad. Then $\theta_d = \frac{180°}{\pi \text{ rad}}\theta_r = \frac{180°}{\pi}(0.4) \approx 22.9°$

(B) $\theta_r = \frac{6\text{ m}}{5.0\text{ m}} = 1.2$ rad. Then $\theta_d = \frac{180°}{\pi \text{ rad}}\theta_r = \frac{180°}{\pi}(1.2) \approx 68.8°$

(C) $\theta_r = \frac{12.5\text{ m}}{5.0\text{ m}} = 2.5$ rad. Then $\theta_d = \frac{180°}{\pi \text{ rad}}\theta_r = \frac{180°}{\pi}(2.5) \approx 143.2°$

(D) $\theta_r = \frac{20\text{ m}}{5.0\text{ m}} = 4$ rad. Then $\theta_d = \frac{180°}{\pi \text{ rad}}\theta_r = \frac{180°}{\pi}(4) \approx 229.2°$

51. (A) $s = r\theta$
$= 7(1.35) = 9.45$ m

(B) $s = \frac{\pi}{180}r\theta$
$= \frac{\pi}{180}(7)(42.0) \approx 5.13$ m

(C) $s = r\theta$
$= 7(0.653) \approx 4.57$ m

(D) $s = \frac{\pi}{180}r\theta$
$= \frac{\pi}{180}(7)(125) \approx 15.27$ m

53. (A) $A = \frac{1}{2}r^2\theta$

$= \frac{1}{2}(12.0)^2(2.00) \approx 144 \text{ cm}^2$

(B) $A = \frac{\pi}{360}r^2\theta$

$= \frac{\pi}{360}(12.0)^2(25.0) = 31.4 \text{ cm}^2$

(C) $A = \frac{1}{2}r^2\theta$

$= \frac{1}{2}(12.0)^2(0.650) \approx 46.8 \text{ cm}^2$

(D) $A = \frac{\pi}{360}r^2\theta$

$= \frac{\pi}{360}(12.0)^2(105) \approx 131.9 \text{ cm}^2$

55. 495° is coterminal with (495 – 360)° or 135°. Since 135° is between 90° and 180°, its terminal side lies in quadrant II.

57. $-\frac{17\pi}{6}$ is coterminal with $-\frac{17\pi}{6} + 2\pi = -\frac{5\pi}{6}$. Since $-\frac{5\pi}{6}$ is between $-\frac{\pi}{2}$ and $-\pi$, its terminal side lies in quadrant III.

59. $\theta_r = \frac{\pi \text{ rad}}{180°}(\theta_d) = \frac{\pi}{180}\left(87 + \frac{39}{60} + \frac{42}{3{,}600}\right) \approx 1.530 \text{ rad}$

61. $\theta_d = \frac{180°}{\pi \text{ rad}}(\theta_r) = \frac{180}{\pi} \cdot \frac{19\pi}{7} \approx 488.5714°$

63. We use $s = r\theta$ to find the central angle that corresponds to the inscribed angle.

$2 = 3\theta$

$\theta = \frac{2}{3} \text{ rad}$

Then the inscribed angle has measure $\frac{\theta}{2} = \frac{1}{2} \cdot \frac{2}{3} \text{ rad} = \frac{1}{3} \text{ rad}$.

65. We use $s = \frac{\pi}{180}r\theta$ to find the central angle that corresponds to the inscribed angle.

$13 = \frac{\pi}{180} \cdot 6\theta$

$\theta = \frac{390°}{\pi}$

Then the inscribed angle has measure $\frac{\theta}{2} = \frac{1}{2} \cdot \frac{390°}{\pi} = \frac{195°}{\pi} \approx 62°$.

67. An inscribed angle of 3° corresponds to a central angle with $\frac{\theta}{2} = 3°$ or $\theta = 6°$. Then

$s = \frac{\pi}{180}r\theta$

$s = \frac{\pi}{180}(85)(6)$

$s = 8.9 \text{ km}$

69. An inscribed angle of measure $\frac{\pi}{8}$ corresponds to a central angle with $\frac{\theta}{2} = \frac{\pi}{8}$ or $\theta = \frac{\pi}{4}$ rad. Then

$$\begin{aligned} s &= r\theta \\ 42 &= r \cdot \frac{\pi}{4} \\ r &= \frac{168}{\pi} \\ r &= 53 \text{ cm} \end{aligned}$$

71. At 2:30, the minute hand has moved $\frac{1}{2}$ of a circumference from its position at the top of the clock. The hour hand has moved $2\frac{1}{2}$ twelfths of a circumference from the same position. Therefore, they form an angle of

$$\frac{1}{2}C - \frac{2\frac{1}{2}}{12}C \text{ radians,}$$

where C = a total circumference or 2π radians. Thus, the desired angle is

$$\frac{1}{2}(2\pi) - \frac{2\frac{1}{2}}{12}(2\pi) = \pi - \frac{5\pi}{12} = \frac{7\pi}{12} \text{ rad} \approx 1.83 \text{ rad}$$

73. We are to find the arc length subtended by a central angle of 32°, in a circle of radius 22 cm.

$$s = \frac{\pi}{180} r\theta = \frac{\pi}{180}(22)(32) = 12 \text{ cm}$$

75. We are to find the angle, in degrees, that subtends an arc of 24 inches, in a circle of radius 72 inches.

$$s = \frac{\pi}{180} r\theta; \quad \theta = \frac{180s}{\pi r} = \frac{180(24)}{\pi(72)} = 19°$$

77. Since $s = r\theta$ we have

$$s = (54.3 \text{ cm})\frac{3\pi}{2} \approx 256 \text{ cm}$$

79. Since $s = r\theta$, we have

$$\begin{aligned} \text{diameter} &\approx s = (1.5 \times 10^8 \text{ km})(9.3 \times 10^{-3} \text{ rad}) \\ &\approx 1.4 \times 10^6 \text{ km} \end{aligned}$$

81. Since $s = \frac{\pi}{180} r\theta$, we have

$$\text{width of field} \approx s = \frac{\pi}{180}(1{,}250 \text{ ft})(8) = 175 \text{ ft}$$

83. $\theta = 5 \times 10^{-7}$ rad, $r = 250$ mi, s

Since $s = r\theta$, we have

width of object $\approx s = (250 \text{ mi})(5 \times 10^{-7} \text{ rad}) = 1.25 \times 10^{-4}$ mi

1.25×10^{-4} mi $= (1.25 \times 10^{-4} \text{ mi})(1{,}609 \text{ m/mi}) \approx 0.2$ m

1.25×10^{-4} mi $= (0.2 \text{ m})(39.37 \text{ in}) \approx 7.9$ in

85. Assuming that an angle corresponding to an entire circumference is swept out in 1 year (52 weeks), and that the amount swept out in one week is proportional to the time, we can write

$$\frac{\text{angle}}{\text{time}} = \frac{\text{angle}}{\text{time}}$$
$$\frac{\theta}{1} = \frac{2\pi}{52}$$
$$\theta = \frac{2\pi}{52} = \frac{\pi}{26}\text{ rad} \approx 0.12 \text{ rad}$$

87. We use the proportion $\dfrac{\text{error in distance}}{\text{error in time}} = \dfrac{\text{actual distance}}{\text{actual time}}$. Let x = error in distance. The actual time,

$$1 \text{ year } = (365 \text{ days})\left(24\frac{\text{hours}}{\text{day}}\right)\left(3{,}600\frac{\text{seconds}}{\text{hour}}\right).$$

Thus,
$$\frac{x}{365 \text{ seconds}} = \frac{2\pi r}{365 \cdot 24 \cdot 3{,}600 \text{ seconds}}$$

$$x = 365 \cdot \frac{2\pi(9.3 \times 10^7 \text{ miles})}{365 \cdot 24 \cdot 3{,}600}$$

$$= \frac{2\pi(9.3 \times 10^7 \text{ miles})}{24 \cdot 3{,}600}$$

$$= 6{,}800 \text{ miles}$$

89. Since $A = \frac{1}{2}r^2\theta$ and $P = s + 2r = r\theta + 2r$, we can eliminate θ between the two equations and write $2A = r^2\theta$, $\theta = \dfrac{2A}{r^2}$.

$$P = r\left(\frac{2A}{r^2}\right) + 2r = \frac{2A}{r} + 2r .$$

Thus, $P = \dfrac{2(52.39)}{10.5} + 2(10.5) \approx 31$ ft

91. (A) Since one revolution corresponds to 2π radians, n revolutions corresponds to $2\pi n$ radians.
(B) No. Radian measure is independent of the size of the circle used; hence, the size of the wheel used.
(C) Since there are $2\pi n$ radians in n revolutions, in 5 revolutions there are $2\pi(5) = 10\pi \approx 31.42$ rad, and in 3.6 revolutions there are $2\pi(3.6) = 7.2\pi \approx 22.62$ rad.

93. Since the two wheels are coupled together, the distance (arc length) that the drive wheel turns is equal to the distance that the shaft turns. Thus,

$$s = r_1\theta_1 \qquad s = r_2\theta_2$$
$$r_1\theta_1 = r_2\theta_2$$
$$\theta_1 = \frac{r_2}{r_1}\theta_2 = \frac{26}{12}(3 \text{ revolutions})$$
$$= 6.5 \text{ revolutions}$$

In Problem 91, we noted that there are $2\pi n$ rad in n revolutions, hence there are $2\pi(6.5) \approx 40.8$ rad in 6.5 revolutions.

95. Since $s = \frac{\pi}{180} r\theta$, we have $\frac{s}{r} = \frac{\pi}{180}\theta$, $\theta = \frac{\pi}{180}\left(\frac{s}{r}\right)$. Here,

$$r = \frac{1}{2}(32) = 16 \text{ in and } s = 20 \text{ ft} = 20 \text{ ft}\left(12\frac{\text{in}}{\text{ft}}\right) = 240 \text{ in}$$

Thus, $\theta = \frac{180°}{\pi}\left(\frac{240}{16}\right) = 859°$.

97. The largest distance will be traveled when the largest front gear (48 teeth) engages the smallest rear gear (11 teeth). Then, one revolution of the pedals will cause $\frac{48}{11}$ of a revolution of the rear wheel. One revolution of the rear wheel is $\pi d = \pi \cdot 28$ in traveled. Therefore, one revolution of the pedals will cause $\frac{48}{11} \cdot \pi \cdot 28 \text{ in} = 384 \text{ in}$ of travel.

EXERCISE 2.2 Linear and Angular Velocity

1. The linear velocity of a point moving on the circumference of a circle is the distance it travels per unit of time.

3. This represents angle per unit of time, that is, angular velocity.

5. $V = r\omega = 12(0.7) = 8.4$ cm/min

7. $V = r\omega = 1.2(200) = 240$ ft/hr

9. $\omega = \frac{V}{r} = \frac{1{,}950}{250} = 7.8$ rad/hr

11. $\omega = \frac{V}{r} = \frac{210}{98} \approx 2.14$ rad/min

13. $\omega = \frac{\theta}{t} = \frac{3\pi}{0.056} \approx 168$ rad/hr

15. $\omega = \frac{\theta}{t} = \frac{9.62}{1.53} \approx 6.29$ rad/min

17. The mean solar day is the time, 24 hr, for the earth to complete one rotation relative to the sun.

19. 1,500 revolutions per second = $1{,}500 \cdot 2\pi$ rad/sec = $3{,}000\pi$ rad/sec

$$V = r\omega = \left(\frac{1}{2}\cdot 16\right)(3{,}000\pi) \text{ mm/sec} = 24{,}000\pi \text{ mm/sec}$$

$$= 24{,}000\pi \frac{\text{mm}}{\text{sec}} \cdot \frac{1}{1000}\frac{\text{m}}{\text{mm}} = 75\frac{\text{m}}{\text{sec}}$$

21. $\omega = \frac{V}{r} = \frac{20{,}000 \text{ mph}}{4{,}300 \text{ mi}} = 4.65$ rad/hr

23. $\omega = \frac{V}{r} = \frac{335.3 \text{ m/sec}}{\frac{1}{2}(3.000 \text{ m})} = 223.5 \text{ rad/sec} = (223.5 \text{ rad/sec})\left(\frac{1}{2\pi}\frac{\text{revolution}}{\text{radian}}\right) = 35.6$ rev/sec

25. The earth travels 1 revolution, or 2π radian, in 1 year, or $24 \cdot 365$ hours. Thus,

$$\omega = \frac{\theta}{t} = \frac{2\pi}{24 \cdot 365} = \frac{\pi}{4{,}380}\frac{\text{rad}}{\text{hr}}$$

$$V = r\omega = \left(\frac{\pi}{4{,}380}\frac{\text{rad}}{\text{hr}}\right)(9.3 \times 10^7 \text{ mi})$$

$$= 6.67 \times 10^4 \text{ mi/hr or } 66{,}700 \text{ mi/hr}$$

27. (A) Jupiter travels 1 revolution, or 2π radian, in 9 hr 55 min, or $\left(9+\frac{55}{60}\right)$hr. Thus,

$$\omega = \frac{\theta}{t} = \frac{2\pi}{9+\frac{55}{60}} = 0.634\frac{\text{rad}}{\text{hr}}$$

(B) Note that $r = \frac{1}{2} \times$ diameter. Thus,

$$V = r\omega = \left(\frac{1}{2}\times 88{,}700\text{ miles}\right)\left(0.634\frac{\text{rad}}{\text{hr}}\right) = 28{,}100\text{ mi/hr}$$

29. The satellite travels 1 revolution, or 2π radian, in 23.93 hr. So,

$$\omega = \frac{\theta}{t} = \frac{2\pi}{23.93} = 0.2626\frac{\text{rad}}{\text{hr}}$$

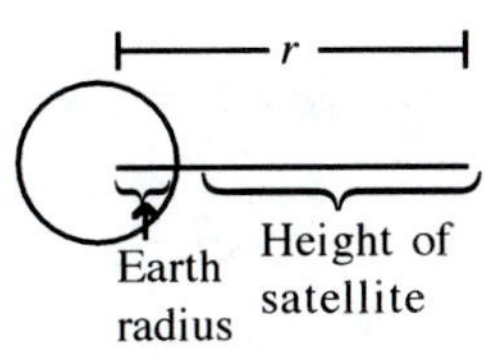

The radius, r, of the satellite's orbit is given by adding the satellite's distance above the earth's surface to the radius of the earth (see sketch). Thus,

$$V = r\omega = (22{,}300 + 3{,}964\text{ miles})\left(0.2626\frac{\text{rad}}{\text{hr}}\right) = 6{,}900\text{ mi/hr}$$

31. Using subscript E to denote quantities associated with the earth, and subscript S to denote quantities associated with the satellite, we can write:

$$\omega_E = \frac{\theta_E}{t} \qquad \omega_S = \frac{\theta_S}{t} \qquad \theta_E = \omega_E t \qquad \theta_S = \omega_S t$$

Using the hint, $2\pi = \theta_S - \theta_E$, hence,

$$2\pi = \omega_S t - \omega_E t = (\omega_S - \omega_E)t$$

$$t = \frac{2\pi}{\omega_S \omega_E}$$

Thus, $$t = \frac{2\pi}{\frac{2\pi}{1.51} - \frac{2\pi}{23.93}} = 1.61\text{ hr}$$

33. (A) 1 rps corresponds to an angular velocity of 2π rad/sec, so that at the end of t sec, $\theta = 2\pi t$.

(B) In triangle ABC, we can write $\tan\theta = \frac{a}{b} = \frac{a}{15}$. Thus, $a = 15\tan\theta$. But $\theta = \omega t$, where ω is given by 1 revolution per second = $2\pi\frac{\text{rad}}{\text{sec}}$. Thus, $\theta = 2\pi t$, and $a = 15\tan 2\pi t$.

(C) The speed of the light spot on the wall increases as t increases from 0.00 to 0.24. When $t = 0.25$, $a = 15\tan(\pi/2)$, which is not defined. The light has made one quarter turn and the spot is no longer on the wall.

t sec	0.00	0.04	0.08	0.12	0.16	0.20	0.24
a ft	0.00	3.85	8.25	14.09	23.64	46.17	238.42

35. The maximum speed will be attained when the largest front gear (53 teeth) engages the smallest rear gear (11 teeth). Then one revolution of the pedals will cause $\frac{53}{11}$ of a revolution of the rear wheel. One revolution of the rear wheel is $\pi d = \pi \cdot 28$in traveled. Therefore, 90 revolutions per minute will result in

$$\left(\frac{53}{11}\cdot\pi\cdot 28\text{ in}\cdot 90\frac{\text{revolutions}}{\text{min}}\cdot 60\frac{\text{min}}{\text{hr}}\right) \div \left(5280\cdot 12\frac{\text{in}}{\text{mi}}\right) = 36.1\frac{\text{mi}}{\text{hr}}$$

EXERCISE 2.3 Trigonometric Functions: Unit Circle Approach

1. $Q(a, b) = Q(0, 1)$

   $a^2 + b^2 = 0^2 + 1^2 = 1$, so Q is indeed on the unit circle.

   $\sin x = b = 1$ $\qquad$ $\csc x = \frac{1}{b} = \frac{1}{1} = 1$

   $\cos x = a = 0$ $\qquad$ $\sec x = \frac{1}{a} = \frac{1}{0}$ = not defined

   $\tan x = \frac{b}{a}$ = not defined $\qquad$ $\cot x = \frac{a}{b} = \frac{0}{1} = 0$

3. $Q(a, b) = Q\left(\frac{3}{5}, \frac{4}{5}\right)$

   $a^2 + b^2 = \left(\frac{3}{5}\right)^2 + \left(\frac{4}{5}\right)^2 = \frac{9}{25} + \frac{16}{25} = 1$, so Q is indeed on the unit circle.

   $\sin x = b = \frac{4}{5}$ $\qquad$ $\csc x = \frac{1}{b} = \frac{5}{4}$

   $\cos x = a = \frac{3}{5}$ $\qquad$ $\sec x = \frac{1}{a} = \frac{5}{3}$

   $\tan x = \frac{b}{a} = \frac{4}{3}$ $\qquad$ $\cot x = \frac{a}{b} = \frac{3}{4}$

5. $Q(a, b) = Q\left(\frac{1}{\sqrt{2}}, \frac{1}{\sqrt{2}}\right)$

   $a^2 + b^2 = \left(\frac{1}{\sqrt{2}}\right)^2 + \left(\frac{1}{\sqrt{2}}\right)^2 = \frac{1}{2} + \frac{1}{2} = 1$, so Q is indeed on the unit circle.

   $\sin x = b = \frac{1}{\sqrt{2}}$ $\qquad$ $\csc x = \frac{1}{b} = \sqrt{2}$

   $\cos x = a = \frac{1}{\sqrt{2}}$ $\qquad$ $\sec x = \frac{1}{a} = \sqrt{2}$

   $\tan x = \frac{b}{a} = 1$ $\qquad$ $\cot x = \frac{a}{b} = 1$

7. $Q(a, b) = Q\left(-\frac{1}{2}, \frac{\sqrt{3}}{2}\right)$

   $a^2 + b^2 = \left(-\frac{1}{2}\right)^2 + \left(\frac{\sqrt{3}}{2}\right)^2 = \frac{1}{4} + \frac{3}{4} = 1$, so Q is indeed on the unit circle.

   $\sin x = b = \frac{\sqrt{3}}{2}$ $\qquad$ $\csc x = \frac{1}{b} = \frac{2}{\sqrt{3}}$

   $\cos x = a = -\frac{1}{2}$ $\qquad$ $\sec x = \frac{1}{a} = -2$

   $\tan x = \frac{b}{a} = -\sqrt{3}$ $\qquad$ $\cot x = \frac{a}{b} = -\frac{1}{\sqrt{3}}$

9. The distance from P to the origin is

$$r = \sqrt{(-10)^2 + 0^2} = \sqrt{100+0} = \sqrt{100} = 10$$

Therefore, the point $\left(-\frac{10}{10}, \frac{0}{10}\right) = (-1, 0)$ lies on the unit circle.

Apply Definition 1 with $a = -1$ and $b = 0$

$\sin x = b = 0$

$\cos x = a = -1$

$\tan x = \frac{b}{a} = \frac{0}{-1} = 0$

$\csc x = \frac{1}{b} = \frac{1}{0} =$ not defined

$\sec x = \frac{1}{a} = \frac{1}{-1} = -1$

$\cot x = \frac{a}{b} = \frac{-1}{0} =$ not defined

11. The distance from P to the origin is

$$r = \sqrt{8^2 + 6^2} = \sqrt{64+36} = \sqrt{100} = 10$$

Therefore, the point $\left(\frac{8}{10}, \frac{6}{10}\right) = \left(\frac{4}{5}, \frac{3}{5}\right)$ lies on the unit circle.

Apply Definition 1 with $a = \frac{4}{5}$ and $b = \frac{3}{5}$

$\sin x = b = \frac{3}{5}$

$\cos x = a = \frac{4}{5}$

$\tan x = \frac{b}{a} = \frac{3}{5} \div \frac{4}{5} = \frac{3}{4}$

$\csc x = \frac{1}{b} = \frac{5}{3}$

$\sec x = \frac{1}{a} = \frac{5}{4}$

$\cot x = \frac{a}{b} = \frac{4}{5} \div \frac{3}{5} = \frac{4}{3}$

13. The distance from P to the origin is

$$r = \sqrt{(-7)^2 + (-24)^2} = \sqrt{49+576} = \sqrt{625} = 25$$

Therefore, the point $\left(-\frac{7}{25}, -\frac{24}{25}\right)$ lies on the unit circle.

Apply Definition 1 with $a = -\frac{7}{25}$ and $b = -\frac{24}{25}$

$\sin x = b = -\frac{24}{25}$

$\cos x = a = -\frac{7}{25}$

$\tan x = \frac{b}{a} = \left(-\frac{24}{25}\right) \div \left(-\frac{7}{25}\right) = \frac{24}{7}$

$\csc x = \frac{1}{b} = -\frac{25}{24}$

$\sec x = \frac{1}{a} = -\frac{25}{7}$

$\cot x = \frac{a}{b} = \left(-\frac{7}{25}\right) \div \left(-\frac{24}{25}\right) = \frac{7}{24}$

15. The distance from P to the origin is

$$r = \sqrt{(-12)^2 + 5^2} = \sqrt{144+25} = \sqrt{169} = 13$$

Therefore, the point $\left(-\frac{12}{13}, \frac{5}{13}\right)$ lies on the unit circle.

Apply Definition 1 with $a = -\frac{12}{13}$ and $b = \frac{5}{13}$

$\sin x = b = \frac{5}{13}$ $\quad\quad$ $\csc x = \frac{1}{b} = \frac{13}{5}$

$\cos x = a = -\frac{12}{13}$ $\quad\quad$ $\sec x = \frac{1}{a} = -\frac{13}{12}$

$\tan x = \frac{b}{a} = \frac{5}{13} \div \left(-\frac{12}{13}\right) = -\frac{5}{12}$ $\quad\quad$ $\cot x = \frac{a}{b} = \left(-\frac{12}{13}\right) \div \frac{5}{13} = -\frac{12}{5}$

17. Because $\sin x = \frac{b}{r} = \frac{3}{5}$, we let $b = 3, r = 5$. We find a so that (a, b) is the point in quadrant I that lies on the circle of radius $r = 5$ with center the origin.

$$\begin{aligned} a^2 + b^2 &= r^2 \\ a^2 + 9 &= 25 \\ a^2 &= 16 \\ a &= 4 \end{aligned}$$

We now apply the third remark following Definition 1 with $a = 4, b = 3$, and $r = 5$:

$\left(\sin x = \frac{b}{r} = \frac{3}{5}\right)$ $\quad\quad$ $\csc x = \frac{r}{b} = \frac{5}{3}$

$\cos x = \frac{a}{r} = \frac{4}{5}$ $\quad\quad$ $\sec x = \frac{r}{a} = \frac{5}{4}$

$\tan x = \frac{b}{a} = \frac{3}{4}$ $\quad\quad$ $\cot x = \frac{a}{b} = \frac{4}{3}$

19. Because $\cos x = \frac{a}{r} = \frac{5}{13}$, we let $a = 5, r = 13$. We find b so that (a, b) is the point in quadrant IV that lies on the circle of radius $r = 13$ with center the origin.

$$\begin{aligned} a^2 + b^2 &= r^2 \\ 25 + b^2 &= 169 \\ b^2 &= 144 \\ b &= -12 \end{aligned}$$

We now apply the third remark following Definition 1 with $a = 5, b = -12$, and $r = 13$:

$\sin x = \frac{b}{r} = -\frac{12}{13}$ $\quad\quad$ $\csc x = \frac{r}{b} = -\frac{13}{12}$

$\left(\cos x = \frac{a}{r} = \frac{5}{13}\right)$ $\quad\quad$ $\sec x = \frac{r}{a} = \frac{13}{5}$

$\tan x = \frac{b}{a} = -\frac{12}{5}$ $\quad\quad$ $\cot x = \frac{a}{b} = -\frac{5}{12}$

21. Because $\tan x = \frac{b}{a} = \frac{3}{2}$ and x is a quadrant III angle, we let $a = -2, b = -3$. The distance from (a, b) to the origin is

$$r = \sqrt{(-2)^2 + (-3)^2} = \sqrt{4+9} = \sqrt{13}$$

We apply the third remark following Definition 1 with $a = -2, b = -3$, and $r = \sqrt{13}$:

$$\sin x = \frac{b}{r} = -\frac{3}{\sqrt{13}} \qquad \csc x = \frac{r}{b} = -\frac{\sqrt{13}}{3}$$

$$\cos x = \frac{a}{r} = -\frac{2}{\sqrt{13}} \qquad \sec x = \frac{r}{a} = -\frac{\sqrt{13}}{2}$$

$$\left(\tan x = \frac{b}{a} = \frac{3}{2}\right) \qquad \cot x = \frac{a}{b} = \frac{2}{3}$$

23. Because $\cot x = \frac{a}{b} = \frac{1}{2}$, we let $a = 1, b = 2$. The distance from (a, b) to the origin is

$$r = \sqrt{1^2 + 2^2} = \sqrt{1+4} = \sqrt{5}$$

We apply the third remark following Definition 1 with $a = 1, b = 2$, and $r = \sqrt{5}$:

$$\sin x = \frac{b}{r} = \frac{2}{\sqrt{5}} \qquad \csc x = \frac{r}{b} = \frac{\sqrt{5}}{2}$$

$$\cos x = \frac{a}{r} = \frac{1}{\sqrt{5}} \qquad \sec x = \frac{r}{a} = \frac{\sqrt{5}}{1} = \sqrt{5}$$

$$\tan x = \frac{b}{a} = \frac{2}{1} = 2 \qquad \left(\cot x = \frac{a}{b} = \frac{1}{2}\right)$$

25. Because $\sec x = \frac{r}{a} = \sqrt{2}$, we let $r = \sqrt{2}, a = 1$. We find b so that (a, b) is the point in quadrant IV that lies on the circle of radius $\sqrt{2}$ with center the origin.

$$a^2 + b^2 = r^2$$
$$1 + b^2 = 2$$
$$b^2 = 1$$
$$b = -1$$

We now apply the third remark following Definition 1 with $a = 1, b = -1$, and $r = \sqrt{2}$:

$$\sin x = \frac{b}{r} = -\frac{1}{\sqrt{2}} \qquad \csc x = \frac{r}{b} = \frac{\sqrt{2}}{-1} = -\sqrt{2}$$

$$\cos x = \frac{a}{r} = \frac{1}{\sqrt{2}} \qquad \left(\sec x = \frac{r}{a} = \sqrt{2}\right)$$

$$\tan x = \frac{b}{a} = -\frac{1}{1} = -1 \qquad \cot x = \frac{a}{b} = \frac{1}{-1} = -1$$

27. Because $\csc x = \frac{r}{b} = -\frac{25}{24}$, we let $r = 25, b = -24$. We find a so that (a, b) is the point in quadrant III that lies on the circle of radius 25 with center the origin.

$$\begin{aligned} a^2 + b^2 &= r^2 \\ a^2 + 576 &= 625 \\ a^2 &= 49 \\ a &= -7 \end{aligned}$$

We now apply the third remark following Definition 1 with $a = -7, b = -24$, and $r = 25$:

$$\sin x = \frac{b}{r} = -\frac{24}{25} \qquad \left(\csc x = \frac{r}{b} = -\frac{25}{24}\right)$$

$$\cos x = \frac{a}{r} = -\frac{7}{25} \qquad \sec x = \frac{r}{a} = -\frac{25}{7}$$

$$\tan x = \frac{b}{a} = \frac{-24}{-7} = \frac{24}{7} \qquad \cot x = \frac{a}{b} = \frac{-7}{-24} = \frac{7}{24}$$

29. Degree mode: $\tan 30° = 0.5774$

31. Degree mode: $\cos(-85°) = 0.0872$

33. Radian mode: $\csc 3 = \frac{1}{\sin 3} = 7.086$

35. Radian mode: $\cot 6 = \frac{1}{\tan 6} = -3.436$

35. Radian mode: $\sin\left(-\frac{\pi}{10}\right) = -0.3090$

39. Degree mode: $\sec(265°) = \frac{1}{\cos(265°)} = -11.47$

41. Since for any point (a, b) with $r^2 = a^2 + b^2$, r is greater than or at least equal to b, $\sin x = \frac{b}{r}$ is at most equal to 1.

Algebraically:

$$\begin{aligned} a^2 &\geq 0 \\ a^2 + b^2 &\geq b^2 \\ r^2 &\geq b^2 \\ 1 &\geq \frac{b^2}{r^2} \\ 1 &\geq \frac{b}{r} = \sin x \end{aligned}$$

43. Since a can take on any arbitrarily large value, $a^2 + b^2 = r^2$ can be arbitrarily larger than b^2. Thus $\frac{r^2}{b^2}$ and $\csc x = \frac{r}{b}$ can be arbitrarily large.

45. $\sin(x + 2\pi) = \sin x$, since the terminal sides of each angle will coincide and therefore the same point $P(a, b) \neq (0, 0)$ can be chosen on the terminal side s. Thus, $\sin(x + 2\pi) = \frac{b}{r} = \sin x$ where $r = \sqrt{a^2 + b^2} \neq 0$.

47. $(a, b) = (\sqrt{3}, 1), r = \sqrt{a^2 + b^2} = \sqrt{(\sqrt{3})^2 + 1^2} = \sqrt{4} = 2$

$\sin x = \frac{b}{r} = \frac{1}{2}$ $\quad \cos x = \frac{a}{r} = \frac{\sqrt{3}}{2}$ $\quad \tan x = \frac{b}{a} = \frac{1}{\sqrt{3}}$

$\csc x = \frac{r}{b} = \frac{2}{1} = 2$ $\quad \sec x = \frac{r}{a} = \frac{2}{\sqrt{3}}$ $\quad \cot x = \frac{a}{b} = \frac{\sqrt{3}}{1} = \sqrt{3}$

49. $(a, b) = (1, -\sqrt{3}), r = \sqrt{a^2 + b^2} = \sqrt{1^2 + (-\sqrt{3})^2} = \sqrt{4} = 2$

$\sin x = \frac{b}{r} = \frac{-\sqrt{3}}{2} = -\frac{\sqrt{3}}{2}$ $\quad \cos x = \frac{a}{r} = \frac{1}{2}$ $\quad \tan x = \frac{b}{a} = \frac{-\sqrt{3}}{1} = -\sqrt{3}$

$\csc x = \frac{r}{b} = \frac{2}{-\sqrt{3}} = -\frac{2}{\sqrt{3}}$ $\quad \sec x = \frac{r}{a} = \frac{2}{1} = 2$ $\quad \cot x = \frac{a}{b} = \frac{1}{-\sqrt{3}} = -\frac{1}{\sqrt{3}}$

51. I, IV

53. I, III

55. I, IV

57. Because $\sin x = \frac{b}{r} = \frac{1}{2}$, we let $b = 1, r = 2$. Since $\sin x$ is positive and $\tan x$ is negative, x is a quadrant II angle. We find a so that (a, b) is the point in quadrant II that lies on the circle of radius 2 with center the origin.

$$a^2 + b^2 = r^2$$
$$a^2 + 1 = 4$$
$$a^2 = 3$$
$$a = -\sqrt{3}$$

We now apply the third remark following Definition 1 with $a = -\sqrt{3}$, $b = 1$, and $r = 2$:

$\left(\sin x = \frac{b}{r} = \frac{1}{2}\right)$ $\quad \csc x = \frac{r}{b} = \frac{2}{1} = 2$

$\cos x = \frac{a}{r} = -\frac{\sqrt{3}}{2}$ $\quad \sec x = \frac{r}{a} = -\frac{2}{\sqrt{3}}$

$\tan x = \frac{b}{a} = -\frac{1}{\sqrt{3}}$ $\quad \cot x = \frac{a}{b} = -\frac{\sqrt{3}}{1} = -\sqrt{3}$

59. Because $\sec x = \frac{r}{a} = \frac{3}{2}$, we let $a = 2, r = 3$. Since $\sec x$ is positive and $\sin x$ is negative, x is a quadrant IV angle. We find b so that (a, b) is the point in quadrant IV that lies on the circle of radius 3 with center the origin.

$$a^2 + b^2 = r^2$$
$$4 + b^2 = 9$$
$$b^2 = 5$$
$$b = -\sqrt{5}$$

We now apply the third remark following Definition 1 with $a = 2$, $b = -\sqrt{5}$, and $r = 3$:

$\sin x = \frac{b}{r} = -\frac{\sqrt{5}}{3}$ $\quad \csc x = \frac{r}{b} = -\frac{3}{\sqrt{5}}$

$\cos x = \frac{a}{r} = \frac{2}{3}$ $\quad \left(\sec x = \frac{r}{a} = \frac{3}{2}\right)$

$\tan x = \frac{b}{a} = -\frac{\sqrt{5}}{2}$ $\quad \cot x = \frac{a}{b} = -\frac{2}{\sqrt{5}}$

61. Because cot x and sin x are negative, x is a quadrant IV angle. Because $\cot x = \frac{a}{b} = -\sqrt{3}$, we let $a = \sqrt{3}, b = -1$. The distance from (a, b) to the origin is

$$r = \sqrt{(\sqrt{3})^2 + (-1)^2} = \sqrt{3+1} = 2$$

We apply the third remark following Definition 1 with $a = \sqrt{3}, b = -1$, and $r = 2$:

$$\sin x = \frac{b}{r} = -\frac{1}{2} \qquad \csc x = \frac{r}{b} = \frac{2}{-1} = -2$$

$$\cos x = \frac{a}{r} = \frac{\sqrt{3}}{2} \qquad \sec x = \frac{r}{a} = \frac{2}{\sqrt{3}}$$

$$\tan x = \frac{b}{a} = -\frac{1}{\sqrt{3}} \qquad \left(\cot x = \frac{a}{b} = -\sqrt{3}\right)$$

63. Radian mode: $\cos 7.215 = 0.5964$

65. Radian mode: $\cot 3.142 = \frac{1}{\tan 3.142} = 2{,}455$

67. Degree mode: $\sec(-85°04'13'') = \frac{1}{\cos\left(-85 - \frac{4}{60} - \frac{13}{3600}\right)^{\circ}} = 11.64$

69. Radian mode: $\sin(-13.28) = -0.6546$

71. Degree mode: $\tan 25°14'54'' = \tan\left(25 + \frac{14}{60} + \frac{54}{3600}\right)^{\circ} = 0.4716$

73. Degree mode: $\csc(-415.3°) = \frac{1}{\sin(-415.3°)} = -1.216$

75. When the terminal side of an angle lies along the vertical axis, the coordinates of any point on the terminal side have the form $(0, b)$, that is, $a = 0$. Therefore, $\tan \theta = \frac{b}{a}$ and $\sec \theta = \frac{1}{a}$ are not defined.

77. (A) Since $x = \frac{s}{r}$, and $s = 6$, and r, the measure of CA, is 5, we have $x = \frac{6}{5} = 1.2$ rad

(B) Since $\cos x = \frac{a}{r}$ and $\sin x = \frac{b}{r}$, we have

$$a = r \cos x = 5 \cos 1.2 \qquad b = r \sin x = 5 = \sin 1.2$$

Thus, $(a, b) = (5 \cos 1.2, 5 \sin 1.2) = (1.81, 4.66)$.

79. (A) Since $x = \frac{s}{r}$, and $s = 2$, and r, the measure of CA, is 1, we have $x = \frac{2}{1} = 2$ rad

(B) Since $\cos x = \frac{a}{r}$ and $\sin x = \frac{b}{r}$, we have

$$a = r \cos x = 1 \cos 2 \qquad b = r \sin x = 1 \sin 2$$

Thus, $(a, b) = (1 \cos 2, 1 \sin 2) = (-0.416, 0.909)$.

81. If $\theta = 0°, I = k\cos 0° = k \cdot 1 = k$
If $\theta = 20°, I = k\cos 20° = k(0.94) = 0.94k$
If $\theta = 40°, I = k\cos 40° = k(0.77) = 0.77k$
If $\theta = 60°, I = k\cos 60° = k(0.50) = 0.50k$
If $\theta = 80°, I = k\cos 80° = k(0.17) = 0.17k$

83. If $\theta = 15°$ (summer solstice), $I = k\cos 15° = k(0.97) = 0.97k$
If $\theta = 63°$ (winter solstice), $I = k\cos 63° = k(0.45) = 0.45k$

85. (A) If $n = 6, A = \frac{n}{2}\sin\left(\frac{360}{n}\right)^\circ = \frac{6}{2}\sin\left(\frac{360}{6}\right)^\circ = 3\sin 60° = 2.59808$

If $n = 10, A = \frac{10}{2}\sin\left(\frac{360}{10}\right)^\circ = 5\sin 36° = 2.93893$

If $n = 100, A = \frac{100}{2}\sin\left(\frac{360}{100}\right)^\circ = 50\sin 3.6° = 3.13953$

If $n = 1000, A = \frac{1000}{2}\sin\left(\frac{360}{1000}\right)^\circ = 500\sin(0.36)° = 3.14157$

If $n = 10{,}000, A = \frac{10{,}000}{2}\sin\left(\frac{360}{10{,}000}\right)^\circ = 5{,}000\sin(0.036)° = 3.14159$

n	6	10	100	1000	10,000
A_n	2.59808	2.93893	3.13953	3.14157	3.14159

(B) The area of the circle is $A = \pi r^2 = \pi(1)^2 = \pi$, and A_n seem to approach π, the area of the circle as n increases.

(C) No. An n sided polygon is always a polygon, no matter the size of n, but the inscribed polygon can be made as close to the circle as desired by taking n sufficiently large.

87. From the diagram, we can see that $x = a + \ell$. To determine a, we note that $\cos\theta = \frac{a}{r}, r = 1$, and $\theta = 20\pi t$. Thus,

$$\cos 20\pi t = \frac{a}{1} \text{ and } a = \cos 20\pi t$$

To determine ℓ, we note that triangle PFL is a right triangle. Thus, from the Pythagorean theorem,

$$b^2 + \ell^2 = 5^2$$
$$\ell^2 = 25 - b^2$$

Since $\sin\theta = \frac{b}{r}, r = 1$, and $\theta = 20\pi t$, we have

$$\sin 20\pi t = \frac{b}{1} \text{ and } b = \sin 20\pi t.$$

Thus,

$$\ell^2 = 25 - (\sin 20\pi t)^2,$$
$$\ell = \sqrt{25 - (\sin 20\pi t)^2}$$
$$x = a + \ell = \cos 20\pi t + \sqrt{25 - (\sin 20\pi t)^2}$$

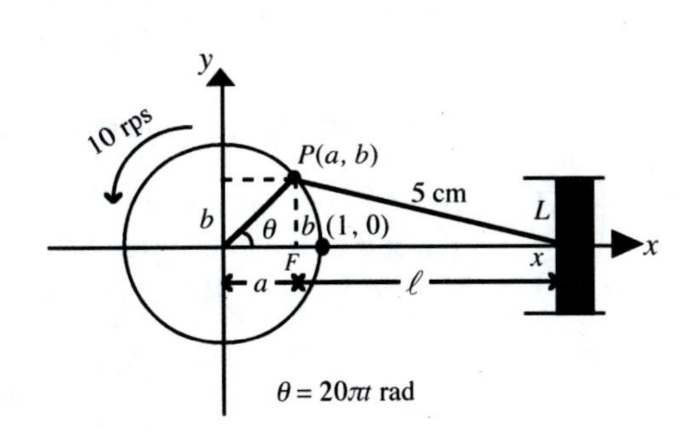

89. Since $I = 35\sin(48\pi t - 12\pi)$ and $t = 0.13$,

$$= 35\sin(48\pi(0.13) - 12\pi) \text{ (calculator in radian mode)}$$
$$\approx 35\sin(-18.095574\ldots) \approx 24 \text{ amperes}$$

91. (A) If the angle of inclination $\theta = 63.5°$, then $m = \tan\theta = \tan 63.5° = 2.01$

If the angle of inclination $\theta = 172°$, then $m = \tan\theta = \tan 172° = -0.14$

(B) If the angle of inclination $\theta = 143°$, then $m = \tan\theta = \tan 143°$. Then the equation of the line is given by

$$y - 6 = \tan 143°(x - (-3))$$
$$y = \tan 143°(x + 3) + 6 = x\tan 143° + 3\tan 143° + 6$$
$$y = -0.75x + 3.74$$

EXERCISE 2.4 Additional Applications

1. Light, in going from one point to another, travels along the path that takes the least time.

3. The angle of incidence is the angle between the incident ray and the line perpendicular to the surface.

5. The angle of incidence equals the angle of reflection.

7. The ratio of the angle of incidence to the angle of refraction is the reciprocal of the ratio of the index of refraction of the medium of incidence to the index of refraction of the medium of refraction.

9. Use $\frac{n_2}{n_1} = \frac{\sin\alpha}{\sin\beta}$,

where $n_2 = 1.33$, $n_1 = 1.00$, and $\alpha = 40.6°$

Solve for β: $\frac{1.33}{1.00} = \frac{\sin 40.6°}{\sin\beta}$

$$\sin\beta = \frac{\sin 40.6°}{1.33}$$
$$\beta = \sin^{-1}\left(\frac{\sin 40.6°}{1.33}\right)$$
$$= 29.3°$$

11. Use $\frac{n_2}{n_1} = \frac{\sin\alpha}{\sin\beta}$,

where $n_2 = 1.66$, $n_1 = 1.33$, and $\alpha = 32.0°$

Solve for β: $\frac{1.66}{1.33} = \frac{\sin 32.0°}{\sin\beta}$

$$\sin\beta = \frac{1.33\sin 32.0°}{1.66}$$
$$\beta = \sin^{-1}\left(\frac{1.33\sin 32.0°}{1.66}\right)$$
$$= 25.1°$$

13. The index of refraction for diamond is $n_1 = 2.42$ and that for air is 1.00. Find the angle of incidence α such that the angle of refraction β is 90°.

$$\frac{\sin\alpha}{\sin\beta} = \frac{n_2}{n_1};\ \sin\alpha = \frac{1.00}{2.42}\sin 90°;\ \alpha = \sin^{-1}\left[\frac{1.00}{2.42}(1)\right] = 24.4°$$

15. See figure (modified from the figure in the text).

The eye tends to assume that light travels straight. Thus, the ball at B will be interpreted as a ball at B'. The ball appears to be above the real ball.

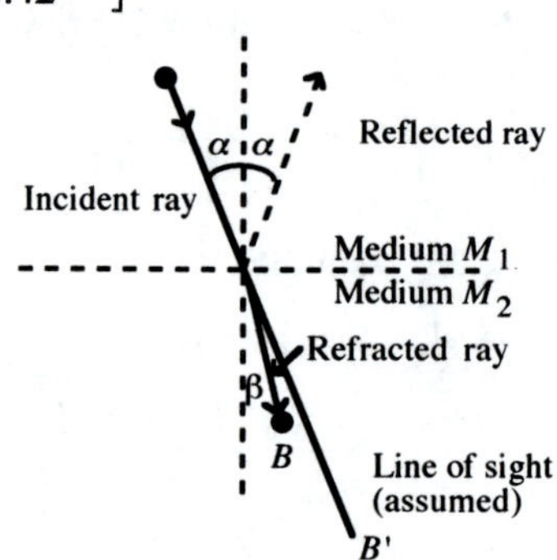

17. Use $\dfrac{n_2}{n_1} = \dfrac{\sin\alpha}{\sin\beta}$,

where $n_1 = 1.00$.

$$\alpha = 90° - 38° = 52°$$

$$\sin\beta = \frac{7.2}{r} = \frac{7.2}{\sqrt{7.2^2 + 12^2}}$$

Solve for n_2:

$$\frac{n_2}{1.00} = \sin 52° \div \frac{7.2}{\sqrt{7.2^2 + 12^2}} \approx 1.5$$

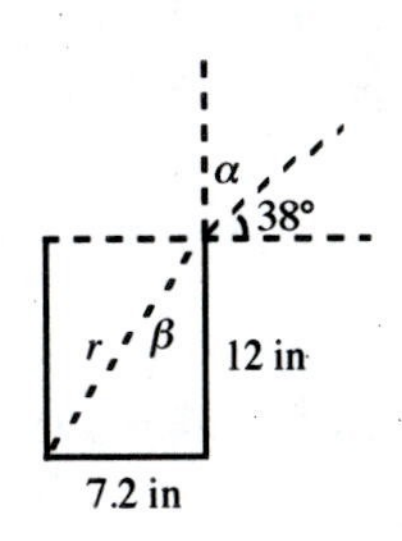

19. A boat moving at a constant rate, faster than the water waves it produces, generates a bow wave that extends back from the bow of the boat.

21. We use $\sin\dfrac{\theta}{2} = \dfrac{S_w}{S_b}$, where $\theta = 60°$ and $S_w = 20$ km/hr. Then we solve for S_b:

$$\sin\frac{60°}{2} = \frac{20}{S_b};\ S_b = \frac{20}{\sin 30°} = 40 \text{ km/hr}$$

23. We use $\sin\dfrac{\theta}{2} = \dfrac{S_s}{S_a}$, where S_a = Mach 1.5 = $1.5S_s$ Then we solve for θ:

$$\sin\frac{\theta}{2} = \frac{S_s}{1.5S_s};\ \sin\frac{\theta}{2} = \frac{1}{1.5}. \text{ Thus } \frac{\theta}{2} = 42°,\ \theta = 84°.$$

25. We use $\sin\dfrac{\theta}{1.97} = \dfrac{S_1}{S_p}$, where $\theta = 92°$ and $S_1 = 1.97 \times 10^{10}$ cm/sec. Then we solve for S_p:

$$\sin\frac{92°}{1.97} = \frac{1.97\times10^{10}}{S_p};\ S_p = \frac{1.97\times10^{10}}{\sin 46°} = 2.74 \times 10^{10} \text{ cm/sec}$$

27. $d = a + b\sin 4\theta = -2.2 + (-4.5)\sin(4\cdot 30°) = -2.2 - 4.5\sin 120° = -6°$.

29.

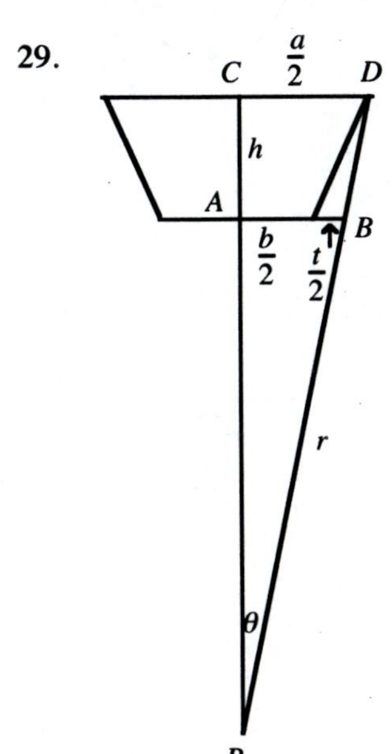

In the text drawing (shown modified), triangle PAB is similar to triangle PCD. Hence $\tan\theta = \dfrac{CD}{CP} = \dfrac{AB}{AP}$.

Thus,

$$\frac{a/2}{CP} = \frac{b/2 + t/2}{CP - h}$$

$$\frac{8}{CP} = \frac{7.5}{CP - 12}$$

$$8CP - 96 = 7.5CP$$

$$CP = 192 \text{ in.}$$

Then r can be calculated as follows:

$$\tan\theta = \frac{a/2}{CP} = \frac{8}{192}$$

$$\theta = \tan^{-1}\frac{8}{192}$$

$$\sin\theta = \frac{a/2}{r}$$

$$r = \frac{a/2}{\sin\theta} = \frac{8}{\sin\left(\tan^{-1} 8/192\right)}$$

r = 192.17 in or 192 in to the nearest inch.

EXERCISE 2.5 Exact Value and Properties of Trigonometric Functions

1. A quadrantal angle is an angle that is an integer multiple of $\frac{\pi}{2}$ or 90°.

3. Drop a perpendicular from any point $P(a, b)$ on the terminal side of θ to the horizontal axis. Let F denote the foot of the perpendicular. The right triangle PFO is a reference triangle for θ.

Note for Problems 5—15: The reference angle α is the angle (always taken positive) between the terminal side of θ and the horizontal axis.

5. $\alpha = \theta = 60°$

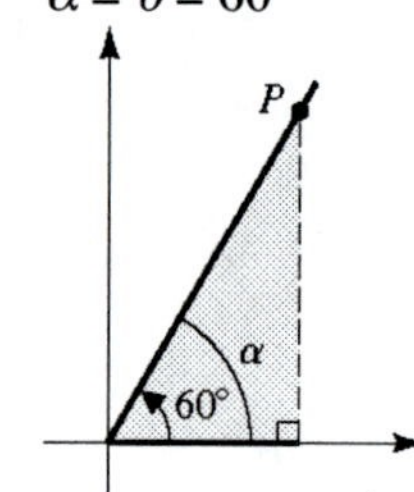

7. $\alpha = |-60°| = 60°$

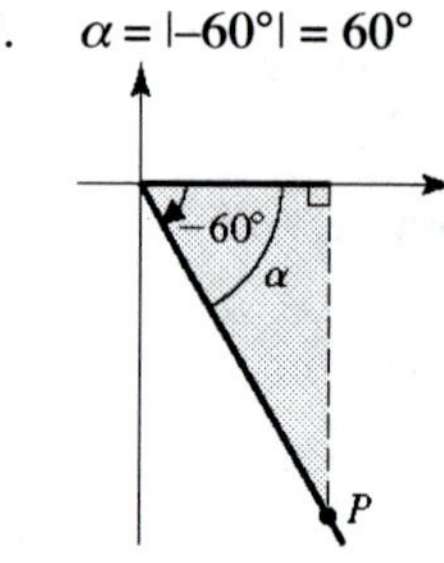

9. $\alpha = |-\pi/3| = \pi/3$

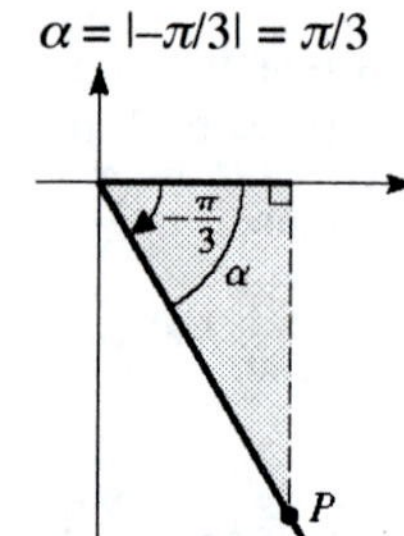

11. $\alpha = \pi - \frac{3\pi}{4} = \frac{\pi}{4}$

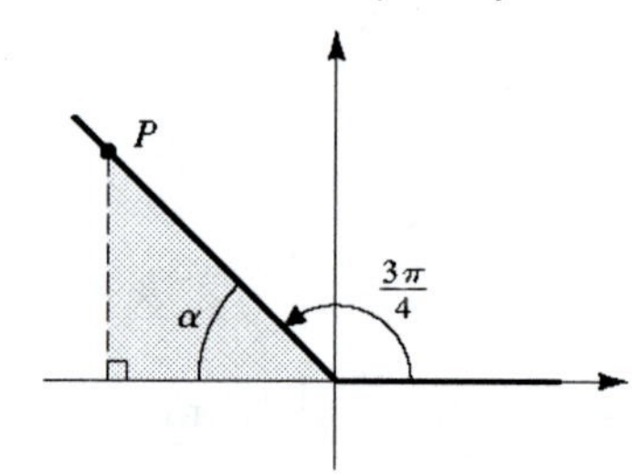

13. $\alpha = 210° - 180° = 30°$

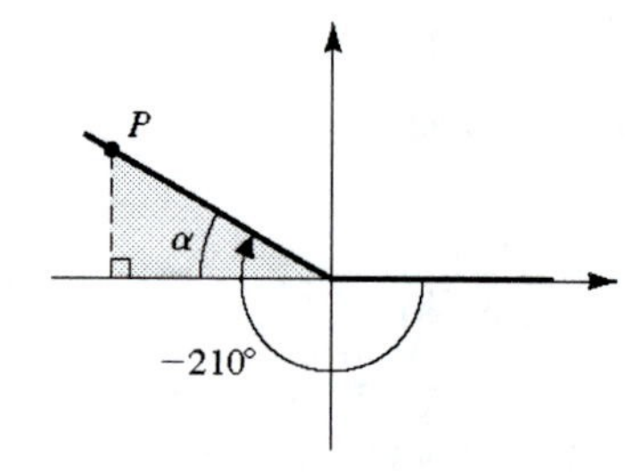

15. $\alpha = \frac{5\pi}{4} - \pi = \frac{\pi}{4}$

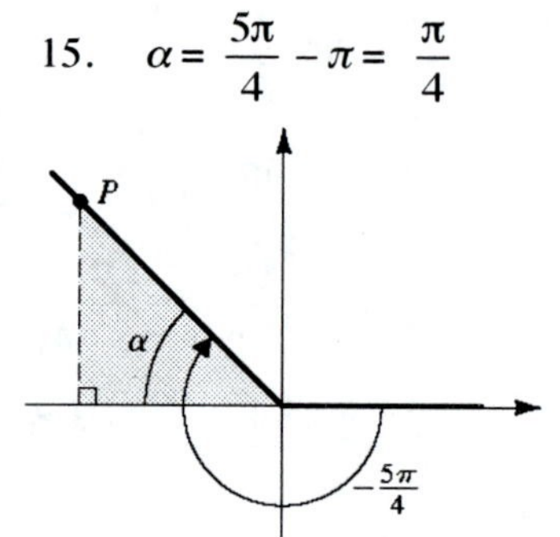

17. $(a, b) = (1, 0), r = 1$

$\sin 0° = \frac{b}{r} = \frac{0}{1} = 0$

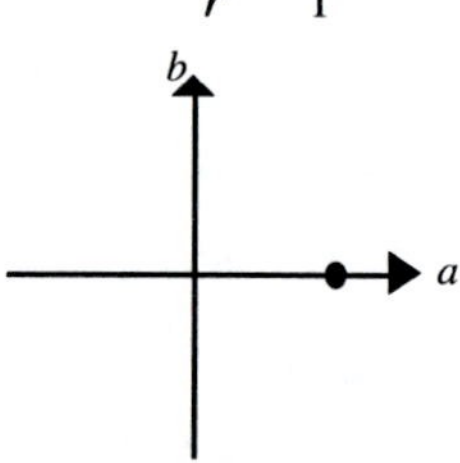

19. $(a, b) = (1, 0), r = 1$

$\tan 0 = \frac{b}{a} = \frac{0}{1} = 0$

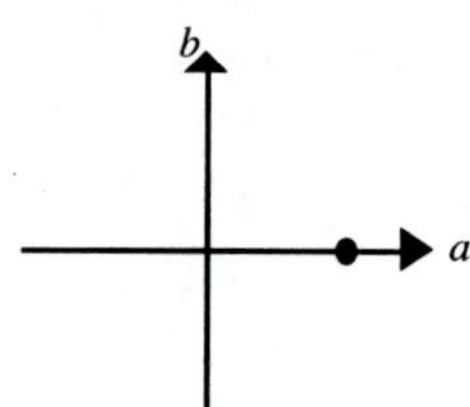

21. Use the special 30°–60° triangle as the reference triangle. Use the sides of the reference triangle to determine $P(a, b)$. Then use Definition 1.

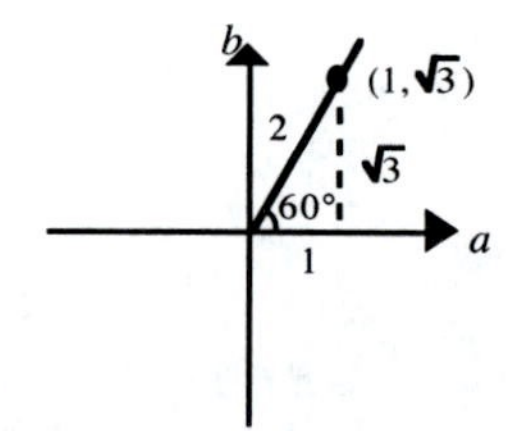

$(a, b) = (1, \sqrt{3}), r = 2$

$\cos(60°) = \frac{a}{r} = \frac{1}{2}$

23. Use the special 45° triangle as the reference triangle. Use the sides of the reference triangle to determine $P(a, b)$ and r. Then use Definition 1.

$(a, b) = (1, 1), r = \sqrt{2}$

$\cot 45° = \frac{a}{b} = \frac{1}{1} = 1$

25. Use the special 30°–60° triangle as the reference triangle. Use the sides of the reference triangle to determine $P(a, b)$. Then use Definition 1.

$(a, b) = (\sqrt{3}, 1), r = 2$

$\sec \frac{\pi}{6} = \frac{r}{a} = \frac{2}{\sqrt{3}}$ or $\frac{2\sqrt{3}}{3}$

27.

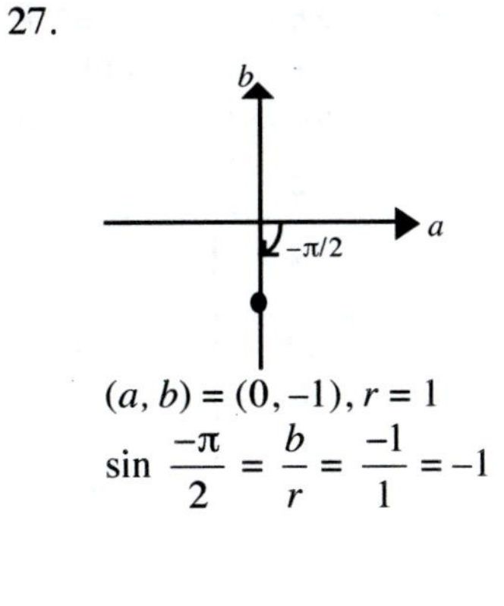

$(a, b) = (0, -1), r = 1$

$\sin \frac{-\pi}{2} = \frac{b}{r} = \frac{-1}{1} = -1$

29.

$(a, b) = (-1, 0), r = 1$

$\cot \pi = \frac{a}{b} = \frac{-1}{0}$

is not defined

31. Locate the 45° reference triangle, determine (a, b) and r, then evaluate.

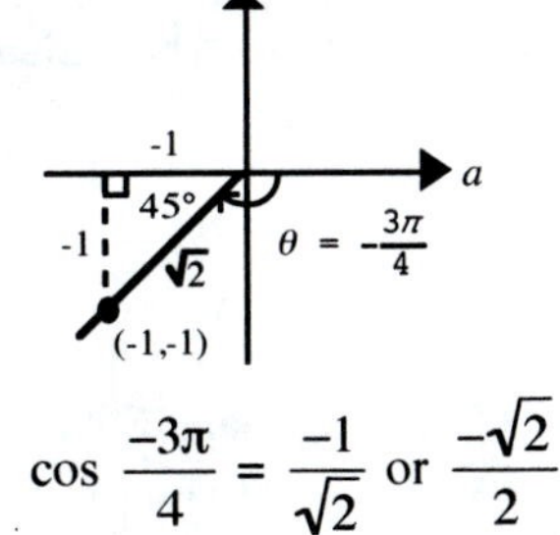

$\cos \frac{-3\pi}{4} = \frac{-1}{\sqrt{2}}$ or $\frac{-\sqrt{2}}{2}$

33. Locate the 30°–60° reference triangle, determine (a, b) and r, then evaluate.

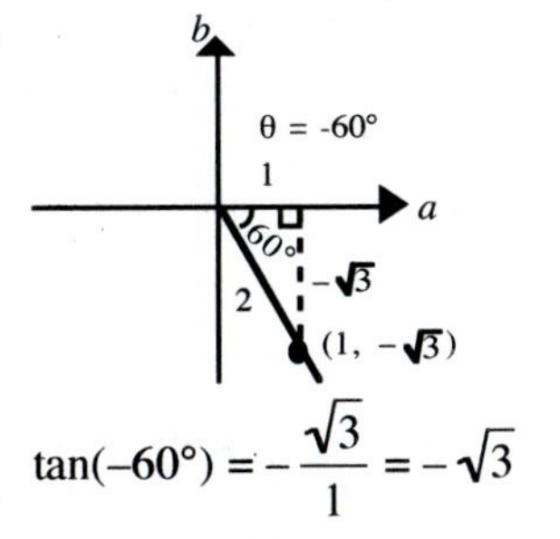

$\tan(-60°) = -\frac{\sqrt{3}}{1} = -\sqrt{3}$

35. Locate the 45° reference triangle, determine (a, b) and r, then evaluate.

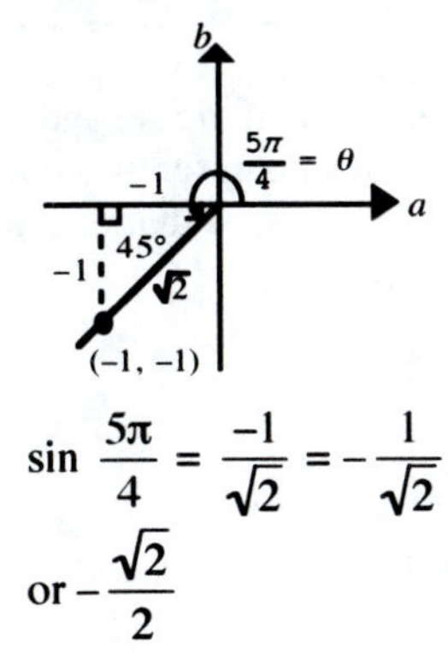

$\sin \frac{5\pi}{4} = \frac{-1}{\sqrt{2}} = -\frac{1}{\sqrt{2}}$

or $-\frac{\sqrt{2}}{2}$

37. Locate the 30°–60° reference triangle, determine (a, b) and r, then evaluate.

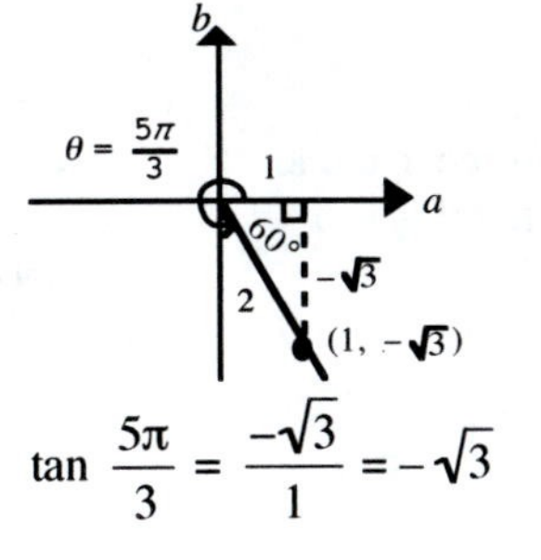

$\tan \frac{5\pi}{3} = \frac{-\sqrt{3}}{1} = -\sqrt{3}$

39.

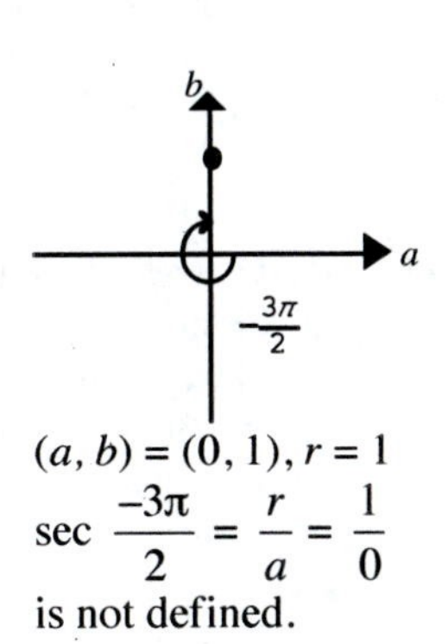

$(a, b) = (0, 1), r = 1$

$\sec \frac{-3\pi}{2} = \frac{r}{a} = \frac{1}{0}$

is not defined.

41. Locate the 30°–60° reference triangle, determine (a, b) and r, then evaluate.

$\cos(-210°) = \dfrac{-\sqrt{3}}{2}$

43. Locate the 45° reference triangle, determine (a, b) and r, then evaluate.

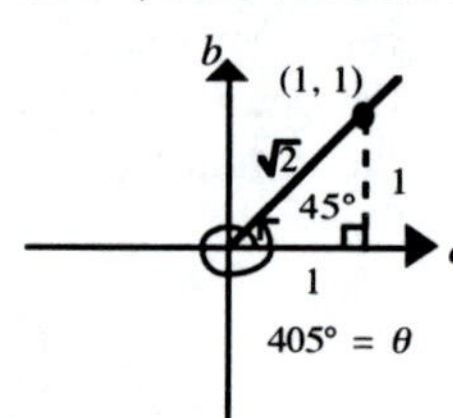

$\cot 405° = \dfrac{1}{1} = 1$

45. Locate the 45° reference triangle, determine (a, b) and r, then evaluate.

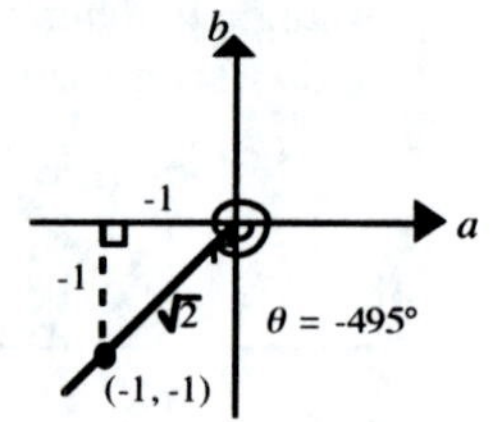

$\csc(-495°) = \dfrac{\sqrt{2}}{-1} = -\sqrt{2}$

47. The tangent function is not defined at $\theta = \dfrac{\pi}{2}$ and $\dfrac{3\pi}{2}$, because $\tan\theta = \dfrac{b}{a}$ and $a = 0$ for any point on the vertical axis.

49. The cosecant function is not defined at $\theta = 0, \pi,$ and 2π, because $\csc\theta = \dfrac{r}{b}$ and $b = 0$ for any point on the horizontal axis.

51. It is known that $\sin\left(\dfrac{\pi}{6}\right) = \dfrac{1}{2} = 0.5000\ldots$ and $\cos(0) = 1 = 1.0000\ldots$, so that the calculator displays exact values of these. Thus $\sin(-45°)$ is not given exactly. To find $\sin(-45°)$, locate the 45° reference triangle, determine (a, b) and r, then evaluate.

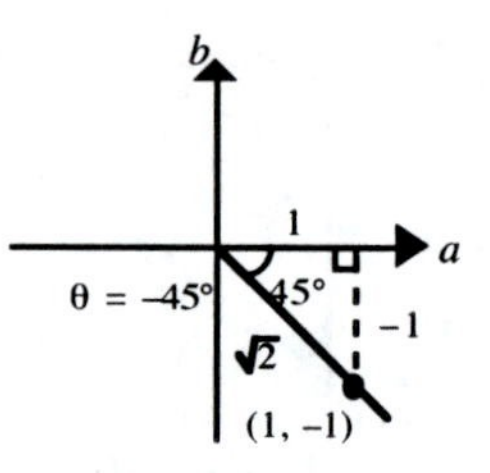

$\sin(-45°) = \dfrac{-1}{\sqrt{2}} = -\dfrac{1}{\sqrt{2}}$ or $-\dfrac{\sqrt{2}}{2}$

53. It is known that $\tan(45°) = 1 = 1.0000\ldots$ and $\tan(180°) = 0 = 0.0000\ldots$, so that the calculator displays exact values for these. Thus $\tan\left(-\dfrac{\pi}{3}\right)$ is not given exactly. To find $\tan\left(-\dfrac{\pi}{3}\right)$, locate the 30°–60° reference triangle, determine (a, b) and r, then evaluate.

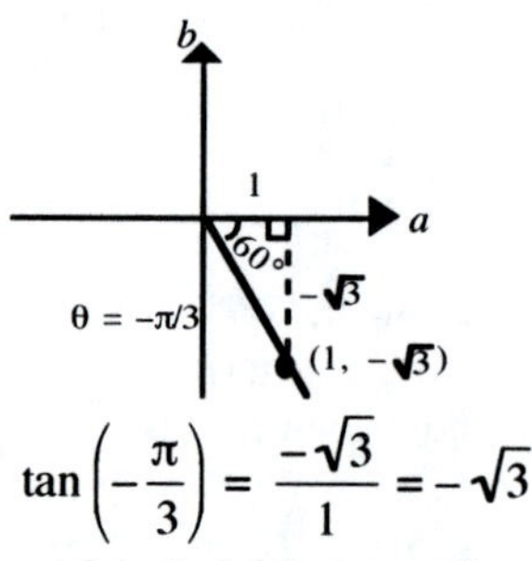

$\tan\left(-\dfrac{\pi}{3}\right) = \dfrac{-\sqrt{3}}{1} = -\sqrt{3}$

55. Draw a reference triangle in the first quadrant with side opposite reference angle 1 and hypotenuse 2. Observe that this is a special 30°–60° triangle.

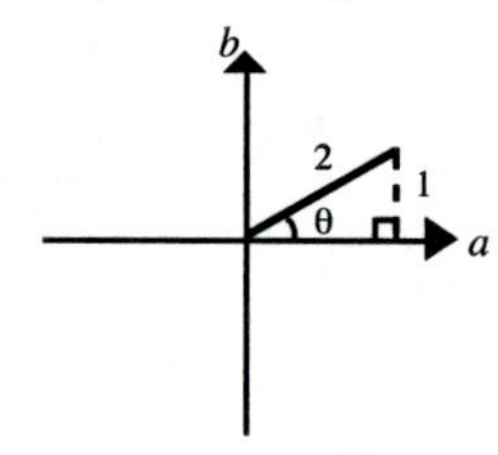

(A) $\theta = 30°$ (B) $\theta = \dfrac{\pi}{6}$

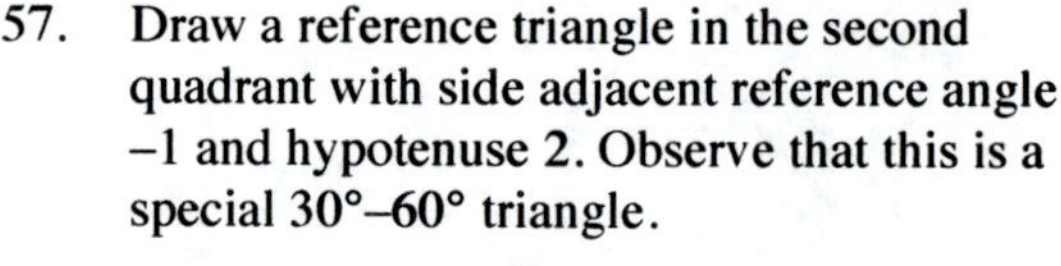

57. Draw a reference triangle in the second quadrant with side adjacent reference angle −1 and hypotenuse 2. Observe that this is a special 30°–60° triangle.

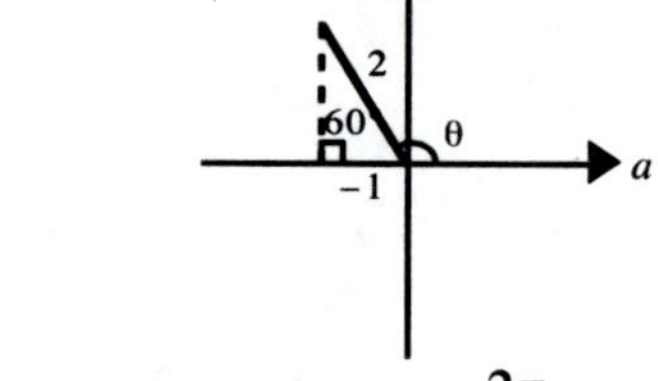

(A) $\theta = 120°$ (B) $\theta = \dfrac{2\pi}{3}$

59. Draw a reference triangle in the second quadrant with side opposite reference angle $\sqrt{3}$ and side adjacent -1. Observe that this is a special 30°–60° triangle.

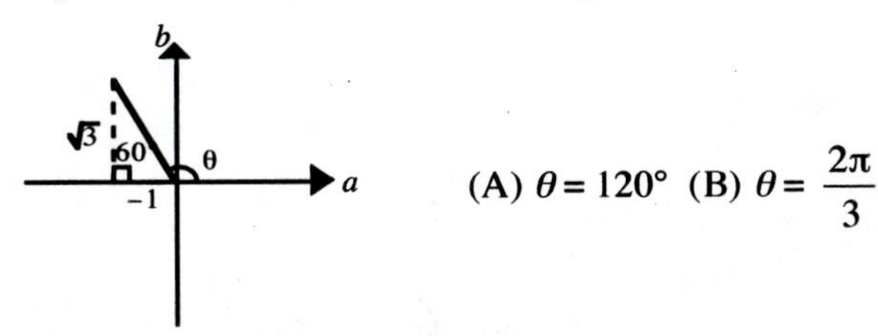

(A) $\theta = 120°$ (B) $\theta = \frac{2\pi}{3}$

61. Yes. By convention, $\sin^2 x = (\sin x)^2$.
$\sin x^2$ means $\sin(x^2)$. These are almost always different in value.

63. (A) – (D) All should equal 0.9525, because the sine function is periodic with period 2π.

65. Calculator in radian mode:

(A) $\tan 1 = 1.6$ $\quad \frac{\sin 1}{\cos 1} = 1.6$

(B) $\tan 5.3 = -1.5$ $\quad \frac{\sin 5.3}{\cos 5.3} = -1.5$

(C) $\tan(-2.376) = 0.96$ $\quad \frac{\sin(-2.376)}{\cos(-2.376)} = 0.96$

67. Calculator in radian mode:

(A) $\sin(-3) = -0.14$ $\quad -\sin 3 = -0.14$

(B) $\sin[-(-12.8)] = 0.23$ $\quad -\sin(-12.8) = 0.23$

(C) $\sin(-407) = 0.99$ $\quad -\sin(407) = 0.99$

Note: Some (very old model) calculators cannot evaluate sin(–407) and, instead, signal an error. If this occurs, use the periodicity of the sine function and evaluate $\sin(-407 + 2\pi \cdot k)$, where k is an appropriate integer.

69. Calculator in radian mode:

(A) $\sin^2 1 + \cos^2 1 = (0.841\ldots)^2 + (0.540\ldots)^2 = 1.0$

(B) $\sin^2(-8.6) + \cos^2(-8.6) = (-0.734\ldots)^2 + (-0.678\ldots)^2 = 1.0$

(C) $\sin^2(263) + \cos^2(263) = (-0.779\ldots)^2 + (0.626\ldots)^2 = 1.0$

71. $\sin x \csc x = \sin x \frac{1}{\sin x}$ Use Identity (1)

$= 1$

73. $\cot x \sec x = \frac{\cos x}{\sin x} \cdot \frac{1}{\cos x}$ Use Identities (5) and (2)

$= \frac{1}{\sin x}$ Use Identity (1)

$= \csc x$

75. $\frac{\sin x}{1 \ \cos^2 x} = \frac{\sin x}{\sin^2 x + \cos^2 x - \cos^2 x}$ Use Identity (9)

$= \frac{\sin x}{\sin^2 x}$

$= \frac{1}{\sin x} = \csc x$ Use Identity (1)

77. $\cot(-x)\ \sin(-x) = \dfrac{\cos(-x)}{\sin(-x)}\sin(-x)$ Use Identity (5)

$= \cos(-x)$

$= \cos x$ Use Identity (7)

79. We can draw reference triangles in both quadrants III and IV with side opposite reference angle $-\sqrt{3}$ and hypotenuse 2. Each triangle is a special 30°–60° triangle.

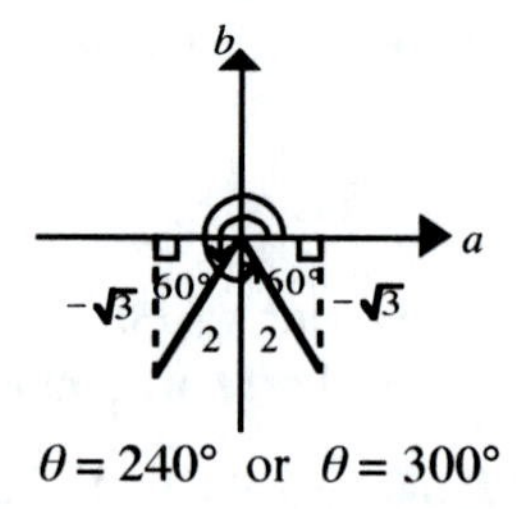

$\theta = 240°$ or $\theta = 300°$

81. We can draw a reference triangle in the second quadrant with side opposite reference angle 1 and side adjacent $-\sqrt{3}$. We can also draw a reference triangle in the fourth quadrant with side opposite reference angle -1 and side adjacent $\sqrt{3}$. Each triangle is a special 30°–60° triangle.

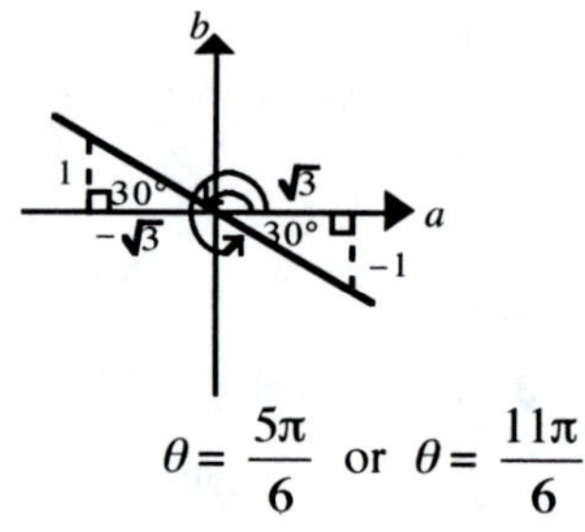

$\theta = \dfrac{5\pi}{6}$ or $\theta = \dfrac{11\pi}{6}$

83. $\cos x = -\dfrac{1}{\sqrt{2}}$

We can draw a reference triangle in the second quadrant with side adjacent reference angle -1 and hypotenuse $\sqrt{2}$. Observe that this is a special 45° triangle.

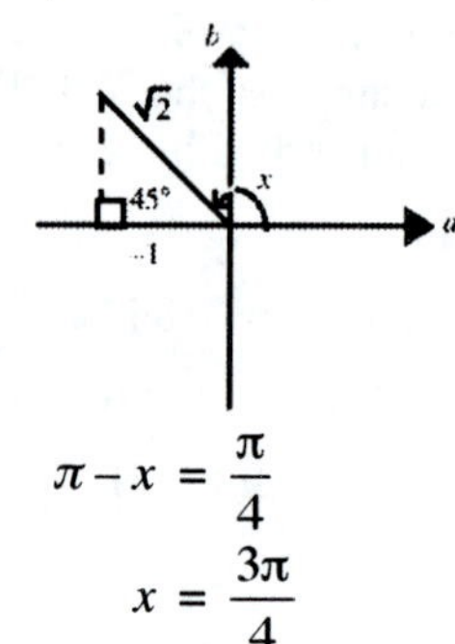

$\pi - x = \dfrac{\pi}{4}$

$x = \dfrac{3\pi}{4}$

85. $\tan x = \dfrac{1}{\sqrt{3}}$

We can draw a reference triangle in the first quadrant with side opposite reference angle 1 and side adjacent $\sqrt{3}$. Then

$$\begin{aligned} a^2 + b^2 &= r^2 \\ 1^2 + (\sqrt{3})^2 &= r^2 \\ 4 &= r^2 \\ r &= 2 \end{aligned}$$

Thus the triangle is a special 30°–60° triangle.

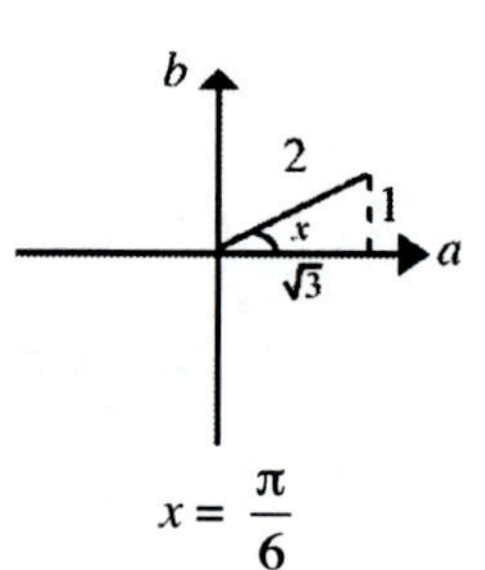

$x = \dfrac{\pi}{6}$

87. (A) Since $\dfrac{7}{x} = \sin 30°$ and $\sin 30° = \dfrac{1}{2}$; $\dfrac{7}{x} = \dfrac{1}{2}$; $x = 14$

Since $\dfrac{7}{y} = \tan 30°$ and $\tan 30° = \dfrac{1}{\sqrt{3}}$; $\dfrac{7}{y} = \dfrac{1}{\sqrt{3}}$; $y = 7\sqrt{3}$

(B) Since $\frac{x}{4} = \sin 45°$ and $\sin 45° = \frac{1}{\sqrt{2}}$; $\frac{x}{4} = \frac{1}{\sqrt{2}}$; $x = \frac{4}{\sqrt{2}}$

Since $\frac{y}{4} = \cos 45°$ and $\cos 45° = \frac{1}{\sqrt{2}}$; $\frac{y}{4} = \frac{1}{\sqrt{2}}$; $y = \frac{4}{\sqrt{2}}$

(C) Since $\frac{5}{x} = \sin 60°$ and $\sin 60° = \frac{\sqrt{3}}{2}$; $\frac{5}{x} = \frac{\sqrt{3}}{2}$; $x = \frac{10}{\sqrt{3}}$

Since $\frac{5}{y} = \tan 60°$ and $\tan 60° = \sqrt{3}$; $\frac{5}{y} = \sqrt{3}$; $y = \frac{5}{\sqrt{3}}$

89. (A) Identity (4) (B) Identity (9) (C) Identity (2)

91. 2π

93. Since for all x in the domain of f (that is, all real numbers) $5\sin(2\pi x + 2\pi) = 5\sin(2\pi x)$, f is periodic. Since $5\sin(2\pi x + 2\pi) = 5\sin 2\pi(x + 1) = f(x + 1)$, the period p of this function is 1.

95. There is no number p such that $\frac{\sin(x+p)}{x+p} = \frac{\sin x}{x}$ for all $x \neq 0$ (the domain of h).

97.
$$
\begin{aligned}
S_1 &= 1\\
S_2 &= S_1 + \cos S_1 = 1 + \cos 1 = 1.540302\\
S_3 &= S_2 + \cos S_2 = 1.540302 + \cos 1.540302 = 1.570792\\
S_4 &= S_3 + \cos S_3 = 1.570792 + \cos 1.570792 = 1.570796\\
S_5 &= S_4 + \cos S_4 = 1.570796 + \cos 1.570796 = 1.570796\\
\frac{\pi}{2} &= 1.570796
\end{aligned}
$$

CHAPTER 2 REVIEW EXERCISE

1. (A) $\theta_r = \frac{\pi \text{ rad}}{180°}\theta_d = \frac{\pi}{180}60 = \frac{\pi}{3}$ (B) $\theta_r = \frac{\pi \text{ rad}}{180°}\theta_d = \frac{\pi}{180}45 = \frac{\pi}{4}$ (C) $\theta_r = \frac{\pi \text{ rad}}{180°}\theta_d = \frac{\pi}{180}90 = \frac{\pi}{2}$

2. (A) $\theta_d = \frac{180°}{\pi \text{ rad}}\theta_r = \frac{180}{\pi}\cdot\frac{\pi}{6} = 30°$ (B) $\theta_d = \frac{180°}{\pi \text{ rad}}\theta_r = \frac{180}{\pi}\cdot\frac{\pi}{2} = 90°$ (C) $\theta_d = \frac{180°}{\pi \text{ rad}}\theta_r = \frac{180}{\pi}\cdot\frac{\pi}{4} = 45°$

3. A central angle of radian measure 2 is an angle subtended by an arc with length twice the length of the radius of the circle.

4. An angle of radian measure 1.5 is larger, since the corresponding degree measure of the angle would be $1.5\left(\frac{180°}{\pi}\right)$ or, approximately, 85.94°.

5. (A) $\theta_d = \dfrac{180°}{\pi \text{ rad}}\theta_r = \dfrac{180}{\pi}\cdot 15.26 = 874.3°$ (B) $\theta_r = \dfrac{\pi \text{ rad}}{180°}\theta_d = \dfrac{\pi}{180}(-389.2) = -6.793 \text{ rad}$

6. $V = r\omega = 25(7.4) = 185$ ft/min

7. $\omega = \dfrac{V}{r} = \dfrac{415}{5.2} = 80$ rad/hr

8. Let $Q(a, b)$ be the point on ray OP that lies on the unit circle (see figure). Segment OP has length $r = \sqrt{(-4)^2 + 3^2} = \sqrt{25} = 5$. The coordinates of $Q(a, b)$ are obtained by dividing the coordinates of P by $r = 5$: $a = -\dfrac{4}{5}$ and $b = \dfrac{3}{5}$.

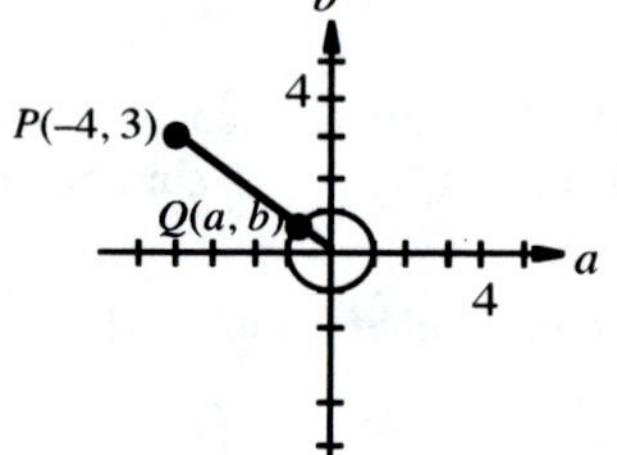

Apply the definition of the trigonometric functions to $Q\left(-\dfrac{4}{5}, \dfrac{3}{5}\right)$.

$\sin\theta = b = \dfrac{3}{5}$ $\tan\theta = \dfrac{b}{a} = -\dfrac{3}{4}$

9. No, since $\csc x = \dfrac{1}{\sin x}$, then one is positive so is the other.

10. (A) Use the reciprocal relationship $\cot\theta = \dfrac{1}{\tan\theta}$.
Degree mode: $\cot 53°40' = \cot(53.666\ldots°)$ Convert to decimal degrees, if necessary.
$= \dfrac{1}{\tan(53.666\ldots°)} = 0.7355$

(B) Use the reciprocal relationship $\csc\theta = \dfrac{1}{\sin\theta}$.
Degree mode: $\csc 67°10' = \csc(67.166\ldots°)$ Convert to decimal degrees, if necessary.
$= \dfrac{1}{\sin(67.1666\ldots°)} = 1.085$

11. (A) Degree mode: $\cos 23.5° = 0.9171$ (B) Degree mode: $\tan 42.3° = 0.9099$

12. (A) Radian mode: $\cos 0.35 = \cos(0.35 \text{ rad}) = 0.9394$ (B) Radian mode: $\tan 1.38 = \tan(1.38 \text{ rad}) = 5.177$

13. The reference angle α is the angle (always taken positive) between the terminal side of θ and the horizontal axis.

(A)

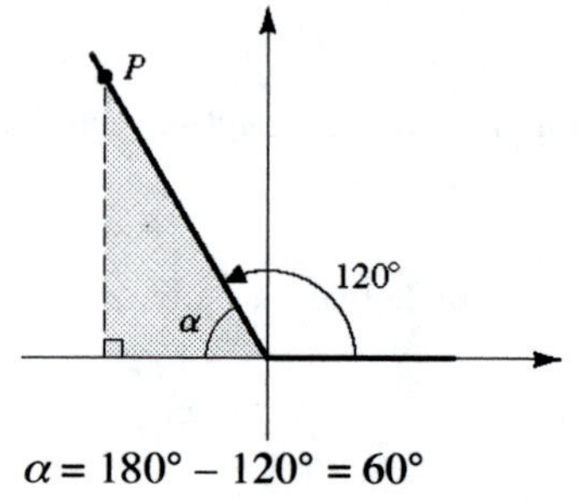

$\alpha = 180° - 120° = 60°$

(B)

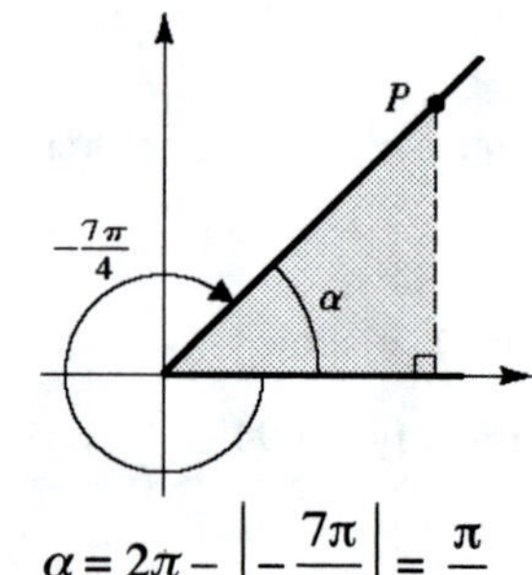

$\alpha = 2\pi - \left|-\dfrac{7\pi}{4}\right| = \dfrac{\pi}{4}$

14. (A) Use the special 30°–60° triangle as the reference triangle. Use the sides of the reference triangle to determine $P(a, b)$ and r. Then use Definition 1.

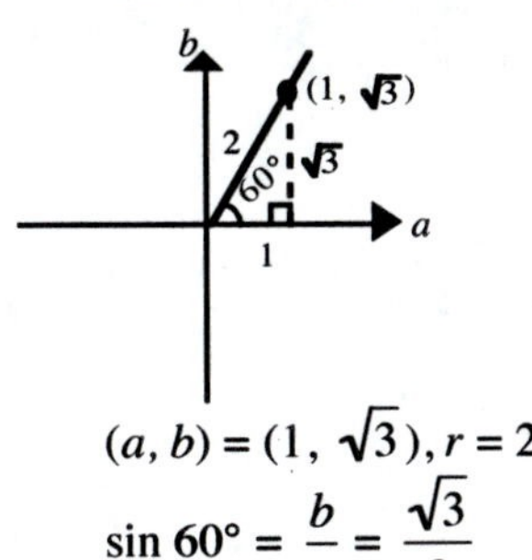

$(a, b) = (1, \sqrt{3}), r = 2$

$\sin 60° = \frac{b}{r} = \frac{\sqrt{3}}{2}$

(B) Use the special 45° triangle as the reference triangle. Use the sides of the reference triangle to determine $P(a, b)$ and r. Then use Definition 1.

$(a, b) = (1, 1), r = \sqrt{2}$

$\cos \frac{\pi}{4} = \frac{a}{r} = \frac{1}{\sqrt{2}}$ or $\frac{\sqrt{2}}{2}$

(C)

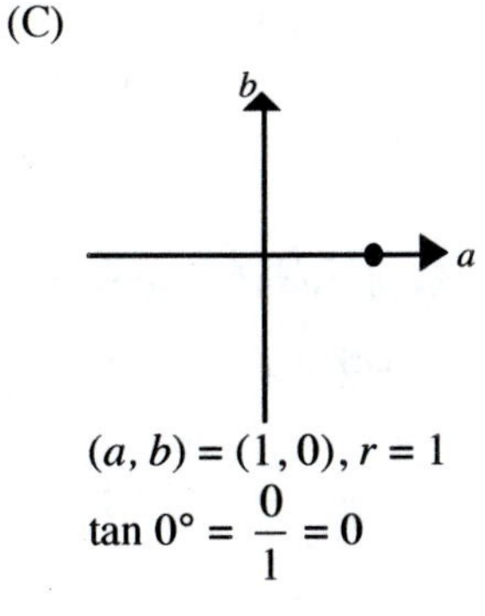

$(a, b) = (1, 0), r = 1$

$\tan 0° = \frac{0}{1} = 0$

15. (A)

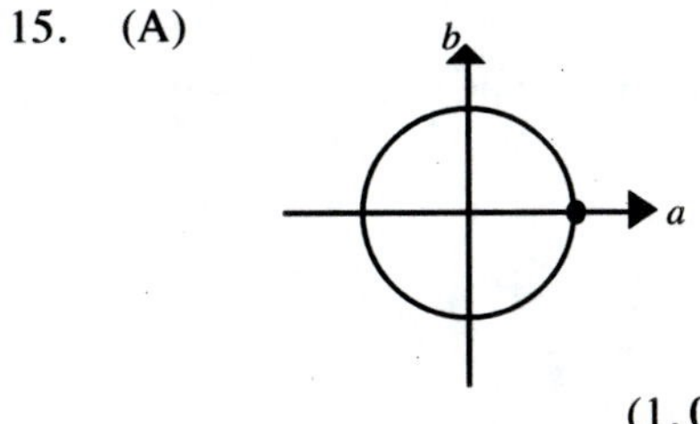

(1, 0)

(B)

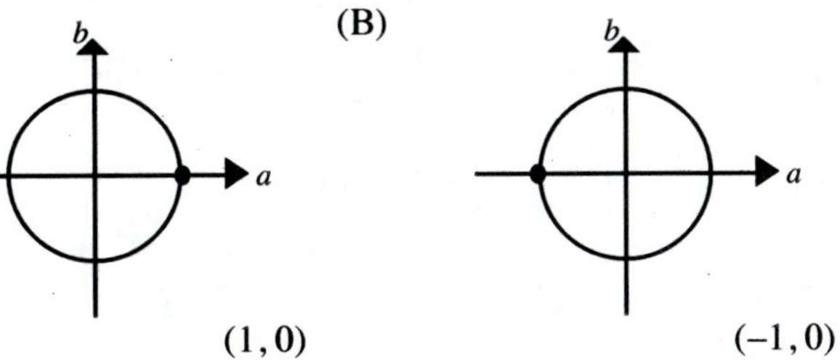

(−1, 0)

(C)

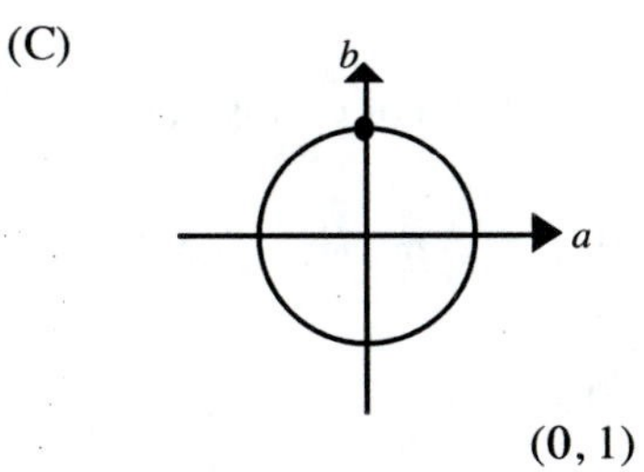

(0, 1)

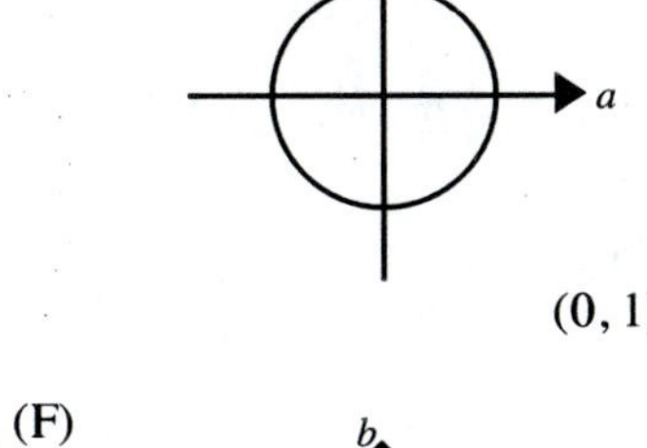

(D)

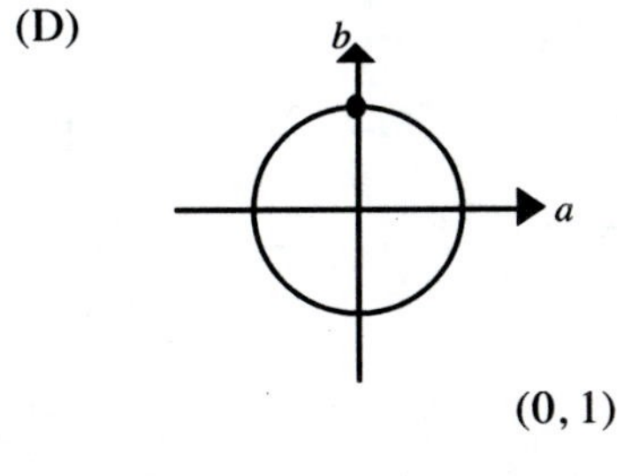

(0, 1)

(E)

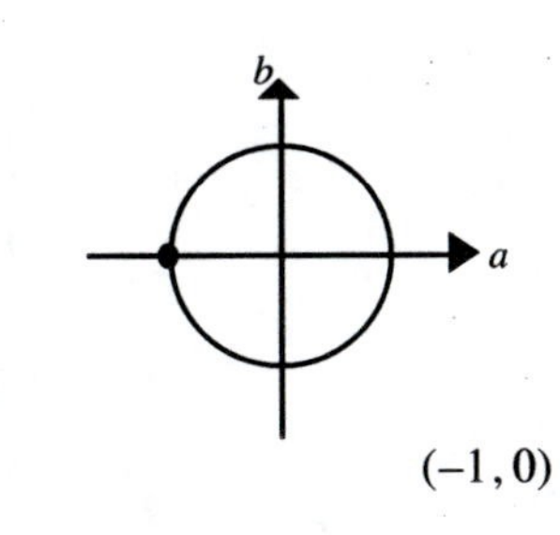

(−1, 0)

(F)

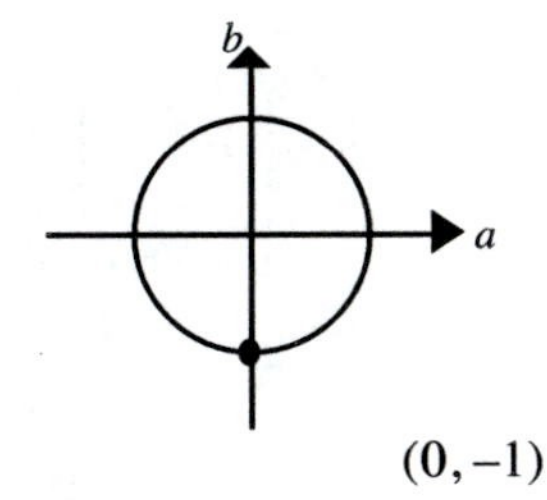

(0, −1)

16. (A) As x varies from 0 to $\frac{\pi}{2}$, $y = \sin x$ varies from 0 to 1.

(B) As x varies from $\frac{\pi}{2}$ to π, $y = \sin x$ varies from 1 to 0.

(C) As x varies from π to $\frac{3\pi}{2}$, $y = \sin x$ varies from 0 to −1.

(D) As x varies from $\frac{3\pi}{2}$ to 2π, $y = \sin x$ varies from −1 to 0.

(E) As x varies from 2π to $\frac{5\pi}{2}$, $y = \sin x$ varies from 0 to 1.

(F) As x varies from $\frac{5\pi}{2}$ to 3π, $y = \sin x$ varies from 1 to 0.

17. When the terminal side of the angle is rotated any multiple of a complete revolution (2π rad), in either direction the resulting angle will be coterminal with the original. In this case, for the restricted interval, this happens for $\frac{\pi}{6} \pm 2\pi$. $\frac{\pi}{6} + 2\pi = \frac{13\pi}{6}$; $\frac{\pi}{6} - 2\pi = \frac{-11\pi}{6}$

18. Since the central angle subtended by a circumference has degree measure 360°, the central angle subtended by an arc $\frac{7}{60}$ of a circumference has degree measure $\frac{7}{60}(360°) = 42°$.

19. $s = r\theta = 4(1.5) = 6$ cm.

20. $\theta_r = \frac{\pi \text{ rad}}{180°}\theta_d = \frac{\pi}{180}212 = \frac{53\pi}{45}$

21. $\theta_d = \frac{180°}{\pi \text{ rad}}\theta_r = \frac{180}{\pi} \cdot \frac{\pi}{12} = 15°$

22. (A) –3.72 (B) 264.71°

23. Yes, since $\theta_d = (180°/\pi \text{ rad})\theta_r$, and if θ_r is tripled, θ_d will also be tripled.

24. No. For example, if $\alpha = \frac{\pi}{6}$ and $\beta = \frac{5\pi}{6}$, α and β are not coterminal, but $\sin\frac{\pi}{6} = \sin\frac{5\pi}{6}$.

25. Use the reciprocal identity: $\csc x = \frac{1}{\sin x} = \frac{1}{0.8594} = 1.1636$.

26. (A)

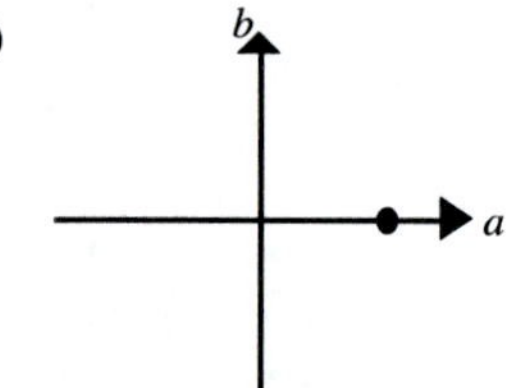

$(a, b) = (1, 0), r = 1$

$\tan 0 = \frac{b}{a} = \frac{0}{1} = 0$

(B)

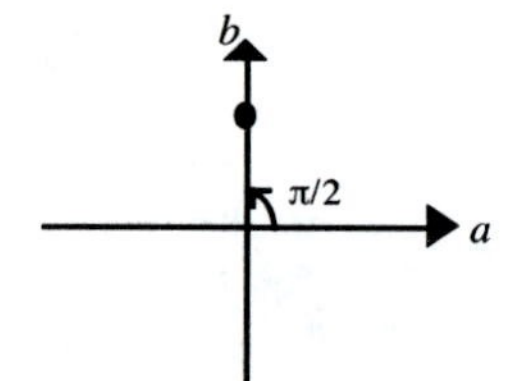

$(a, b) = (0, 1), r = 1$

$\tan\frac{\pi}{2} = \frac{b}{a} = \frac{1}{0}$

Not defined.

(C)

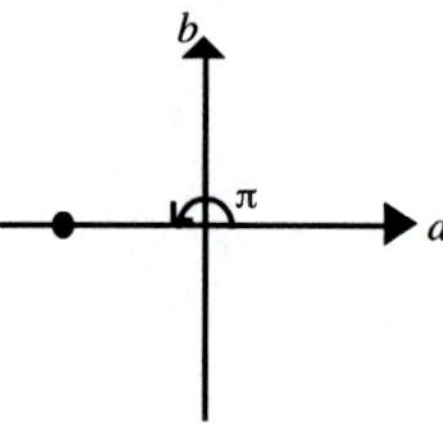

$(a, b) = (-1, 0), r = 1$

$\tan \pi = \frac{b}{a} = \frac{0}{-1} = 0$

(D)

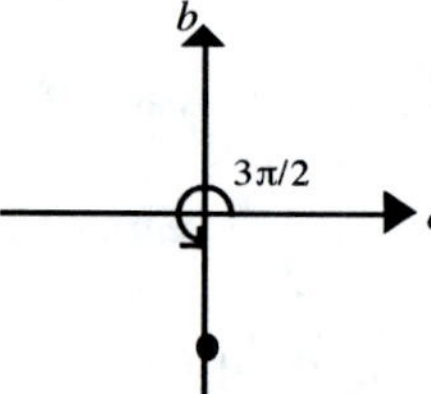

$(a, b) = (0, -1), r = 1$

$\tan\frac{3\pi}{2} = \frac{b}{a} = \frac{-1}{0}$

Not defined.

27. (A) 732° is coterminal with (732 – 360°) = 372°, and thus with (372 – 360)° = 12°. Since 12° is between 0° and 90°, its terminal side lies in quadrant I.

(B) –7 rad is coterminal with $(-7 + 2\pi)$rad ≈ –0.72 rad. Since –0.72 rad is between 0 and $-\frac{\pi}{2}$, its terminal side lies in quadrant IV.

28. Degree mode: $\csc(-3.2°) = \dfrac{1}{\sin(-3.2°)} = -17.9$

29. Degree mode: $\cot 183.5° = \dfrac{1}{\tan 183.5°} = 16.3$

30. Degree mode: $\tan 45°15' = \tan\left(45+\dfrac{15}{60}\right)^{\circ} = 1.01$

31. Radian mode: $\sin 2.8\pi = 0.588$

32. Radian mode: $\cos 11.5 = 0.483$

33. Radian mode: $\sec(-4.7) = \dfrac{1}{\cos(-4.7)} = -80.7$

34. Locate the 30°–60° reference triangle, determine (a, b) and r, then evaluate.

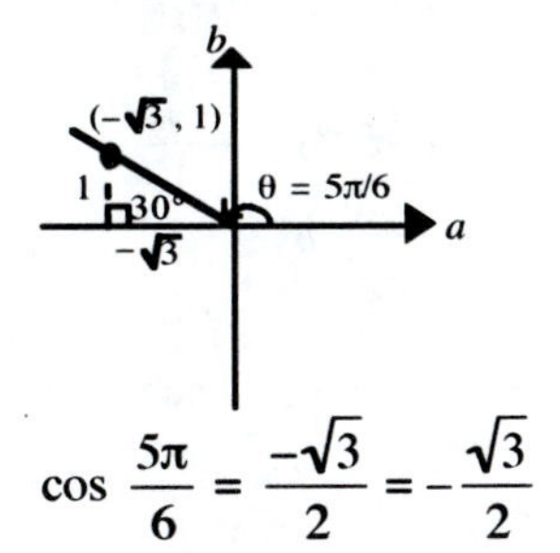

$$\cos\frac{5\pi}{6} = \frac{-\sqrt{3}}{2} = -\frac{\sqrt{3}}{2}$$

35. Locate the 45° reference triangle, determine (a, b) and r, then evaluate.

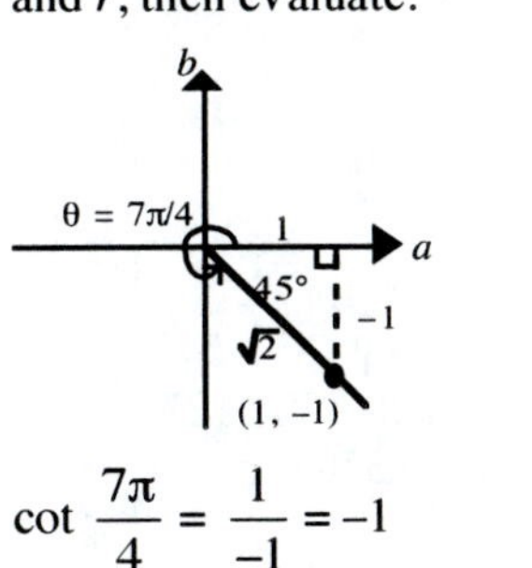

$$\cot\frac{7\pi}{4} = \frac{1}{-1} = -1$$

36.

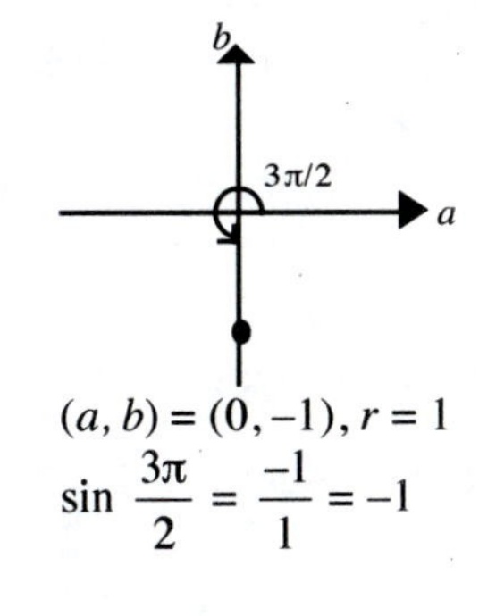

$(a, b) = (0, -1)$, $r = 1$

$$\sin\frac{3\pi}{2} = \frac{-1}{1} = -1$$

37. See Problem 36.

$$\cos\frac{3\pi}{2} = \frac{0}{1} = 0$$

38. Locate the 30°–60° reference triangle, determine (a, b) and r, then evaluate.

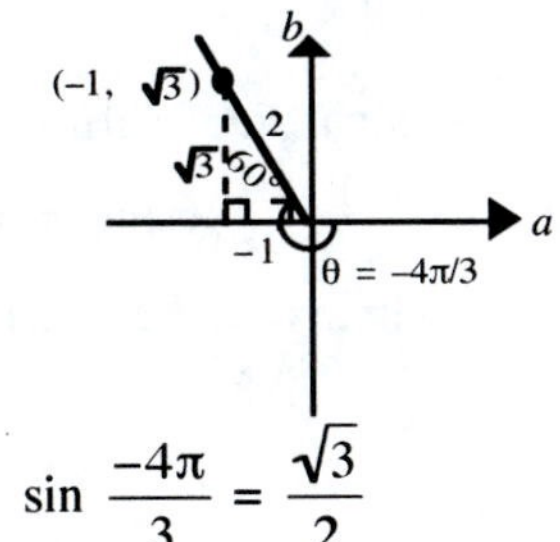

$$\sin\frac{-4\pi}{3} = \frac{\sqrt{3}}{2}$$

39. See Problem 38.

$$\sec\frac{-4\pi}{3} = \frac{2}{-1} = -2$$

40. $(a, b) = (-1, 0), r = 1$

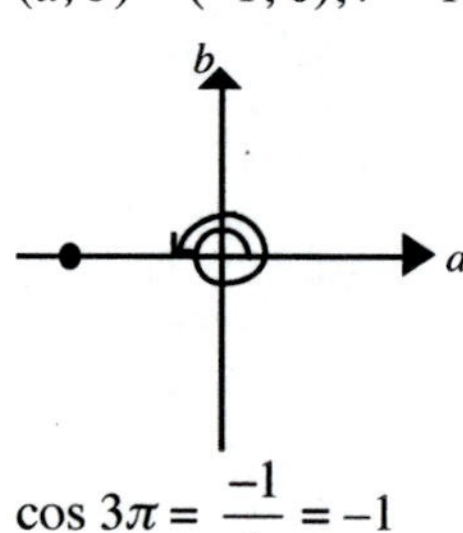

$\cos 3\pi = \frac{-1}{1} = -1$

41. See Problem 40.
$\cot 3\pi = \frac{-1}{0}$
Not defined.

42. Locate the 30°–60° reference triangle, determine (a, b) and r, then evaluate.

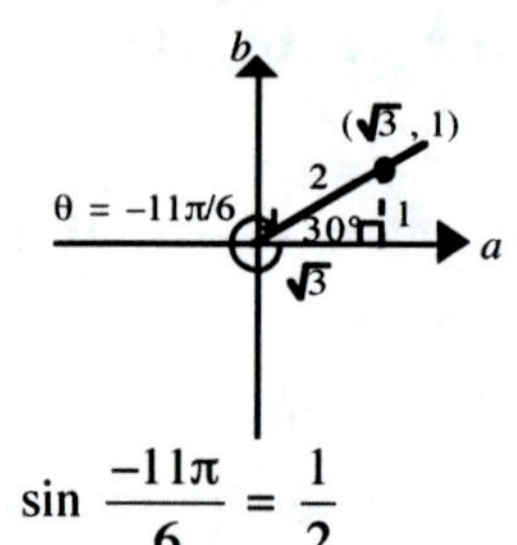

$\sin \frac{-11\pi}{6} = \frac{1}{2}$

43. It is known that $\cos\left(\frac{\pi}{3}\right) = \frac{1}{2} = 0.5000\ldots$ and $\sin(180°) = 0 = 0.0000\ldots$, so that the calculator displays exact values for these. Thus $\tan(-60°)$ is not given exactly. To find $\tan(-60°)$, locate the 30°–60° reference triangle, determine (a, b) and r, then evaluate.

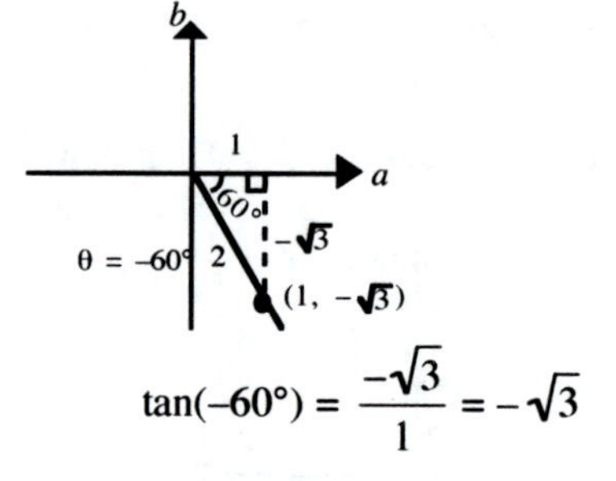

$\tan(-60°) = \frac{-\sqrt{3}}{1} = -\sqrt{3}$

44. Degree mode: $\sin 384.0314° = 0.40724$

45. Degree mode: $\tan(-198°43'6'') = \tan(-198.71833\ldots°)$ Convert to decimal degrees, if necessary.
$= -0.33884$

46. Radian mode: $\cos 26 = 0.64692$

47. Use the reciprocal relationship $\cot \theta = \frac{1}{\tan \theta}$.

Radian mode: $\cot(-68.005) = \frac{1}{\tan(-68.005)}$
$= 0.49639$

48. Because $\sin \theta$ is negative and the terminal side of θ does not lie in the third quadrant, it must lie in the fourth quadrant. Because $\sin \theta = \frac{b}{r} = -\frac{4}{5}$, we let $b = -4, r = 5$. We find a so that (a, b) is the point in quadrant IV that lies on the circle of radius 5 with center the origin.

$$a^2 + b^2 = r^2$$
$$a^2 + 16 = 25$$
$$a^2 = 9$$
$$a = 3$$

We now apply the third remark following Definition 1, Section 2.3, with $a = 3, b = -4$, and $r = 5$.

$\cos \theta = \frac{a}{r} = \frac{3}{5}$ $\qquad$ $\tan \theta = \frac{b}{a} = -\frac{4}{3}$

49. Draw a reference triangle in the third quadrant with side opposite reference angle –1 and hypotenuse 2. Observe that this is a special 30°–60° triangle.

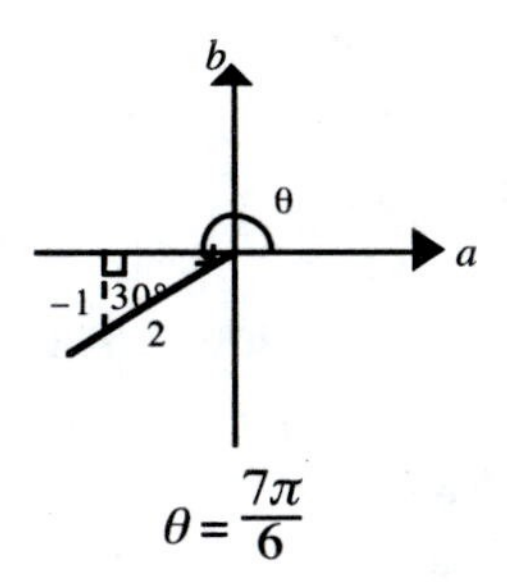

$\theta = \frac{7\pi}{6}$

50. Because sin θ and tan θ are negative, θ is a quadrant IV angle. Because $\sin\theta = \frac{b}{r} = -\frac{2}{5}$, we let $b = -2$, $r = 5$. We find a so that (a, b) is the point in quadrant IV that lies on the circle of radius 5 with center the origin.

$$\begin{aligned} a^2 + b^2 &= r^2 \\ a^2 + 4 &= 25 \\ a^2 &= 21 \\ a &= \sqrt{21} \end{aligned}$$

We now apply the third remark following Definition 1, Section 2.3, with $a = \sqrt{21}$, $b = -2$, and $r = 5$:

$\left(\sin\theta = \frac{b}{r} = -\frac{2}{5}\right)$ $\qquad$ $\csc\theta = \frac{r}{b} = -\frac{5}{2}$

$\cos\theta = \frac{a}{r} = \frac{\sqrt{21}}{5}$ $\qquad$ $\sec\theta = \frac{r}{a} = \frac{5}{\sqrt{21}}$

$\tan\theta = \frac{b}{a} = -\frac{2}{\sqrt{21}}$ $\qquad$ $\cot\theta = \frac{a}{b} = -\frac{\sqrt{21}}{2}$

51. $\tan\theta = \frac{b}{a} = -1 = \frac{1}{1}$ or $\frac{1}{1}$

We can draw a reference triangle in the second quadrant with side opposite reference angle 1 and side adjacent –1. We can also draw a reference triangle in the fourth quadrant with side opposite reference angle –1 and side adjacent 1. Each triangle is a special 45° triangle.

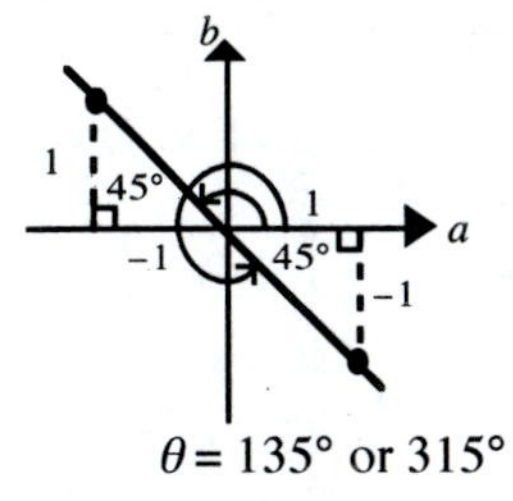

$\theta = 135°$ or $315°$

52. We can draw reference triangles in both quadrants II and III with sides adjacent reference angle $-\sqrt{3}$ and hypotenuse 2. Each triangle is a special 30°–60° triangle.

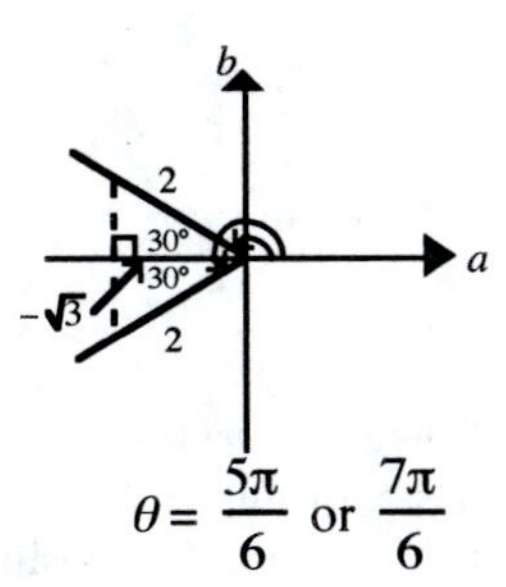

$\theta = \frac{5\pi}{6}$ or $\frac{7\pi}{6}$

53. (A) $s = r\theta = (12.0)(1.69) \approx 20.3$ cm $\qquad$ (B) $s = \frac{\pi}{180} r\theta = \frac{\pi}{180}(12.0)(22.5) \approx 4.71$ cm

54. (A) $A = \frac{1}{2}r^2\theta$

$\left[r = \frac{1}{2}D = \frac{1}{2}(80) = 40 \text{ ft}\right]$

$A = \frac{1}{2}(40)^2(0.773) \approx 618 \text{ ft}^2$

(B) $A = \frac{\pi}{360}r^2\theta = \frac{\pi}{360}(40)^2(135) \approx 1{,}880 \text{ ft}^2$

55. Since the cities have the same longitude, θ is given by their difference in latitude.

$\theta = 41°28' - 38°21' = 3°7' = \left(3 + \frac{7}{60}\right)°$

$s = \frac{\pi}{180}r\theta = \frac{\pi}{180}(3{,}694 \text{ miles})\left(3 + \frac{7}{60}\right)$

$= 215.6$ miles

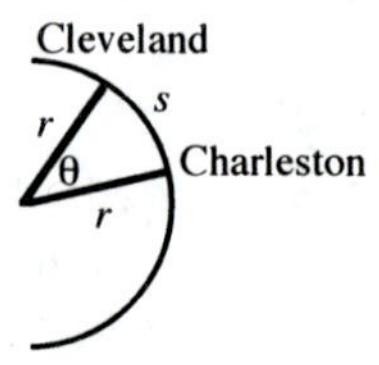

56. $\omega = \frac{\theta}{t} = \frac{6.43}{15.24} = 0.422$ rad/sec

57. A radial line from the axis of rotation sweeps out at an angle at the rate of 12π rad per sec.

58. (A) Calculator in radian mode: $\cos 7 = 0.754$

(B), (C) Both should equal 0.754, because the cosine function is periodic with period 2π.

59. Calculator in radian mode:

(A) $\tan(-7) = -0.871$
$-\tan 7 = -0.871$

(B) $\tan[-(-17.9)] = -1.40$
$-\tan(-17.9) = -1.40$

(C) $\tan[-(-2{,}135)] = -3.38$
$-\tan(-2{,}135) = -3.38$

Note: Some (very old model) calculators cannot evaluate tan(–2,135) and, instead, signal an error. If this occurs, use the periodicity of the tangent function and evaluate $\tan(2{,}135 - 2\pi \cdot k)$, where k is an appropriate integer.

60. (D) is not an identity, since it is not true for all values of the variable x.

61. $(\csc x)(\cot x)(1 - \cos^2 x) = \frac{1}{\sin x}\frac{\cos x}{\sin x}(1 - \cos^2 x)$ Use Identities (1) and (5)

$= \frac{\cos x}{\sin^2 x}(\sin^2 x + \cos^2 x - \cos^2 x)$ Use Identity (9)

$= \frac{\cos x}{\sin^2 x} \cdot \sin^2 x = \cos x$

62. $\cot(-x)\sin(-x) = \frac{\cos(-x)}{\sin(-x)}\sin(-x)$ Use Identity (5)

$= \cos(-x) = \cos x$ Use Identity (7)

63. Since $P(a, b)$ is moving clockwise, $x = -29.37$. By the definition of the circular functions, the point has coordinates $P(\cos(-29.37), \sin(-29.37)) = P(-0.4575, 0.8892)$. P lies in quadrant II, since a is negative and b is positive.

64. The radian measure of a central angle θ subtended by an arc of length s is $\theta = \frac{s}{r}$, where r is the radius of the circle. In this case $\theta = \frac{1.3}{1} = 1.3$ rad.

65. Since $\cot x = \dfrac{1}{\tan x}$ and $\csc x = \dfrac{1}{\sin x}$, and $\sin(k\pi) = \tan(k\pi) = 0$ for all integers k, $\cot x$ and $\sec x$ are not defined for these values.

66. $\sin x = -\dfrac{\sqrt{3}}{2}$

We can draw a reference triangle in the third quadrant with side opposite reference angle $-\sqrt{3}$ and hypotenuse 2. Observe that this is a special 30°–60° triangle.

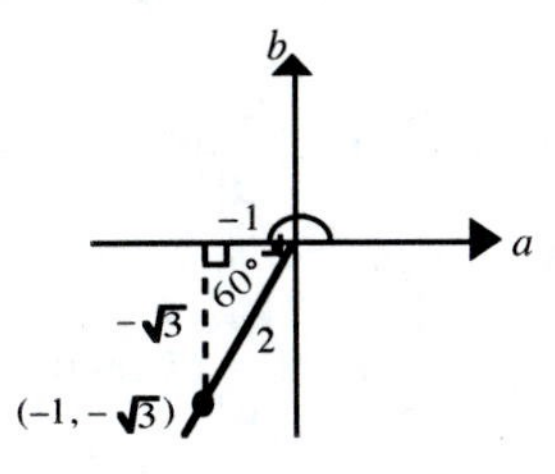

$$x - \pi = \frac{\pi}{3}$$
$$x = \frac{4\pi}{3}$$

67. Since $A = \dfrac{1}{2}r^2\theta$ and $s = r\theta$, we can eliminate θ between these two equations as follows:

$$\theta = \frac{s}{r}$$
$$A = \frac{1}{2}r^2\left(\frac{s}{r}\right) = \frac{1}{2}rs.$$

Then, $s = \dfrac{2A}{r} = \dfrac{2(342.5)}{12} \approx 57$ m.

68. From the figure, we note $s = r\theta$.

$$r = \sqrt{a^2 + b^2} = \sqrt{4^2 + 5^2} = \sqrt{41}$$
$$\tan\theta = \frac{b}{a} = \frac{5}{4}$$
$$\theta = \tan^{-1}\frac{5}{4}$$

Thus, $s = \sqrt{41}\,\tan^{-1}\dfrac{5}{4} \approx 5.74$ units.

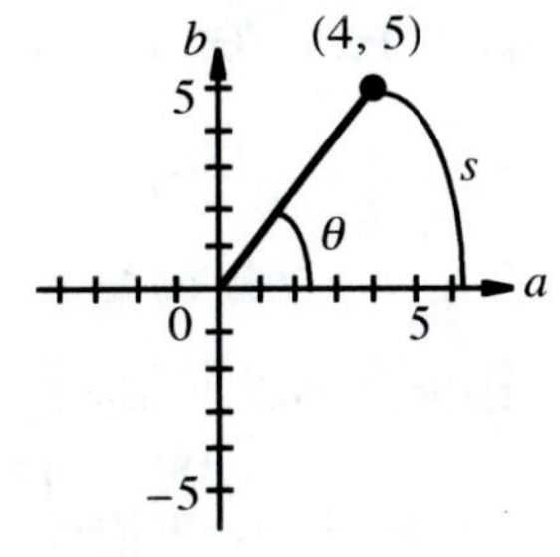

(Calculator in radian mode.)

69. We use $s = r\theta$ with $r = \dfrac{1}{2}d = 5$ cm and $s = 10$ m $= 1000$ cm.

Then, $\theta = \dfrac{s}{r} = \dfrac{1000}{5} = 200$ rad. Since 1 revolution corresponds to 1 circumference $= 2\pi$ radians, 200 radians corresponds to $\dfrac{200}{2\pi} = \dfrac{100}{\pi} \approx 31.8$ revolutions.

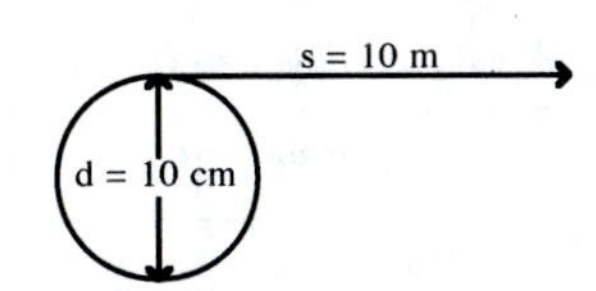

70. Since the three gear wheels are coupled together, each must turn through the same distance (arc length). Thus,

$s = r_1\theta_1 \quad s = r_2\theta_2 \quad s = r_3\theta_3 \quad r_1 = 30 \text{ cm} \quad r_2 = 20 \text{ cm} \quad r_3 = 10 \text{ cm}$

$$r_1\theta_1 = r_2\theta_2$$

$$\theta_2 = \frac{r_1}{r_2}\theta_1 = \frac{30}{20}(5 \text{ revolutions}) = 7.5 \text{ revolutions}$$

$$r_3\theta_3 = r_1\theta_1$$

$$\theta_3 = \frac{r_1}{r_3}\theta_1 = \frac{30}{10}(5 \text{ revolutions}) = 15 \text{ revolutions}$$

71. We use $\omega = \frac{V}{r}$ with $V = 70$ ft/sec and $r = \frac{27 \text{ in.}}{2} = \frac{27 \text{ in.}}{2} \cdot \frac{1 \text{ in.}}{12 \text{ ft}} = \frac{27}{24}\text{ft.}$

Then $\omega = \frac{V}{r} = \frac{70}{27/24} = 62$ rad/sec.

72. We use $V = r\omega$.

$$r = \text{radius of orbit} = 3{,}964 + 1{,}000 = 4{,}964 \text{ mi}$$

$$\omega = \frac{\theta}{t} = \frac{2\pi}{114/60 \text{ hr}} = \frac{120\pi}{114}\text{rad/hr}$$

$$V = (4{,}964 \text{ mi})\left(\frac{120\pi}{114}\text{rad/hr}\right) = 16{,}400 \text{ mi/hr}$$

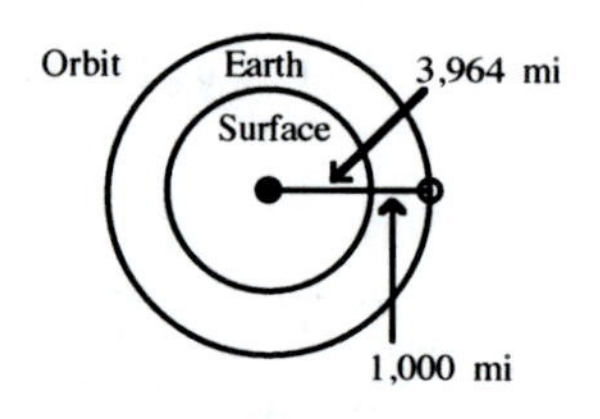

73. $I = 30\sin(120\pi t - 60\pi) = 30\sin[120\pi(0.015) - 60\pi] = -17.6$ amp

74. (A) We note:

The length of the ladder $AC = AB + BC$

In triangle ABE, $\csc\theta = \frac{AB}{BE} = \frac{AB}{10 \text{ ft}}$

In triangle BCD, $\sec\theta = \frac{BC}{BD} = \frac{BC}{2 \text{ ft}}$

Then $AB = 10\csc\theta$, $BC = 2\sec\theta$, hence length of ladder $AC = 10\csc\theta + 2\sec\theta$

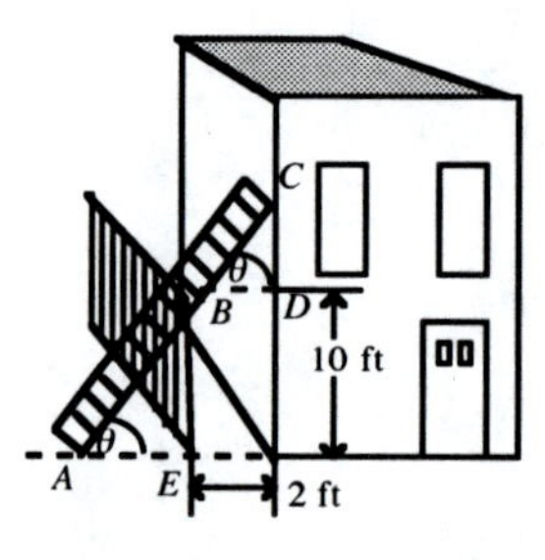

(B) As θ decreases to 0 rad, L will increase without bound; as θ increases to $\frac{\pi}{2}$, L increases without bound. Between these extremes, there appears to be a value of θ that produces a minimum L.

(C)

θ rad	0.70	0.80	0.90	1.00	1.10	1.20	1.30
L ft	18.14	16.81	15.98	15.59	15.63	16.25	17.85

(D) $L = 15.59$ ft for $\theta = 1.00$ rad

75. Use $\frac{n_2}{n_1} = \frac{\sin\alpha}{\sin\beta}$, where $n_2 = 1.33, n_1 = 1.00$, and $\alpha = 31.7°$

Solve for β:
$$\frac{1.33}{1.00} = \frac{\sin 31.7°}{\sin\beta}$$
$$\sin\beta = \frac{\sin 31.7°}{1.33}$$
$$\beta = \sin^{-1}\left(\frac{\sin 31.7°}{1.33}\right)$$
$$= 23.3°$$

76. Find the angle of incidence α such that the angle of refraction is 90°.

$$\frac{\sin\alpha}{\sin\beta} = \frac{n_2}{n_1} \quad \sin\alpha = \frac{n_2}{n_1}\sin\beta = \frac{1.00}{1.52}\sin 90° \quad \alpha = \sin^{-1}\left[\frac{1.00}{1.52}(1)\right] = 41.1°$$

77. We use $\sin\frac{\theta}{2} = \frac{S_w}{S_b}$, where $\theta = 51°$ and $S_b = 25$ mph. Then we solve for S_w :

$$\sin\frac{51°}{2} = \frac{S_w}{25} \qquad S_w = 25\sin 25.5° = 11 \text{ mph}$$

Chapter 3 Graphing Trigonometric Functions

EXERCISE 3.1 Basic Graphs

1. $2\pi, 2\pi, \pi$

3. Since $\sin x$ represents the y coordinate of a point moving on the unit circle from $(1, 0)$, the intercepts of the graph of $y = \sin x$ occur when the point is on the x axis, that is, when the point has moved $0, \pi, 2\pi, \ldots$ and $-\pi, -2\pi, \ldots$
Since $\cos x$ represents the x coordinate of the point, the intercepts of its graph occur when the point is on the y axis, that is, when the point has moved $\frac{\pi}{2}, \frac{3\pi}{2}, \frac{5\pi}{2}, \ldots$ and $-\frac{\pi}{2}, -\frac{3\pi}{2}, \ldots$.

5. Draw vertical asymptotes through the x intercepts of the graph of $y = \sin x$ (at $x = n\pi$, n any integer). Note the points $x = \frac{\pi}{2} + 2n\pi, y = 1$ and $x = -\frac{\pi}{2} + 2n\pi, y = -1$. Take reciprocal values for a few points between $-\pi$ and 0, and between 0 and π and sketch the graph between the asymptotes.

7. (A) 1 unit (B) Indefinitely far (C) Indefinitely far

9. $-\frac{3\pi}{2}, -\frac{\pi}{2}, \frac{\pi}{2}, \frac{3\pi}{2}$ 11. $-2\pi, -\pi, 0, \pi, 2\pi$ 13. The graph has no x intercepts; $\sec x$ is never 0.

15. (A) None; $\sin x$ is always defined. (B) $-2\pi, -\pi, 0, \pi, 2\pi$ (C) $-\frac{3\pi}{2}, -\frac{\pi}{2}, \frac{\pi}{2}, \frac{3\pi}{2}$

17.

x	0	0.1	0.2	0.3	0.4	0.5	0.6	0.7	0.8
$\cos x$	1	1.0	0.98	0.96	0.92	0.88	0.83	0.76	0.70

x	0.9	1.0	1.1	1.2	1.3	1.4	1.5	1.6
$\cos x$	0.62	0.54	0.45	0.36	0.27	0.17	0.07	−0.03

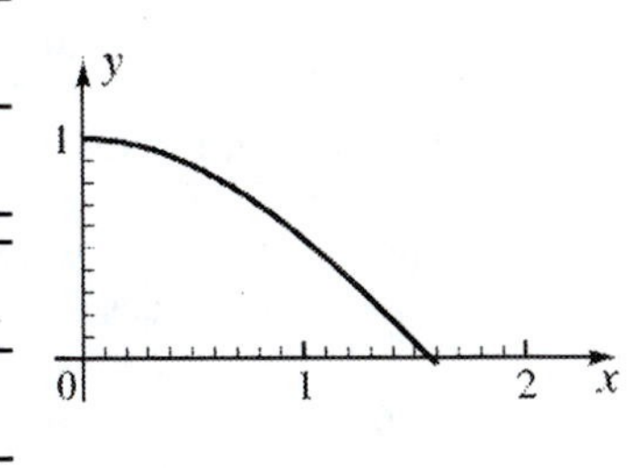

19.

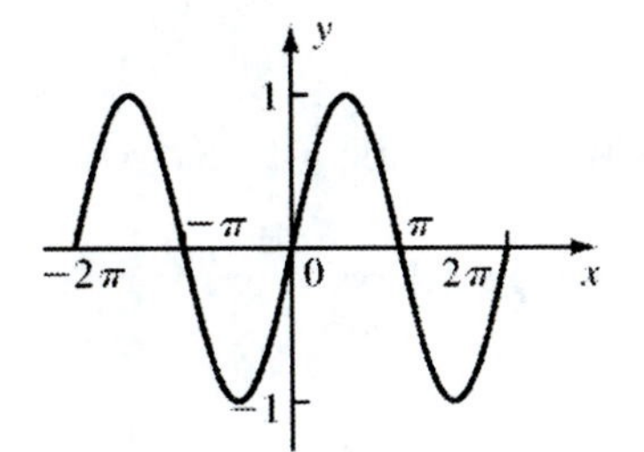

21.

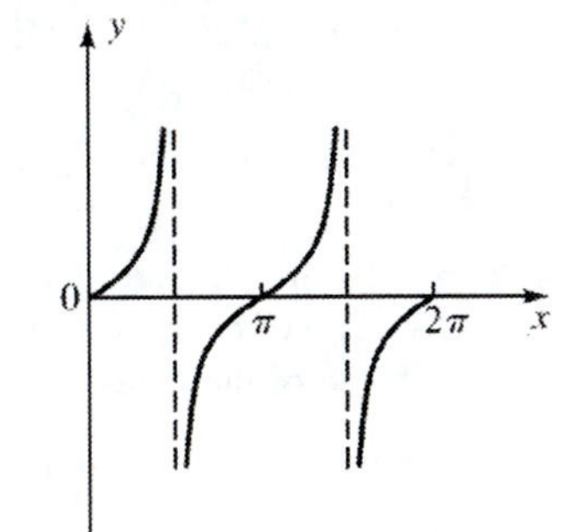

23. The dashed line shows $y = \sin x$ in this interval. The solid line is $y = \csc x$.

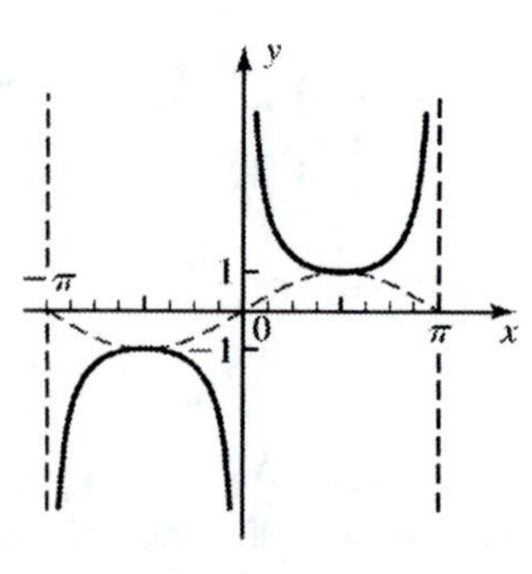

25. (A)

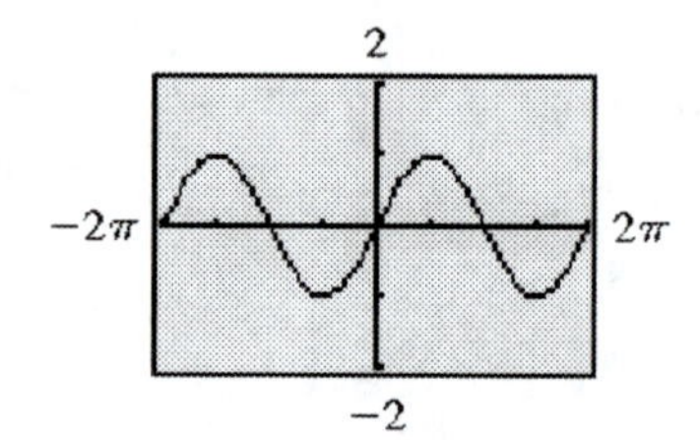

(B)

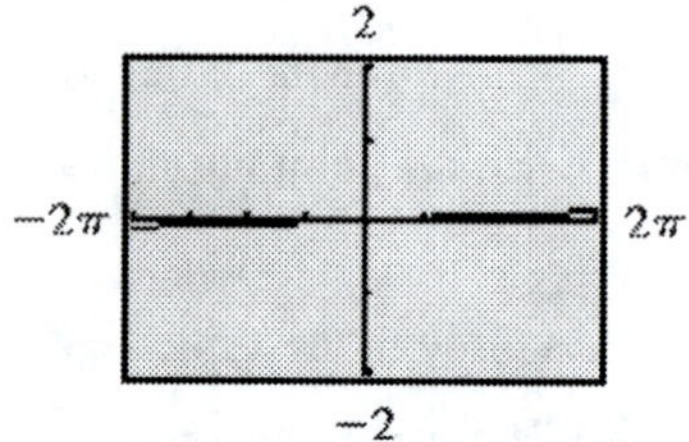

(C) The mode setting is crucial. Degree mode will make the graph totally different.

27. The range of the cosine function is the set of all real numbers between –1 and 1:

$-1 \le \cos x \le 1$ Multiply both inequalities by 5

$-5 \le 5 \cos x \le 5$

Therefore, $-5 \le y \le 5$ and Max $y = 5$, Min $y = -5$.

29. The range of the sine function is the set of all real numbers between –1 and 1:

$-1 \le \sin x \le 1$ Add 9 to both inequalities

$9 - 1 \le 9 + \sin x \le 9 + 1$ Simplify

$8 \le 9 + \sin x \le 10$

Therefore, $8 \le y \le 10$ and Max $y = 10$, Min $y = 8$.

31. The cosecant function has no maximum and no minimum, so Max y and Min y do not exist.

33. The range of the cosine function is the set of all real numbers between –1 and 1:

$-1 \le \cos x \le 1$ Multiply both inequalities by 7

$-7 \le 7 \cos x \le 7$ Add –6 to both inequalities

$-6 - 7 \le -6 + 7 \cos x \le -6 + 7$ Simplify

$-13 \le -6 + 7 \cos x \le 1$

Therefore, $-13 \le y \le 1$ and Max $y = 1$, Min $y = -13$.

35. The tangent function has no maximum and no minimum, so Max y and Min y do not exist.

37. Depending on the particular calculator used, either an error message will occur or some very large number will occur because of round-off error, because:

(A) 0 is not in the domain of the cotangent function.

(B) $\frac{\pi}{2}$ is not in the domain of the tangent function.

(C) π is not in the domain of the cosecant function.

39. (A) Both graphs are almost indistinguishable the closer x is to the origin.

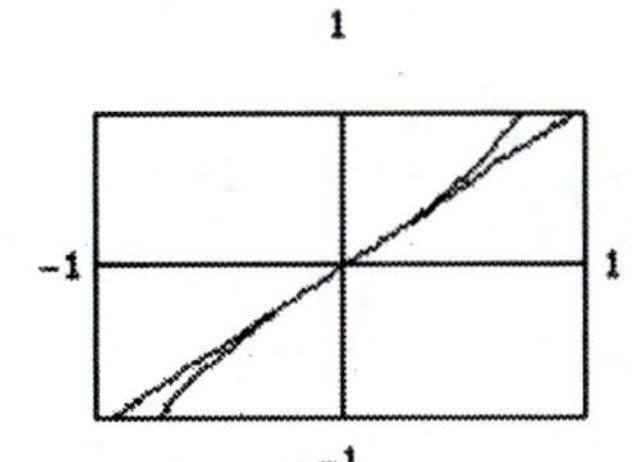

(B)

x	−0.3	−0.2	−0.1	0.0	0.1	0.2	0.3
$\tan x$	−0.309	−0.203	−0.100	0.000	0.100	0.203	0.309

(C) It is not valid to replace $\tan x$ with x for small x if x is in degrees, as is clear from the graph.

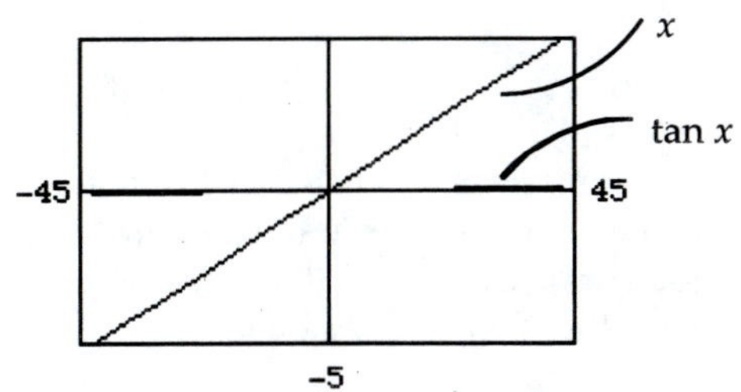

41. For a given value of T, the y value on the unit circle and the corresponding y value on the sine curve are the same. This is a graphing utility illustration of how the sine function is defined as a circular function. See Figure 3 in this section of the text.

43. Comparing the graph of $y = \sin x$, which has x intercepts at $x = 0$ and $x = \pi$ and a maximum at $x = \frac{\pi}{2}$, with the graph of $y = \cos x$, which has x intercepts at $x = \frac{3\pi}{2}$ and $x = \frac{5\pi}{2}$ and a maximum at $x = 2\pi$, the graph of $y = \sin x$ could be shifted $\frac{3\pi}{2}$ units to the right to coincide with the graph of $y = \cos x$.

45. Comparing the graph of $y = \sec x$, which has a minimum at $x = 0$ and asymptotes at $x = -\frac{\pi}{2}$ and $x = \frac{\pi}{2}$, with the graph of $y = \csc x$, which has a minimum at $x = -\frac{3\pi}{2}$ and asymptotes at $x = -2\pi$ and $x = -\pi$, the graph of $y = \sec x$ could be shifted $\frac{3\pi}{2}$ units to the left to coincide with the graph of $y = \csc x$.

47.

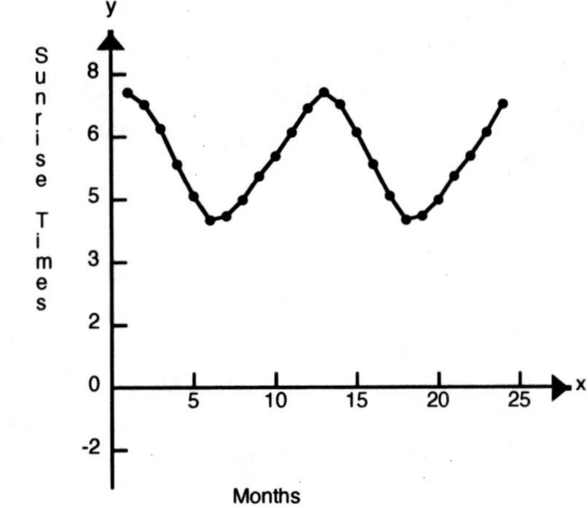

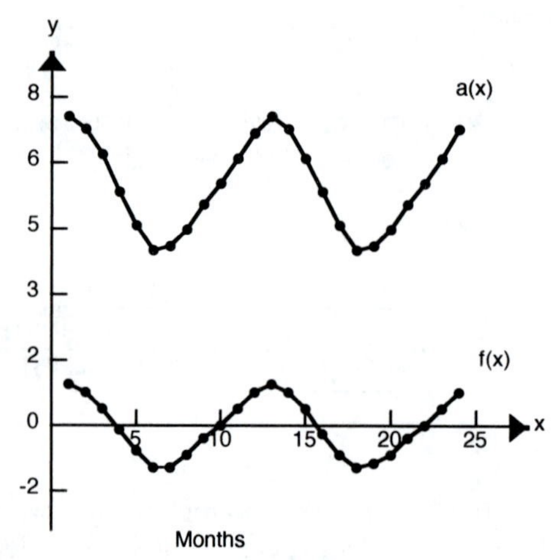

49. The maximum value shown in the table is $y_{max} = 7.4$. The minimum value is $y_{min} = 4.2$.
Therefore, set amplitude $= |A| = \dfrac{7.4 - 4.2}{2} = 1.6$
Half-way mark $= k = |A| + y_{min} = 1.6 + 4.2 = 5.8$
Then $f(x) = \dfrac{a(x) - 5.8}{1.6}$ will be a function transforming $a(x)$ to have minimum $f(6) = -1$ and maximum $f(1) = -1$.

51. $d(x)$ represents the changing difference, over a 24-month period, between sunrise and sunset times, and therefore can be best interpreted as the changing length of the daylight over this period. Since the seasons and their attendant daylight lengths repeat every year, the three functions are expected to be (approximately) periodic.

EXERCISE 3.2 Graphing $y = k + A \sin Bx$ and $y = k + A \cos Bx$

1. The amplitude of the graph of a sine or cosine function represents one-half the distance between the maximum and the minimum values of the function.

3. For any periodic phenomenon, if P is the period and f is the frequency, each quantity is the reciprocal of the other, thus,

$$P = \frac{1}{f} \text{ and } f = \frac{1}{p}$$

5.

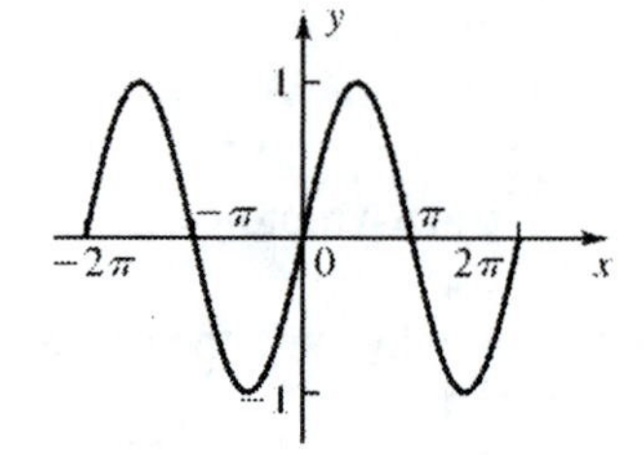

7. $y = -2 \sin x$. Amplitude $= |-2| = 2$. Period $= \dfrac{2\pi}{B} = \dfrac{2\pi}{1} = 2\pi$.
Since $A = -2$ is negative, the basic curve for $y = \sin x$ is turned upside down. One full cycle of the graph is completed as x goes from 0 to 2π. Block out this interval, divide it into four equal parts, locate high and low points, and locate x intercepts. Then complete the graph.

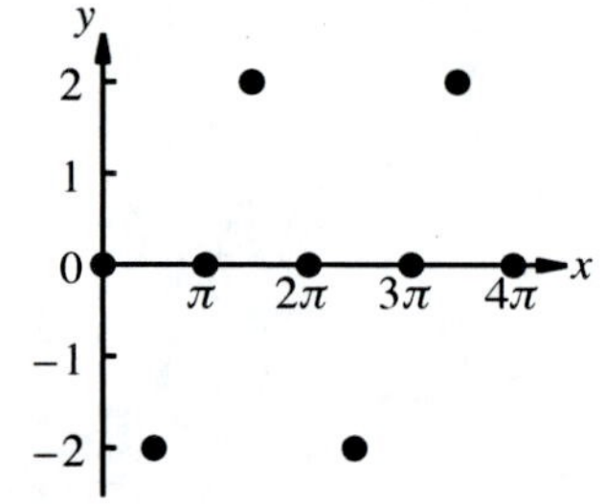

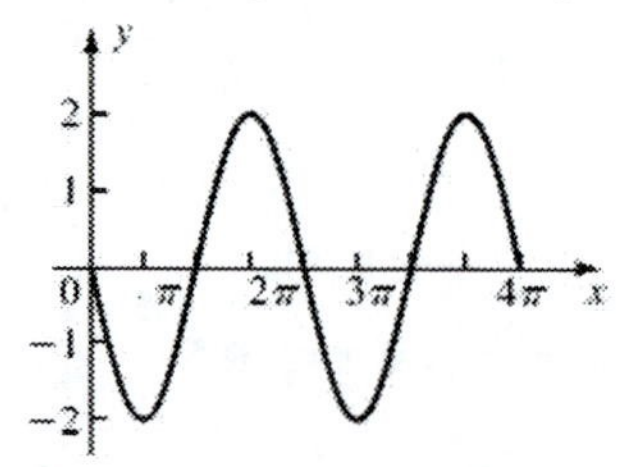

9. $y = \frac{1}{2}\sin x$. Amplitude $= \left|\frac{1}{2}\right| = \frac{1}{2}$.

Period $= \frac{2\pi}{B} = \frac{2\pi}{1} = 2\pi$.

One full cycle of the graph is completed as x goes from 0 to 2π. Block out this interval, divide it into four equal parts, locate high and low points, and locate x intercepts. Then complete the graph.

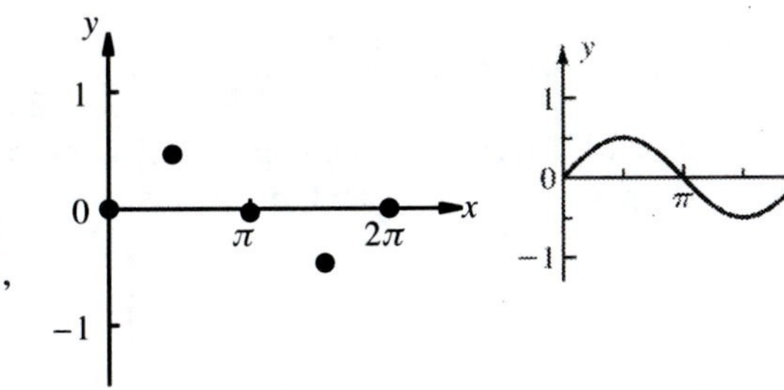

11. $y = \sin 2\pi x$. Amplitude $= |A| = |1| = 1$. Period $= \frac{2\pi}{2\pi} = 1$.

One full cycle of this graph is completed as x goes from 0 to 1. Block out this interval, divide it into four equal parts, locate high and low points, and locate x intercepts. Then complete the graph.

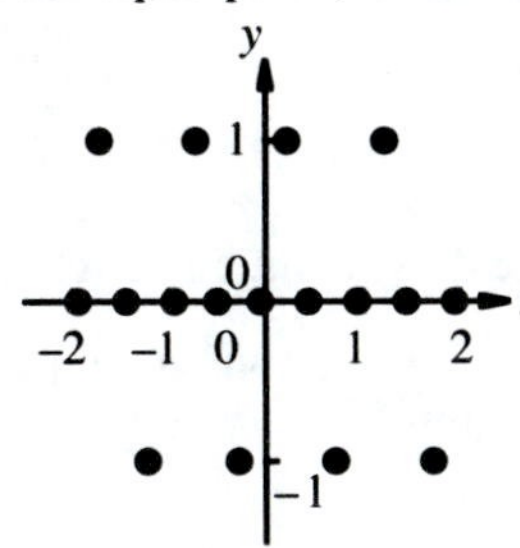

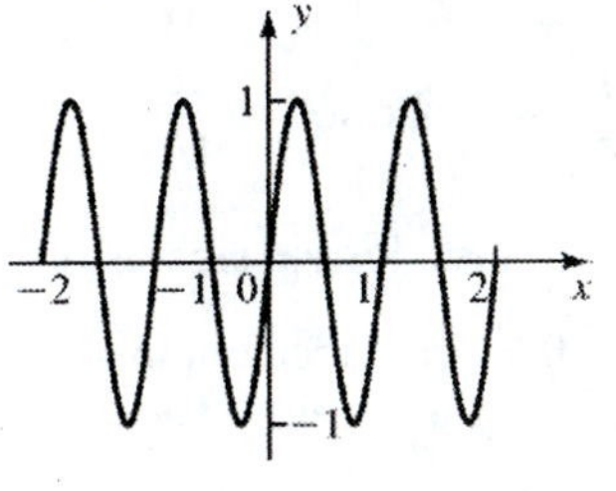

13. $y = \cos\frac{x}{4}$. Amplitude $= |A| = |1| = 1$. Period $= \frac{2\pi}{1/4} = 8\pi$.

One full cycle of this graph is completed as x goes from 0 to 8π. Block out this interval, divide it into four equal parts, locate high and low points, and locate x intercepts. Then complete the graph.

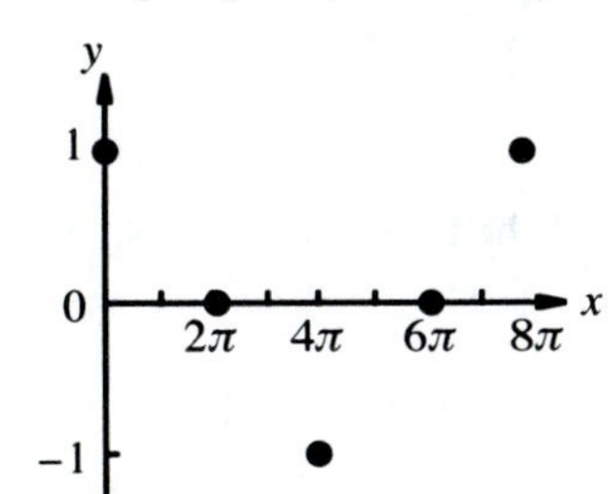

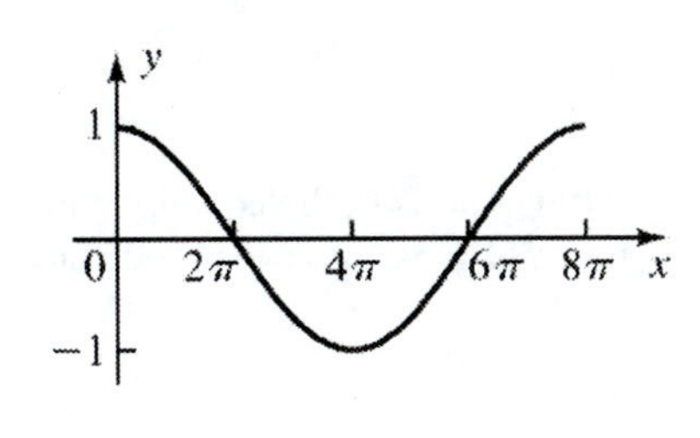

15. $y = 2\sin 4x$. Amplitude $= |2| = 2$. Period $= \frac{2\pi}{4} = \frac{\pi}{2}$.

One full cycle of this graph is completed as x goes from 0 to $\frac{\pi}{2}$. Block out this interval, divide it into four equal parts, locate high and low points, and locate x intercepts. Then complete the graph.

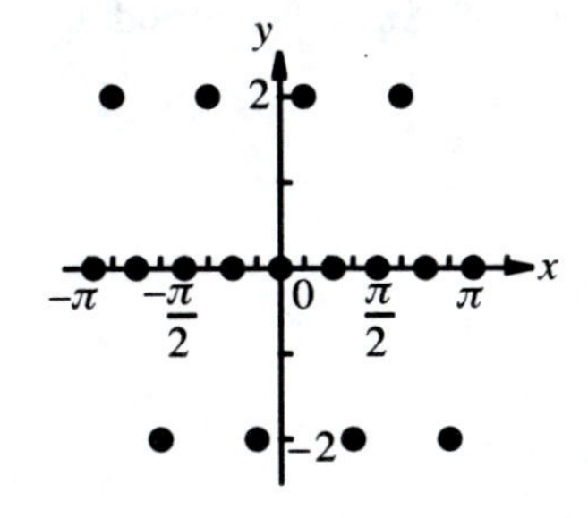

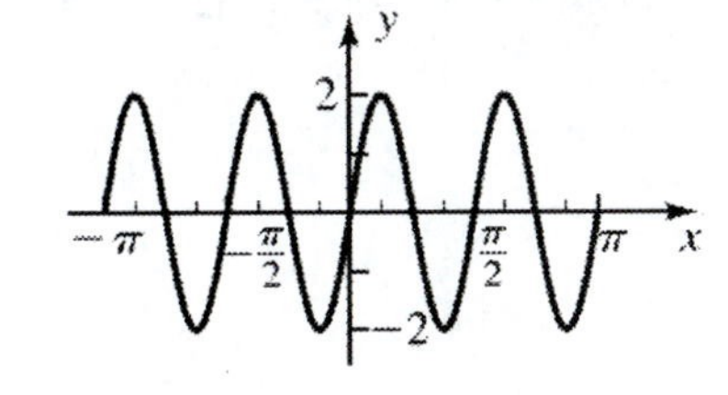

17. $y = \frac{1}{3} \cos 2\pi x$. Amplitude $= \left|\frac{1}{3}\right| = \frac{1}{3}$. Period $= \frac{2\pi}{2\pi} = 1$.

One full cycle of this graph is completed as x goes from 0 to 1. Block out this interval, divide it into four equal parts, locate high and low points, and locate x intercepts. Then complete the graph.

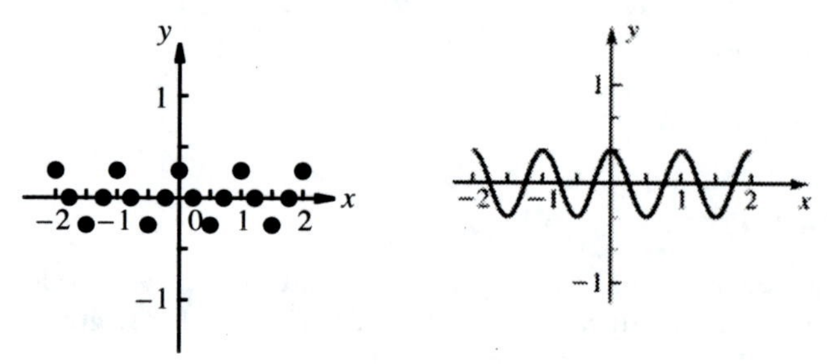

19. $y = -\frac{1}{4} \sin \frac{x}{2}$. Amplitude $= \left|-\frac{1}{4}\right| = \frac{1}{4}$. Period $= \frac{2\pi}{1/2} = 4\pi$.

Since $A = -\frac{1}{4}$ is negative, the basic curve for $y = \sin x$ is turned upside down. One full cycle of this graph is completed as x goes from 0 to 4π. Block out this interval, divide it into four equal parts, locate high and low points, and locate x intercepts. Then complete the graph.

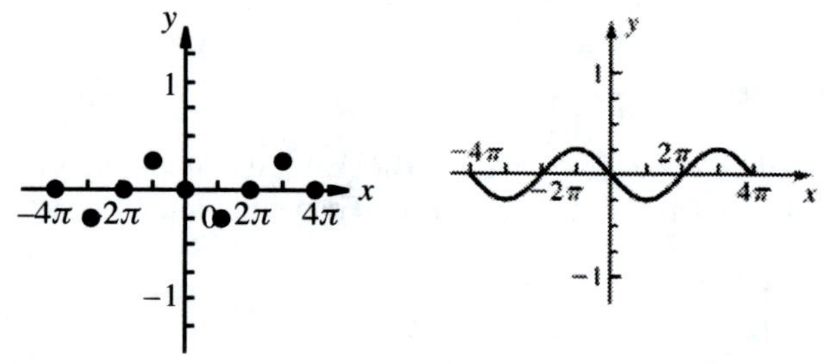

21. Since the displacement is 0 ft when t is 0, the equation should be of the form $y = A \sin Bt$. Since the amplitude is 3 ft, $|A| = 3$, hence $A = 3$ or $A = -3$. Since the period is 4 sec, write

$$\frac{2\pi}{B} = 4$$
$$2\pi = 4B$$
$$B = \frac{\pi}{2}$$

Thus, the required equation is $y = 3 \sin\left(\frac{\pi t}{2}\right)$ or $y = -3 \sin\left(\frac{\pi t}{2}\right)$.

23. Since the displacement is equal to the amplitude when t is 0, the equation should be of the form $y = A \cos Bt$. Since the amplitude is 9 m, $|A| = 9$, hence $A = 9$ or $A = -9$. Since the period is 0.2 sec, write

$$\frac{2\pi}{B} = 0.2$$
$$2\pi = 0.2B$$
$$B = \frac{2\pi}{0.2}$$
$$B = 10\pi$$

Thus, the required equation is $y = 9 \cos(10\pi t)$ or $y = -9 \cos(10\pi t)$.
Since the displacement is 9 in when $t = 0$, the correct equation is $y = 9 \cos(10\pi t)$.

25. Since $P = \frac{2\pi}{B}$, P tends to zero as B increases without bound.

27. $y = -1 + \frac{1}{3}\cos 2\pi x$. Amplitude $= \left|\frac{1}{3}\right| = \frac{1}{3}$. Period $= \frac{2\pi}{2\pi} = 1$.

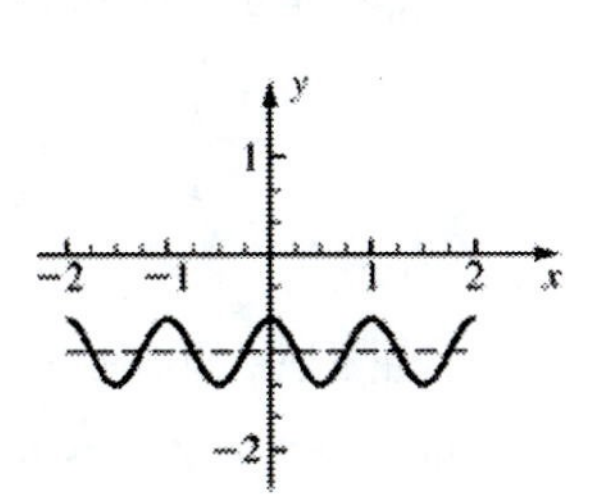

$y = \frac{1}{3}\cos 2\pi x$ was graphed in Problem 17. This graph is the graph of $y = \frac{1}{3}\cos 2\pi x$ moved down $|k| = |-1| = 1$ unit. We start by drawing a horizontal broken line 1 unit below the x axis, then graph $y = \frac{1}{3}\cos 2\pi x$ relative to the broken line and the original y axis.

29. $y = 2 - \frac{1}{4}\sin\frac{x}{2}$. Amplitude $= \left|-\frac{1}{4}\right| = \frac{1}{4}$. Period $= \frac{2\pi}{1/2} = 4\pi$.

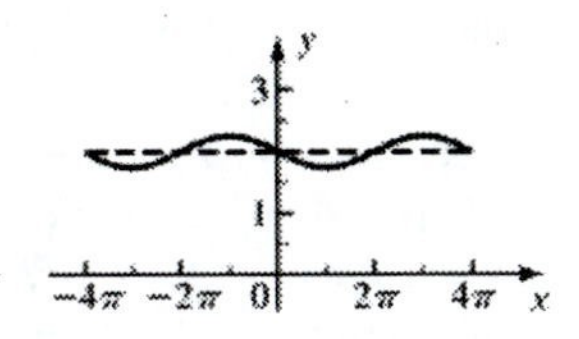

$y = -\frac{1}{4}\sin\frac{x}{2}$ was graphed in Problem 19. This graph is the graph of $y = -\frac{1}{4}\sin\frac{x}{2}$ moved up $k = 2$ units. We start by drawing a horizontal broken line 2 units above the x axis, then graph $y = -\frac{1}{4}\sin\frac{x}{2}$ relative to the broken line and the original y axis.

31. Amplitude $= 5 = |A|$. Period $= \frac{2\pi}{B} = \pi$. Thus, $B = 2$. The form of the graph is that of the basic sine curve. Thus, $y = |A|\sin Bx = 5\sin 2x$.

33. Amplitude $= 4 = |A|$. Period $= \frac{2\pi}{B} = 4$. Thus, $B = \frac{2\pi}{4} = \frac{\pi}{2}$. The form of the graph is that of the basic sine curve turned upside down. Thus, $y = -|A|\sin Bx = -4\sin\left(\frac{\pi x}{2}\right)$.

35. Amplitude $= 8 = |A|$. Period $= \frac{2\pi}{B} = 8\pi$. Thus, $B = \frac{2\pi}{8\pi} = \frac{1}{4}$. The form of the graph is that of the basic cosine curve. Thus, $y = |A|\cos Bx = 8\cos\left(\frac{1}{4}x\right)$.

37. Amplitude $= 1 = |A|$. Period $= \frac{2\pi}{B} = 6$. Thus, $B = \frac{2\pi}{6} = \frac{\pi}{3}$. The form of the graph is that of the basic cosine curve turned upside down. Thus, $y = -|A|\cos Bx = -\cos\left(\frac{\pi x}{3}\right)$.

39. The graph of $y = \sin x\cos x$ is shown in the figure.
Amplitude $= 0.5 = |A|$. Period $= \frac{2\pi}{B} = \pi$. Thus, $B = \frac{2\pi}{\pi} = 2$.
The form of the graph is that of the basic sine curve.
Thus, $y = |A|\sin Bx = 0.5\sin 2x$.

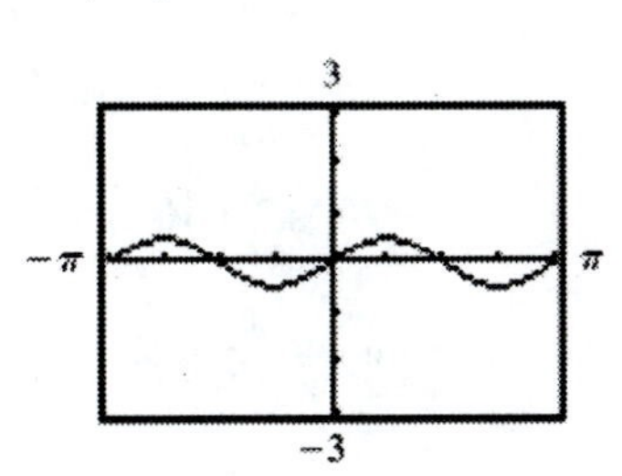

41. The graph of $y = 2 \cos^2 x$ is shown in the figure.

$$\text{Amplitude} = \frac{1}{2}(y \text{ coordinate of highest point} - y \text{ coordinate of lowest point})$$

$$= \frac{1}{2}(2 - 0) = 1 = |A|; \text{ Period} = \frac{2\pi}{B} = \pi.$$

Thus, $B = \dfrac{2\pi}{\pi} = 2$. The form of the graph is that of the basic cosine curve shifted up 1 unit.

Thus, $y = 1 + |A| \cos Bx = 1 + \cos 2x$.

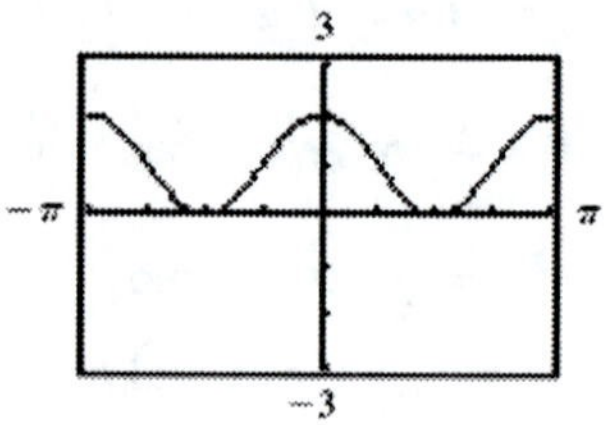

43. The graph of $y = 2 - 4 \sin^2 2x$ is shown in the figure.

Amplitude $= 2 = |A|$. Period $= \dfrac{2\pi}{B} = \dfrac{\pi}{2}$.

Thus, $B = 2\pi \div \dfrac{\pi}{2} = 4$. The form of the graph is that of the basic cosine curve. Thus, $y = |A| \cos Bx = 2 \cos 4x$.

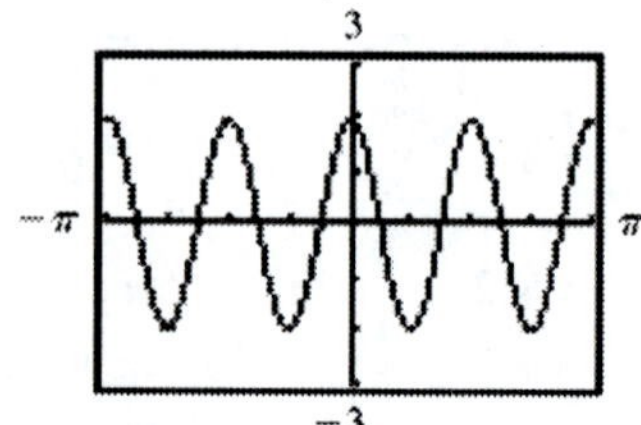

45. The range of $y = \sin Bx$ and the range of $y = \sin x$ are the same. $-1 \le \sin x \le 1$ and $-1 \le \sin Bx \le 1$. Therefore, for positive A, $-A \le A \sin x \le A$ and $-A \le A \sin Bx \le A$ and $k - A \le k + A \sin x \le k + A$ and $k - A \le k + A \sin Bx \le k + A$. Thus, both functions have the same maximum, $k + A$, and the same minimum, $k - A$. Similarly for negative A.

47. The range of the sine function is the set of all real numbers between -1 and 1:

$$-1 \le \sin 4x \le 1 \qquad \text{Multiply both inequalities by 9}$$

$$-9 \le 9 \sin 4x \le 9$$

Therefore, $-9 \le y \le 9$ and Max $y = 9$, Min $y = -9$.

49. The range of the cosine function is the set of all real numbers between -1 and 1:

$$-1 \le \cos \pi x \le 1 \qquad \text{Multiply both inequalities by } -25, \text{ reversing the sense.}$$

$$25 \ge -25 \cos \pi x \ge -25 \qquad \text{Add } -5 \text{ to both inequalities}$$

$$-5 + 25 \ge -5 - 25 \cos \pi x \ge -5 - 25 \qquad \text{Simplify}$$

$$20 \ge -5 - 25 \cos \pi x \ge -30$$

Therefore, $-30 \le y \le 20$ and Max $y = 20$, Min $y = -30$.

51. The secant function has no maximum and no minimum, so Max y and Min y do not exist.

53. The range of the cosine function is the set of all real numbers between -1 and 1:

$$-1 \le \cos \frac{x}{4} \le 1 \qquad \text{Multiply both inequalities by 10}$$

$$-10 \le 10 \cos \frac{x}{4} \le 10 \qquad \text{Add 2 to both inequalities}$$

$$2 - 10 \le 2 + 10 \cos \frac{x}{4} \le 2 + 10 \qquad \text{Simplify}$$

$$-8 \le 2 + 10 \cos \frac{x}{4} \le 12$$

Therefore, $-8 \le y \le 12$ and Max $y = 12$, Min $y = -8$.

55. The graph of $f(x) = \cos^2 x$, $-2\pi \le x \le 2\pi$, is shown in the figure.

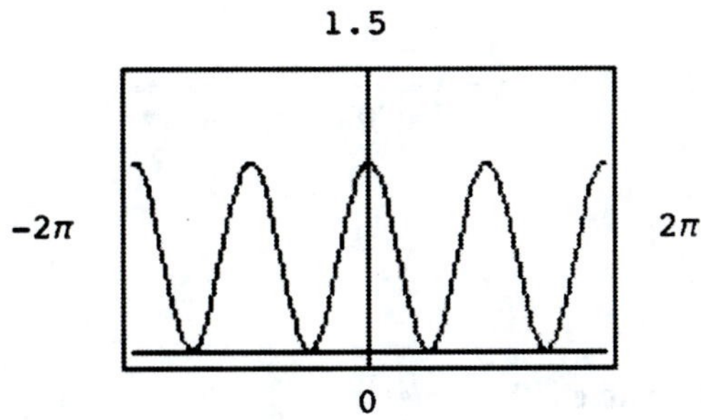

It appears that four full periods of the graph occur in the interval from -2π to 2π, thus, the period is the length of this interval divided by 4.

$$P = \frac{2\pi - (-2\pi)}{4} = \frac{4\pi}{4} = \pi.$$

57. Since the maximum value occurs at the end points of the interval, it would appear that a function of the form $y = A \cos Bx$ would be required, with A positive.
Since the maximum value of the function seems to be 2, and the minimum value seems to be -2, $A = \frac{2-(-2)}{2} = \frac{4}{2} = 2$. Since the maximum value is achieved at $x = 0$ and $x = 3$, the period of the function is 3. Hence $\frac{2\pi}{B} = 3$ and $B = \frac{2\pi}{3}$. Thus the required function is $y = 2\cos\frac{2\pi x}{3}$.

59. Given that the frequency, f, is 20 Hz, then the period is found using the reciprocal formula:

$$P = \frac{1}{f} = \frac{1}{20} \text{ sec or } 0.05 \text{ sec}$$

61. Given that the frequency, f, is 500 Hz, then the period is found using the reciprocal formula:

$$P = \frac{1}{f} = \frac{1}{500} \text{ sec or } 0.002 \text{ sec}$$

63. $E = 110 \sin 120\pi t$. Amplitude $= |110| = 110$. Period $= \frac{2\pi}{120\pi} = \frac{1}{60}$ sec.

Frequency $f = \frac{1}{p} = \frac{1}{1/60} = 60$ Hz

One full cycle of this graph is completed as t goes from 0 to $\frac{1}{60}$. Block out this interval, divide it into four equal parts, locate high and low points, and locate t intercepts. Then complete the graph.

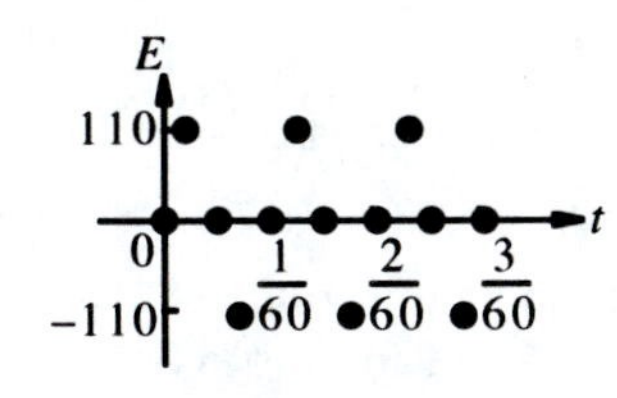

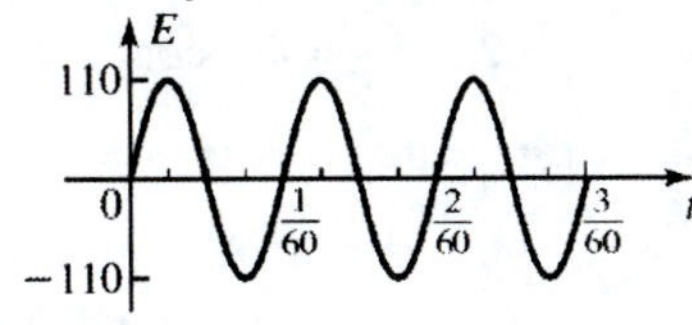

65. *Find A:* The amplitude $|A|$ is given to be 12. Since $E = 12$ when $t = 0$, $A = 12$ (and not -12).
Find B: We are given that the frequency, f, is 40 Hz. Hence, the period is found using the reciprocal formula:

$$P = \frac{1}{f} = \frac{1}{40} \text{ sec. But, } P = \frac{2\pi}{B}. \text{ Thus, } B = \frac{2\pi}{P} = \frac{2\pi}{1/40} = 80\pi.$$

Write the equation: $E = 12 \cos 80\pi t$

67. We use the formula: Period $= \dfrac{2\pi}{\sqrt{1{,}000gA/M}}$ with $g = 9.75$ m/sec^2,

$A = 0.4$ m^2, and Period $= \dfrac{1}{2}$ sec, and solve for M. Thus, $\dfrac{1}{2} = \dfrac{2\pi}{\sqrt{(1{,}000)(9.75)(0.4)\ M}}$;

$M = \dfrac{(1{,}000)(9.75)(0.4)}{4(4\pi^2)} = 24.7$ kg.

No, the containers are not within the 20 kg limit.

69. (A) We use the formula: Period $= \dfrac{2\pi}{\sqrt{1{,}000gA/M}}$ with $g = 9.75$ m/sec^2,

$A = 3\text{ m} \times 3\text{ m} = 9\text{ m}^2$, and Period = 1 sec, and solve for M. Thus, $1 = \dfrac{2\pi}{\sqrt{(1{,}000)(9.75)(9)/M}}$;

$M = \dfrac{(1{,}000)(9.75)(9)}{4\pi^2} \approx 2{,}220$ kg

(B) $D = 0.2$, $B = \dfrac{2\pi}{\text{Period}}$, $y = 0.2 \sin 2\pi t$

(C) One full cycle of this graph is completed as t goes from 0 to 1. Block out this interval, divide it into four equal parts, locate high and low points, and locate t intercepts. Then complete the graph.

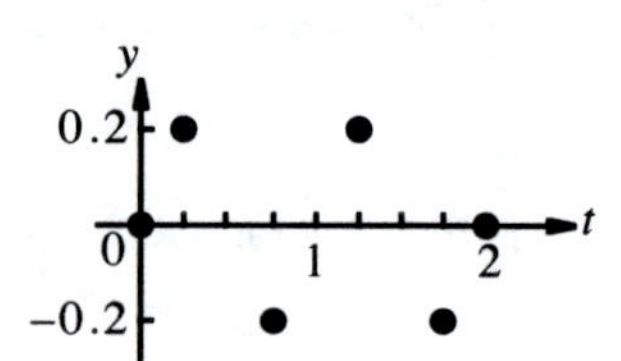

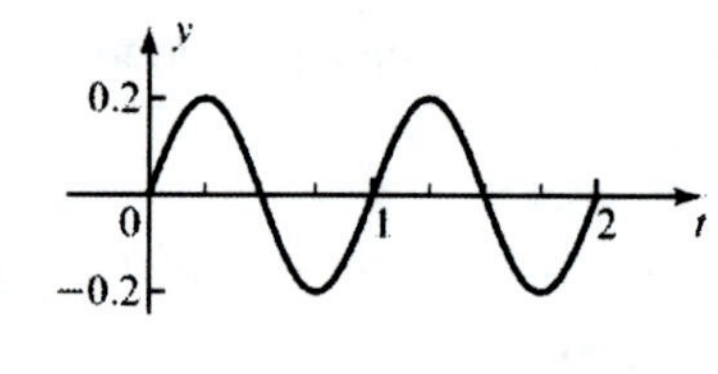

71. (A) max vol = 85 ℓ, min vol = 0.05 ℓ; $0.40 \cos \dfrac{\pi t}{2}$ is maximum when $\cos \dfrac{\pi t}{2}$ is 1 and is minimum when $\cos \dfrac{\pi t}{2}$ is -1.

Therefore, max vol = 0.45 + 0.40 = 0.85 ℓ and min vol = 0.45 − 0.40 = 0.05 ℓ.

(B) The period $= \dfrac{2\pi}{B} = 2\pi \div \left(\dfrac{\pi}{2}\right) = 2\pi \cdot \dfrac{2}{\pi} = 4$ seconds

(C) A breath is taken every 4 sec. Since there are 60 seconds per minute, there are $\dfrac{60}{4} = 15$ breaths per minute.

(D)

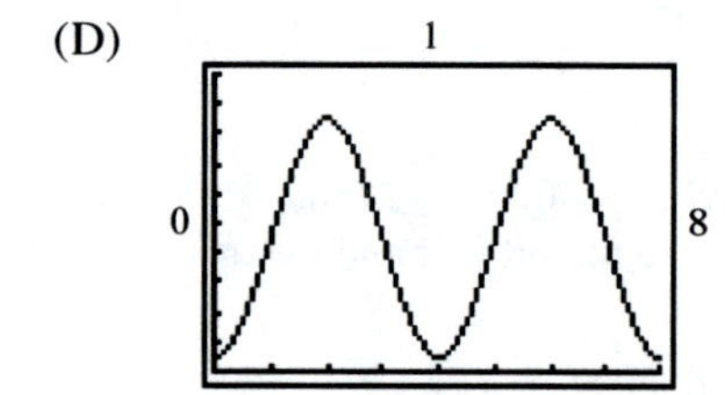

From the graph, the maximum volume occurs at $t = 2$ and $t = 6$ and appears to be 0.85ℓ. The minimum volume occurs at $t = 0, 4, 8$, and appears to be 0.05ℓ.

73. Since the rotation is at 4 revolutions per minute, in 1 minute it covers 4 revolutions, or 8π radians. In t minutes, it covers $8\pi t$ radians, thus $\theta = 8\pi t$. To see that $x = 20 \sin 8\pi t$, it might help to look sideways at the wheel. Below, we have indicated a new coordinate system in which θ is in standard position. Then, $\sin \theta = \frac{b}{20}$, so $b = 20 \sin \theta = 20 \sin 8\pi t$. Thus, the coordinate of the shadow $= b = x$ in the author's coordinate system. That is, $x = 20 \sin 8\pi t$.

To graph this, note:

Amplitude $= |A| = |20| = 20$

Period $= \frac{2\pi}{8\pi} = \frac{1}{4}$.

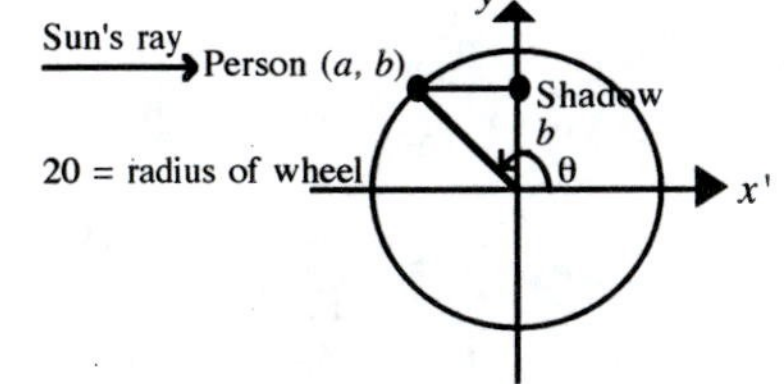

One full cycle of the graph is completed as t goes from 0 to $\frac{1}{4}$.

Block out this interval, divide it into four equal parts, locate high and low points, and locate t intercepts. Then complete the graph.

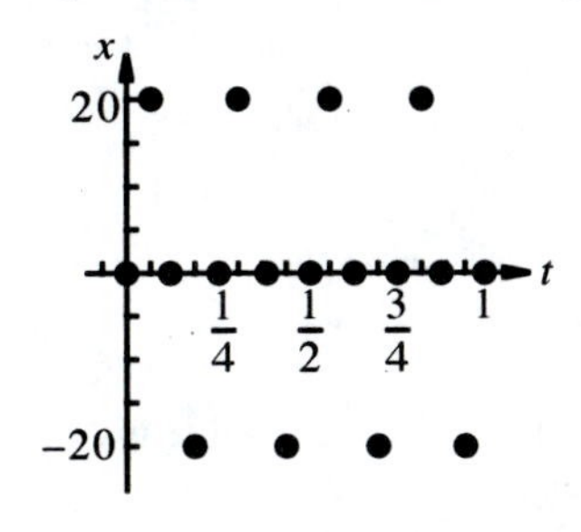

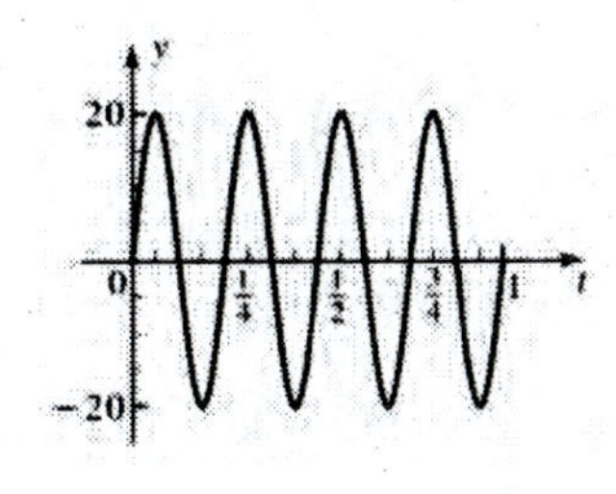

75. (A) The data for θ repeats every 2 seconds, so the period is $P = 2$ sec. The angle θ deviates from 0 by 25° in each direction, so the amplitude is $|A| = 25°$.

(B) $\theta = A \sin Bt$ is not suitable, because, for example, for $t = 0$, $A \sin Bt = 0$ no matter what the choice of A and B. $\theta = A \cos Bt$ appears suitable, because, for example if $t = 0$ and $A = -25$, then we can get the first value in the table, a good start. Choose $A = -25$ and $B = \frac{2\pi}{P} = \pi$, which yields $\theta = -25 \cos \pi t$. A calculator can be used to check that this equation produces (or comes close to producing) all the values in the table.

(C) One full cycle of this graph is completed as t goes from 0 to 2. Block out this interval, divide it into four equal parts, locate high and low points, and locate t intercepts. Then complete the graph.

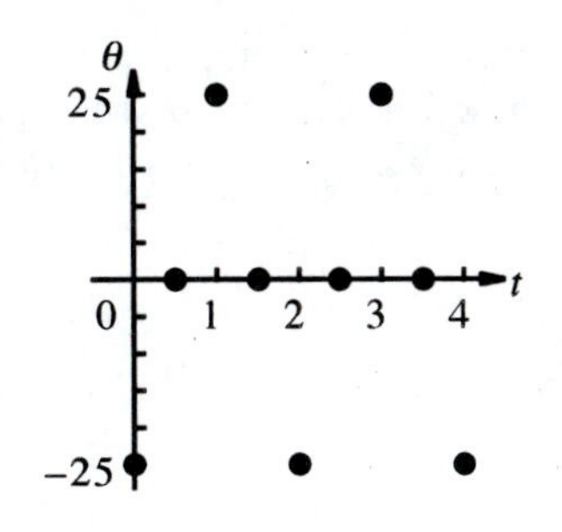

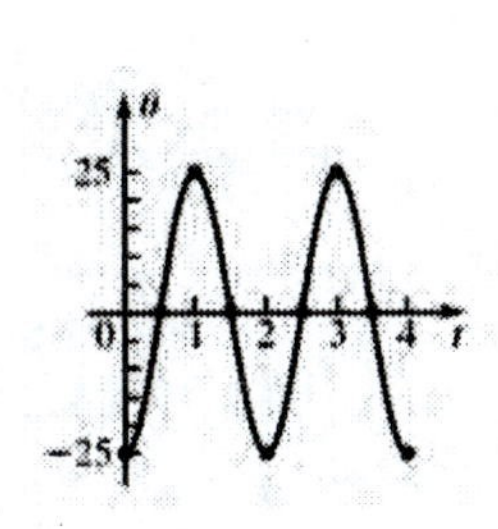

77. (A)

t sec	0	3	6	9	12	15	18	21	24
h ft	28	13	−2	13	28	13	−2	13	28

(B)

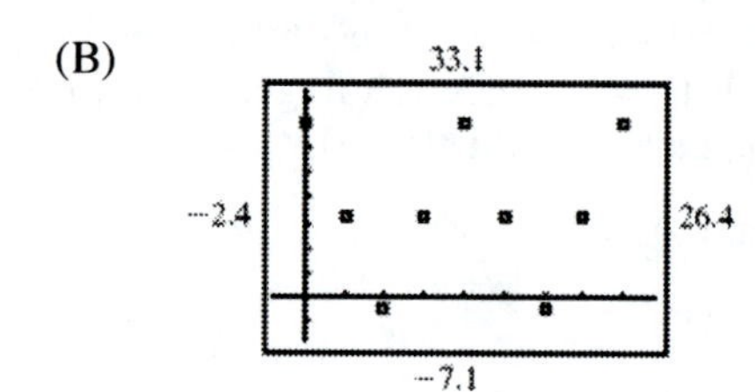

(C) Since the maximum value of h is repeated at intervals of 12 sec, the period = 12 sec. Because the maximum value of $h = k + A \cos Bt$ occurs when $t = 0$ (assuming A is positive), and this corresponds to a maximum value in the table 1 when $t = 0$, a function of this form appears to be a better model for the data.

(D) $|A| = \dfrac{\max h - \min h}{2} = \dfrac{28-(-2)}{2} = 15$

$B = \dfrac{2\pi}{\text{Period}} = \dfrac{2\pi}{12} = \dfrac{\pi}{6}$

$k = |A| + \min h = 15 + (-2) = 13$

Thus, $h = 13 + 15 \cos \dfrac{\pi}{6}t$ or $h = 13 - 15 \cos \dfrac{\pi}{6}t$.

Since $h = 28$ when $t = 0$, $h = 13 + 15 \cos \dfrac{\pi}{6} \cdot 0 = 28$ indicates that

$h = 13 + 15 \cos \dfrac{\pi}{6}t$ is the correct equation.

(E)

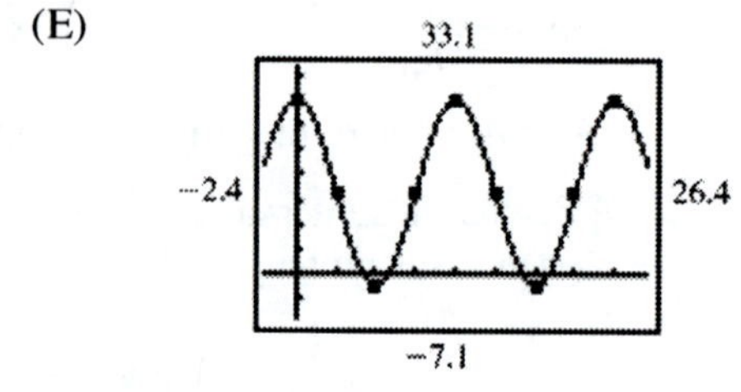

79. The maximum value shown in the table is $t_{max} = 27.0$. The minimum value is $t_{min} = 19.5$. Therefore, set

$|A| = \dfrac{t_{max} - t_{min}}{2} = \dfrac{27.0 - 19.5}{2} = 3.75$

$B = \dfrac{2\pi}{\text{Period}} = \dfrac{2\pi}{12} = \dfrac{\pi}{6}$

$k = |A| + t_{min} = 3.75 + 19.5 = 23.25$

Thus, $t = 23.25 + 3.75 \sin \dfrac{\pi}{6}x$ or $t = 23.25 - 3.75 \sin \dfrac{\pi}{6}x$.

Since $t = 20.4$ when $x = 0$, $t = 23.25 - 3.75 \sin \dfrac{\pi}{6}x$ is the better model.

One full cycle of this graph is completed as x goes from 0 to 12. Block out this interval (0 to 1 is not part of the domain, so is not shown in the final graph), divide it into four equal parts, locate high and low points and the halfway marks. Then complete the graph.

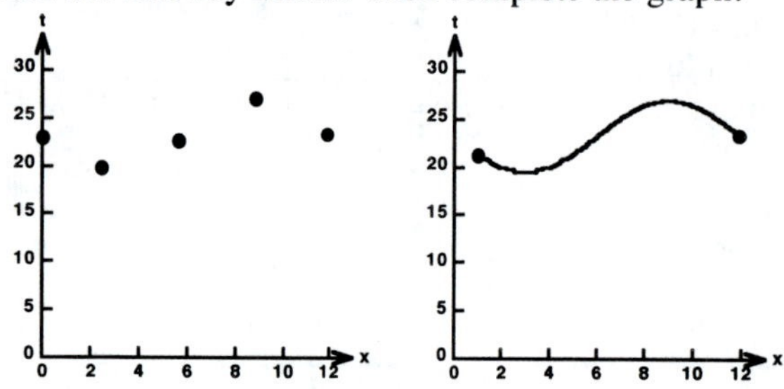

81. The maximum value shown in the table is $g_{max} = 1{,}963$. The minimum value is $g_{min} = 715$. Therefore, set

$$|A| = \frac{g_{max} - g_{min}}{2} = \frac{1{,}963 - 715}{2} = 624$$
$$B = \frac{2\pi}{\text{Period}} = \frac{2\pi}{12} = \frac{\pi}{6}$$
$$k = |A| + g_{min} = 624 + 715 = 1{,}339$$

Thus, $g = 1{,}339 + 624 \sin \frac{\pi x}{6}$ or $g = 1{,}339 - 624 \sin \frac{\pi x}{6}$

Since g increases as x increases from 1 to 4, $g = 1{,}339 + 624 \sin \frac{\pi x}{6}$ is the better model.

One full cycle of this graph is completed as x goes from 0 to 12. Block out this interval (0 to 1 is not part of the domain, so is not shown in the final graph), divide it into four equal parts, locate high and low points and the halfway marks. Then complete the graph.

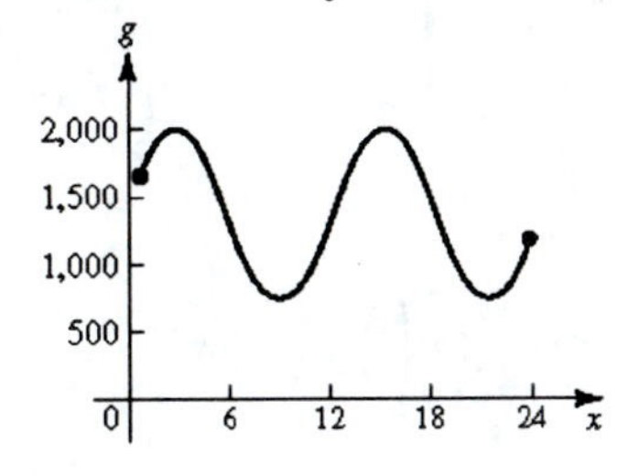

EXERCISE 3.3 Graphing $y = k + A \sin(Bx + C)$ and $y = k + A \cos(Bx + C)$

1. A function of the form $y = A \sin(Bx + C)$ or $y = A \cos(Bx + C)$ is called a simple harmonic.

3. The phase shift of a simple harmonic $y = A \sin(Bx + C)$ or $y = A \cos(Bx + C)$ is given by $-C/B$.

5. The graph of y_1 is the graph of y_2 translated horizontally to the right by the number of units in the phase shift.

7. Amplitude $= |A| = |-5| = 5$.

9. Phase shift $= -C/B = -18/3 = -6$.

11. Period $= 2\pi/B = 2\pi/\pi = 2$.

13. The amplitude of $y_1 = -10 \cos(2\pi x + \pi)$ is $|-10|$ or 10. The graph of $y = 8 - 10 \cos(2\pi x + \pi)$ is the graph of $y_1 = -10 \cos(2\pi x + \pi)$ translated 8 units up. Since the maximum value of y_1 is 10, the maximum value of y is $10 + 8$ or 18.

15. The amplitude of $y_1 = -30 \sin 4x$ is $|-30|$ or 30. The graph of $y = 11 - 30 \sin 4x$ is the graph of $y_1 = -30 \sin 4x$ translated 11 units up. Since the minimum value of y_1 is -30, the minimum value of y is $-30 + 11$ or -19.

17. Amplitude = $|A| = |1| = 1$
    Phase Shift and Period: Solve

$$Bx + C = 0 \quad \text{and} \quad Bx + C = 2\pi$$

$$x + \frac{\pi}{2} = 0 \qquad x + \frac{\pi}{2} = 2\pi$$

$$x = -\frac{\pi}{2} \qquad x = -\frac{\pi}{2} + 2\pi$$

↑ Phase Shift ↑ Period = 2π

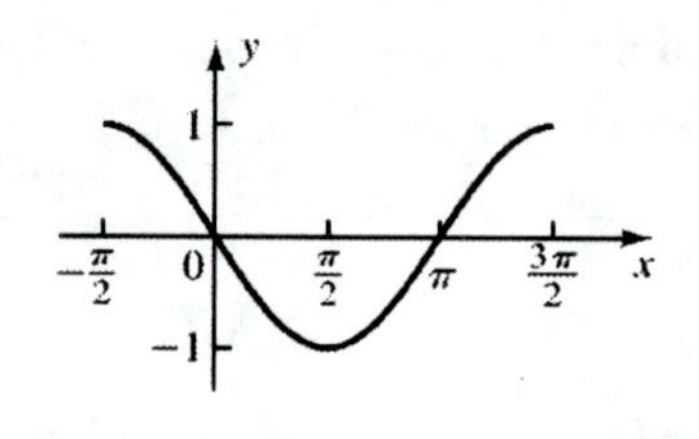

Phase Shift = $-\frac{\pi}{2}$. Graph one cycle over the interval from $-\frac{\pi}{2}$ to $\left(-\frac{\pi}{2} + 2\pi\right) = \frac{3\pi}{2}$.

19. Amplitude = $|A| = |1| = 1$
    Phase Shift and Period: Solve

$$Bx + C = 0 \quad \text{and} \quad Bx + C = 2\pi$$

$$x - \frac{\pi}{4} = 0 \qquad x - \frac{\pi}{4} = 2\pi$$

$$x = \frac{\pi}{4} \qquad x = \frac{\pi}{4} + 2\pi$$

↑ Phase Shift ↑ Period = 2π

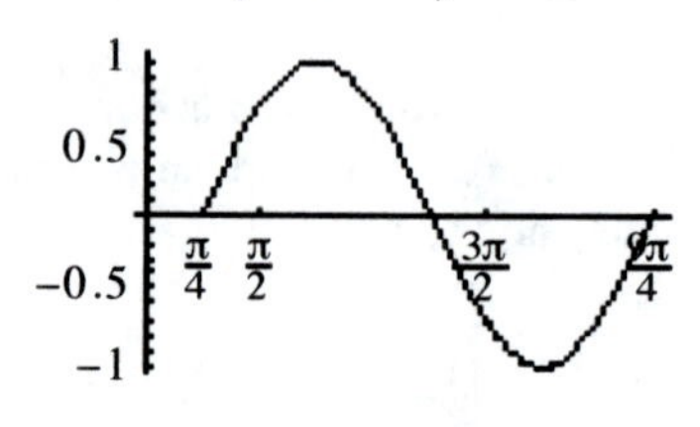

Phase Shift = $\frac{\pi}{4}$. Graph one cycle over the interval from $\frac{\pi}{4}$ to $\left(\frac{\pi}{4} + 2\pi\right) = \frac{9\pi}{4}$.

Extend the graph from $-\pi$ to $\frac{\pi}{4}$ and delete the portion of the graph from 2π to $\frac{9\pi}{4}$, since this was not required.

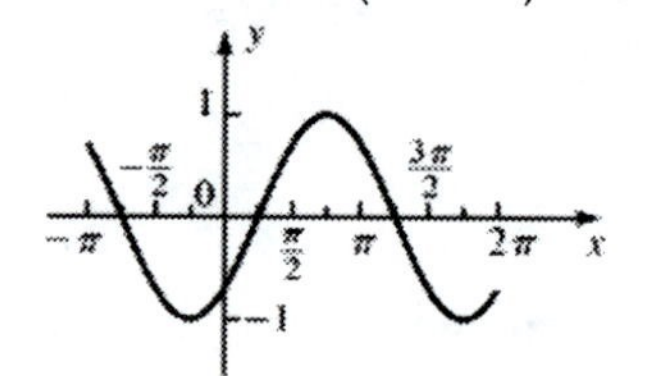

21. Amplitude = $|A| = |4| = 4$
    Phase Shift and Period: Solve

$$Bx + C = 0 \quad \text{and} \quad Bx + C = 2\pi$$

$$\pi x + \frac{\pi}{4} = 0 \qquad \pi x + \frac{\pi}{4} = 2\pi$$

$$x = -\frac{1}{4} \qquad x = -\frac{1}{4} + 2$$

↑ Phase Shift ↑ Period = 2

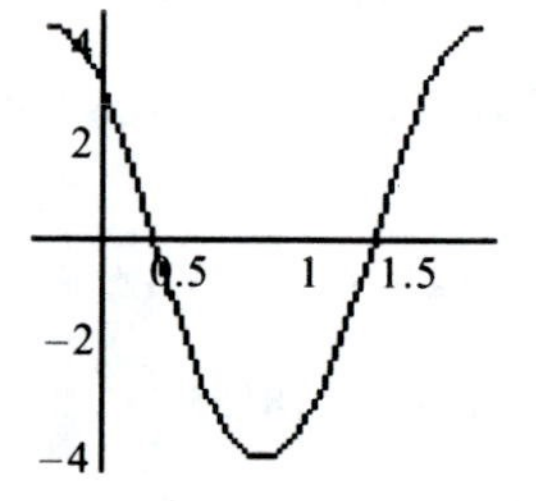

Phase Shift = $-\frac{1}{4}$. Graph one cycle over the interval from $-\frac{1}{4}$ to $\left(-\frac{1}{4} + 2\right) = \frac{7}{4}$.
Extend the graph from −1 to 3.

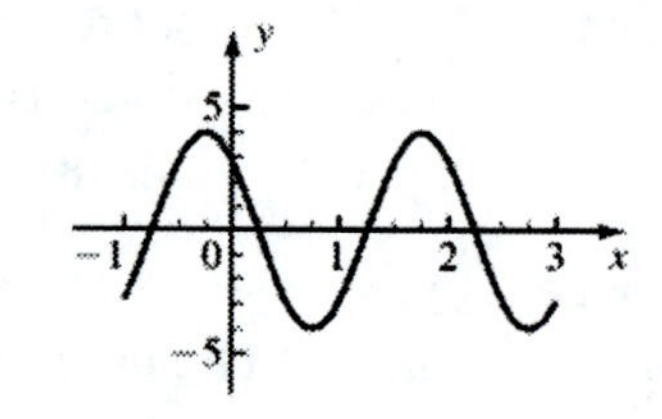

23. Amplitude = $|A| = |-2| = 2$

Phase Shift and Period: Solve

$$\begin{aligned} Bx + C &= 0 \quad \text{and} \quad & Bx + C &= 2\pi \\ 2x + \pi &= 0 & 2x + \pi &= 2\pi \\ x &= -\frac{\pi}{2} & x &= -\frac{\pi}{2} + \pi \end{aligned}$$

$\uparrow$ Phase Shift $\quad \uparrow$ Period $= \pi$

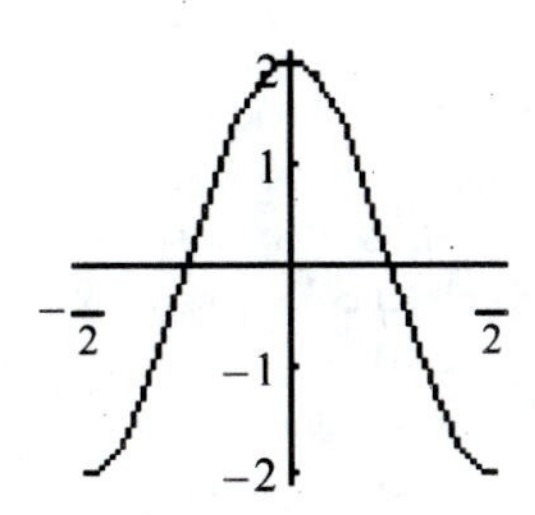

Phase Shift $= -\frac{\pi}{2}$. Graph one cycle (the basic cosine curve turned upside down) over the interval from $-\frac{\pi}{2}$ to $\left(-\frac{\pi}{2} + \pi\right) = \frac{\pi}{2}$.

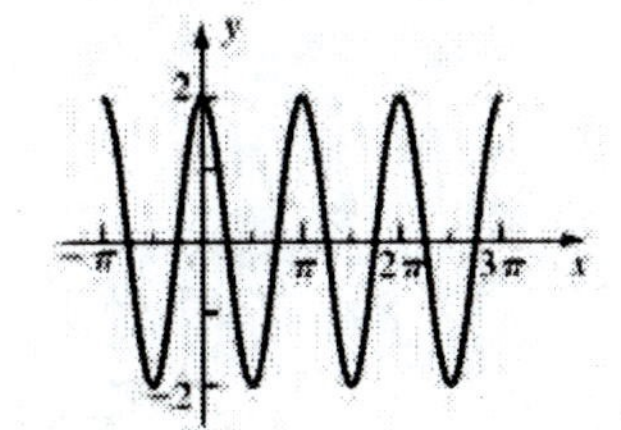

Extend the graph from $-\pi$ to 3π.

25. We sketch one period of the graph of $y = \sin x$ below. It has amplitude 1, period 2π, and phase shift 0. To graph $y = \cos\left(x - \frac{\pi}{2}\right)$ we compute its amplitude, period, and phase shift:

Amplitude = $|A| = |1| = 1$

Phase Shift and Period: Solve

$$\begin{aligned} Bx + C &= 0 \quad \text{and} \quad & Bx + C &= 2\pi \\ x - \frac{\pi}{2} &= 0 & x - \frac{\pi}{2} &= 2\pi \\ x &= \frac{\pi}{2} & x &= \frac{\pi}{2} + 2\pi \end{aligned}$$

$\uparrow$ Phase Shift $\quad \uparrow$ Period $= 2\pi$

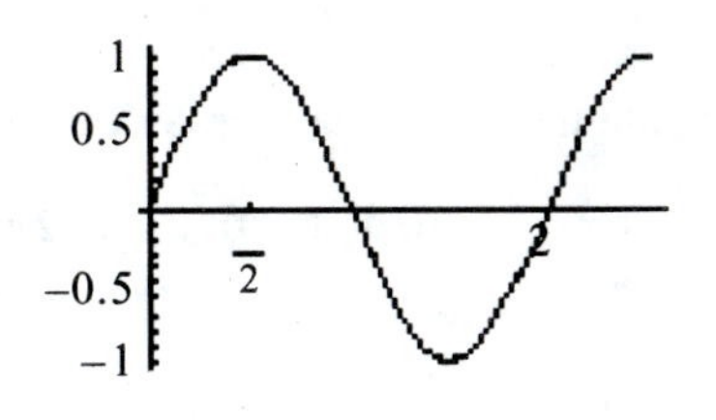

If we sketch one period of the graph starting at $x = \frac{\pi}{2}$ (the phase shift) and ending at $x = \frac{\pi}{2} + 2\pi = \frac{5\pi}{2}$ (the phase shift plus one period), we see that it is the same as that of $y = \sin x$ over the interval from $x = \frac{\pi}{2}$ to $x = 2\pi$. Since both curves can be extended indefinitely in both directions, we conclude that the graphs are the same, thus $\cos\left(x - \frac{\pi}{2}\right) = \sin x$ for all x.

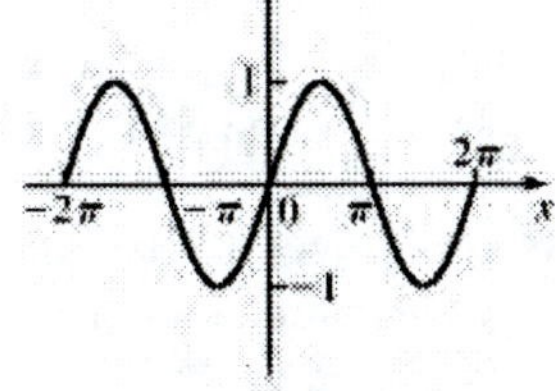

27. $y = -2 + 4\cos\left(\pi x + \frac{\pi}{4}\right)$. $y = 4\cos\left(\pi x + \frac{\pi}{4}\right)$ was graphed in Problem 21. This graph is the graph of $y = 4\cos\left(\pi x + \frac{\pi}{4}\right)$ moved down $|k| = |-2| = 2$ units. We start by drawing a horizontal broken line 2 units below the x axis, then graph $y = 4\cos\left(\pi x + \frac{\pi}{4}\right)$ relative to the broken line and the original y axis.

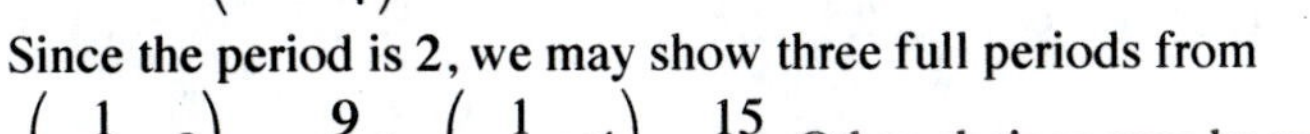
Since the period is 2, we may show three full periods from $\left(-\frac{1}{4} - 2\right) = -\frac{9}{4}$ to $\left(-\frac{1}{4} + 4\right) = \frac{15}{4}$. Other choices may be made.

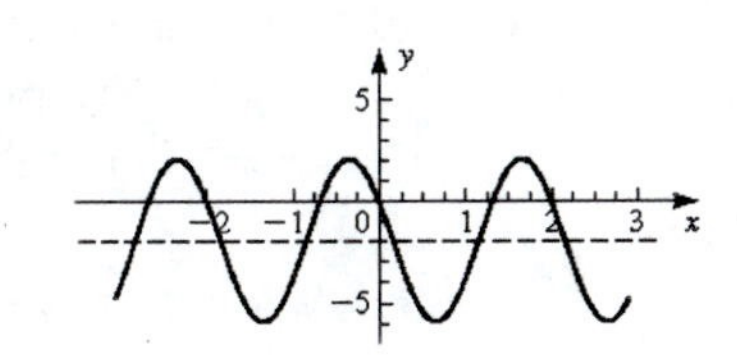

29. $y = 3 - 2\cos(2x + \pi)$.
$y = -2\cos(2x + \pi)$ was graphed in Problem 23.
This graph is the graph of $y = -2\cos(2x + \pi)$ moved up $k = 3$ units. We start by drawing a horizontal broken line 3 units above the x axis, then graph $y = -2\cos(2x + \pi)$ relative to the broken line and the original y axis.
Since the period is π, we may show three full periods from $-\pi$ to $(-\pi + 3\pi) = 2\pi$. Other choices may be made.

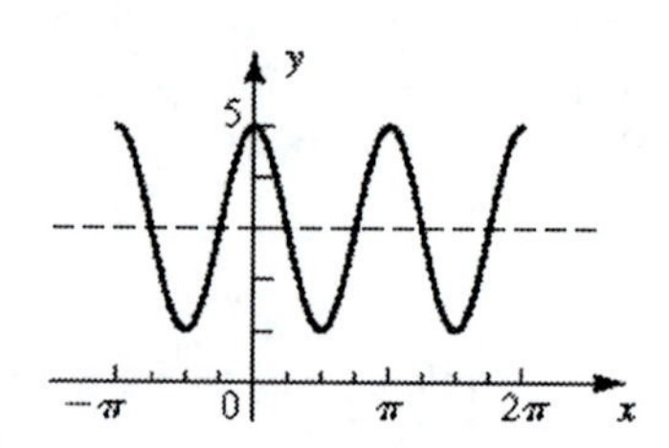

31. Compute the period and phase shift for $y = 2\sin\left(\pi x - \frac{\pi}{2}\right)$:
Solve $Bx + C = 0$ and $Bx + C = 2\pi$

$$\pi x - \frac{\pi}{2} = 0 \qquad \pi x - \frac{\pi}{2} = 2\pi$$
$$\pi x = \frac{\pi}{2} \qquad \pi x = \frac{\pi}{2} + 2\pi$$
$$x = \frac{1}{2} \qquad x = \frac{1}{2} + 2$$

Thus, the graph of the equation is a sine curve with a period of 2 and a phase shift of $\frac{1}{2}$, which means the sine curve is shifted $\frac{1}{2}$ unit to the right. This matches (b).

33. Compute the period and phase shift for $y = 2\cos\left(2x + \frac{\pi}{2}\right)$:

Solve $Bx + C = 0$ and $Bx + C = 2\pi$

$$2x + \frac{\pi}{2} = 0 \qquad 2x + \frac{\pi}{2} = 2\pi$$
$$2x = -\frac{\pi}{2} \qquad 2x = -\frac{\pi}{2} + 2\pi$$
$$x = -\frac{\pi}{4} \qquad x = -\frac{\pi}{4} + \pi$$

↑ Phase Shift ↑ Period

Thus, the graph of the equation is a cosine curve with a period of π and a phase shift of $-\frac{\pi}{4}$, which means the cosine curve is shifted $-\frac{\pi}{4}$ unit to the left. This matches (a).

35. Since the maximum deviation from the x axis is 5, we can write:
Amplitude $= |A| = 5$. Thus, $A = 5$ or -5. Since the period is $3 - (-1) = 4$, we can write:
Period $= \frac{2\pi}{B} = 4$. Thus, $B = \frac{2\pi}{4} = \frac{\pi}{2}$. Since we are instructed to choose the phase shift between 0 and 2, we can regard this graph as containing the basic sine curve with a phase shift of 1. This requires us to choose A positive, since the graph shows that as x increases from 1 to 2, y *increases* like the basic sine curve (not the upside down sine curve). So $A = 5$. Then, $-\frac{C}{B} = 1$. Thus,

$$C = -B = -\frac{\pi}{2} \text{ and } y = A\sin(Bx + C) = 5\sin\left(\frac{\pi}{2}x - \frac{\pi}{2}\right).$$

Check: When $x = 0$, $y = 5\sin\left(\frac{\pi}{2}\cdot 0 - \frac{\pi}{2}\right) = 5\sin\left(-\frac{\pi}{2}\right) = -5$

When $x = 1$, $y = 5\sin\left(\frac{\pi}{2}\cdot 1 - \frac{\pi}{2}\right) = 5\sin 0 = 0$

37. Since the maximum deviation from the x axis is 2, we can write:

Amplitude $= |A| = 2$. Thus, $A = 2$ or -2. Since the period is $\frac{5\pi}{2} - \left(-\frac{3\pi}{2}\right) = 4\pi$, we can write:

Period $= \frac{2\pi}{B} = 4\pi$. Thus, $B = \frac{2\pi}{4\pi} = \frac{1}{2}$. Since we are instructed to choose $-2\pi < -\frac{C}{B} < 0$, that is, the phase shift to the left, we regard this graph as containing the upside down cosine curve, with a phase shift of $-\frac{\pi}{2}$. This requires us to choose A negative, since the graph shows that as x increases from $-\frac{\pi}{2}$ to 0, y *increases* like the upside down cosine curve (not the basic cosine curve).

So, $A = -2$. Then, $-\frac{C}{B} = -\frac{\pi}{2}$.

Thus, $C = \frac{\pi}{2}B = \frac{\pi}{2} \cdot \frac{1}{2} = \frac{\pi}{4}$ and $y = A\cos(Bx + C) = -2\cos\left(\frac{1}{2}x + \frac{\pi}{4}\right)$.

Check: When $x = 0$, $y = -2\cos\left(\frac{1}{2}\cdot 0 + \frac{\pi}{4}\right) = -2\cos\frac{\pi}{4} = -\sqrt{2}$

When $x = \frac{\pi}{2}$, $y = -2\cos\left(\frac{1}{2}\cdot\frac{\pi}{2} + \frac{\pi}{4}\right) = -2\cos\frac{\pi}{2} = 0$

39. Amplitude $= |A| = |2| = 2$

Phase Shift and Period: Solve

$Bx + C = 0$ and $Bx + C = 2\pi$

$3x - \frac{\pi}{2} = 0 \qquad 3x - \frac{\pi}{2} = 2\pi$

$x = \frac{\pi}{6} \qquad x = \frac{\pi}{6} + \frac{2\pi}{3}$

↑ Phase Shift ↑ Period $= \frac{2\pi}{3}$

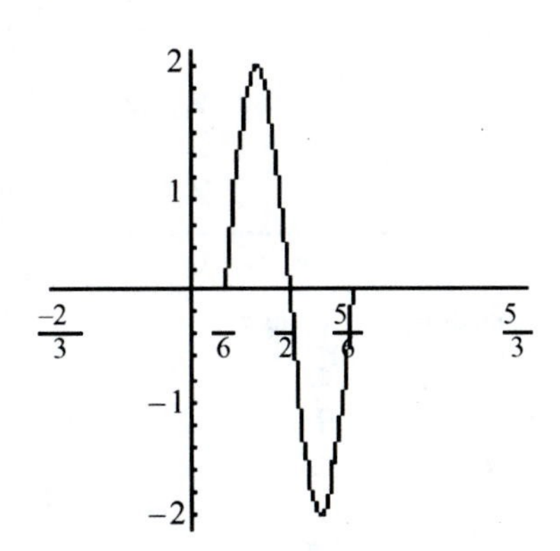

Phase Shift $= \frac{\pi}{6}$. Graph one cycle over the interval from $\frac{\pi}{6}$ to $\left(\frac{\pi}{6} + \frac{2\pi}{3}\right) = \frac{5\pi}{6}$.

Extend the graph from $-\frac{2\pi}{3}$ to $\frac{5\pi}{3}$.

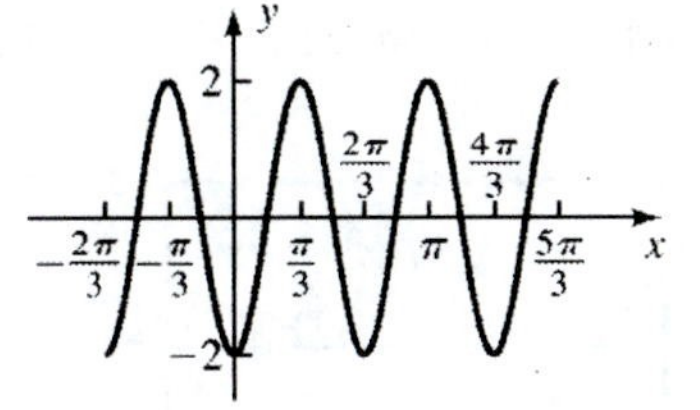

41. $y = 4 + 2\sin\left(3x - \frac{\pi}{2}\right)$.

$y = 2\sin\left(3x - \frac{\pi}{2}\right)$ was graphed in Problem 39. This graph is the graph of $y = 2\sin\left(3x - \frac{\pi}{2}\right)$ moved up $k = 4$ units. We start by drawing a horizontal broken line 4 units above the x axis, then graph $y = 2\sin\left(3x - \frac{\pi}{2}\right)$ relative to the broken line and the original y axis.

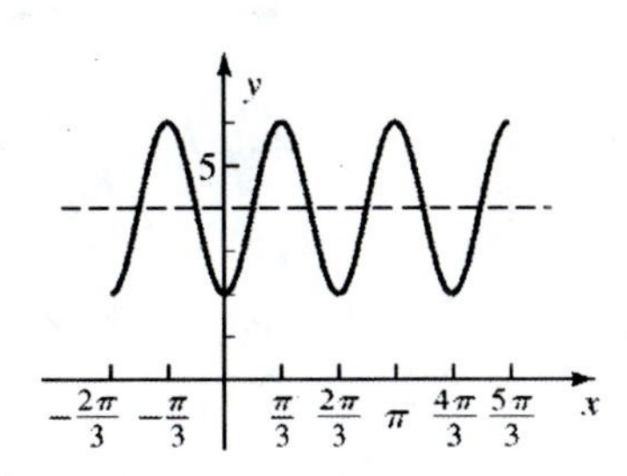

43. Amplitude = $|A| = |2.3| = 2.3$
Phase Shift and Period: Solve

$$\frac{\pi}{1.5}(x-2) = 0 \qquad \frac{\pi}{1.5}(x-2) = 2\pi$$
$$x - 2 = 0 \qquad x - 2 = 3$$
$$x = 2 \qquad x = 2 + 3$$

↑ Phase Shift ↑ Period

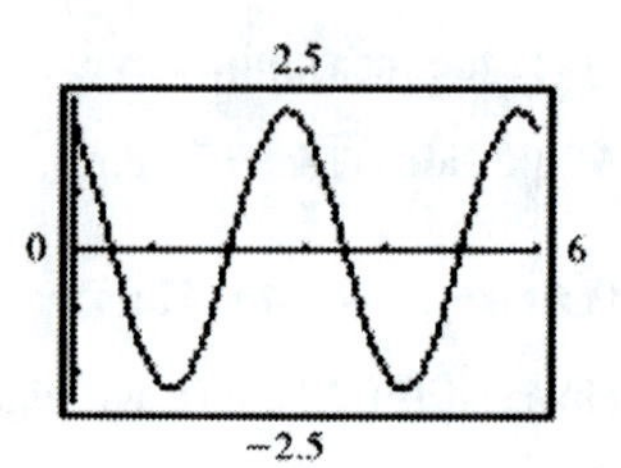

45. Amplitude = $|A| = |18| = 18$
Phase Shift and Period: Solve

$$4\pi(x + 0.137) = 0 \qquad 4\pi(x + 0.137) = 2\pi$$
$$x + 0.137 = 0 \qquad x + 0.137 = \frac{1}{2}$$
$$x = -0.137 \qquad x = -0.137 + \frac{1}{2}$$

↑ Phase Shift ↑ Period

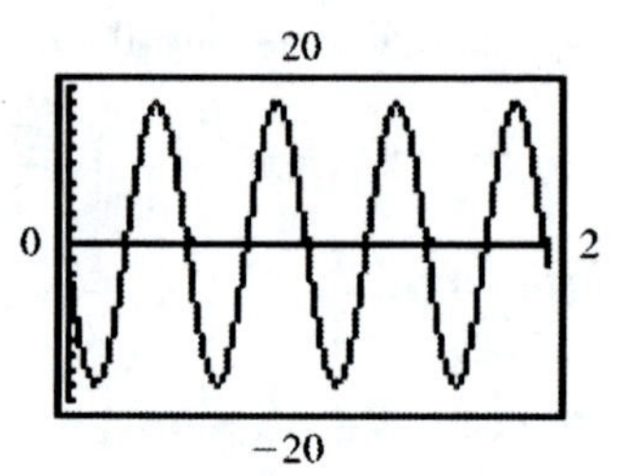

47. 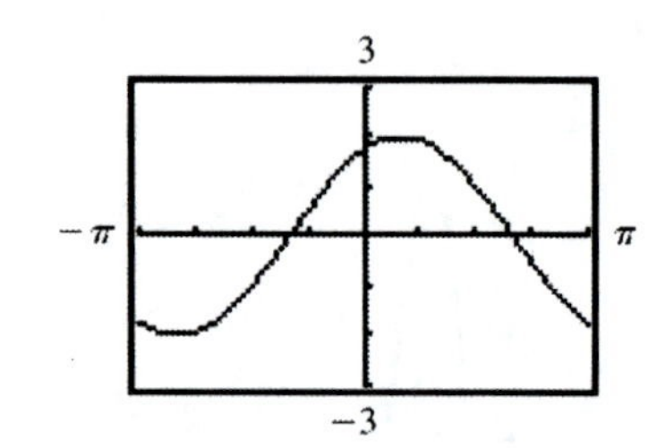

The graph of $y = \sin x + \sqrt{3}\cos x$ is shown in the figure. This graph appears to be a sine wave with amplitude 2 and period 2π that has been shifted to the left. Thus, we conclude that $A = 2$ and $B = \frac{2\pi}{2\pi} = 1$. To determine C, we use the zoom feature or the built-in approximation routine to locate the x intercept closest to the origin at $x = -1.047$. This is the phase-shift for the graph.

Substitute $B = 1$ and $x = -1.047$ into the phase-shift equation

$$x = -\frac{C}{B}$$
$$-1.047 = -\frac{C}{1}$$
$$C = 1.047$$

Thus, the equation required is $y = 2\sin(x + 1.047)$.

49. 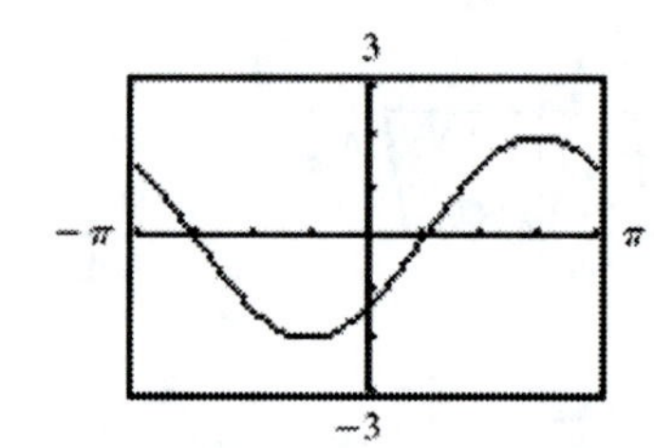

The graph of $y = \sqrt{2}\sin x - \sqrt{2}\cos x$ is shown in the figure. This graph appears to be a sine wave with amplitude 2 and period 2π that has been shifted to the right. Thus, we conclude that $A = 2$ and $B = \frac{2\pi}{2\pi} = 1$. To determine C, we use the zoom feature or the built-in approximation routine to locate the x intercept closest to the origin at $x = 0.785$. This is the phase-shift for the graph.

Substitute $B = 1$ and $x = 0.785$ into the phase-shift equation

$$x = -\frac{C}{B}$$
$$0.785 = -\frac{C}{1}$$
$$C = -0.785$$

Thus, the equation required is $y = 2\sin(x - 0.785)$.

51. 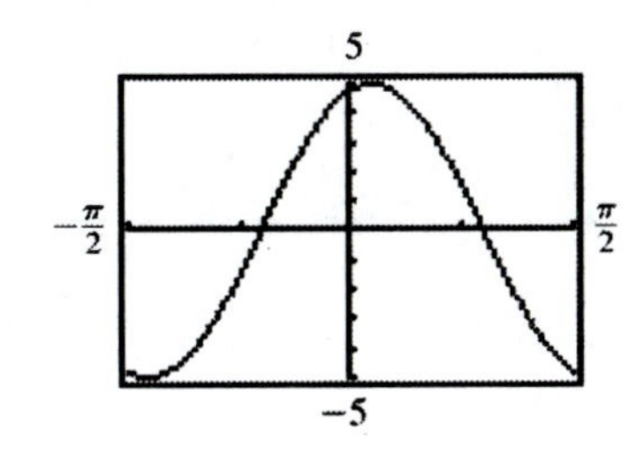

The graph of $y = 1.4 \sin 2x + 4.8 \cos 2x$ is shown in the figure. This graph appears to be a sine wave with amplitude 5 and period π that has been shifted to the left. Thus, we conclude that $A = 5$ and $B = \frac{2\pi}{\pi} = 2$. To determine C, we use the zoom feature or the built-in approximation routine to locate the x intercept closest to the origin at $x = -0.644$. This is the phase-shift for the graph.

Substitute $B = 2$ and $x = -0.644$ into the phase-shift equation

$$x = -\frac{C}{B}$$
$$-0.644 = -\frac{C}{2}$$
$$C = 1.288$$

Thus, the equation required is $y = 5 \sin(x + 1.288)$.

53. 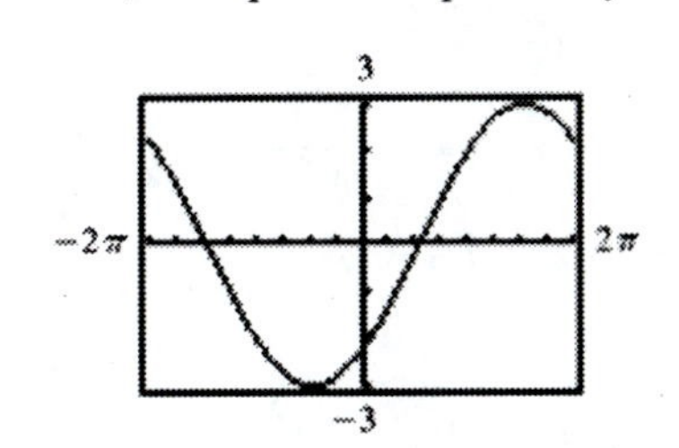

The graph of $y = 2 \sin \frac{x}{2} - \sqrt{5} \cos \frac{x}{2}$ is shown in the figure. This graph appears to be a sine wave with amplitude 3 and period 4π that has been shifted to the right. Thus, we conclude that $A = 3$ and $B = \frac{2\pi}{4\pi} = \frac{1}{2}$. To determine C, we use the zoom feature or the built-in approximation routine to locate the x intercept closest to the origin at $x = 1.682$. This is the phase-shift for the graph.

Substitute $B = \frac{1}{2}$ and $x = 1.682$ into the phase-shift equation
$$x = -\frac{C}{B}$$
$$1.682 = -\frac{C}{1/2}$$
$$C = -\frac{1}{2}(1.682) = -0.841$$

Thus, the equation required is $y = 3 \sin\left(\frac{x}{2} - 0.841\right)$.

55. We look for a function of the form $y = A \sin(Bx + C)$ with the required properties (although a function of the form $y = A \cos(Bx + C)$ would do equally well).
Since amplitude = 10, we set $A = 10$.
Since period = 12, we set $2\pi/B = 12$, thus $2\pi = 12B$ and $B = \pi/6$.
Since phase shift = −2, we set $-C/B = -2$, thus $C = 2B = \pi/3$.
Therefore, $y = 10 \sin\left(\frac{\pi}{6}x + \frac{\pi}{3}\right)$ is one of several possible answers.

57. We look for a function of the form $y = A \sin(Bx + C)$ with the required properties (although a function of the form $y = A \cos(Bx + C)$ would do equally well).
Since amplitude = 1, we set $A = 1$.
Since period = π, we set $2\pi/B = \pi$, thus $2\pi = \pi B$ and $B = 2$.
Since phase shift = $\frac{\pi}{8}$, we set $-C/B = \frac{\pi}{8}$, thus $C = -\frac{\pi}{8}B = -\frac{\pi}{8} \cdot 2 = -\frac{\pi}{4}$.
Therefore, $y = 1 \sin\left(2x - \frac{\pi}{4}\right)$ or $y = \sin\left(2x - \frac{\pi}{4}\right)$ is one of several possible answers.

59. Amplitude $= |A| = |5| = 5$

Phase Shift and Period: Solve

$$Bx + C = 0 \quad \text{and} \quad Bx + C = 2\pi$$

$$\frac{\pi}{6}(t+3) = 0 \qquad \frac{\pi}{6}(t+3) = 2\pi$$

$$t = -3 \qquad t = -3 + 12$$

(↑ Phase Shift, ↑ Period)

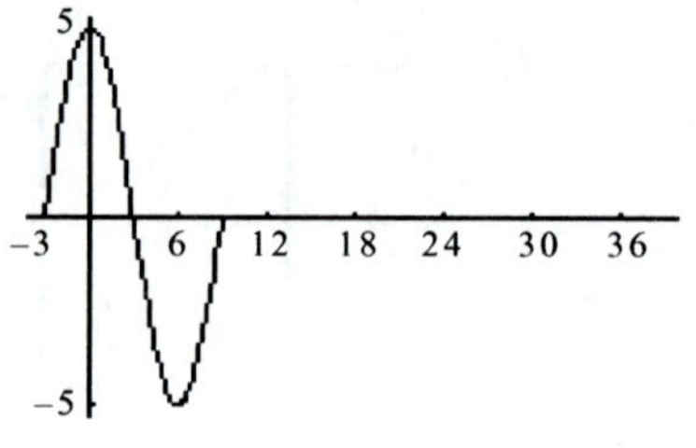

Phase Shift $= -3$. Period $= 12$ sec

Graph one cycle over the interval from -3 to $(-3 + 12) = 9$.

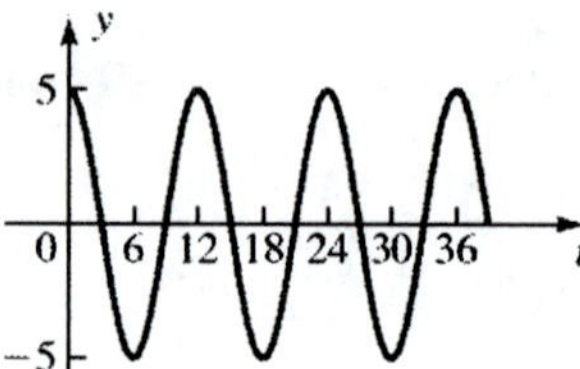

Extend the graph from 9 to 39, and delete the portion of the graph from -3 to 0, since this was not required.

61. Amplitude $= |A| = |30| = 30$

Phase Shift and Period: Solve

$$Bx + C = 0 \quad \text{and} \quad Bx + C = 2\pi$$

$$120\pi t - \pi = 0 \qquad 120\pi t - \pi = 2\pi$$

$$t = \frac{1}{120} \qquad t = \frac{1}{120} + \frac{1}{60}$$

(↑ Phase Shift, ↑ Period)

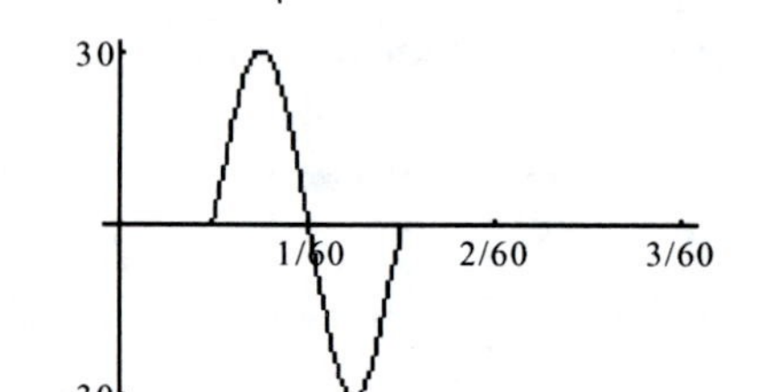

Phase Shift $= \frac{1}{120}$. Period $= \frac{1}{60}$. Frequency $= \frac{1}{\text{Period}} = \frac{1}{1/60} = 60$ Hz

Graph one cycle over the interval from $\frac{1}{120}$ to $\left(\frac{1}{120} + \frac{1}{60}\right) = \frac{3}{120}$.

Extend the graph from 0 to $\frac{3}{60}$.

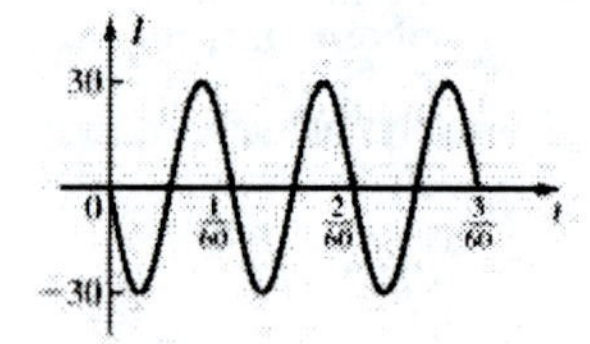

63. (A)

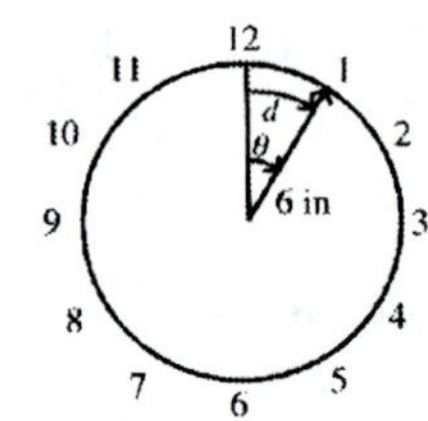

Since the second hand sweeps out 2π radians in 60 seconds, it sweeps out an angle of $\frac{2\pi}{12}$ or $\frac{\pi}{6}$ radians every 5 seconds. Since the required distance d satisfies $\frac{d}{6} = \sin\theta$ or $d = 6\sin\theta$, we can complete the table by adding $\frac{\pi}{6}$ radians every 5 seconds to θ starting at $\theta = 0$ when $t = 0$ and the hand points to 12.

Thus:

$t = 5$	$d = 6\sin\frac{\pi}{6} = 3.0$	$t = 35$	$d = 6\sin\frac{7\pi}{6} = -3$
$t = 10$	$d = 6\sin\frac{2\pi}{6} = 5.2$	$t = 40$	$d = 6\sin\frac{8\pi}{6} = -5.2$
$t = 15$	$d = 6\sin\frac{3\pi}{6} = 6$	$t = 45$	$d = 6\sin\frac{9\pi}{6} = -6$
$t = 20$	$d = 6\sin\frac{4\pi}{6} = 5.2$	$t = 50$	$d = 6\sin\frac{10\pi}{6} = -5.2$
$t = 25$	$d = 6\sin\frac{5\pi}{6} = 3$	$t = 55$	$d = 6\sin\frac{11\pi}{6} = -3.0$
$t = 30$	$d = 6\sin\frac{6\pi}{6} = 0$	$t = 60$	$d = 6\sin\frac{12\pi}{6} = 0$

Thus, we can complete the table 1 as follows:

TABLE 1 (Distance d, t sec after the second hand points to 12.)

t sec	0	5	10	15	20	25	30	35	40	45	50	55	60
d in	0.0	3.0	5.2	6.0	5.2	3.0	0.0	–3.0	–5.2	–6.0	–5.2	–3.0	0.0

TABLE 2 (has the same values for d, but starting with $d = 6 \sin \frac{9\pi}{6} = -6$)

t sec	0	5	10	15	20	25	30	35	40	45	50	55	60
d in	–6.0	–5.2	–3.0	0.0	3.0	5.2	6.0	5.2	3.0	0.0	–3.0	–5.2	–6.0

(B) From the table values and the position of the hand on the clock, since the same values are found in Table 2 15 seconds later than in Table 1, we see that relation (2) is 15 seconds out of phase with relation (1).

(C) Clearly the relations repeat the values every 60 seconds, hence this is the period. Since the largest value is 6.0 and the smallest value is –6.0, the amplitude is 6.0 inches.

(D) For relation (1), $|A| = 6.0$

$$\text{Period} = 60\text{, thus } \frac{2\pi}{B} = 60\text{, hence, } B = \frac{2\pi}{60} = \frac{\pi}{30}.$$

Since $d = 0$ when $t = 0$, there is no phase shift and $C = 0$. Thus, $y = 6.0 \sin \frac{\pi}{30}t$ or $y = -6.0 \sin \frac{\pi}{30}t$.

Since $d = 15$ when $t = 6.0$, the equation must be

$$y = 6.0 \sin \frac{\pi}{30}t$$

For relation (2), $A = 6.0$ and $B = \frac{\pi}{30}$ again. Since the phase shift is 15, $-\frac{C}{B} = 15$ and

$C = -15B = -15\left(\frac{\pi}{30}\right) = -\frac{\pi}{2}$. Hence the equation must be

$$y = 6.0 \sin\left(\frac{\pi}{30}t - \frac{\pi}{2}\right)$$

The student should check that these equations give the values in the tables above.

(E) Here the equation will be again of the form $y = 6.0 \sin\left(\frac{\pi}{30}t + C\right)$. The second hand will point to 3 after 15 sec, hence the phase shift will be –15. $-\frac{C}{B} = -15$, hence $C = +15B = 15\left(\frac{\pi}{30}\right) = \frac{\pi}{2}$.

Hence the equation must be

$$y = 6.0 \sin\left(\frac{\pi}{30}t + \frac{\pi}{2}\right)$$

(F)

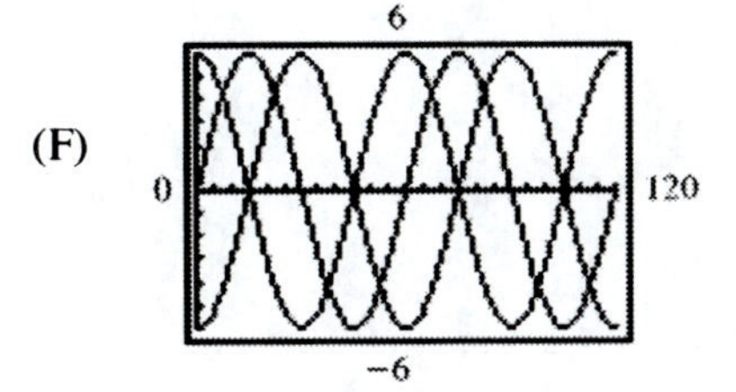

65. (A)

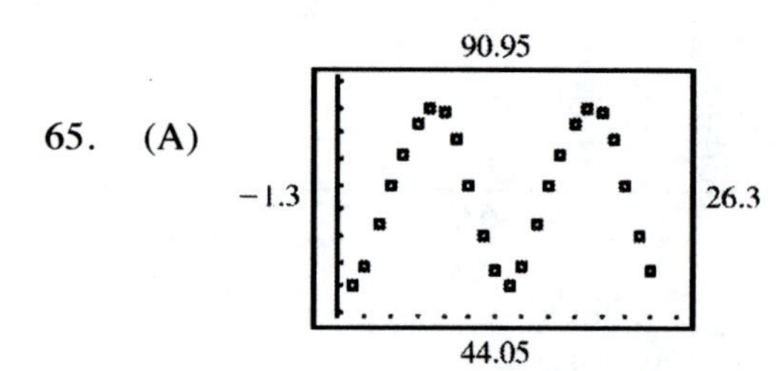

(B) $|A| = \dfrac{\max y - \min y}{2} = \dfrac{85-50}{2} = 17.5$; Period = 12 months, therefore, $B = \dfrac{2\pi}{P} = \dfrac{\pi}{6}$;
$k = |A| + \min y = 17.5 + 50 = 67.5$. From the scatter plot it appears that if we use $A = 17.5$, then the phase shift will be about 3.0--which can be adjusted for a better visual fit later, if necessary. Thus, using $PS = -\dfrac{C}{B}$, $C = -\dfrac{\pi}{6}(3.0) = -1.6$, and $y = 67.5 + 17.5\sin\left(\dfrac{\pi t}{6} - 1.6\right)$.

Graphing this equation in the same viewing window as the scatter plot, we see that adjusting C to -2.1 produces a little better visual fit. Thus, with this adjustment, the equation and graph are:

$$y = 67.5 + 17.5\sin\left(\frac{\pi t}{6} - 2.1\right)$$

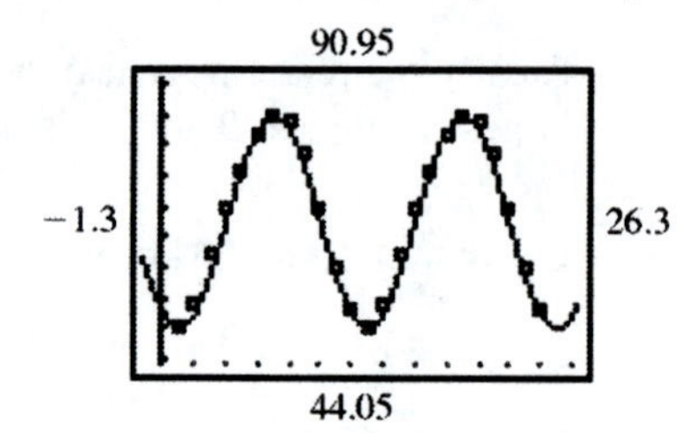

(C) $y = 68.7 + 17.1\sin(0.5t - 2.1)$

90.95

−1.3 26.3

44.05

(D) The regression equation differs slightly in k, A, and B, but not in C. Both equations appear to fit the data very well.

67. No answer is provided here. The steps are as in the previous problem, but the data should be collected by the student for the relevant city.

EXERCISE 3.4 Additional Applications

1. (A) Amplitude $|A| = |10| = 10$ amperes. Phase Shift: Solve $Bx + C = 0$. $120\pi t - \dfrac{\pi}{2} = 0$ $\quad t = \dfrac{1}{240}$

Phase Shift $= \dfrac{1}{240}$ seconds. Frequency $= \dfrac{B}{2\pi} = \dfrac{120\pi}{2\pi} = 60$ Hz

(B) The maximum current is the amplitude, which is 10 amperes.

(C)

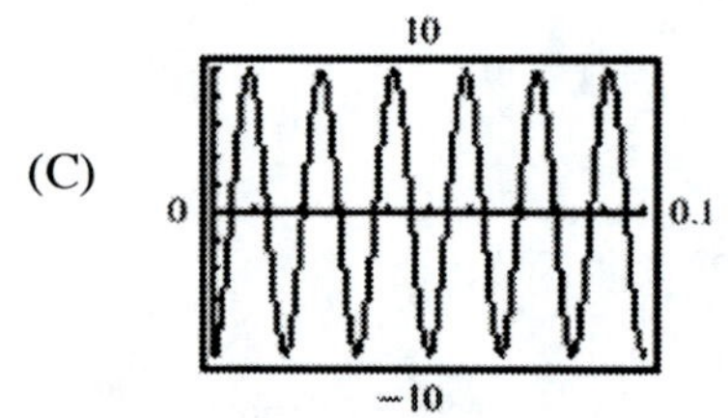

There are 60 cycles per second, hence there are $60(0.1) = 6$ cycles in 0.1 second, as shown.

3. *Find A:* The amplitude $|A|$ is given to be 20. Since $A > 0, A = 20$.

   Find B: We are given that the frequency $f = 30$ Hz. But $f = \dfrac{B}{2\pi}$. Thus, $\dfrac{B}{2\pi} = 30, B = 60\pi$

   Write the equation: $I = 20 \cos 60\pi t$

5. The height of the wave from trough to crest is the difference in height between the crest (height A) and the trough (height $-A$). In this case, $A = 15$ ft. $A - (-A) = 2A = 2(15 \text{ ft}) = 30$ ft. To find the wavelength γ, we note: $\gamma = 5.12T^2, T = \dfrac{2\pi}{B}, B = \dfrac{\pi}{8}$. Thus, $T = \dfrac{2\pi}{\pi/8} = 16$ sec,

   $\gamma = 5.12(16)^2 \approx 1311$ ft. To find the speed S, we use

$$
\begin{aligned}
S &= \sqrt{\frac{g\gamma}{2\pi}} \qquad g = 32 \text{ ft/sec}^2 \\
&= \sqrt{\frac{32(1311)}{2\pi}} \approx 82 \text{ ft/sec}
\end{aligned}
$$

7. To graph $y = 15 \sin \dfrac{\pi}{8}t$, we note: Amplitude $= |A| = 15$ ft. Period $= \dfrac{2\pi}{B} = 16$ sec. One full cycle of the graph is completed as t goes from 0 to 16. Block out this interval, divide it into four equal parts, locate high and low points, and locate t intercepts. Then complete the graph.

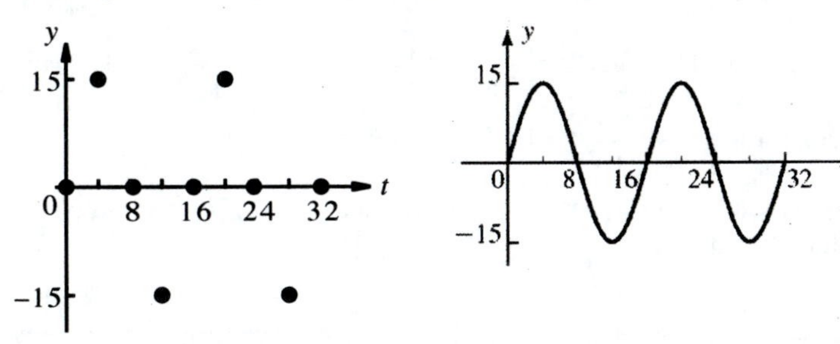

9. (A) *Find A:* The amplitude $|A|$ is given to be 2. Although, on the basis of the given information, A could be either 2 or -2, it is natural to choose $A = 2$.

   Find B: Since the variable in this problem is r, the distance from the source, the length of one cycle = wavelength = $\lambda = \dfrac{2\pi}{B}$. Thus, $B = \dfrac{2\pi}{\lambda} = \dfrac{2\pi}{150} = \dfrac{\pi}{75}$

   Write the equation: $y = 2 \sin \dfrac{\pi}{75}r$

   (B) To find the period T, we use: $\lambda = 5.12T^2$ (λ in feet), $T = \sqrt{\dfrac{\lambda}{5.12}}$

   Substituting $\lambda = 150$ miles $= (150)(5280)$ feet, we have $T = \sqrt{\dfrac{(150)(5280)}{5.12}} \approx 393$ sec

11. (A) The equation $y = 25 \sin 2\pi\left(\dfrac{t}{10} + \dfrac{1024}{512}\right) = 25 \sin 2\pi\left(\dfrac{t}{10} + 2\right)$ models the vertical motion of the wave at the fixed point $r = 1{,}024$ ft from the source relative to time in seconds.

   (B) The appropriate choice is period, since period is defined in terms of time and wavelength is defined in terms of distance. The period of $25 \sin 2\pi\left(\dfrac{t}{10} + 2\right) = 25 \sin\left(\dfrac{2\pi}{10}t + 4\pi\right)$ is $\dfrac{2\pi}{B}$,

   where $B = \dfrac{2\pi}{10}$. Thus, period $= 2\pi \div \dfrac{2\pi}{10} = 10$ seconds.

(C)

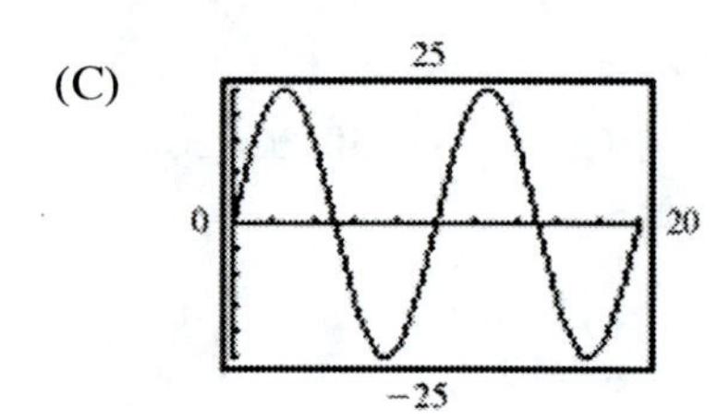

13. Period = $\frac{1}{\nu} = \frac{1}{10^8} = 10^{-8}$ sec. To find the wavelength λ, we use the formula $\lambda\nu = c$

with $c \approx 3 \times 10^8$ m/sec, $\lambda = \frac{c}{\nu} = \frac{3 \times 10^8 \text{ m/sec}}{10^8 \text{ Hz}} = 3$ m

15. We first use $\lambda\nu = c$ to find the frequency and then use $\nu = \frac{B}{2\pi}$ to find B:

$\nu = \frac{c}{\lambda} = \frac{3 \times 10^8 \text{ m/sec}}{3 \times 10^{-10}} = 10^{18}$ Hz; $B = 2\pi\nu = 2\pi \times 10^{18}$

17. If $y = A \sin 2\pi \times 10^6\, t$, then $B = 2\pi \times 10^6$. Since $\nu = \frac{B}{2\pi}$, we have $\nu = \frac{2\pi \times 10^6}{2\pi} = 10^6$ Hz

Since Period = $\frac{2\pi}{B}$, we have Period = $\frac{2\pi}{2\pi \times 10^6} = 10^{-6}$ sec

Figure 2 (text) shows atmospheric adsorption, for waves of frequency 10^6 Hz, as total. No, such waves cannot pass through the atmosphere.

19. (A)

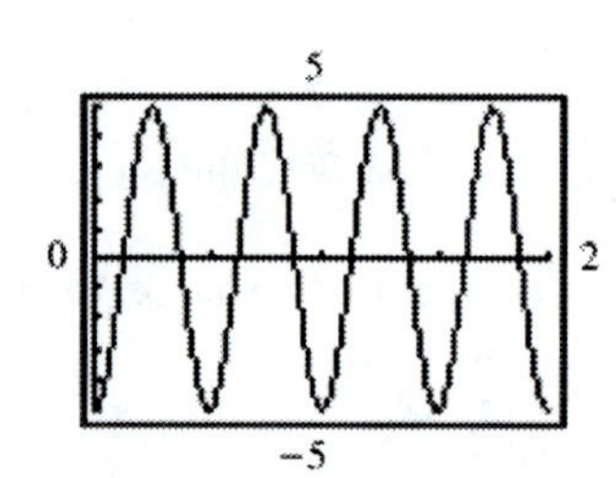

Since the amplitude remains constant over time, this represents simple harmonic motion.

(B)

Since the amplitude decreases to 0 as time increases, this represents damped harmonic motion.

21. (A) The lowest point is set at $t = 0$, $y = 42 - \frac{36}{0+2}\cos(0.989 \cdot 0)$ = 24 ft. Using the built-in maximum routine, the highest point is found to be $y = 49$ ft when $t = 2.97$ sec.

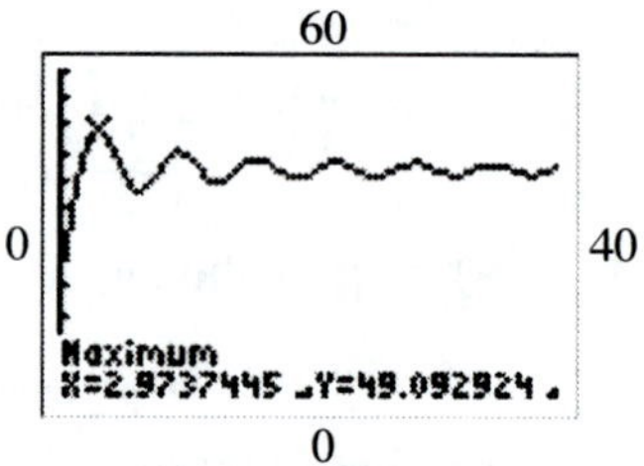

(B) Using the trace feature of the calculator, 5 minimum height times may be found between $t = 0$ and $t = 30$ sec.

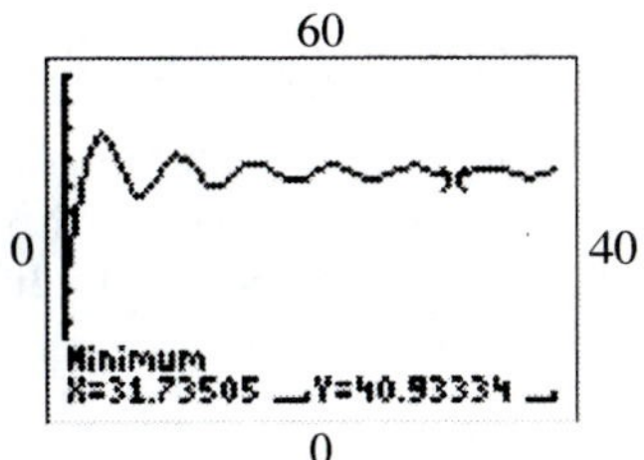

(C) The last minimum not within 1 foot of the resting height of 42 feet is found at $y = 40.9$ ft when $t = 31.7$ sec. Thus, after 32 sec the jump is considered over.

23. For men, to find $m(x) = k + A \sin Bx$, set

$$|A| = \frac{37.7 - 35.7}{2} = 1$$

$$B = \frac{2\pi}{\text{Period}} = \frac{2\pi}{24} = \frac{\pi}{12}$$

$$k = |A| + 35.7 = 36.7$$

Since we are told that the temperature decreases to a minimum at about 6 AM, set $A = -1$.

Thus, $m(x) = 36.7 - \sin \frac{\pi}{12}x$.

For women, to find $w(x) = k + A \sin Bx$, set

$$|A| = \frac{38.1 - 33.2}{2} = 2.45$$

$$B = \frac{2\pi}{\text{Period}} = \frac{2\pi}{24} = \frac{\pi}{12}$$

$$k = |A| + 33.2 = 35.65$$

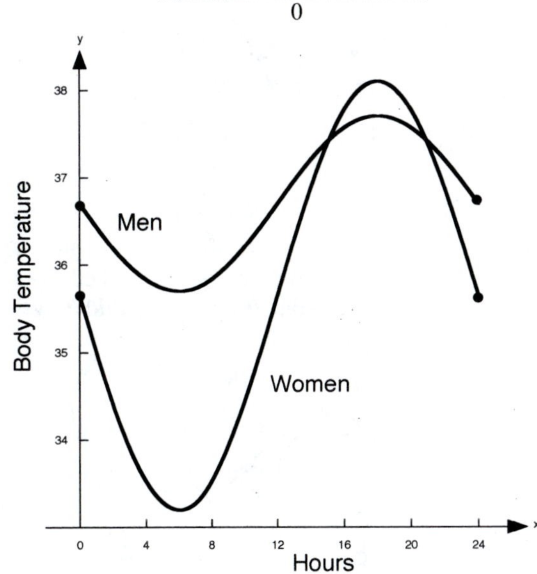

As before, set $A = -2.45$. Thus, $w(x) = 35.65 - 2.45 \sin \frac{\pi}{12}x$.

EXERCISE 3.5 Graphing the Sum of Functions

1. Draw the graph of each equation on the same axes, then use a compass, ruler, or eye to add the y coordinates to build the graph of the sum.

3. The frequencies of overtones are integer multiples of the frequency of the fundamental tone.

5. We form $y_1 = 1$ and $y_2 = \sin x$. We sketch the graph of each equation in the same coordinate system (dashed lines), then add the ordinates $y_1 + y_2$ (solid curve).

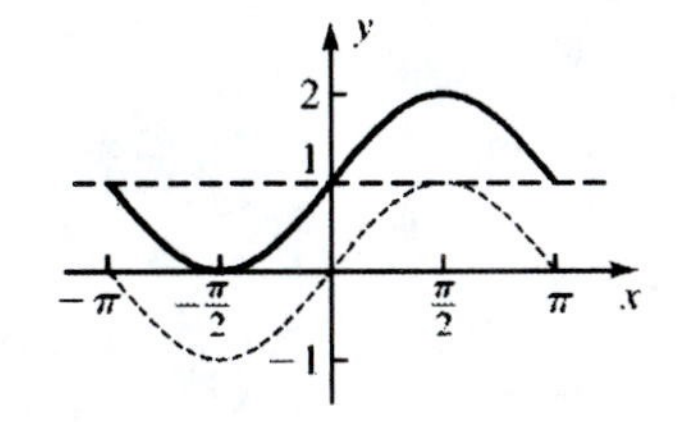

7. We form $y_1 = x$ and $y_2 = \cos x$. We sketch the graph of each equation in the same coordinate system (dashed lines), then add the ordinates $y_1 + y_2$ (solid curve).

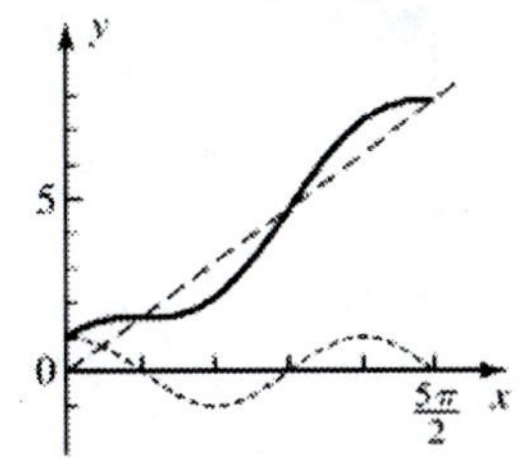

9. We form $y_1 = \frac{x}{2}$ and $y_2 = \cos \pi x$. We sketch the graph of each equation in the same coordinate system (dashed lines), then add the ordinates $y_1 + y_2$ (solid curve).

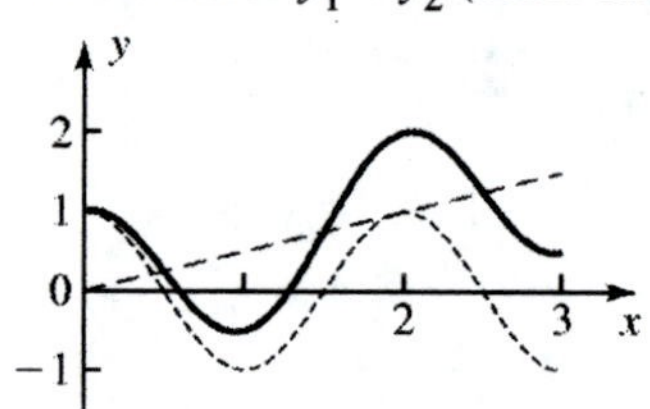

11. We form $y_1 = x$ and $y_2 = -\cos 2\pi x$. We sketch the graph of each equation in the same coordinate system (dashed lines), then add the ordinates $y_1 + y_2$ (solid curve).

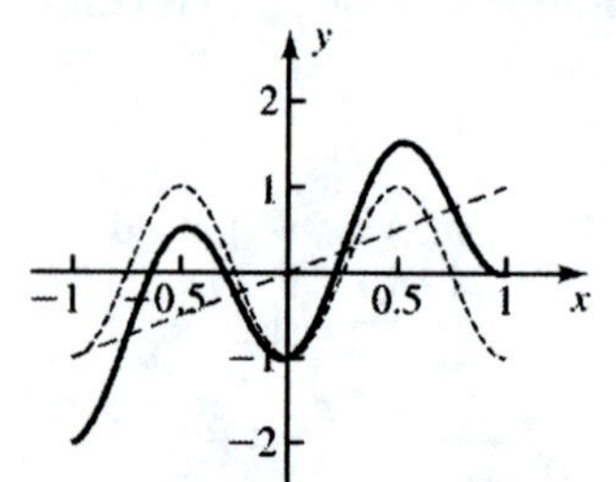

13. We form $y_1 = \sin x$ and $y_2 = \cos x$. We sketch the graph of each equation in the same coordinate system (dashed lines), then add the ordinates $y_1 + y_2$ (solid curve).

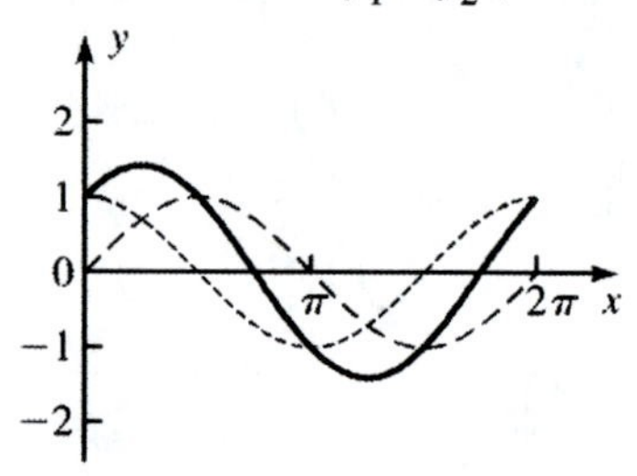

15. We form $y_1 = 3 \sin x$ and $y_2 = \cos x$. We sketch the graph of each equation in the same coordinate system (dashed lines), then add the ordinates $y_1 + y_2$ (solid curve).

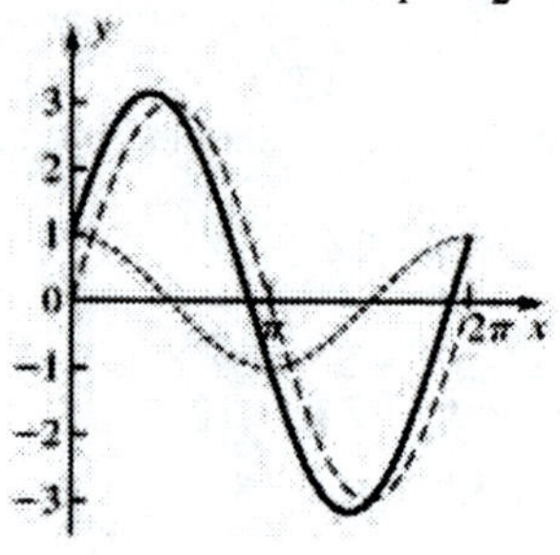

17. We form $y_1 = 3 \cos x$ and $y_2 = \sin 2x$. We sketch the graph of each equation in the same coordinate system (dashed curves), then add the ordinates $y_1 + y_2$ (solid curve).

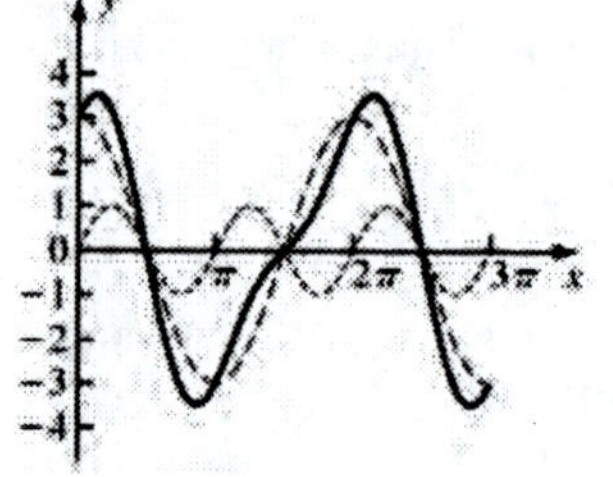

19. We form y1 = sin x and y2 = 2 cos 2x. We sketch the graph of each equation in the same coordinate system (dashed curves), then add the ordinates y1 + y2 (solid curve).

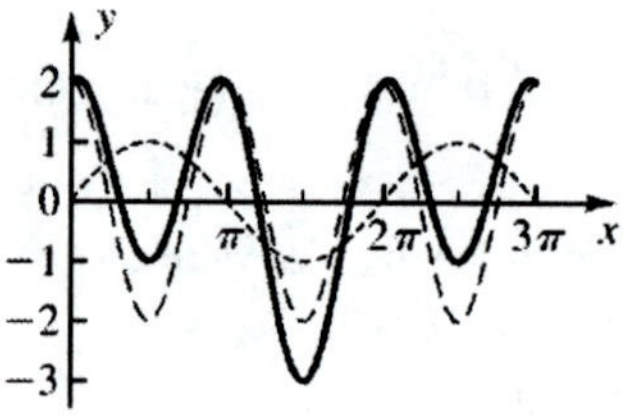

21.

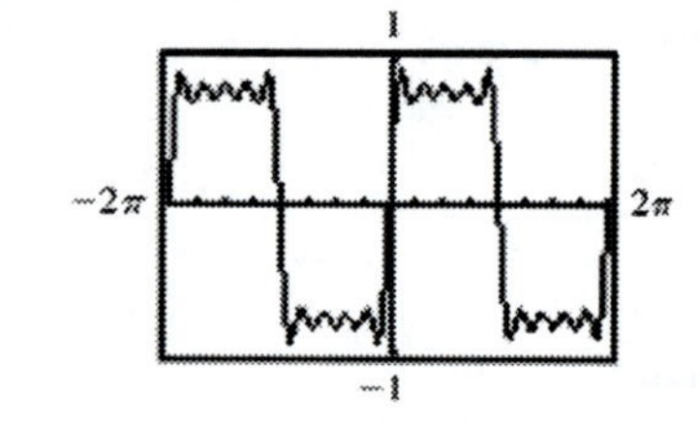

23.

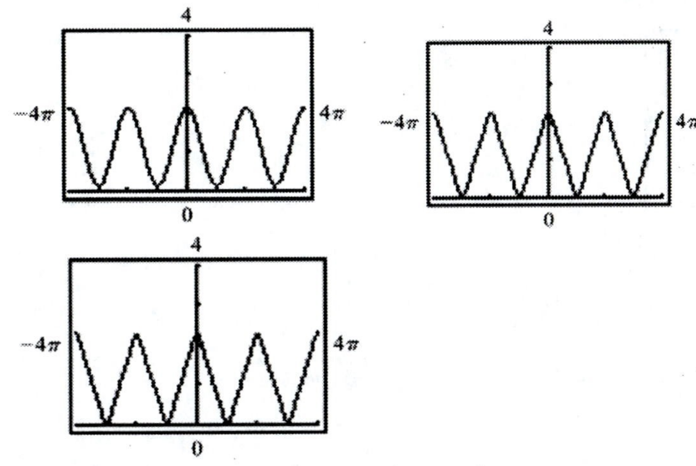

(B) The graphs approach a saw-tooth wave form.

(C)

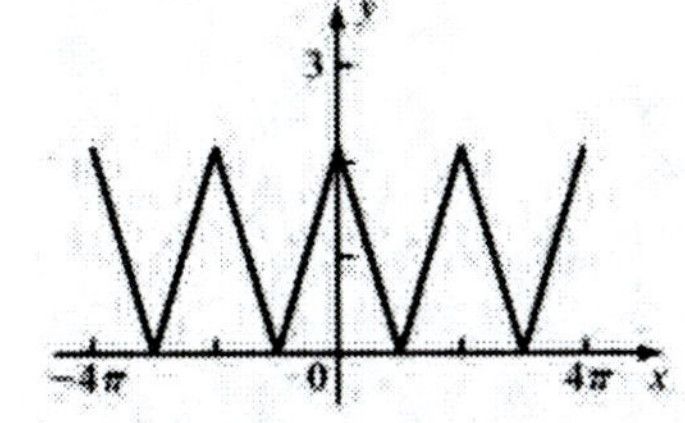

25. *Determine k:* The dashed line indicates that high and low points of the curve are equal distances from the line $V = 0.45$. Hence, $k = 0.45$.

Determine A: The maximum deviation from the line $V = 0.45$ is seen at points such as $t = 2$, $V = 0.8$ or $t = 4$, $V = 0.1$. Thus, $|A| = 0.8 - 0.45$ or $|A| = 0.45 - 0.1$. From either statement, we see, $|A| = 0.35$. So, $A = 0.35$ or $A = -0.35$. Since the portion of the curve as t increases from 0 to 1 has the form of an upside down basic cosine curve (V increasing), A must be negative. $A = -0.35$.

Determine B: One full cycle of the curve is completed as t varies from 0 to 4 seconds. Hence, the period $P = 4$ sec. Since

$P = \frac{2\pi}{B}$, we have $4 = \frac{2\pi}{B}$, or $B = \frac{2\pi}{4} = \frac{\pi}{2}$.

$$V = k + A\cos Bt = 0.45 - 0.35\cos\frac{\pi}{2}t.$$

27. (A) We form $S_1 = 5 + \frac{t}{52}$ and $S_2 = -4\cos\frac{\pi t}{26}$.
We sketch the graph of each equation in the same coordinate system (dashed lines), then add the ordinates $S_1 + S_2$ (solid curve).

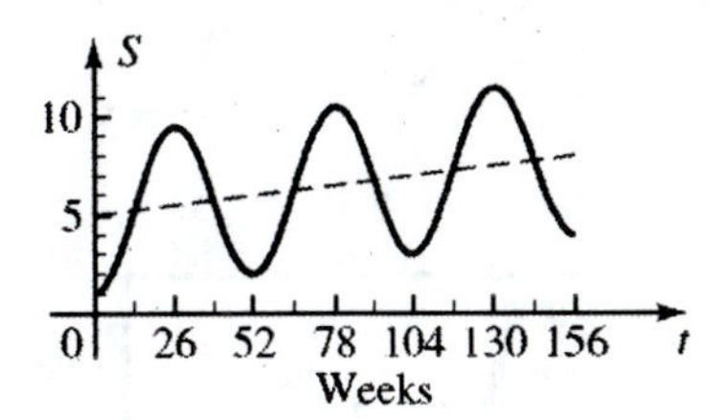

(B) In the 26th week of the 3rd year,
$t = 2 \cdot 52 + 26 = 130$. The sales are given by

$$S = 5 + \frac{130}{52} - 4\cos\frac{\pi \cdot 130}{26}$$
$$= 5 + 2.5 - 4\cos 5\pi = 7.5 - 4(-1) = \$11.5 \text{ million}$$

(C) In the 52nd week of the 3rd year, $t = 3 \cdot 52 = 156$. The sales are given by

$$S = 5 + \frac{156}{52} - 4\cos\frac{\pi \cdot 156}{26} = 5 + 3 - 4\cos 6\pi = 8 - 4(1) = \$4 \text{ million}$$

29. (A)

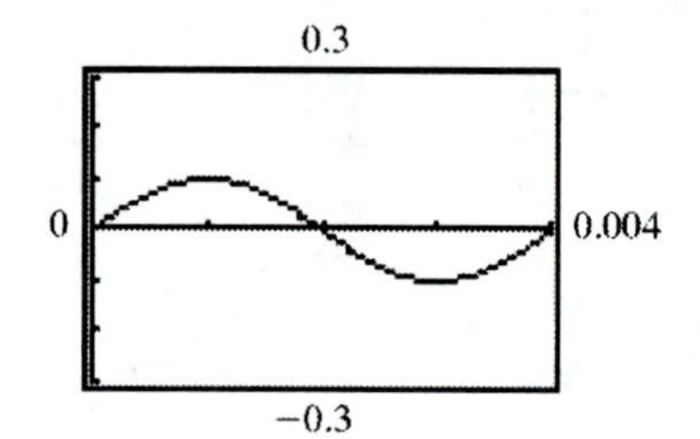

(B)

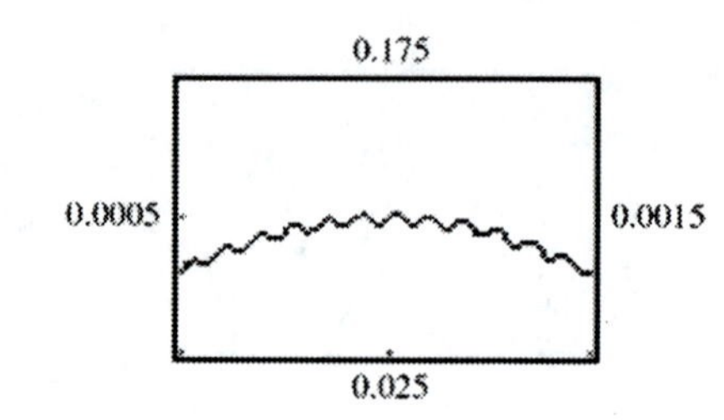

31.

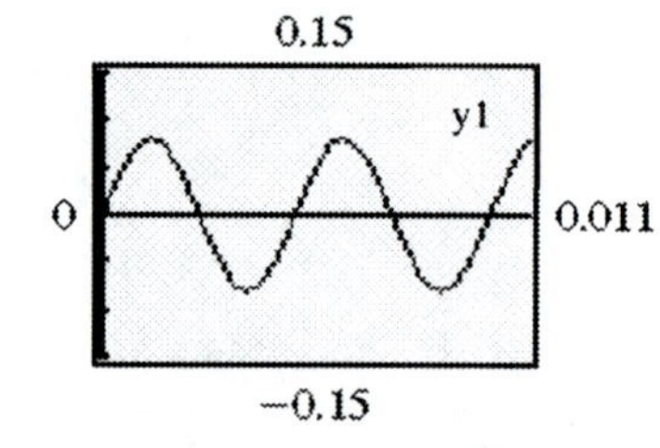

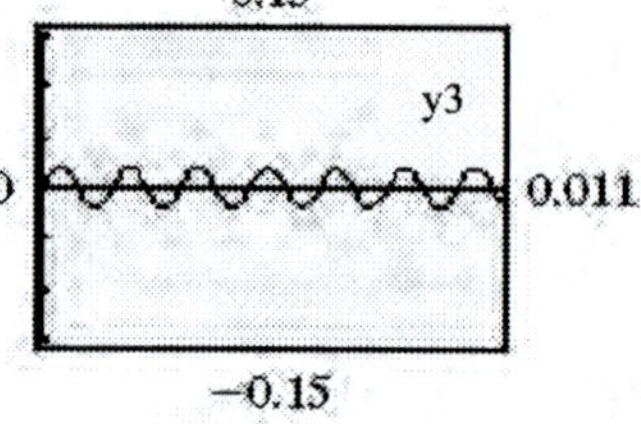

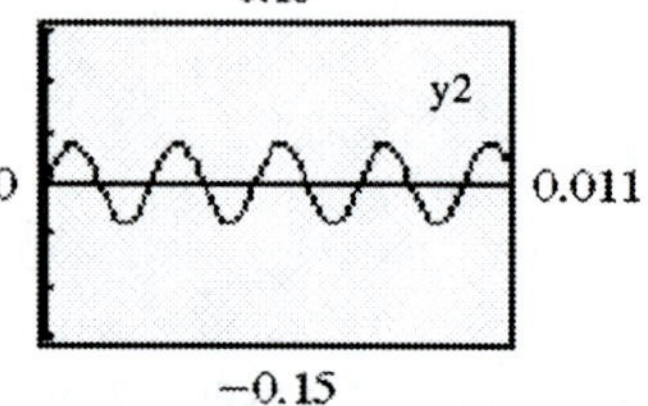

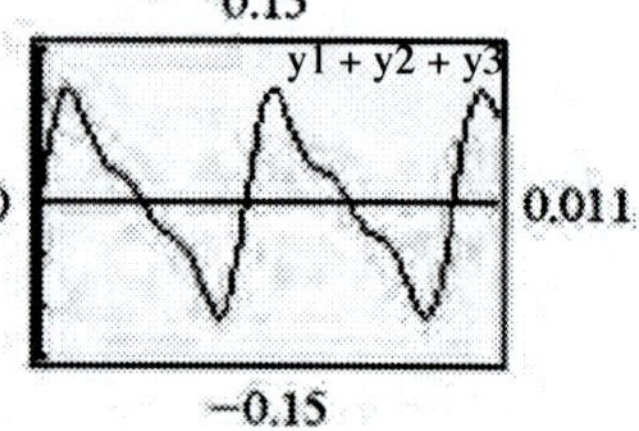

The period of the combined tone is the same as the period of the fundamental tone, that is, $\dfrac{2\pi}{400\pi} = 0.005$ seconds.

33. (A)

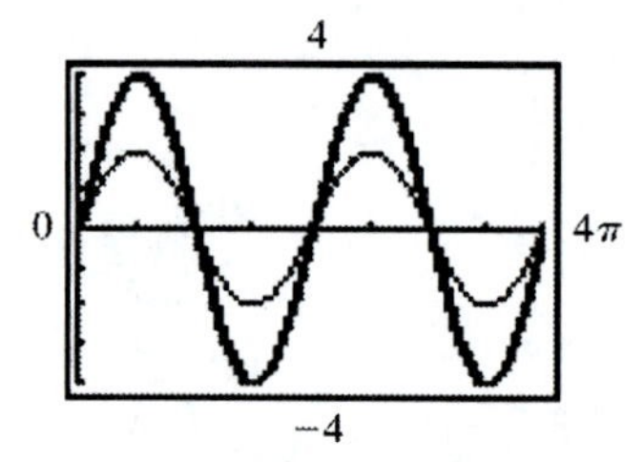

y1 = y2 = 2 sin t
y3 = 4 sin t
Since the amplitude of y3 (darker line) is greater than that of y1 and y2, this represents constructive interference.

(B)

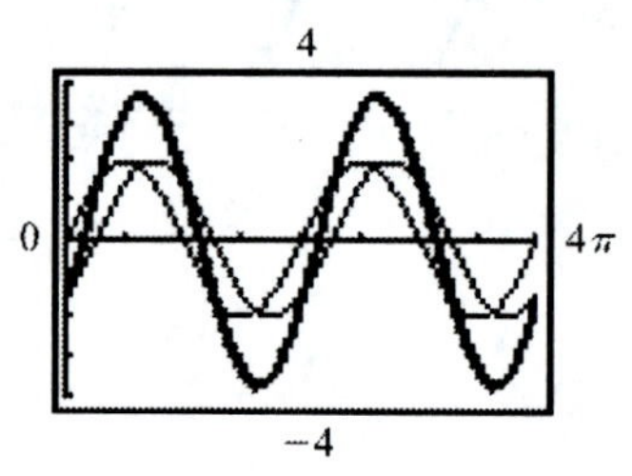

$y1 = 2 \sin t,\ y2 = 2 \sin\left(t - \dfrac{\pi}{4}\right)$

$B = y1 + y2 = 2 \sin t + 2 \sin\left(t - \dfrac{\pi}{4}\right)$

Since the amplitude of y3 (darker line) is greater than that of y1 and y2, this represents constructive interference.

(C)

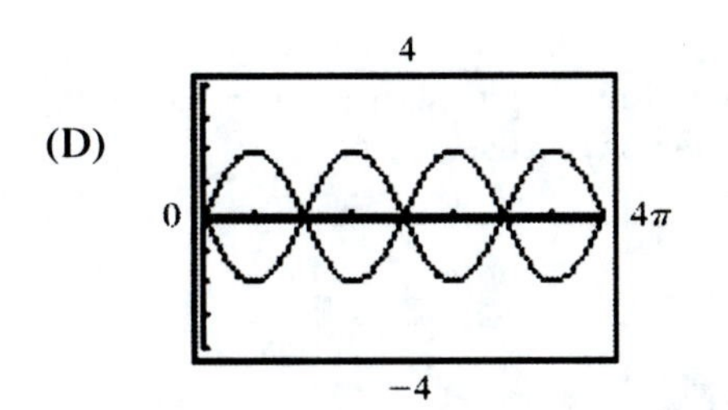

$y1 = 2 \sin t$, $y2 = 2 \sin\left(t - \frac{3\pi}{4}\right)$

$y3 = 2 \sin t + 2 \sin\left(t - \frac{3\pi}{4}\right)$

Since the amplitude of y3 (darker line) is less than that of y1 and y2, this represents destructive interference.

(D)

4

0 4π

−4

$y1 = 2 \sin t$, $y2 = 2 \sin(t - \pi)$
$y3 = 2 \sin t + 2 \sin(t - \pi)$
Since the amplitude of y3 (dark line along x axis) is less than that of y1 and y2 (in fact, 0), this represents destructive interference.

35. (A) y2 must have the same amplitude and period as y1, but must be completely out of phase with y2. Hence $C = \pi$ or $-\pi$. Since C is required to be negative, $y2 = 65 \sin(400\pi t - \pi)$.

(B) y2 added to y1 produces a sound wave of zero amplitude--no noise.

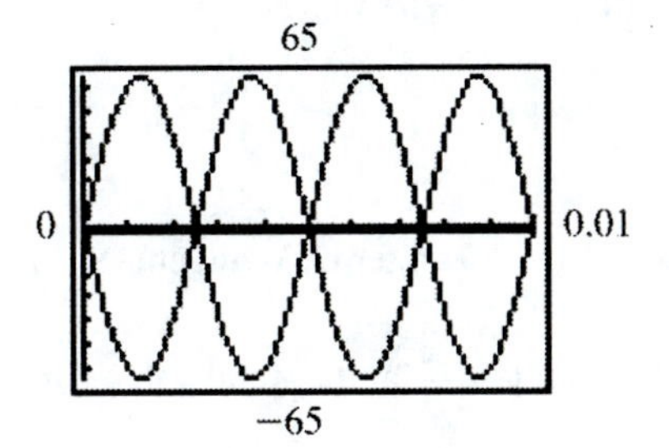

37. (A)

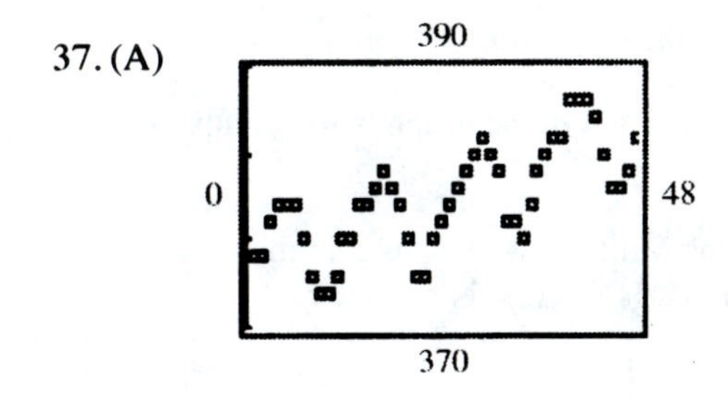

(B) Using the LinReg feature of the TI-83 on (6.5, 376), (18.5, 378), (30.5, 380), and (42.5, 382) yields $f(x) = 0.167x + 375$.

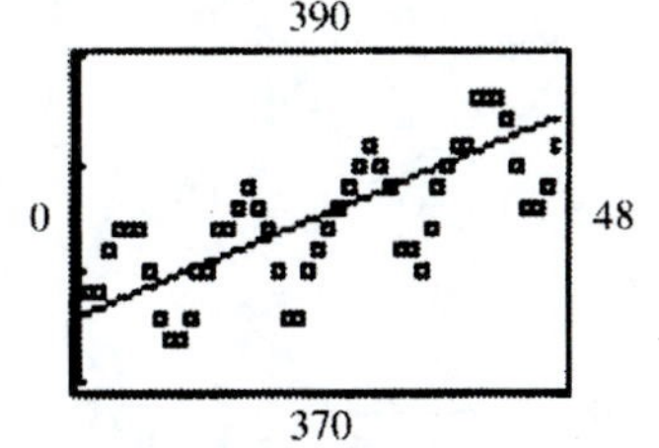

(C) Subtraction yields the following table of values:

	J	F	M	A	M	J	J	A	S	O	N	D
2003	0.17	1.00	1.83	2.67	3.50	2.33	1.17	−2.00	−3.17	−3.33	−1.50	−0.67
2004	0.17	1.00	1.83	2.67	3.50	2.33	−0.83	−2.00	−4.17	−4.33	−2.50	−0.67
2005	−0.83	1.00	1.83	2.67	2.50	2.33	1.17	−1.00	−3.17	−3.33	−2.50	−0.67
2006	0.17	1.00	1.83	3.67	3.50	2.33	0.17	−2.00	−3.17	−3.33	−2.50	−0.67

The values from the table are shown as plotted on the TI-83.

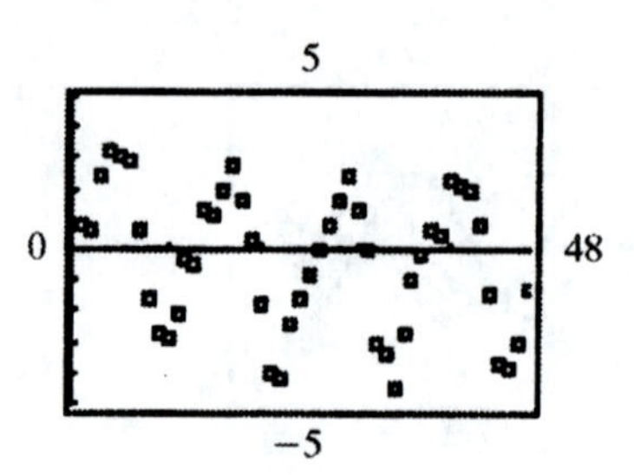

(D) Applying the SinReg feature yields $s(x) = 3.12 \sin(0.526x - 0.618) + 0.004$.

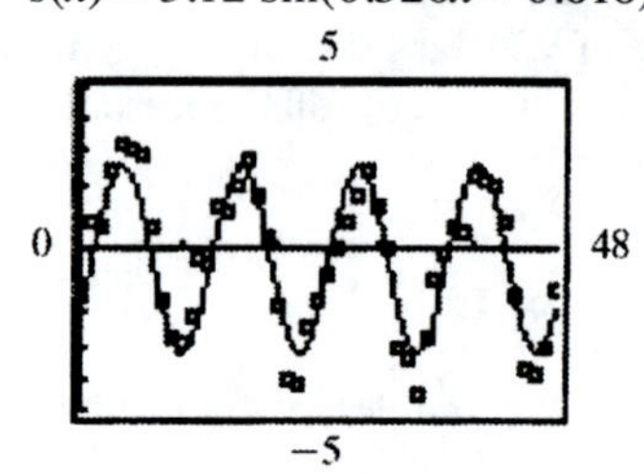

(E) The following plot shows $g(x) = f(x) + s(x)$ and the original scatter plot.

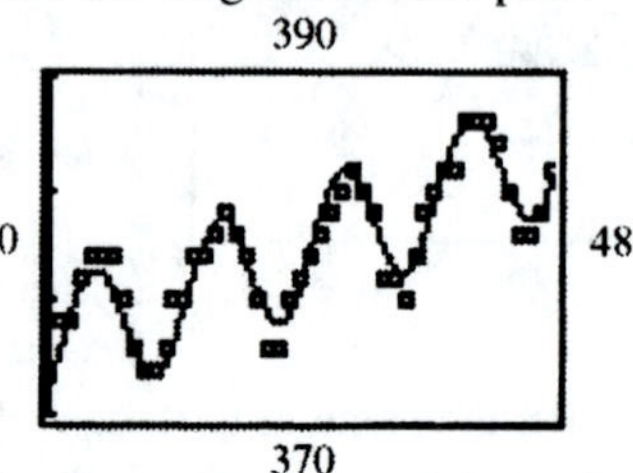

The function $g(x)$ appears to be a good model (despite occasional deviations) for the data for these years.

EXERCISE 3.6 Tangent, Cotangent, Secant, and Cosecant Functions Revisited

1. Set $Bx + C$ equal to $-\frac{\pi}{2}$ and $\frac{\pi}{2}$ and solve for x.

3. The secant and tangent functions are undefined for exactly the same input values.

5. A makes the graph steeper if $|A| > 1$ and less steep if $|A| < 1$. If $A < 0$, the graph is turned upside down.

7.

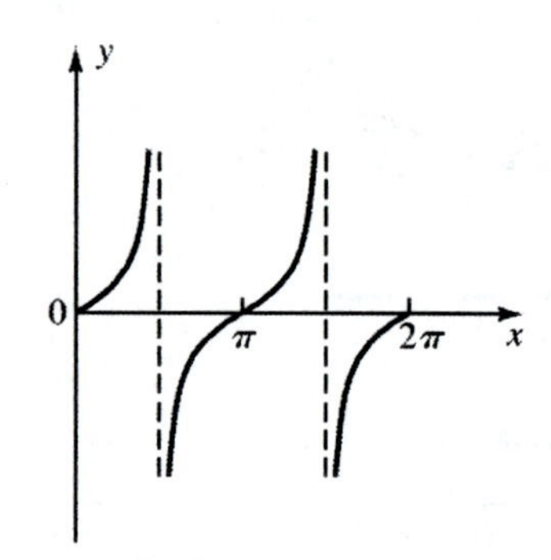

9. The dashed line shows $y = \sin x$ in this interval. The solid line is $y = \csc x$.

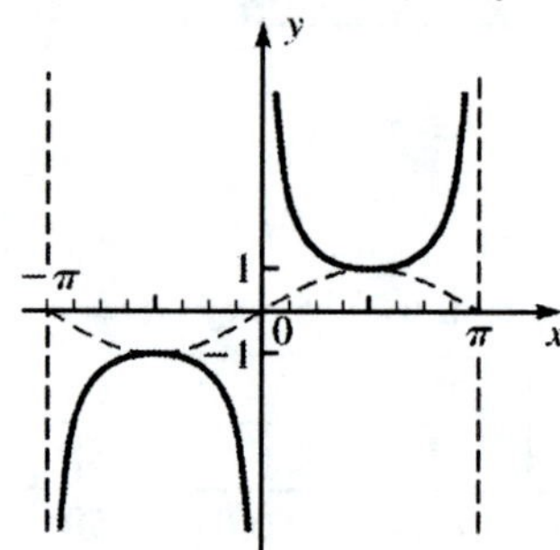

11. Period $= \dfrac{\pi}{B} = \dfrac{\pi}{2}$

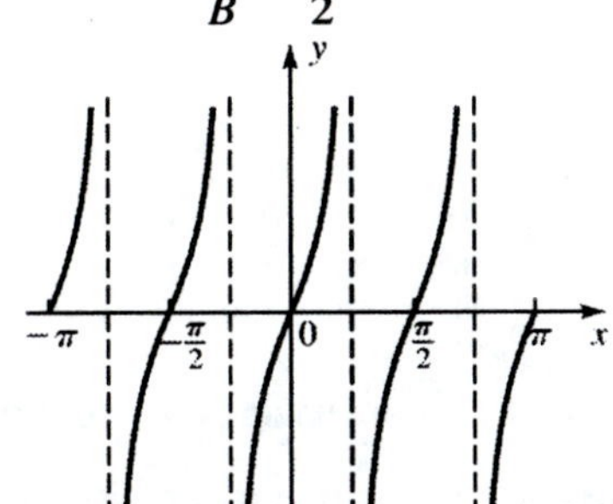

13. Period $= \dfrac{\pi}{1/2} = 2\pi$

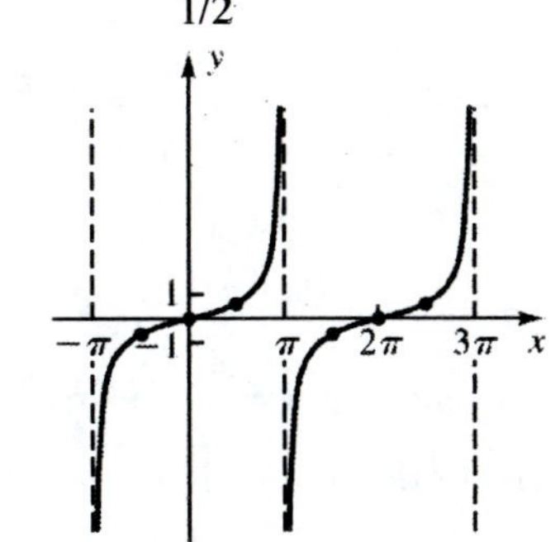

15. Period $= \dfrac{2\pi}{1/2} = 4\pi$

The dashed line shows $y = \dfrac{1}{2}\sin\left(\dfrac{x}{2}\right)$ in this interval. The solid line is $y = 2\csc\left(\dfrac{x}{2}\right)$.

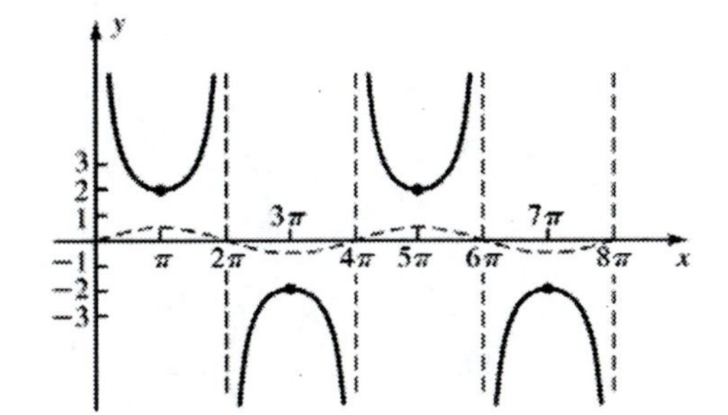

17. Period $= \dfrac{\pi}{\pi} = 1$

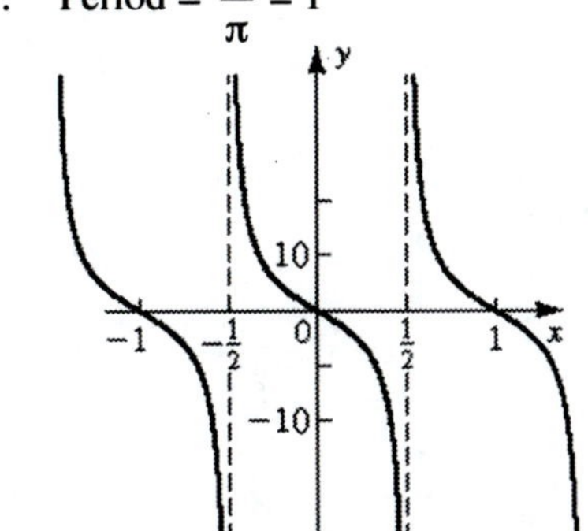

19. Period $= \dfrac{2\pi}{1/4} = 8\pi$

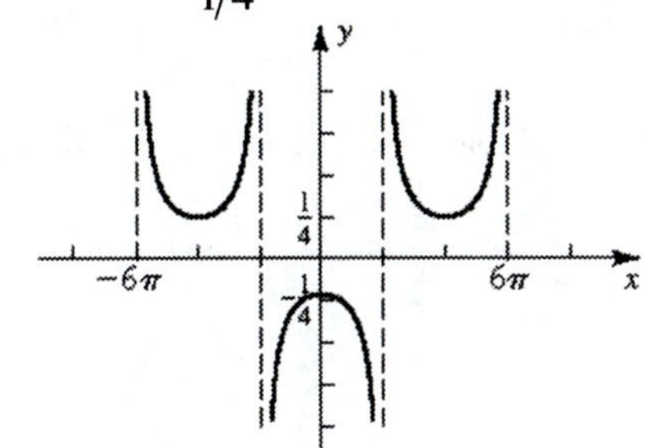

21. To find two asymptotes, set $2x - \pi$ equal to 0 and π and solve for x.

$$2x - \pi = 0 \qquad\qquad 2x - \pi = \pi$$
$$2x = \pi \qquad\qquad 2x = \pi + \pi$$
$$x = \frac{\pi}{2} = \text{Phase shift} \qquad\qquad x = \frac{\pi}{2} + \frac{\pi}{2} = \pi$$

The two asymptotes, $x = \dfrac{\pi}{2}$ and $x = \pi$, are $\dfrac{\pi}{2}$ units apart, so the period is $\dfrac{\pi}{2}$. The phase shift is $\dfrac{\pi}{2}$. Thus the asymptotes for the required region are $x = \dfrac{\pi}{2}$, $x = \dfrac{\pi}{2} - \dfrac{\pi}{2} = 0$, and $x = 0 - \dfrac{\pi}{2} = -\dfrac{\pi}{2}$.
Sketch in these asymptotes and fill in the curve.

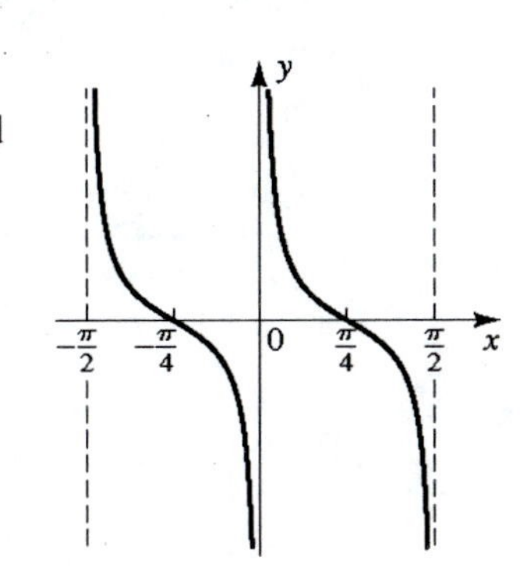

23. To find two asymptotes, set $\pi x - \frac{\pi}{2}$ equal to 0 and π and solve for x.

$$\pi x - \frac{\pi}{2} = 0 \qquad\qquad \pi x - \frac{\pi}{2} = \pi$$

$$\pi x = \frac{\pi}{2} \qquad\qquad \pi x = \pi + \frac{\pi}{2}$$

$$x = \frac{1}{2} = \text{Phase shift} \qquad\qquad x = 1 + \frac{1}{2} = \frac{3}{2}$$

There are asymptotes at $x = \frac{1}{2}$ and $x = \frac{3}{2}$, and a portion of the graph that opens up is between them. The low point of this portion is at height 1. The remaining asymptotes for the required region are at $x = -\frac{1}{2}$ and $x = \frac{5}{2}$. The period is 2 (twice the distance between successive asymptotes) and the phase shift is $\frac{1}{2}\left(= -\frac{C}{B}\right)$.

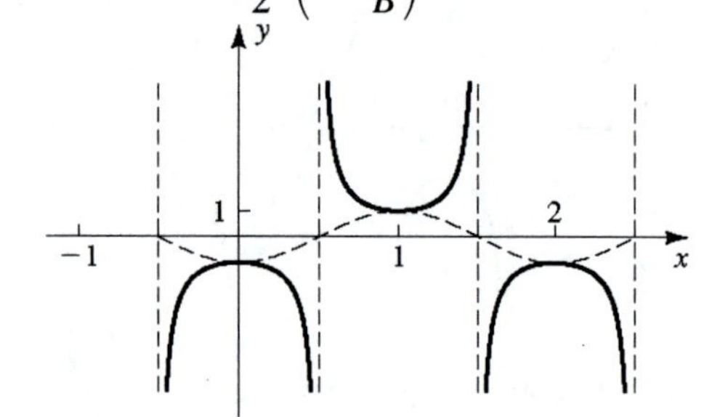

25. To find two asymptotes, set $x - \frac{\pi}{4}$ equal to $-\frac{\pi}{2}$ and $\frac{\pi}{2}$ and solve for x.

$$x - \frac{\pi}{4} = -\frac{\pi}{2} \qquad\qquad x - \frac{\pi}{4} = \frac{\pi}{2}$$

$$x = -\frac{\pi}{2} + \frac{\pi}{4} \qquad\qquad x = \frac{\pi}{2} + \frac{\pi}{4}$$

$$x = -\frac{\pi}{4} \qquad\qquad x = \frac{3\pi}{4}$$

The two asymptotes, $x = -\frac{\pi}{4}$ and $x = \frac{3\pi}{4}$, are π units apart, so the period is π. The phase shift is $\frac{\pi}{4}$.

Thus the asymptotes in the required region are $x = -\frac{\pi}{4}$ and $x = -\frac{\pi}{4} - \pi = -\frac{5\pi}{4}$. Sketch these asymptotes and fill in the curve, noting that the coefficient of -1 turns the tangent curve upside down.

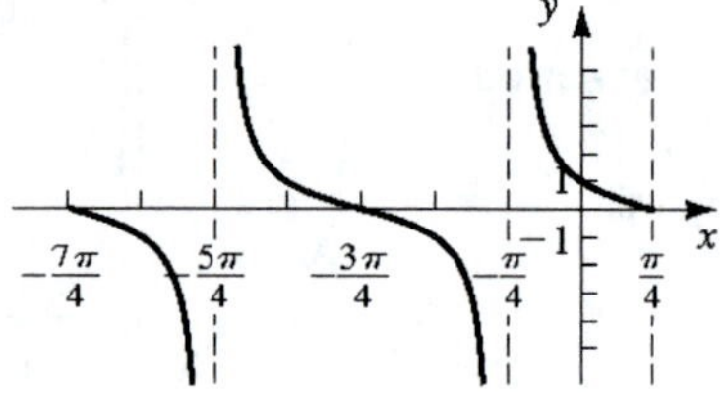

27. The graph of $y = \csc x - \cot x$ is shown in the figure. This graph appears to have vertical asymptotes at $x = -\pi$ and $x = \pi$, and period 2π. It appears, therefore, to be the same as the graph of $y = \tan Bx$, with $\frac{\pi}{B} = 2\pi$, that is $B = \frac{1}{2}$. The required equation is $y = \tan \frac{1}{2}x$.

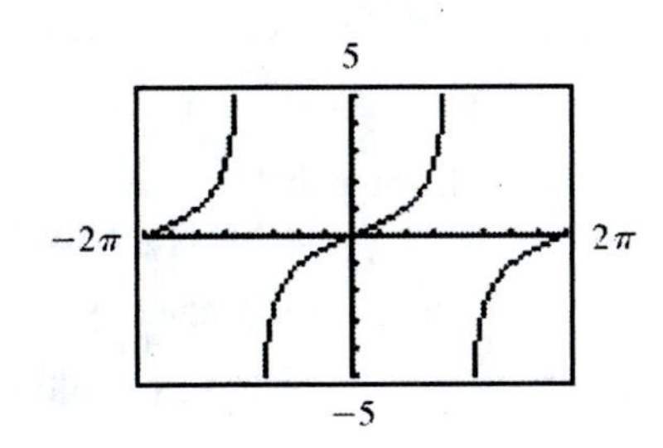

29. The graph of $y = \cot x + \tan x$ is shown in the figure. This graph appears to have vertical asymptotes at $x = -\pi$ and $x = -\frac{\pi}{2}, x = 0, x = \frac{\pi}{2}$, and $x = \pi$, and period π. Its high and low points appear to have y coordinates of -2 and 2, respectively. It appears, therefore, to be the same as the graph of $y = 2 \csc Bx$, with $\frac{2\pi}{B} = \pi$, that is $B = 2$. The required equation is $y = 2 \csc 2x$.

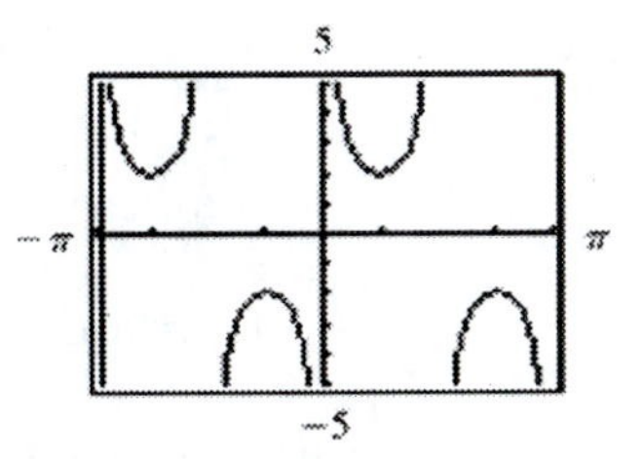

31. To find two asymptotes, set $\pi x - \pi$ equal to 0 and π and solve for x.

$$\pi x - \pi = 0 \qquad\qquad \pi x - \pi = \pi$$
$$\pi x = \pi \qquad\qquad \pi x = \pi + \pi$$
$$x = 1 = \text{Phase shift} \qquad\qquad x = 1 + 1 = 2$$

The two asymptotes, $x = 1$ and $x = 2$, are 1 unit apart, so the period is 1. The phase shift is 1. Thus the asymptotes for the required region are $x = 2, x = 1, x = 0, x = -1$, and $x = -2$. Sketch in these asymptotes and fill in the curve, noting that the coefficient of -3 turns the cotangent curve upside down and makes it steeper.

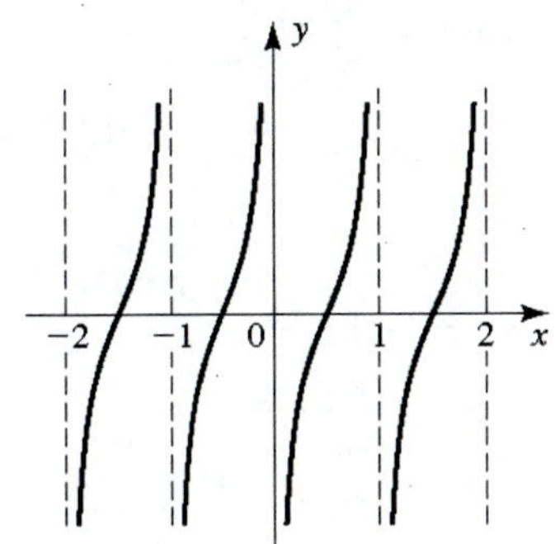

33. To find two asymptotes, set $\pi x - \frac{\pi}{2}$ equal to $-\frac{\pi}{2}$ and $\frac{\pi}{2}$ and solve for x.

$$\pi x - \frac{\pi}{2} = -\frac{\pi}{2} \qquad\qquad \pi x - \frac{\pi}{2} = \frac{\pi}{2}$$
$$\pi x = -\frac{\pi}{2} + \frac{\pi}{2} \qquad\qquad \pi x = \frac{\pi}{2} + \frac{\pi}{2}$$
$$\pi x = 0 \qquad\qquad \pi x = \pi$$
$$x = 0 \qquad\qquad x = 1$$

There are asymptotes at $x = 0$ and $x = 1$, and a portion of the graph that opens up is between them. The low point of this portion is at height 2. The remaining asymptotes in the required region are at $x = -1, x = 2$, and $x = 3$. The period is 2 (twice the distance between successive asymptotes) and the phase shift is $\frac{1}{2}\left(= -\frac{C}{B}\right)$

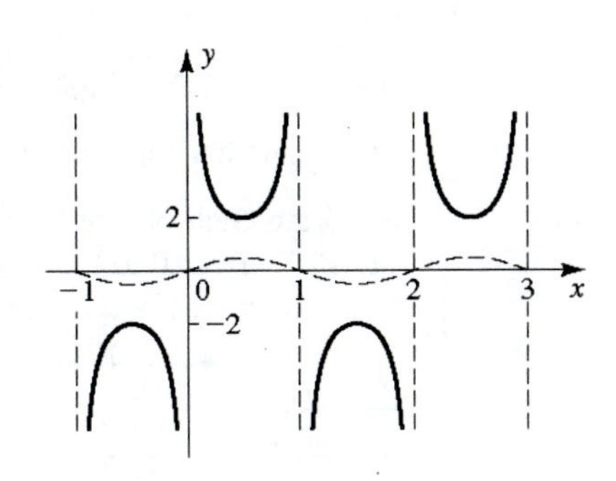

35. The graph of $y = \cos 2x + \sin 2x \tan 2x$ is shown in the figure. This graph appears to have vertical asymptotes at $x = -\frac{3\pi}{4}, x = -\frac{\pi}{4}, x = \frac{\pi}{4}$, and $x = \frac{3\pi}{4}$, and period π. Its high and low points appear to have y coordinates of -1 and 1, respectively. It appears, therefore, to be the same as the graph of $y = \sec Bx$, $\frac{2\pi}{B} = \pi$, that is $B = 2$. The required equation is $y = \sec 2x$.

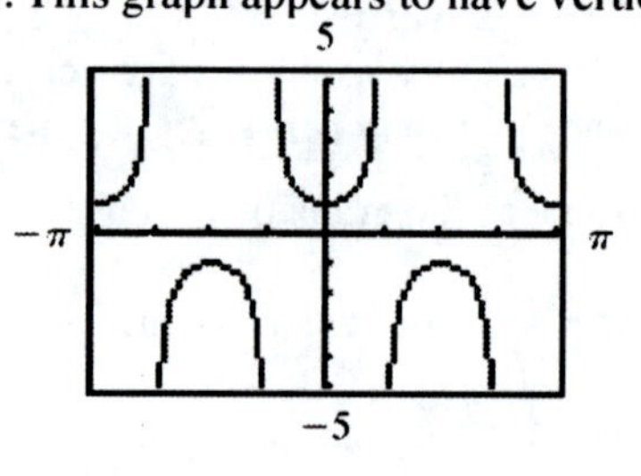

37. The graph of $y = \frac{\sin 6x}{1 - \cos 6x}$ is shown in the figure. This graph appears to have vertical asymptotes at $x = -\pi, -\frac{2\pi}{3}, -\frac{\pi}{3}, 0, \frac{\pi}{3}$, and π, and period $\frac{\pi}{3}$. It appears, therefore, to be the same as the graph of $y = \cot Bx$, $\frac{\pi}{B} = \frac{\pi}{3}$, that is $B = 3$. The required equation is $y = \cot 3x$.

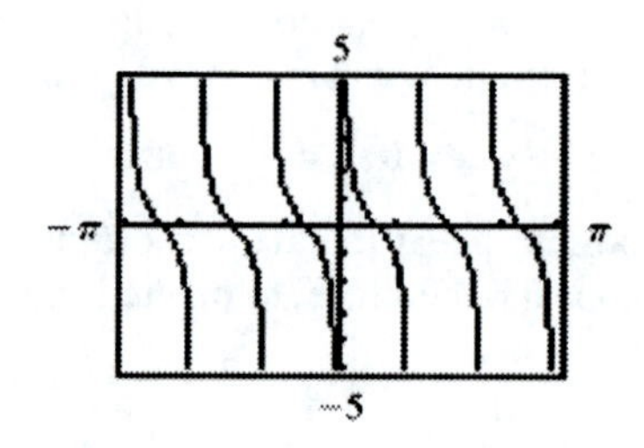

39. Amplitude $= |A| = |-5| = 5$.

41. Amplitude is not defined for the secant function.

43. Period $= \frac{\pi}{B} = \frac{\pi}{4\pi} = \frac{1}{4}$.

45. Period $= \frac{2\pi}{B} = \frac{2\pi}{4} = \frac{\pi}{2}$.

47. Phase shift $= -\frac{C}{B} = -\frac{-0.1}{\pi} = \frac{0.1}{\pi}$

49. There is no minimum value for the cotangent function.

51. There is no minimum value for the cosecant function itself. However, since $0 \le |\sin \pi x| \le 1$, we can write

$$\frac{1}{|\sin \pi x|} \ge 1$$
$$|\csc \pi x| \ge 1$$
$$|12 \csc \pi x| \ge 12$$

Therefore, the minimum value for this function is 12.

53. (A) In triangle ABC, we can write

$$\tan \theta = \frac{a}{b} = \frac{a}{15}$$

Thus, $a = 15 \tan \theta$, or $a = 15 \tan 2\pi t$.

(B) Period $= \frac{\pi}{2\pi} = \frac{1}{2}$

One period of the graph would therefore extend from 0 to $\frac{1}{2}$, with a vertical asymptote at $t = \frac{1}{4}$, or 0.25. We sketch half of one period, since the required interval is from 0 to 0.25 only. Ordinates can be determined from a calculator, thus:

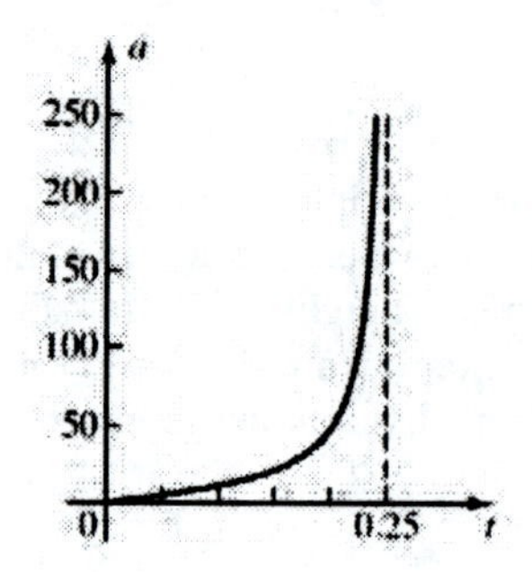

t	0	0.05	0.10	0.15	0.20	0.24
$15 \tan 2\pi t$	0	4.9	11	21	46	240

(C) a increases without bound as t approaches 0.25 (the graph has a vertical asymptote at $t = 0.25$.)

CHAPTER 3 REVIEW EXERCISE

1.

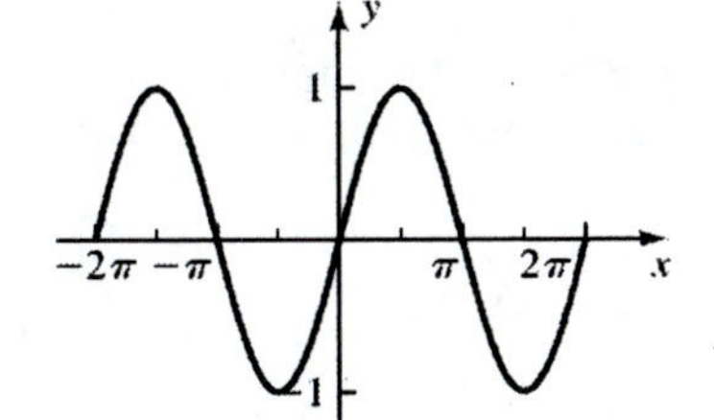

2.

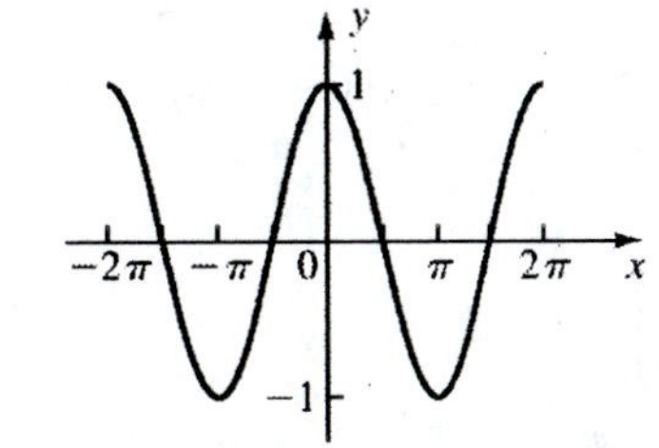

3.

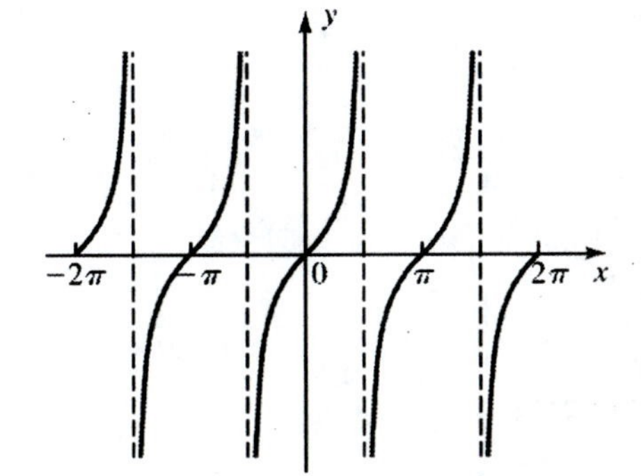

4.

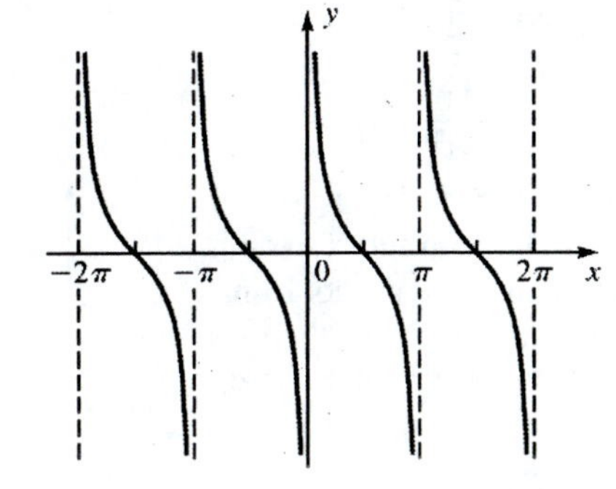

5.

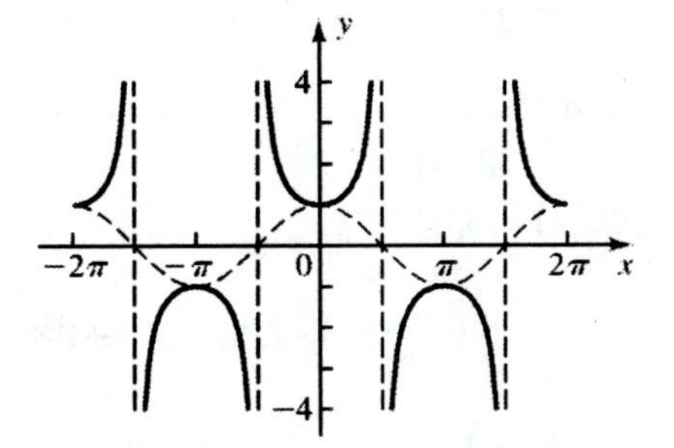

6.

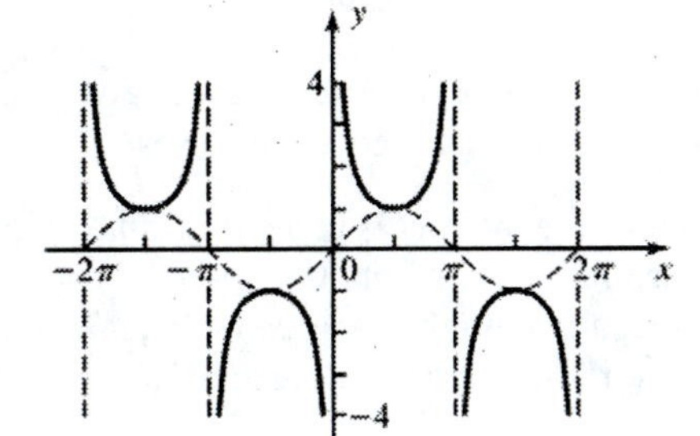

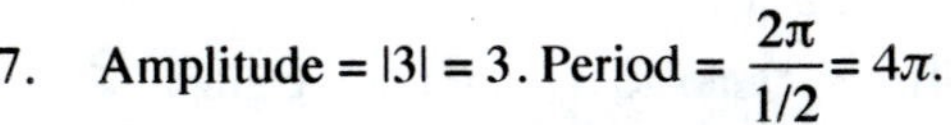

7. Amplitude = $|3| = 3$. Period = $\dfrac{2\pi}{1/2} = 4\pi$.

One full cycle of this graph is completed as x goes from 0 to 4π. Block out this interval, divide it into four equal parts, locate high and low points, and locate x intercepts. Then complete the graph.

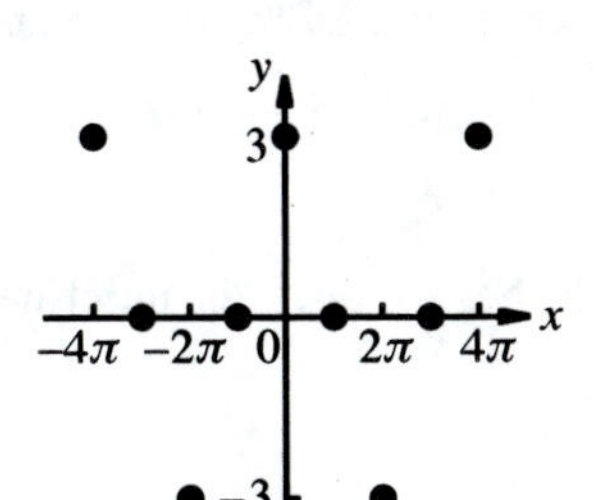

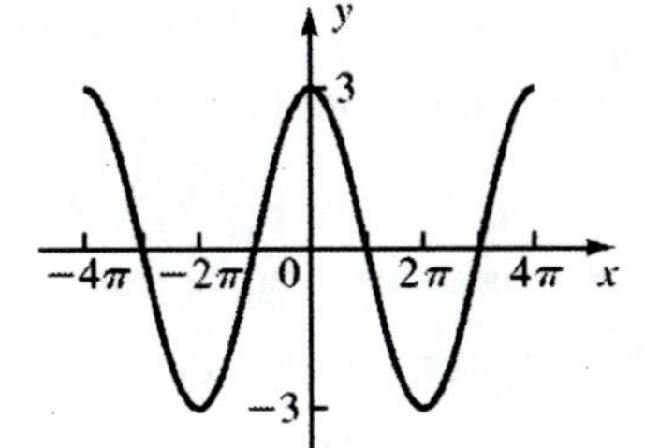

8. Amplitude $= \left|\frac{1}{2}\right| = \frac{1}{2}$. Period $= \frac{2\pi}{2} = \pi$.

One full cycle of this graph is completed as x goes from 0 to π. Block out this interval, divide it into four equal parts, locate high and low points, and locate x intercepts, then complete the graph.

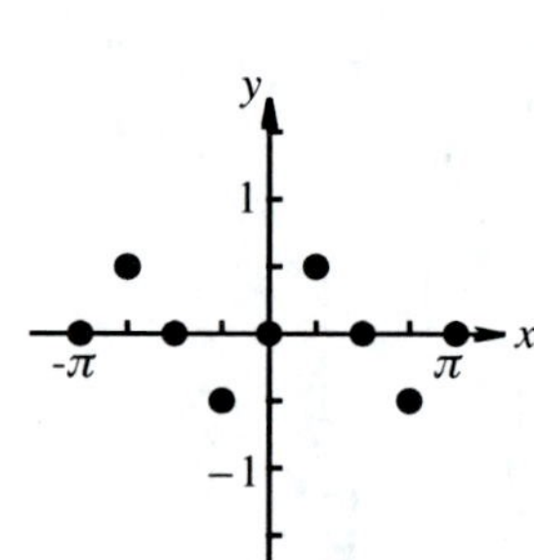

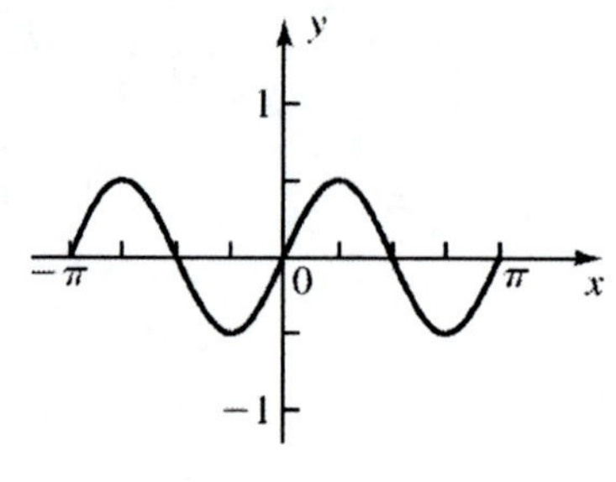

9. We form $y_1 = 4$ and $y_2 = \cos x$. We sketch the graph of each equation in the same coordinate system (dashed lines), then add the ordinates $y_1 + y_2$ (solid curve).

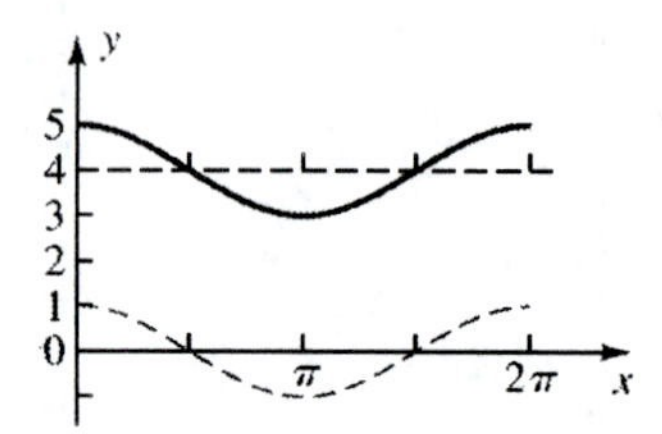

10. All six trigonometric functions are periodic. This is the key property shared by all.

11. Since the period of $y = A\cos Bx$ is given by $\frac{2\pi}{B}$ = period, if B is increased, the period decreases; if B is decreased, the period increases.

12. If a positive number is added to C, the graph is moved to the left, since the graph of $y = f(x + h)$ is shifted left h units from the graph of $y = f(x)$.
If a negative number is added to C, the graph is moved to the right, since the graph of $y = f(x - h)$ is shifted right h units from the graph of $y = f(x)$.

Thus, for example, the graph of $y = A\sin(3x + \pi)$ is shifted $\frac{\pi}{3}$ units to the left of the graph of $y = A\sin 3x$, while the graph of $y = A\sin(3x - \pi)$ is shifted $\frac{\pi}{3}$ units to the right of the graph of $y = A\sin 3x$.

13. (A) The basic functions with period 2π are sine, cosine, secant and cosecant. Of these, sine and cosine are always defined, and secant is undefined at $x = \frac{\pi}{2} + n\pi$, n an integer. The correct match is with cosecant.

(B) The basic functions with period π are tangent and cotangent. Of these, tangent is undefined at $x = \frac{\pi}{2} + n\pi$, n an integer. The correct match is with cotangent.

(C) The basic functions with amplitude 1 are sine and cosine. Since $\sin 0 = 0$ and $\cos 0 = 1$, only the graph of sine passes through $(0, 0)$.

14. Amplitude $= |A| = |3| = 3$. Period $= \frac{2\pi}{B} = \frac{2\pi}{3\pi} = \frac{2}{3}$. There is no phase shift.

15. Amplitude = $|A| = \left|\frac{1}{4}\right|$. To find the period and phase shift, solve $Bx + C = 0$ and $Bx + C = 2\pi$.

$$\frac{1}{2}x - 2\pi = 0 \qquad \frac{1}{2}x - 2\pi = 2\pi$$
$$x = 4\pi \qquad x = 4\pi + 4\pi$$
$$\text{Phase shift} = 4\pi \qquad \text{Period} = 4\pi$$

16. Amplitude is not defined for the tangent function. One cycle of $y = -5\tan\left(\frac{\pi}{3}x\right)$ is completed as $\frac{\pi}{3}x$ varies from 0 to π.

Solve for x:
$$\frac{\pi}{3}x = 0 \qquad \frac{\pi}{3}x = \pi$$
$$x = 0 \qquad x = 0 + 3$$
$$\text{Period} = 3 \qquad \text{Phase shift} = 0 \text{ (there is no phase shift)}$$

17. Amplitude is not defined for the cotangent function. One cycle of $y = 4\cot(x + 7)$ is completed as $x + 7$ varies from 0 to π. Solve for x:
$$x + 7 = 0 \qquad x + 7 = \pi$$
$$x = -7 \qquad x = -7 + \pi$$
$$\text{Period} = \pi \qquad \text{Phase shift} = -7$$

18. Amplitude is not defined for the secant function. One cycle of $y = -\frac{1}{2}\sec(2\pi x - 4\pi)$ is completed as $2\pi x - 4\pi$ varies from 0 to 2π. Solve for x:
$$2\pi x - 4\pi = 0 \qquad 2\pi x - 4\pi = 2\pi$$
$$2\pi x = 4\pi \qquad 2\pi x = 4\pi + 2\pi$$
$$x = 2 \qquad x = 2 + 1$$
$$\text{Period} = 1 \qquad \text{Phase shift} = 2$$

19. Amplitude is not defined for the cosecant function. One cycle of $y = 2\csc 5x$ is completed as $5x$ varies from 0 to 2π. Solve for x:
$$5x = 0 \qquad 5x = 2\pi$$
$$x = 0 \qquad x = \frac{2\pi}{5}$$
$$\text{Period} = \frac{2\pi}{5} \qquad \text{Phase shift} = 0 \text{ (there is no phase shift)}$$

20. Amplitude = $\left|-\frac{1}{3}\right| = \frac{1}{3}$. Period = $\frac{2\pi}{2\pi} = 1$.

Since $A = -\frac{1}{3}$ is negative, the basic curve for $y = \cos x$ is turned upside down. One full cycle of the graph is completed as x goes from 0 to 1. Block out this interval, divide it into four equal parts, locate high and low points, and locate x intercepts, then complete the graph.

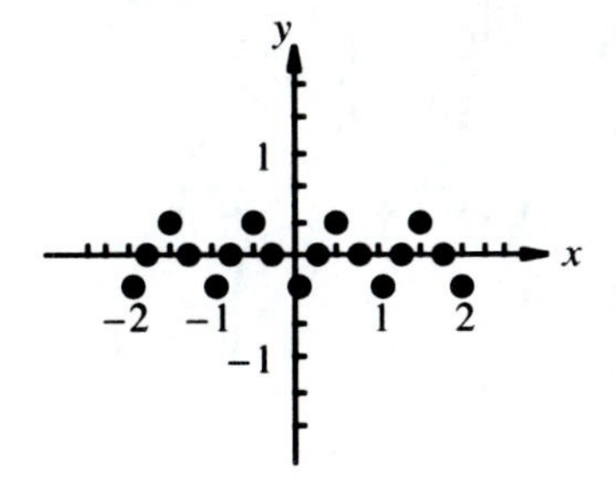

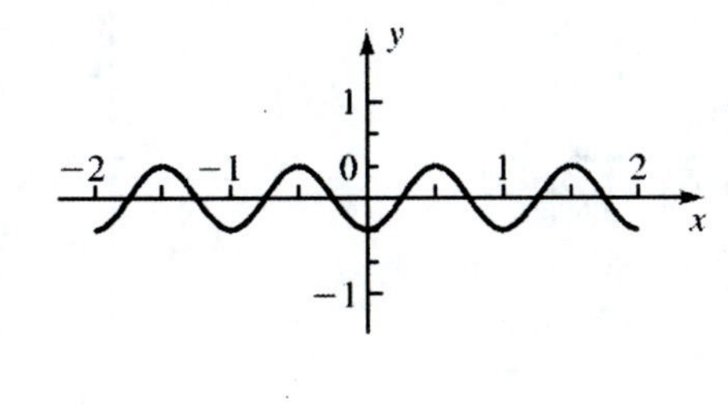

21. $y = -1 + \frac{1}{2}\sin 2x$. Amplitude $= \left|\frac{1}{2}\right| = \frac{1}{2}$. Period $= \frac{2\pi}{2} = \pi$,

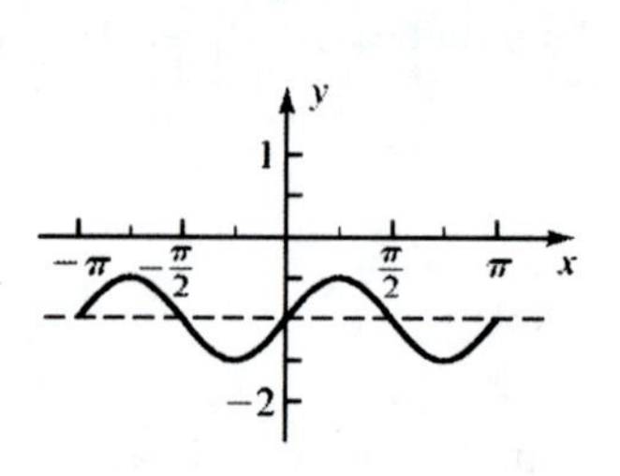

$y = \frac{1}{2}\sin 2x$ was graphed in Problem 8. This graph is the graph of $y = \frac{1}{2}\sin 2x$ moved down $|k| = |-1| = 1$ unit.
We start by drawing a horizontal broken line 1 unit below the x axis, then graph $y = \frac{1}{2}\sin 2x$ relative to the broken line and the original y axis.

22. This graph is the graph of $y = -2\sin(\pi x - \pi)$ moved up 4 units. To graph $y = -2\sin(\pi x - \pi)$, we work as follows:
Amplitude $= |A| = |-2| = 2$. Phase Shift and Period: Solve

$$\begin{aligned} Bx + C &= 0 \\ \pi x - \pi &= 0 \\ x &= 1 \\ \text{Phase Shift} &= 1 \end{aligned} \quad \text{and} \quad \begin{aligned} Bx + C &= 2\pi \\ \pi x - \pi &= 2\pi \\ x &= 1 + 2 \\ \text{Period} &= 2 \end{aligned}$$

Graph one cycle (the basic sine curve turned upside down) over the interval from 1 to $(1 + 2) = 3$. Then extend the graph from 0 to 1 and delete the portion of the graph from 2 to 3 since this was not required.

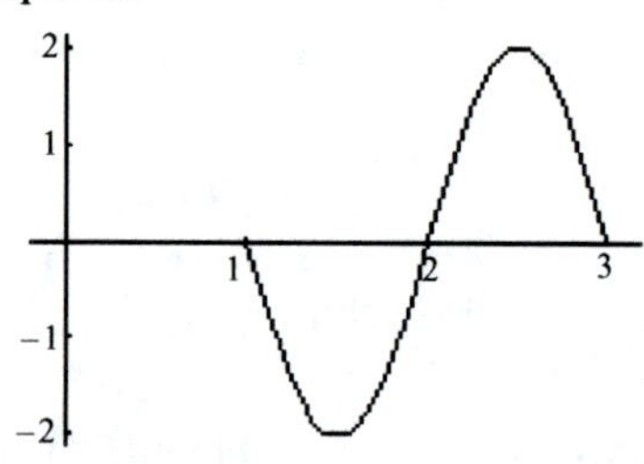

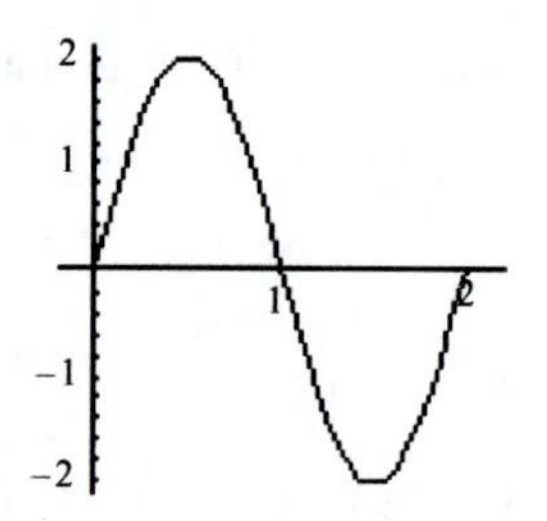

We then shift this graph up 4 units by drawing a horizontal broken line 4 units above the x axis, then graphing $y = -2\sin(\pi x - \pi)$ relative to the broken line and the original y axis, $y = 4 - 2\sin(\pi x - \pi)$

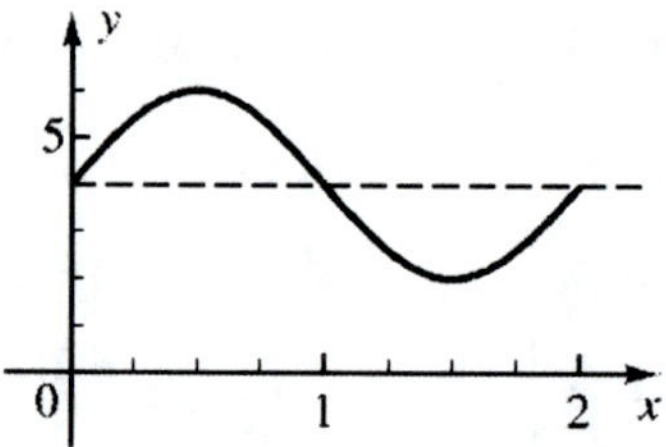

23. Period $= \frac{\pi}{2}$

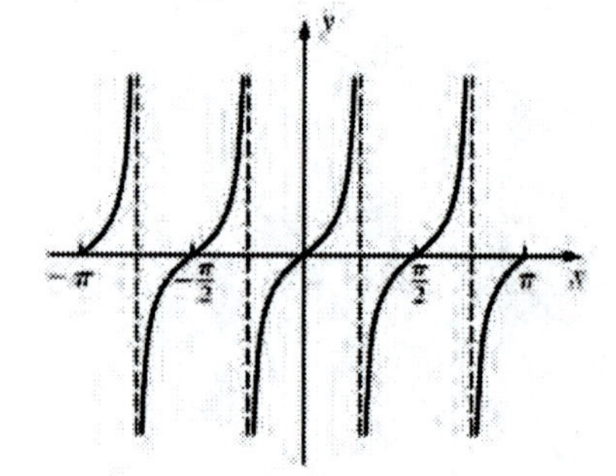

24. Period $= \frac{\pi}{\pi} = 1$

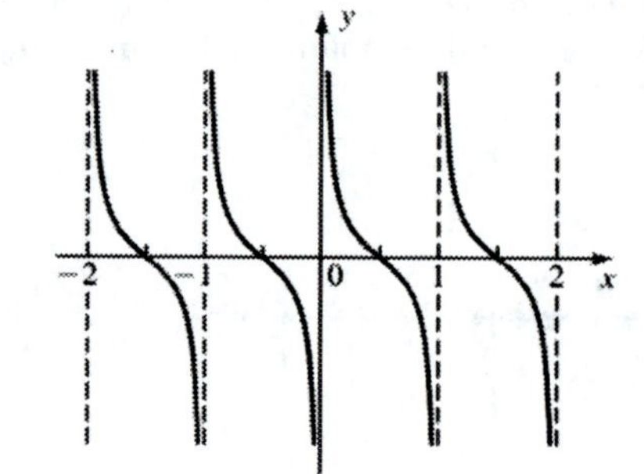

25. Period $= \dfrac{2\pi}{\pi} = 2$. We first sketch a graph of $y = \dfrac{1}{3}\sin \pi x$ from -1 to 2, which has amplitude $\dfrac{1}{3}$ and period 2. This curve (dashed curve) can serve as a guide for $y = 3\csc \pi x = \dfrac{1}{(1/3)\sin \pi x}$ by taking reciprocals of ordinates.

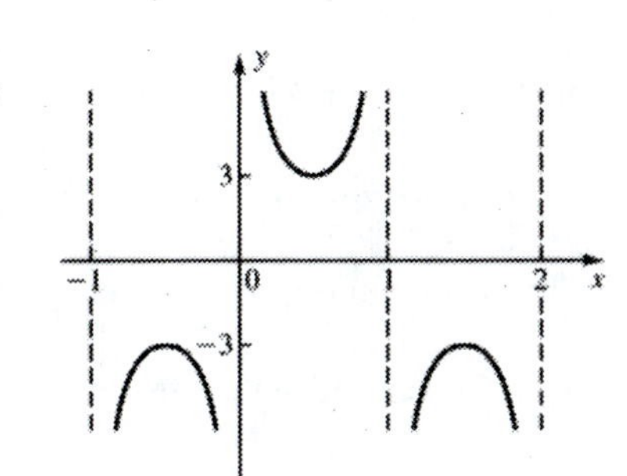

26. Period $= \dfrac{2\pi}{1/2} = 4\pi$. We first sketch a graph of $y = \dfrac{1}{2}\cos \dfrac{x}{2}$ from $-\pi$ to 3π, which has amplitude $\dfrac{1}{2}$ and period 4π. This curve (dashed curve) can serve as a guide for $y = 2\sec \dfrac{x}{2} = \dfrac{1}{(1/2)\cos(x/2)}$ by taking reciprocals of ordinates.

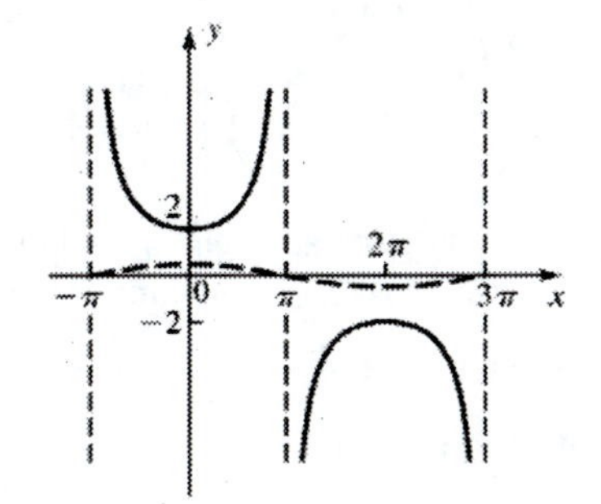

27. To find two asymptotes, set $x + \dfrac{\pi}{2}$ equal to $-\dfrac{\pi}{2}$ and $\dfrac{\pi}{2}$ and solve for x.

$$x + \frac{\pi}{2} = -\frac{\pi}{2} \qquad\qquad x + \frac{\pi}{2} = \frac{\pi}{2}$$
$$x = -\frac{\pi}{2} - \frac{\pi}{2} \qquad\qquad x = \frac{\pi}{2} - \frac{\pi}{2}$$
$$x = -\pi \qquad\qquad x = 0$$

The two asymptotes, $x = -\pi$ and $x = 0$, are π units apart, so the period is π. Thus, the asymptotes for the required region are $x = -\pi$, $x = 0$, and $x = \pi$. Sketch in these asymptotes and fill in the curve, a standard tangent curve.

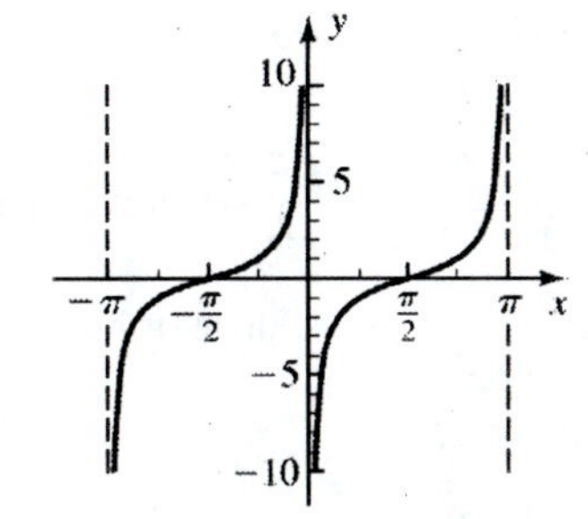

28. (A) y2 is y1 with half the period.

(B) y2 is y1 reflected across the x axis with twice the amplitude.

(C) The amplitudes of y1 and y2 are both 1. Hence $|A| = 1$. Since both y1 and y2 have the form of a basic sine curve, and not an upside down sine curve, $A = 1$. The period of y1 is 2, hence $\dfrac{2\pi}{B} = 2$ and $B = \dfrac{2\pi}{2} = \pi$. Hence y1 = $\sin \pi x$. The period of y2 is 1, hence $\dfrac{2\pi}{B} = 1$ and $B = 2\pi$. Hence y2 = $\sin 2\pi x$.

(D) The amplitude of y1 is 1. Hence $|A| = 1$. Since y1 has the form of a basic cosine curve and not an upside down cosine curve, $A = 1$. The period of y1 is 2, hence $\dfrac{2\pi}{B} = 2$ and $B = \dfrac{2\pi}{2} = \pi$. Hence y1 = $\cos \pi x$. Since y2 is y1 reflected across the x axis with twice the amplitude, but the same period, $A = -2$ and y2 = $-2\cos \pi x$.

29. Since the period is 4, we can write $\frac{2\pi}{B} = 4$, hence $B = \frac{2\pi}{4} = \frac{\pi}{2}$. Since the $y_{max} = 7$ and $y_{min} = -1$, we can write amplitude $= |A| = \frac{y_{max} - y_{min}}{2} = \frac{7-(-1)}{2} = 4$ and $k = \frac{y_{max} + y_{min}}{2} = \frac{7+(-1)}{2} = 3$. Thus $y = 3 + 4\sin\frac{\pi}{2}x$ or $y = 3 - 4\sin\frac{\pi}{2}x$. Since the portion of the graph for $0 \le x \le 4$ has the form of the basic sine curve, and not the upside down sine curve, we have $y = 3 + 4\sin\frac{\pi}{2}x$.

Check: When $x = 0$, $y = 3 + 4\sin\left(\frac{\pi}{2}\cdot 0\right) = 3 + 4 \cdot 0 = 3$

When $x = 1$, $y = 3 + 4\sin\left(\frac{\pi}{2}\cdot 1\right) = 3 + 4 \cdot 1 = 7$

30. Since the displacement is 0 when $t = 0$, the equation has the form $y = A \sin Bt$, and not $y = A\cos Bt$. Since the amplitude is 65, we can choose $A = 65$ or $A = -65$. Since the period is 0.01, we can write $\frac{2\pi}{B} = 0.01$, hence $B = \frac{2\pi}{0.01} = 200\pi$. We choose A positive for the sake of simplicity, thus $y = 65\sin 200\pi t$.

31. Since, for each value of x, y2 is the sum of the ordinate values for y1 and y3, y2 = y1 + y3.

32.

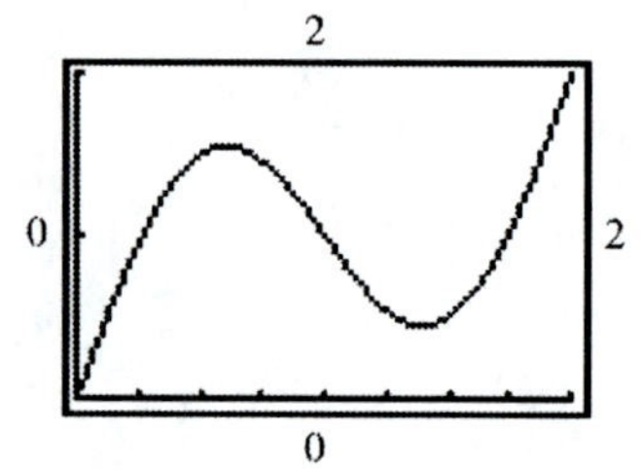

33.

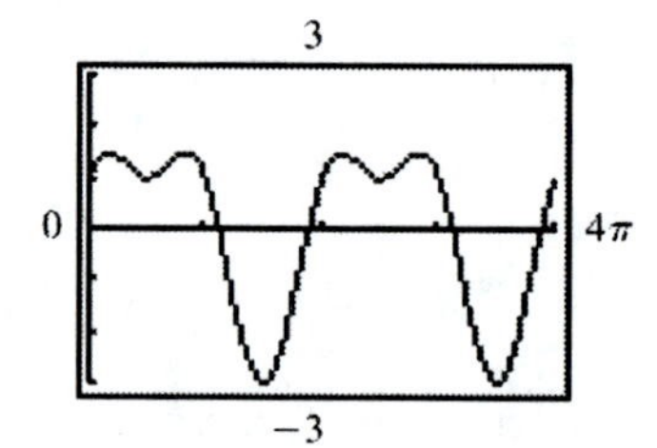

34. The graph of $y = \frac{1}{1 + \tan^2 x}$ is shown in the figure.

Amplitude $= \frac{1}{2}$ (y coordinate of highest point $-y$ coordinate of lowest point) $= \frac{1}{2}(1-0) = \frac{1}{2} = |A|$

Period $= \frac{2\pi}{2} = \pi$.

Thus, $B = \frac{2\pi}{\pi} = 2$. The form of the graph is that of the basic cosine curve shifted up $\frac{1}{2}$ unit.

Thus, $y = \frac{1}{2} + |A|\cos Bx = \frac{1}{2} + \frac{1}{2}\cos 2x$.

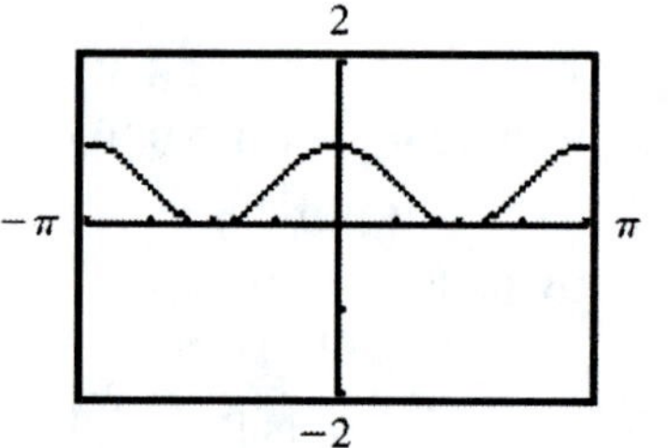

35.

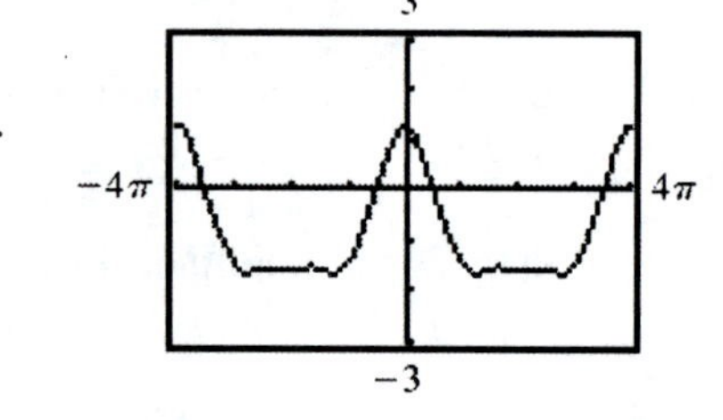

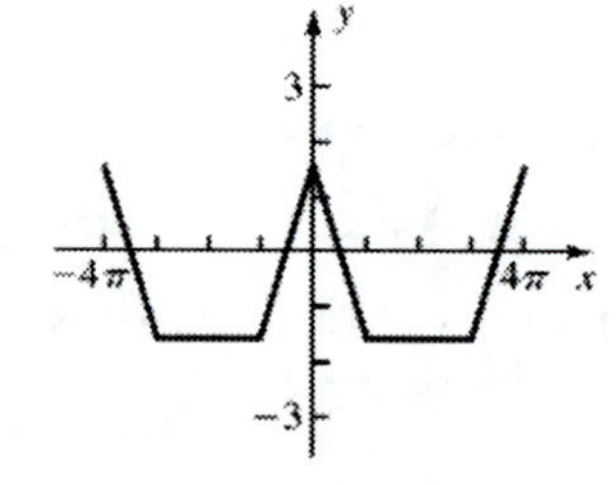

36. (A) The graph of $y = \dfrac{2\sin x}{\sin 2x}$ is shown in the figure. This graph appears to have vertical asymptotes at $x = -\dfrac{3\pi}{4}, x = -\dfrac{\pi}{4}, x = \dfrac{\pi}{4}$, and $x = \dfrac{3\pi}{4}$, and period 2π. Its high and low points appear to have y coordinates of -1 and 1, respectively. It appears, therefore, to be the same as the graph of $y = \sec Bx$, $\dfrac{2\pi}{B} = 2\pi$, that is, $B = 1$. The required equation is $y = \sec x$.

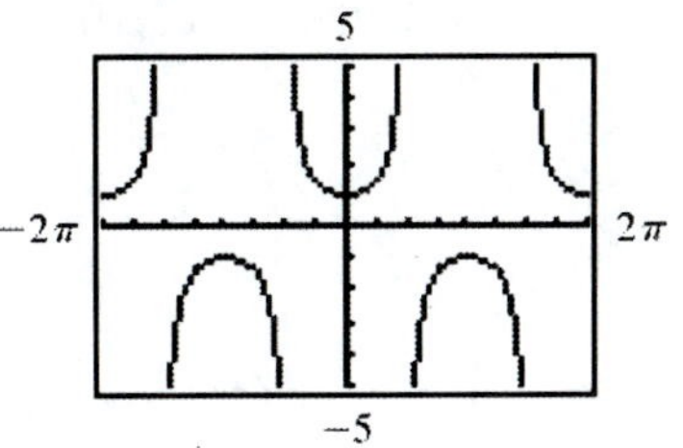

(B) The graph of $y = \dfrac{2\cos x}{\sin 2x}$ is shown in the figure. This graph appears to have vertical asymptotes at $x = -2\pi$, $x = -\pi, x = 0, x = \pi$, and $x = 2\pi$, and period 2π. Its high and low points appear to have y coordinates of 1 and -1, respectively. It appears, therefore, to be the same as the graph of $y = \csc Bx$, $\dfrac{2\pi}{B} = 2\pi$, that is, $B = 1$.
The required equation is $y = \csc x$.

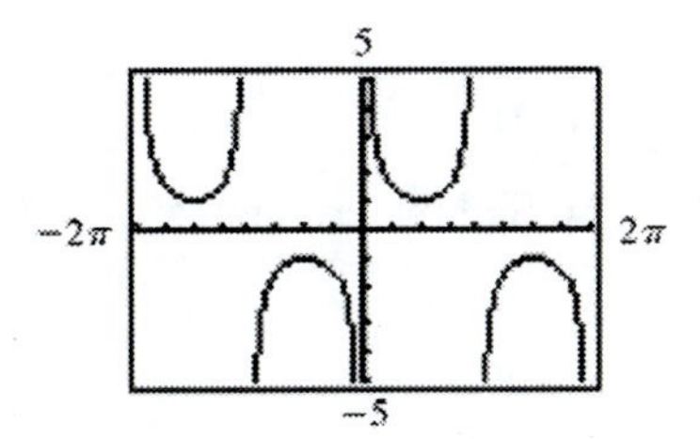

(C) The graph of $y = \dfrac{2\cos^2 x}{\sin 2x}$ is shown in the figure. This graph appears to have vertical asymptotes at $x = -2\pi$, $x = -\pi, x = 0, x = \pi$, and $x = 2\pi$, and period π. It appears, therefore, to be the same as the graph of $y = \cot Bx$, $\dfrac{\pi}{B} = \pi$, that is, $B = 1$. The required equation is $y = \cot x$.

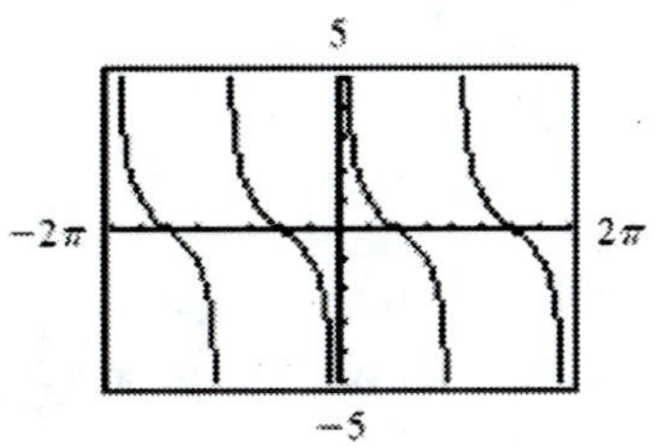

(D) The graph of $y = \dfrac{2\sin^2 x}{\sin 2x}$ is shown in the figure. This graph appears to have vertical asymptotes at $x = -\dfrac{3\pi}{2}$, $x = -\dfrac{\pi}{2}, x = \dfrac{\pi}{2}$, and $x = \dfrac{3\pi}{2}$, and period π. It appears, therefore, to be the same as the graph of $y = \tan Bx$, with $\dfrac{\pi}{B} = \pi$, that is, $B = 1$.
The required equation is $y = \tan x$.

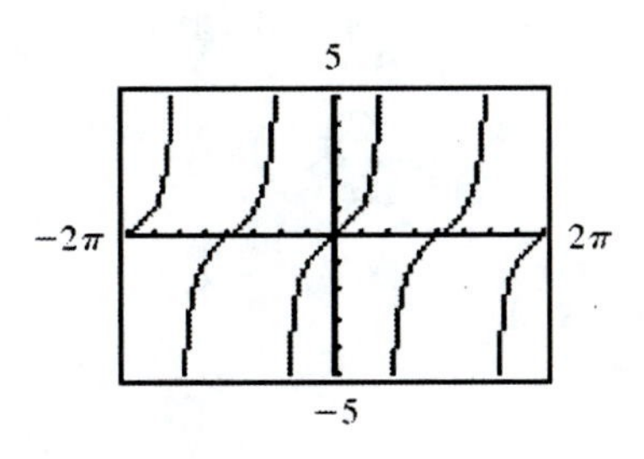

37. A horizontal shift of $\dfrac{\pi}{2}$ to the left (or right) combined with a reflection in the x axis transforms the graph of $y = \cot x$ into the graph of $y = \tan x$.

38. To find two asymptotes, set $\pi x + \dfrac{\pi}{2}$ equal to $-\dfrac{\pi}{2}$ and $\dfrac{\pi}{2}$ and solve for x.

$$\begin{aligned} \pi x + \frac{\pi}{2} &= -\frac{\pi}{2} \\ \pi x &= -\frac{\pi}{2} - \frac{\pi}{2} \\ x &= -\frac{1}{2} - \frac{1}{2} \\ x &= -1 \end{aligned} \qquad \begin{aligned} \pi x + \frac{\pi}{2} &= \frac{\pi}{2} \\ \pi x &= \frac{\pi}{2} - \frac{\pi}{2} \\ x &= \frac{1}{2} - \frac{1}{2} \\ x &= 0 \end{aligned}$$

The two asymptotes, $x = -1$ and $x = 0$, are 1 unit apart, so the period is 1. Thus, the asymptotes for the required region are $x = -1$, $x = 0$, and $x = 1$. Sketch in these asymptotes and fill in the curve, noting that the coefficient of 2 makes the graph steeper.

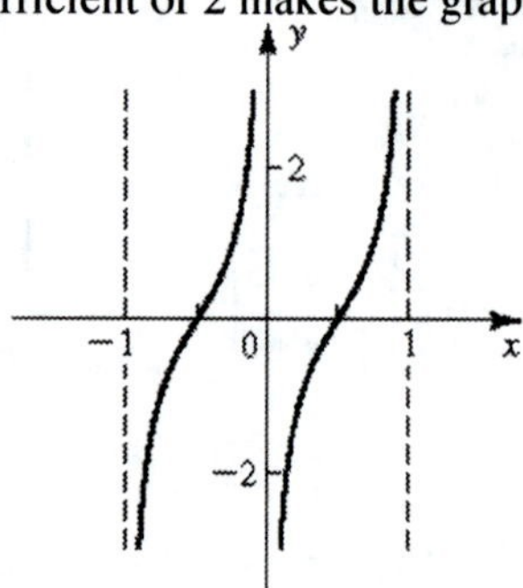

39. To find two asymptotes, set $2x - \pi$ equal to $-\frac{\pi}{2}$ and $\frac{\pi}{2}$ and solve for x.

$$2x - \pi = -\frac{\pi}{2} \qquad\qquad 2x - \pi = \frac{\pi}{2}$$
$$2x = -\frac{\pi}{2} + \pi \qquad\qquad 2x = \frac{\pi}{2} + \pi$$
$$x = -\frac{\pi}{4} + \frac{\pi}{2} \qquad\qquad x = \frac{\pi}{4} + \frac{\pi}{2}$$
$$x = \frac{\pi}{4} \qquad\qquad x = \frac{3\pi}{4}$$

There are asymptotes at $x = \frac{\pi}{4}$ and $x = \frac{3\pi}{4}$, and a portion of the graph that opens up is between them. The low point of this portion is at height 2. The remaining asymptote in the required region is at $x = \frac{5\pi}{4}$. The period is π (twice the distance between successive asymptotes) and the phase shift is

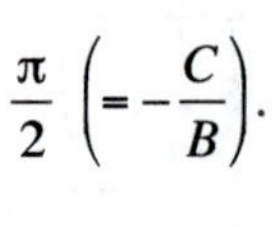

$\frac{\pi}{2}\left(= -\frac{C}{B}\right)$.

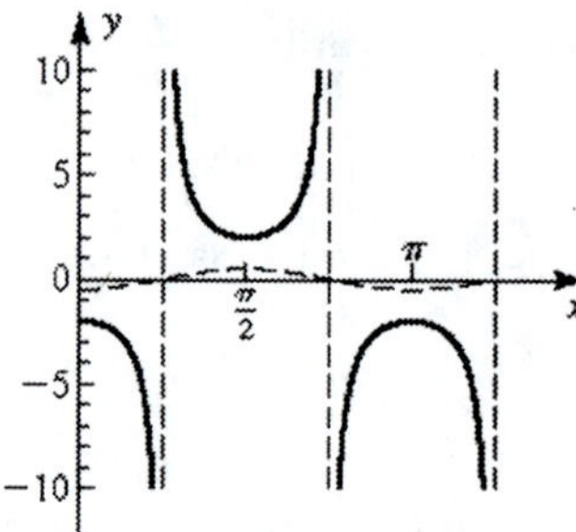

40. Since the function values follow the pattern 0, maximum, 0, minimum, 0, it would appear that a function of the form $y = A \sin Bx$ would be required, with A positive.
Since the maximum value of the function seems to be 2, and the minimum value seems to be −2, $A = \frac{2-(-2)}{2} = \frac{4}{2} = 2$.
Since the maximum value is achieved at $x = 0.25$ and $x = 1.25$, the period of the function is 1. Hence $\frac{2\pi}{B} = 1$ and $B = 2\pi$. Thus, the required function is $y = 2 \sin 2\pi x$.

41. Since the maximum deviation from the x axis is 1, we can write: Amplitude $= |A| = 1$.
Thus, $A = 1$ or -1.
Since the period is $\frac{5}{4} - \left(-\frac{3}{4}\right) = 2$, we can write: Period $= \frac{2\pi}{B} = 2$. Thus, $B = \frac{2\pi}{2} = \pi$.
Since we are instructed to choose the phase shift between 0 and 1, we can regard this graph as containing the upside down sine curve, with a phase shift of $\frac{1}{4}$. This requires us to choose A negative, since the graph shows that as x increases from $\frac{1}{4}$ to $\frac{3}{4}$, y *decreases* like the upside down sine curve. So, $A = -1$. Then, $-\frac{C}{B} = \frac{1}{4}$. Thus, $C = -\frac{1}{4}B = -\frac{\pi}{4}$.
$y = A\sin(Bx + C) = -\sin\left(\pi x - \frac{\pi}{4}\right)$.

Check: When $x = 0$, $y = -\sin\left(\pi \cdot 0 - \frac{\pi}{4}\right) = -\sin\left(-\frac{\pi}{4}\right) = \frac{\sqrt{2}}{2}$

When $x = \frac{1}{4}$, $y = -\sin\left(\pi \cdot \frac{1}{4} - \frac{\pi}{4}\right) = -\sin 0 = 0$.

42. The graph of $y = 1.2\sin 2x + 1.6\cos 2x$ is shown in the figure. This graph appears to be a sine wave with amplitude 2 and period π that has been shifted to the left. Thus, we conclude that $A = 2$ and $B = \frac{2\pi}{\pi} = 2$. To determine C, we use the zoom feature or the built-in approximation routine to locate the x intercept closest to the origin at $x = -0.464$. This is the phase-shift for the graph. Substitute $B = 2$ and $x = -0.464$ into the phase-shift equation $x = -\frac{C}{B}$; $-0.464 = -\frac{C}{2}$;
$C = 0.928$.
Thus, the equation required is $y = 2\sin(2x + 0.928)$.

2

$-\pi$ π

-2

43. (A)

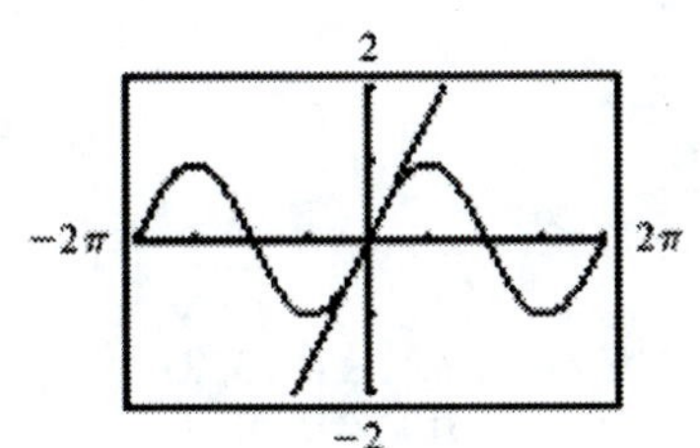

(B)

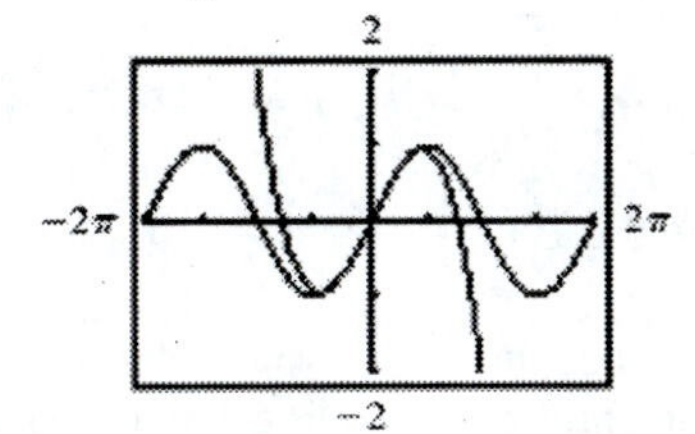

(C)

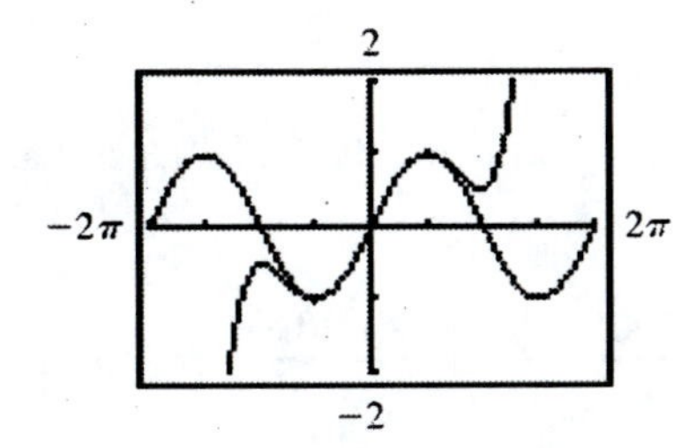

(D) As more terms of the series are used, the resulting approximation of $\sin x$ improves over a wider interval.

44. Here is a calculator graph of $f(x) = |\sin x|, -2\pi \le x \le 2\pi$

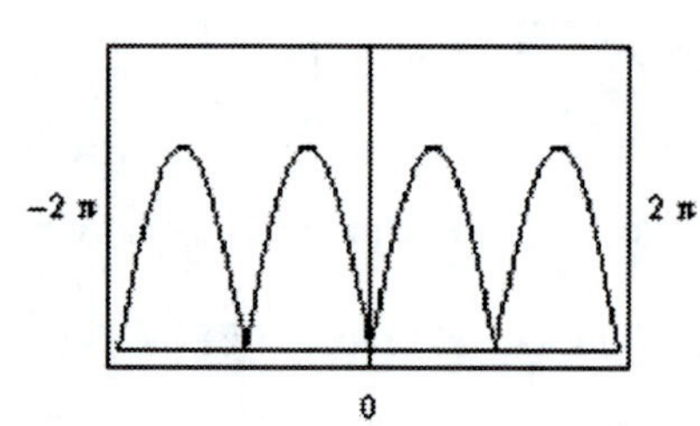

From the graph, it appears that $f(x)$ repeats its value after an interval of π, thus, $|\sin(x + \pi)| = |\sin x|$, and the period of $f(x)$ is π.

45. *Find A:* The amplitude $|A|$ is given to be 4. Since $y = -4$ (4 cm *below* position at rest) when $t = 0$, $A = -4$ (and not 4).
Find B: We are given that the frequency, f, is 8 Hz. Hence, the period is found using the reciprocal formula: $P = \frac{1}{f} = \frac{1}{8}$ sec. But $P = \frac{2\pi}{B}$. Thus, $B = \frac{2\pi}{P} = \frac{2\pi}{1/8} = 16\pi$.
Write the equation: $y = -4 \cos 16\pi t$.
An equation of the form $y = A \sin Bt$ cannot be used to model the motion, because when $t = 0$, y cannot equal -4 for any values of A and B.

46. *Determine K:* The dashed line indicates that high and low points of the curve are equal distances from the line $P = 1$. Hence, $K = 1$.

Determine A: The maximum deviation from the line $P = 1$ is seen at points such as $n = 0, P = 2$ or $n = 26, P = 0$. Thus, $|A| = 2 - 1$ or $|A| = 1 - 0$. From either statement, we see $|A| = 1$. So, $A = 1$ or $A = -1$. Since the portion of the curve as n increases from 0 to 26 has the form of the basic cosine curve (P decreasing)--and not the upside down cosine curve--A must be positive. $A = 1$.

Determine B: one full cycle of the curve is completed as n varies from 0 to 52 weeks. Hence, the Period = 52 weeks.

Since Period $= \frac{2\pi}{B}$, we have $52 = \frac{2\pi}{B}$, or $B = \frac{2\pi}{52} = \frac{\pi}{26}$.

$P = K + A \cos Bn = 1 + \cos \frac{\pi n}{26}, 0 \le n \le 104.$

An equation of the form $P = k + A \sin Bn$ will not work: The curve starts at (0, 2) and oscillates 1 unit above and below the line $P = 1$; a sine curve would have to start at (0, 1).

47. (A) We use the formula:

$$\text{Period} = \frac{2\pi}{\sqrt{1000gA/M}}$$ with $g = 9.75$ m/sec^2, $A = \pi\left(\frac{1.2}{2}\text{m}\right)^2$, and Period = 0.8 sec, and solve for M. Thus,

$$0.8 = \frac{2\pi}{\sqrt{(1000)(9.75)\pi(0.6)^2/M}} \qquad M = \frac{(1000)(9.75)\pi(0.6)^2(0.8)^2}{4\pi^2} \approx 179 \text{ kg}$$

(B) $D =$ amplitude $= 0.6$ $\qquad B = \dfrac{2\pi}{\text{Period}} = \dfrac{2\pi}{0.8} = 2.5\pi$ $\qquad y = 0.6 \sin(2.5\pi t)$

(C) One full cycle of this graph is completed as t goes from 0 to 0.8. block out this interval, divide it into four equal parts, locate high and low points, and locate t intercepts. Then complete the graph.

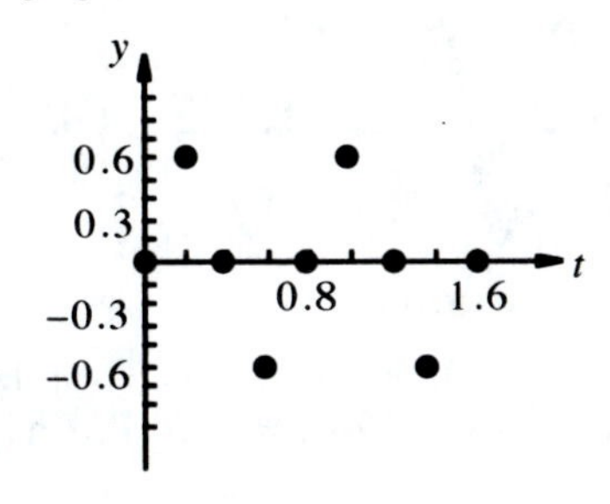

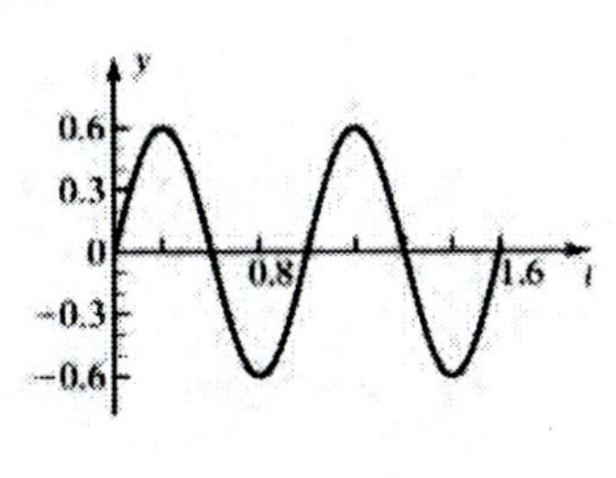

48. (A) Use $\text{Period} = \dfrac{1}{v}$ and $B = \dfrac{2\pi}{\text{Period}}$

$\text{Period} = \dfrac{1}{280}$ sec $\qquad B = \dfrac{2\pi}{1/280} = 560\pi$

(B) Use $v = \dfrac{1}{\text{Period}}$ and $B = \dfrac{2\pi}{\text{Period}}$

$v = \dfrac{1}{.0025\,\text{sec}} = 400$ Hz $\qquad B = \dfrac{2\pi}{.0025} = 800\pi$

(C) Use $\text{Period} = \dfrac{2\pi}{B}$ and $v = \dfrac{1}{\text{Period}}$

$\text{Period} = \dfrac{2\pi}{700\pi} = \dfrac{1}{350}$ sec $\qquad v = \dfrac{1}{1/350\,\text{sec}} = 350$ Hz

49. *Find A:* The amplitude $|A|$ is given to be 18. Since $E = 18$ when $t = 0$, $A = 18$ (and not -18).
Find B: We are given that the frequency, v, is 30 Hz. Hence, the period is found using the reciprocal formula:

$P = \dfrac{1}{v} = \dfrac{1}{30}$ sec. But, $P = \dfrac{2\pi}{B}$. Thus, $B = \dfrac{2\pi}{P} = \dfrac{2\pi}{1/30} = 60\pi$.

Write the equation: $y = 18 \cos 60\pi t$

50. $y = 6 \cos \dfrac{\pi}{10}(t - 5)$

We compute amplitude, period, and phase shift as follows:
Amplitude $= |A| = |6| = 6$
Phase Shift and Period: Solve

$Bx + C = 0$ and $Bx + C = 2\pi$

$\dfrac{\pi}{10}(t-5) = 0 \qquad \dfrac{\pi}{10}(t-5) = 2\pi$

$t = 5 \qquad t = 5 + 20$

↑ Phase Shift ↑ Period

The phase shift is 5 seconds and the period is 20 seconds.

Graph one cycle over the interval from 5 to 25.

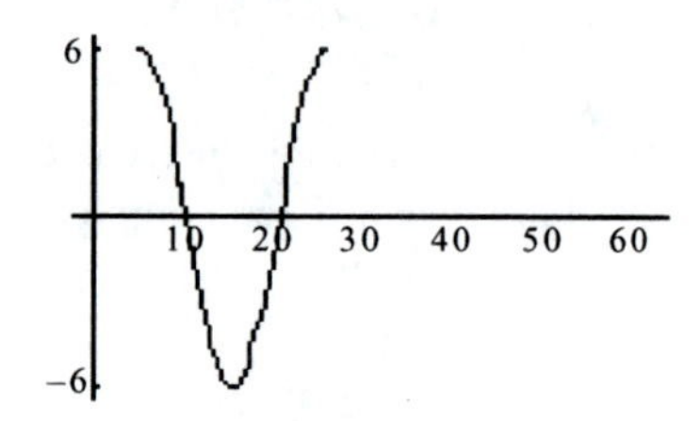

Extend the graph from 0 to 60.

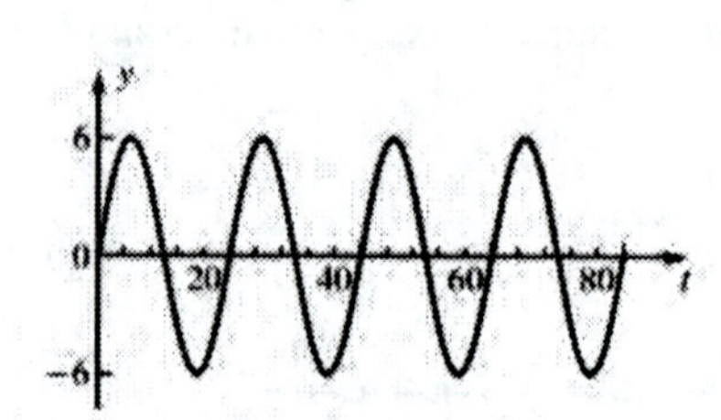

51. The height of the wave from trough to crest is the difference in height between the crest (height A) and the trough (height $-A$). In this case, $A = 12$ ft. $A - (-A) = 2A = 2(12 \text{ ft}) = 24$ ft. To find the wavelength λ, we note: $\lambda = 5.12T^2$,
$T = \dfrac{2\pi}{B}$, $B = \dfrac{\pi}{3}$. Thus, $T = \dfrac{2\pi}{\pi/3} = 6$ sec, $\lambda = 5.12(6)^2 \approx 184$ ft. To find the speed S, we use

$$S = \sqrt{\frac{g\lambda}{2\pi}} \qquad g = 32 \text{ ft/sec}^2 = \sqrt{\frac{32(184)}{2\pi}} \approx 31 \text{ ft/sec}$$

52. Period $= \dfrac{1}{\nu} = \dfrac{1}{10^{15}} = 10^{-15}$ sec. To find the wavelength λ, we use the formula $\lambda\nu = c$ with $c \approx 3 \times 10^8$ m/sec, $\lambda = \dfrac{c}{\nu} = \dfrac{3 \times 10^8 \text{ m/sec}}{10^{15} \text{ Hz}} = 3 \times 10^{-7}$ m

53. (A)

3

0 2

−3

Since the amplitude remains constant over time, this represents simple harmonic motion.

(B)

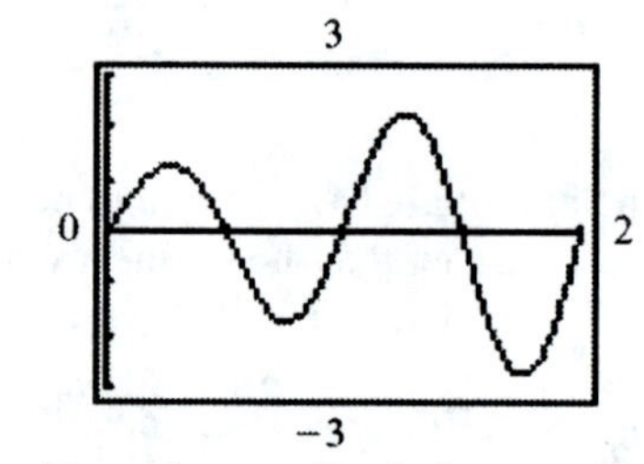

Since the amplitude increases as time increases, this represents resonance.

(C)

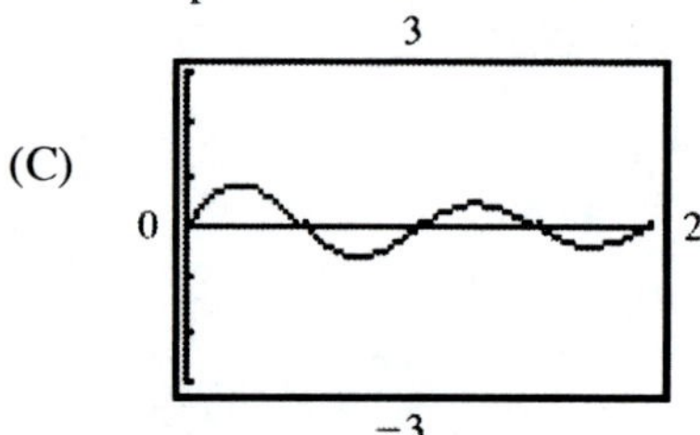

Since the amplitude decreases to 0 as time increases, this represents damped harmonic motion.

54. (A)

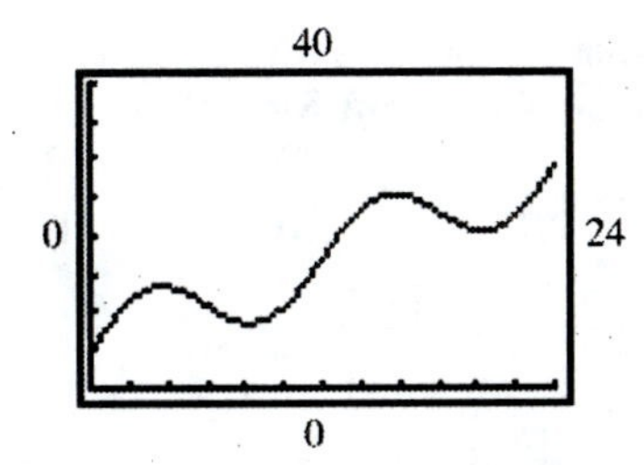

(B) It shows that sales have an overall upward trend with seasonal variations.

55. (A) The triangle shown in the text figure is a right triangle. Hence,

$$\tan\theta = \frac{a}{b} = \frac{h}{1000}.$$

Thus, $h = 1000\tan\theta$.

(B)

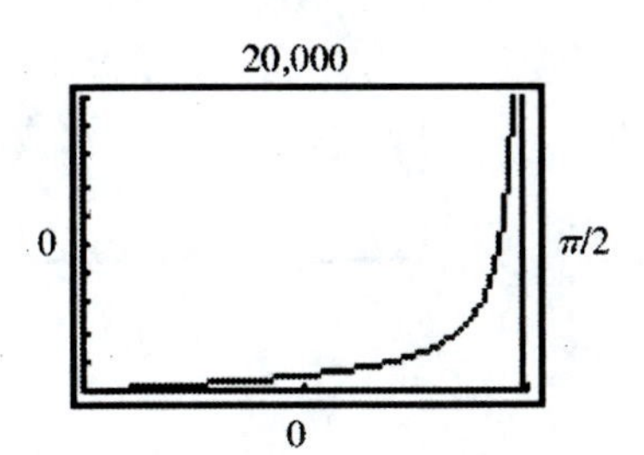

(C) As θ approaches $\frac{\pi}{2}$, h increases without bound.

56. (A) The data for θ repeats every second, so the period is $P = 1$ sec. The angle θ deviates from 0 by 36° in each direction, so the amplitude is $|A| = 36°$.

(B) $\theta = A\cos Bt$ is not suitable, because, for example, for $t = 0$, $A\cos Bt = A$, which will be 36° and not 0°. $\theta = A\sin Bt$ appears suitable, because, for example, if $t = 0$, then $\theta = 0°$, a good start. Choose $A = 36$ and $B = \frac{2\pi}{P} = 2\pi$, which yields $\theta = 36\sin 2\pi t$. The student should check that this equation produces (or comes close to producing) all the values in the table.

(C) Amplitude = 36. Period = 1.
One full cycle of this graph is completed as x goes from 0 to 1. Block out this interval, divide it into four equal parts, locate high and low points, and locate t intercepts. The points that result are precisely the points given in the table. Then complete the graph.

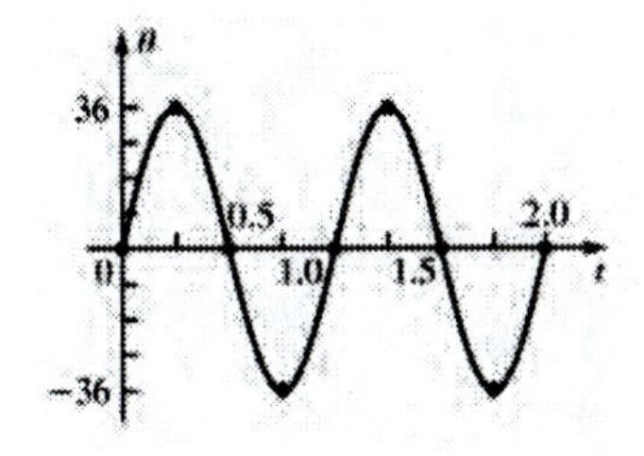

57. (A)

x(months)	1, 13	2, 14	3, 15	4, 16	5, 17	6, 18	7, 19	8, 20	9, 21
y (decimal hrs)	17.08	17.63	18.12	18.60	19.07	19.48	19.58	19.25	18.57

x	10, 22	11, 23	12, 24
y	17.78	17.12	16.85

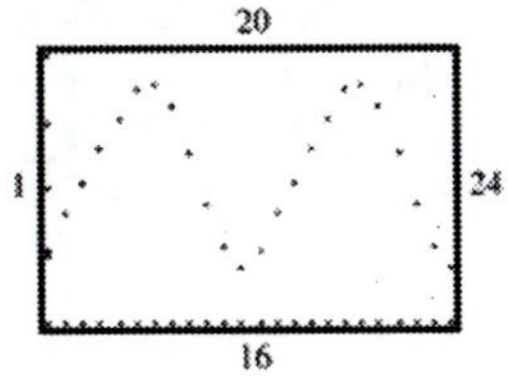

(B) From the table, Max $y = 19.58$ and Min $y = 16.85$. Then,

$$A = \frac{(\text{Max } y - \text{Min } y)}{2} = \frac{(19.58 - 16.85)}{2} = 1.37$$

$$B = \frac{2\pi}{\text{Period}} = \frac{2\pi}{12} = \frac{\pi}{6}$$

$$k = \text{Min } y + A = 16.85 + 1.37 = 18.22$$

From the plot in (A) or the table, we estimate the smallest positive value of x for which $y = k = 18.22$ to be approximately 3.2. Then this is the phase-shift for the graph. Substitute $B = \frac{\pi}{6}$ and $x = 3.2$ into the phase-shift equation $x = -\frac{C}{B}$; $3.2 = -\frac{C}{\pi/6}$; $C = -\frac{3.2\pi}{6} \approx -1.7$.

Thus, the equation required is $y = 18.22 + 1.37 \sin\left(\frac{\pi}{6}x - 1.7\right)$.

(C)

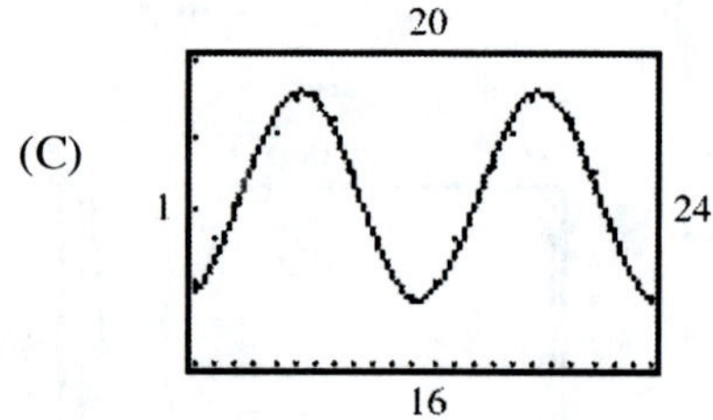

CUMULATIVE REVIEW EXERCISE CHAPTERS 1—3

1. Since $32' = \frac{32°}{60}$, then $43°32' = \left(43+\frac{32}{60}\right)^{\circ} \approx 43.53°$

2. $\theta_r = \frac{\pi \text{ rad}}{180°}(\theta_d) = \frac{\pi}{180}(88.73) = 1.55 \text{ rad}$

3. $\theta_r = \frac{\pi \text{ rad}}{180°}(\theta_d) = \frac{\pi}{180}\left(122+\frac{17}{60}\right) = 2.13 \text{ rad}$

4. $\theta_d = \frac{180°}{\pi \text{ rad}}(\theta_r) = \frac{180}{\pi}(2.75) = 157.56°$

5. $\theta_d = \frac{180°}{\pi \text{ rad}}(\theta_r) = \frac{180}{\pi}(-1.45) = -83.08° = -83°(0.08 \times 60)' = -83°5'$

6. The distance from P to the origin is

   $$r = \sqrt{7^2+24^2} = \sqrt{49+576} = \sqrt{625} = 25$$

   Therefore, the point $\left(\frac{7}{25}, \frac{24}{25}\right)$ lies on the unit circle.

   Apply Definition 1, Section 2.3, with $a = \frac{7}{25}$ and $b = \frac{24}{25}$.

   $\sin\theta = b = \frac{24}{25}$ $\qquad$ $\csc\theta = \frac{1}{b} = \frac{25}{24}$

   $\cos\theta = a = \frac{7}{25}$ $\qquad$ $\sec\theta = \frac{1}{a} = \frac{25}{7}$

   $\tan\theta = \frac{b}{a} = \frac{24}{25} \div \frac{7}{25} = \frac{24}{7}$ $\qquad$ $\cot\theta = \frac{a}{b} = \frac{7}{25} \div \frac{24}{25} = \frac{7}{24}$

7. We set $c = 13$ and $a = 11$

   Solve for b:
   $$\begin{aligned} c^2 &= a^2 + b^2 \\ b^2 &= c^2 - a^2 \\ b &= \sqrt{c^2 - a^2} \\ &= \sqrt{13^2 - 11^2} \\ &= \sqrt{48} \\ &= 6.9 \text{ in} \end{aligned}$$

   Solve for θ: We will use the cosine. Thus, $\cos\theta = \frac{a}{c} = \frac{11 \text{ in}}{13 \text{ in}} \approx 0.8462$

   $$\theta = \cos^{-1} 0.8462 = 32°$$

   Solve for complementary angle: $90° - \theta = 90° - 32° = 58°$

8. (A) Degree mode: $\sin 72.5° = 0.9537$

   (B) Degree mode: $\cos 104°52' = \cos\left(104+\frac{52}{60}\right)^{\circ} = -0.2566$

   (C) Radian mode: $\tan 2.41 = -0.8978$

9. (A) Use the reciprocal relationship $\sec\theta = \dfrac{1}{\cos\theta}$.

   Degree mode: $\sec 246.8° = \dfrac{1}{\cos 246.8°} = -2.538$

   (B) Use the reciprocal relationship $\cot\theta = \dfrac{1}{\tan\theta}$

   Degree mode: $\cot 23°15' = \dfrac{1}{\tan\left(23 + {}^{15}\!/_{60}\right)°} = 2.328$

   (C) Use the reciprocal relationship $\csc\theta = \dfrac{1}{\sin\theta}$

   Radian mode: $\csc 1.83 = \dfrac{1}{\sin 1.83} = 1.035$

10. The reference angle α is the angle (always taken positive) between the terminal side of θ and the horizontal axis.

    (A) $\alpha = 2\pi - \dfrac{11\pi}{6} = \dfrac{\pi}{6}$

    (B) $\alpha = |-225°| - 180° = 45°$

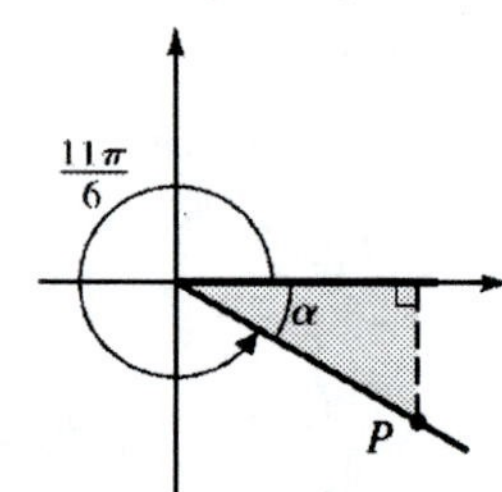

11. (A)

y 1 −1 −2π −π π 2π x

(B)

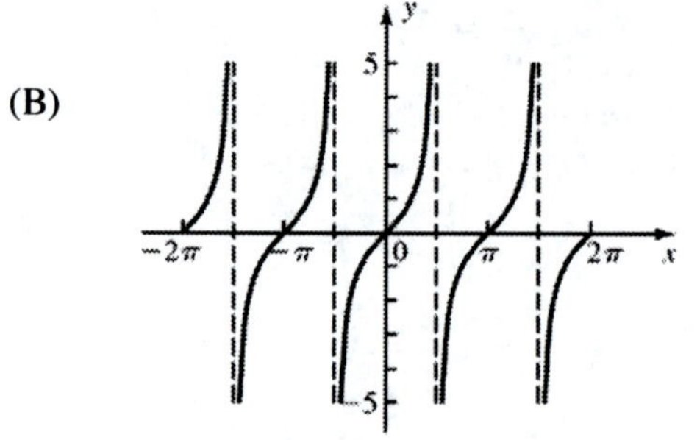

(C)

y 5 −5 −2π −π π 2π x

12. The central angle of a circle subtended by an arc that is one and one-half the length of the radius of the circle.

13. Yes. For example, for any x such that $\frac{\pi}{2} < x < \pi$, $\cos x$ is negative and $\csc x = \frac{1}{\sin x}$ is positive.

14. No. The sum of all three angles in any triangle is 180°. An obtuse angle is one that has a measure between 90° and 180°, and a triangle with more than one obtuse angle would have angles whose measure would add up to more than 180°, which would be a contradiction to the first statement.

15. If $\tan\theta = 0.9465$, then

$$\theta = \tan^{-1} 0.9465 = 43°30'$$

16. The tip of the second hand travels 1 revolution, or 2π radian, in 60 seconds. In 40 seconds it travels $\frac{40}{60}$ revolution, or $\frac{40}{60}\cdot 2\pi$ radian, that is, $\frac{4\pi}{3}$ radian. Since the distance traveled $s = R\theta$, we have

$$s = R\theta = (5.00\text{ cm})\left(\frac{4\pi}{3}\text{ rad}\right) = 20.94\text{ cm}$$

17. The tip of the second hand travels 1 circumference in 1 minute. Thus, its speed is given by

$$V = \frac{d}{t} = \frac{2\pi r}{t} = \frac{2\pi(5.00\text{ cm})}{1\text{ min}} = 31.4\text{ cm/min}$$

18. Label the sides of the triangle as shown at the right. Since triangles ABC and ADE are similar, we have

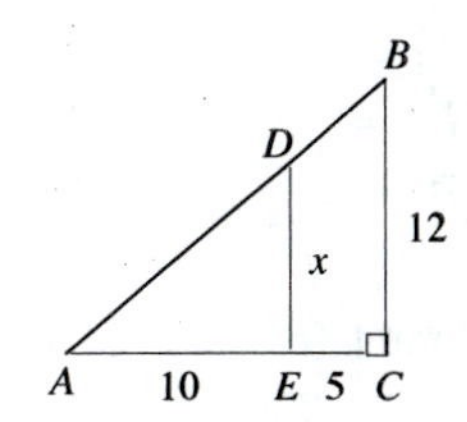

$$\frac{DE}{AE} = \frac{BC}{AC}$$

$$\frac{x}{10} = \frac{12}{10+5}$$

$$x = 10\cdot\frac{12}{15} = 8$$

19. $$\theta_r = \frac{\pi\text{ rad}}{180°}\theta_d$$

$$= \frac{\pi}{180}(48) = \frac{4\pi}{15}\text{ rad}$$

20. Since $\sin(3.78°) = 0.0659$ and $\tan(-76.25°) = -4.0867$, the first display (a) is the result of the calculator being set in degree mode. Since $\sin(3.78\text{ rad}) = -0.5959$ and $\tan(-76.25\text{ rad}) = -1.1424$, the second display (b) is the result of the calculator being set in radian mode.

21. Yes. Each is a different measure of the same angle, and if one is doubled the other must be doubled. this can be seen using the conversion formula $\theta_r = \frac{\pi}{180}\theta_d$: double θ_d, then θ_r is doubled.

22. Use the identity: $\cot x = \frac{1}{\tan x}$. Thus $\cot x = \frac{1}{0.5453} = 1.8339$

23. (A) Locate the 45° reference triangle, determine (a, b) and r, then evaluate.

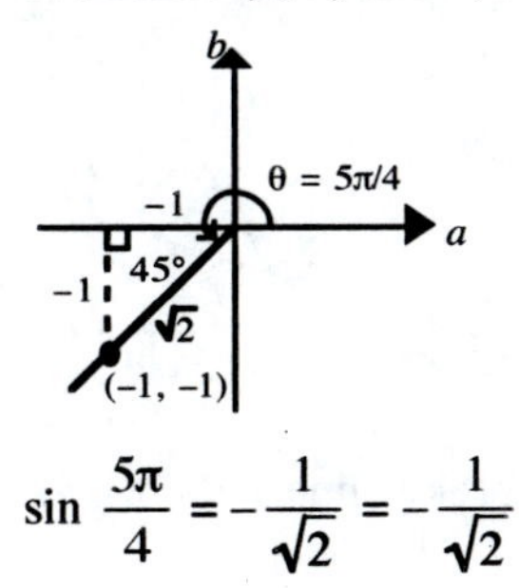

$$\sin\frac{5\pi}{4} = -\frac{1}{\sqrt{2}} = -\frac{1}{\sqrt{2}}$$

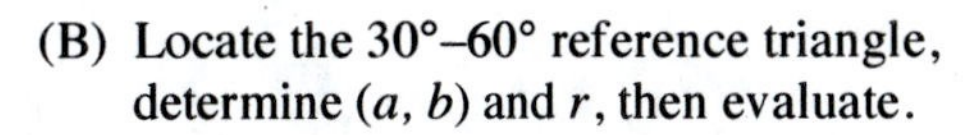

(B) Locate the 30°–60° reference triangle, determine (a, b) and r, then evaluate.

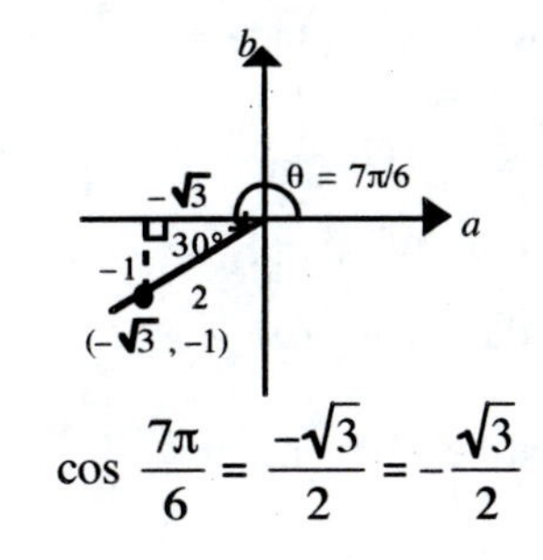

$$\cos\frac{7\pi}{6} = \frac{-\sqrt{3}}{2} = -\frac{\sqrt{3}}{2}$$

(C) Locate the 30°–60° reference triangle, determine (a, b) and r, then evaluate.

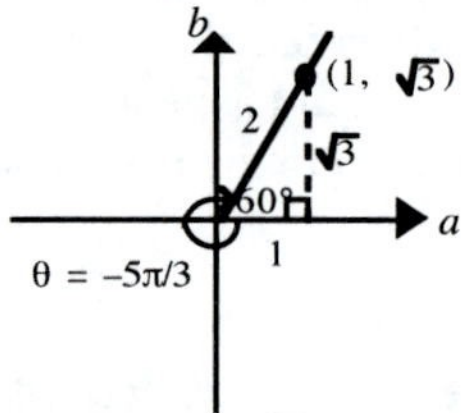

$$\tan \frac{5\pi}{3} = \frac{\sqrt{3}}{1} = \sqrt{3}$$

(D) $(a, b) = (-1, 0), r = 1$

$$\csc 3\pi = \frac{1}{0}$$

Not defined

24. Because $\cos\theta = \frac{a}{r} = -\frac{2}{3}$, we let $a = -2, r = 3$. Since $\cos\theta$ and $\tan\theta$ are negative, θ is a quadrant II angle. We find b so that (a, b) is the point in quadrant II that lies on the circle of radius 3 with center the origin.

$$a^2 + b^2 = r^2$$
$$(-2)^2 + b^2 = 3^2$$
$$4 + b^2 = 9$$
$$b^2 = 5$$
$$b = \sqrt{5}$$

We now apply the third remark following Definition 1, Section 2.3, with $a = -2, b = \sqrt{5}$ and $r = 3$.

$$\sin\theta = \frac{b}{r} = \frac{\sqrt{5}}{3} \qquad \csc\theta = \frac{r}{b} = \frac{3}{\sqrt{5}}$$

$$\left(\cos\theta = \frac{a}{r} = -\frac{2}{3}\right) \qquad \sec\theta = \frac{r}{a} = -\frac{3}{2}$$

$$\tan\theta = \frac{b}{a} = -\frac{\sqrt{5}}{2} \qquad \cot\theta = \frac{a}{b} = -\frac{2}{\sqrt{5}}$$

25. It is known that $\cos(0°) = 1 = 1.0000\ldots$ and $\tan(135°) = -1 = -1.0000\ldots$, so that the calculator displays exact values for these. Thus $\sin\left(-\frac{\pi}{3}\right)$ is not given exactly. To find $\sin\left(-\frac{\pi}{3}\right)$, locate the 30°–60° reference triangle, determine (a, b) and r, then evaluate.

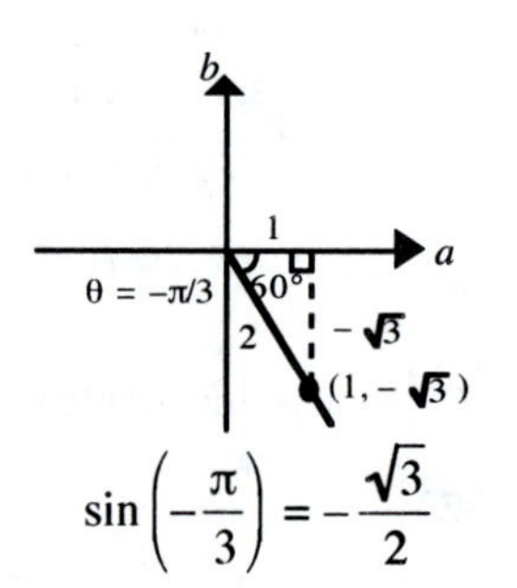

$$\sin\left(-\frac{\pi}{3}\right) = -\frac{\sqrt{3}}{2}$$

26. $y = 1 - \frac{1}{2}\cos 2x$. Amplitude $= \left|-\frac{1}{2}\right| = \frac{1}{2}$.

Period $= \frac{2\pi}{2} = \pi$. This graph is the graph of $y = -\frac{1}{2}\cos 2x$ moved up 1 unit. We start by drawing a horizontal broken line 1 unit above the x axis, then graph $y = -\frac{1}{2}\cos 2x$ (an upside down cosine curve with amplitude $\frac{1}{2}$ and period π) relative to the broken line and the original y axis.

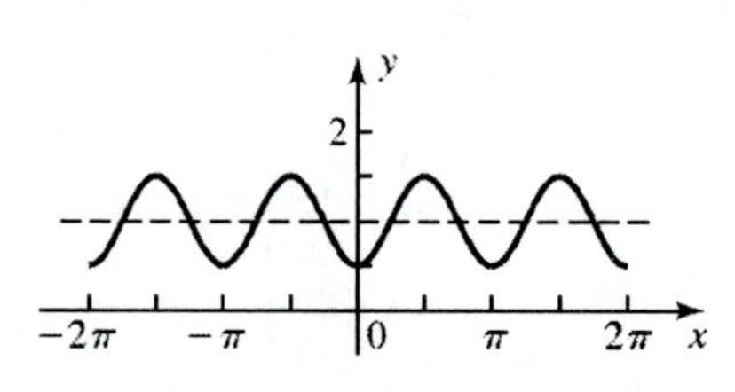

27. Amplitude $= |A| = |2| = 2$. Phase Shift and Period: Solve

$$Bx + C = 0 \qquad \text{and} \qquad Bx + C = 2\pi$$
$$x - \frac{\pi}{4} = 0 \qquad\qquad x - \frac{\pi}{4} = 2\pi$$
$$x = \frac{\pi}{4} \qquad\qquad x = \frac{\pi}{4} + 2\pi$$
$$\text{Phase Shift} = \frac{\pi}{4} \qquad\qquad \text{Period} = 2\pi$$

Graph one cycle over the interval from $\frac{\pi}{4}$ to $\left(\frac{\pi}{4} + 2\pi\right) = \frac{9\pi}{4}$. Then extend the graph from $-\pi$ to 3π.

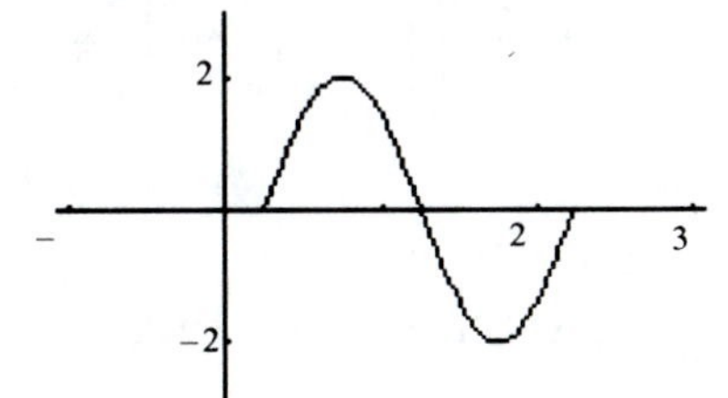

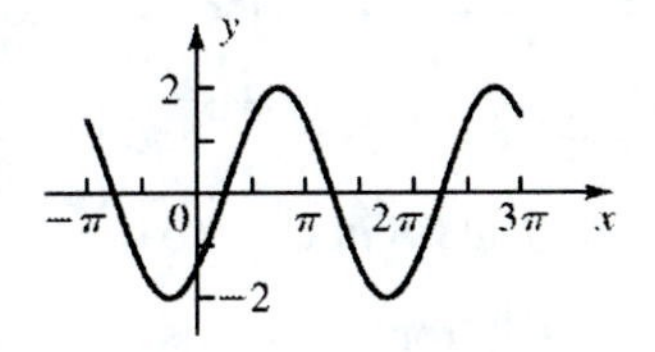

28. Period $= \frac{\pi}{4}$

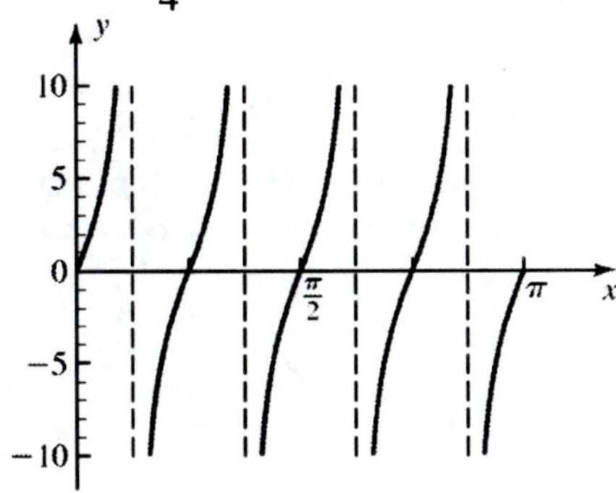

29. Period $= \frac{2\pi}{1/2} = 4\pi$

The dashed line shows $y = \sin\frac{x}{2}$ in this interval. The solid line is $y = \csc\frac{x}{2}$.

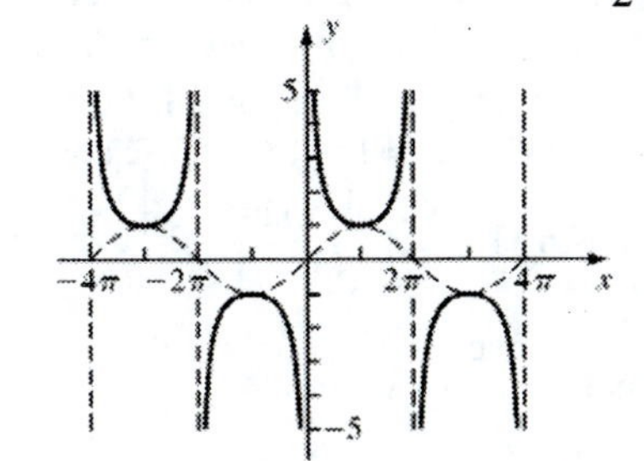

30. Period $= \frac{2\pi}{\pi} = 2$.
The standard secant curve is reflected through the x axis. High and low points are at heights 2 and –2 respectively.

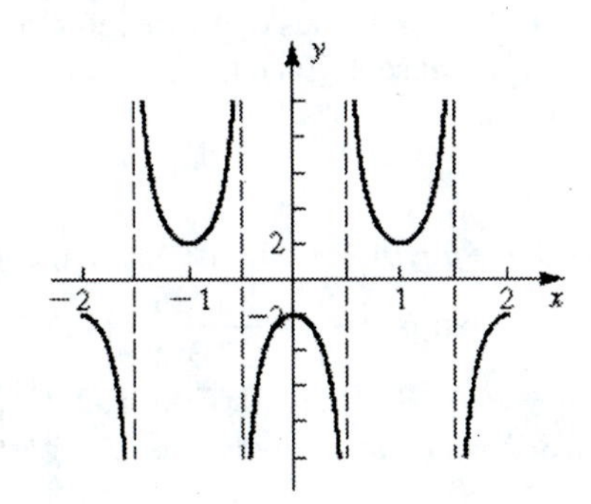

31. To find two asymptotes, set $\pi x + \frac{\pi}{2}$ equal to 0 and π and solve for x.

$$\pi x + \frac{\pi}{2} = 0 \qquad\qquad \pi x + \frac{\pi}{2} = \pi$$

$$\pi x = -\frac{\pi}{2} \qquad\qquad \pi x = \pi - \frac{\pi}{2}$$

$$x = -\frac{1}{2} = \text{Phase shift} \qquad\qquad x = 1 - \frac{1}{2} = \frac{1}{2}$$

The two asymptotes, $x = -\frac{1}{2}$ and $x = \frac{1}{2}$, are 1 unit apart, so the period is 1. The phase shift is $-\frac{1}{2}$. Thus the asymptotes for the required region are $x = -\frac{1}{2}$, $x = \frac{1}{2}$, $x = \frac{3}{2}$, and $x = \frac{5}{2}$. Sketch in these asymptotes and fill in the curve.

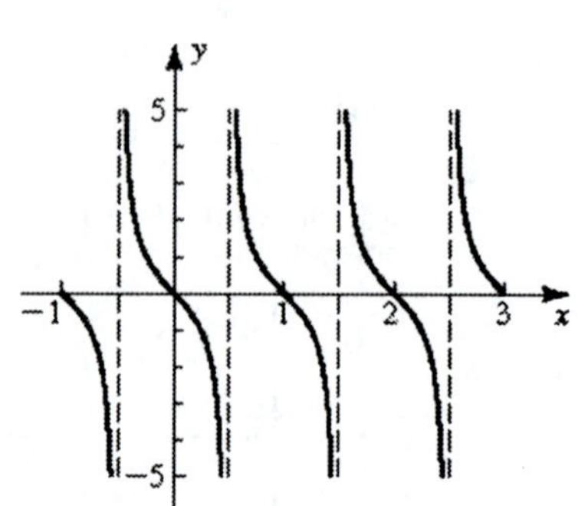

32. Since the maximum deviation from the x axis is 2, we can write: Amplitude = $|A| = 2$. Thus, $A = 2$ or -2. Since the period is 1, we can write: Period = $\frac{2\pi}{B} = 1$. Thus, $B = 2\pi$. As x increases from 0 to 0.25, y *decreases* like the upside down sine curve. So, A is negative. $A = -2$; $y = -2\sin(2\pi x)$.

33. Since the period is 2, we can write $\frac{2\pi}{B} = 2$, hence $B = \frac{2\pi}{B} = \pi$. Since the $y_{max} = 3$ and $y_{min} = -1$, we can write amplitude $|A| = \frac{y_{max} - y_{min}}{2} = \frac{3-(-1)}{2} = 2$, and $k = \frac{y_{max} + y_{min}}{2} = \frac{3+(-1)}{2} = 1$.
Thus, $y = 1 + 2\cos \pi x$ or $y = 1 - 2\cos \pi x$. Since the portion of the graph for $0 \le x \le 2$ has the form of the basic cosine curve, and not the upside-down cosine curve, we have $y = 1 + 2\cos \pi x$.
Check: When $x = 0$ $\quad y = 1 + 2\cos \pi \cdot 0 = 1 + 2 \cdot 1 = 3$
When $x = 1$ $\quad y = 1 + 2\cos \pi(1) = 1 + 2(-1) = -1$

34. $(\tan x)(\sin x) + \cos x = \frac{\sin x}{\cos x}\sin x + \cos x = \frac{\sin^2 x}{\cos x} + \frac{\cos x}{1} = \frac{\sin^2 x}{\cos x} + \frac{\cos^2 x}{\cos x} = \frac{\sin^2 x + \cos^2 x}{\cos x}$
$= \frac{1}{\cos x} = \sec x$

35. We can draw reference triangles in both quadrants III and IV with sides opposite reference angle −1 and hypotenuse 2. Each triangle is a special 30°–60° triangle.

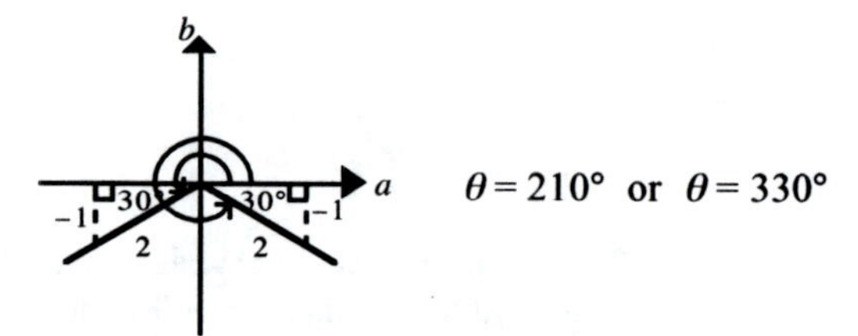

36. *Solve for θ:* We will use the tangent. Thus,

$$\tan\theta = \frac{b}{a} = \frac{23.5\text{ in.}}{37.3\text{ in.}} = 0.6300$$

$$\theta = \tan^{-1} 0.6300 = 32.2°$$

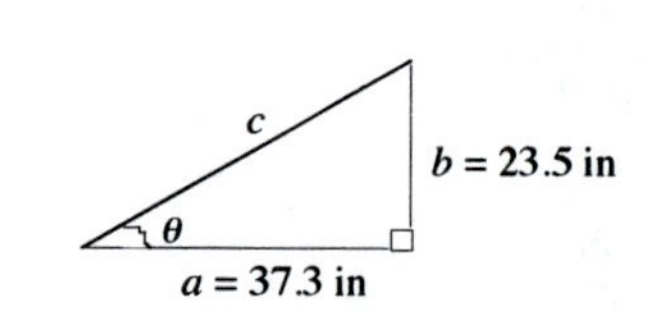

Solve for the complementary angle:
$90° - \theta = 90° - 32.2° = 57.8°$
Solve for c: We use the Pythagorean theorem.

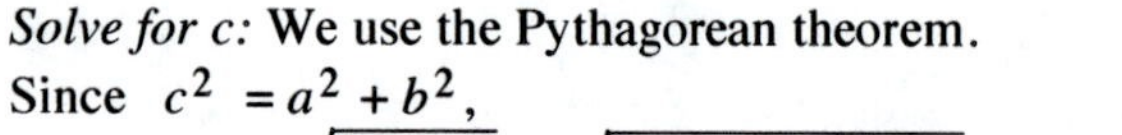
Since $c^2 = a^2 + b^2$,

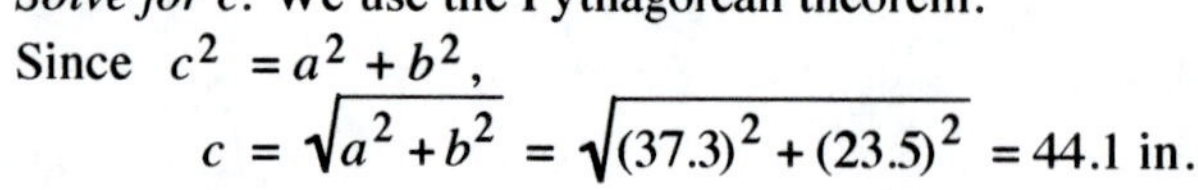

$$c = \sqrt{a^2 + b^2} = \sqrt{(37.3)^2 + (23.5)^2} = 44.1 \text{ in.}$$

37. 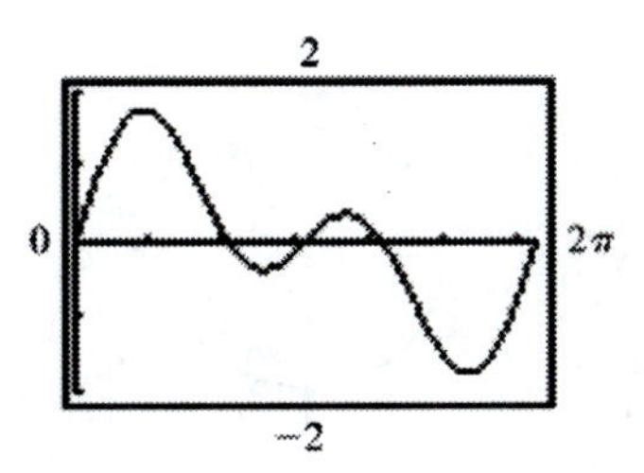

38. The graph of $y = \dfrac{\tan^2 x}{1 + \tan^2 x}$ is shown in the figure.

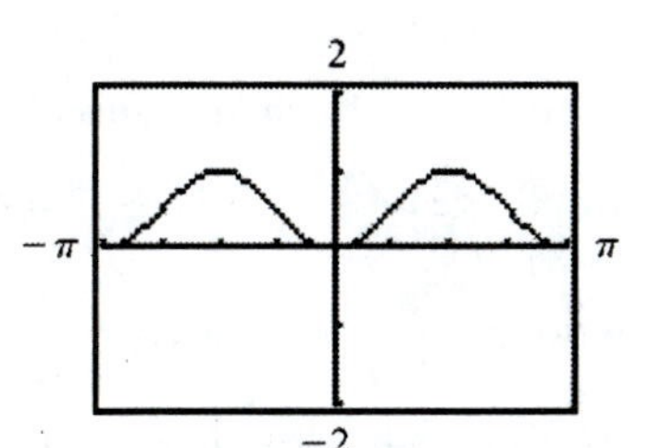

$$\text{Amplitude} = \frac{1}{2}\,(y \text{ coordinate of highest point} - y \text{ coordinate of lowest point})$$
$$= \frac{1}{2}(1 - 0) = \frac{1}{2} = |A|$$
$$\text{Period} = \frac{2\pi}{2} = \pi.$$

Thus, $B = \dfrac{2\pi}{\pi} = 2$. The form of the graph is that of the upside down cosine curve shifted up $\dfrac{1}{2}$ unit.

Thus, $y = \dfrac{1}{2} - |A| \cos Bx = \dfrac{1}{2} - \dfrac{1}{2} \cos 2x$.

39. 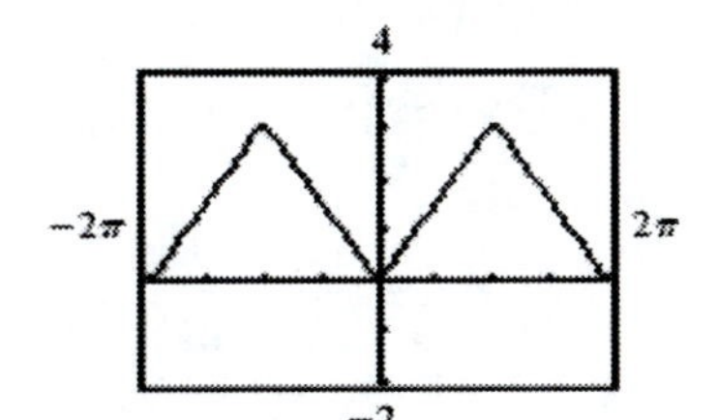

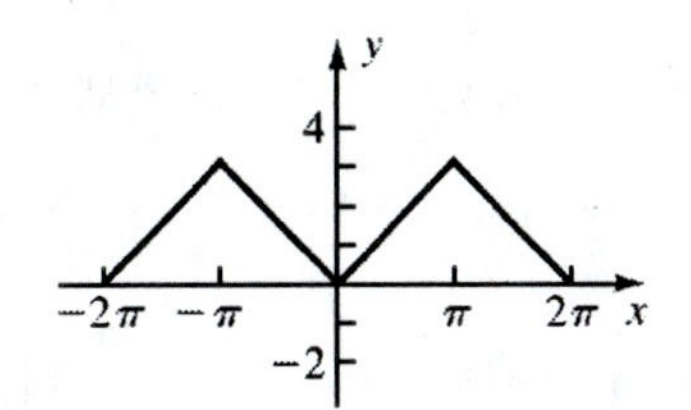

40. Using the fundamental identities, we can write $\tan\theta = a$, hence, $\dfrac{\sin\theta}{\cos\theta} = a$, or $\sin\theta = a\cos\theta$. Also, $\sin^2\theta + \cos^2\theta = 1$. Substituting $a\cos\theta$ for $\sin\theta$ and solving for $\cos\theta$, we obtain:

$$(a\cos\theta)^2 + \cos^2\theta = 1$$
$$(a^2 + 1)\cos^2\theta = 1$$
$$\cos^2\theta = \frac{1}{1 + a^2}$$

Since θ is a first quadrant angle, $\cos\theta$ is positive. Hence, $\cos\theta = \dfrac{1}{\sqrt{1 + a^2}}$. It follows that

$\sin\theta = a\cos\theta = \dfrac{a}{\sqrt{1 + a^2}}$. Using the reciprocal identities, we can write

$$\cot\theta = \frac{1}{\tan\theta} = \frac{1}{a}$$

$$\sec\theta = \frac{1}{\cos\theta} = \frac{1}{1/\sqrt{1 + a^2}} = \sqrt{1 + a^2} \qquad \csc\theta = \frac{1}{\sin\theta} = \frac{1}{a/\sqrt{1 + a^2}} = \frac{\sqrt{1 + a^2}}{a}$$

41. In the figure, we note, $s = r\theta$

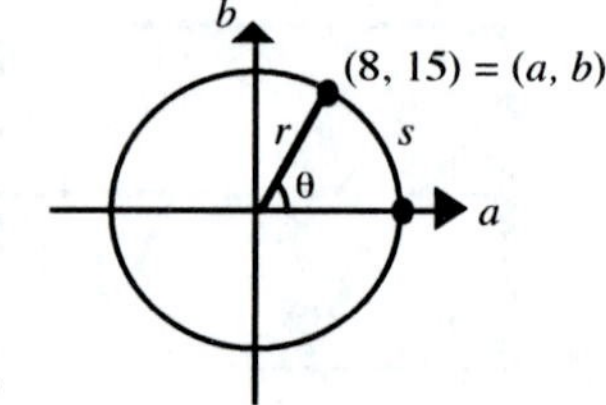

$$r = \sqrt{a^2 + b^2} = \sqrt{8^2 + 15^2} = 17;$$

$$\tan\theta = \frac{b}{a} = \frac{15}{8}; \ \theta = \tan^{-1}\frac{15}{8} \text{ radians}$$

Therefore,

$$s = r\theta = 17\tan^{-1}\frac{15}{8} \approx 18.37 \text{ units (calculator in radian mode).}$$

42. Since the point moves clockwise, x is negative, and the coordinates of the point are: $P(\cos(-53.077), \sin(-53.077)) = (-0.9460, -0.3241)$. The quadrant in which P lies is determined by the signs of the coordinates. In this case, P lies in quadrant III, because both coordinates are negative.

43. Since (a, b) is on a unit circle with $(a, b) = (0.5796, 0.8149) = (\cos s, \sin s)$, we can solve $\cos s = 0.5796$ or $\sin s = 0.8149$. Then $s = \cos^{-1}(0.5796) = 0.9526$ or $s = \sin^{-1}(0.8149) = 0.9526$.

44. Since the maximum deviation from the x axis is 4, we can write:
Amplitude $= |A| = 4$. Thus, $A = 4$ or -4. Since the period is $\frac{4}{3} - \left(-\frac{2}{3}\right) = 2$, we can write:
Period $= \frac{2\pi}{B} = 2$. Thus, $B = \frac{2\pi}{2} = \pi$. Since we are instructed to choose the phase shift between 0 and 1, we can regard this graph as containing the basic sine curve with a phase shift of $\frac{1}{3}$. This requires us to choose A positive, since the graph shows that as x increases from $\frac{1}{3}$ to $\frac{5}{6}$, *y increases* like the basic sine curve (not the upside down sine curve). So $A = 4$. Then, $-\frac{C}{B} = \frac{1}{3}$. Thus,

$$C = -\frac{1}{3}B = -\frac{1}{3}\pi \qquad y = A\sin(Bx + C) = 4\sin\left(\pi x - \frac{1}{3}\pi\right).$$

Check: When $x = 0$, $y = 4\sin\left(\pi \cdot 0 - \frac{1}{3}\pi\right) = 4\sin\left(-\frac{1}{3}\pi\right) = -2\sqrt{3}$

When $x = \frac{1}{3}$, $y = 4\sin\left(\pi \cdot \frac{1}{3} - \frac{1}{3}\pi\right) = 4\sin 0 = 0$

45. The graph of $y = 2.4\sin\frac{x}{2} - 1.8\cos\frac{x}{2}$ is shown in the figure. This graph appears to be a sine wave with amplitude 3 and period 4π that has been shifted to the right. Thus, we conclude that $A = 3$ and $B = \frac{2\pi}{4\pi} = \frac{1}{2}$. To determine C, we use the zoom feature or the built-in approximation routine to locate the x intercept closest to the origin at $x = 1.287$. This is the phase-shift for the graph.
Substitute $B = \frac{1}{2}$ and $x = 1.287$ into the phase-shift equation $x = -\frac{C}{B}$;

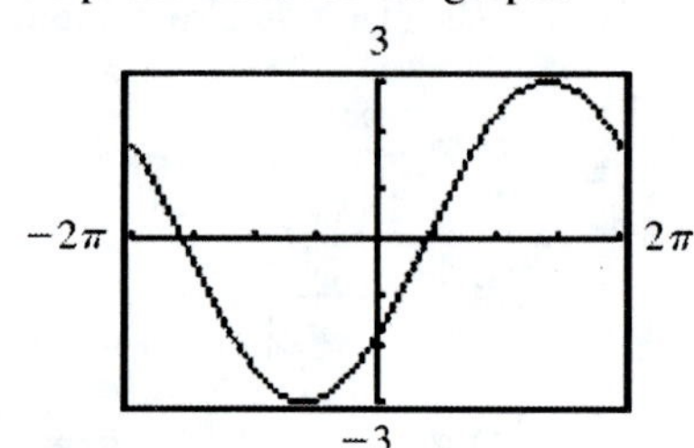

$1.287 = -\frac{C}{1/2}$; $C = -0.6435$.

Thus, the equation required is $y = 3\sin\left(\frac{x}{2} - 0.6435\right)$.

46. (A) The graph of $y = \dfrac{\sin 2x}{1 + \cos 2x}$ is shown in the figure. The graph appears to have vertical asymptotes at $x = -\dfrac{\pi}{2}$ and $x = \dfrac{\pi}{2}$ and period π.
It appears, therefore, to be the same as the graph of $y = \tan Bx$, with $\dfrac{\pi}{B} = \pi$, that is, $B = 1$. The required equation is $y = \tan x$.

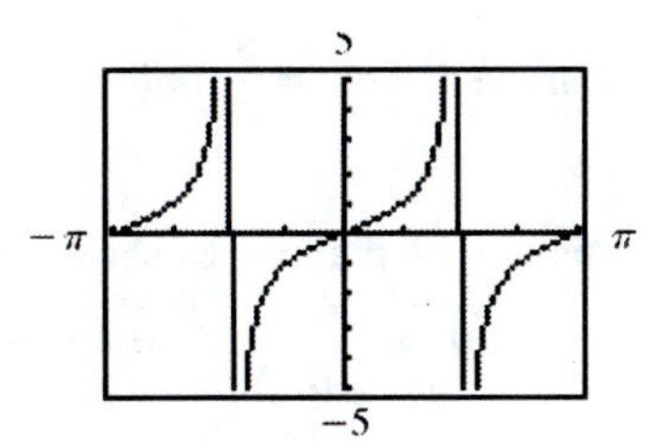

(B) The graph of $y = \dfrac{2\cos x}{1 + \cos 2x}$ is shown in the figure.
The graph appears to have vertical asymptotes at $x = -\dfrac{\pi}{2}$ and $x = \dfrac{\pi}{2}$ and period 2π. Its high and low points appear to have y coordinates of -1 and 1, respectively. It appears, therefore, to be the same as the graph of $y = \sec Bx$, $\dfrac{2\pi}{B} = 2\pi$, that is, $B = 1$.
The required equation is $y = \sec x$.

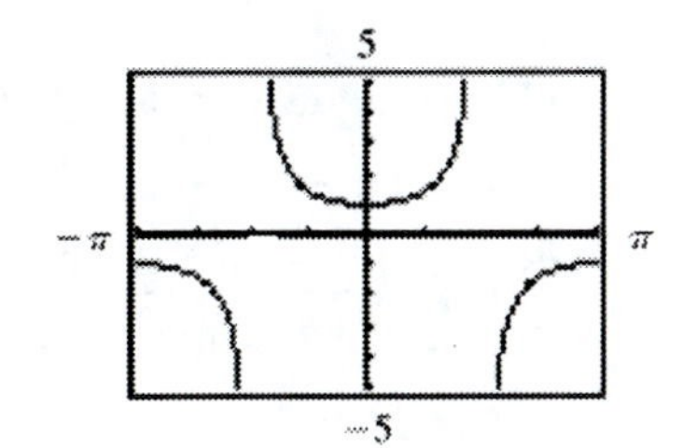

(C) The graph of $y = \dfrac{2\sin x}{1 \cos 2x}$ is shown in the figure.
The graph appears to have vertical asymptotes at $x = -\pi, x = 0$, and $x = \pi$, and period 2π. Its high and low points appear to have y coordinates of -1 and 1, respectively. It appears, therefore, to be the same as the graph of $y = \csc Bx$, $\dfrac{2\pi}{B} = 2\pi$, that is, $B = 1$.
The required equation is $y = \csc x$.

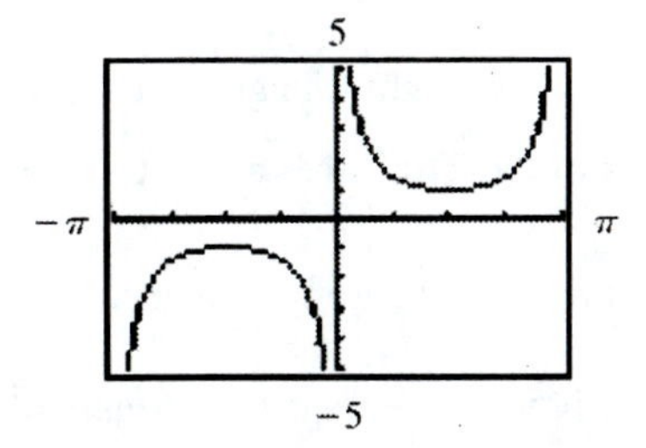

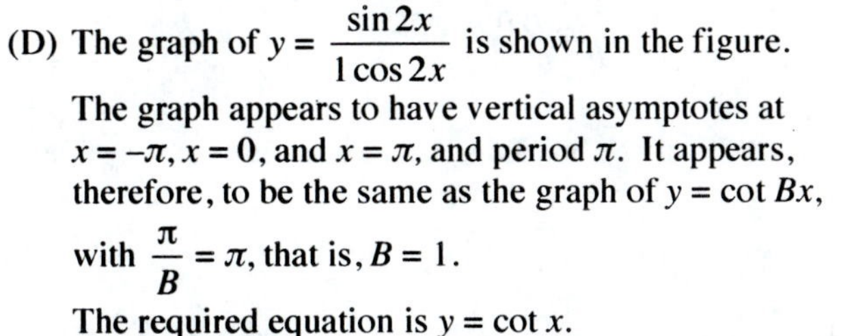
(D) The graph of $y = \dfrac{\sin 2x}{1 \cos 2x}$ is shown in the figure.
The graph appears to have vertical asymptotes at $x = -\pi, x = 0$, and $x = \pi$, and period π. It appears, therefore, to be the same as the graph of $y = \cot Bx$, with $\dfrac{\pi}{B} = \pi$, that is, $B = 1$.
The required equation is $y = \cot x$.

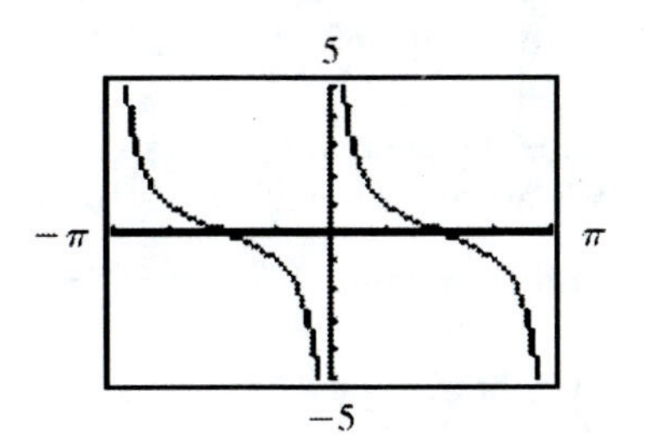

47. We use the diagram and reason as follows: Since the cities have the same longitude, θ is given by their difference in latitude $\theta = 41°36' - 30°25' = 11°11' = \left(11 + \dfrac{11}{60}\right)^\circ$.

Since $\dfrac{s}{C} = \dfrac{\theta}{360°}$ and $C = 2\pi r$, then $\dfrac{s}{2\pi r} = \dfrac{\theta}{360°}$.

$$s = 2\pi r \cdot \frac{\theta}{360°} \approx 2(\pi)(3960 \text{ mi}) \frac{11 + \frac{11}{60}}{360} \approx 773 \text{ mi}.$$

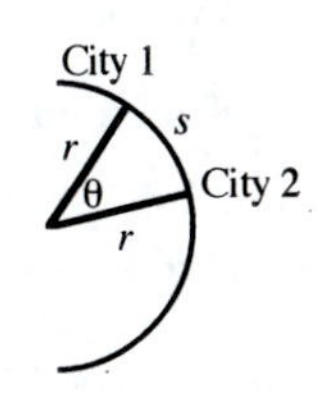

48. Since the two right triangles shown in the figure are similar, we can write $\dfrac{r}{4} = \dfrac{6}{9}$, thus, $r = \dfrac{8}{3}$ cm.

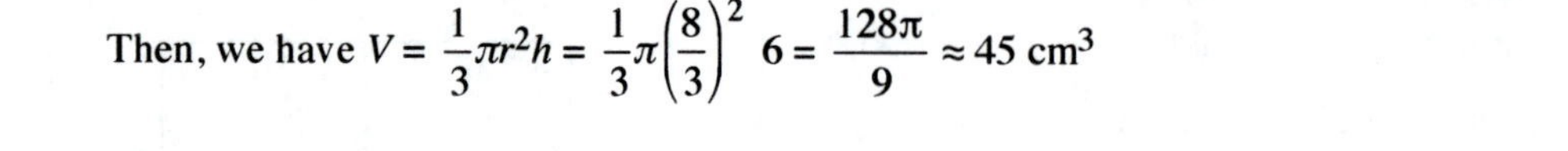

Then, we have $V = \dfrac{1}{3}\pi r^2 h = \dfrac{1}{3}\pi\left(\dfrac{8}{3}\right)^2 6 = \dfrac{128\pi}{9} \approx 45 \text{ cm}^3$

49. Labeling the diagram as shown, we can write:

$$\cos\theta = \frac{AC}{AB} \qquad \theta = \text{angle of elevation}$$

$$\theta = \cos^{-1}\frac{AC}{AB} = \cos^{-1}\frac{30}{40} = 41°$$

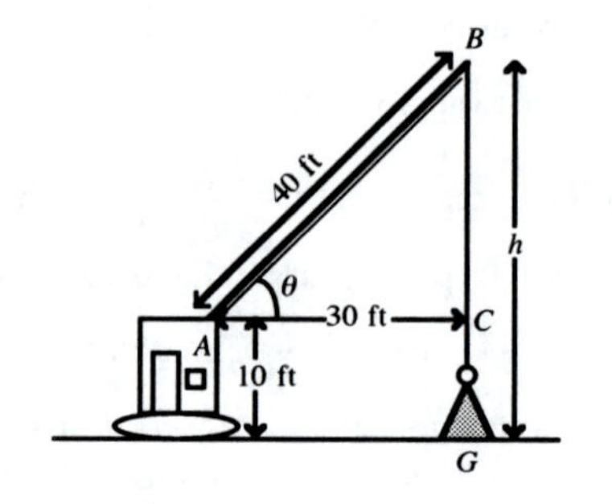

To find h, the altitude of the tip, we note

$$h = BC + CG$$

$$BC^2 = AB^2 - AC^2 \text{ (Pythagorean theorem)}$$

$$BC = \sqrt{AB^2 - AC^2} = \sqrt{40^2 - 30^2}$$

$$= 26 \text{ ft}$$

Then, $h = 26 + 10 = 36$ ft

50. The two right triangles shown in the text figure have corresponding angles equal, hence they are similar. Thus, we can write $\dfrac{x}{2} = \dfrac{4-x}{4}$; $2x = 4 - x$; $x = \dfrac{4}{3}$ ft.

51. From the figure it is clear that $\tan\theta = \dfrac{a}{b}$.

Given: percentage of inclination $\dfrac{a}{b} = 3\% = 0.03$, then $\tan\theta = 0.03$; $\theta = 1.7°$

Given: angle of inclination $\theta = 3°$, then $\dfrac{a}{b} = \tan 3°$; $\dfrac{a}{b} = 0.05$ or 5%

52. Labeling the text figure as shown, we note:
We are asked for $BT = x + 10$, the height of the office building, and d, the width of the street.

In right triangle SCT, $\cot 68° = \dfrac{d}{x}$

In right triangle ABT, $\cot 72° = \dfrac{d}{x+10}$

We solve the system of equations

$$\cot 68° = \frac{d}{x} \qquad \cot 72° = \frac{d}{x+10}$$

by clearing of fractions, then eliminating d.

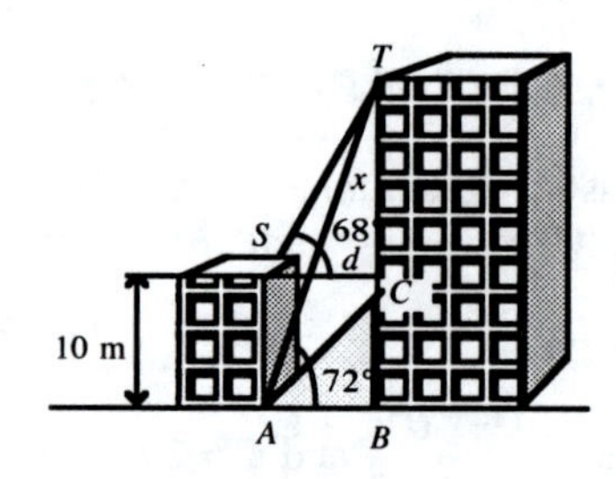

(1) $d = x\cot 68° \quad d = (x + 10)\cot 72°$

$$x\cot 68° = (x + 10)\cot 72°$$

$$= x\cot 72° + 10\cot 72°$$

$$x\cot 68° - x\cot 72° = 10\cot 72°$$

$$x = \frac{10\cot 72°}{\cot 68° - \cot 72°} = 41 \text{ m}$$

Then the height of the office building $= x + 10 = 51$ m. Substituting in (1),

$$d = x\cot 68° = (41 \text{ m})\cot 68° = 17 \text{ m}$$

53. We redraw and label the figure in the text.

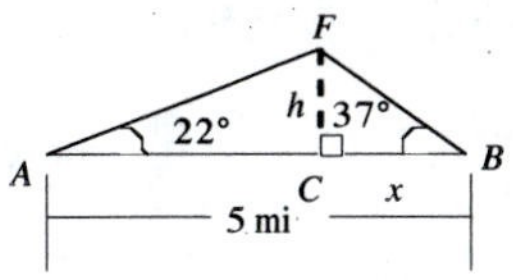

(A) We note: $AC = AB - BC = 5 - x$. We are to find x.

In right triangle BCF, $\tan 37° = \dfrac{h}{x}$

In right triangle ACF, $\tan 22° = \dfrac{h}{5-x}$

We solve the system of equations

$$\tan 37° = \frac{h}{x} \qquad \tan 22° = \frac{h}{5-x}$$

by clearing of fractions, then eliminating h.

(1) $h = x \tan 37°$; $h = (5 - x) \tan 22°$;

$x \tan 37° = (5 - x) \tan 22° = 5 \tan 22° - x \tan 22°$;

$x \tan 37° + x \tan 22° = 5 \tan 22°$; $x = \dfrac{5\tan 22°}{\tan 37° + \tan 22°} = 1.7$ mi

(B) We are to find h. From (1) in part (A), $h = x \tan 37° = 1.7 \tan 37° = 1.3$ mi

54. (A) For the drive motor, we note: 300 rpm $= 300 \cdot 2\pi$ rad/min $= 600\pi$ rad/min.

$$V = r\omega = (15 \text{ in.})(600\pi \text{ rad/min}) = 9000\pi \text{ in./min}$$

This is the linear velocity of the chain, hence the linear velocity of the smaller wheel of the saw.

Then, for the saw,

$$\omega = \frac{V}{r} = \frac{9000\pi \text{ in./min}}{30 \text{ in.}} = 300\pi \text{ rad/min} \approx 942 \text{ rad/min}$$

(B) For the saw itself, ω is also 300π rad/min

$$V = r\omega = (68 \text{ in.})(300\pi \text{ rad/min}) = 20{,}400\pi \text{ in./min} \approx 64{,}088 \text{ in./min}$$

55. Use $\dfrac{n_2}{n_1} = \dfrac{\sin\alpha}{\sin\beta}$, where $n_2 = 1.33$, $n_1 = 1.00$, and $\alpha = 38.4°$.

Solve for β: $\dfrac{1.33}{1.00} = \dfrac{\sin 38.4°}{\sin\beta}$; $\sin\beta = \dfrac{\sin 38.4°}{1.33}$; $\beta = \sin^{-1}\left(\dfrac{\sin 38.4°}{1.33}\right) = 27.8°$

56. We use $\sin\dfrac{\theta}{2} = \dfrac{S_s}{S_p}$, where S_s is the speed of sound and S_p = speed of the plane = $1.5S_s$. Thus,

$$\sin\frac{\theta}{2} = \frac{S_s}{1.5S_s} = \frac{1}{1.5}; \quad \frac{\theta}{2} \approx 42°; \quad \theta \approx 84°$$

57. (A) $A_1 = \dfrac{1}{2}(\text{base})(\text{height}) = \dfrac{1}{2}(1)(\sin x) = \dfrac{1}{2}\sin x$ (the height is the perpendicular distance from P to the x axis, thus, $\sin x$).

$A_2 = \dfrac{1}{2}(\text{radius})^2(\text{angle}) = \dfrac{1}{2}(1)^2 x = \dfrac{1}{2}x$

$A_3 = \dfrac{1}{2}(\text{base})(\text{height}) = \dfrac{1}{2}(1)$ h. In triangle OAB, $\tan x = \dfrac{h}{1}$, hence, $h = \tan x$. Thus, $A_3 = \dfrac{1}{2}\tan x$.

(B) Since $A_1 < A_2 < A_3$, we can write $\dfrac{1}{2}\sin x < \dfrac{1}{2}x < \dfrac{1}{2}\tan x$

Multiplying by 2, we can write $\sin x < x < \tan x$

Applying a fundamental identity, we can write $\sin x < x < \dfrac{\sin x}{\cos x}$

As long as $0 < x < \dfrac{\pi}{2}$, these quantities are positive. For positive quantities, $a < b < c$ is equivalent to $\dfrac{1}{c} < \dfrac{1}{b} < \dfrac{1}{a}$.

Hence, $\dfrac{\cos x}{\sin x} < \dfrac{1}{x} < \dfrac{1}{\sin x}$. If $0 < x < \dfrac{\pi}{2}$, $\sin x > 0$, and we can multiply all parts of this double inequality by $\sin x$ without altering the sense of inequalities. Thus,

$$\sin x \cdot \frac{\cos x}{\sin x} < \sin x \cdot \frac{1}{x} < \sin x \cdot \frac{1}{\sin x}$$

$$\cos x < \frac{\sin x}{x} < 1, \quad x > 0$$

(C) $\cos x$ approaches 1 as x approaches 0, and $\dfrac{\sin x}{x}$ is between $\cos x$ and 1, therefore, $\dfrac{\sin x}{x}$ must approach 1 as x approaches 0.

58. (A)

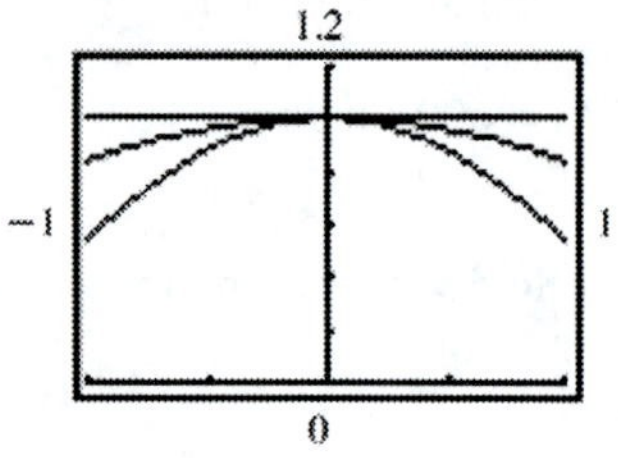

(B) $y1 < y2 < y3$

(C) $\cos x$ approaches 1 as x approaches 0, and $\dfrac{\sin x}{x}$ is between $\cos x$ and 1, therefore $\dfrac{\sin x}{x}$ must approach 1 as x approaches 0.

This can also be observed using TRACE.

59. Since the amplitude is 3.6 cm, $|A| = 3.6$, and since the position when $t = 0$ sec is taken positive, $A = 3.6$. Since the frequency is 6 Hz, write

$$\text{Frequency} = \frac{1}{\text{Period}} = \frac{1}{2\pi/B} = \frac{B}{2\pi}$$

$$6 = \frac{B}{2\pi}$$

$$12\pi = B$$

Thus, the required equation is $y = 3.6 \cos 12\pi t$. An equation of the form $y = A \sin Bt$ cannot be used to model the motion, because then, for $t = 0$, y will be 0 no matter what values are assigned to A and B.

60. (A) $I = 12 \sin(60\pi t - \pi)$

We compute amplitude, period, frequency, and phase shift as follows:

Amplitude $= |A| = |12| = 12$ amperes.

Phase Shift and Period: Solve

$$\begin{aligned} Bx + C &= 0 & \text{and} \quad Bx + C &= 2\pi \\ 60\pi t - \pi &= 0 & 60\pi t - \pi &= 2\pi \\ 60\pi t &= \pi & 60\pi t &= \pi + 2\pi \\ t &= \frac{1}{60} & t &= \frac{1}{60} + \frac{1}{30} \end{aligned}$$

$\frac{1}{60}$ ↑ Phase Shift; $\frac{1}{30}$ ↑ Period in seconds

$$\text{Frequency} = \frac{1}{\text{Period}} = \frac{1}{1/30} = 30 \text{ Hz}$$

(B)

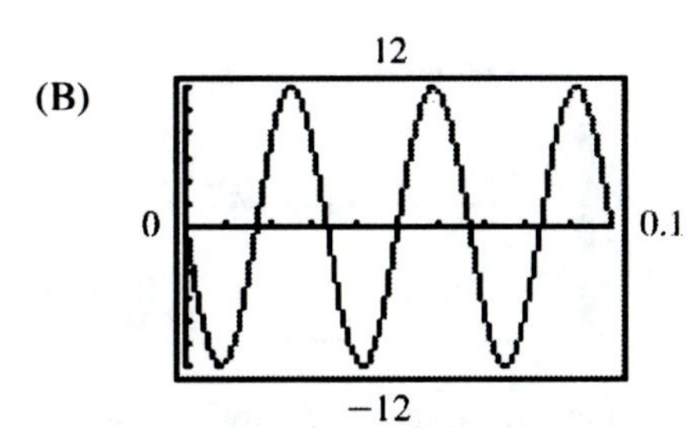

61. To find the period T, we use the formula $\lambda v = c$, with $v = \dfrac{1}{\text{Period}} = \dfrac{1}{T}$ and solve for T.

$\lambda \dfrac{1}{T} = c$; $\lambda = Tc$; $T = \dfrac{\lambda}{c}$. Hence, $T = \dfrac{6 \times 10^{-5} m}{3 \times 10^{8} m/\text{sec}} = 2 \times 10^{-13}$ sec.

62. (A) In the figure, note that ABC and CDE are right triangles, and that $AE = AC + CE = a + b$. Then,

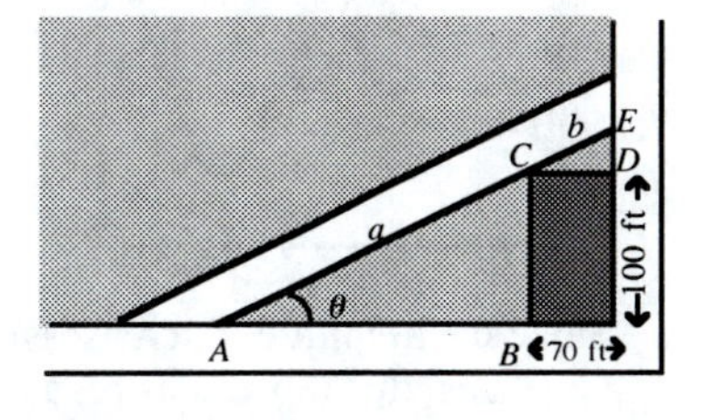

$\csc\theta = \dfrac{a}{100}$ from triangle ABC, so $a = 100 \csc\theta$,

$\sec\theta = \dfrac{b}{70}$ from triangle CDE, so $b = 70 \sec\theta$.

Thus, $AE = 100 \csc\theta + 70 \sec\theta$.

(B) For small θ (near 0°), L is extremely large. As θ increases from 0°, L decreases to some minimum value, then increases again beyond all bounds as θ approaches 90°.

(C)

θ rad	0.50	0.60	0.70	0.80	0.90	1.00	1.10
L ft	288.3	261.9	246.7	239.9	240.3	248.4	266.5

According to the table, the minimum value is 239.9 ft when $\theta = 0.80$ rad.

(D) According to the graph, the minimum $L = 239.16$ ft when $\theta = 0.84$ rad.

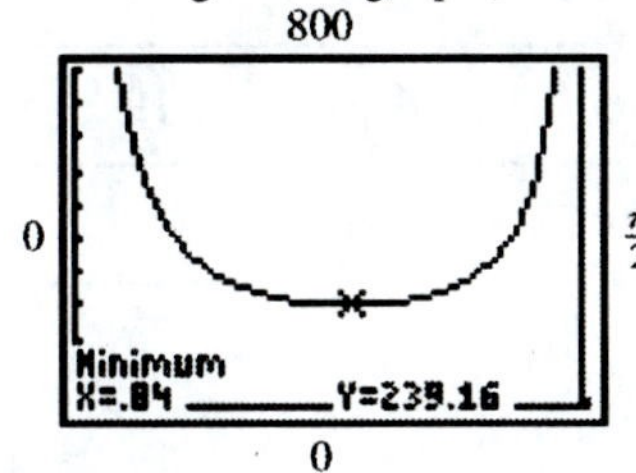

63. (A) In the right triangle in the figure, $\tan\theta = \dfrac{d}{50}$. Hence, $d = 50 \tan\theta$.

(B) 20 rpm = $20(2\pi)$ rad/min = 40π rad/min. Since $\theta = \omega t$, and $\omega = 40\pi$ rad/min, $\theta = 40\pi t$.

(C) Substituting the expression for θ from part (B) into $d = 50 \tan\theta$, we obtain $d = 50 \tan 40\pi t$.

(D)

800

0 $\quad \dfrac{1}{80}$

0

d increases without bound as t approaches $\dfrac{1}{80}$ min.

64. (A)

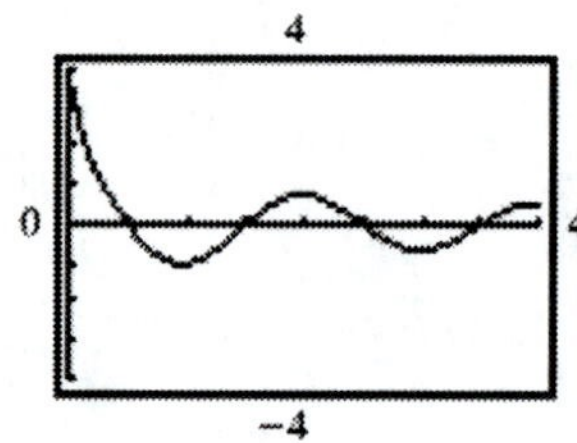

Since the amplitude decreases to 0 as time increases, this represents damped harmonic motion.

(B)

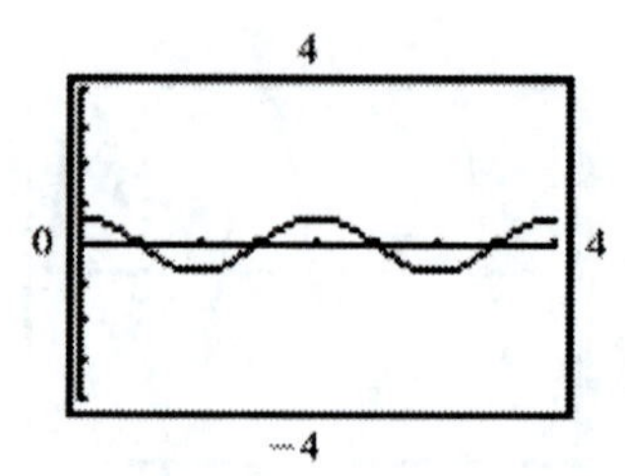

Since the amplitude remains constant over time, this represents simple harmonic motion.

(C)

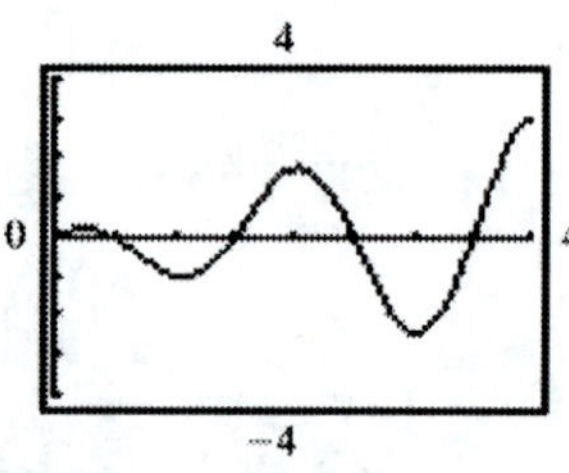

Since the amplitude increases as time increases, this represents resonance.

65. (A)

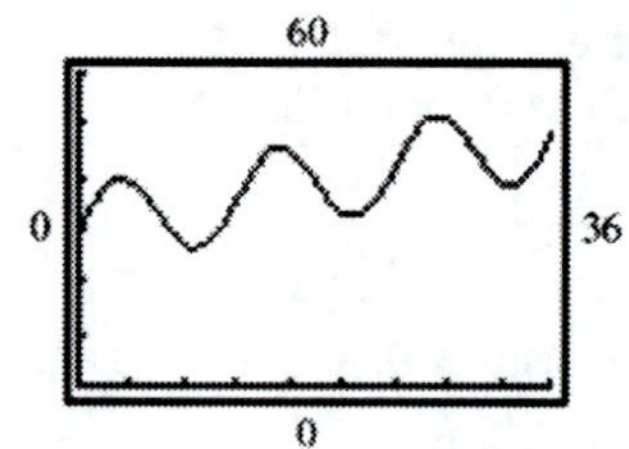

(B) The sales trend is up but with the expected seasonal variations.

66. (A)

x(months)	1, 13	2, 14	3, 15	4, 16	5, 17	6, 18	7, 19	8, 20	9, 21	10, 22	11, 23	12, 24
y (temperatures)	19	23	32	45	55	65	71	69	62	51	37	25

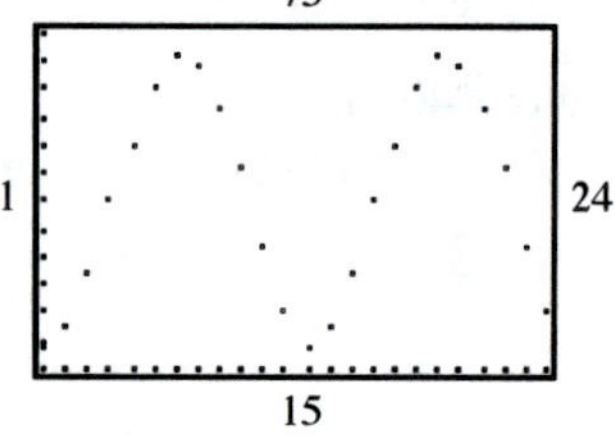

(B) From the table, Max $y = 71$ and Min $y = 19$. Then,

$$A = \frac{(\text{Max } y - \text{Min } y)}{2} = \frac{(71-19)}{2} = 26$$

$$B = \frac{2\pi}{\text{Period}} = \frac{2\pi}{12} = \frac{\pi}{6}$$

$$k = \text{Min } y + A = 19 + 26 = 45$$

From the plot in (A) or the table, we estimate the smallest positive value of x for which $y = k = 45$ to be approximately 4.2. Then this is the phase-shift for the graph. Substitute $B = \frac{\pi}{6}$ and $x = 4.2$ into the phase-shift equation $x = -\frac{C}{B}$; $4.2 = \frac{-C}{\pi/6}$; $C = -\frac{4.2\pi}{6} \approx -2.2$.

Thus, the equation required is $y = 45 + 26 \sin\left(\frac{\pi x}{6} x - 2.2\right)$.

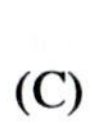
(C)

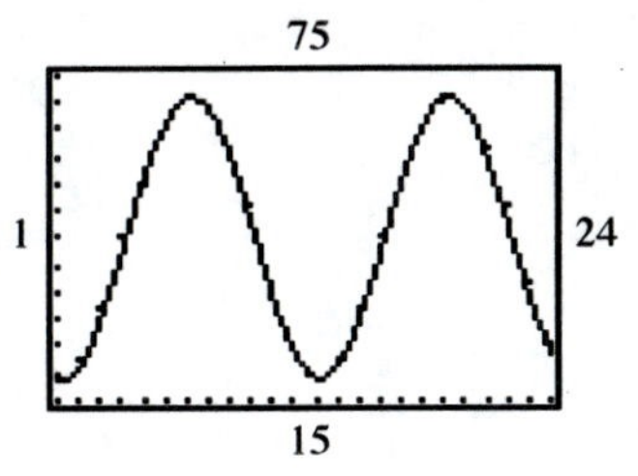

Chapter 4 Identities

EXERCISE 4.1 Fundamental Identities and Their Use

1. An identity is an equation in one or more variables in which the left side is equal to the right side for all replacements of the variables for which both sides are defined.

3. $\sin^2 x + \cos^2 x = 1$, $\tan^2 x + 1 = \sec^2 x$, $1 + \cot^2 x = \csc^2 x$

5. *Find tan x*: $\tan x = \frac{\sin x}{\cos x} = \frac{2/\sqrt{5}}{1/\sqrt{5}} = 2$ *Find cot x:* $\cot x = \frac{1}{\tan x} = \frac{1}{2}$

   Find csc x: $\csc x = \frac{1}{\sin x} = \frac{1}{2/\sqrt{5}} = \frac{\sqrt{5}}{2}$ *Find sec x:* $\sec x = \frac{1}{\cos x} = \frac{1}{1/\sqrt{5}} = \sqrt{5}$

7. *Find sin x:* Since $\csc x = \frac{1}{\sin x}$, we can write $\sin x = \frac{1}{\csc x}$; $\sin x = \frac{1}{\csc x} = \frac{1}{-\sqrt{10}/3} = -\frac{3}{\sqrt{10}}$

   Find tan x: $\tan x = \frac{\sin x}{\cos x} = \frac{-3/\sqrt{10}}{1/\sqrt{10}} = -3$

   Find cot x: $\cot x = \frac{1}{\tan x} = \frac{1}{-3} = -\frac{1}{3}$

   Find sec x: $\sec x = \frac{1}{\cos x} = \frac{1}{1/\sqrt{10}} = \sqrt{10}$

9. *Find cos x*: Since $\sec x = \frac{1}{\cos x}$, we can write $\cos x = \frac{1}{\sec x}$; $\cos x = \frac{1}{\sec x} = \frac{1}{-4/\sqrt{15}} = -\frac{\sqrt{15}}{4}$

   Find sin x: Since $\tan x = \frac{\sin x}{\cos x}$, we can write $\sin x = \tan x \cos x$;

   $\sin x = \tan x \cos x = \left(\frac{1}{\sqrt{15}}\right)\left(-\frac{\sqrt{15}}{4}\right) = -\frac{1}{4}$

   Find cot x: $\cot x = \frac{1}{\tan x} = \frac{1}{1/\sqrt{15}} = \sqrt{15}$

   Find csc x: $\csc x = \frac{1}{\sin x} = \frac{1}{-1/4} = -4$

11. $\tan u \cot u = \tan u \frac{1}{\tan u}$ Reciprocal identity

    $= 1$ Algebra

13. $\tan x \csc x = \frac{\sin x}{\cos x}\frac{1}{\sin x}$ Quotient and reciprocal identities

    $= \frac{1}{\cos x}$ Algebra

    $= \sec x$ Reciprocal identity

15. $\dfrac{\sec^2 x - 1}{\tan x} = \dfrac{\tan^2 x + 1 - 1}{\tan x}$ Pythagorean identity

$= \dfrac{\tan^2 x}{\tan x}$ Algebra

$= \tan x$ Algebra

17. $\dfrac{\sin^2\theta}{\cos\theta} + \cos\theta = \dfrac{\sin^2\theta}{\cos\theta} + \dfrac{\cos\theta}{1}$ Algebra

$= \dfrac{\sin^2\theta}{\cos\theta} + \dfrac{\cos^2\theta}{\cos\theta}$ Algebra

$= \dfrac{\sin^2\theta + \cos^2\theta}{\cos\theta}$ Algebra

$= \dfrac{1}{\cos\theta}$ Pythagorean identity

$= \sec\theta$ Reciprocal identity

Key algebraic steps: $\dfrac{a^2}{b} + b = \dfrac{a^2}{b} + \dfrac{b}{1} = \dfrac{a^2}{b} + \dfrac{b^2}{b} = \dfrac{a^2 + b^2}{b}$

19. $\dfrac{1}{\sin^2\beta} - 1 = \left(\dfrac{1}{\sin\beta}\right)^2 - 1$ Algebra

$= \csc^2\beta - 1$ Reciprocal identity

$= 1 + \cot^2\beta - 1$ Pythagorean identity

$= \cot^2\beta$ Algebra

21. $\dfrac{(1 - \cos x)^2 + \sin^2 x}{1 - \cos x} = \dfrac{1 - 2\cos x + \cos^2 x + \sin^2 x}{1 - \cos x}$ Algebra

$= \dfrac{1 - 2\cos x + 1}{1 - \cos x}$ Pythagorean identity

$= \dfrac{2 - 2\cos x}{1 - \cos x}$ Algebra

$= \dfrac{2(1 - \cos x)}{1 - \cos x}$ Algebra

$= 2$ Algebra

Key algebraic step: $(1 - a)^2 = 1 - 2a + a^2$

23. There are many possible answers. We choose $x = 1$.
Then left side $= x^2 + 1 = 1^2 + 1 = 2$
right side $= (x + 1)^2 = (1 + 1)^2 = 4$
Since the left side and the right side are defined, but not equal, the equation is not an identity.

25. There are many possible answers. We choose $x = 0$.
then left side $= \cos(\pi x) = \cos(\pi \cdot 0) = 1$
right side $= \pi \cos x = \pi \cos 0 = \pi$
Since the left side and the right side are defined, but not equal, the equation is not an identity.

27. There are many possible answers. We choose $x = \frac{\pi}{2}$.
Then left side $= \sin(x + \pi) = \sin\left(\frac{\pi}{2} + \pi\right) = -1$
right side $= \sin x + \sin \pi = \sin \frac{\pi}{2} + \sin \pi = 1$
Since the left side and the right side are defined, but not equal, the equation is not an identity.

29. Not necessarily. For example, $\sin x = 0$ for infinitely many values ($x = k\pi$, k any integer), but the equation is not an identity. The left side is not equal to the right side for any other value of x than $x = k\pi$, k an integer; for example, if $x = \frac{\pi}{2}$, $\sin \frac{\pi}{2} = 1$, hence $\sin \frac{\pi}{2} = 0$ is false.

31. *Find cos x:* We start with the Pythagorean identity $\sin^2 x + \cos^2 x = 1$ and solve for $\cos x$.

$$\cos x = \pm\sqrt{1 - \sin^2 x}$$

Since $\cos x$ is specified negative, we choose

$$\cos x = -\sqrt{1 - \sin^2 x} = -\sqrt{1 - (2/5)^2} = -\sqrt{\frac{21}{25}} = -\frac{\sqrt{21}}{5}$$

Find sec x: $\sec x = \frac{1}{\cos x} = \frac{1}{-\sqrt{21}/5} = -\frac{5}{\sqrt{21}}$ *Find csc x:* $\csc x = \frac{1}{\sin x} = \frac{1}{2/5} = \frac{5}{2}$

Find tan x: $\tan x = \frac{\sin x}{\cos x} = \frac{2/5}{-\sqrt{21}/5} = -\frac{2}{\sqrt{21}}$ *Find cot x:* $\cot x = \frac{1}{\tan x} = \frac{1}{-2/\sqrt{21}} = -\frac{\sqrt{21}}{2}$

33. *Find sec x:* We start with the Pythagorean identity $\tan^2 x + 1 = \sec^2 x$ and solve for $\sec x$.

$$\sec x = \pm\sqrt{\tan^2 x + 1}$$

Since $\sin x$ is positive and $\tan x$ is negative, x is associated with the second quadrant, where $\sec x$ is negative; hence, $\sec x = -\sqrt{\tan^2 x + 1} = -\sqrt{\left(-\frac{1}{2}\right)^2 + 1} = -\sqrt{\frac{5}{4}} = -\frac{\sqrt{5}}{2}$

Find cos x: $\cos x = \frac{1}{\sec x} = \frac{1}{-\sqrt{5}/2} = -\frac{2}{\sqrt{5}}$

Find sin x: Since $\tan x = \frac{\sin x}{\cos x}$, we can write $\sin x = \cos x \tan x = \left(-\frac{2}{\sqrt{5}}\right)\left(-\frac{1}{2}\right) = \frac{1}{\sqrt{5}}$

Find cot x: $\cot x = \frac{1}{\tan x} = \frac{1}{-1/2} = -2$

Find csc x: $\csc x = \frac{1}{\sin x} = \frac{1}{1/\sqrt{5}} = \sqrt{5}$

35. *Find tan x:* We start with the Pythagorean identity $\tan^2 x + 1 = \sec^2 x$ and solve for $\tan x$.

$$\tan x = \pm\sqrt{\sec^2 x - 1}$$

Since $\sec x$ and $\cot x$ are both positive, x is associated with the first quadrant, where $\tan x$ is positive; hence, $\tan x = \sqrt{\sec^2 x - 1} = \sqrt{4^2 - 1} = \sqrt{15}$

Find cos x: $\cos x = \dfrac{1}{\sec x} = \dfrac{1}{4}$

Find sin x: Since $\tan x = \dfrac{\sin x}{\cos x}$, we can write $\sin x = \cos x \tan x = \left(\dfrac{1}{4}\right)\left(\sqrt{15}\right) = \dfrac{\sqrt{15}}{4}$

Find cot x: $\cot x = \dfrac{1}{\tan x} = \dfrac{1}{\sqrt{15}}$

Find csc x: $\csc x = \dfrac{1}{\sin x} = \dfrac{1}{\sqrt{15}/4} = \dfrac{4}{\sqrt{15}}$

37. Given $\sin x = \dfrac{1}{3}$, then $\csc x = \dfrac{1}{\sin x} = \dfrac{1}{1/3} = 3$. However, to find exact values of the other trigonometric functions, it is necessary to use a Pythagorean identity. For example, $\cos x = \pm\sqrt{1 - \sin^2 x}$. However, on the basis of the given information, x could be associated either with quadrants I or II; there is no way to decide which, hence $\cos x$ has an undetermined sign. No, it is not possible to find exact values of all the remaining trigonometric functions.

39. Given $\cot x = -\sqrt{3}$, then $\tan x = \dfrac{1}{\cot x} = \dfrac{1}{-\sqrt{3}}$. However, to find exact values of the other trigonometric functions, it is necessary to use a Pythagorean identity. For example, $\csc x = \pm\sqrt{\cot^2 x + 1}$. However, on the basis of the given information, x could be associated either with quadrants II or IV; there is no way to decide which, hence $\csc x$ has an undetermined sign. No, it is not possible to find exact values of all the remaining trigonometric functions.

41. (A) From the identities for negatives,

$$\sin(-x) = -\sin x$$

Hence

$$\sin(-x) = -0.4350$$

(B) From the Pythagorean identity, $\cos^2 x + \sin^2 x = 1$, thus

$$\cos^2 x = 1 - \sin^2 x$$

Hence

$$(\cos x)^2 = 1 - 0.1892 = 0.8108$$

43.

$\csc(-y)\cos(-y) = \dfrac{1}{\sin(-y)}\cos(-y)$	Reciprocal identity
$= -\dfrac{1}{\sin y}\cos y$	Identities for negatives
$= -\dfrac{\cos y}{\sin y}$	Algebra
$= -\cot y$	Quotient identity

45. $\cot x \cos x + \sin x = \dfrac{\cos x}{\sin x}\cos x + \sin x$ Quotient identity

$= \dfrac{\cos^2 x}{\sin x} + \dfrac{\sin x}{1}$ Algebra

$= \dfrac{\cos^2 x}{\sin x} + \dfrac{\sin^2 x}{\sin x}$ Algebra

$= \dfrac{\cos^2 x + \sin^2 x}{\sin x}$ Algebra

$= \dfrac{1}{\sin x}$ Pythagorean identity

$= \csc x$ Reciprocal identity

Key algebraic steps: $\dfrac{a}{b}a + b = \dfrac{a}{b}\cdot\dfrac{a}{1} + \dfrac{b}{1} = \dfrac{a^2}{b} + \dfrac{b}{1} = \dfrac{a^2}{b} + \dfrac{b^2}{b} = \dfrac{a^2+b^2}{b}$

47. $\dfrac{\cot(-\theta)}{\csc\theta} + \cos\theta = \dfrac{\dfrac{\cos(-\theta)}{\sin(-\theta)}}{\dfrac{1}{\sin\theta}} + \cos\theta$ Quotient and reciprocal identities

$= \dfrac{\dfrac{\cos\theta}{-\sin\theta}}{\dfrac{1}{\sin\theta}} + \cos\theta$ Identities for negatives

$= \dfrac{\cos\theta}{-\sin\theta}\cdot\dfrac{\sin\theta}{1} + \cos\theta$ Algebra

$= -\cos\theta + \cos\theta$ Algebra

$= 0$ Algebra

49. $\dfrac{\cot x}{\tan x} + 1 = \cot x \div \tan x + 1$ Algebra

$= \cot x \div \dfrac{1}{\cot x} + 1$ Reciprocal identity

$= \cot x \cdot \dfrac{\cot x}{1} + 1$ Algebra

$= \cot^2 x + 1$ Algebra

$= \csc^2 x$ Pythagorean identity

51. $\sec w \csc w - \sec w \sin w = \dfrac{1}{\cos w}\cdot\dfrac{1}{\sin w} - \dfrac{1}{\cos w}\cdot \sin w$ Reciprocal identities

$= \dfrac{1}{\cos w \sin w} - \dfrac{\sin w}{\cos w}$ Algebra

$= \dfrac{1}{\cos w \sin w} - \dfrac{\sin^2 w}{\cos w \sin w}$ Algebra

$= \dfrac{1-\sin^2 w}{\cos w \sin w}$ Algebra

$= \dfrac{\cos^2 w}{\cos w \sin w}$ Pythagorean identity (solved for $1 - \sin^2 x = \cos^2 x$)

$= \dfrac{\cos w}{\sin w}$ Algebra

$= \cot w$ Quotient identity

53. Yes. The left side, $\sin x \cot x = \sin x \cdot \dfrac{\cos x}{\sin x} = \cos x$, the right side, for all values of x for which both are defined.

55. Yes. The left side, $\sin(-x) + \sin x = -\sin x + \sin x = 0$, the right side, for all values of x.

57. No. For example, if $x = \dfrac{\pi}{4}$, $\tan^2 x + \sec^2 x = \left(\tan\dfrac{\pi}{4}\right)^2 + \left(\sec\dfrac{\pi}{4}\right)^2 = 1^2 + (\sqrt{2})^2 = 3$.

59. (A) By the Pythagorean identity, $\sin^2 \alpha + \cos^2 \alpha = 1$ for any α. Therefore,
$\sin^2 \dfrac{x}{2} + \cos^2 \dfrac{x}{2} = 1$ (this is independent of x).
(B) By the Pythagorean identity, $\csc^2 \alpha = 1 + \cot^2 \alpha$ for any α. Therefore,
$\csc^2 (2x) - \cot^2 (2x) = 1 + \cot^2 (2x) - \cot^2 (2x) = 1$ (this is independent of x).

61. $\sqrt{1-\cos^2 x} = \sqrt{\sin^2 x}$ by the Pythagorean identity. Therefore, $\sqrt{1-\cos^2 x} = \sin x$ when, and only when, $\sqrt{\sin^2 x} = \sin x$. The latter statement is true whenever $\sin x$ is positive, that is, when x is in quadrant I or II.

63. $\sqrt{1-\sin^2 x} = \sqrt{\cos^2 x}$ by the Pythagorean identity. Therefore, $\sqrt{1-\sin^2 x} = -\cos x$ when, and only when, $\sqrt{\cos^2 x} = -\cos x$. The latter statement is true whenever $\cos x$ is negative, that is, when x is in quadrant II or III.

65. $\sqrt{1-\sin^2 x} = \sqrt{\cos^2 x}$ by the Pythagorean identity. $\sqrt{\cos^2 x} = |\cos x|$ is always true. Hence, $\sqrt{1-\sin^2 x} = |\cos x|$ in all quadrants.

67. $\dfrac{\sin x}{\sqrt{1-\sin^2 x}} = \dfrac{\sin x}{\sqrt{\cos^2 x}}$ by the Pythagorean identity. $\tan x = \dfrac{\sin x}{\cos x}$ by the quotient identity.
Therefore, $\dfrac{\sin x}{\sqrt{1-\sin^2 x}} = \dfrac{\sin x}{\sqrt{\cos^2 x}} = \dfrac{\sin x}{\cos x} = \tan x$ will be true whenever the middle two quantities are equal, that is, when, and only when, $\sqrt{\cos^2 x} = \cos x$. This statement is true whenever $\cos x$ is positive, that is, when x is in quadrant I or IV.

69. $\sqrt{a^2 - u^2} = \sqrt{a^2 - (a\sin x)^2}$ using the given substitution

$= \sqrt{a^2 - a^2\sin^2 x}$ Algebra

$= \sqrt{a^2(1 - \sin^2 x)}$ Algebra

$= \sqrt{a^2\cos^2 x}$ Pythagorean identity

$= |a||\cos x|$ Algebra

$= a\cos x$ since $a > 0$ and x is in quadrant I or IV $\left(\text{given } -\frac{\pi}{2} < x < \frac{\pi}{2}\right)$, thus, $\cos x > 0$.

71. $\sqrt{a^2 + u^2} = \sqrt{a^2 + (a\tan x)^2}$ using the given substitution

$= \sqrt{a^2 + a^2\tan^2 x}$ Algebra

$= \sqrt{a^2(1 + \tan^2 x)}$ Algebra

$= \sqrt{a^2\sec^2 x}$ Pythagorean identity

$= |a||\sec x|$ Algebra

$= a\sec x$ since $a > 0$ and x is in quadrant I $\left(\text{given } 0 < x < \frac{\pi}{2}\right)$, thus, $\sec x > 0$.

73. (A) Since $\frac{x}{5} = \cos t$ and $\frac{y}{4} = \sin t$, write

$$\left(\frac{x}{5}\right)^2 + \left(\frac{y}{4}\right)^2 = \cos^2 t + \sin^2 t$$

$$\frac{x^2}{25} + \frac{y^2}{16} = 1$$

(B) The graph is as shown. The orbit is elliptical.

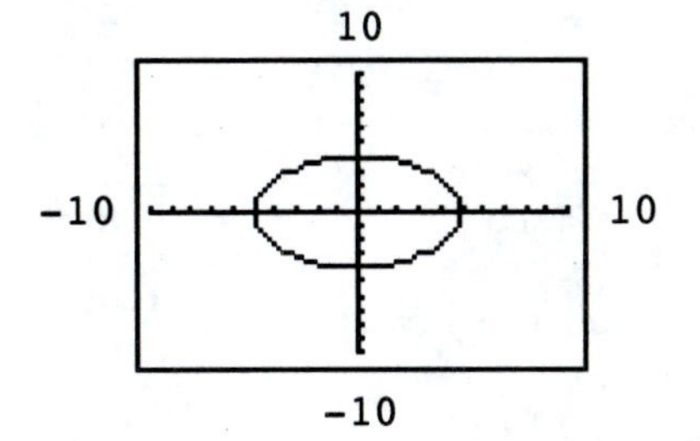

75. (A) Since $\frac{y}{-3} = \csc t$ and $\frac{x}{5} = \cot t$, write

$$\left(\frac{y}{-3}\right)^2 - \left(\frac{x}{5}\right)^2 = \csc^2 t - \cot^2 t$$

$$\frac{y^2}{9} - \frac{x^2}{25} = 1$$

(B) The graph is as shown. The orbit is hyperbolic.

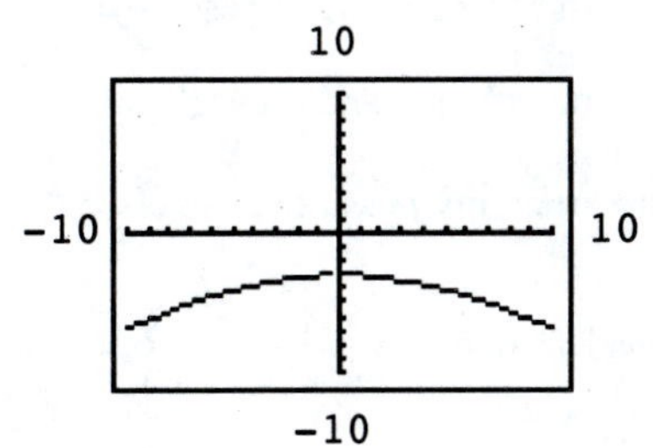

EXERCISE 4.2 Verifying Trigonometric Identities

1. To verify that an equation is an identity is to prove that the equation is satisfied by all values of the variables for which both sides of the equation are defined.

3. No, a calculator cannot prove that an equation is an identity. It is possible that the graph of the left-hand side of a conditional equation is indistinguishable from the graph of the right-hand side in certain viewing windows.

5. $\frac{\tan x}{\sec x} = \tan x \div \sec x$ Algebra

$= \frac{\sin x}{\cos x} \div \frac{1}{\cos x}$ Reciprocal and Quotient identities

$= \frac{\sin x}{\cos x} \cdot \frac{\cos x}{1}$ Algebra

$= \sin x$ Algebra

7. $\frac{\sin x}{\tan x} = \sin x \div \tan x$ Algebra

$= \sin x \div \frac{\sin x}{\cos x}$ Quotient identity

$= \sin x \cdot \frac{\cos x}{\sin x}$ Algebra

$= \cos x$ Algebra

9. $\frac{\csc x}{\cot x} = \csc x \div \cot x$ Algebra

$= \frac{1}{\sin x} \div \frac{\cos x}{\sin x}$ Reciprocal and Quotient identities

$= \frac{1}{\sin x} \cdot \frac{\sin x}{\cos x}$ Algebra

$= \frac{1}{\cos x}$ Algebra

$= \sec x$ Reciprocal identity

11. There are many possible answers. We choose $x = \frac{\pi}{4}$.

Then left side $= \frac{\cos x}{\tan x} = \frac{\cos\frac{\pi}{4}}{\tan\frac{\pi}{4}} = \frac{1}{\sqrt{2}}$

right side $= \csc x = \csc \frac{\pi}{4} = \sqrt{2}$

Since the left side and the right side are defined, but not equal, the equation is not an identity.

13. There are many possible answers. We choose $x = \frac{\pi}{3}$.

Then left side $= \frac{\csc x}{\sec x} = \frac{\csc\frac{\pi}{3}}{\sec\frac{\pi}{3}} = \frac{\frac{2}{\sqrt{3}}}{2} = \frac{1}{\sqrt{3}}$

right side $= \tan x = \tan \frac{\pi}{3} = \sqrt{3}$

15. $\tan x = \dfrac{\sin x}{\cos x}$ Quotient identity

$= \sin x \cdot \dfrac{1}{\cos x}$ Algebra

$= \sin x \sec x$ Reciprocal identity

17. $\csc(-x) = \dfrac{1}{\sin(-x)}$ Reciprocal identity

$= \dfrac{1}{-\sin x}$ Identities for negatives

$= -\dfrac{1}{\sin x}$ Algebra

$= -\csc x$ Reciprocal identity

19. $\dfrac{\sin\alpha}{\cos\alpha\tan\alpha} = \dfrac{\sin\alpha}{\cos\alpha\dfrac{\sin\alpha}{\cos\alpha}}$ Quotient identity

$= \dfrac{\sin\alpha}{\sin\alpha}$ Algebra

$= 1$ Algebra

21. $\dfrac{\cos\beta\sec\beta}{\tan\beta} = \dfrac{\cos\beta\dfrac{1}{\cos\beta}}{\tan\beta}$ Reciprocal identity

$= \dfrac{1}{\tan\beta}$ Algebra

$= \cot\beta$ Reciprocal identity

23. $\sec\theta(\sin\theta + \cos\theta) = \sec\theta\sin\theta + \sec\theta\cos\theta$ Algebra

$= \dfrac{1}{\cos\theta}\sin\theta + \dfrac{1}{\cos\theta}\cos\theta$ Reciprocal identity

$= \dfrac{\sin\theta}{\cos\theta} + 1$ Algebra

$= \tan\theta + 1$ Quotient identity

25. $\dfrac{\cos^2 t - \sin^2 t}{\sin t\cos t} = \dfrac{\cos^2 t}{\sin t\cos t} - \dfrac{\sin^2 t}{\sin t\cos t}$ Algebra

$= \dfrac{\cos t}{\sin t} - \dfrac{\sin t}{\cos t}$ Algebra

$= \cot t - \tan t$ Quotient identity

Key algebraic steps: $\dfrac{b^2 - a^2}{ab} = \dfrac{b^2}{ab} - \dfrac{a^2}{ab} = \dfrac{b}{a} - \dfrac{a}{b}$

27. $\dfrac{\cos\beta}{\cot\beta}+\dfrac{\sin\beta}{\tan\beta}=\cos\beta\div\cot\beta+\sin\beta\div\tan\beta$ Algebra

$=\cos\beta\div\dfrac{\cos\beta}{\sin\beta}+\sin\beta\div\dfrac{\sin\beta}{\cos\beta}$ Quotient identity

$=\dfrac{\cos\beta}{1}\cdot\dfrac{\sin\beta}{\cos\beta}+\dfrac{\sin\beta}{1}\cdot\dfrac{\cos\beta}{\sin\beta}$ Algebra

$=\sin\beta+\cos\beta$ Algebra

29. $\sec^2\theta-\tan^2\theta=\tan^2\theta+1-\tan^2\theta$ Pythagorean identity

$=1$ Algebra

31. $\sin^2 x(1+\cot^2 x)=\sin^2 x\csc^2 x$ Pythagorean identity

$=\sin^2 x\left(\dfrac{1}{\sin x}\right)^2$ Reciprocal identity

$=\dfrac{\sin^2 x}{1}\cdot\dfrac{1}{\sin^2 x}$ Algebra

$=1$ Algebra

33. $(\csc\alpha+1)(\csc\alpha-1)=\csc^2\alpha-1$ Algebra

$=1+\cot^2\alpha-1$ Pythagorean identity

$=\cot^2\alpha$ Algebra

Key algebraic step: $(x+1)(x-1)=x^2-1$

35. $\dfrac{\sin t}{\csc t}+\dfrac{\cos t}{\sec t}=\dfrac{\sin t}{1/\sin t}+\dfrac{\cos t}{1/\cos t}$ Reciprocal identities

$=\sin t\div\dfrac{1}{\sin t}+\cos t\div\dfrac{1}{\cos t}$ Algebra

$=\sin t\cdot\dfrac{\sin t}{1}+\cos t\cdot\dfrac{\cos t}{1}$ Algebra

$=\sin^2 t+\cos^2 t$ Algebra

$=1$ Pythagorean identity

37. Yes. The left side, $\dfrac{x^2+5x+6}{x+2}=\dfrac{(x+2)(x+3)}{x+2}=x+3$, the right side, for all values of x for which both are defined ($x\neq-2$).

39. No. For example, if $x=1$, $\dfrac{x^2}{x+1}=\dfrac{1^2}{1+1}=\dfrac{1}{2}$, but $x+x^2=1+1^2=2$.

41. No. For example, if $x=\dfrac{\pi}{4}$, $\sin^2 x+\csc^2 x=\left(\sin\dfrac{\pi}{4}\right)^2+\left(\csc\dfrac{\pi}{4}\right)^2=\left(\dfrac{1}{\sqrt{2}}\right)^2+(\sqrt{2})^2=\dfrac{5}{2}$.

43. No. For example, if $x=\dfrac{\pi}{4}$, $(\sin x+\cos x)^2=\left(\sin\dfrac{\pi}{4}+\cos\dfrac{\pi}{4}\right)^2=\left(\dfrac{1}{\sqrt{2}}+\dfrac{1}{\sqrt{2}}\right)^2=2$.

45. No. For example, if $x=\dfrac{\pi}{4}$, $\sin^4 x+\cos^4 x=\left(\sin\dfrac{\pi}{4}\right)^4+\left(\cos\dfrac{\pi}{4}\right)^4=\left(\dfrac{1}{\sqrt{2}}\right)^4+\left(\dfrac{1}{\sqrt{2}}\right)^4=\dfrac{1}{2}$.

47. Yes. The left side, $\cos(-x)\sec x = \cos x \sec x = \cos x \cdot \dfrac{1}{\cos x} = 1$, the right side, for all values of x for which both are defined.

49. $\dfrac{1-(\cos\theta-\sin\theta)^2}{\cos\theta} = \dfrac{1-(\cos^2\theta-2\sin\theta\cos\theta+\sin^2\theta)}{\cos\theta}$ Algebra

$= \dfrac{1-\cos^2\theta+2\sin\theta\cos\theta-\sin^2\theta}{\cos\theta}$ Algebra

$= \dfrac{\sin^2\theta+2\sin\theta\cos\theta-\sin^2\theta}{\cos\theta}$ Pythagorean identity

$= \dfrac{2\sin\theta\cos\theta}{\cos\theta}$ Algebra

$= 2\sin\theta$ Algebra

Key algebraic steps: $1-(b-a)^2 = 1-(b^2-2ab+a^2) = 1-b^2+2ab-a^2$

$$\frac{a^2+2ab-a^2}{b} = \frac{2ab}{b} = 2a$$

51. $\dfrac{\tan w+1}{\sec w} = \dfrac{\tan w}{\sec w} + \dfrac{1}{\sec w}$ Algebra

$= \dfrac{\sin w/\cos w}{1/\cos w} + \cos w$ Quotient and reciprocal identities

$= \dfrac{\sin w}{\cos w} \cdot \dfrac{\cos w}{1} + \cos w$ Algebra

$= \sin w + \cos w$ Algebra

Key algebraic steps: $\dfrac{a/b}{1/b} + b = \dfrac{a}{b} \div \dfrac{1}{b} + b = \dfrac{a}{b} \cdot \dfrac{b}{1} + b = a + b$

53. $\dfrac{1}{1-\cos^2\theta} = \dfrac{1}{\sin^2\theta}$ Pythagorean identity

$= \left(\dfrac{1}{\sin\theta}\right)^2$ Algebra

$= \csc^2\theta$ Reciprocal identity

$= 1+\cot^2\theta$ Pythagorean identity

55. $\dfrac{\sin^2\beta}{1-\cos\beta} = \dfrac{1-\cos^2\beta}{1-\cos\beta}$ Pythagorean identity

$= \dfrac{(1-\cos\beta)(1+\cos\beta)}{1-\cos\beta}$ Algebra

$= 1+\cos\beta$ Algebra

Key algebraic steps: $\dfrac{1-b^2}{1-b} = \dfrac{(1-b)(1+b)}{1-b} = 1+b$

57. $\dfrac{2-\cos^2\theta}{\sin\theta} = \dfrac{2-(1-\sin^2\theta)}{\sin\theta}$ Pythagorean identity

$= \dfrac{2-1+\sin^2\theta}{\sin\theta}$ Algebra

$= \dfrac{1+\sin^2\theta}{\sin\theta}$ Algebra

$= \dfrac{1}{\sin\theta} + \dfrac{\sin^2\theta}{\sin\theta}$ Algebra

$= \dfrac{1}{\sin\theta} + \sin\theta$ Algebra

$= \csc\theta + \sin\theta$ Reciprocal identity

Key algebraic steps: $\dfrac{2-(1-a^2)}{a} = \dfrac{2-1+a^2}{a} = \dfrac{1+a^2}{a} = \dfrac{1}{a} + \dfrac{a^2}{a} = \dfrac{1}{a} + a$

59. $\tan x + \cot x = \dfrac{\sin x}{\cos x} + \dfrac{\cos x}{\sin x}$ Quotient identity

$= \dfrac{\sin^2 x}{\sin x\cos x} + \dfrac{\cos^2 x}{\sin x\cos x}$ Algebra

$= \dfrac{\sin^2 x+\cos^2 x}{\sin x\cos x}$ Algebra

$= \dfrac{1}{\sin x\cos x}$ Pythagorean identity

$= \dfrac{1}{\sin x} \cdot \dfrac{1}{\cos x}$ Algebra

$= \sec x \csc x$ Reciprocal identities

Key algebraic steps: $\dfrac{a}{b} + \dfrac{b}{a} = \dfrac{a^2}{ab} + \dfrac{b^2}{ab} = \dfrac{a^2+b^2}{ab}$

61. $\dfrac{1-\csc x}{1+\csc x} = \dfrac{1-\dfrac{1}{\sin x}}{1+\dfrac{1}{\sin x}}$ Reciprocal identity

$= \dfrac{\sin x\cdot 1-\sin x\cdot\dfrac{1}{\sin x}}{\sin x\cdot 1+\sin x\cdot\dfrac{1}{\sin x}}$ Algebra

$= \dfrac{\sin x-1}{\sin x+1}$ Algebra

63. $\csc^2 \alpha - \cos^2 \alpha - \sin^2 \alpha = \csc^2 \alpha - (\cos^2 \alpha + \sin^2 \alpha)$ Algebra

$= \csc^2 \alpha - 1$ Pythagorean identity

$= \cot^2 \alpha$ Pythagorean identity

65. $(\sin x + \cos x)^2 - 1 = \sin^2 x + 2 \sin x \cos x + \cos^2 x - 1$ Algebra

$= 2 \sin x \cos x + \sin^2 x + \cos^2 x - 1$ Algebra

$= 2 \sin x \cos x + 1 - 1$ Pythagorean identity

$= 2 \sin x \cos x$ Algebra

67. $(\sin u - \cos u)^2 + (\sin u + \cos u)^2$

$= \sin^2 u - 2 \sin u \cos u + \cos^2 u + \sin^2 u + 2 \sin u \cos u + \cos^2 u$ Algebra

$= 2 \sin^2 u + 2 \cos^2 u$ Algebra

$= 2(\sin^2 u + \cos^2 u)$ Algebra

$= 2 \cdot 1$ or 2 Pythagorean identity

Key algebraic steps: $(a - b)^2 + (a + b)^2 = a^2 - 2ab + b^2 + a^2 + 2ab + b^2 = 2a^2 + 2b^2 = 2(a^2 + b^2)$

69. $\sin^4 x - \cos^4 x = (\sin^2 x)^2 - (\cos^2 x)^2$ Algebra

$= (\sin^2 x - \cos^2 x)(\sin^2 x + \cos^2 x)$ Algebra

$= (\sin^2 x - \cos^2 x)(1)$ Pythagorean identity

$= \sin^2 x - \cos^2 x$ Algebra

$= (1 - \cos^2 x) - \cos^2 x$ Pythagorean identity

$= 1 - 2 \cos^2 x$ Algebra

Key algebraic steps: $a^4 - b^4 = (a^2)^2 - (b^2)^2 = (a^2 - b^2)(a^2 + b^2)$

71. $\dfrac{\sin\alpha}{1 - \cos\alpha} - \dfrac{1 + \cos\alpha}{\sin\alpha} = \dfrac{\sin\alpha \cdot \sin\alpha}{(1 - \cos\alpha)\sin\alpha} - \dfrac{(1 - \cos\alpha)(1 + \cos\alpha)}{(1 - \cos\alpha)\sin\alpha}$ Algebra

$= \dfrac{\sin\alpha\sin\alpha - (1 - \cos\alpha)(1 + \cos\alpha)}{(1 - \cos\alpha)\sin\alpha}$ Algebra

$= \dfrac{\sin^2\alpha - (1 - \cos^2\alpha)}{(1 - \cos\alpha)\sin\alpha}$ Algebra

$= \dfrac{\sin^2\alpha - \sin^2\alpha}{(1 - \cos\alpha)\sin\alpha}$ Pythagorean identity

$= \dfrac{0}{(1 - \cos\alpha)\sin\alpha}$ Algebra

$= 0$ Algebra

Key algebraic steps: $\dfrac{a}{1-b} - \dfrac{1+b}{a} = \dfrac{a \cdot a}{a(1-b)} - \dfrac{(1-b)(1+b)}{a(1-b)} = \dfrac{a^2 - (1-b^2)}{a(1-b)}$

73.
$$\frac{\cos^2 n - 3\cos n + 2}{\sin^2 n} = \frac{(\cos n - 2)(\cos n - 1)}{\sin^2 n} \qquad \text{Algebra}$$
$$= \frac{(\cos n - 2)(\cos n - 1)}{1 - \cos^2 n} \qquad \text{Pythagorean identity}$$
$$= \frac{(\cos n - 2)(\cos n - 1)}{(1 - \cos n)(1 + \cos n)} \qquad \text{Algebra}$$
$$= \frac{(\cos n - 2)(-1)}{1 + \cos n} \qquad \text{Algebra}$$
$$= \frac{2 - \cos n}{1 + \cos n} \qquad \text{Algebra}$$

75.
$$\frac{1 - \cot^2 x}{\tan^2 x - 1} = \frac{1 - \cot^2 x}{\dfrac{1}{\cot^2 x} - 1} \qquad \text{Reciprocal identity}$$
$$= \frac{\cot^2 x(1 - \cot^2 x)}{\cot^2 x \cdot \dfrac{1}{\cot^2 x} - \cot^2 x \cdot 1} \qquad \text{Algebra}$$
$$= \frac{\cot^2 x(1 - \cot^2 x)}{1 - \cot^2 x} \qquad \text{Algebra}$$
$$= \cot^2 x \qquad \text{Algebra}$$

77.
$$\sec^2 x + \csc^2 x = \frac{1}{\cos^2 x} + \frac{1}{\sin^2 x} \qquad \text{Reciprocal identities}$$
$$= \frac{\sin^2 x}{\cos^2 x \sin^2 x} + \frac{\cos^2 x}{\cos^2 x \sin^2 x} \qquad \text{Algebra}$$
$$= \frac{\sin^2 x + \cos^2 x}{\cos^2 x \sin^2 x} \qquad \text{Algebra}$$
$$= \frac{1}{\cos^2 x \sin^2 x} \qquad \text{Pythagorean identity}$$
$$= \frac{1}{\cos^2 x} \cdot \frac{1}{\sin^2 x} \qquad \text{Algebra}$$
$$= \sec^2 x \csc^2 x \qquad \text{Reciprocal identities}$$

79. $\dfrac{1+\sin t}{\cos t} = \dfrac{(1+\sin t)\cos t}{\cos t \cos t}$ Algebra

$= \dfrac{(1+\sin t)\cos t}{\cos^2 t}$ Algebra

$= \dfrac{(1+\sin t)\cos t}{1-\sin^2 t}$ Pythagorean identity

$= \dfrac{(1+\sin t)\cos t}{(1+\sin t)(1-\sin t)}$ Algebra

$= \dfrac{\cos t}{1-\sin t}$ Algebra

81. (A)

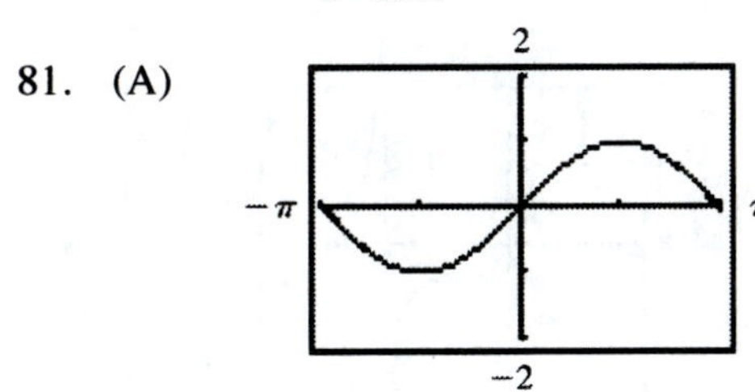

The graphs appear almost identical, but if the trace feature of the calculator is used to move from one graph to the other, different y values will arise for the same value of x. Although close, the graphs are not the same, and the equation is not an identity over the interval.

(B)

2
-2π 2π
-2

Outside the interval $[-\pi, \pi]$ the graphs differ widely.

83. Graph both sides of the equation in the same viewing window.

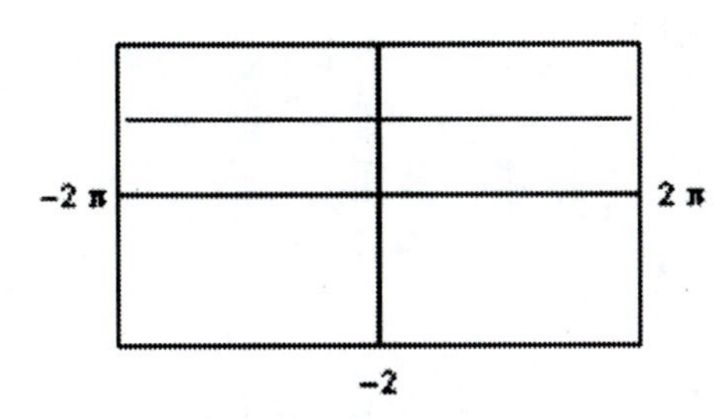

$\dfrac{\cos x}{\sin(-x)\cot(-x)} = 1$ appears to be an identity, which we verify.

$\dfrac{\cos x}{\sin(-x)\cot(-x)} = \dfrac{\cos x}{\sin(-x)\dfrac{\cos(-x)}{\sin(-x)}}$ Quotient identity

$= \dfrac{\cos x}{\cos(-x)}$ Algebra

$= \dfrac{\cos x}{\cos x}$ Identities for negatives

$= 1$ Algebra

85. Graph both sides of the equation in the same viewing window.

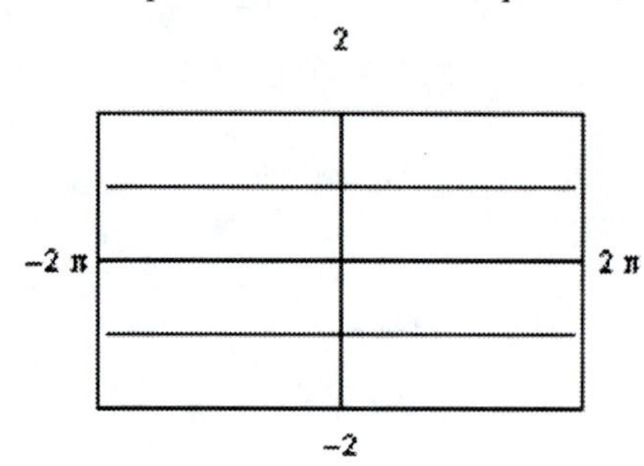

$\dfrac{\cos(-x)}{\sin x\cot(-x)} = 1$ is not an identity, since the graphs do not match. Try $x = \dfrac{\pi}{4}$.

Left side: $\dfrac{\cos(-\pi/4)}{\sin(\pi/4)\cot(-\pi/4)} = \dfrac{1/\sqrt{2}}{(1/\sqrt{2})(-1)} = -1$

Right side: 1

This verifies that the equation is not an identity.

87. Graph both sides of the equation in the same viewing window.

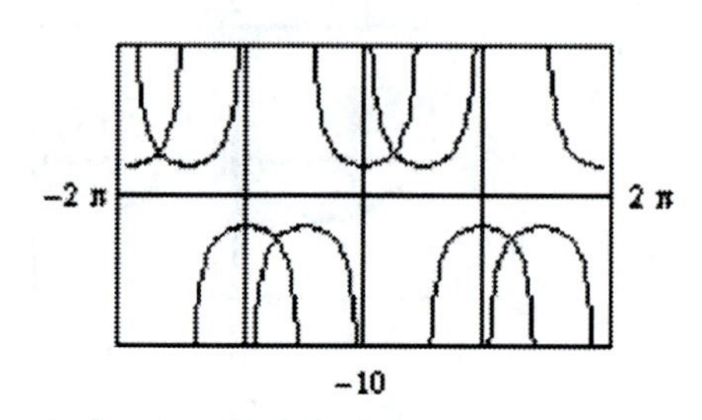

$\dfrac{\cos x}{\sin x+1} - \dfrac{\cos x}{\sin x-1} = 2\csc x$

is not an identity, since the graphs do not match.

Try $x = -\dfrac{\pi}{4}$.

Left side: $\dfrac{\cos(-\pi/4)}{\sin(-\pi/4)+1} - \dfrac{\cos(-\pi/4)}{\sin(-\pi/4)-1} = \dfrac{1/\sqrt{2}}{(-1/\sqrt{2})+1} - \dfrac{1/\sqrt{2}}{(-1/\sqrt{2})-1} = \dfrac{1}{-1+\sqrt{2}} - \dfrac{1}{-1-\sqrt{2}}$

$$= \frac{(-1-\sqrt{2})-(-1+\sqrt{2})}{(-1+\sqrt{2})(-1-\sqrt{2})} = \frac{-2\sqrt{2}}{-1} = 2\sqrt{2}$$

Right side: $2\csc\left(-\dfrac{\pi}{4}\right) = 2(-\sqrt{2}) = -2\sqrt{2}$. This verifies that the equation is not an identity.

89. Graph both sides of the equation in the same viewing window.

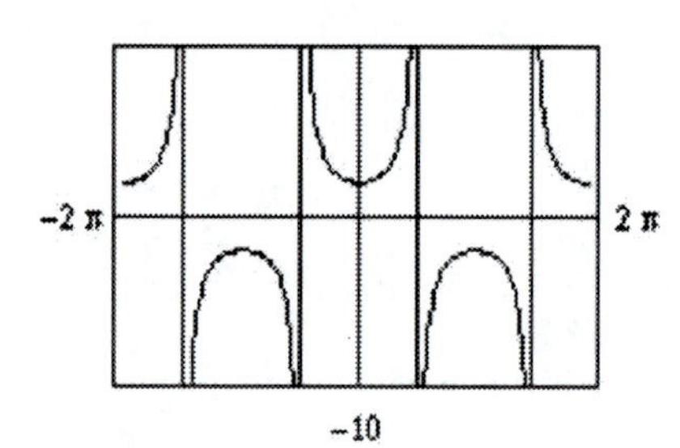

$\dfrac{\cos x}{1-\sin x} + \dfrac{\cos x}{1+\sin x} = 2\sec x$ appears to be an identity, which we verify.

$$\frac{\cos x}{1-\sin x} + \frac{\cos x}{1+\sin x} = \frac{\cos x(1+\sin x)}{(1-\sin x)(1+\sin x)} + \frac{\cos x(1-\sin x)}{(1+\sin x)(1-\sin x)} \qquad \text{Algebra}$$

$$= \frac{\cos x(1+\sin x)+\cos x(1-\sin x)}{(1-\sin x)(1+\sin x)} \qquad \text{Algebra}$$

$$= \frac{\cos x+\cos x\sin x+\cos x-\cos x\sin x}{1-\sin^2 x} \qquad \text{Algebra}$$

$= \dfrac{2\cos x}{1-\sin^2 x}$ Algebra

$= \dfrac{2\cos x}{\cos^2 x}$ Pythagorean identity

$= \dfrac{2}{\cos x}$ Algebra

$= 2\sec x$ Reciprocal identity

91. $\dfrac{\sin x}{1-\cos x} - \cot x = \dfrac{\sin x}{1-\cos x} - \dfrac{\cos x}{\sin x}$ Quotient identity

$= \dfrac{(\sin x)(\sin x)}{\sin x(1-\cos x)} - \dfrac{(1-\cos x)(\cos x)}{(1-\cos x)(\sin x)}$ Algebra

$= \dfrac{(\sin x)(\sin x) - (1-\cos x)(\cos x)}{(1-\cos x)\sin x}$ Algebra

$= \dfrac{\sin^2 x - \cos x + \cos^2 x}{(1-\cos x)\sin x}$ Algebra

$= \dfrac{\sin^2 x + \cos^2 x - \cos x}{(1-\cos x)\sin x}$ Algebra

$= \dfrac{1-\cos x}{(1-\cos x)\sin x}$ Pythagorean identity

$= \dfrac{1}{\sin x}$ Algebra

$= \csc x$ Reciprocal identity

Key algebraic steps: $\dfrac{a}{1-b} - \dfrac{b}{a} = \dfrac{aa}{a(1-b)} - \dfrac{b(1-b)}{a(1-b)} = \dfrac{a^2 - b(1-b)}{a(1-b)}$

$= \dfrac{a^2 - b + b^2}{a(1-b)} = \dfrac{a^2 + b^2 - b}{a(1-b)}$

93. $\dfrac{\cot\beta}{\csc\beta + 1} = \dfrac{\cot\beta(\csc\beta - 1)}{(\csc\beta + 1)(\csc\beta - 1)}$ Algebra

$= \dfrac{\cot\beta(\csc\beta - 1)}{\csc^2\beta - 1}$ Algebra

$= \dfrac{\cot\beta(\csc\beta - 1)}{1 + \cot^2\beta - 1}$ Pythagorean identity

$= \dfrac{\cot\beta(\csc\beta - 1)}{\cot^2\beta}$ Algebra

$= \dfrac{\csc\beta - 1}{\cot\beta}$ Algebra

95.

$$\frac{3\cos^2 m + 5\sin m - 5}{\cos^2 m} = \frac{3(1 - \sin^2 m) + 5\sin m - 5}{1 - \sin^2 m} \quad \text{Pythagorean identity}$$

$$= \frac{3 - 3\sin^2 m + 5\sin m - 5}{1 - \sin^2 m} \quad \text{Algebra}$$

$$= \frac{-3\sin^2 m + 5\sin m - 2}{1 - \sin^2 m} \quad \text{Algebra}$$

$$= \frac{(-\sin m + 1)(3\sin m - 2)}{(1 - \sin m)(1 + \sin m)} \quad \text{Algebra}$$

$$= \frac{(1 - \sin m)(3\sin m - 2)}{(1 - \sin m)(1 + \sin m)} \quad \text{Algebra}$$

$$= \frac{3\sin m - 2}{1 + \sin m} \quad \text{Algebra}$$

97. In this problem, it is more straightforward to start with the right-hand side of the identity to be verified. The student can confirm that the steps would be valid if reversed.

$$\frac{\tan x + \tan y}{1 - \tan x \tan y} = \frac{\frac{\sin x}{\cos x} + \frac{\sin y}{\cos y}}{1 - \frac{\sin x}{\cos x}\frac{\sin y}{\cos y}} \quad \text{Quotient identity}$$

$$= \frac{\cos x \cos y\left(\frac{\sin x}{\cos x} + \frac{\sin y}{\cos y}\right)}{\cos x \cos y\left(1 - \frac{\sin x \sin y}{\cos x \cos y}\right)} \quad \text{Algebra}$$

$$= \frac{\cos x \cos y \frac{\sin x}{\cos x} + \cos x \cos y \frac{\sin y}{\cos y}}{\cos x \cos y - \cos x \cos y \frac{\sin x \sin y}{\cos x \cos y}} \quad \text{Algebra}$$

$$= \frac{\sin x \cos y + \cos x \sin y}{\cos x \cos y - \sin x \sin y} \quad \text{Algebra}$$

EXERCISE 4.3 Sum, Difference, and Cofunction Identities

1. To show that an equation in two variables is not an identity, it is sufficient to find one value of x and one value of y for which the left and right sides are defined but not equal.

3. Not necessarily. It is necessary that the left and right sides are equal for every value of x and every value of y for which both sides are defined, not just values for which $x = y$. See Problem 5 below.

5. There are many possible answers. We choose $x = 1, y = -1$.
Then left side $= (x - y)^2 = [1 - (-1)]^2 = 4$
right side $= x^2 - y^2 = 1^2 - (-1)^2 = 0$
Since the left side and the right side are defined, but not equal, the equation is not an identity.

7. There are many possible answers. We choose $x = \frac{\pi}{2}, y = \frac{\pi}{2}$.

   Then left side $= \sin(x + y) = \sin\left(\frac{\pi}{2} + \frac{\pi}{2}\right) = 0$

   right side $= \sin x + \sin y = \sin \frac{\pi}{2} + \sin \frac{\pi}{2} = 2$

   Since the left side and the right side are defined, but not equal, the equation is not an identity.

9. There are many possible answers. We choose $x = \frac{\pi}{6}, y = \frac{\pi}{6}$.

   Then left side $= \tan(x + y) = \tan\left(\frac{\pi}{6} + \frac{\pi}{6}\right) = \sqrt{3}$

   right side $= \tan x + \tan y = \tan \frac{\pi}{6} + \tan \frac{\pi}{6} = \frac{2}{\sqrt{3}}$

   Since the left side and the right side are defined, but not equal, the equation is not an identity.

11. There are many possible answers. We choose $x = \frac{\pi}{4}, y = -\frac{\pi}{4}$.

   Then left side $= \csc(x - y) = \csc\left[\frac{\pi}{4} - \left(-\frac{\pi}{4}\right)\right] = 1$

   right side $= \csc x - \csc y = \csc \frac{\pi}{4} - \csc\left(-\frac{\pi}{4}\right) = 2\sqrt{2}$

   Since the left side and the right side are defined, but not equal, the equation is not an identity.

13. $\tan\left(\frac{\pi}{2} - 2\right) = \frac{\sin\left(\frac{\pi}{2} - x\right)}{\cos\left(\frac{\pi}{2} - x\right)}$ Quotient identity

   $= \frac{\cos x}{\sin x}$ Cofunction identities

   $= \cot x$ Quotient identity

15. $\sec\left(\frac{\pi}{2} - x\right) = \frac{1}{\cos\left(\frac{\pi}{2} - x\right)}$ Reciprocal identity

   $= \frac{1}{\sin x}$ Cofunction identity

   $= \csc x$ Reciprocal identity

17. Use the sum identity for cosine, replacing y with 30°.

   $\sin(x + y) = \sin x \cos y + \cos x \sin y$

   $\sin(x + 30°) = \sin x \cos 30° + \cos x \sin 30° = \sin x \frac{\sqrt{3}}{2} + \cos x \frac{1}{2} = \frac{1}{2}(\sqrt{3} \sin x + \cos x)$

19. Use the sum identity for cosine, replacing y with 60°.

   $\cos(x + y) = \cos x \cos y - \sin x \sin y$

   $\cos(x + 60°) = \cos x \cos 60° - \sin x \sin 60° = \cos x \frac{1}{2} - \sin x \frac{\sqrt{3}}{2} = \frac{1}{2}(\cos x - \sqrt{3} \sin x)$

21. Use the sum identity for tangent, replacing y with $\frac{\pi}{4}$.

$$\tan(x+y) = \frac{\tan x + \tan y}{1 - \tan x \tan y} = \frac{\tan x + \tan\frac{\pi}{4}}{1 - \tan x \tan\frac{\pi}{4}} = \frac{\tan x + 1}{1 - \tan x(1)} = \frac{\tan x + 1}{1 - \tan x}$$

23. Since $\cot a = \frac{1}{\tan a}$, write $\cot\left(\frac{\pi}{6} - x\right) = \frac{1}{\tan\left(\frac{\pi}{6} - x\right)}$ and use the difference identity for tangent, replacing x with $\frac{\pi}{6}$ and y with x.

$$\cot\left(\frac{\pi}{6} - x\right) = \frac{1}{\tan\left(\frac{\pi}{6} - x\right)} = \frac{1}{\frac{\tan\frac{\pi}{6} - \tan x}{1 + \tan\frac{\pi}{6}\tan x}} = \frac{1 + \tan\frac{\pi}{6}\tan x}{\tan\frac{\pi}{6} - \tan x}$$

$$= \frac{1 + \frac{1}{\sqrt{3}}\tan x}{\frac{1}{\sqrt{3}} - \tan x} = \frac{\sqrt{3} + \tan x}{1 - \sqrt{3}\tan x}$$

25. Since we can write $15° = 45° - 30°$, the difference of two special angles, we can use the difference identity for tangent, with $x = 45°$ and $y = 30°$.

$$\tan(x-y) = \frac{\tan x - \tan y}{1 + \tan x \tan y}$$

$$\tan 15° = \tan(45° - 30°)$$

$$= \frac{\tan 45° - \tan 30°}{1 + \tan 45° \tan 30°}$$

$$= \frac{1 - \frac{1}{\sqrt{3}}}{1 + 1\left(\frac{1}{\sqrt{3}}\right)}$$

$$= \frac{\sqrt{3} - 1}{\sqrt{3} + 1}$$

27. Since we can write $90° = 102° - 12°$, we can use the difference identity for cosine, with $x = 102°$ and $y = 12°$.

$$\cos x \cos y + \sin x \sin y = \cos(x-y)$$

$$\cos 102° \cos 12° + \sin 102° \sin 12° = \cos(102° - 12°) = \cos 90° = 0$$

29. Since we can write $30° = 50° - 20°$, we can use the difference identity for sine with $x = 50°$ and $y = 20°$.

$$\sin x \cos y - \cos x \sin y = \sin(x-y)$$

$$\sin 50° \cos 20° - \cos 50° \sin 20° = \sin(50° - 20°) = \sin 30° = \frac{1}{2}$$

31. Since we can write 60° = 75° – 15°, we can use the difference identity for tangent, with $x = 75°$ and $y = 15°$.

$$\frac{\tan x - \tan y}{1 + \tan x \tan y} = \tan(x - y)$$

$$\frac{\tan 75° - \tan 15°}{1 + \tan 75° \tan 15°} = \tan(75° - 15°) = \tan 60° = \sqrt{3}$$

33. Since we can write $\frac{5\pi}{12} = \frac{\pi}{6} + \frac{\pi}{4}$, the sum of two special angles, we can use the sum identity for cosine, with $x = \frac{\pi}{6}$ and $y = \frac{\pi}{4}$.

$$\cos(x + y) = \cos x \cos y - \sin x \sin y$$

$$\cos \frac{5\pi}{12} = \cos\left(\frac{\pi}{6} + \frac{\pi}{4}\right)$$

$$= \cos \frac{\pi}{6} \cos \frac{\pi}{4} - \sin \frac{\pi}{6} \sin \frac{\pi}{4}$$

$$= \frac{\sqrt{3}}{2} \cdot \frac{1}{\sqrt{2}} - \frac{1}{2} \cdot \frac{1}{\sqrt{2}}$$

$$= \frac{\sqrt{3} - 1}{2\sqrt{2}}$$

35. Since we can write $\frac{\pi}{4} = \frac{6\pi}{24} = \frac{\pi}{24} + \frac{5\pi}{24}$, we can use the sum identity for tangent with $x = \frac{\pi}{24}$ and $y = \frac{5\pi}{24}$.

$$\frac{\tan x + \tan y}{1 - \tan x \tan y} = \tan(x + y)$$

$$\frac{\tan \frac{\pi}{24} + \tan \frac{5\pi}{24}}{1 - \tan \frac{\pi}{24} \tan \frac{5\pi}{24}} = \tan\left(\frac{\pi}{24} + \frac{5\pi}{24}\right) = \tan \frac{\pi}{4} = 1$$

37. Since we can write $\pi = \frac{12\pi}{12} = \frac{5\pi}{12} + \frac{7\pi}{12}$, we can use the sum identity for sine with $x = \frac{5\pi}{12}$ and $y = \frac{7\pi}{12}$.

$$\sin x \cos y + \cos x \sin y = \sin(x + y)$$

$$\sin \frac{5\pi}{12} \cos \frac{7\pi}{12} + \cos \frac{5\pi}{12} \sin \frac{7\pi}{12} = \sin\left(\frac{5\pi}{12} + \frac{7\pi}{12}\right) = \sin \pi = 0$$

39. Since we can write $\frac{\pi}{3} = \frac{8\pi}{24} = \frac{11\pi}{24} - \frac{3\pi}{24} = \frac{11\pi}{24} - \frac{\pi}{8}$, we can use the difference identity for sine with $x = \frac{11\pi}{24}$ and $y = \frac{\pi}{8}$.

$$\sin x \cos y - \cos x \sin y = \sin(x - y)$$

$$\sin \frac{11\pi}{24} \cos \frac{\pi}{8} - \cos \frac{11\pi}{24} \sin \frac{\pi}{8} = \sin\left(\frac{11\pi}{24} - \frac{\pi}{8}\right) = \sin \frac{\pi}{3} = \frac{\sqrt{3}}{2}$$

41. To find $\sin(x + y)$, we start with the sum identity for sine:

$$\sin(x + y) = \sin x \cos y + \cos x \sin y.$$

We know $\sin x$ and $\cos y$, but not $\cos x$ and $\sin y$. We find the latter two values by using reference triangles and the Pythagorean theorem:

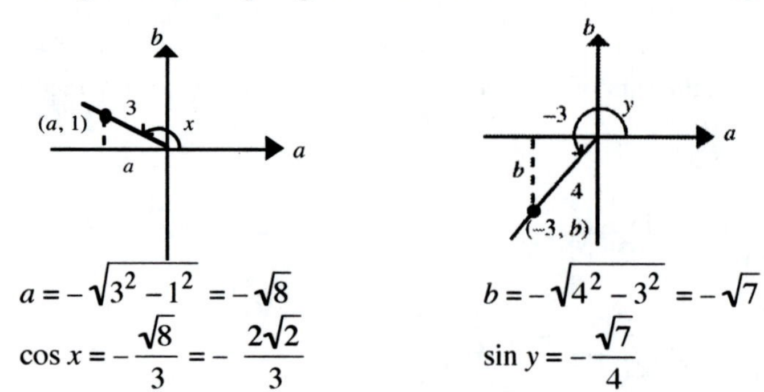

$$a = -\sqrt{3^2 - 1^2} = -\sqrt{8} \qquad b = -\sqrt{4^2 - 3^2} = -\sqrt{7}$$

$$\cos x = -\frac{\sqrt{8}}{3} = -\frac{2\sqrt{2}}{3} \qquad \sin y = -\frac{\sqrt{7}}{4}$$

Thus,
$$\begin{aligned}\sin(x + y) &= \sin x \cos y + \cos x \sin y \\ &= \frac{1}{3}\left(-\frac{3}{4}\right) + \left(-\frac{2\sqrt{2}}{3}\right)\left(-\frac{\sqrt{7}}{4}\right) \\ &= -\frac{1}{4} + \frac{\sqrt{14}}{6}\end{aligned}$$

43. To find $\cos(x - y)$, we start with the difference identity for cosine:

$$\cos(x - y) = \cos x \cos y + \sin x \sin y.$$

To find the four values that we need, we use reference triangles and the Pythagorean theorem.

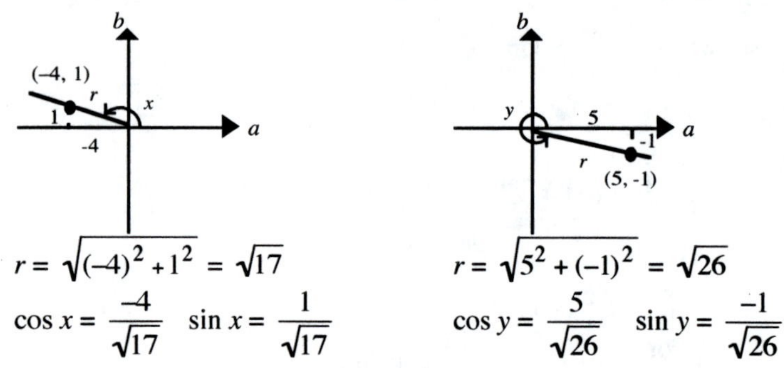

$$r = \sqrt{(-4)^2 + 1^2} = \sqrt{17} \qquad r = \sqrt{5^2 + (-1)^2} = \sqrt{26}$$

$$\cos x = \frac{-4}{\sqrt{17}} \quad \sin x = \frac{1}{\sqrt{17}} \qquad \cos y = \frac{5}{\sqrt{26}} \quad \sin y = \frac{-1}{\sqrt{26}}$$

Thus,
$$\begin{aligned}\cos(x - y) &= \cos x \cos y + \sin x \sin y \\ &= \left(\frac{-4}{\sqrt{17}}\right)\left(\frac{5}{\sqrt{26}}\right) + \left(\frac{1}{\sqrt{17}}\right)\left(\frac{-1}{\sqrt{26}}\right) \\ &= \frac{-20}{\sqrt{442}} + \frac{-1}{\sqrt{442}} \\ &= \frac{-21}{\sqrt{442}}\end{aligned}$$

45. To find $\tan(x + y)$, we start with the sum identity for tangent:

$$\tan(x + y) = \frac{\tan x + \tan y}{1 - \tan x \tan y}$$

To find $\tan x$ and $\tan y$, we use the given values and fundamental identities.

Find cos x: Since $\cos^2 x + \sin^2 x = 1$, $\cos x = \pm\sqrt{1 - \sin^2 x}$. Since $\cos x < 0$,

$$\cos x = -\sqrt{1 - \sin^2 x} = -\sqrt{1 - \left(-\frac{1}{4}\right)^2} = -\sqrt{\frac{15}{16}} = -\frac{\sqrt{15}}{4}$$

Find tan x: $\tan x = \dfrac{\sin x}{\cos x} = \dfrac{-1/4}{-\sqrt{15}/4} = \dfrac{1}{\sqrt{15}}$

Find sin y: Since $\cos^2 y + \sin^2 y = 1$, $\sin y = \pm\sqrt{1 - \cos^2 y}$. Since $\sin y < 0$,

$$\sin y = -\sqrt{1 - \cos^2 y} = -\sqrt{1 - \left(-\frac{1}{3}\right)^2} = -\sqrt{\frac{8}{9}} = -\frac{\sqrt{8}}{3}$$

Find tan y: $\tan y = \dfrac{\sin y}{\cos y} = \dfrac{-\sqrt{8}/3}{-1/3} = \sqrt{8}$

Thus,
$$\tan(x + y) = \frac{\tan x + \tan y}{1 - \tan x \tan y} = \frac{\frac{1}{\sqrt{15}} + \sqrt{8}}{1 - \frac{1}{\sqrt{15}}\sqrt{8}} = \frac{1 + \sqrt{120}}{\sqrt{15} - \sqrt{8}}$$

47. No. For example, if $x = \dfrac{\pi}{4}$, $\tan(x + \pi) = \tan\left(\dfrac{\pi}{4} + \pi\right) = \tan\dfrac{5\pi}{4} = 1$, but $-\tan x = -\tan\dfrac{\pi}{4} = -1$.

49. Yes. The left side, $\cos\left(x - \dfrac{\pi}{2}\right) = \cos x \cos\dfrac{\pi}{2} + \sin x \sin\dfrac{\pi}{2} = \cos x(0) + \sin x(1) = \sin x$, the right side, for all values of x.

51. No. For example, if $x = \dfrac{3\pi}{2}$, $\sin(2\pi - x) = \sin\left(2\pi - \dfrac{3\pi}{2}\right) = \sin\dfrac{\pi}{2} = 1$, but $\sin x = \sin\dfrac{3\pi}{2} = -1$.

53. Yes. The left side, $\csc(\pi - x) = \dfrac{1}{\sin(\pi - x)} = \dfrac{1}{\sin\pi\cos x - \cos\pi\sin x} = \dfrac{1}{(0)\cos x - (-1)\sin x} = \dfrac{1}{\sin x}$ $= \csc x$, the right side, for all values of x for which both are defined.

55.

$\sin 2x = \sin(x + x)$	Algebra
$= \sin x \cos x + \cos x \sin x$	Sum identity for sine
$= \sin x \cos x + \sin x \cos x$	Algebra
$= 2 \sin x \cos x$	Algebra

57. $\cot(x-y) = \dfrac{\cos(x-y)}{\sin(x-y)}$ Quotient identity

$= \dfrac{\cos x \cos y + \sin x \sin y}{\sin x \cos y - \cos x \sin y}$ Difference identities for sine and cosine

$= \dfrac{\dfrac{\cos x \cos y}{\sin x \sin y} + \dfrac{\sin x \sin y}{\sin x \sin y}}{\dfrac{\sin x \cos y}{\sin x \sin y} - \dfrac{\cos x \sin y}{\sin x \sin y}}$ Algebra

$= \dfrac{\dfrac{\cos x}{\sin x}\dfrac{\cos y}{\sin y} + 1}{\dfrac{\cos y}{\sin y} - \dfrac{\cos x}{\sin x}}$ Algebra

$= \dfrac{\cot x \cot y + 1}{\cot y - \cot x}$ Quotient identity

59. $\cot 2x = \dfrac{\cos 2x}{\sin 2x}$ Quotient identity

$= \dfrac{\cos(x+x)}{\sin(x+x)}$ Algebra

$= \dfrac{\cos x \cos x - \sin x \sin x}{\sin x \cos x + \cos x \sin x}$ Sum identities for sine and cosine

$= \dfrac{\cos x \cos x - \sin x \sin x}{2 \sin x \cos x}$ Algebra

$= \dfrac{\dfrac{\cos x \cos x}{\sin x \sin x} - \dfrac{\sin x \sin x}{\sin x \sin x}}{\dfrac{2 \sin x \cos x}{\sin x \sin x}}$ Algebra

$= \dfrac{\dfrac{\cos x}{\sin x}\dfrac{\cos x}{\sin x} - 1}{2\dfrac{\cos x}{\sin x}}$ Algebra

$= \dfrac{\cot x \cot x - 1}{2 \cot x}$ Quotient identity

$= \dfrac{\cot^2 x - 1}{2 \cot x}$ Algebra

61. $\dfrac{\tan\alpha+\tan\beta}{\tan\alpha-\tan\beta} = \dfrac{\dfrac{\sin\alpha}{\cos\alpha}+\dfrac{\sin\beta}{\cos\beta}}{\dfrac{\sin\alpha}{\cos\alpha}-\dfrac{\sin\beta}{\cos\beta}}$ Quotient identity

$= \dfrac{\dfrac{\cos\alpha\cos\beta}{1}\cdot\dfrac{\sin\alpha}{\cos\alpha}+\dfrac{\cos\alpha\cos\beta}{1}\cdot\dfrac{\sin\beta}{\cos\beta}}{\dfrac{\cos\alpha\cos\beta}{1}\cdot\dfrac{\sin\alpha}{\cos\alpha}-\dfrac{\cos\alpha\cos\beta}{1}\cdot\dfrac{\sin\beta}{\cos\beta}}$ Algebra

$= \dfrac{\sin\alpha\cos\beta+\cos\alpha\sin\beta}{\sin\alpha\cos\beta-\cos\alpha\sin\beta}$ Algebra

$= \dfrac{\sin(\alpha+\beta)}{\sin(\alpha-\beta)}$ Sum and difference identities for sine

63. $\dfrac{\sin(x-y)}{\cos x\cos y} = \dfrac{\sin x\cos y-\cos x\sin y}{\cos x\cos y}$ Difference identity for sine

$= \dfrac{\sin x\cos y}{\cos x\cos y}-\dfrac{\cos x\sin y}{\cos x\cos y}$ Algebra

$= \dfrac{\sin x}{\cos x}-\dfrac{\sin y}{\cos y}$ Algebra

$= \tan x-\tan y$ Quotient identities

65. $\tan(x+y) = \dfrac{\sin(x+y)}{\cos(x+y)}$ Quotient identity

$= \dfrac{\sin x\cos y+\cos x\sin y}{\cos x\cos y-\sin x\sin y}$ Sum identities for sine and cosine

$= \dfrac{\dfrac{\sin x\cos y}{\sin x\sin y}+\dfrac{\cos x\sin y}{\sin x\sin y}}{\dfrac{\cos x\cos y}{\sin x\sin y}-\dfrac{\sin x\sin y}{\sin x\sin y}}$ Algebra

$= \dfrac{\dfrac{\cos y}{\sin y}+\dfrac{\cos x}{\sin x}}{\dfrac{\cos x}{\sin x}\,\dfrac{\cos y}{\sin y}-1}$ Algebra

$= \dfrac{\cot y+\cot x}{\cot y\cot x-1}$ Quotient identities

$= \dfrac{\cot x+\cot y}{\cot x\cot y-1}$ Algebra

67. $\dfrac{\sin(x+h)-\sin x}{h} = \dfrac{\sin x\cos h + \cos x\sin h - \sin x}{h}$ Sum identity for sine

$= \dfrac{\sin x\cos h - \sin x + \cos x\sin h}{h}$ Algebra

$= \dfrac{\sin x(\cos h - 1) + \cos x\sin h}{h}$ Algebra

$= \sin x\ \dfrac{\cos h - 1}{h} + \cos x\ \dfrac{\sin h}{h}$ Algebra

69. Graph y1 = $\sin(x-2)$ and y2 = $\sin x - \sin 2$ in the same viewing window and observe that the graphs are not the same:

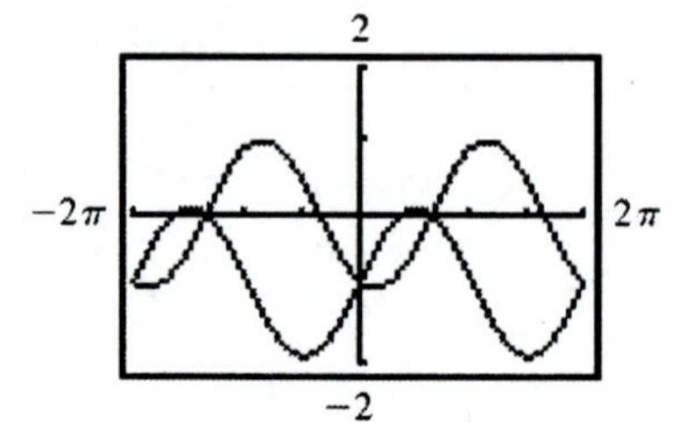

71. $\cos\left(x+\dfrac{5\pi}{6}\right) = \cos x\cos\dfrac{5\pi}{6} - \sin x\sin\dfrac{5\pi}{6}$ Sum identity for cosine

$= \cos x\left(-\dfrac{\sqrt{3}}{2}\right) - \sin x\left(\dfrac{1}{2}\right)$ Known values

$= -\dfrac{\sqrt{3}}{2}\cos x - \dfrac{1}{2}\sin x$

Graph y1 = $\cos\left(x+\dfrac{5\pi}{6}\right)$ and

y2 = $-\dfrac{\sqrt{3}}{2}\cos x - \dfrac{1}{2}\sin x$.

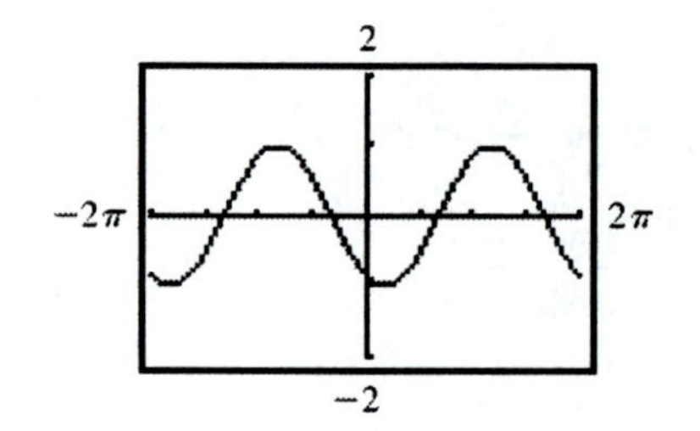

73. $\tan\left(x-\dfrac{\pi}{4}\right) = \dfrac{\tan x - \tan\dfrac{\pi}{4}}{1+\tan x\tan\dfrac{\pi}{4}}$ Difference identity for tangent

$= \dfrac{\tan x - 1}{1+\tan x\cdot 1}$ Known values

$= \dfrac{\tan x - 1}{1+\tan x}$

Graph y1 = $\tan\left(x-\dfrac{\pi}{4}\right)$ and y2 = $\dfrac{\tan x - 1}{1+\tan x}$.

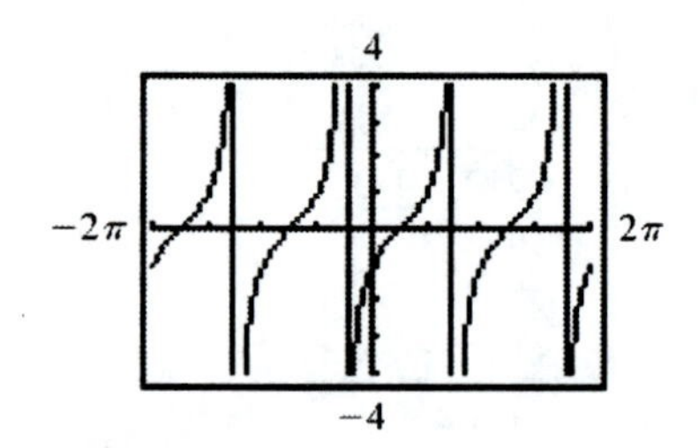

75. $\sin 3x \cos x - \cos 3x \sin x = \sin(3x - x)$ — Difference identity for sine

$= \sin 2x$ — Algebra

Graph $y1 = \sin 3x \cos x - \cos 3x \sin x$ and $y2 = \sin 2x$.

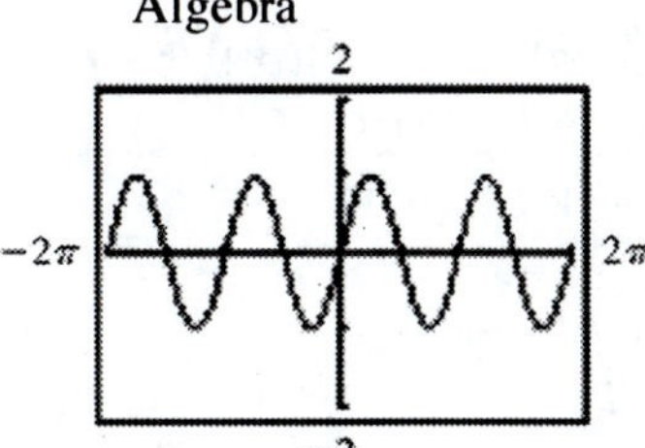

77. $\sin \dfrac{\pi x}{4} \cos \dfrac{3\pi x}{4} + \cos \dfrac{\pi x}{4} \sin \dfrac{3\pi x}{4} = \sin\left(\dfrac{\pi x}{4} + \dfrac{3\pi x}{4}\right)$ — Sum identity for sine

$= \sin \pi x$ — Algebra

Graph $y1 = \sin \dfrac{\pi x}{4} \cos \dfrac{3\pi x}{4} + \cos \dfrac{\pi x}{4} \sin \dfrac{3\pi x}{4}$ and $y2 = \sin \pi x$.

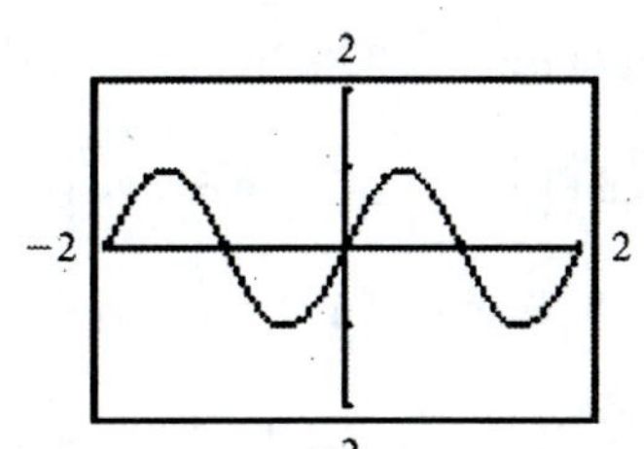

79. $\sin(x + y + z) = \sin[(x + y) + z]$ — Algebra

$= \sin(x + y) \cos z + \cos(x + y) \sin z$ — Sum identity for sine

$= (\sin x \cos y + \cos x \sin y) \cos z + (\cos x \cos y - \sin x \sin y) \sin z$ — Sum identities for sine and cosine

$= \sin x \cos y \cos z + \cos x \sin y \cos z + \cos x \cos y \sin z - \sin x \sin y \sin z$ — Algebra

81. $\tan(\theta_2 - \theta_1) = \dfrac{\tan\theta_2 - \tan\theta_1}{1 + \tan\theta_2 \tan\theta_1}$ — Difference identity for tangent

$= \dfrac{m_2 - m_1}{1 + m_2 m_1}$ — Given $m_1 = \tan \theta_1$ and $m_2 = \tan \theta_2$

$= \dfrac{m_2 - m_1}{1 + m_1 m_2}$ — Algebra

83. Following the hint, label the text diagram as follows:

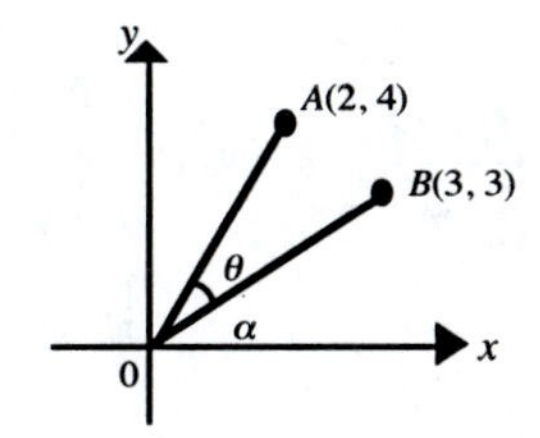

Then $\tan\theta = \frac{y_B}{x_B} = \frac{3}{3} = 1 \qquad \tan(\alpha+\theta) = \frac{y_A}{x_A} = \frac{4}{2} = 2$

$$\begin{aligned}\tan\theta &= \tan(\alpha+\theta-\alpha) = \tan[(\alpha+\theta)-\alpha]\\ &= \frac{\tan(\alpha+\theta)-\tan\alpha}{1+\tan(\alpha+\theta)\tan\alpha}\\ &= \frac{2-1}{1+2\cdot 1}\\ &= \frac{1}{3}\end{aligned}$$

Therefore, $\theta = \tan^{-1}\frac{1}{3} = 0.322$ rad.

85. (A) In right triangle ABE, we have (1) $\cot\alpha = \frac{AB}{AE} = \frac{AB}{h}$

In right triangle BCD, we have (2) $\cot\alpha = \frac{BC}{CD} = \frac{BC}{H}$

In right triangle $EE'D$, we have (3) $\tan\beta = \frac{E'D}{EE'} = \frac{H-h}{AC} = \frac{H-h}{AB+BC}$

From (3), $H-h = (AB+BC)\tan\beta$. From (1) and (2), $AB = h\cot\alpha$ and $BC = H\cot\alpha$. Hence, substituting, we have (4) $H-h = (h\cot\alpha + H\cot\alpha)\tan\beta$,

or $\quad = (h+H)\cot\alpha\tan\beta$

Solving (4) for H, we have

$$\begin{aligned}H-h &= h\cot\alpha\tan\beta + H\cot\alpha\tan\beta\\ H - H\cot\alpha\tan\beta &= h + h\cot\alpha\tan\beta\\ H(1-\cot\alpha\tan\beta) &= h(1+\cot\alpha\tan\beta)\\ H &= h\left(\frac{1+\cot\alpha\tan\beta}{1-\cot\alpha\tan\beta}\right)\end{aligned}$$

(B) Start with $H = h\left(\frac{1+\cot\alpha\tan\beta}{1-\cot\alpha\tan\beta}\right)$

Apply the quotient identities: $H = h\left(\frac{1+\frac{\cos\alpha}{\sin\alpha}\frac{\sin\beta}{\cos\beta}}{1-\frac{\cos\alpha}{\sin\alpha}\frac{\sin\beta}{\cos\beta}}\right)$

Reduce the complex fraction to a simple one: $H = h\left(\frac{\sin\alpha\cos\beta+\cos\alpha\sin\beta}{\sin\alpha\cos\beta-\cos\alpha\sin\beta}\right)$

Apply the sum and difference identities for sine: $H = h\left(\frac{\sin(\alpha+\beta)}{\sin(\alpha-\beta)}\right)$

(C) Substituting the given values, we have: $H = (5.50\text{ ft})\frac{\sin(45.00°+44.92°)}{\sin(45.00°-44.92°)} = 3{,}940$ ft

EXERCISE 4.4 Double-Angle and Half-Angle Identities

1. $\dfrac{\sin 2x}{2\sin x} = \dfrac{2\sin x\cos x}{2\sin x}$ Double-angle identity

$= \cos x$ Algebra

3. $1 - \sin x \tan \dfrac{x}{2} = 1 - \sin x\, \dfrac{1-\cos x}{\sin x}$ Half-angle identity

$= 1 - (1 - \cos x)$ Algebra

$= 1 - 1 + \cos x$ Algebra

$= \cos x$ Algebra

5. $\dfrac{1-\cos 2x}{2\sin x} = \dfrac{1-(1-2\sin^2 x)}{2\sin x}$ Double-angle identity

$= \dfrac{1-1+2\sin^2 x}{2\sin x}$ Algebra

$= \dfrac{2\sin^2 x}{2\sin x}$ Algebra

$= \sin x$ Algebra

7. $\sin 105° = \sin \dfrac{210°}{2} = \sqrt{\dfrac{1-\cos 210°}{2}}$

The positive sign is used since 105° is in the second quadrant and sine is positive there. We note that the reference triangle for 210° is a 30°–60° triangle in the third quadrant. Thus,

$$\sin 105° = \sqrt{\frac{1-(-\sqrt{3}/2)}{2}} = \sqrt{\frac{2+\sqrt{3}}{4}} = \frac{\sqrt{2+\sqrt{3}}}{2}$$

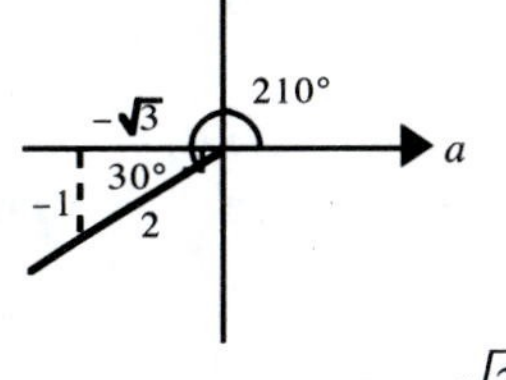

$\cos 210° = -\cos 30° = -\dfrac{\sqrt{3}}{2}$

9. $$\tan 15° = \tan \frac{30°}{2} = \frac{1-\cos 30°}{\sin 30°} = \frac{1-\sqrt{3}/2}{1/2} = \frac{2\left(1-\frac{\sqrt{3}}{2}\right)}{2\left(\frac{1}{2}\right)} = \frac{2-\sqrt{3}}{1} = 2-\sqrt{3}$$

11.

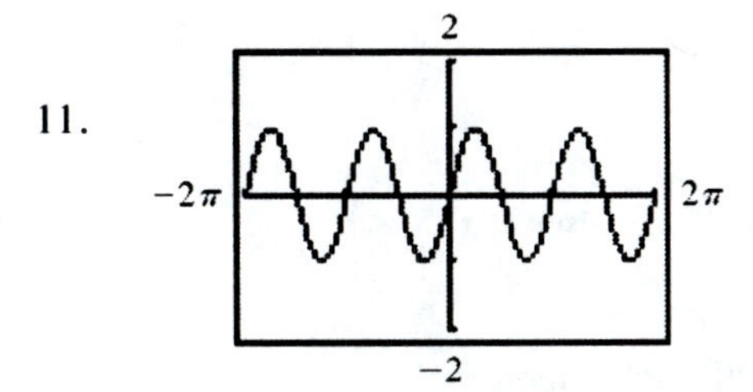

The graphs are identical, as can be seen from the trace function, or the double-angle identity for sine.

13.

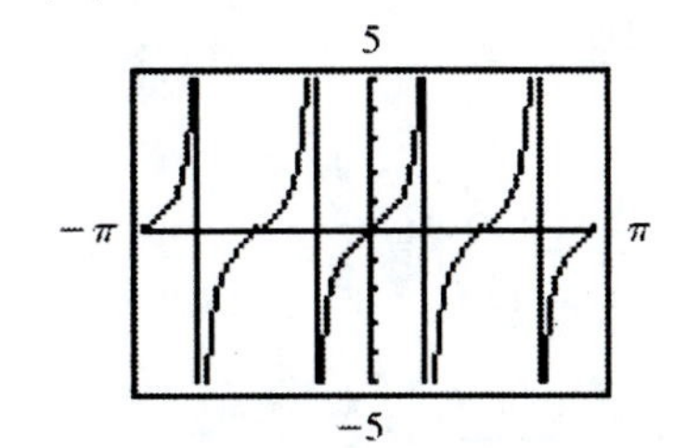

The graphs are identical, as can be seen from the trace function, or the double-angle identity for tangent.

15. $(\tan x)(1 + \cos 2x) = (\tan x)(1 + 2\cos^2 x - 1)$ — Double-angle identity

$= (\tan x)\ 2\cos^2 x$ — Algebra

$= \dfrac{\sin x}{\cos x}\ 2\cos^2 x$ — Quotient identity

$= 2\sin x \cos x$ — Algebra

$= \sin 2x$ — Double-angle identity

17. $2\sin^2 \frac{x}{2} = 2\left(\pm\sqrt{\dfrac{1-\cos x}{2}}\right)^2$ — Half-angle identity

$= 2\left(\dfrac{1-\cos x}{2}\right)$ — Algebra

$= \dfrac{1-\cos x}{1}$ — Algebra

$= \dfrac{1-\cos x}{1} \cdot \dfrac{1+\cos x}{1+\cos x}$ — Algebra

$= \dfrac{1-\cos^2 x}{1+\cos x}$ — Algebra

$= \dfrac{\sin^2 x}{1+\cos x}$ — Pythagorean identity

19. $(\sin\theta - \cos\theta)^2 = \sin^2\theta - 2\sin\theta\cos\theta + \cos^2\theta$ — Algebra

$= \sin^2\theta + \cos^2\theta - 2\sin\theta\cos\theta$ — Algebra

$= 1 - 2\sin\theta\cos\theta$ — Pythagorean identity

$= 1 - \sin 2\theta$ — Double-angle identity

21. $\cos^2 \frac{w}{2} = \left(\pm\sqrt{\dfrac{1+\cos w}{2}}\right)^2$ — Half-angle identity

$= \dfrac{1+\cos w}{2}$ — Algebra

23. $\cot\frac{\alpha}{2} = \dfrac{1}{\tan\frac{\alpha}{2}}$ — Reciprocal identity

$= \dfrac{1}{\dfrac{\sin\alpha}{1+\cos\alpha}}$ — Half-angle identity

$= \dfrac{1+\cos\alpha}{\sin\alpha}$ — Algebra

25. $\dfrac{\cos 2t}{1-\sin 2t} = \dfrac{\cos^2 t - \sin^2 t}{1 - 2\sin t \cos t}$ Double-angle identity

$= \dfrac{\cos^2 t - \sin^2 t}{\cos^2 t + \sin^2 t - 2 \sin t \cos t}$ Pythagorean identity

$= \dfrac{\cos^2 t - \sin^2 t}{\cos^2 t - 2 \sin t \cos t + \sin^2 t}$ Algebra

$= \dfrac{(\cos t - \sin t)(\cos t + \sin t)}{(\cos t - \sin t)(\cos t - \sin t)}$ Algebra

$= \dfrac{\cos t + \sin t}{\cos t - \sin t}$ Algebra

$= \dfrac{\dfrac{\cos t}{\cos t} + \dfrac{\sin t}{\cos t}}{\dfrac{\cos t}{\cos t} - \dfrac{\sin t}{\cos t}}$ Algebra

$= \dfrac{1 + \tan t}{1 - \tan t}$ Quotient identity

27. $\tan 2x = \tan(x + x)$ Algebra

$= \dfrac{\tan x + \tan x}{1 - \tan x \tan x}$ Sum identity for tangent

$= \dfrac{2 \tan x}{1 - \tan^2 x}$ Algebra

29. $\tan \dfrac{x}{2} = \dfrac{\sin \frac{x}{2}}{\cos \frac{x}{2}}$ Quotient identity

$= \dfrac{\pm\sqrt{\dfrac{1-\cos x}{2}}}{\pm\sqrt{\dfrac{1+\cos x}{2}}}$ Half-angle identities

$= \pm\sqrt{\dfrac{1-\cos x}{1+\cos x}}$ Algebra

$\left|\tan \dfrac{x}{2}\right| = \sqrt{\dfrac{1-\cos x}{1+\cos x}}$ Algebra

$= \sqrt{\dfrac{1-\cos x}{1+\cos x} \cdot \dfrac{1+\cos x}{1+\cos x}}$ Algebra

$$= \sqrt{\frac{1-\cos^2 x}{(1+\cos x)^2}}$$ Algebra

$$= \sqrt{\frac{\sin^2 x}{(1+\cos x)^2}}$$ Pythagorean identity

$$= \left|\frac{\sin x}{1+\cos x}\right|$$ Algebra

Since $1 + \cos x \geq 0$ and $\sin x$ has the same sign as $\tan \frac{x}{2}$, we may drop the absolute value signs to obtain $\tan \frac{x}{2} = \frac{\sin x}{1+\cos x}$

To show that $\sin x$ has the same sign as $\tan \frac{x}{2}$, we note the following cases:

If $0 < x < \pi$, $\sin x > 0$, then $0 < \frac{x}{2} < \frac{\pi}{2}$, $\tan \frac{x}{2} > 0$.

If $\pi < x < 2\pi$, $\sin x < 0$, then $\frac{\pi}{2} < \frac{x}{2} < \pi$, $\tan \frac{x}{2} < 0$.

The truth of the statement for other values of x follows since $\sin(x + 2k\pi) = \sin x$ and $\tan \frac{x+2k\pi}{2} = \tan \frac{x}{2}$ by the periodic properties of sine and tangent.

(Note: if $x = 0$ both sides of the proposed identity are 0, if $x = \pi$ both sides are meaningless.)

Alternative proof:

$$\tan \frac{x}{2} = \frac{\sin \frac{x}{2}}{\cos \frac{x}{2}}$$ Quotient identity

$$= \frac{2 \sin \frac{x}{2} \cos \frac{x}{2}}{2 \cos \frac{x}{2} \cos \frac{x}{2}}$$ Algebra

$$= \frac{2 \sin \frac{x}{2} \cos \frac{x}{2}}{2 \cos^2 \frac{x}{2}}$$ Algebra

$$= \frac{2 \sin \frac{x}{2} \cos \frac{x}{2}}{1 + 2 \cos^2 \frac{x}{2} - 1}$$ Algebra

$$= \frac{\sin 2(x/2)}{1 + \cos 2(x/2)}$$ Double-angle identities

$$= \frac{\sin x}{1+\cos x}$$ Algebra

31. $(\sec 2x)(2 - \sec^2 x) = \dfrac{1}{\cos 2x}\left(2 - \dfrac{1}{\cos^2 x}\right)$ Reciprocal identity

$= \dfrac{1}{\cos 2x}\left(\dfrac{2}{1} - \dfrac{1}{\cos^2 x}\right)$ Algebra

$= \dfrac{1}{\cos 2x}\left(\dfrac{2\cos^2 x}{\cos^2 x} - \dfrac{1}{\cos^2 x}\right)$ Algebra

$= \dfrac{1}{\cos 2x}\left(\dfrac{2\cos^2 x - 1}{\cos^2 x}\right)$ Algebra

$= \dfrac{1}{\cos 2x}\,\dfrac{\cos 2x}{\cos^2 x}$ Double-angle identity

$= \dfrac{1}{\cos^2 x}$ Algebra

$= \sec^2 x$ Reciprocal identity

33. No. For example, if $x = -4$, $\sqrt{x^2 + 6x + 9} = \sqrt{(-4)^2 + 6(-4) + 9} = \sqrt{1} = 1$, but $x + 3 = -4 + 3 = -1$.

35. Yes. The right side, $\dfrac{x-1}{\sqrt{x-1}} = \dfrac{\left(\sqrt{x-1}\right)^2}{\sqrt{x-1}} = \sqrt{x-1}$, the left side, for all values of x for which both are defined $(x > 1)$.

37. No. For example, if $x = \dfrac{\pi}{2}$, $\sin 3x = \sin 3\left(\dfrac{\pi}{2}\right) = -1$, but $3 \sin x \cos x = 3 \sin \dfrac{\pi}{2} \cos \dfrac{\pi}{2} = 3(1)(0) = 0$.

39. Yes. The right side can be shown to equal the left side for all values of x for which both are defined, as follows:

$\dfrac{2}{\tan x - \cot x} = \dfrac{2}{\tan x - \dfrac{1}{\tan x}}$ Reciprocal identity

$= \dfrac{2\tan x}{\tan^2 x - 1}$ Algebra

$= -\dfrac{2\tan x}{1 - \tan^2 x}$ Algebra

$= -\tan 2x$ Double-angle identity

$= \tan(-2x)$ Identities for negatives

41. No. For example, if $x = \dfrac{5\pi}{2}$, $\sin \dfrac{x}{2} = \sin \dfrac{5\pi/2}{2} = \sin \dfrac{5\pi}{4} = -\dfrac{1}{\sqrt{2}}$,

but $\sqrt{\dfrac{1 - \cos x}{2}} = \sqrt{\dfrac{1 - \cos 5\pi/2}{2}} = \sqrt{\dfrac{1}{2}} = \dfrac{1}{\sqrt{2}}$.

43. No. For example, if $x = 0$, $\left|\cos\frac{x}{4}\right| = \left|\cos\frac{0}{4}\right| = |1| = 1$, but $\sqrt{\frac{1+\cos x}{4}} = \sqrt{\frac{1+\cos 0}{4}} = \sqrt{\frac{2}{4}} = \frac{\sqrt{2}}{2}$.

45. First draw a reference triangle in the second quadrant and find cos x and tan x: $a = -\sqrt{25^2 - 7^2} = -24$.
$\sin x = \frac{7}{25}$; $\cos x = -\frac{24}{25}$; $\tan x = -\frac{7}{24}$. Now use the double-angle identities.

$$\sin 2x = 2\sin x\cos x = 2\left(\frac{7}{25}\right)\left(-\frac{24}{25}\right) = -\frac{336}{625}$$

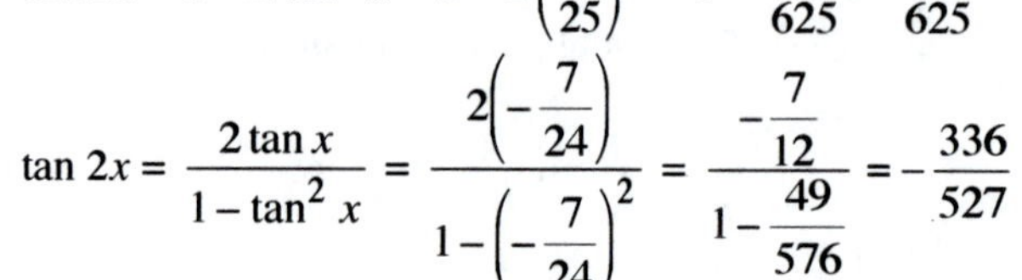

$$\cos 2x = 1 - 2\sin^2 x = 1 - 2\left(\frac{7}{25}\right)^2 = 1 - \frac{98}{625} = \frac{527}{625}$$

$$\tan 2x = \frac{2\tan x}{1-\tan^2 x} = \frac{2\left(-\frac{7}{24}\right)}{1-\left(-\frac{7}{24}\right)^2} = \frac{-\frac{7}{12}}{1-\frac{49}{576}} = -\frac{336}{527}$$

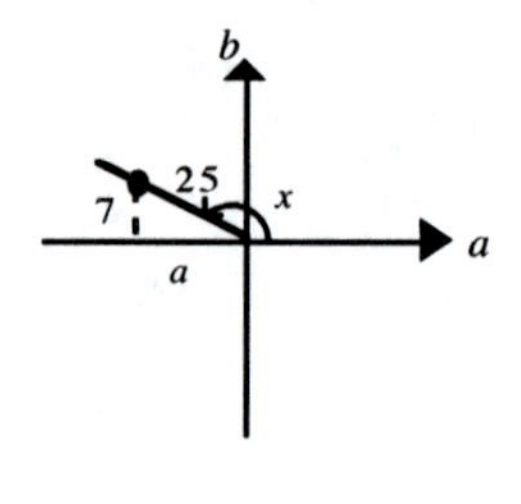

Alternatively, $\tan 2x = \frac{\sin 2x}{\cos 2x} = \frac{-336/625}{527/625} = -\frac{336}{527}$

47. First draw a reference triangle in the fourth quadrant and find sin x, cos x, and tan x:
$r = \sqrt{12^2 + (-35)^2} = 37$
$\sin x = -\frac{35}{37}$; $\cos x = \frac{12}{37}$; $\tan x = -\frac{35}{12}$. Now use the double-angle identities.

$$\sin 2x = 2\sin x\cos x = 2\left(-\frac{35}{37}\right)\left(\frac{12}{37}\right) = -\frac{840}{1{,}369}$$

$$\cos 2x = 2\cos^2 x - 1 = 2\left(\frac{12}{37}\right)^2 - 1 = \frac{288}{1{,}369} - 1 = -\frac{1{,}081}{1{,}369}$$

$$\tan 2x = \frac{2\tan x}{1-\tan^2 x} = \frac{2\left(-\frac{35}{12}\right)}{1-\left(-\frac{35}{12}\right)^2} = \frac{-\frac{35}{6}}{1-\frac{1{,}225}{144}} = \frac{840}{1{,}081}$$

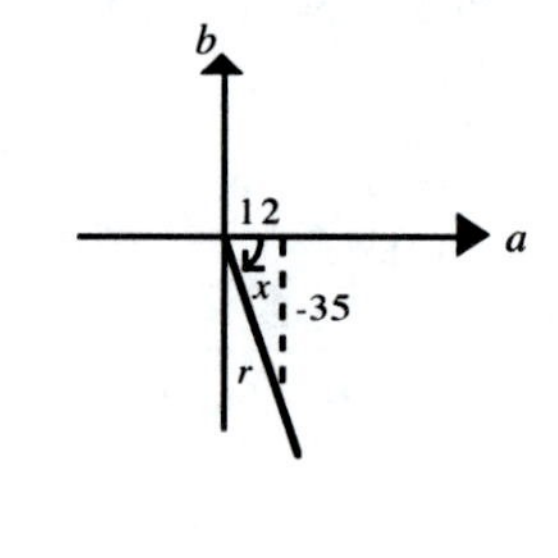

Alternatively, $\tan 2x = \frac{\sin 2x}{\cos 2x} = \frac{-840/1{,}369}{-1{,}081/1{,}369} = \frac{840}{1{,}081}$

49. We are given cos x. We can find $\sin\frac{x}{2}$ and $\cos\frac{x}{2}$ from the half-angle identities, after determining their sign, as follows:
If $0° < x < 90°$, then $0° < \frac{x}{2} < 45°$. Thus, $\frac{x}{2}$ is in the first quadrant, where sine and cosine are positive. Using half-angle identities, we obtain:

$$\sin\frac{x}{2} = \sqrt{\frac{1-\cos x}{2}} = \sqrt{\frac{1-(1/4)}{2}} = \sqrt{\frac{3}{8}} \qquad \cos\frac{x}{2} = \sqrt{\frac{1+\cos x}{2}} = \sqrt{\frac{1+(1/4)}{2}} = \sqrt{\frac{5}{8}}$$

51. Draw a reference triangle in the second quadrant and find cos x.

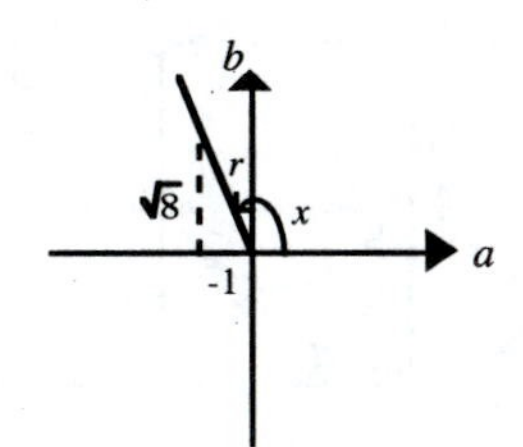

$r = \sqrt{(-1)^2 + (\sqrt{8})^2} = 3$

$\cos x = -\frac{1}{3}$

If $90° < x < 180°$, then $45° < \frac{x}{2} < 90°$. Thus, $\frac{x}{2}$ is in the first quadrant, where sine and cosine are positive. Using half-angle identities, we obtain:

$$\sin\frac{x}{2} = \sqrt{\frac{1-\cos x}{2}} = \sqrt{\frac{1-(-1/3)}{2}} = \sqrt{\frac{2}{3}}$$

$$\cos\frac{x}{2} = \sqrt{\frac{1+\cos x}{2}} = \sqrt{\frac{1+(-1/3)}{2}} = \sqrt{\frac{1}{3}}$$

53. Draw a reference triangle in the fourth quadrant and find cos x.

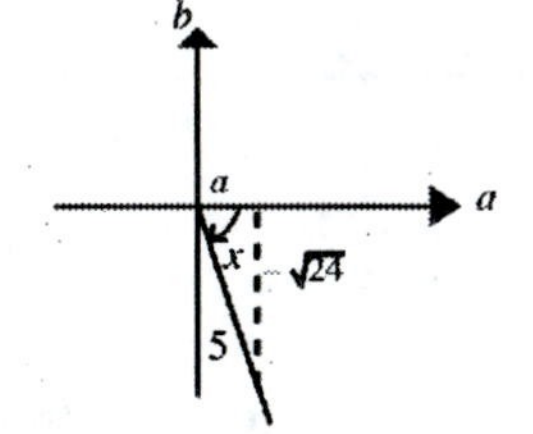

$a = \sqrt{5^2 - (-\sqrt{24})^2} = 1$

$\cos x = \frac{1}{5}$

If $-90° < x < 0°$, then $-45° < \frac{x}{2} < 0°$. Thus, $\frac{x}{2}$ is in the fourth quadrant, where sine is negative and cosine is positive. Using half-angle identities, we obtain:

$$\sin\frac{x}{2} = -\sqrt{\frac{1-\cos x}{2}} = -\sqrt{\frac{1-(1/5)}{2}} = -\sqrt{\frac{2}{5}}$$

$$\cos\frac{x}{2} = \sqrt{\frac{1+\cos x}{2}} = \sqrt{\frac{1+(1/5)}{2}} = \sqrt{\frac{3}{5}}$$

55. (A) Since θ is a first quadrant angle and sec 2θ is negative for 2θ in the second quadrant and not for 2θ in the first, 2θ is a second quadrant angle.

(B) Construct a reference triangle for 2θ in the second quadrant with $a = -4$ and $r = 5$. Use the Pythagorean theorem to find $b = 3$. Thus, $\sin 2\theta = \frac{3}{5}$ and $\cos 2\theta = -\frac{4}{5}$.

(C) The double angle identities $\cos 2\theta = 1 - 2\sin^2\theta$ and $\cos 2\theta = 2\cos^2\theta - 1$.

(D) Use the identities in part (C) in the form

$$\sin\theta = \sqrt{\frac{1-\cos 2\theta}{2}} \text{ and } \cos\theta = \sqrt{\frac{1+\cos 2\theta}{2}}$$

The positive radicals are used because θ is in quadrant one.

(E) $$\sin\theta = \sqrt{\frac{1-\cos 2\theta}{2}} = \sqrt{\frac{1-(-4/5)}{2}} = \sqrt{\frac{9}{10}} = \frac{3}{\sqrt{10}} \text{ or } \frac{3\sqrt{10}}{10}$$

$$\cos\theta = \sqrt{\frac{1+\cos 2\theta}{2}} = \sqrt{\frac{1+(-4/5)}{2}} = \sqrt{\frac{1}{10}} = \frac{1}{\sqrt{10}} \text{ or } \frac{\sqrt{10}}{10}$$

55. (E)

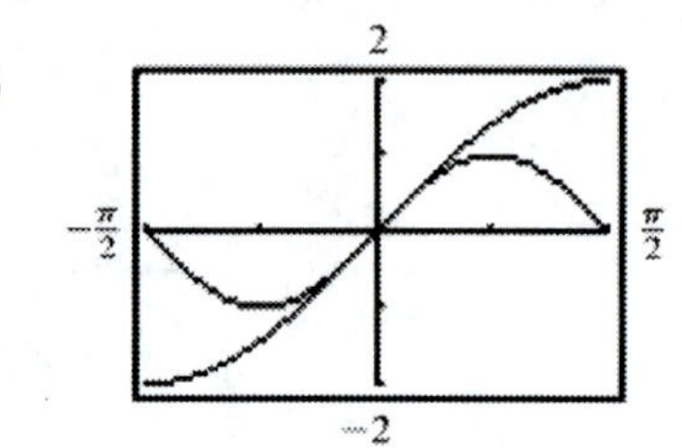

57. (A)

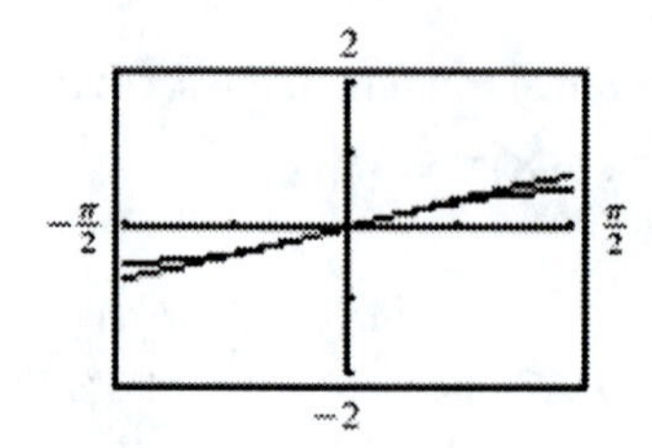

In each case, as x gets closer to zero, the two curves get closer together; the approximation improves.

57. (B)

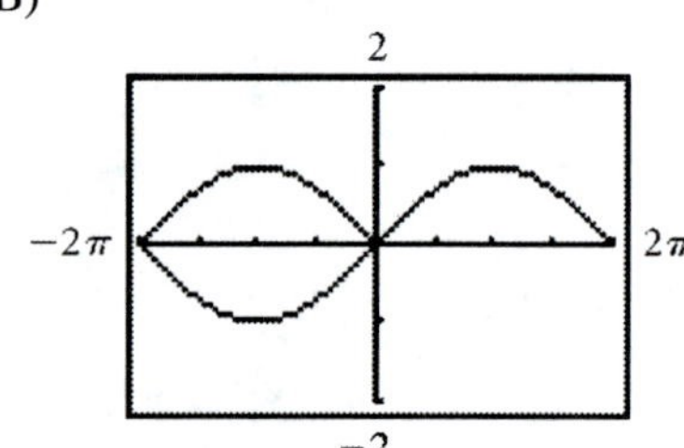

59. $0 \le x \le 2\pi$

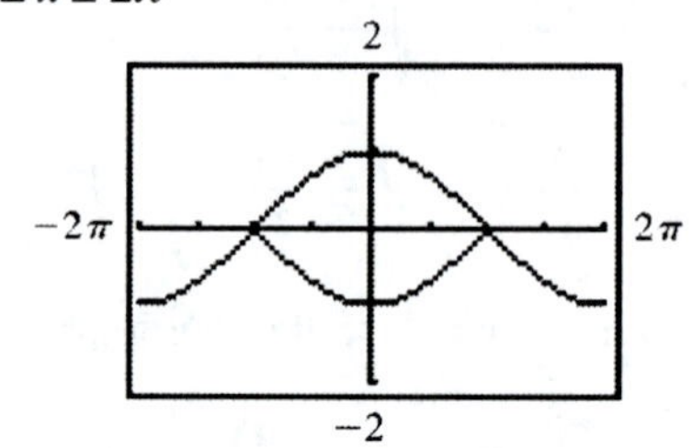

61. $-2\pi \le x \le -\pi, \pi \le x \le 2\pi$

63. To obtain $\sin x$ and $\cos x$ from $\sin 2x$, we use the half-angle identities with x replaced by $2x$. Thus,

$\sin \frac{x}{2} = \pm\sqrt{\frac{1-\cos x}{2}}$ becomes $\sin \frac{2x}{2} = \pm\sqrt{\frac{1-\cos 2x}{2}}$ or $\sin x = \pm\sqrt{\frac{1-\cos 2x}{2}}$

$\cos \frac{x}{2} = \pm\sqrt{\frac{1+\cos x}{2}}$ becomes $\cos \frac{2x}{2} = \pm\sqrt{\frac{1+\cos 2x}{2}}$ or $\cos x = \pm\sqrt{\frac{1+\cos 2x}{2}}$

Since $0 < x < \frac{\pi}{4}, 0 < 2x < \frac{\pi}{2}$. Hence $2x$ is in the first quadrant.

To obtain $\cos 2x$ from $\sin 2x$, we draw a reference triangle for $2x$ in the first quadrant.

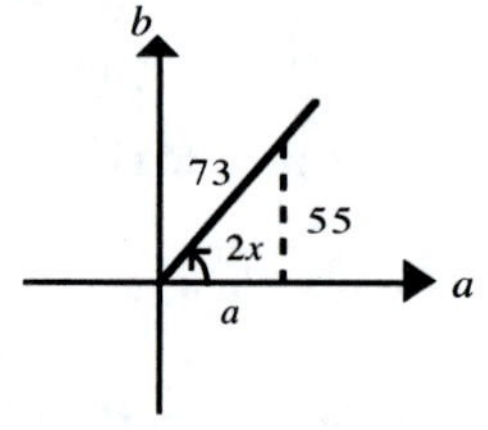

$a = \sqrt{73^2 - 55^2} = 48,$

$\cos 2x = \frac{48}{73}.$

Since $0 < x < \frac{\pi}{4}$, $\sin x$, $\cos x$, and $\tan x$ are positive.

Thus,

$$\sin x = \sqrt{\frac{1-\cos 2x}{2}} = \sqrt{\frac{1-(48/73)}{2}} = \sqrt{\frac{25}{146}} \text{ or } \frac{5}{\sqrt{146}}$$

$$\cos x = \sqrt{\frac{1+\cos 2x}{2}} = \sqrt{\frac{1+(48/73)}{2}} = \sqrt{\frac{121}{146}} \text{ or } \frac{11}{\sqrt{146}}$$

$$\tan x = \frac{\sin x}{\cos x} = \frac{5/\sqrt{146}}{11/\sqrt{146}} = \frac{5}{11}$$

65. In Problem 63, we derived the identities:

$$\sin x = \pm\sqrt{\frac{1-\cos 2x}{2}} \text{ and } \cos x = \pm\sqrt{\frac{1+\cos 2x}{2}}$$

Since $\frac{\pi}{4} < x < \frac{\pi}{2}$, $\frac{\pi}{2} < 2x < \pi$. Hence $2x$ is in the second quadrant. To obtain $\cos 2x$ from $\tan 2x$, we draw a reference triangle for $2x$ in the second quadrant.

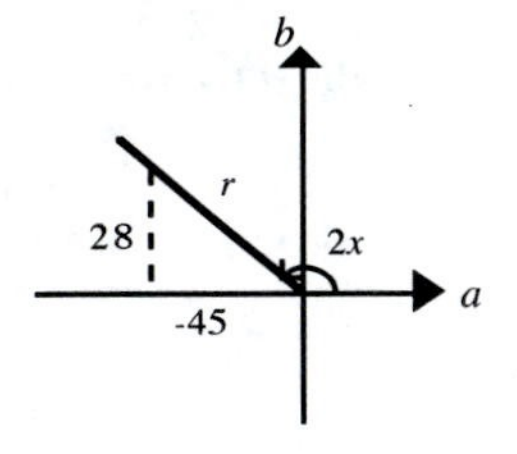

$r = \sqrt{(-45)^2 + 28^2} = 53$

$\cos 2x = -\frac{45}{53}$

Since $\frac{\pi}{4} < x < \frac{\pi}{2}$, $\sin x$, $\cos x$, and $\tan x$ are positive.

Thus,

$$\sin x = \sqrt{\frac{1-\cos 2x}{2}} = \sqrt{\frac{1-(-45/53)}{2}} = \sqrt{\frac{49}{53}} \text{ or } \frac{7}{\sqrt{53}}$$

$$\cos x = \sqrt{\frac{1+\cos 2x}{2}} = \sqrt{\frac{1+(-45/53)}{2}} = \sqrt{\frac{4}{53}} \text{ or } \frac{2}{\sqrt{53}}$$

$$\tan x = \frac{\sin x}{\cos x} = \frac{7/\sqrt{53}}{2/\sqrt{53}} = \frac{7}{2}$$

67. In Problem 63, we derived the identities:

$$\sin x = \pm\sqrt{\frac{1-\cos 2x}{2}} \text{ and } \cos x = \pm\sqrt{\frac{1+\cos 2x}{2}}$$

To obtain $\cos 2x$ from $\sec 2x$, we use the reciprocal identity:

$$\sec 2x = \frac{1}{\cos 2x} \text{ and } \cos 2x = \frac{1}{\sec 2x} = \frac{1}{65/33} = \frac{33}{65}$$

Since $-\frac{\pi}{4} < x < 0$, x is in the fourth quadrant, where sine is negative and cosine is positive. Thus,

$$\sin x = -\sqrt{\frac{1-\cos 2x}{2}} = -\sqrt{\frac{1-(33/65)}{2}} = -\sqrt{\frac{16}{65}} \text{ or } -\frac{4}{\sqrt{65}}$$

$$\cos x = \sqrt{\frac{1+\cos 2x}{2}} = \sqrt{\frac{1+(33/65)}{2}} = \sqrt{\frac{49}{65}} \text{ or } \frac{7}{\sqrt{65}}$$

$$\tan x = \frac{\sin x}{\cos x} = \frac{-4/\sqrt{65}}{7/\sqrt{65}} = -\frac{4}{7}$$

69. $\sin 3x = \sin(2x + x)$ — Algebra

$= \sin 2x \cos x + \cos 2x \sin x$ — Sum identity

$= 2 \sin x \cos x \cos x + (1 - 2 \sin^2 x) \sin x$ — Double-angle identities

$= 2 \sin x \cos^2 x + \sin x - 2 \sin^3 x$ — Algebra

$= 2 \sin x(1 - \sin^2 x) + \sin x - 2 \sin^3 x$ — Pythagorean identity

$= 2 \sin x - 2 \sin^3 x + \sin x - 2 \sin^3 x$ — Algebra

$= 3 \sin x - 4 \sin^3 x$ — Algebra

71. $\sin 4x = \sin 2(2x)$ — Algebra

$= 2 \sin 2x \cos 2x$ — Double-angle identity

$= 2(2 \sin x \cos x)(1 - 2 \sin^2 x)$ — Double-angle identity

$= \cos x(4 \sin x)(1 - 2 \sin^2 x)$ — Algebra

$= \cos x(4 \sin x - 8 \sin^3 x)$ — Algebra

73. $\tan 3x = \tan(2x + x)$ — Algebra

$= \dfrac{\tan 2x + \tan x}{1 - \tan 2x \tan x}$ — Sum identity

$= \dfrac{\dfrac{2 \tan x}{1 - \tan^2 x} + \tan x}{1 - \dfrac{2 \tan x}{1 - \tan^2 x} \tan x}$ — Double-angle identity

$= \dfrac{(1 - \tan^2 x)\dfrac{2 \tan x}{1 - \tan^2 x} + (1 - \tan^2 x) \tan x}{(1 - \tan^2 x) \cdot 1 - (1 - \tan^2 x)\dfrac{2 \tan x}{1 - \tan^2 x} \tan x}$ — Algebra

$= \dfrac{2 \tan x + (1 - \tan^2 x) \tan x}{1 - \tan^2 x - 2 \tan x \tan x}$ — Algebra

$= \dfrac{2 \tan x + \tan x - \tan^3 x}{1 - \tan^2 x - 2 \tan^2 x}$ — Algebra

$= \dfrac{3 \tan x - \tan^3 x}{1 - 3 \tan^2 x}$ — Algebra

75. The graph of $f(x)$ is shown in the figure. The graph appears to have vertical asymptotes $x = -2\pi$, $x = 0$, and $x = 2\pi$, x intercepts $-\pi$ and π, and period 2π. It appears that $g(x) = \cot \frac{x}{2}$ would be an appropriate choice. We verify $f(x) = g(x)$ as follows:

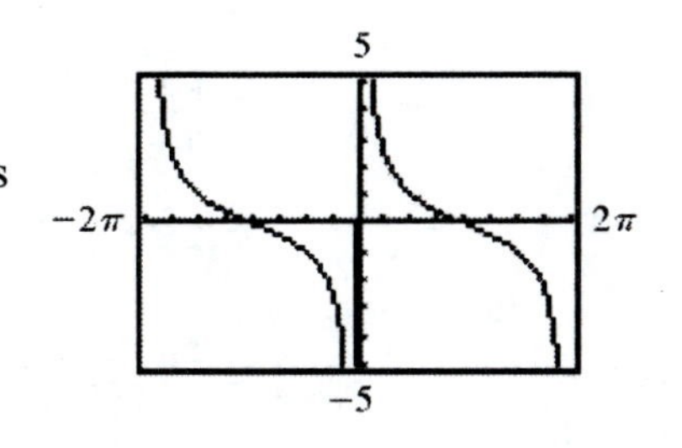

$f(x) = \csc x + \cot x$	
$= \frac{1}{\sin x} + \frac{\cos x}{\sin x}$	Reciprocal and quotient identities
$= \frac{1 + \cos x}{\sin x}$	Algebra
$= 1 \div \frac{\sin x}{1 + \cos x}$	Algebra
$= 1 \div \tan \frac{x}{2}$	Half-angle identity
$= \cot \frac{x}{2} = g(x)$	Reciprocal identity

77. The graph of $f(x)$ is shown in the figure. The graph appears to have vertical asymptotes $x = -\pi, -\frac{\pi}{2}, 0, \frac{\pi}{2}$, and π, and period π. It appears to have high and low points with y coordinate -1 and 1 respectively. It appears that $g(x) = \csc 2x$ would be an appropriate choice. We verify $f(x) = g(x)$ as follows:

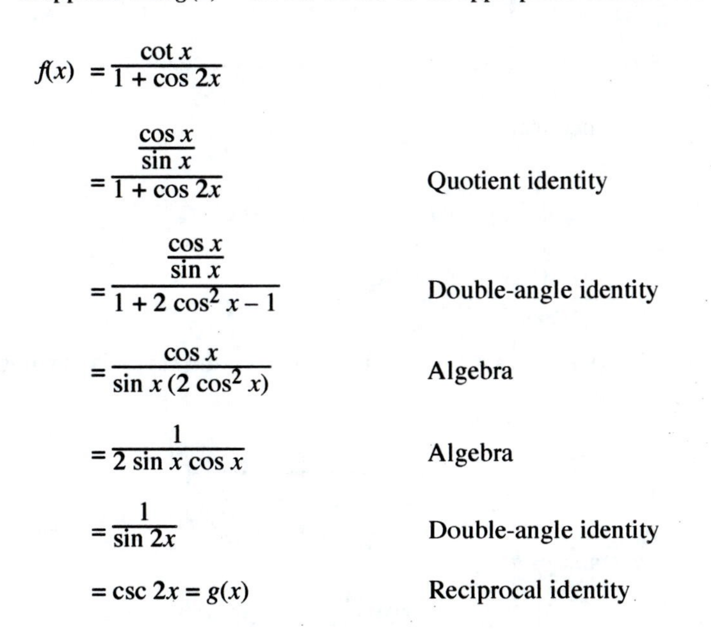

$f(x) = \frac{\cot x}{1 + \cos 2x}$	
$= \frac{\frac{\cos x}{\sin x}}{1 + \cos 2x}$	Quotient identity
$= \frac{\frac{\cos x}{\sin x}}{1 + 2\cos^2 x - 1}$	Double-angle identity
$= \frac{\cos x}{\sin x\,(2\cos^2 x)}$	Algebra
$= \frac{1}{2 \sin x \cos x}$	Algebra
$= \frac{1}{\sin 2x}$	Double-angle identity
$= \csc 2x = g(x)$	Reciprocal identity

79. The graph of $f(x)$ is shown in the figure. The graph appears to be a basic cosine curve with period 2π, amplitude $= \frac{1}{2}(y\text{max} - y\text{ min}) = \frac{1}{2}[1 - (-3)] = 2$, displaced downward by $|k| = 1$ unit. It appears that $g(x) = 2 \cos x - 1$ would be an appropriate choice. We verify $f(x) = g(x)$ as follows:

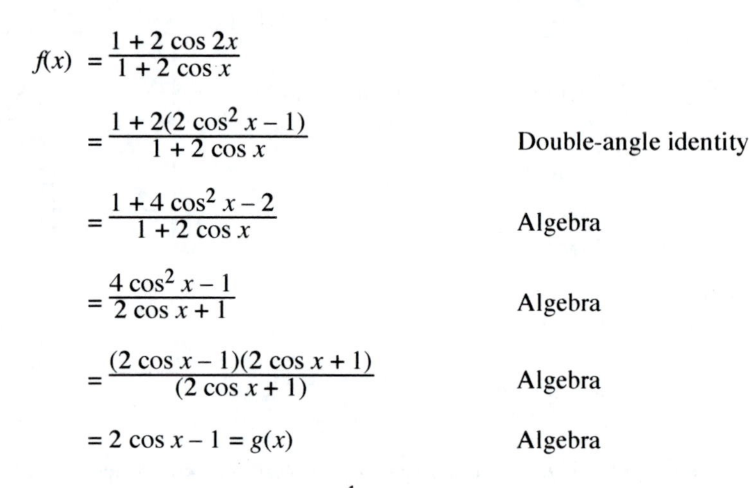

$$f(x) = \frac{1 + 2\cos 2x}{1 + 2\cos x}$$

$$= \frac{1 + 2(2\cos^2 x - 1)}{1 + 2\cos x} \quad \text{Double-angle identity}$$

$$= \frac{1 + 4\cos^2 x - 2}{1 + 2\cos x} \quad \text{Algebra}$$

$$= \frac{4\cos^2 x - 1}{2\cos x + 1} \quad \text{Algebra}$$

$$= \frac{(2\cos x - 1)(2\cos x + 1)}{(2\cos x + 1)} \quad \text{Algebra}$$

$$= 2\cos x - 1 = g(x) \quad \text{Algebra}$$

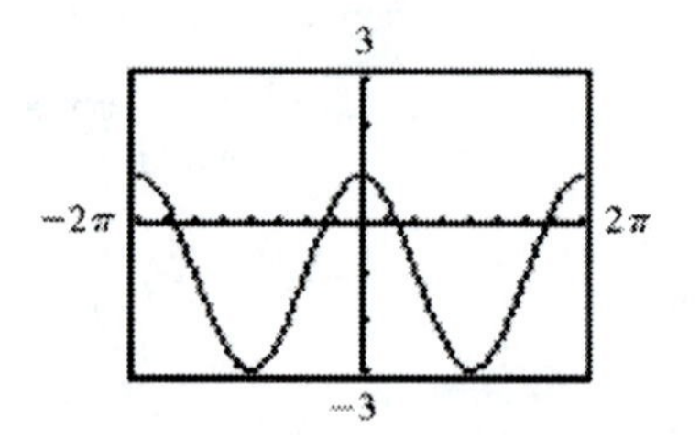

81. For $n = 2$, the left side is $y1 = \frac{1}{2} + \cos x + \cos 2x$ and the right side is

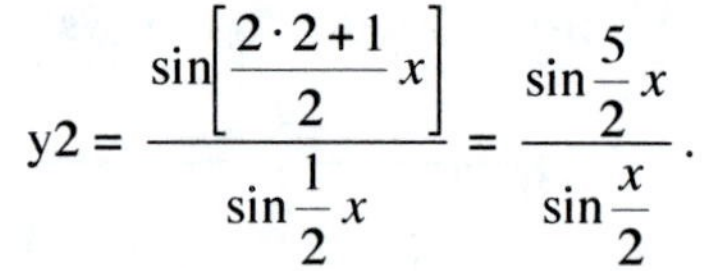

$$y2 = \frac{\sin\left[\frac{2\cdot 2+1}{2}x\right]}{\sin\frac{1}{2}x} = \frac{\sin\frac{5}{2}x}{\sin\frac{x}{2}}.$$

Since the identity holds, the graphs are identical.

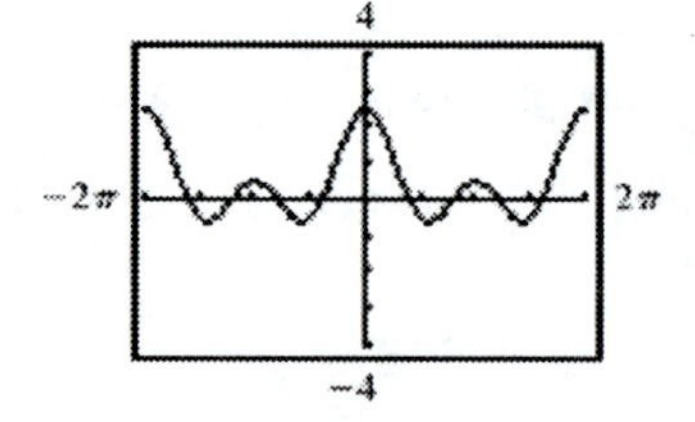

83. (A) Since $2 \sin\theta \cos\theta = \sin 2\theta$ by the double-angle identity, we can write

$$d = \frac{2v_0^2 \sin\theta\cos\theta}{32 \text{ ft/sec}^2} = \frac{v_0^2 (2\sin\theta\cos\theta)}{32 \text{ ft/sec}^2} = \frac{v_0^2 \sin 2\theta}{32 \text{ ft/sec}^2}$$

(B) Since v_0 is a given constant, d is maximum when $\sin 2\theta$ is maximum, and $\sin 2\theta$ is maximum when $2\theta = 90°$, that is, when $\theta = 45°$.

(C) Graph $d = \frac{100^2 \sin 2\theta}{32} = \frac{10{,}000 \sin 2\theta}{32}$.

As θ increases from 0° to 90°, d increases to a maximum of 312.5 feet when $\theta = 45°$, then decreases.

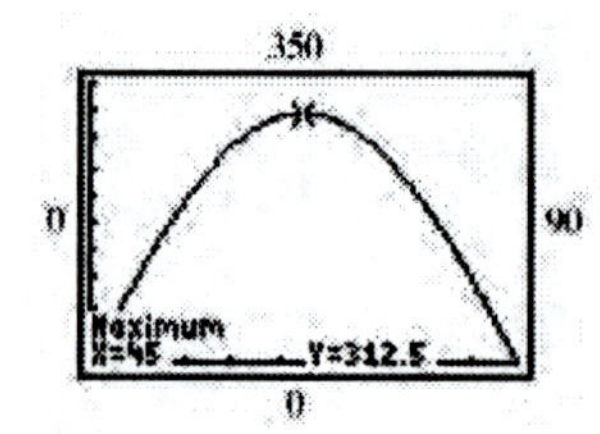

85. (A) Label the figure in the text as shown at the right. Then since triangle ABC is isosceles, the altitude BD bisects since AC and $AD = DC = \frac{b}{2}$.

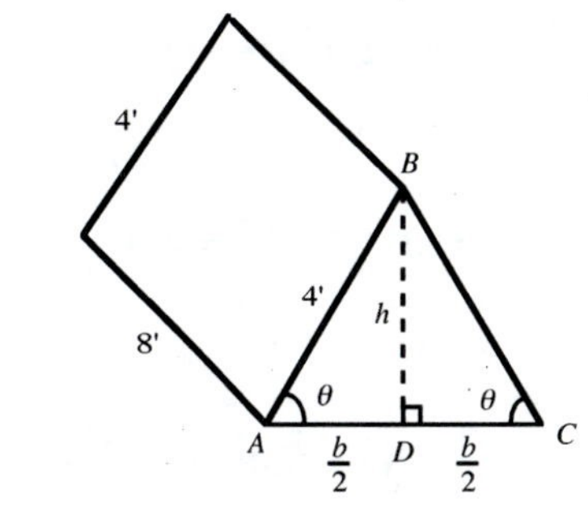

In triangle ABD,

$$\sin\theta = \frac{BD}{AB} = \frac{h}{4}\text{, thus } h = 4\sin\theta$$

$$\cos\theta = \frac{AD}{AB} = \frac{b/2}{4} = \frac{b}{8}\text{, thus } b = 8\cos\theta$$

Since the volume $V = 8 \cdot \frac{bh}{2}$, we have

$$V = 8 \cdot \frac{8\cos\theta \cdot 4\sin\theta}{2} = 128\cos\theta\sin\theta$$

Hence, $V = 64 \cdot 2\sin\theta\cos\theta = 64\sin 2\theta$

(B) $\sin 2\theta$ has a maximum value of 1 when $2\theta = 90°$, that is, $\theta = 45°$. When $\theta = 45°$, $V = 64\sin 2(45°) = 64\text{ ft}^3$

(C)

$\theta°$	30	35	40	45	50	55	60
V ft^3	55.4	60.1	63.0	64.0	63.0	60.1	55.4

(D)

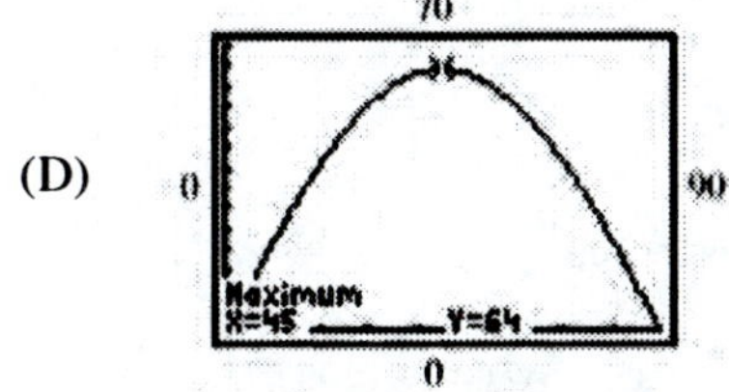

From the graphing calculator, the maximum value of V is 64 when $\theta = 45°$.

87. We note that $\tan\theta = \frac{2}{x}$ and $\tan 2\theta = \frac{6}{x}$ (see figure). Using the hint, we have $\tan 2\theta = \frac{2\tan\theta}{1 - \tan^2\theta}$

$$\frac{6}{x} = \frac{2\left(\frac{2}{x}\right)}{1 - \left(\frac{2}{x}\right)^2} = \frac{\frac{4}{x}}{1 - \left(\frac{4}{x^2}\right)} = \frac{x^2 \cdot \left(\frac{4}{x}\right)}{x^2 \cdot 1 - x^2 \cdot \left(\frac{4}{x^2}\right)} \quad x \neq 0$$

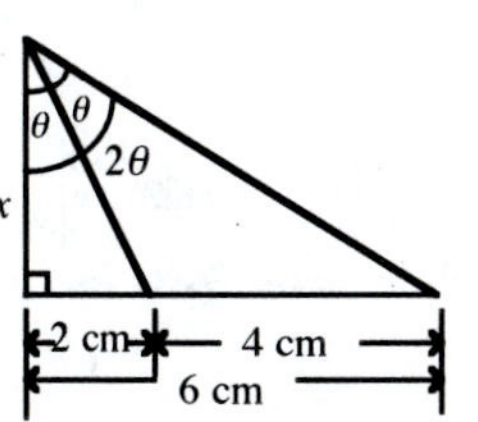

$$\frac{6}{x} = \frac{4x}{x^2 - 4}$$

$$x(x^2 - 4) \cdot \frac{6}{x} = x(x^2 - 4) \cdot \frac{4x}{x^2 - 4} \quad x \neq 2, -2$$

$$6(x^2 - 4) = x \cdot 4x$$

$$6x^2 - 24 = 4x^2$$

$$2x^2 = 24$$

$$x = \sqrt{12} \text{ or } 2\sqrt{3} \text{ (We discard the negative solution.)}$$

Then, $\frac{2}{x} = \tan\theta$, so $\tan\theta = \frac{2}{2\sqrt{3}} = \frac{1}{\sqrt{3}}$. To three decimal places $x \approx 3.464$ cm, $\theta = 30.000°$.

89. We label the figure as shown. From the Pythagorean theorem:

D N C E M θ/2 A s B

Since $AB = s$ and $MB = \frac{s}{2}$;

$AM^2 = AB^2 + MB^2 = s^2 + \frac{s^2}{4} = \frac{5s^2}{4}$; $AM = s\,\frac{\sqrt{5}}{2}$

Since $CN = CM = \frac{s}{2}$;

$NM^2 = CN^2 + CM^2 = \left(\frac{s}{2}\right)^2 + \left(\frac{s}{2}\right)^2 = \frac{2s^2}{4}$; $NM = s\,\frac{\sqrt{2}}{2}$

From the fact that $NA = MA$, thus triangle AMN is isosceles:

AE bisects NM, hence $ME = \frac{1}{2}MN = \frac{1}{2}\cdot s\,\frac{\sqrt{2}}{2} = s\,\frac{\sqrt{2}}{4}$.

From the definition of sine: $\sin\frac{\theta}{2} = \frac{ME}{MA} = s\,\frac{\sqrt{2}}{4} \div s\,\frac{\sqrt{5}}{2} = \frac{1}{2}\frac{\sqrt{2}}{\sqrt{5}}$

From the half-angle identity: $\sin\frac{\theta}{2} = \sqrt{\frac{1-\cos\theta}{2}}$.

Hence, $\sqrt{\frac{1-\cos\theta}{2}} = \frac{1}{2}\frac{\sqrt{2}}{\sqrt{5}}$; $\frac{1-\cos\theta}{2} = \frac{1}{10}$; $1 - \cos\theta = \frac{1}{5}$; $\cos\theta = \frac{4}{5}$.

(The student may wish to compare with Exercise 1.4, Problem 53.)

EXERCISE 4.5 Product-Sum and Sum-Product Identities

1. $\cos x \cos y = \frac{1}{2}[\cos(x + y) + \cos(x - y)]$ Let $x = 4w$ and $y = w$.

$\cos 4w \cos w = \frac{1}{2}[\cos(4w + w) + \cos(4w - w)] = \frac{1}{2}(\cos 5w + \cos 3w) = \frac{1}{2}\cos 5w + \frac{1}{2}\cos 3w$

3. $\cos x \sin y = \frac{1}{2}[\sin(x + y) - \sin(x - y)]$ Let $x = 2u$ and $y = u$.

$\cos 2u \sin u = \frac{1}{2}[\sin(2u + u) - \sin(2u - u)] = \frac{1}{2}(\sin 3u - \sin u) = \frac{1}{2}\sin 3u - \frac{1}{2}\sin u$

5. $\sin x \cos y = \frac{1}{2}[\sin(x + y) + \sin(x - y)]$ Let $x = 2B$ and $y = 5B$.

$\sin 2B \cos 5B = \frac{1}{2}[\sin(2B + 5B) + \sin(2B - 5B)] = \frac{1}{2}[\sin 7B + \sin(-3B)]$

$= \frac{1}{2}(\sin 7B - \sin 3B) = \frac{1}{2}\sin 7B - \frac{1}{2}\sin 3B$

7. $\sin x \sin y = \frac{1}{2}[\cos(x - y) - \cos(x + y)]$ Let $x = 3m$ and $y = 4m$.

$\sin 3m \sin 4m = \frac{1}{2}[\cos(3m - 4m) - \cos(3m + 4m)] = \frac{1}{2}[\cos(-m) - \cos 7m]$

$= \frac{1}{2}(\cos m - \cos 7m) = \frac{1}{2}\cos m - \frac{1}{2}\cos 7m$

9. $\cos x + \cos y = 2\cos\dfrac{x+y}{2}\cos\dfrac{x-y}{2}$ Let $x = 5\theta$ and $y = 3\theta$.

$$\cos 5\theta + \cos 3\theta = 2\cos\frac{5\theta+3\theta}{2}\cos\frac{5\theta-3\theta}{2} = 2\cos 4\theta\cos\theta$$

11. $\sin x - \sin y = 2\cos\dfrac{x+y}{2}\sin\dfrac{x-y}{2}$ Let $x = 6u$ and $y = 2u$.

$$\sin 6u - \sin 2u = 2\cos\frac{6u+2u}{2}\sin\frac{6u-2u}{2} = 2\cos 4u\sin 2u$$

13. $\sin x + \sin y = 2\sin\dfrac{x+y}{2}\cos\dfrac{x-y}{2}$ Let $x = 3B$ and $y = 5B$.

$$\sin 3B + \sin 5B = 2\sin\frac{3B+5B}{2}\cos\frac{3B-5B}{2} = 2\sin 4B\cos(-B) = 2\sin 4B\cos B$$

15. $\cos x - \cos y = -2\sin\dfrac{x+y}{2}\sin\dfrac{x-y}{2}$ Let $x = w$ and $y = 5w$.

$$\cos w - \cos 5w = -2\sin\frac{w+5w}{2}\sin\frac{w-5w}{2} = -2\sin 3w\sin(-2w) = 2\sin 3w\sin 2w$$

17. $\cos x\sin y = \dfrac{1}{2}[\sin(x+y) - \sin(x-y)]$ Let $x = 75°$ and $y = 15°$.

$$\cos 75°\sin 15° = \frac{1}{2}[\sin(75°+15°) - \sin(75°-15°)] = \frac{1}{2}[\sin 90° - \sin 60°]$$

$$= \frac{1}{2}\left(1 - \frac{\sqrt{3}}{2}\right) = \frac{1}{2}\left(\frac{2-\sqrt{3}}{2}\right) = \frac{2-\sqrt{3}}{4}$$

19. $\sin x\sin y = \dfrac{1}{2}[\cos(x-y) - \cos(x+y)]$ Let $x = 105°$ and $y = 165°$.

$$\sin 105°\sin 165° = \frac{1}{2}[\cos(105°-165°) - \cos(105°+165°)]$$

$$= \frac{1}{2}[\cos(-60°) - \cos 270°] = \frac{1}{2}\left(\frac{1}{2} - 0\right) = \frac{1}{4}$$

21. $\sin x + \sin y = 2\sin\dfrac{x+y}{2}\cos\dfrac{x-y}{2}$ Let $x = 195°$ and $y = 105°$.

$$\sin 195° + \sin 105° = 2\sin\frac{195°+105°}{2}\cos\frac{195°-105°}{2} = 2\sin 150°\cos 45°$$

$$= 2\left(\frac{1}{2}\right)\left(\frac{\sqrt{2}}{2}\right) = \frac{\sqrt{2}}{2}$$

23. $\sin x - \sin y = 2\cos\frac{x+y}{2}\sin\frac{x-y}{2}$ Let $x = 75°$ and $y = 165°$.

$$\sin 75° - \sin 165° = 2\cos\frac{75°+165°}{2}\sin\frac{75°-165°}{2} = 2\cos 120°\sin(-45°)$$
$$= 2\left(-\frac{1}{2}\right)\left(-\frac{\sqrt{2}}{2}\right) = \frac{\sqrt{2}}{2}$$

25.

$$\cos(x-y) = \cos x\cos y + \sin x\sin y$$
$$\cos(x+y) = \cos x\cos y - \sin x\sin y$$

$$\cos(x-y) - \cos(x+y) = 2\sin x\sin y \qquad \text{subtracting the above}$$
$$\sin x\sin y = \frac{1}{2}[\cos(x-y) - \cos(x+y)]$$

27. Let $x = u + v$ and $y = u - v$ and solve the resulting system for u and v in terms of x and y to obtain $u = \frac{x+y}{2}$ and $v = \frac{x-y}{2}$. Substituting into the product-sum identity yields

$$\cos\frac{x+y}{2}\cos\frac{x-y}{2} = \frac{1}{2}[\cos x + \cos y]$$

or

$$\cos x + \cos y = 2\cos\frac{x+y}{2}\cos\frac{x-y}{2}$$

29.

$$\frac{\cos t - \cos 3t}{\sin t + \sin 3t} = \frac{-2\sin\frac{t+3t}{2}\sin\frac{t-3t}{2}}{2\sin\frac{t+3t}{2}\cos\frac{t-3t}{2}} \qquad \text{Sum-product identities}$$

$$= \frac{-\sin 2t\sin(-t)}{\sin 2t\cos(-t)} \qquad \text{Algebra}$$

$$= \frac{\sin 2t\sin t}{\sin 2t\cos t} \qquad \text{Identities for negatives}$$

$$= \frac{\sin t}{\cos t} \qquad \text{Algebra}$$

$$= \tan t \qquad \text{Quotient identity}$$

31.

$$\frac{\sin x + \sin y}{\cos x + \cos y} = \frac{2\sin\frac{x+y}{2}\cos\frac{x-y}{2}}{2\cos\frac{x+y}{2}\cos\frac{x-y}{2}} \qquad \text{Sum-product identities}$$

$$= \frac{\sin\frac{x+y}{2}}{\cos\frac{x+y}{2}} \qquad \text{Algebra}$$

$$= \tan\frac{x+y}{2} \qquad \text{Quotient identity}$$

33. $\dfrac{\cos x - \cos y}{\sin x + \sin y} = \dfrac{-2\sin\frac{x+y}{2}\sin\frac{x-y}{2}}{2\sin\frac{x+y}{2}\cos\frac{x-y}{2}}$ Sum-product identities

$= \dfrac{-\sin\frac{x-y}{2}}{\cos\frac{x-y}{2}}$ Algebra

$= -\tan\frac{x-y}{2}$ Quotient identity

35. $\dfrac{\sin x + \sin y}{\sin x - \sin y} = \dfrac{2\sin\frac{x+y}{2}\cos\frac{x-y}{2}}{2\cos\frac{x+y}{2}\sin\frac{x-y}{2}}$ Sum-product identities

$= \dfrac{\sin\frac{x+y}{2}\cos\frac{x-y}{2}}{\cos\frac{x+y}{2}\sin\frac{x-y}{2}}$ Algebra

$= \tan\frac{x+y}{2}\cot\frac{x-y}{2}$ Quotient identities

$= \tan\frac{x+y}{2}\dfrac{1}{\tan\frac{x-y}{2}}$ Reciprocal identity

$= \dfrac{\tan\frac{x+y}{2}}{\tan\frac{x-y}{2}}$ Algebra

37. No. For example, if $x = \frac{\pi}{2}$, $\cos 2x \sin x = \cos\left(2\cdot\frac{\pi}{2}\right)\sin\left(\frac{\pi}{2}\right) = (-1)(1) = -1$,

but $\frac{1}{2}[\sin 3x + \sin x] = \frac{1}{2}\left[\sin\left(3\cdot\frac{\pi}{2}\right) + \sin\frac{\pi}{2}\right] = \frac{1}{2}[(-1) + 1] = 0$.

39. Yes. The right side, $\frac{1}{2}\sin 4x = \frac{1}{2}\sin 2(2x) = \frac{1}{2}(2\sin 2x\cos 2x) = \sin 2x\cos 2x$, the left side, for all values of x.

41. Yes. The left side, $\sin 5x + \sin x = 2\sin\frac{5x+x}{2}\cos\frac{5x-x}{2} = 2\sin 3x\cos 2x$, the right side, for all values of x.

43. No. For example, if $x = \frac{\pi}{4}$, $\cos x - \cos 3x = \cos\frac{\pi}{4} - \cos\left(3\cdot\frac{\pi}{4}\right) = \frac{1}{\sqrt{2}} - \left(-\frac{1}{\sqrt{2}}\right) = \sqrt{2}$,

but $-2\sin 2x\sin x = -2\sin\left(2\cdot\frac{\pi}{4}\right)\sin\frac{\pi}{4} = -2\cdot 1\cdot\frac{1}{\sqrt{2}} = -\sqrt{2}$.

45. $$\cos u \cos v = \frac{1}{2}[\cos(u+v) + \cos(u-v)] \qquad \text{Let } u = 5x \text{ and } v = 3x.$$

$$\cos 5x \cos 3x = \frac{1}{2}[\cos(5x+3x) + \cos(5x-3x)] = \frac{1}{2}(\cos 8x + \cos 2x) = y2$$

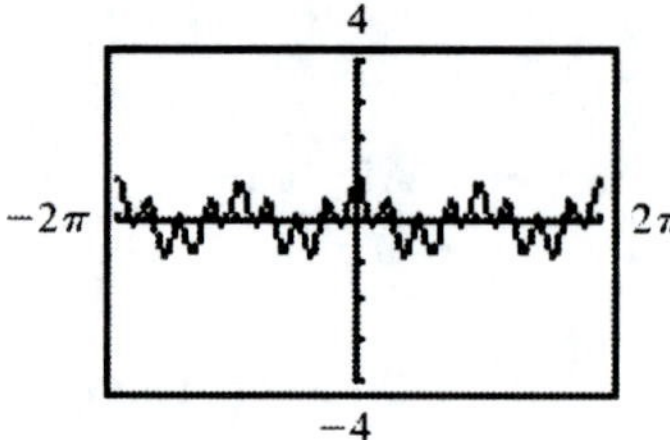

47. $$\cos u \sin v = \frac{1}{2}[\sin(u+v) - \sin(u-v)] \qquad \text{Let } u = 1.9x \text{ and } v = 3.5x.$$

$$\cos 1.9x \sin 0.5x = \frac{1}{2}[\sin(1.9x+0.5x) - \sin(1.9x-0.5x)] = \frac{1}{2}(\sin 2.4x - \sin 1.4x) = y2$$

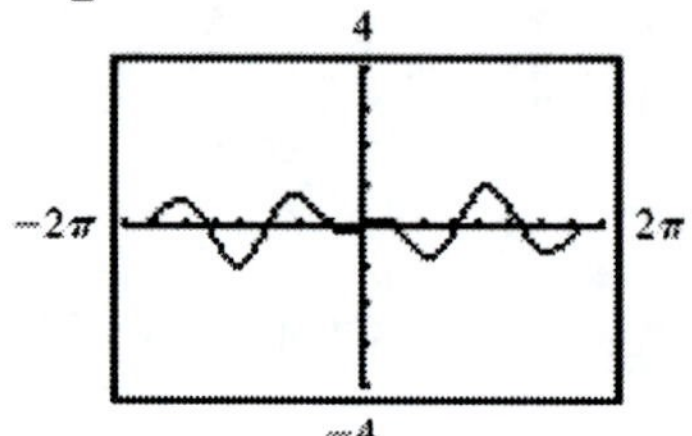

49. $$\cos u + \cos v = 2\cos\frac{u+v}{2}\cos\frac{u-v}{2} \qquad \text{Let } u = 3x \text{ and } v = x.$$

$$\cos 3x + \cos x = 2\cos\frac{3x+x}{2}\cos\frac{3x-x}{2} = 2\cos 2x \cos x = y2$$

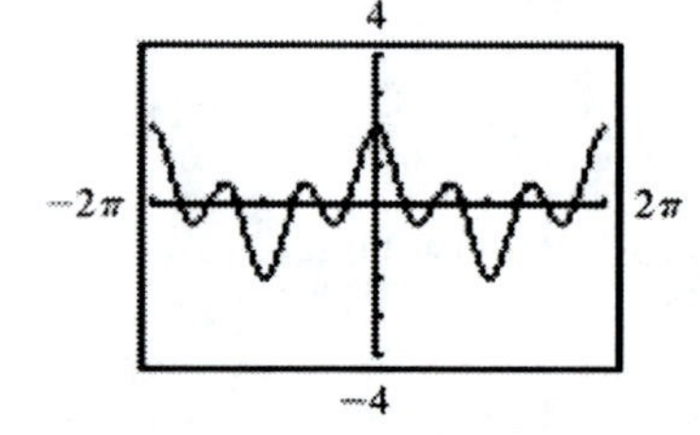

51. $$\sin u - \sin y = 2\cos\frac{u+v}{2}\sin\frac{u-v}{2} \qquad \text{Let } u = 2.1x \text{ and } v = 0.5x.$$

$$\sin 2.1x - \sin 0.5x = 2\cos\frac{2.1x+0.5x}{2}\sin\frac{2.1x-0.5x}{2} = 2\cos 1.3x \sin 0.8x = y2$$

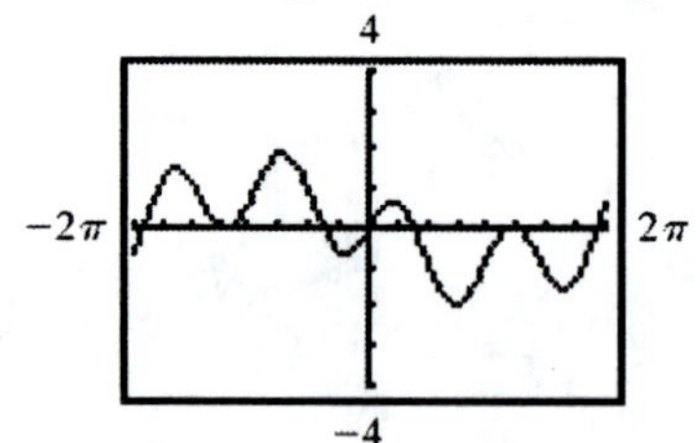

53.
$$\begin{aligned}
\sin x \sin y \sin z &= \sin x \,\frac{1}{2}[\cos(y-z) - \cos(y+z)] && \text{Product-sum identity}\\
&= \frac{1}{2}\sin x \cos(y-z) - \frac{1}{2}\sin x \cos(y+z) && \text{Algebra}\\
&= \frac{1}{2}\left\{\frac{1}{2}[\sin(x+y-z) + \sin(x-\{y-z\})]\right\}\\
&\quad - \frac{1}{2}\left\{\frac{1}{2}[\sin(x+y+z) + \sin(x-\{y+z\})]\right\} && \text{Product-sum identities}\\
&= \frac{1}{4}\sin(x+y-z) + \frac{1}{4}\sin(x-y+z) - \frac{1}{4}\sin(x+y+z)\\
&\quad - \frac{1}{4}\sin(x-y-z) && \text{Algebra}\\
&= \frac{1}{4}[\sin(x+y-z) - \sin(x-y-z) + \sin(z+x-y)\\
&\quad - \sin(x+y+z)] && \text{Algebra}\\
&= \frac{1}{4}[\sin(x+y-z) + \sin\{-(x-y-z)\} + \sin(z+x-y)\\
&\quad - \sin(x+y+z)] && \text{Identity for negatives}\\
&= \frac{1}{4}[\sin(x+y-z) + \sin(y+z-x) + \sin(z+x-y)\\
&\quad - \sin(x+y+z)] && \text{Algebra}
\end{aligned}$$

55. (A)

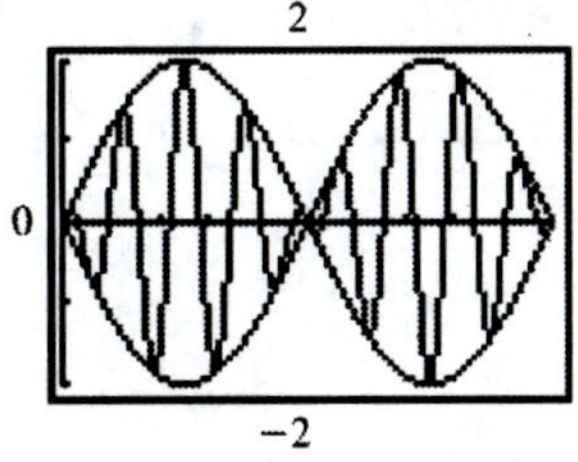

(B) $\cos u \sin v = \frac{1}{2}[\sin(u+v) - \sin(u-v)]$ Let $u = 16\pi x$ and $v = 2\pi x$.

$$\begin{aligned}
2\cos 16\pi x \sin 2\pi x &= 2\left(\frac{1}{2}\right)[\sin(16\pi x + 2\pi x) - \sin(16\pi x - 2\pi x)]\\
&= \sin(18\pi x) - \sin(14\pi x)
\end{aligned}$$

The graph is the same as in part (A).

57. (A)

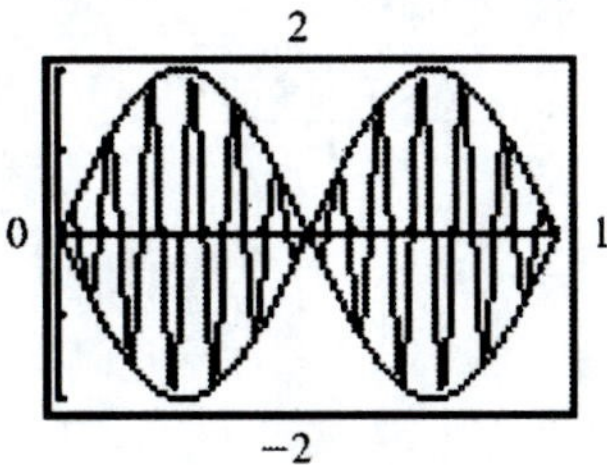

(B) $\sin u \sin v = \frac{1}{2}[\cos(u-v) - \cos(u+v)]$ Let $u = 24\pi x$ and $v = 2\pi x$.

$$\begin{aligned}
2\sin(24\pi x)\sin(2\pi x) &= 2\left(\frac{1}{2}\right)[\cos(24\pi x - 2\pi x) - \cos(24\pi x + 2\pi x)]\\
&= \cos(22\pi x) - \cos(26\pi x)
\end{aligned}$$

The graph is the same as in part (A).

59. Assume that t has units of seconds. The sum of the two tones is

$$y = k \sin 522\pi t + k \sin 512\pi t = k(\sin 522\pi t + \sin 512\pi t)$$

$$= k\left(2\sin\frac{522\pi t + 512\pi t}{2}\cos\frac{522\pi t - 512\pi t}{2}\right) \qquad \text{Sum-product identity}$$

This simplifies to $y = 2k \sin 517\pi t \cos 5\pi t$ Algebra

To find the beat frequency, we note

$$\text{Period of first tone} = \frac{2\pi}{B_1} = \frac{2\pi}{522\pi} = \frac{1}{261}$$

$$\text{Frequency of first tone} = \frac{1}{\text{Period}} = \frac{1}{1/261} = 261 \text{ Hz}$$

$$\text{Period of second tone} = \frac{2\pi}{B_2} = \frac{2\pi}{512\pi} = \frac{1}{256}$$

$$\text{Frequency of second tone} = \frac{1}{\text{Period}} = \frac{1}{1/256} = 256 \text{ Hz}$$

Beat frequency = Frequency of first tone – Frequency of second tone

$f_b = 261 \text{ Hz} - 256 \text{ Hz} = 5 \text{ Hz}$

61. (A)

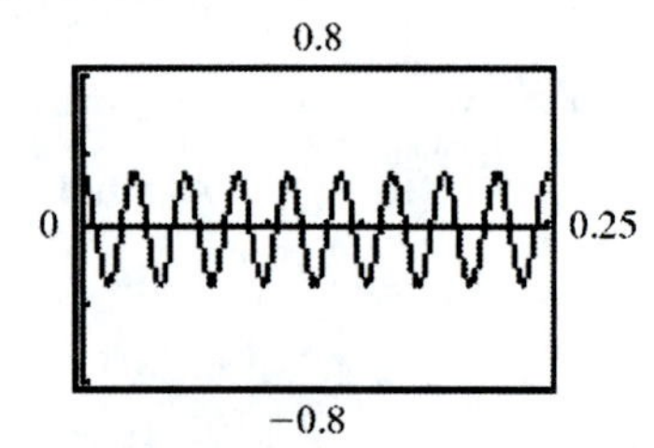

(B)

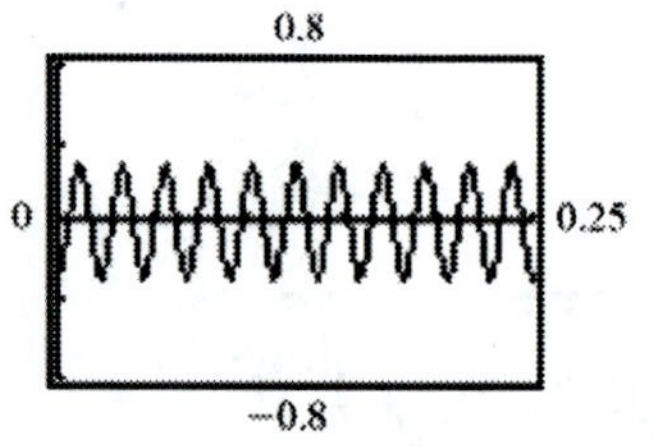

(C)

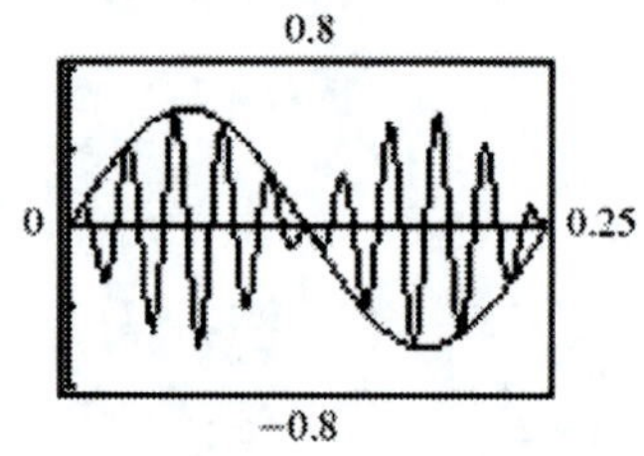

(D) $\cos x - \cos y = -2 \sin\frac{x+y}{2} \sin\frac{x-y}{2}$

Let $x = 72\pi t$ and $y = 88\pi t$

$$0.3(\cos 72\pi t - \cos 88\pi t) = (0.3)(-2) \sin\frac{72\pi t + 88\pi t}{2} \sin\frac{72\pi t - 88\pi t}{2}$$

$$= -0.6 \sin 80\pi t \sin(-8\pi t)$$

$$= 0.6 \sin 80\pi t \sin 8\pi t$$

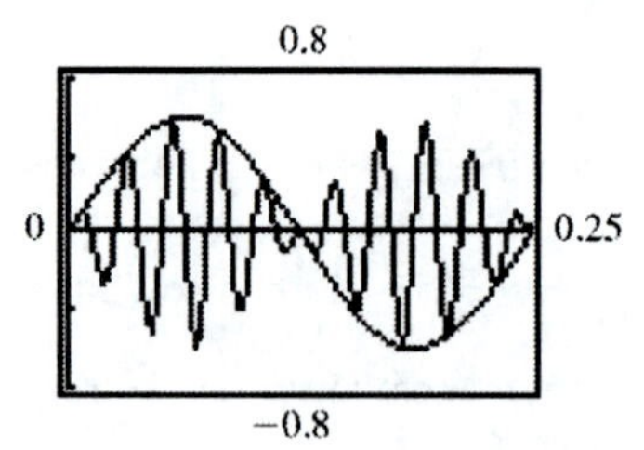

CHAPTER 4 REVIEW EXERCISE

1. Equation (1) is an identity, because it is true for all replacements of x by real numbers for which both sides are defined. Equation (2) is a conditional equation, because it is only true for $x = -2$ and $x = 3$, but not true for $x = 0$, for example.

2. There are many possible answers. We choose $x = 3$.
   Then left side $= x(x - 1)(x + 1)(x - 2)(x + 2) = 3(3 - 1)(3 + 1)(3 - 2)(3 + 2) = 120$
   right side $= 0$
   Since the left side and the right side are defined, but not equal, the equation is not an identity.

3. There are many possible answers. We choose $x = 0$.
   Then left side $= \sec(x + \pi) = \sec(0 + \pi) = -1$
   right side $= \sec x = \sec 0 = 1$
   Since the left side and the right side are defined, but not equal, the equation is not an identity.

4. There are many possible answers. We choose $x = \frac{\pi}{4}$.
   Then left side $= \tan x = \tan \frac{\pi}{4} = 1$
   right side $= \sin x \cos x = \sin \frac{\pi}{4} \cos \frac{\pi}{4} = \frac{1}{2}$
   Since the left side and the right side are defined, but not equal, the equation is not an identity.

5. There are many possible answers. We choose $x = \frac{\pi}{6}$ and $y = 0$.
   Then left side $= \sin 2(x - y) = \sin\left[2\left(\frac{\pi}{6} - 0\right)\right] = \frac{\sqrt{3}}{2}$
   right side $= 2 \sin x - 2 \sin y = 2 \sin \frac{\pi}{6} - 2 \sin 0 = 1$
   Since the left side and the right side are defined, but not equal, the equation is not an identity.

6. There are many possible answers. We choose $x = 0$ and $y = \frac{\pi}{4}$.
   Then left side $= \tan x \cot y = \tan 0 \cot \frac{\pi}{4} = 0$
   right side $= 1$
   Since the left side and the right side are defined, but not equal, the equation is not an identity.

7. $\cos^2 x (1 + \cot^2 x) = \cos^2 x \csc^2 x$ — Pythagorean identity
   $= \cos^2 x \dfrac{1}{\sin^2 x}$ — Reciprocal identity
   $= \dfrac{\cos^2 x}{\sin^2 x}$ — Algebra
   $= \cot^2 x$ — Quotient identity

8. $$\frac{\sin\alpha\csc\alpha}{\cot\alpha} = \frac{\sin\alpha\cdot\frac{1}{\sin\alpha}}{\cot\alpha}$$ Reciprocal identity

$$= \frac{1}{\cot\alpha}$$ Algebra

$$= \tan\alpha$$ Reciprocal identity

9. $$\frac{\sin^2 u-\cos^2 u}{\sin u\cos u} = \frac{\sin^2 u}{\sin u\cos u} - \frac{\cos^2 u}{\sin u\cos u}$$ Algebra

$$= \frac{\sin u}{\cos u} - \frac{\cos u}{\sin u}$$ Algebra

$$= \tan u - \cot u$$ Quotient identities

10. $$\frac{\sec\theta-\csc\theta}{\sec\theta\csc\theta} = \frac{\sec\theta}{\sec\theta\csc\theta} - \frac{\csc\theta}{\sec\theta\csc\theta}$$ Algebra

$$= \frac{1}{\csc\theta} - \frac{1}{\sec\theta}$$ Algebra

$$= \sin\theta - \cos\theta$$ Reciprocal identities

11. $$\begin{aligned}\cos(x+y) &= \cos x\cos y - \sin x\sin y\\ \cos(x+2\pi) &= \cos x\cos 2\pi - \sin x\sin 2\pi\\ &= \cos x(1) - \sin x(0)\\ &= \cos x\end{aligned}$$

12. $$\begin{aligned}\sin(x+y) &= \sin x\cos y + \cos x\sin y\\ \sin(x+y) &= \sin x\cos\pi + \cos x\sin\pi\\ &= \sin x(-1) + \cos x(0)\\ &= -\sin x\end{aligned}$$

13. $$\cos 2x = \cos 2(30^\circ) = \cos 60^\circ = \frac{1}{2}$$

$$\begin{aligned}1-2\sin^2 x &= 1-2\sin^2(30^\circ)\\ &= 1-2(\sin 30^\circ)^2\\ &= 1-2\left(\frac{1}{2}\right)^2 = 1-\frac{1}{2}\\ &= \frac{1}{2}\end{aligned}$$

14. $$\sin\frac{x}{2} = \sin\frac{\pi/2}{2} = \sin\frac{\pi}{4} = \frac{1}{\sqrt{2}}$$

Since $\frac{\pi}{4}$ is in the first quadrant, the sign of the square root is chosen to be positive.

$$\sqrt{\frac{1-\cos x}{2}} = \sqrt{\frac{1-\cos\pi/2}{2}} = \sqrt{\frac{1-0}{2}} = \sqrt{\frac{1}{2}} = \frac{1}{\sqrt{2}}$$

15. $\sin x \sin y = \frac{1}{2}[\cos(x-y) - \cos(x+y)]$ Let $x = 8t$ and $y = 5t$

$\sin 8t \sin 5t = \frac{1}{2}[\cos(8t-5t) - \cos(8t+5t)] = \frac{1}{2}(\cos 3t - \cos 13t) = \frac{1}{2}\cos 3t - \frac{1}{2}\cos 13t$

16. $\sin x + \sin y = 2\sin\frac{x+y}{2}\cos\frac{x-y}{2}$ Let $x = w$ and $y = 5w$

$\sin w + \sin 5w = 2\sin\frac{w+5w}{2}\cos\frac{w-5w}{2} = 2\sin 3w\cos(-2w) = 2\sin 3w\cos 2w$

17. $\frac{1-\cos^2 t}{\sin^3 t} = \frac{\sin^2 t}{\sin^3 t}$ Pythagorean identity

$= \frac{1}{\sin t}$ Algebra

$= \csc t$ Reciprocal identity

18. $\frac{(\cos\alpha - 1)^2}{\sin^2\alpha} = \frac{(\cos\alpha - 1)^2}{1-\cos^2\alpha}$ Pythagorean identity

$= \frac{(\cos\alpha-1)(\cos\alpha-1)}{(1-\cos\alpha)(1+\cos\alpha)}$ Algebra

$= \frac{(-1)(\cos\alpha-1)}{1+\cos\alpha}$ Algebra

$= \frac{1-\cos\alpha}{1+\cos\alpha}$ Algebra

Key algebraic steps: $\frac{(b-1)^2}{1-b^2} = \frac{(b-1)(b-1)}{(1-b)(1+b)} = \frac{(-1)(b-1)}{1+b} = \frac{1-b}{1+b}$

19. $\frac{1-\tan^2 x}{1-\tan^4 x} = \frac{1-\tan^2 x}{(1)^2 - (\tan^2 x)^2}$ Algebra

$= \frac{1-\tan^2 x}{(1-\tan^2 x)(1+\tan^2 x)}$ Algebra

$= \frac{1}{1+\tan^2 x}$ Algebra

$= \frac{1}{\sec^2 x}$ Pythagorean identity

$= \left(\frac{1}{\sec x}\right)^2$ Algebra

$= \cos^2 x$ Reciprocal identity

Key algebraic steps: $\frac{1-c^2}{1-c^4} = \frac{1-c^2}{(1)^2-(c^2)^2} = \frac{1-c^2}{(1-c^2)(1+c^2)} = \frac{1}{1+c^2}$

20. $\cot^2 x \cos^2 x = (\csc^2 x - 1)\cos^2 x$ Pythagorean identity
$= \csc^2 x \cos^2 x - \cos^2 x$ Algebra
$= \left(\dfrac{1}{\sin x}\right)^2 \cos^2 x - \cos^2 x$ Reciprocal identity
$= \left(\dfrac{\cos x}{\sin x}\right)^2 - \cos^2 x$ Algebra
$= \cot^2 x - \cos^2 x$ Quotient identity

21. The equation is not an identity. For example, let $x = \dfrac{\pi}{2}$. Then both sides are defined, however, the left side is $\sin \dfrac{\pi}{2}$ or 1 while the right side is 0, hence the equation is not true for all values of x for which both sides are defined.

22. Graph each side of the equation in the same viewing window and observe that the graphs are not the same, except where the graph of y1 = sin x crosses the x axis. (Note that the graph of y2 = 0 is the x axis.)

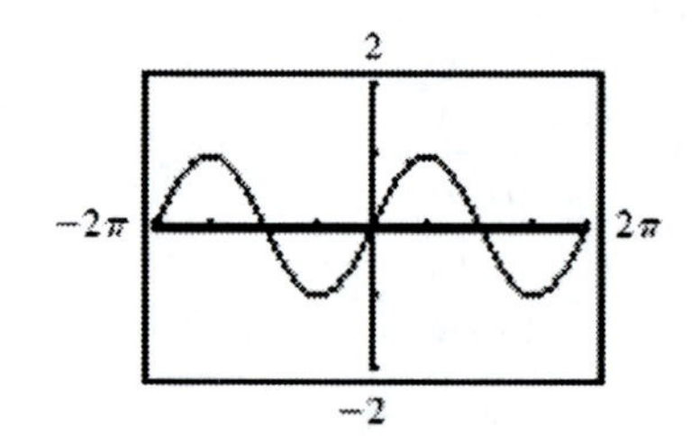

23. No. For all values of x, the left side equals 1 by the Pythagorean identity.

24. No. For example, if $x = \dfrac{\pi}{8}$, $\dfrac{\tan x}{\cot x} = \dfrac{\tan \frac{\pi}{8}}{\cot \frac{\pi}{8}} = \tan^2 \dfrac{\pi}{8} \approx 0.172 \neq 1$.

25. Yes. The left side, $\tan\left(x - \dfrac{\pi}{2}\right) = \dfrac{\sin\left(x - \frac{\pi}{2}\right)}{\cos\left(x - \frac{\pi}{2}\right)} = \dfrac{-\cos x}{\sin x} = -\cot x$ for all values of x for which both are defined.

26. No. For example, if $x = \dfrac{\pi}{4}$, $\sec\left(\dfrac{\pi}{2} - x\right) = \sec\left(\dfrac{\pi}{2} - \dfrac{\pi}{4}\right) = \sec \dfrac{\pi}{4} = \sqrt{2}$, but $\cos x = \cos \dfrac{\pi}{4} = \dfrac{1}{\sqrt{2}}$.

27. No. For example, if $x = 0$, $\cos 2x = \cos(2 \cdot 0) = 1$, but $2 \sin^2 x - 1 = 2(\sin 0)^2 - 1 = 0 - 1 = -1$.

28. Yes. The left side can be shown to equal the right side for all values of x for which both are defined as follows:

$\dfrac{\tan 2x}{\tan x} = \tan 2x \div \tan x$	Algebra
$= \dfrac{2\tan x}{1-\tan^2 x} \div \tan x$	Double-angle identity
$= \dfrac{2\tan x}{1-\tan^2 x} \cdot \dfrac{1}{\tan x}$	Algebra
$= \dfrac{2}{1-\tan^2 x}$	Algebra
$= \dfrac{2}{(1-\tan x)(1+\tan x)}$	Algebra

29. No. For example, if $x = 0$, $\cos^2 x = (\cos 0)^2 = 1^2 = 1$, but $\dfrac{1}{2}\cos 2x = \dfrac{1}{2}\cos(2 \cdot 0) = \dfrac{1}{2}\cos 0 = \dfrac{1}{2}$.

30. Yes. The left side, $2 \sin x \cos 5x = 2 \cdot \dfrac{1}{2}[\sin(x + 5x) + \sin(x - 5x)] = \sin 6x + \sin(-4x)$ $= \sin 6x - \sin 4x$ for all values of x.

31.

$\dfrac{\sin x}{1-\cos x} = \dfrac{\sin x}{1-\cos x}\dfrac{1+\cos x}{1+\cos x}$	Algebra
$= \dfrac{\sin x\,(1+\cos x)}{1-\cos^2 x}$	Algebra
$= \dfrac{\sin x\,(1+\cos x)}{\sin^2 x}$	Pythagorean identity
$= \dfrac{\sin x}{\sin^2 x}(1+\cos x)$	Algebra
$= \dfrac{1}{\sin x}(1+\cos x)$	Algebra
$= \csc x\,(1+\cos x)$	Reciprocal identity

32.

$\dfrac{1-\tan^2 x}{1-\cot^2 x} = \dfrac{1-\tan^2 x}{1-\dfrac{1}{\tan^2 x}}$	Reciprocal identity
$= \dfrac{\tan^2 x(1-\tan^2 x)}{\tan^2 x\left(1-\dfrac{1}{\tan^2 x}\right)}$	Algebra
$= \dfrac{\tan^2 x\,(1-\tan^2 x)}{\tan^2 x - 1}$	Algebra

$= \frac{-\tan^2 x(\tan^2 x - 1)}{\tan^2 x - 1}$ Algebra

$= -\tan^2 x$ Algebra

$= -(\sec^2 x - 1)$ Pythagorean identity

$= 1 - \sec^2 x$ Algebra

Key algebraic steps: $\frac{1 - a^2}{1 - \frac{1}{a^2}} = \frac{a^2(1-a^2)}{a^2\left(1 - \frac{1}{a^2}\right)} = \frac{a^2(1-a^2)}{a^2 - 1} = \frac{-a^2(a^2-1)}{a^2 - 1} = -a^2$

33. $\tan(x + \pi) = \frac{\tan x + \tan \pi}{1 - \tan x \tan \pi}$ Sum identity

$= \frac{\tan x + 0}{1 - \tan x \cdot 0}$ Known values

$= \tan x$ Algebra

34. $1 - (\cos\beta - \sin\beta)^2 = 1 - (\cos^2\beta - 2\sin\beta\cos\beta + \sin^2\beta)$ Algebra

$= 1 - (\cos^2\beta + \sin^2\beta - 2\sin\beta\cos\beta)$ Algebra

$= 1 - (1 - 2\sin\beta\cos\beta)$ Pythagorean identity

$= 1 - 1 + 2\sin\beta\cos\beta$ Algebra

$= 2\sin\beta\cos\beta$ Algebra

$= \sin 2\beta$ Double-angle identity

35. $\frac{\sin 2x}{\cot x} = \frac{2\sin x\cos x}{\cot x}$ Double-angle identity

$= \frac{2\sin x\cos x}{\frac{\cos x}{\sin x}}$ Quotient identity

$= 2\sin x\cos x \div \frac{\cos x}{\sin x}$ Algebra

$= 2\sin x\cos x \cdot \frac{\sin x}{\cos x}$ Algebra

$= 2\sin^2 x$ Algebra

$= 2\sin^2 x - 1 + 1$ Algebra

$= 1 + (2\sin^2 x - 1)$ Algebra

$= 1 - (1 - 2\sin^2 x)$ Algebra

$= 1 - \cos 2x$ Double-angle identity

36. $\dfrac{2\tan x}{1+\tan^2 x} = \dfrac{2\dfrac{\sin x}{\cos x}}{1+\dfrac{\sin^2 x}{\cos^2 x}}$ Quotient identity

$= \dfrac{\cos^2 x \cdot 2\dfrac{\sin x}{\cos x}}{\cos^2 x \cdot 1 + \cos^2 x \cdot \dfrac{\sin^2 x}{\cos^2 x}}$ Algebra

$= \dfrac{2\sin x\cos x}{\cos^2 x + \sin^2 x}$ Algebra

$= \dfrac{2\sin x\cos x}{1}$ Pythagorean identity

$= 2\sin x\cos x$ Algebra

$= \sin 2x$ Double-angle identity

37. $2\csc 2x = \dfrac{2}{\sin 2x}$ Reciprocal identity

$= \dfrac{2}{2\sin x\cos x}$ Double-angle identity

$= \dfrac{1}{\sin x\cos x}$ Algebra

$= \dfrac{\sin^2 x + \cos^2 x}{\sin x\cos x}$ Pythagorean identity

$= \dfrac{\sin^2 x}{\sin x\cos x} + \dfrac{\cos^2 x}{\sin x\cos x}$ Algebra

$= \dfrac{\sin x}{\cos x} + \dfrac{\cos x}{\sin x}$ Algebra

$= \tan x + \cot x$ Quotient identities

38. $\dfrac{\cot\frac{x}{2}}{1+\cos x} = \cot\dfrac{x}{2} \cdot \dfrac{1}{1+\cos x}$ Algebra

$= \dfrac{1}{\tan\frac{x}{2}} \cdot \dfrac{1}{1+\cos x}$ Reciprocal identity

$= \dfrac{1}{\dfrac{\sin x}{1+\cos x}} \cdot \dfrac{1}{1+\cos x}$ Half-angle identity

$= \dfrac{1+\cos x}{\sin x} \cdot \dfrac{1}{1+\cos x}$ Algebra

$= \dfrac{1}{\sin x}$ Algebra

$= \csc x$ Reciprocal identity

39. $\frac{\sin(x-y)}{\sin(x+y)} = \frac{\sin x \cos y - \cos x \sin y}{\sin x \cos y + \cos x \sin y}$ Sum and difference identities

$= \frac{\frac{\sin x \cos y}{\cos x \cos y} - \frac{\cos x \sin y}{\cos x \cos y}}{\frac{\sin x \cos y}{\cos x \cos y} + \frac{\cos x \sin y}{\cos x \cos y}}$ Algebra

$= \frac{\frac{\sin x}{\cos x} - \frac{\sin y}{\cos y}}{\frac{\sin x}{\cos x} + \frac{\sin y}{\cos y}}$ Algebra

$= \frac{\tan x - \tan y}{\tan x + \tan y}$ Quotient identity

40. $\csc 2x = \frac{1}{\sin 2x}$ Reciprocal identity

$= \frac{1}{2 \sin x \cos x}$ Double-angle identity

$= \frac{\sin^2 x + \cos^2 x}{2 \sin x \cos x}$ Pythagorean identity

$= \frac{\sin^2 x}{2 \sin x \cos x} + \frac{\cos^2 x}{2 \sin x \cos x}$ Algebra

$= \frac{\sin x}{2 \cos x} + \frac{\cos x}{2 \sin x}$ Algebra

$= \frac{1}{2}\frac{\sin x}{\cos x} + \frac{1}{2}\frac{\cos x}{\sin x}$ Algebra

$= \frac{1}{2} \tan x + \frac{1}{2} \cot x$ Quotient identities

$= \frac{\tan x + \cot x}{2}$ Algebra

41. $\frac{2 - \sec^2 x}{\sec^2 x} = \frac{1 - \frac{1}{\cos^2 x}}{\frac{1}{\cos^2 x}}$ Reciprocal identity

$= \frac{\cos^2 x \cdot 2 - \cos^2 x \cdot \frac{1}{\cos^2 x}}{\cos^2 x \cdot \frac{1}{\cos^2 x}}$ Algebra

$= \frac{2 \cos^2 x - 1}{1}$ Algebra

$= 2 \cos^2 x - 1$ Algebra

$= \cos 2x$ Double-angle identity

42. $\tan\frac{x}{2} = \frac{1-\cos x}{\sin x}$ Half-angle identity

$= \dfrac{\frac{1}{\cos x} - \frac{\cos x}{\cos x}}{\frac{\sin x}{\cos x}}$ Algebra

$= \dfrac{\sec x - \frac{\cos x}{\cos x}}{\frac{\sin x}{\cos x}}$ Reciprocal identity

$= \dfrac{\sec x - 1}{\frac{\sin x}{\cos x}}$ Algebra

$= \dfrac{\sec x - 1}{\tan x}$ Quotient identity

43. $\dfrac{\sin t + \sin 5t}{\cos t + \cos 5t} = \dfrac{2\sin\frac{t+5t}{2}\cos\frac{t-5t}{2}}{2\cos\frac{t+5t}{2}\cos\frac{t-5t}{2}}$ Sum-product identities

$= \dfrac{2\sin 3t\cos(-2t)}{2\cos 3t\cos(-2t)}$ Algebra

$= \dfrac{\sin 3t}{\cos 3t}$ Algebra

$= \tan 3t$ Quotient identity

44. $\dfrac{\sin x + \sin y}{\cos x - \cos y} = \dfrac{2\sin\frac{x+y}{2}\cos\frac{x-y}{2}}{-2\sin\frac{x+y}{2}\sin\frac{x-y}{2}}$ Sum-product identities

$= -\dfrac{\cos\frac{x-y}{2}}{\sin\frac{x-y}{2}}$ Algebra

$= -\cot\frac{x-y}{2}$ Quotient identity

45. $\dfrac{\cos x - \cos y}{\cos x + \cos y} = \dfrac{-2\sin\frac{x+y}{2}\sin\frac{x-y}{2}}{2\cos\frac{x+y}{2}\cos\frac{x-y}{2}}$ Sum-product identities

$= -\dfrac{\sin\frac{x+y}{2}}{\cos\frac{x+y}{2}}\dfrac{\sin\frac{x-y}{2}}{\cos\frac{x-y}{2}}$ Algebra

$= -\tan\frac{x+y}{2}\tan\frac{x-y}{2}$ Quotient identity

46. $$\sin x \sin y = \frac{1}{2}[\cos(x-y) - \cos(x+y)] \qquad \text{Let } x = 165° \text{ and } y = 15°.$$

$$\sin 165° \sin 15° = \frac{1}{2}[\cos(165° - 15°) - \cos(165° + 15°)] = \frac{1}{2}[\cos 150° - \cos 180°]$$

$$= \frac{1}{2}\left[-\frac{\sqrt{3}}{2} - (-1)\right] = -\frac{\sqrt{3}}{4} + \frac{1}{2} \text{ or } \frac{1}{2} - \frac{\sqrt{3}}{4}$$

47. $$\cos x - \cos y = -2 \sin \frac{x+y}{2} \sin \frac{x-y}{2} \qquad \text{Let } x = 165° \text{ and } y = 75°.$$

$$\cos 165° - \cos 75° = -2 \sin \frac{165° + 75°}{2} \sin \frac{165° - 75°}{2} = -2 \sin 120° \sin 45°$$

$$= -2\left(\frac{\sqrt{3}}{2}\right)\left(\frac{\sqrt{2}}{2}\right) = -\frac{\sqrt{6}}{2}$$

48. *Find sin x:* We start with the Pythagorean identity $\sin^2 x + \cos^2 x = 1$ and solve for $\sin x$:

$$\sin x = \pm\sqrt{1 - \cos^2 x}$$

Since $\cos x$ and $\tan x$ are negative, x is associated with the second quadrant, where $\sin x$ is positive; hence,

$$\sin x = \sqrt{1 - \cos^2 x} = \sqrt{1 - \left(-\frac{2}{3}\right)^2} = \sqrt{\frac{5}{9}} = \frac{\sqrt{5}}{3}$$

Find tan x: $\tan x = \dfrac{\sin x}{\cos x} = \dfrac{\sqrt{5}/3}{-2/3} = -\dfrac{\sqrt{5}}{2}$ *Find cot x:* $\cot x = \dfrac{1}{\tan x} = \dfrac{1}{-\sqrt{5}/2} = -\dfrac{2}{\sqrt{5}}$

Find sec x: $\sec x = \dfrac{1}{\cos x} = \dfrac{1}{-2/3} = -\dfrac{3}{2}$ *Find csc x:* $\csc x = \dfrac{1}{\sin x} = \dfrac{1}{\sqrt{5}/3} = \dfrac{3}{\sqrt{5}}$

49. First draw a reference triangle in the first quadrant and find $\sin x$ and $\cos x$: $r = \sqrt{3^2 + 4^2} = 5$; $\sin x = \dfrac{4}{5}$, $\cos x = \dfrac{3}{5}$. Now use the double-angle identities:

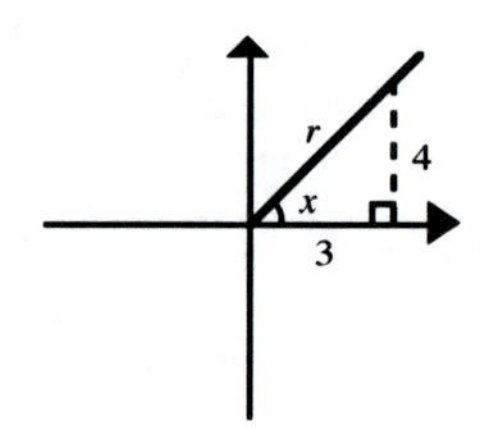

$$\sin 2x = 2 \sin x \cos x = 2\left(\frac{4}{5}\right)\left(\frac{3}{5}\right) = \frac{24}{25}$$

$$\cos 2x = 2\cos^2 x - 1 = 2\left(\frac{3}{5}\right)^2 - 1 = \frac{18}{25} - 1 = -\frac{7}{25}$$

$$\tan 2x = \frac{2\tan x}{1 - \tan^2 x} = \frac{2(4/3)}{1 - (4/3)^2} = \frac{(8/3)}{1 - (16/9)} = -\frac{24}{7}$$

50. We are given $\cos x$. We can find $\sin \dfrac{x}{2}$ and $\cos \dfrac{x}{2}$ from the half-angle identities, after determining their sign, as follows: if $-\pi < x < -\dfrac{\pi}{2}$, than $-\dfrac{\pi}{2} < \dfrac{x}{2} < -\dfrac{\pi}{4}$. Thus, $\dfrac{x}{2}$ is in the fourth quadrant, where sine is negative and cosine is positive. Using half-angle identities, we obtain:

$$\sin \frac{x}{2} = -\sqrt{\frac{1 - \cos x}{2}} = -\sqrt{\frac{1 - (-5/13)}{2}} = -\sqrt{\frac{9}{13}} \text{ or } -\frac{3}{\sqrt{13}}$$

$$\cos\frac{x}{2} = \sqrt{\frac{1+\cos x}{2}} = \sqrt{\frac{1+(-5/13)}{2}} = \sqrt{\frac{4}{13}} \text{ or } \frac{2}{\sqrt{13}}$$

$$\tan\frac{x}{2} = \frac{\sin\left(\frac{x}{2}\right)}{\cos\left(\frac{x}{2}\right)} = \frac{-3/\sqrt{13}}{2/\sqrt{13}} = -\frac{3}{2}$$

51. $\tan\left(x+\frac{\pi}{4}\right) = \dfrac{\tan x + \tan\frac{\pi}{4}}{1-\tan x\tan\frac{\pi}{4}}$ Sum identity

$= \dfrac{\tan x + 1}{1-\tan x\cdot 1}$ Known values

$= \dfrac{\tan x + 1}{1-\tan x}$ Algebra

Graph y1 = $\tan\left(x+\frac{\pi}{4}\right)$ and y2 = $\dfrac{\tan x + 1}{1-\tan x}$.

Use the trace function to confirm that the graphs are identical.

52. $\cos 1.5x\cos 0.3x - \sin 1.5x\sin 0.3x$ $= \cos(1.5x + 0.3x)$ Sum identity

$= \cos(1.8x)$ Algebra

Graph y1 = $\cos 1.5x\cos 0.3x - \sin 1.5x\sin 0.3x$ and y2 = $\cos(1.8x)$. Use the trace function to confirm that the graphs are identical.

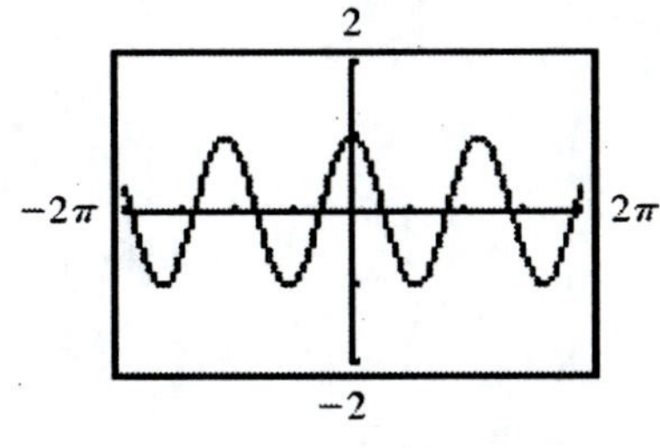

53.

The two graphs coincide on the interval $-2\pi \le x \le 0$. $\sin\frac{x}{2} = -\sqrt{\frac{1-\cos x}{2}}$ is an identity on this interval.

54. (A) Graph both sides of the equation in the same viewing window.

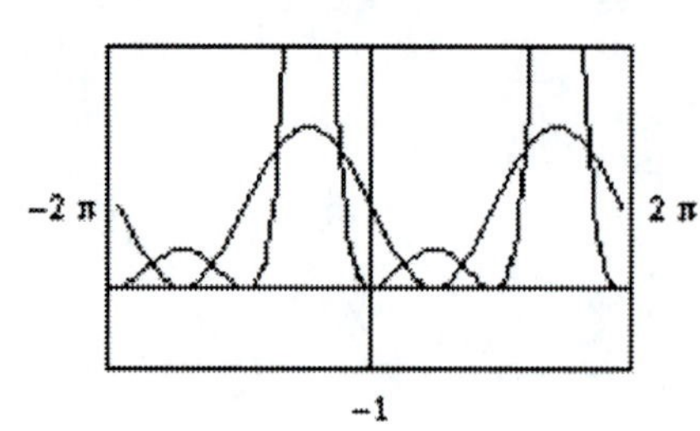

$\frac{\sin^2 x}{1+\sin x} = 1 - \sin x$ is not an identity, since the graphs do not match.

Try $x = 0$

Left side: $\frac{\sin^2 0}{1+\sin 0} = \frac{0}{1+0} = 0$

Right side: $1 - \sin 0 = 1 - 0 = 1$

This verifies that the equation is not an identity.

(B) Graph both sides of the equation in the same viewing window.

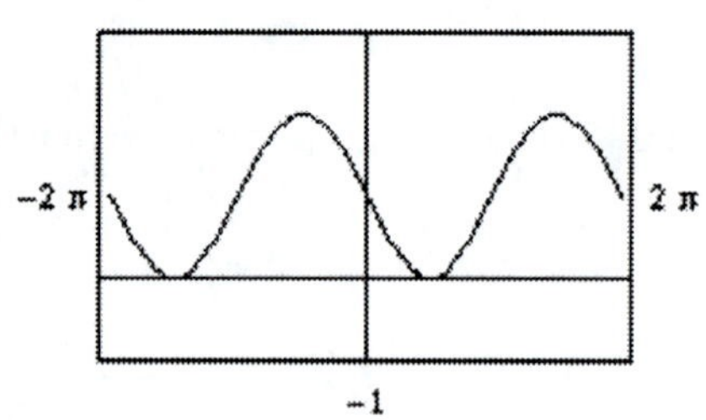

$\frac{\cos^2 x}{1+\sin x} = 1 - \sin x$ appears to be an identity which we verify.

$\frac{\cos^2 x}{1+\sin x} = \frac{1-\sin^2 x}{1+\sin x}$ Pythagorean identity

$= \frac{(1+\sin x)(1-\sin x)}{1+\sin x}$ Algebra

$= 1 - \sin x$ Algebra

55. To obtain $\sin x$ and $\cos x$ from $\sec 2x$, we use the half-angle identities with x replaced by $2x$. Thus,

$$\sin\frac{x}{2} = \pm\sqrt{\frac{1-\cos x}{2}} \quad \text{becomes} \quad \sin\frac{2x}{2} = \pm\sqrt{\frac{1-\cos 2x}{2}} \quad \text{or } \sin x = \pm\sqrt{\frac{1-\cos 2x}{2}}$$

$$\cos\frac{x}{2} = \pm\sqrt{\frac{1+\cos x}{2}} \quad \text{becomes} \quad \cos\frac{2x}{2} = \pm\sqrt{\frac{1+\cos x}{2}} \quad \text{or } \cos x = \pm\sqrt{\frac{1+\cos x}{2}}$$

To obtain $\cos 2x$ from $\sec 2x$, we use the reciprocal identity

$$\sec 2x = \frac{1}{\cos 2x} \qquad \cos 2x = \frac{1}{\sec 2x} = \frac{1}{-\frac{13}{12}} = -\frac{12}{13}$$

Since $-\frac{\pi}{2} < x < 0$, x is in the fourth quadrant, where sine is negative and cosine is positive. Thus,

$$\sin x = -\sqrt{\frac{1-\cos 2x}{2}} = -\sqrt{\frac{1-(-12/13)}{2}} = -\sqrt{\frac{25}{26}} \text{ or } -\frac{5}{\sqrt{26}}$$

$$\cos x = \sqrt{\frac{1+\cos 2x}{2}} = \sqrt{\frac{1+(-12/13)}{2}} = \sqrt{\frac{1}{26}} \text{ or } \frac{1}{\sqrt{26}}$$

$$\tan x = \frac{\sin x}{\cos x} = \frac{-5/\sqrt{26}}{1/\sqrt{26}} = -5$$

56.
$$\frac{\cot x}{\csc x + 1} = \frac{(\csc x - 1)\cot x}{(\csc x - 1)(\csc x + 1)} \qquad \text{Algebra}$$

$$= \frac{(\csc x - 1)\cot x}{\csc^2 x - 1} \qquad \text{Algebra}$$

$$= \frac{(\csc x - 1)\cot x}{\cot^2 x} \qquad \text{Pythagorean identity}$$

$$= \frac{\csc x - 1}{\cot x} \qquad \text{Algebra}$$

57.
$$\cot 3x = \frac{1}{\tan 3x} \qquad \text{Reciprocal identity}$$

$$= 1 \div \tan 3x \qquad \text{Algebra}$$

$$= 1 \div \tan(2x + x) \qquad \text{Algebra}$$

$$= 1 \div \frac{\tan 2x + \tan x}{1 - \tan 2x \tan x} \qquad \text{Sum identity for tangent}$$

$$= \frac{1 - \tan 2x \tan x}{\tan 2x + \tan x} \qquad \text{Algebra}$$

$$= \frac{1 - \frac{2\tan x}{1-\tan^2 x}\cdot \tan x}{\frac{2\tan x}{1-\tan^2 x} + \tan x} \qquad \text{Double-angle identity}$$

$$= \frac{(1 - \tan^2 x) \cdot 1 - \frac{(1 - \tan^2 x)}{1} \cdot \frac{2 \tan x}{1 - \tan^2 x} \cdot \tan x}{\frac{2 \tan x}{1 - \tan^2 x} \cdot \frac{(1 - \tan^2 x)}{1} + \tan x (1 - \tan^2 x)} \quad \text{Algebra}$$

$$= \frac{1 - \tan^2 x - 2 \tan x \cdot \tan x}{2 \tan x + \tan x - \tan^3 x} \quad \text{Algebra}$$

$$= \frac{1 - \tan^2 x - 2 \tan^2 x}{3 \tan x - \tan^3 x} \quad \text{Algebra}$$

$$= \frac{1 - 3 \tan^2 x}{3 \tan x - \tan^3 x} \quad \text{Algebra}$$

$$= \frac{(-1)(1 - 3 \tan^2 x)}{(-1)(3 \tan x - \tan^3 x)} \quad \text{Algebra}$$

$$= \frac{3 \tan^2 x - 1}{\tan^3 x - 3 \tan x} \quad \text{Algebra}$$

58. The definitions of the functions involved a point (a, b) on a unit circle. Recall:

$\sin x = b \qquad \cos x = a \qquad \tan x = \frac{b}{a} \qquad a \neq 0$

Thus, $\tan x = \frac{b}{a} = \frac{\sin x}{\cos x}$

59. We are to show $\cos(x + 2k\pi) = \cos x$. But, $\cos(x + 2k\pi) = \cos x \cos 2k\pi - \sin x \sin 2k\pi$ by the sum identity for cosine. A quick sketch shows $(a, b) = (1, 0)$, $r = 1$, $\cos 2k\pi = 1$, and $\sin 2k\pi = 0$. Hence,

$$\cos(x + 2k\pi) = \cos x \cdot 1 - \sin x \cdot 0$$
$$\cos(x + 2k\pi) = \cos x$$

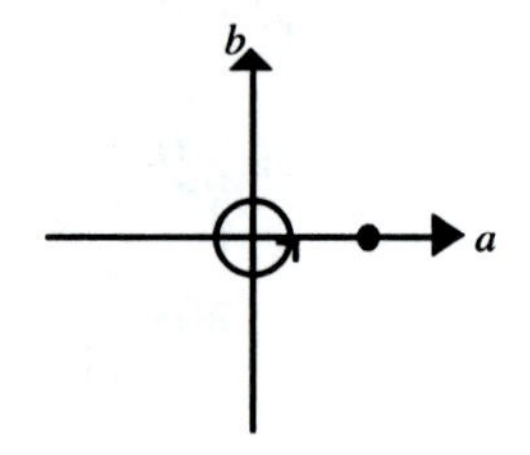

60. We are to show $\cot(x + k\pi) = \cot x$. But, $\cot(x + k\pi) = \frac{1}{\tan(x + k\pi)}$

$$= \frac{1}{\frac{\tan x + \tan k\pi}{1 - \tan x \tan k\pi}} \quad \text{Sum identity}$$

$$= \frac{1 - \tan x \tan k\pi}{\tan x + \tan k\pi} \quad \text{Algebra}$$

A quick sketch shows $(a, b) = (\pm 1, 0)$, $r = 1$.

$\tan k\pi = \frac{0}{\pm 1} = 0$. Hence, $\cot(x + k\pi) = \frac{1 - \tan x \cdot 0}{\tan x + 0}$ Known values

$= \frac{1}{\tan x}$ Algebra

$= \cot x$ Reciprocal identity

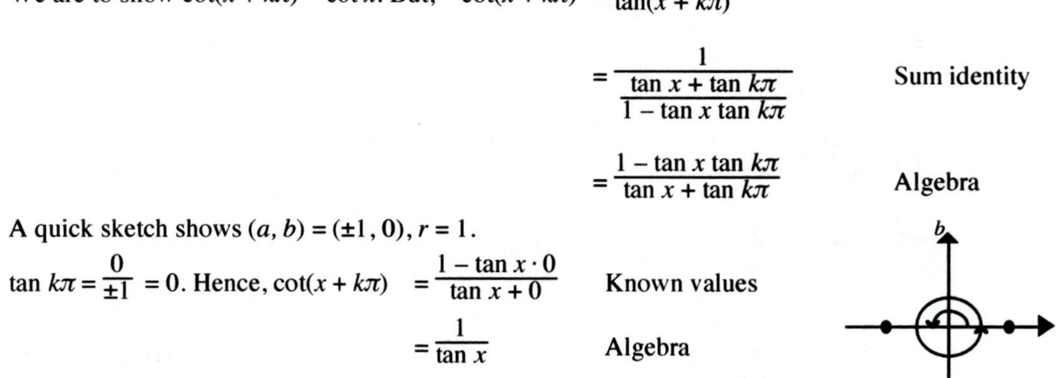

61. If we let $x + y = u, x - y = v$, and solve the system, we obtain $2x = u + v, 2y = u - v$; that is, $x = \frac{u+v}{2}, y = \frac{u-v}{2}$. Hence,

$$\sin x \sin y = \frac{1}{2}[\cos(x-y) - \cos(x+y)] \text{ becomes } \sin\frac{u+v}{2}\sin\frac{u-v}{2} = \frac{1}{2}[\cos v - \cos u]$$

upon substitution. Solving for the quantity in the brackets, we obtain

$$\frac{1}{2}[\cos v - \cos u] = \sin\frac{u+v}{2}\sin\frac{u-v}{2};\quad \cos v - \cos u = 2\sin\frac{u+v}{2}\sin\frac{u-v}{2}$$

62. The graph of $f(x)$ is shown in the figure. The graph appears to be a basic cosine curve with period 2π, amplitude $= \frac{1}{2}(y\text{ max} - y\text{ min}) = \frac{1}{2}(6-2) = 2$, displaced upward by $k = 4$ units. It appears that $g(x) = 4 + 2\cos x$ would be an appropriate choice. We verify $f(x) = g(x)$ as follows:

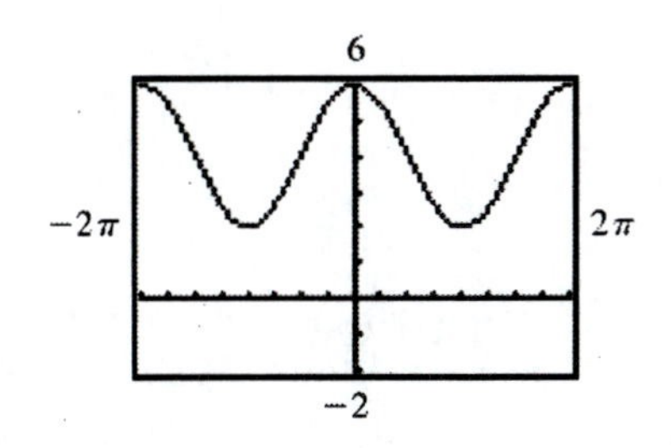

$$f(x) = \frac{3\sin^2 x}{1-\cos x} + \frac{\tan^2 x\cos^2 x}{1+\cos x}$$

$$= \frac{3\sin^2 x}{1-\cos x} + \frac{\frac{\sin^2 x}{\cos^2 x}\cos^2 x}{1+\cos x} \qquad \text{Quotient identity}$$

$$= \frac{3\sin^2 x}{1-\cos x} + \frac{\sin^2 x}{1+\cos x} \qquad \text{Algebra}$$

$$= \frac{3(1-\cos^2 x)}{1-\cos x} + \frac{1-\cos^2 x}{1+\cos x} \qquad \text{Pythagorean identity}$$

$$= \frac{3(1+\cos x)(1-\cos x)}{1-\cos x} + \frac{(1-\cos x)(1+\cos x)}{1+\cos x} \qquad \text{Algebra}$$

$$= 3 + 3\cos x + 1 - \cos x \qquad \text{Algebra}$$

$$= 4 + 2\cos x = g(x)$$

63. The graph of $f(x)$ is shown in the figure. The graph appears to have vertical asymptotes $x = -\frac{\pi}{4}$ and $x = \frac{\pi}{4}$, x intercepts $-\frac{\pi}{2}$, 0, and $\frac{\pi}{2}$, and period $\frac{\pi}{2}$. It appears that $g(x) = \tan 2x$ would be an appropriate choice. We verify $f(x) = g(x)$ as follows:

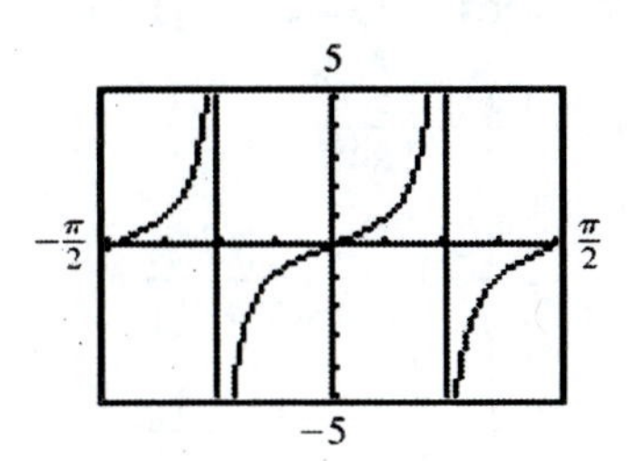

$$f(x) = \frac{\sin x}{\cos x - \sin x} + \frac{\sin x}{\cos x + \sin x}$$

$$= \frac{\sin x(\cos x + \sin x)}{(\cos x - \sin x)(\cos x + \sin x)} + \frac{\sin x(\cos x - \sin x)}{(\cos x + \sin x)(\cos x - \sin x)} \qquad \text{Algebra}$$

$$= \frac{\sin x(\cos x + \sin x) + \sin x(\cos x - \sin x)}{(\cos x - \sin x)(\cos x + \sin x)} \qquad \text{Algebra}$$

$= \dfrac{\sin x \cos x + \sin^2 x + \sin x \cos x - \sin^2 x}{\cos^2 x - \sin^2 x}$ Algebra

$= \dfrac{2 \sin x \cos x}{\cos^2 x - \sin^2 x}$ Algebra

$= \dfrac{\sin 2x}{\cos 2x}$ Double-angle identities

$= \tan 2x = g(x)$ Quotient identity

Key algebraic steps: $\dfrac{a}{b-a} + \dfrac{a}{b+a} = \dfrac{a(b+a)}{(b-a)(b+a)} + \dfrac{a(b-a)}{(b+a)(b-a)} = \dfrac{a(b+a)+a(b-a)}{(b-a)(b+a)}$

$= \dfrac{ab + a^2 + ab - a^2}{b^2 - a^2} = \dfrac{2ab}{b^2 - a^2}$

64. The graph of $f(x)$ is shown in the figure. The graph appears to be an upside down cosine curve with period π, amplitude $= \frac{1}{2}(y \text{ max} - y \text{ min}) = \frac{1}{2}(3 - 1) = 1$, displaced upward by $k = 2$ units. It appears that $g(x) = 2 - \cos 2x$ would be an appropriate choice. We verify $f(x) = g(x)$ as follows:

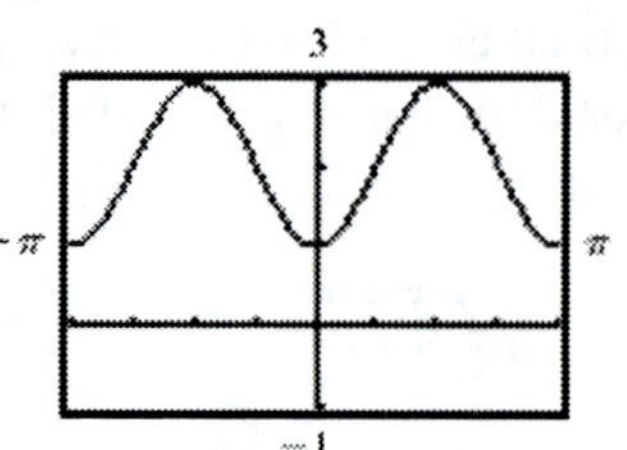

$f(x) = 3 \sin^2 x + \cos^2 x$

$= 3 \sin^2 x + 1 - \sin^2 x$ Pythagorean identity

$= 2 \sin^2 x + 1$ Algebra

$= 2 - (1 - 2 \sin^2 x)$ Algebra

$= 2 - \cos 2x$ Double-angle identity

65. The graph of $f(x)$ is shown in the figure. The graph appears to have vertical asymptotes $x = -\frac{3\pi}{4}, -\frac{\pi}{4}, \frac{\pi}{4}$, and $\frac{3\pi}{4}$, and period π. It appears to have high and low points with y coordinates -3 and -1, respectively. It appears that $g(x) = -2 + \sec 2x$ would be an appropriate choice. We verify $f(x) = g(x)$ as follows:

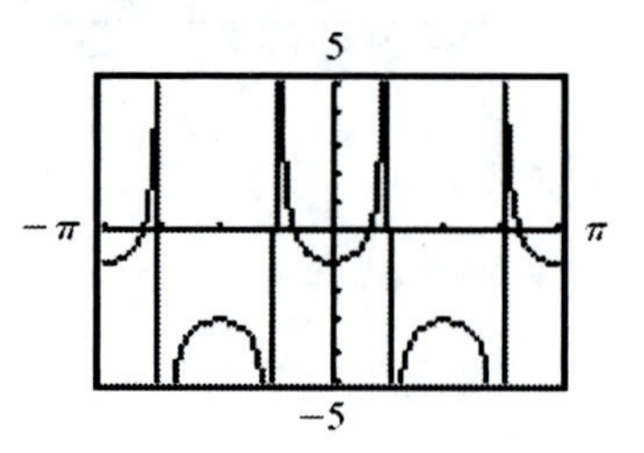

$f(x) = \dfrac{3 - 4\cos^2 x}{1 - 2\sin^2 x}$

$= \dfrac{1 - (-2 + 4\cos^2 x)}{1 - 2\sin^2 x}$ Algebra

$= \dfrac{1 - 2(2\cos^2 x - 1)}{1 - 2\sin^2 x}$ Algebra

$= \dfrac{1 - 2\cos 2x}{\cos 2x}$ Double-angle identities

$= \dfrac{1}{\cos 2x} - \dfrac{2\cos 2x}{\cos 2x}$ Algebra

$= \sec 2x - 2$ or $-2 + \sec 2x = g(x)$ Reciprocal identity, Algebra

66. The graph of $f(x)$ is shown in the figure. The graph appears to have vertical asymptotes $x = -2\pi$, $x = 0$, and $x = 2\pi$, and period 2π. Its x intercepts are difficult to determine, but since there appears to be symmetry with respect to the points where the curve crosses the line $y = 2$, it appears to be a cotangent curve displaced upward by $k = 2$ units. It appears that

$g(x) = 2 + \cot\frac{x}{2}$ would be an appropriate choice. We verify $f(x) = g(x)$ as follows:

$$f(x) = \frac{2 + \sin x - 2\cos x}{1 - \cos x}$$

$$= \frac{2 - 2\cos x + \sin x}{1 - \cos x} \qquad \text{Algebra}$$

$$= \frac{2 - 2\cos x}{1 - \cos x} + \frac{\sin x}{1 - \cos x} \qquad \text{Algebra}$$

$$= 2 + 1 \div \frac{1 - \cos x}{\sin x} \qquad \text{Algebra}$$

$$= 2 + 1 \div \tan\frac{x}{2} \qquad \text{Half-angle identity}$$

$$= 2 + \cot\frac{x}{2} \qquad \text{Reciprocal identity}$$

Key algebraic steps: $\frac{2 + a - 2b}{1 - b} = \frac{2 - 2b + a}{1 - b} = \frac{2 - 2b}{1 - b} + \frac{a}{1 - b} = 2 + 1 \div \frac{1 - b}{a}$

67. $-2\pi \le x < -\pi,\ 0 \le x < \pi$

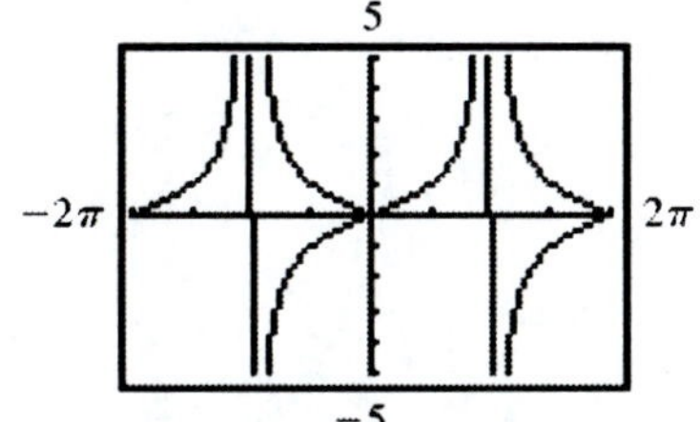

68. 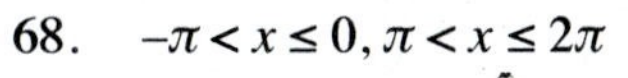
$-\pi < x \le 0,\ \pi < x \le 2\pi$

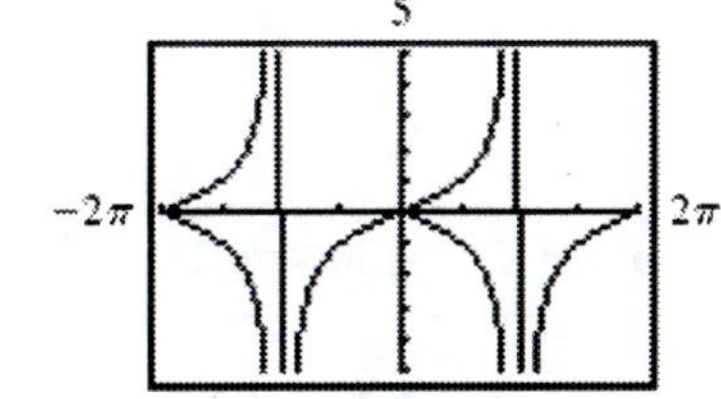

69.

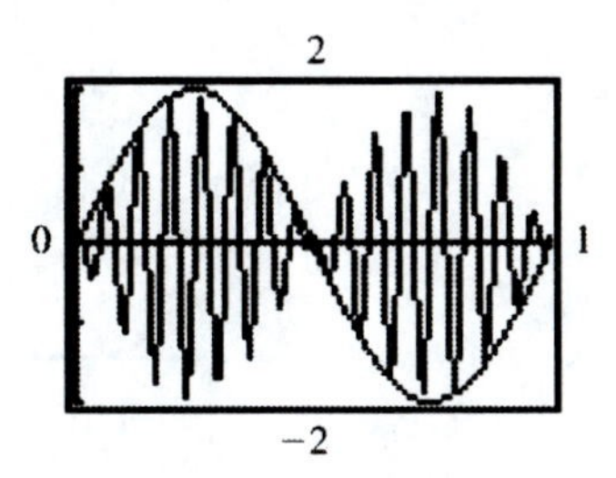

70. $\cos x \sin y = \frac{1}{2}[\sin(x + y) - \sin(x - y)]$

Let $x = 30\pi X$ and $y = 2\pi X$

$$2\cos 30\pi X \sin 2\pi X = 2\left(\frac{1}{2}\right)[\sin(30\pi X + 2\pi X) - \sin(30\pi X - 2\pi X)]$$

$$= \sin 32\pi X - \sin 28\pi X$$

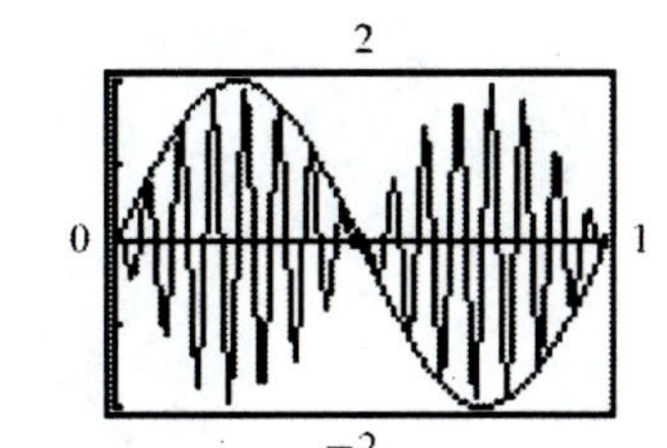

71.

$\sqrt{u^2 - a^2} = \sqrt{(a\sec x)^2 - a^2}$	Using the given substitution
$= \sqrt{a^2 \sec^2 x - a^2}$	Algebra
$= \sqrt{a^2(\sec^2 x - 1)}$	Algebra
$= \sqrt{a^2 \tan^2 x}$	Pythagorean identity
$= \lvert a \rvert \lvert \tan x \rvert$	Algebra
$= a \tan x$	Since $a > 0$ and x is in quadrant I $\left(\text{given } 0 < x < \frac{\pi}{2}\right)$, thus, $\tan x > 0$.

72. In Problem 81, Exercise 4.3, the formula

$$\tan(\theta_2 - \theta_1) = \frac{m_2 - m_1}{1 + m_1 m_2}$$

was derived. Since the given lines have slopes $4 = m_2$ and $\frac{1}{3} = m_1$, we can write

$$\tan(\theta_2 - \theta_1) = \frac{4 - \frac{1}{3}}{1 + \left(\frac{1}{3}\right)(4)} = \frac{3 \cdot 4 - 3 \cdot \frac{1}{3}}{3 + 3\left(\frac{1}{3}\right)4} = \frac{12 - 1}{3 + 4} = \frac{11}{7}$$

$$\theta_2 - \theta_1 = \tan^{-1}\left(\frac{11}{7}\right)$$

$$\approx 57.5°$$

73. We note that $\tan\theta = \frac{5}{8}$ and $\tan 2\theta = \frac{5 + x}{8}$ (see figure). Then,

$$\tan 2\theta = \frac{2\tan\theta}{1 - \tan^2\theta}, \theta = \tan^{-1}\frac{5}{8} \approx 32.005°$$

$$\frac{5 + x}{8} = \frac{2\left(\frac{5}{8}\right)}{1 - \left(\frac{5}{8}\right)^2} = \frac{\frac{5}{4}}{1 - \frac{25}{64}} = \frac{\frac{5}{4}}{\frac{39}{64}} = \frac{80}{39}$$

$$8 \cdot \frac{5 + x}{8} = 8 \cdot \frac{80}{39}$$

$$5 + x = \frac{640}{39}$$

$$x = \frac{445}{39} \approx 11.410$$

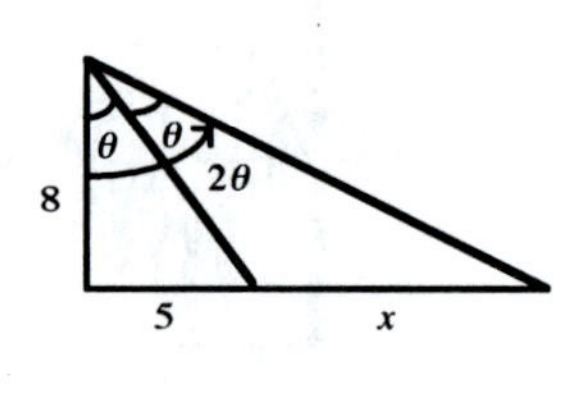

74. Following the hint, label the text diagram as follows:

Then $\tan \alpha = \dfrac{y_B}{x_B} = \dfrac{4}{4} = 1 \qquad \tan(\alpha + \theta) = \dfrac{y_A}{x_A} = \dfrac{6}{2} = 3$

$$\begin{aligned}\tan \theta &= \tan(\alpha + \theta - \alpha) = \tan[(\alpha + \theta) - \alpha] \\ &= \frac{\tan(\alpha + \theta) - \tan \alpha}{1 + \tan(\alpha + \theta)\tan \alpha} \\ &= \frac{3 - 1}{1 + 3 \cdot 1} \\ &= \frac{1}{2}\end{aligned}$$

Therefore, $\theta = \tan^{-1}\frac{1}{2} = 0.464$ rad.

75. (A) Redrawing the front of the trim and labeling, we have:

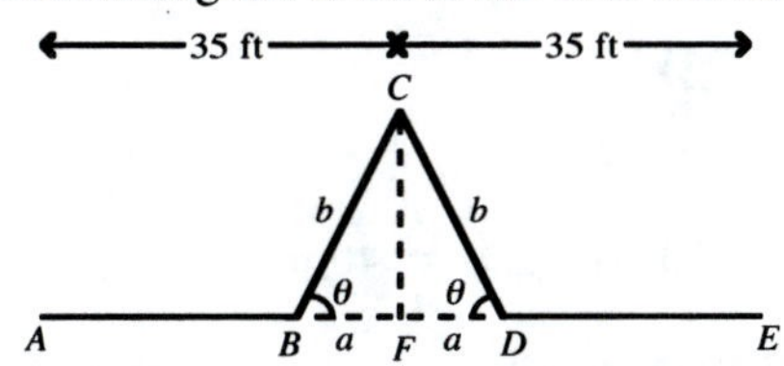

The length of this edge is given by

$$\begin{aligned}\ell = AB + BC + CD + DE &= AF - BF + BC + CD + FE - FD \\ &= 35 - a + b + b + 35 - a \\ &= 70 + 2b - 2a\end{aligned}$$

The length b of the entire trim is then $2\ell + 2 \cdot 50 = 2(70 + 2b - 2a) + 100$

Thus, $L = 140 + 4b - 4a + 100 = 240 + 4b - 4a$.

To express L in terms of θ, we note:

In triangle BCF, $\sin \theta = \dfrac{10}{b}$, hence $b \sin \theta = 10$ and $b = \dfrac{10}{\sin\theta}$

Also, $\cos \theta = \dfrac{a}{b}$, hence $a = b \cos \theta = \dfrac{10}{\sin\theta} \cos \theta = \dfrac{10\cos\theta}{\sin\theta}$

$$\begin{aligned}\text{Hence, } L &= 240 + 4b - 4a \\ &= 240 + 4\left(\frac{10}{\sin\theta}\right) - 4\left(\frac{10\cos\theta}{\sin\theta}\right)\end{aligned}$$

$$= 240 + 40\left(\frac{1}{\sin\theta} - \frac{\cos\theta}{\sin\theta}\right)$$
$$= 240 + 40\ \frac{1-\cos\theta}{\sin\theta}$$
$$= 240 + 40 \tan\frac{\theta}{2} \text{ as required}$$

(B) As θ varies from 30° to 60 °, $\frac{\theta}{2}$ varies from 15° to 30°; tan $\frac{\theta}{2}$ and therefore L should increase steadily on this interval.

(C)

θ°	30	35	40	45	50	55	60
L ft	250.7	252.6	254.6	256.6	258.7	260.8	263.1

(D) From the table, we find:
Min L = 250.7 ft; Max L = 263.1 ft

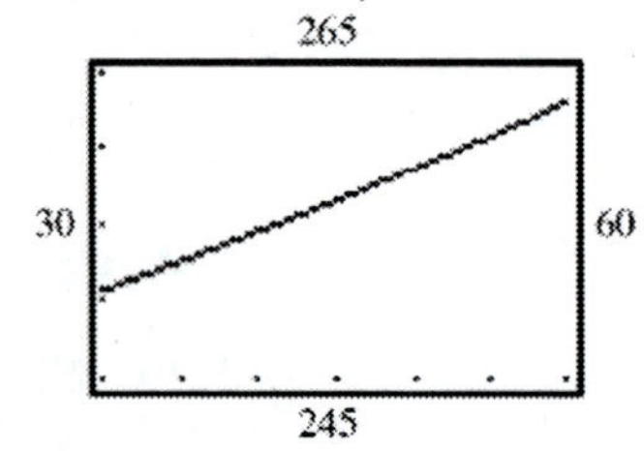

76. Assume that t is measured in seconds. Sum = 0.3 cos $120\pi t$ – 0.3 cos $140\pi t$.

$\cos x - \cos y = -2 \sin\frac{x+y}{2} \sin\frac{x-y}{2}$. Let $x = 120\pi t$ and $y = 140\pi t$.

$$0.3\cos 120\pi t - 0.3\cos 140\pi t = 0.3(\cos 120\pi t - \cos 140\pi t)$$
$$= 0.3(-2)\sin\frac{120\pi t + 140\pi t}{2}\sin\frac{120\pi t - 140\pi t}{2}$$
$$= -0.6\sin 130\pi t\sin(-10\pi t)$$
$$= 0.6\sin 130\pi t\sin 10\pi t$$

To find the beat frequency, we note:

Period of first tone = $\frac{2\pi}{B_1} = \frac{2\pi}{120\pi} = \frac{1}{60}$; Frequency of first tone = $\frac{1}{\text{Period}} = \frac{1}{\frac{1}{60}}$ = 60 Hz

Period of second tone = $\frac{2\pi}{B_2} = \frac{2\pi}{140\pi} = \frac{1}{70}$; Frequency of first tone = $\frac{1}{\text{Period}} = \frac{1}{\frac{1}{70}}$ = 70 Hz

Beat frequency = Frequency of second tone – Frequency of first tone
f_b = 70 Hz – 60 Hz = 10 Hz

77. (A)

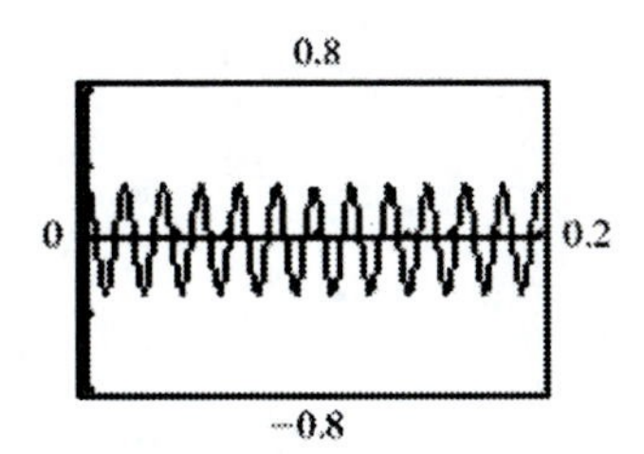

(B)

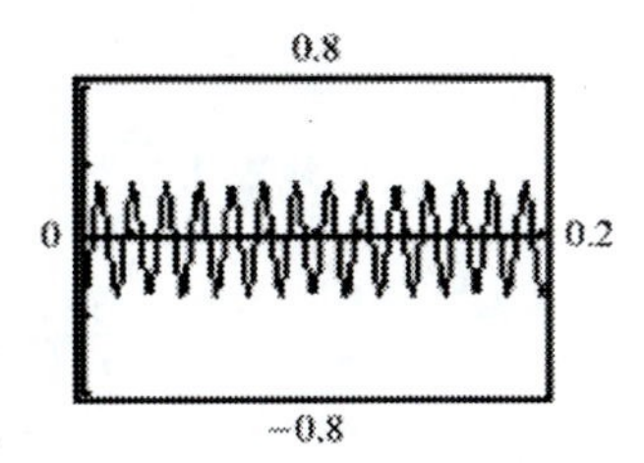

(C) $y3 = 0.3 \cos 120\pi t - 0.3 \cos 140\pi t$

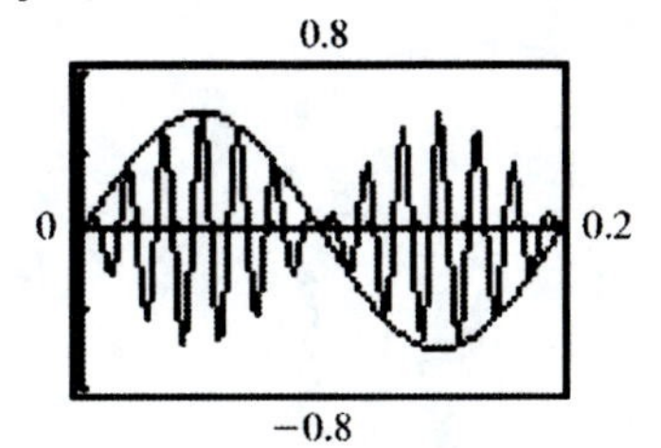

(D) $y3 = 0.6 \sin 130\pi t \sin 10\pi t$

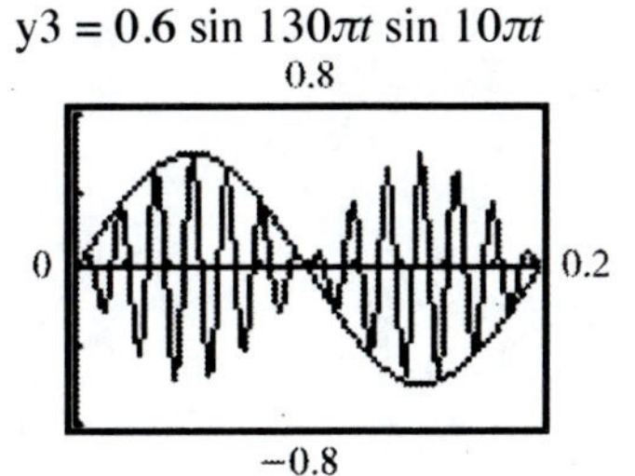

78. In the formula $M \sin Bt + N \cos Bt = A \sin(Bt + C)$, we have $M = -8$, $N = -6$, and $B = 3$.

Locate $P(M, N) = P(-8, -6)$ to determine C: $r = \sqrt{(-8)^2 + (-6)^2} = 10$,

$\sin C = \dfrac{-6}{10} = -0.6$, $\tan C = \dfrac{-6}{-8} = 0.75$. Find the reference angle α.

Then $C = \pi + \alpha$. $\tan \alpha = 0.75$, $\alpha = \tan^{-1}(0.75) = 0.6435$.

Thus, $C \approx 3.79$. We can now write:

$y = -8 \sin 3t - 6 \cos 3t = 10 \sin(3t + 3.79)$

Amplitude $= |10| = 10$

Period and Phase Shift:

$$3t + 3.79 = 0 \qquad\qquad 3t + 3.79 = 2\pi$$

$$t = -1.26 \qquad\qquad t = -1.26 + \frac{2\pi}{3}$$

Period $= \dfrac{2\pi}{3}$ Phase Shift ≈ -1.26

$$\text{Frequency} = \frac{1}{\text{Period}} = \frac{1}{2\pi/3} = \frac{3}{2\pi}$$

79. The graph of $y = -8 \sin 3t - 6 \cos 3t$ is shown in the figure. We use the zoom feature or the built-in approximation routine to locate the t intercepts in this interval at $t = -1.26$, -0.21, 0.83, and 1.88.

The phase shift for $y = 10 \sin(3t + 3.79)$ is -1.26.

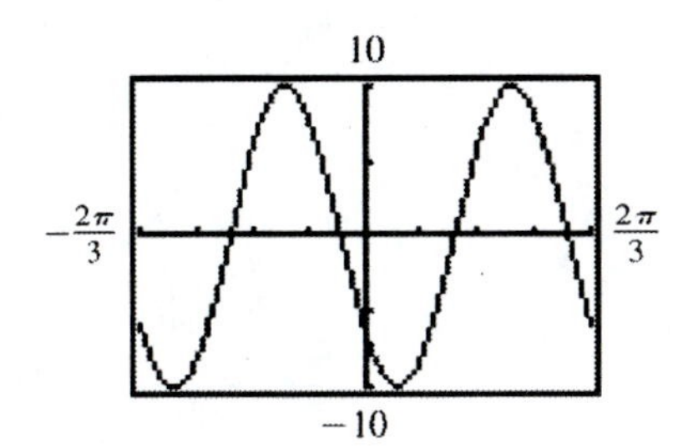

Chapter 5 Inverse Trigonometric Functions; Trigonometric Equations

EXERCISE 5.1 Inverse Sine, Cosine, and Tangent Functions

1. For a function to be one-to-one, each element of the range must correspond to a single element of the domain. Since there are, for example, an infinite number of domain elements ($x = n\pi$, n any integer) corresponding to range element 0, $y = \sin x$ is not a one-to-one function.

3. No. For example, $\sin \frac{\pi}{6} = \frac{1}{2}$ and $\sin \frac{5\pi}{6} = \frac{1}{2}$, so more than one element of the domain $[0, \pi]$ corresponds to the element $\frac{1}{2}$ of the range.

5. Make a careful drawing.

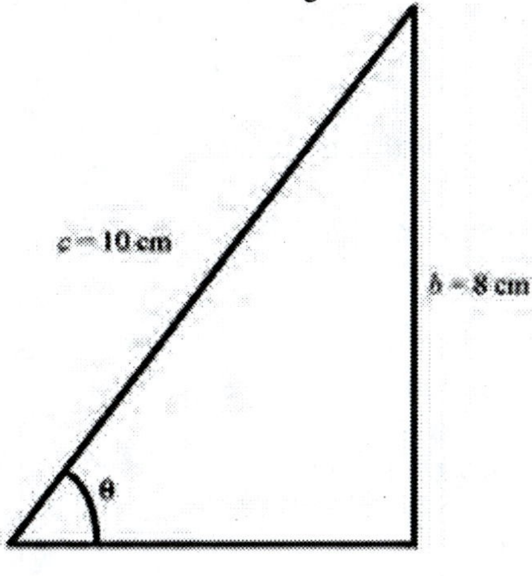

Measure $\theta \approx 53°$.

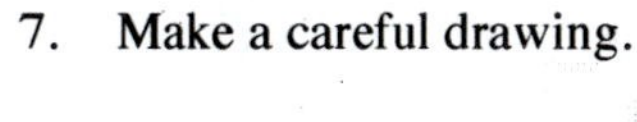

7. Make a careful drawing.

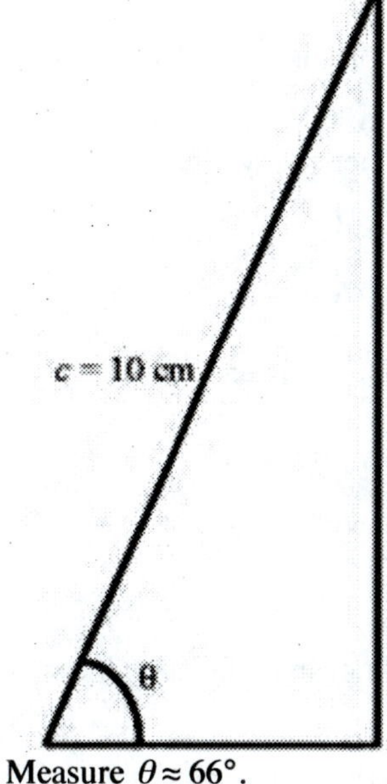

Measure $\theta \approx 66°$.

9. Make a careful drawing.

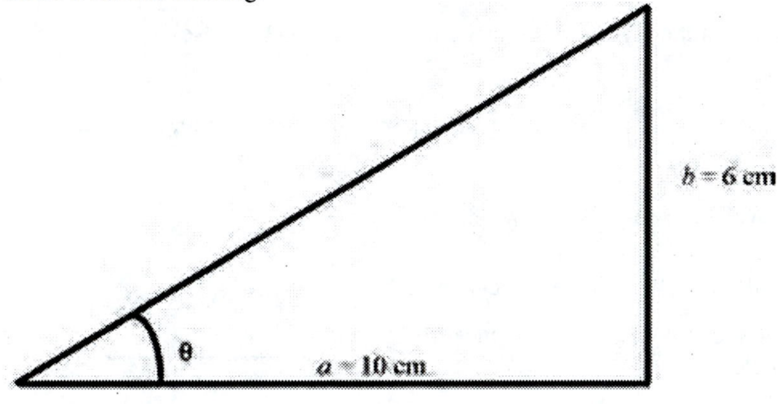

Measure $\theta \approx 31°$.

11. $y = \sin^{-1} 0$ is equivalent to $\sin y = 0$. No reference triangle can be drawn, but the only y between $-\frac{\pi}{2}$ and $\frac{\pi}{2}$ which has sine equal to 0 is $y = 0$. Thus, $\sin^{-1} 0 = 0$.

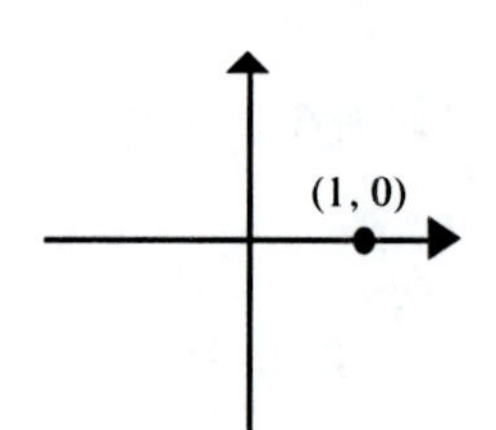

13. $y = \arccos \frac{\sqrt{3}}{2}$ is equivalent to $\cos y = \frac{\sqrt{3}}{2}$. What y between 0 and π has cosine equal to $\frac{\sqrt{3}}{2}$? y must be associated with a reference triangle in the first quadrant. Reference triangle is a special 30°–60° triangle.
$y = \frac{\pi}{6}$, $\arccos \frac{\sqrt{3}}{2} = \frac{\pi}{6}$

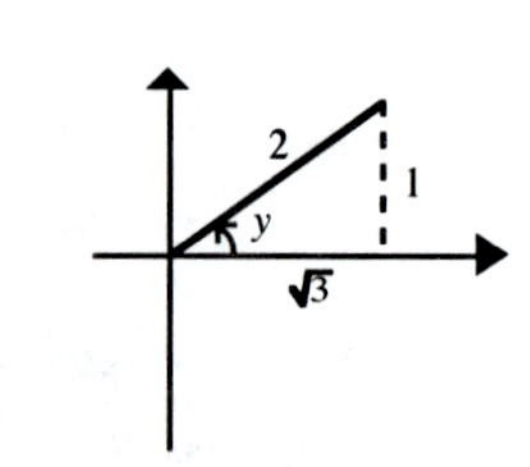

15. $y = \tan^{-1} 1$ is equivalent to $\tan y = 1$. What y between $-\frac{\pi}{2}$ and $\frac{\pi}{2}$ has tangent equal to 1? y must be associated with a reference triangle in the first quadrant. Reference triangle is a special 45° triangle.
$y = \frac{\pi}{4}$, $\tan^{-1} 1 = \frac{\pi}{4}$

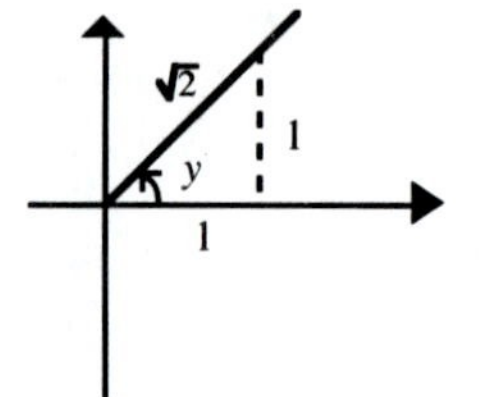

17. $y = \cos^{-1} \frac{1}{2}$ is equivalent to $\cos y = \frac{1}{2}$. What y between 0 and π has cosine equal to $\frac{1}{2}$? y must be associated with a reference triangle in the first quadrant. Reference triangle is a special 30°–60° triangle.

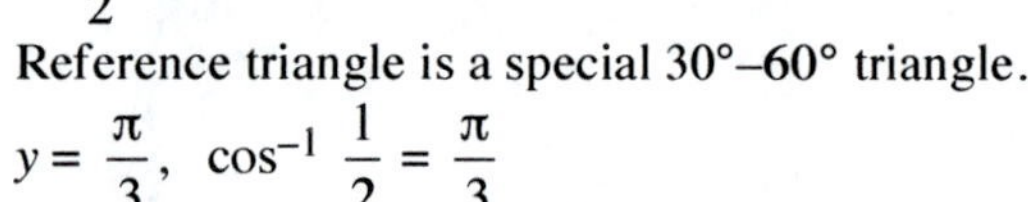
$y = \frac{\pi}{3}$, $\cos^{-1} \frac{1}{2} = \frac{\pi}{3}$

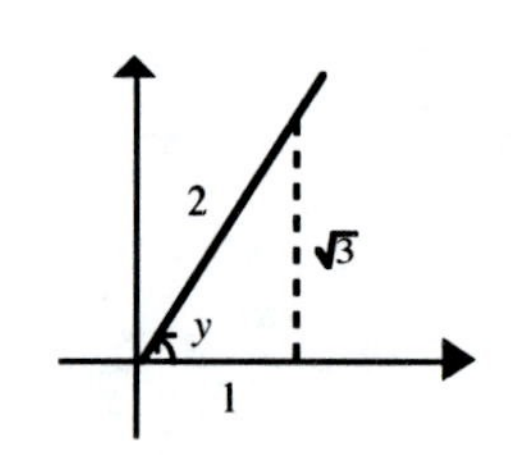

19. Calculator in radian mode: $\cos^{-1}(-0.9999) = 3.127$

21. Calculator in radian mode: $\tan^{-1} 4.056 = 1.329$

23. 3.142 is not in the domain of the inverse sine function. $-1 \le 3.142 \le 1$ is false. arcsin 3.142 is not defined.

25. $\sin^{-1} x = 37$ is equivalent to $\sin 37 = x$. Since the calculator is in degree mode, calculate $x = \sin 37° = 0.601815$.

27. $y = \arccos\left(-\frac{1}{2}\right)$ is equivalent to $\cos y = -\frac{1}{2}$. What y between 0 and π has cosine equal to $-\frac{1}{2}$? y must be associated with a reference triangle in the second quadrant. Reference triangle is a special 30°–60° triangle.
$y = \frac{2\pi}{3}$, $\arccos\left(-\frac{1}{2}\right) = \frac{2\pi}{3}$

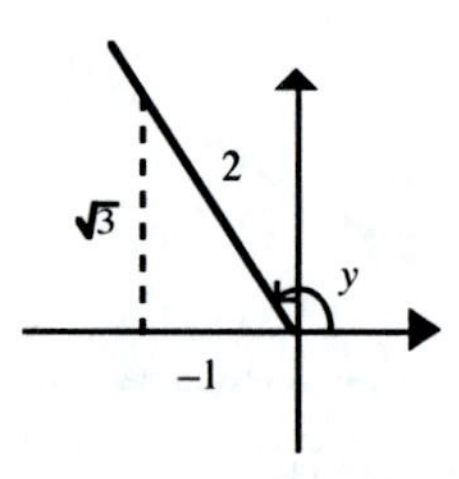

29. $y = \tan^{-1}(-1)$ is equivalent to $\tan y = -1$. What y between $-\frac{\pi}{2}$ and $\frac{\pi}{2}$ has tangent equal to -1?

y must be negative and associated with a reference triangle in the fourth quadrant. Reference triangle is a special 45° triangle.

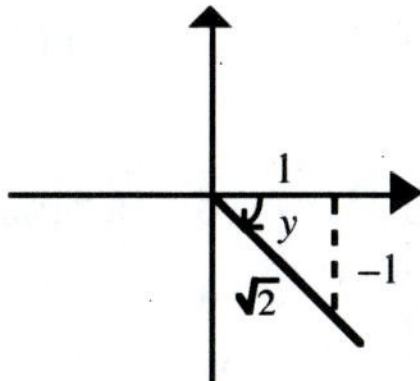

$y = -\frac{\pi}{4}$, $\tan^{-1}(-1) = -\frac{\pi}{4}$

31. $y = \sin^{-1}\left(-\frac{\sqrt{3}}{2}\right)$ is equivalent to $\sin y = -\frac{\sqrt{3}}{2}$. What y between $-\frac{\pi}{2}$ and $\frac{\pi}{2}$ has sine equal to $-\frac{\sqrt{3}}{2}$?

y must be negative and associated with a reference triangle in the fourth quadrant. Reference triangle is a special 30°–60° triangle.

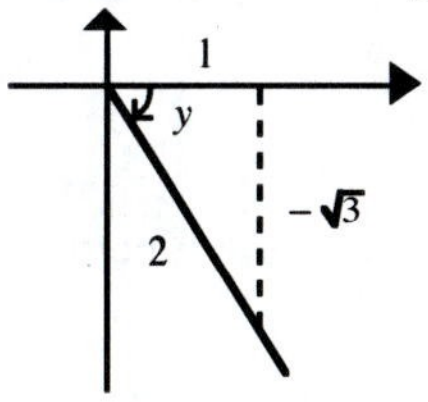

$y = -\frac{\pi}{3}$, $\sin^{-1}\left(-\frac{\sqrt{3}}{2}\right) = -\frac{\pi}{3}$

33. $\sin[\sin^{-1}(-0.6)] = -0.6$ from the sine-inverse sine identity.

35. Let $y = \sin^{-1}\left(-\frac{\sqrt{2}}{2}\right)$, then $\sin y = -\frac{\sqrt{2}}{2}$, $-\frac{\pi}{2} \le y \le \frac{\pi}{2}$. Draw the reference triangle associated with y.

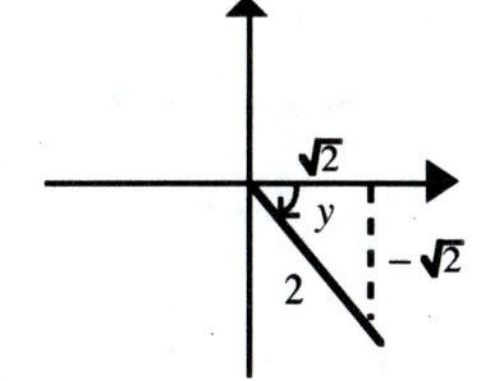

Then $\cos y = \cos\left[\sin^{-1}\left(-\frac{\sqrt{2}}{2}\right)\right]$ can be determined directly from the triangle or by recognizing that $y = -\frac{\pi}{4}$.

$$\cos\left[\sin^{-1}\left(-\frac{\sqrt{2}}{2}\right)\right] = \cos y = \frac{\sqrt{2}}{2} \text{ or } \cos\left(-\frac{\pi}{4}\right) = \frac{\sqrt{2}}{2}$$

37. Let $y = \sin^{-1}\frac{2}{3}$, then $\sin y = \frac{2}{3}$, $-\frac{\pi}{2} \le y \le \frac{\pi}{2}$. Draw the reference triangle associated with y; then $\tan y = \tan\left(\sin^{-1}\frac{2}{3}\right)$ can be determined directly from the triangle (after finding the third side).

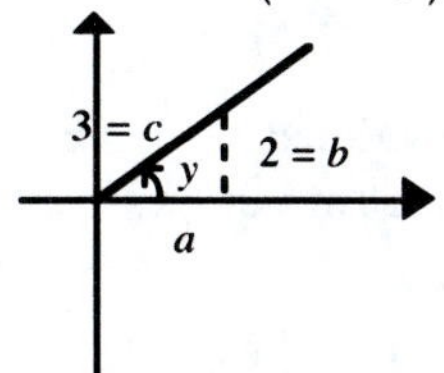

$$a^2 + b^2 = c^2$$
$$a = \sqrt{3^2 - 2^2} = \sqrt{5}$$

Thus, $\tan\left(\sin^{-1}\frac{2}{3}\right) = \tan y = \frac{b}{a} = \frac{2}{\sqrt{5}}$.

39. Let $y = \tan^{-1}(-2)$, then $\tan y = -2$, $-\frac{\pi}{2} < y < \frac{\pi}{2}$. Draw the reference triangle associated with y; then $\cos y = \cos[\tan^{-1}(-2)]$ can be determined directly from the triangle (after finding the third side).

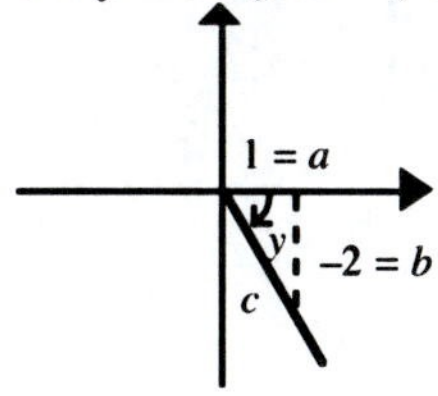

$$a^2 + b^2 = c^2$$
$$c = \sqrt{1^2 + (-2)^2} = \sqrt{5}$$

Thus, $\cos[\tan^{-1}(-2)] = \cos y = \frac{a}{c} = \frac{1}{\sqrt{5}}$.

41. Calculator in radian mode: $\tan^{-1}(-4.038) = -1.328$

43. Calculator in radian mode: $\sec[\sin^{-1}(-0.0399)] = \dfrac{1}{\cos[\sin^{-1}(-0.0399)]} = 1.001$

45. Calculator in radian mode: $\tan^{-1}(\tan 3) = -0.1416$

47. The graph of $y = \sin x$ shows that the function is one-to-one on the intervals $\left[-\dfrac{\pi}{2}, \dfrac{\pi}{2}\right]$ and $\left[-\dfrac{\pi}{2} + n\pi, \dfrac{\pi}{2} + n\pi\right]$ for any integer n. If $n = 1$, the interval $\left[\dfrac{\pi}{2}, \dfrac{3\pi}{2}\right]$ contains $x = \pi$, so this is the required interval.

49. The graph of $y = \tan x$ shows that the function is one-to-one on the intervals $\left(-\dfrac{\pi}{2}, \dfrac{\pi}{2}\right)$ and $\left(-\dfrac{\pi}{2} + n\pi, \dfrac{\pi}{2} + n\pi\right)$ for any integer n. If $n = -1$, the interval $\left(-\dfrac{3\pi}{2}, -\dfrac{\pi}{2}\right)$ contains $x = -\pi$, so this is the required interval.

51. Since $\tan 60° = \sqrt{3}$, $\arcsin(\tan 60°) = \arcsin(\sqrt{3})$ is not defined. $\sqrt{3}$ is not in the domain of the inverse sine function. $-1 \le \sqrt{3} \le 1$ is false.

53. Since $\cos(-60°) = \dfrac{1}{2}$, $\theta = \cos^{-1}[\cos(-60°)] = \cos^{-1}\dfrac{1}{2}$ is equivalent to $\cos\theta = \dfrac{1}{2}$, $0° \le \theta \le 180°$. Thus, $\theta = 60°$.

55. Calculator in degree mode: $\theta = \tan^{-1} 3.0413 = 71.80°$

57. Calculator in degree mode: $\theta = \arcsin(-0.8107) = -54.16°$

59. Calculator in degree mode: $\theta = \arctan(-17.305) = -86.69°$

61. $\cos^{-1}[\cos(-0.3)] = 0.3$. This does not illustrate a cosine-inverse cosine identity because $\cos^{-1}(\cos x) = x$ only if $0 \le x \le \pi$.

63. (A)

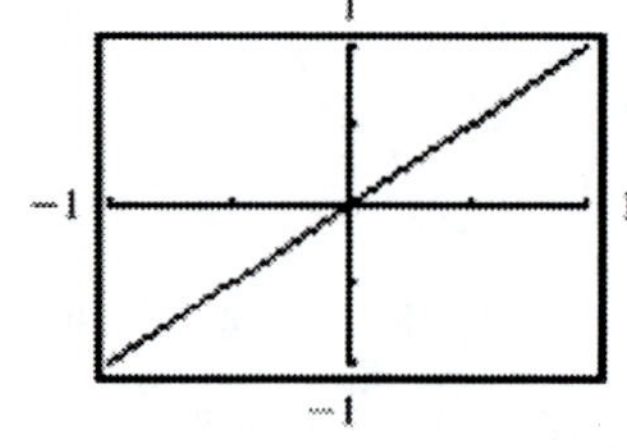

(B) The domain of $\sin^{-1}$ is restricted to $-1 \le x \le 1$; hence no graph will appear for other x.

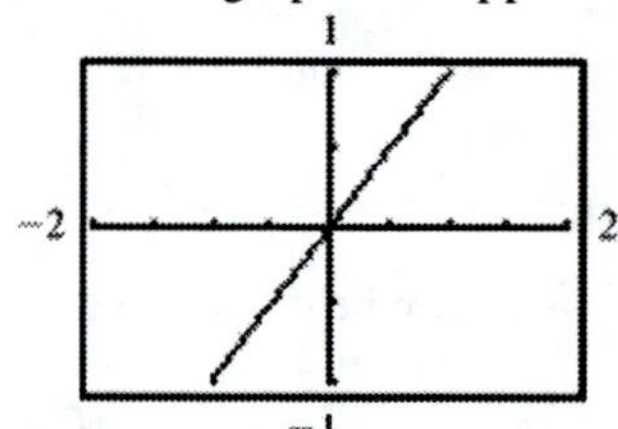

65. Since we recognize $\arccos \frac{1}{2} = \frac{\pi}{3}$ and $\arcsin(-1) = -\frac{\pi}{2}$, we can write

$$\sin\left[\arccos\frac{1}{2} + \arcsin(-1)\right] = \sin\left(\frac{\pi}{3} + \left(-\frac{\pi}{2}\right)\right) = \sin\frac{\pi}{3}\cos\left(-\frac{\pi}{2}\right) + \cos\frac{\pi}{3}\sin\left(-\frac{\pi}{2}\right)$$

$$= \frac{\sqrt{3}}{2}(0) + \left(\frac{1}{2}\right)(-1) = -\frac{1}{2}$$

67. Let $y = \sin^{-1}\left(-\frac{4}{5}\right)$. Then, $\sin y = -\frac{4}{5}$. We are asked to evaluate $\sin(2y)$, which is $2 \sin y \cos y$ by the double-angle identity. Draw a reference triangle associated with y; then, $\cos y = \cos\left[\sin^{-1}\left(-\frac{4}{5}\right)\right]$ can be determined directly from the triangle.

$\sin y = \sin\left[\sin^{-1}\left(-\frac{4}{5}\right)\right] = -\frac{4}{5}$ by the sine-inverse sine identity.

Note: $-\frac{\pi}{2} \le y \le \frac{\pi}{2}$ $\quad a = \sqrt{5^2 - (-4)^2} = 3 \quad \cos y = \frac{3}{5}$.

Thus, $\sin\left[2\sin^{-1}\left(-\frac{4}{5}\right)\right] = 2\sin\left[\sin^{-1}\left(-\frac{4}{5}\right)\right]\cos\left[\sin^{-1}\left(-\frac{4}{5}\right)\right] = 2\left(-\frac{4}{5}\right)\left(\frac{3}{5}\right) = -\frac{24}{25}$

69. Let $u = \arctan 2$ and $v = \arcsin\frac{1}{3}$. Then, $\tan u = 2$ and $\sin v = \frac{1}{3}$. We are asked to evaluate $\cos(u + v)$, which is $\cos u \cos v - \sin u \sin v$ by the sum identity.

Draw reference triangles associated with u and v; then $\cos u$, $\cos v$, and $\sin u$ can be determined directly from the triangles.

$c = \sqrt{1^2 + 2^2} = \sqrt{5}$ $\qquad a = \sqrt{3^2 - 1^2} = \sqrt{8} = 2\sqrt{2}$

$\sin u = \frac{2}{\sqrt{5}}$ $\qquad \sin v = \frac{1}{3}$

$\cos u = \frac{1}{\sqrt{5}}$ $\qquad \cos v = \frac{2\sqrt{2}}{3}$

Then, $\cos\left(\arctan 2 + \arcsin\frac{1}{3}\right) = \cos(u + v) = \cos u \cos v - \sin u \sin v$

$$= \left(\frac{1}{\sqrt{5}}\right)\left(\frac{2\sqrt{2}}{3}\right) - \left(\frac{2}{\sqrt{5}}\right)\left(\frac{1}{3}\right)$$

$$= \frac{2\sqrt{2} - 2}{3\sqrt{5}}$$

71. Let $u = \sin^{-1}\dfrac{1}{4}$ and $v = \cos^{-1}\dfrac{1}{4}$. Then, $\sin u = \dfrac{1}{4}$ and $\cos v = \dfrac{1}{4}$. We are asked to evaluate $\tan(u - v)$, which is $\dfrac{\tan u - \tan v}{1 + \tan u \tan v}$ by the difference identity. Draw reference triangles associated with u and v; then $\tan u$ and $\tan v$ can be determined directly from the triangles.

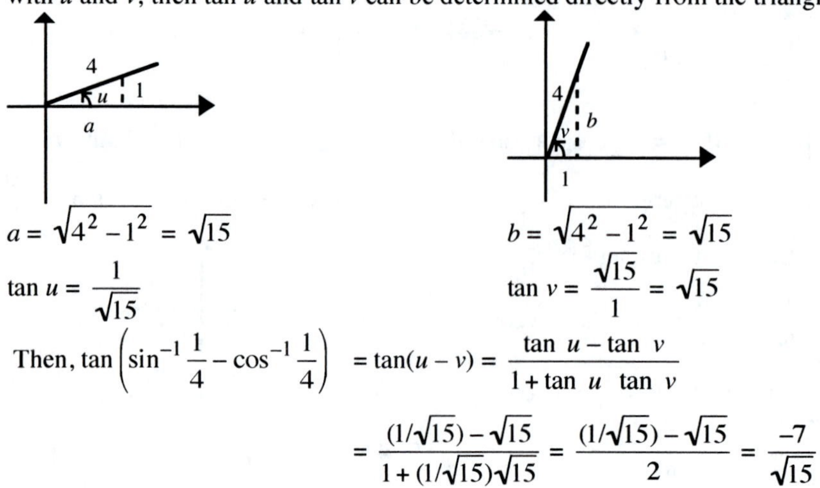

$a = \sqrt{4^2 - 1^2} = \sqrt{15}$ $\qquad$ $b = \sqrt{4^2 - 1^2} = \sqrt{15}$

$\tan u = \dfrac{1}{\sqrt{15}}$ $\qquad$ $\tan v = \dfrac{\sqrt{15}}{1} = \sqrt{15}$

Then, $\tan\left(\sin^{-1}\dfrac{1}{4} - \cos^{-1}\dfrac{1}{4}\right) = \tan(u - v) = \dfrac{\tan u - \tan v}{1 + \tan u \tan v}$

$$= \frac{(1/\sqrt{15}) - \sqrt{15}}{1 + (1/\sqrt{15})\sqrt{15}} = \frac{(1/\sqrt{15}) - \sqrt{15}}{2} = \frac{-7}{\sqrt{15}}$$

73. Let $y = \arctan(-2)$. Then $\tan y = -2, -\dfrac{\pi}{2} < y < 0$. We are asked to evaluate $\sin\dfrac{y}{2}$, which is $\sqrt{\dfrac{1 - \cos y}{2}}$ or $-\sqrt{\dfrac{1 - \cos y}{2}}$. In this case, since $-\dfrac{\pi}{2} < y < 0, -\dfrac{\pi}{4} < \dfrac{y}{2} < 0$; hence $\dfrac{y}{2}$ is in the fourth quadrant, and $\sin\dfrac{y}{2} = -\sqrt{\dfrac{1 - \cos y}{2}}$. Draw a reference triangle associated with y, then $\cos y$ can be determined directly from the triangle.

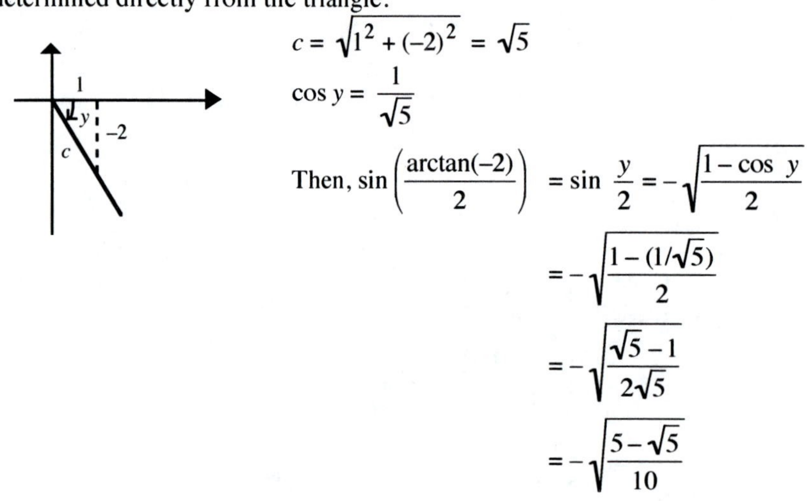

$c = \sqrt{1^2 + (-2)^2} = \sqrt{5}$

$\cos y = \dfrac{1}{\sqrt{5}}$

Then, $\sin\left(\dfrac{\arctan(-2)}{2}\right) = \sin\dfrac{y}{2} = -\sqrt{\dfrac{1 - \cos y}{2}}$

$$= -\sqrt{\frac{1 - (1/\sqrt{5})}{2}}$$

$$= -\sqrt{\frac{\sqrt{5} - 1}{2\sqrt{5}}}$$

$$= -\sqrt{\frac{5 - \sqrt{5}}{10}}$$

75. Let $y = \cos^{-1} x$ $(-1 \le x \le 1$ corresponds to $0 \le y \le \pi)$
or, equivalently, $\cos y = x$ $\quad 0 \le y \le \pi$
Geometrically,

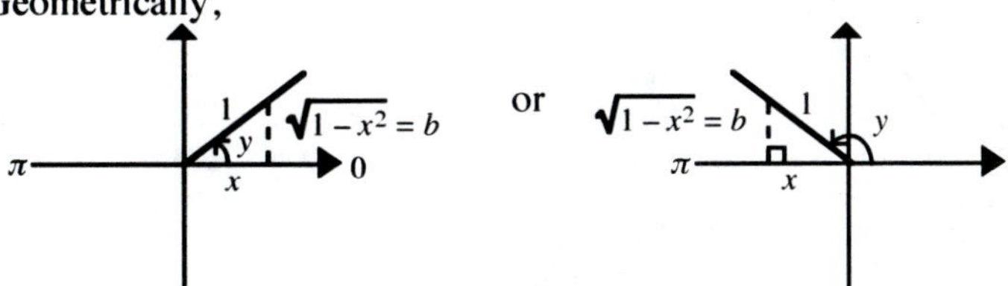

In either case, $b = \sqrt{1-x^2}$. Thus, $\sin(\cos^{-1} x) = \sin y = \dfrac{b}{r} = \sqrt{1-x^2}$

77. Let $y = \arcsin x$ $\left(-1 \le x \le 1 \text{ corresponds to } -\dfrac{\pi}{2} \le y \le \dfrac{\pi}{2}\right)$

or, equivalently, $\sin y = x$ $\quad -\dfrac{\pi}{2} \le y \le \dfrac{\pi}{2}$

Geometrically,

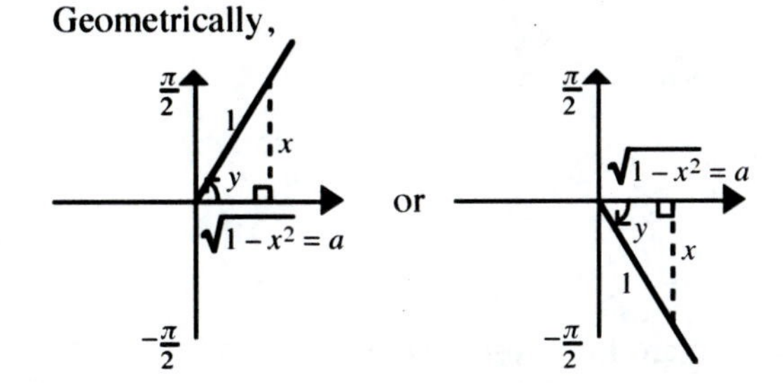

In either case, $a = \sqrt{1-x^2}$. Thus, $\tan(\arcsin x) = \tan y = \dfrac{b}{a} = \dfrac{x}{\sqrt{1-x^2}}$.

79. Let $y = \tan^{-1}(-x)$.
This is equivalent to $-x = \tan y$ $\left(-\dfrac{\pi}{2} < y < \dfrac{\pi}{2}\right)$ by the definition of the inverse tangent function.
By algebra, this is equivalent to $x = -\tan y$, which in turn is equivalent to $x = \tan(-y)$ by the identities for negatives. This equation is equivalent to $-y = \tan^{-1} x$ $\left(-\dfrac{\pi}{2} < -y < \dfrac{\pi}{2}\right)$ or
$y = -\tan^{-1} x$.

81. (A)

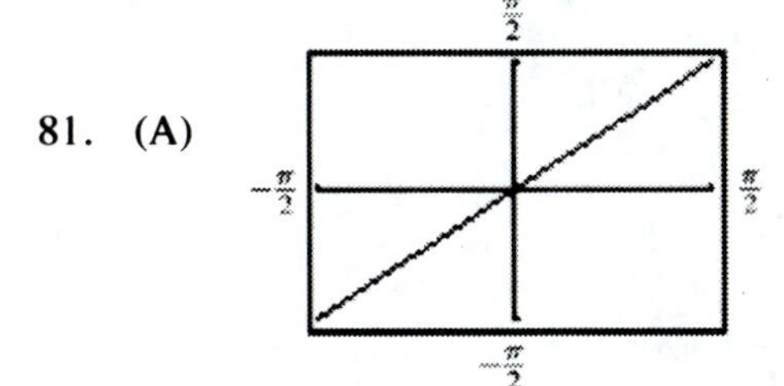

(B) The domain for $\sin x$ is $(-\infty, \infty)$ and the range is $[-1, 1]$, which is the domain for $\sin^{-1} x$.
Thus, $y = \sin^{-1}(\sin x)$ has a graph over the interval $(-\infty, \infty)$, but $\sin^{-1}(\sin x) = x$ only on the restricted domain of $\sin x$, $\left[-\dfrac{\pi}{2}, \dfrac{\pi}{2}\right]$.

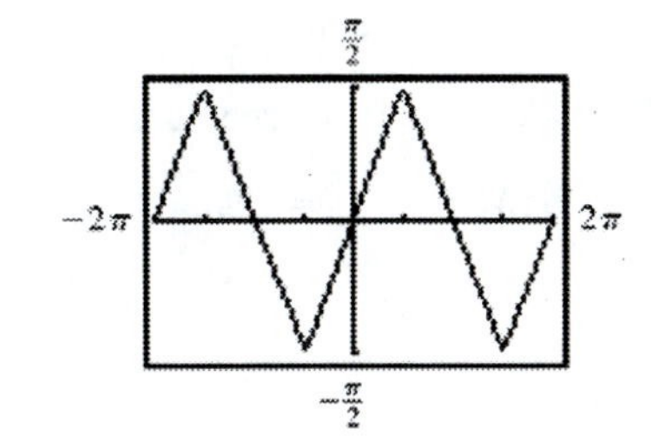

83. Let $y = \sin^{-1} x$ $\left(-1 \le x \le 1 \text{ corresponds to } -\frac{\pi}{2} \le y \le \frac{\pi}{2}\right)$
or, equivalently, $\sin y = x$ $\quad -\frac{\pi}{2} \le y \le \frac{\pi}{2}$.
Geometrically,

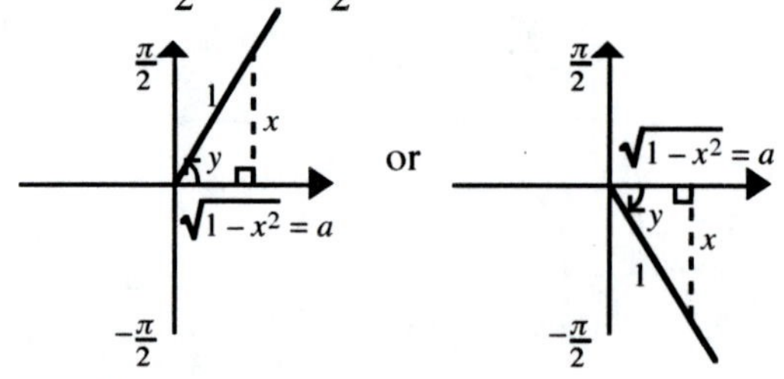

In either case, $a = \sqrt{1-x^2}$. Thus, $\tan(\sin^{-1} x) = \tan y = \frac{b}{a} = \frac{x}{\sqrt{1-x^2}}$. If $-1 < x < 1$,
$-\frac{\pi}{2} < y < \frac{\pi}{2}$, this is equivalent to $y = \tan^{-1} \frac{x}{\sqrt{1-x^2}}$. If $x = -1$ or $x = 1$, this expression is undefined.

Thus, $\sin^{-1} x = \tan^{-1} \frac{x}{\sqrt{1-x^2}}$ as long as both sides are defined.

85. Let $y = \arcsin x$ $\left(-1 \le x \le 1 \text{ corresponds to } -\frac{\pi}{2} \le y \le \frac{\pi}{2}\right)$
or, equivalently, $\sin y = x$ $\quad -\frac{\pi}{2} \le y \le \frac{\pi}{2}$.
By the double-angle identity, $\sin(2 \arcsin x) = 2 \sin(\arcsin x) \cos(\arcsin x)$.
Geometrically,

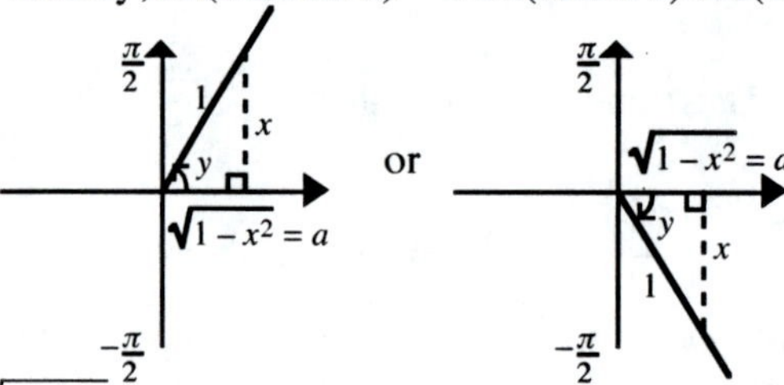

In either case, $a = \sqrt{1-x^2}$. Thus, $\sin(\arcsin x) = \sin y = x$, $\cos(\arcsin x) = \cos y = \frac{a}{r} = \frac{\sqrt{1-x^2}}{1}$ or $\sqrt{1-x^2}$. Finally, $\sin(2 \arcsin x) = 2 \sin(\arcsin x) \cos(\arcsin x) = 2x\sqrt{1-x^2}$.

87. Substitute $x = 3$ into $\theta = \tan^{-1}\left(\frac{x}{5.2}\right)$ to obtain $\theta = \tan^{-1}\left(\frac{3}{5.2}\right) = 30°$.

89. (A) If $\sin \frac{\theta}{2} = \frac{1}{M}$, then $\frac{\theta}{2} = \sin^{-1}\left(\frac{1}{M}\right)$ $\quad -1 \le \frac{1}{M} \le 1$.

Thus, $\theta = 2 \sin^{-1}\left(\frac{1}{M}\right)$. Since M must be positive, $\frac{1}{M} \le 1$, or $M \ge 1$.

(B) Calculator in degree mode:

For $M = 1.7$, $\theta = 2 \sin^{-1}\left(\frac{1}{1.7}\right) = 72°$. For $M = 2.3$, $\theta = 2 \sin^{-1}\left(\frac{1}{2.3}\right) = 52°$

91. (A) By comparison of θ_1, θ_2, θ_3, drawn for various values of x, it appears that θ increases with increasing x, then decreases somewhat.

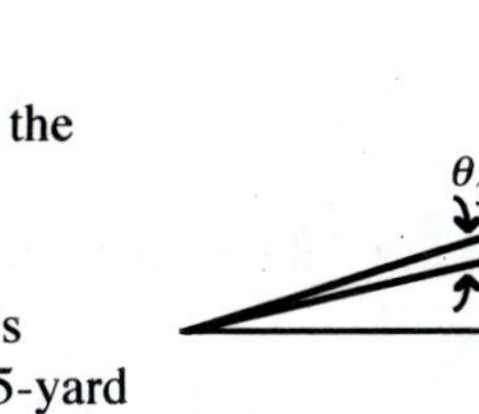

(B) Take an arbitrary value of θ (shown as θ_2 in the diagram.)
Then $\theta = \angle CPA - \angle BPA$
$AB + BC + CD = 60 - 2 \cdot 5 = 50$ yd since this distance is the width of the field minus two 5-yard lengths.

Let $y = AB = CD$, $BC = 8$ yd (width of the goal)

Then $2y + 8 = 50$, hence $y = 21$

$$\tan BPA = \frac{y}{x} = \frac{21}{x} \qquad \tan CPA = \frac{AB+BC}{x} = \frac{21+8}{x} =$$

$$\tan\theta = \tan(\angle CPA - \angle BPA) = \frac{\tan\angle CPA - \tan\angle BPA}{1+(\tan\angle CPA)(\tan\angle BPA)}$$

$$\tan\theta = \frac{\frac{29}{x} - \frac{21}{x}}{1+\frac{29}{x}\cdot\frac{21}{x}} = \frac{\frac{8}{x}}{1+\frac{609}{x^2}} = \frac{\frac{8}{x}\cdot x^2}{x^2+\frac{609}{x^2}\cdot x^2} = \frac{8x}{x^2+609}$$

Hence $\qquad \theta = \tan^{-1}\dfrac{8x}{x^2+609}$

(C) From the table the max $\theta = 9.21°$ when $x = 25$ yd.

x yd	10	15	20	25	30	35
$\theta°$	6.44	8.19	9.01	9.21	9.04	8.68

(D)

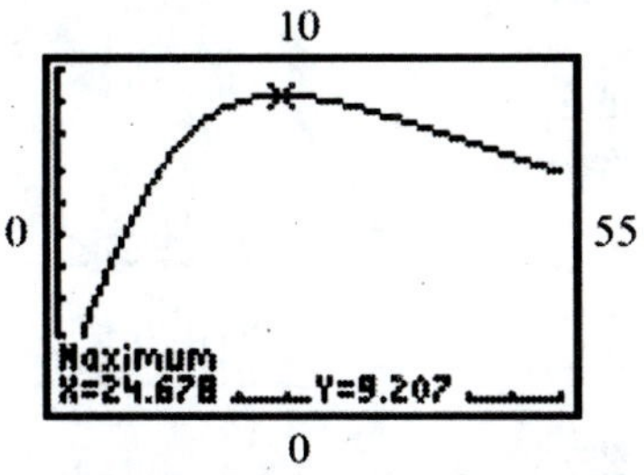

The angle θ increases rapidly until a maximum is reached then declines more slowly. The maximum $\theta = 9.21°$ when $x = 24.68$ yd, which is about 7 yd before from the left corner of the penalty area (shown in the text diagram.)

93. (A) The volume of the fuel is clearly given by Volume = (height)(cross-sectional area) with L = height. To determine the cross-sectional area (see figure), we reason that

Area of segment $AEBF$ = Area of sector $ACBF$ – Area of triangle ABC

Area of sector $ACBF$ = 2(area of sector CFB) = $2\left(\frac{1}{2}R^2\theta\right) = R^2\theta$

Since triangle CEB is a right triangle, we have

$$\cos\theta = \frac{EC}{BC} = \frac{R-x}{R} \qquad \theta = \cos^{-1}\frac{R-x}{R}$$

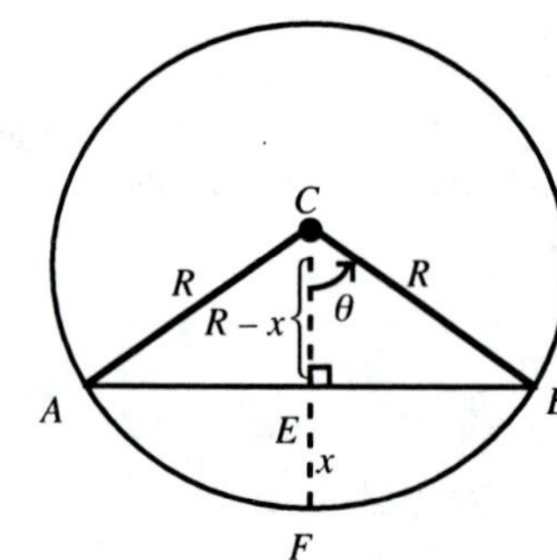

Therefore,

$$\text{Area of sector } ACBF = R^2\cos^{-1}\frac{R-x}{R}$$

Area of triangle $ABC = \frac{1}{2}(\text{base})(\text{altitude}) = \frac{1}{2}(AB)(CE) = \frac{1}{2}(2\cdot EB)(CE) = (EB)(R-x)$

By the Pythagorean theorem applied to triangle CEB, we have $EB^2 + (R-x)^2 = R^2$.

$EB = \sqrt{R^2-(R-x)^2}$. Therefore, Area of triangle $ABC = (R-x)\sqrt{R^2-(R-x)^2}$.

Finally,

$$\text{Area of segment } AEBF = R^2\cos^{-1}\frac{R-x}{R} - (R-x)\sqrt{R^2-(R-x)^2}$$

$$\text{Volume} = \left[R^2\cos^{-1}\frac{R-x}{R} - (R-x)\sqrt{R^2-(R-x)^2}\right]L$$

(B) Substituting the given values, we have $R = 3, x = 2, R - x = 1, L = 30$

$$\text{Volume} = [3^2\cos^{-1}\frac{1}{3} - 1\sqrt{3^2-1^2}\,]30$$

$$= [9\cos^{-1}\frac{1}{3} - \sqrt{8}\,]30 \qquad \text{(Calculator in radian mode)}$$

$$= 248 \text{ ft}^3$$

(C) The graphs of y1 and y2 are shown in the figure. We use the zoom feature or the built-in approximation routine to locate the x coordinate of the point of intersection at $x = 2.6$.

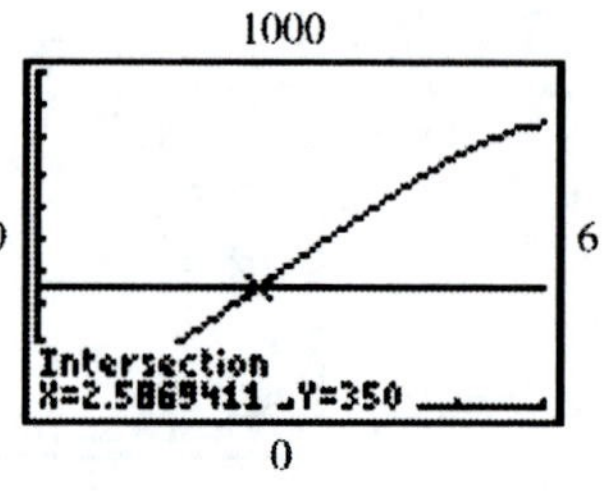

The depth is 2.6 feet.

95. To find the inverse function to $y = \arcsin\sqrt{x}$, interchange x and y, then solve for y in terms of x.

$$y = \arcsin\sqrt{x} \qquad 0 \le x \le 1 \qquad 0 \le y \le \frac{\pi}{2}$$

Inverse function:

$$x = \arcsin\sqrt{y} \qquad 0 \le x \le \frac{\pi}{2} \qquad 0 \le y \le 1$$

Solve for y in terms of x:

$$\sin x = \sqrt{y}$$

$$y = \sin^2 x$$

97. (A)

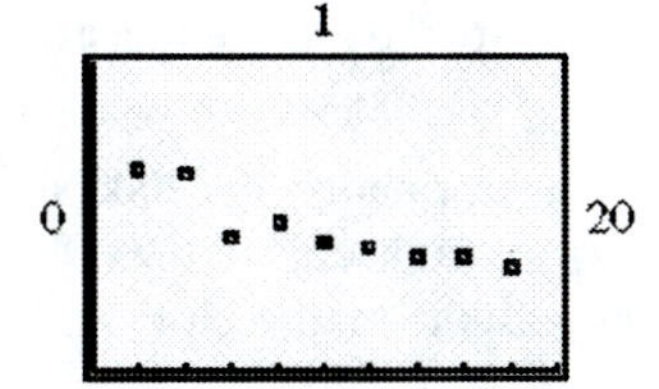

(B)

y	0.657	0.642	0.437	0.481	0.425	0.399	0.374	0.370	0.341
$\sin^{-1}\sqrt{y}$	0.945	0.929	0.722	0.766	0.710	0.684	0.658	0.654	0.624

(C)

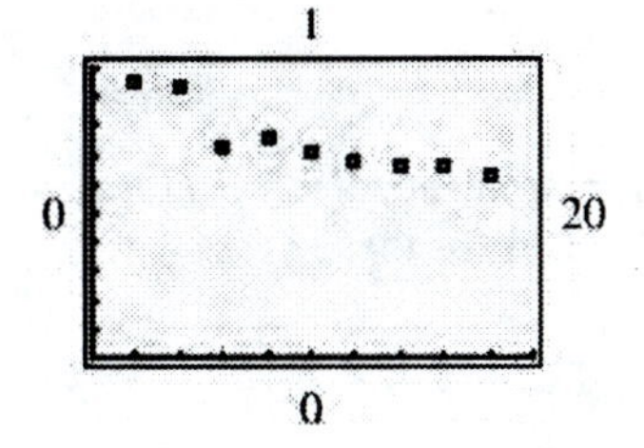

(D) Applying the LinReg feature yields the regression line $y = -0.0193x + 0.937$.

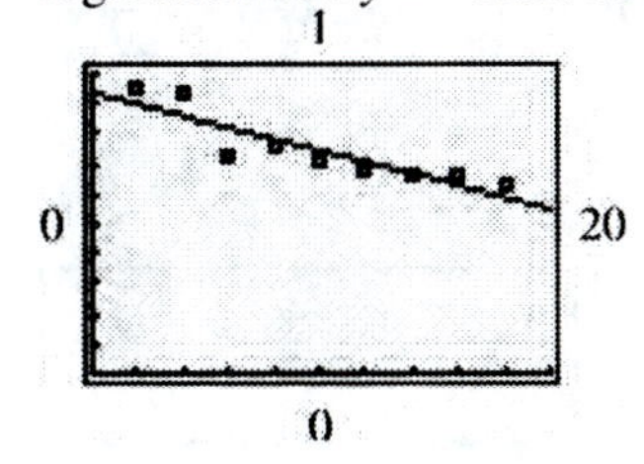

(E) Since $f(x) = \sin^2 x$, $f(ax + b) = \sin^2(-0.0193x + 0.937)$.

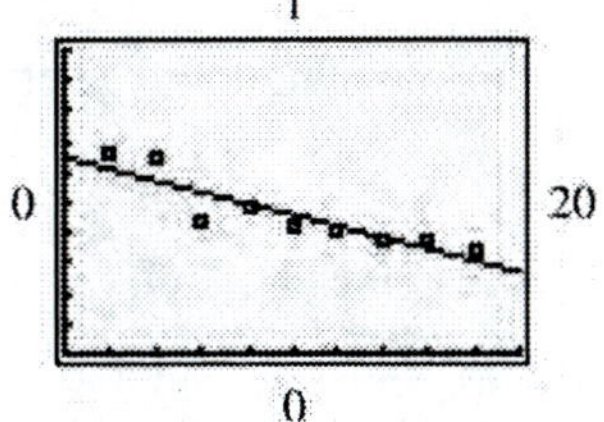

The model appears to fit the data fairly well, especially for the more recent years.

99. (A)

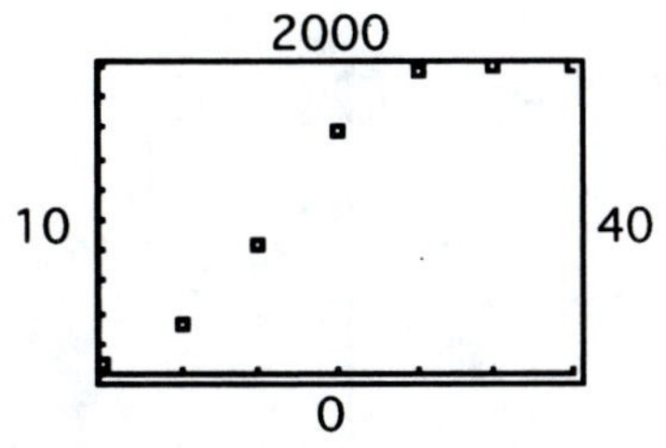

(B)

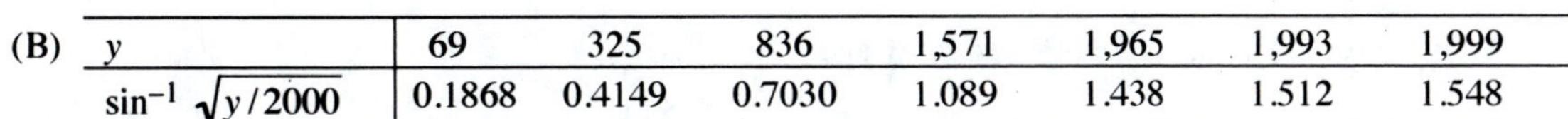

y	69	325	836	1,571	1,965	1,993	1,999
$\sin^{-1}\sqrt{y/2000}$	0.1868	0.4149	0.7030	1.089	1.438	1.512	1.548

(C)

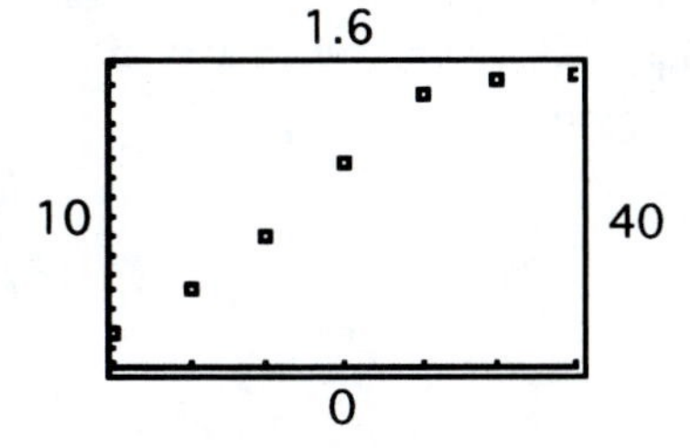

(D) Applying the LinReg feature yields the regression line
$y = 0.05009x - 0.2678$.

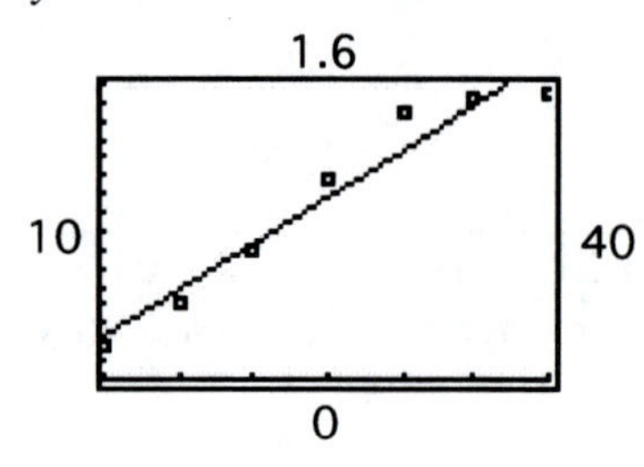

(E) In Problem 96, the arcsine transformation $y = \arcsin\sqrt{x/2000}$ was shown to have the inverse transformation $y = 2000\sin^2 x$. Graphing $y = 2000\sin^2(0.05009x - 0.2678)$ together with the scatter plot yields:

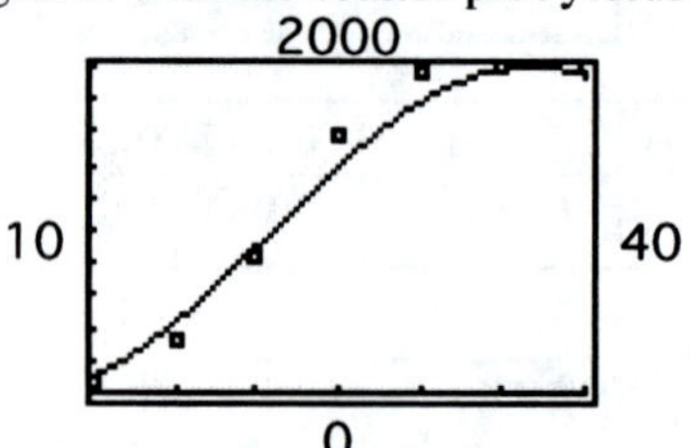

The function seems to be a fairly good model for the data for these years.

EXERCISE 5.2 Inverse Cotangent, Secant, and Cosecant Functions

1. For a function to be one-to-one, each element of the range must correspond to a single element of the domain. Since there are, for example, an infinite number of domain elements, $\left(x = \frac{\pi}{2} + n\pi\right)$, n any integer, corresponding to range element 0, $y = \cot x$ is not a one-to-one function.

3. Yes. A glance at the graph of $y = \frac{1}{\tan x} = \cot x$ on the interval $(\pi, 2\pi)$ suggests, but does not prove, that this statement is true.

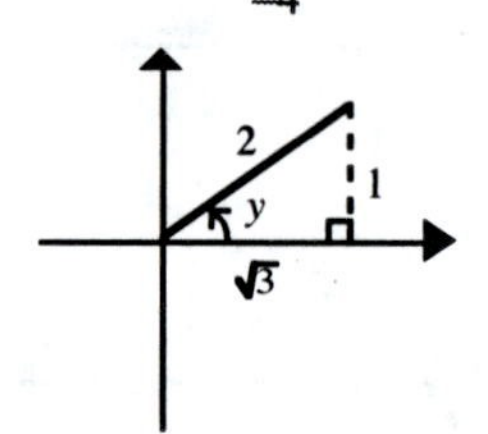

5. $y = \cot^{-1}\sqrt{3}$ is equivalent to $\cot y = \sqrt{3}$ and $0 < y < \pi$. What number between 0 and π has cotangent equal to $\sqrt{3}$?

y must be in the first quadrant. $\cot y = \sqrt{3} = \frac{\sqrt{3}}{1}$, $y = \frac{\pi}{6}$.

Thus, $\cot^{-1}\sqrt{3} = \frac{\pi}{6}$

7. $y = \operatorname{arccsc} 1$ is equivalent to $\csc y = 1$ and $-\frac{\pi}{2} \le y \le \frac{\pi}{2}$, $y \ne 0$. What number between $-\frac{\pi}{2}$ and $\frac{\pi}{2}$ has cosecant equal to 1? No reference triangle can be drawn, but from the diagram we see

$(a, b) = (0, 1)$, $r = 1$, $\csc y = \frac{1}{1} = 1$, $y = \frac{\pi}{2}$

Thus, $\operatorname{arccsc} 1 = \frac{\pi}{2}$

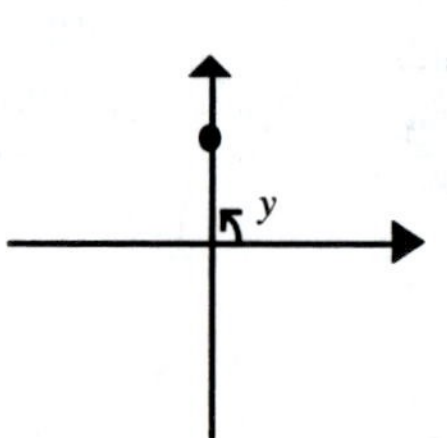

9. $y = \sec^{-1}\sqrt{2}$ is equivalent to $\sec y = \sqrt{2}$ and $0 \le y \le \pi$, $y \ne \frac{\pi}{2}$. What number between 0 and π has secant equal to $\sqrt{2}$? y must be in the first quadrant.

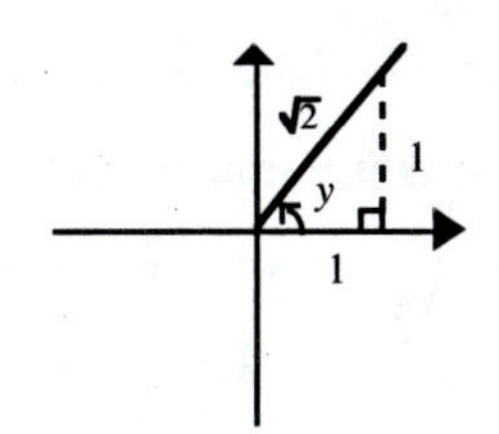

$\sec y = \sqrt{2} = \frac{\sqrt{2}}{1}$, $y = \frac{\pi}{4}$

Thus, $\sec^{-1}\sqrt{2} = \frac{\pi}{4}$

11. Let $y = \cot^{-1} 0$, then $\cot y = 0, 0 < y < \pi$. No reference triangle can be drawn, but from the diagram we see $(a, b) = (0, 1)$, $r = 1$, $\cot y = \frac{0}{1} = 0$, $y = \frac{\pi}{2}$

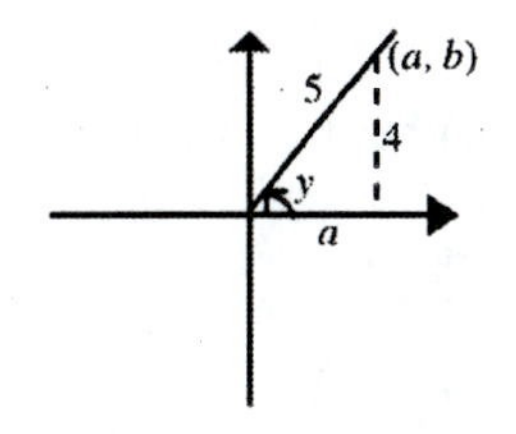

Thus, $\sin(\cot^{-1} 0) = \sin\frac{\pi}{2} = 1$

13. Let $y = \csc^{-1}\frac{5}{4}$, then $\csc y = \frac{5}{4}, -\frac{\pi}{2} \le y \le \frac{\pi}{2}, y \ne 0$. y is in the first quadrant. Draw a reference triangle, find the third side, then determine $\tan y$ from the triangle.

$a = \sqrt{5^2 - 4^2} = 3$

Thus, $\tan\left(\csc^{-1}\frac{5}{4}\right) = \tan y = \frac{b}{a} = \frac{4}{3}$

15. The expression is defined. The domain of the inverse cotangent function is all real numbers.

17. The expression is not defined. The domain of the inverse secant function excludes all numbers between –1 and 1.

19. The expression is not defined. The domain of the inverse cosecant function excludes all numbers between –1 and 1.

21. The expression is defined. –1.2 is in the domain of the inverse secant function.

23. The expression is not defined. $-1 < \sin\frac{23\pi}{12} < 1$ and therefore $\sin\frac{23\pi}{12}$ is not in the domain of the inverse cosecant function.

25. $y = \cot^{-1}(-1)$ is equivalent to $\cot y = -1$ and $0 < y < \pi$. What number between 0 and π has cotangent equal to –1? y must be positive and in the second quadrant.

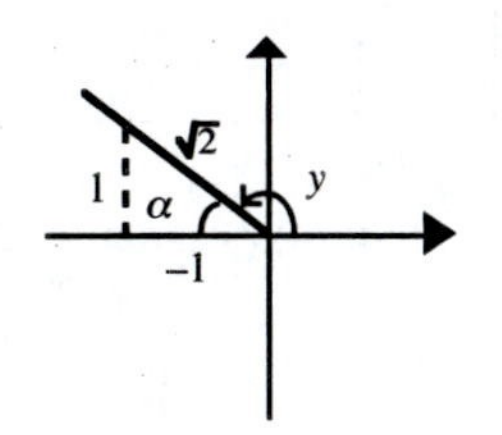

$\cot y = -1 = \frac{-1}{1}$, $\alpha = \frac{\pi}{4}$, $y = \frac{3\pi}{4}$

Thus, $\cot^{-1}(-1) = \frac{3\pi}{4}$

27. $y = \text{arcsec}(-2)$ is equivalent to $\sec y = -2$ and $0 \le y \le \pi$, $y \ne \frac{\pi}{2}$. What number between 0 and π has secant equal to -2? y must be positive and in the second quadrant.
$\sec y = -2 = \frac{2}{-1}$, $\alpha = \frac{\pi}{3}$, $y = \frac{2\pi}{3}$
Thus, $\text{arcsec}(-2) = \frac{2\pi}{3}$.

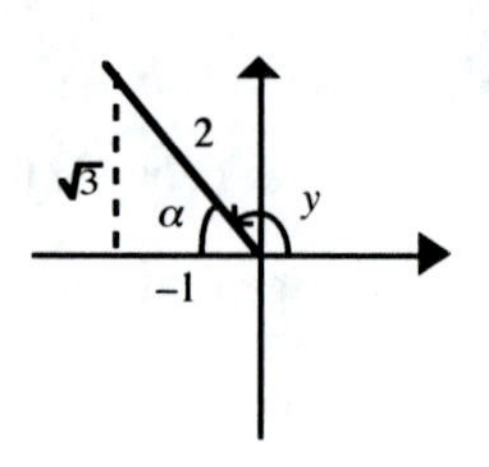

29. $y = \text{arccsc}(-2)$ is equivalent to $\csc y = -2$ and $-\frac{\pi}{2} \le y \le \frac{\pi}{2}$, $y \ne 0$. What number between $-\frac{\pi}{2}$ and $\frac{\pi}{2}$ has cosecant equal to -2? y must be negative and in the fourth quadrant.
$\csc y = -2 = \frac{2}{-1}$, $\alpha = \frac{\pi}{6}$, $y = -\frac{\pi}{6}$
Thus, $\text{arccsc}(-2) = -\frac{\pi}{6}$

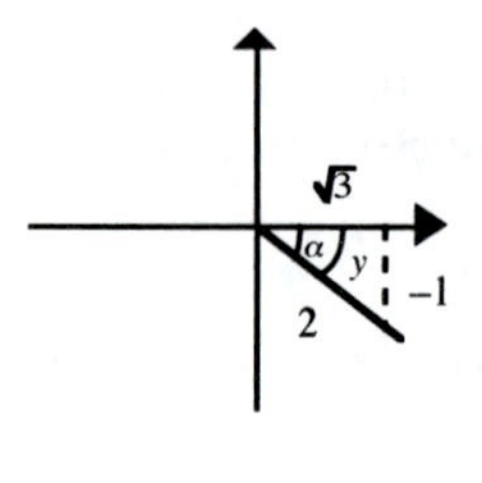

31. If $x = \frac{1}{2}$, neither $x \le -1$ nor $x \ge 1$ is a true statement. Thus, x is not in the domain of the inverse cosecant function. $\csc^{-1} \frac{1}{2}$ is not defined.

33. Let $y = \csc^{-1}\left(-\frac{5}{3}\right)$; then, $\csc y = -\frac{5}{3}, -\frac{\pi}{2} \le y \le \frac{\pi}{2}, y \ne 0$. y is negative and in the fourth quadrant. Draw a reference triangle, find the third side, and then determine $\cos y$ from the triangle.
$a = \sqrt{5^2 - (-3)^2} = 4$ $\left(\csc y = -\frac{5}{3} = \frac{5}{-3}\right)$
Thus, $\cos\left[\csc^{-1}\left(-\frac{5}{3}\right)\right] = \cos y = \frac{a}{r} = \frac{4}{5}$

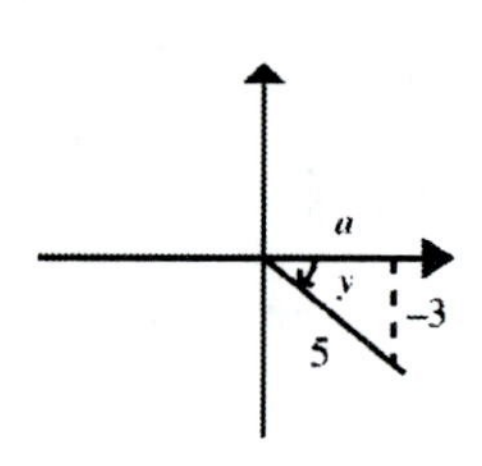

35. Let $y = \sec^{-1}\left(-\frac{5}{4}\right)$; then $\sec y = -\frac{5}{4}, 0 \le y \le \pi, y \ne \frac{\pi}{2}$.
y is positive and in the second quadrant. Draw a reference triangle, find the third side, and then determine $\cot y$ from the triangle.
$b = \sqrt{5^2 - (-4)^2} = 3$ $\left(\sec y = -\frac{5}{4} = \frac{5}{-4}\right)$
Thus, $\cot\left[\sec^{-1}\left(-\frac{5}{4}\right)\right] = \cot y = \frac{a}{b} = -\frac{4}{3}$

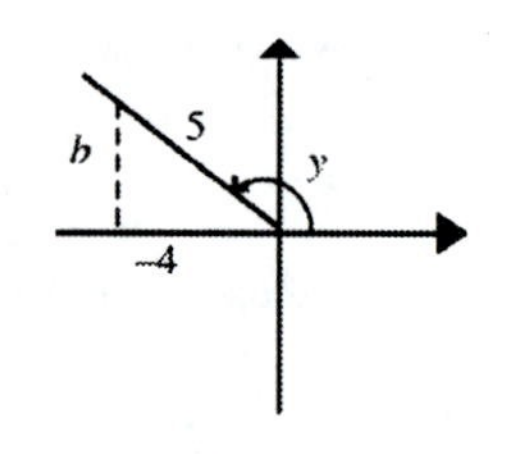

37. From the inverse trigonometric identities, we see $\sec^{-1} x = \cos^{-1} \frac{1}{x}, x \ge 1$ or $x \le -1$.
Thus, $\cos(\sec^{-1} x) = \cos\left(\cos^{-1}\frac{1}{x}\right) = \frac{1}{x}$ if $\frac{1}{x}$ is in the domain of the inverse cosine function.
Hence, $\cos[\sec^{-1}(-2)] = \cos\left[\cos^{-1}\left(\frac{1}{-2}\right)\right] = \frac{1}{-2}$ or $-\frac{1}{2}$, since $-\frac{1}{2}$ is in the domain of the inverse cosine function.

39. We note: $\cot^{-1} x = \tan^{-1} \frac{1}{x}$ for $x > 0$.

Thus,

$$\begin{aligned}
\cot(\cot^{-1} x) &= \cot\left(\tan^{-1}\frac{1}{x}\right) \\
&= \frac{1}{\tan\left(\tan^{-1}\frac{1}{x}\right)} && \text{Reciprocal identity} \\
&= \frac{1}{\frac{1}{x}} && \text{Tangent-inverse tangent identity} \\
&= x && \text{Algebra}
\end{aligned}$$

Since $33.4 > 0$, the above reasoning applies; thus, $\cot(\cot^{-1} 33.4) = 33.4$

41. We note: $\csc^{-1} x = \sin^{-1} \frac{1}{x}$ for $x \geq 1$ or $x \leq -1$.

Thus,

$$\begin{aligned}
\csc(\csc^{-1} x) &= \csc\left(\sin^{-1}\frac{1}{x}\right) \\
&= \frac{1}{\sin\left(\sin^{-1}\frac{1}{x}\right)} && \text{Reciprocal identity} \\
&= \frac{1}{\frac{1}{x}} && \text{Sine-inverse sine identity. [If } x \geq 1 \text{ or } x \leq -1 \text{ then } 1/x \text{ will be in the domain of the inverse sine function.]} \\
&= x && \text{Algebra}
\end{aligned}$$

Since $-4 \leq -1$, the above reasoning applies; thus $\csc[\csc^{-1}(-4)] = -4$

43. $\sec^{-1} 5.821 = \cos^{-1} \frac{1}{5.821}$ (Calculator in radian mode)
$= 1.398$

45. $\cot^{-1} 0.035 = \tan^{-1} \frac{1}{0.035}$ (Calculator in radian mode)
$= 1.536$

47. 0.847 is not in the domain of the inverse cosecant function. Neither $0.847 \leq -1$ nor $0.847 \geq 1$ is true. $\csc^{-1} 0.847$ is not defined.

49. We note: $\cot^{-1} x = \pi + \tan^{-1} \frac{1}{x}$ if $x < 0$.

Thus,

$$\begin{aligned}
\text{arccot}(-3.667) &= \cot^{-1}(-3.667) \\
&= \pi + \tan^{-1}\frac{1}{-3.667} && \text{(Calculator in radian mode)} \\
&= 2.875
\end{aligned}$$

51. $\text{arcsec}(-15.025) = \sec^{-1}(-15.025)$

$= \cos^{-1} \dfrac{1}{-15.025}$ (Calculator in radian mode)

$= 1.637$

53. $\theta = \text{arcsec}(-2)$ is equivalent to $\sec\theta = -2 \quad 0 \le \theta \le 180°, \theta \ne 90°$

$\cos\theta = -\dfrac{1}{2} \quad 0 \le \theta \le 180°$

Thus, $\theta = 120°$

55. $\theta = \cot^{-1}(-1)$ is equivalent to $\cot\theta = -1 \quad 0 < \theta < 180°$

$\tan\theta = \dfrac{1}{-1} = -1 \quad 0 < \theta < 180°$

Thus, $\theta = 135°$

57. $\theta = \csc^{-1}\left(-\dfrac{2}{\sqrt{3}}\right)$ is equivalent to $\csc\theta = -\dfrac{2}{\sqrt{3}} \quad -90° \le \theta \le 90°, \theta \ne 0°$

$\sin\theta = -\dfrac{\sqrt{3}}{2} \quad -90° \le \theta \le 90°$

Thus, $\theta = -60°$

59. Since $\sin 60° = \dfrac{\sqrt{3}}{2}$, $\text{arcsec}(\sin 60°) = \text{arcsec}\left(\dfrac{\sqrt{3}}{2}\right)$ is not defined. Neither $\dfrac{\sqrt{3}}{2} \le -1$ nor $\dfrac{\sqrt{3}}{2} \ge 1$ is true. $\dfrac{\sqrt{3}}{2}$ is not in the domain of the inverse secant function.

61. $\theta = \text{arccsc}(\sec 135°) = \text{arccsc}(-\sqrt{2})$ is equivalent to $\csc\theta = -\sqrt{2} \quad -90° \le \theta \le 90°$, $\theta \ne 0°$

$\sin\theta = -\dfrac{1}{\sqrt{2}} \quad -90° \le \theta \le 90°$

Thus, $\theta = -45°$

63. We note: $\cot^{-1} x = 180° + \tan^{-1} \dfrac{1}{x}$ for $x < 0$, values in degrees.

Thus, $\cot^{-1}(\cot(-15°)) = 180° + \tan^{-1} \dfrac{1}{\cot(-15°)}$

$= 180° + \tan^{-1}[\tan(-15°)]$

$= 180° + (-15°)$

$= 165°$

65. Calculator in degree mode: $\theta = \cot^{-1} 0.3288 = \tan^{-1} \dfrac{1}{0.3288} = 71.80°$

67. Calculator in degree mode: $\theta = \text{arccsc}(-1.2336) = \arcsin \dfrac{1}{-1.2336} = -54.16°$

69. Calculator in degree mode: $\theta = \text{arccot}(-0.0578) = 180° + \tan^{-1}\left(\dfrac{1}{-0.0578}\right) = 93.31°$

71. Let $u = \csc^{-1}\left(-\frac{5}{3}\right)$ and $v = \tan^{-1}\frac{1}{4}$. Then we are asked to evaluate tan($u + v$), which is $\frac{\tan u + \tan v}{1 + \tan u \tan v}$ by the sum identity for tangent. We know $\tan v = \tan\left(\tan^{-1}\frac{1}{4}\right) = \frac{1}{4}$ by the tangent-inverse tangent identity. It remains to find $\tan u = \tan\left[\csc^{-1}\left(-\frac{5}{3}\right)\right]$.

See sketch, Problem 33. $\tan\left[\csc^{-1}\left(-\frac{5}{3}\right)\right] = \frac{b}{a} = \frac{-3}{4}$ from the reference triangle. Hence,

$$\tan\left[\csc^{-1}\left(-\frac{5}{3}\right) + \tan^{-1}\frac{1}{4}\right] = \tan(u + v) = \frac{\tan u + \tan v}{1 - \tan u \tan v} = \frac{-\frac{3}{4} + \frac{1}{4}}{1 - \left(-\frac{3}{4}\right)\left(\frac{1}{4}\right)} = \frac{-\frac{2}{4}}{1 + \frac{3}{16}} = \frac{-8}{19}$$

73. Let $y = \cot^{-1}\left(-\frac{3}{4}\right)$. Then we are asked to evaluate tan(2y), which is $\frac{2\tan y}{1 - \tan^2 y}$ from the double-angle identity.

Draw a reference triangle associated with y. Then, $\tan y = \tan\left[2\cot^{-1}\left(-\frac{3}{4}\right)\right]$ can be determined directly from the triangle.

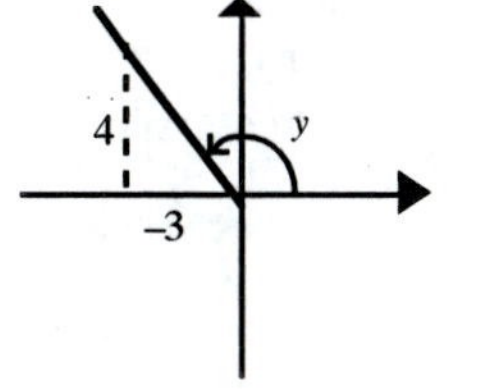

y is positive and in the second quadrant. $\tan y = \frac{4}{-3} = -\frac{4}{3}$.

$$\text{Thus, } \tan\left[2\cot^{-1}\left(-\frac{3}{4}\right)\right] = \tan 2y = \frac{2\left(-\frac{4}{3}\right)}{1 - \left(-\frac{4}{3}\right)^2} = \frac{-\frac{8}{3}}{1 - \left(\frac{16}{9}\right)} = \frac{24}{7}$$

75. Let $y = \cot^{-1} x \quad 0 < y < \pi$ or, equivalently,
$x = \cot y \quad 0 < y < \pi$

Geometrically,

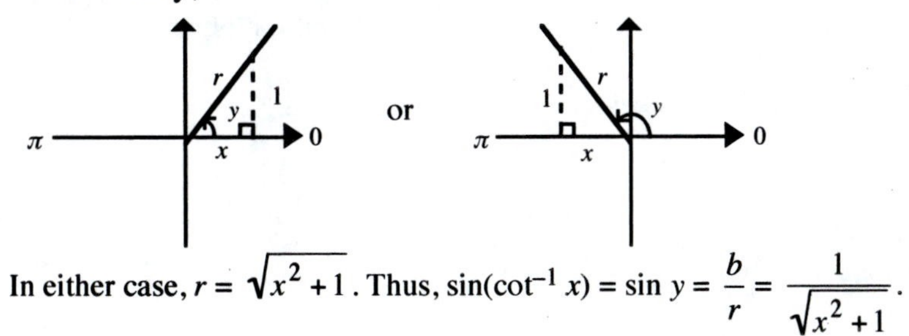

In either case, $r = \sqrt{x^2 + 1}$. Thus, $\sin(\cot^{-1} x) = \sin y = \frac{b}{r} = \frac{1}{\sqrt{x^2 + 1}}$.

77. Let $y = \sec^{-1} x$ $\quad 0 \le y \le \pi \quad y \ne \frac{\pi}{2}$ $\quad$ or, equivalently,

$x = \sec y \quad 0 \le y \le \pi$

Geometrically,

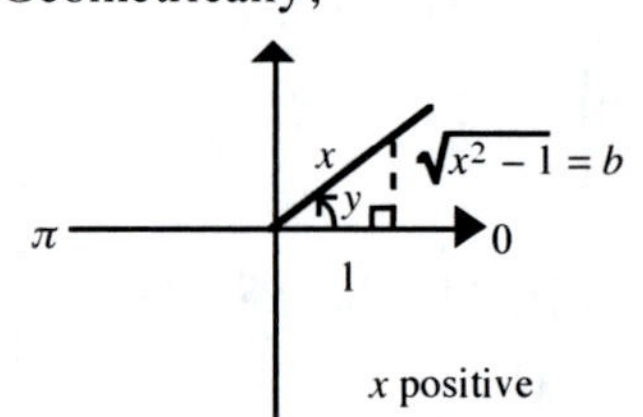

or

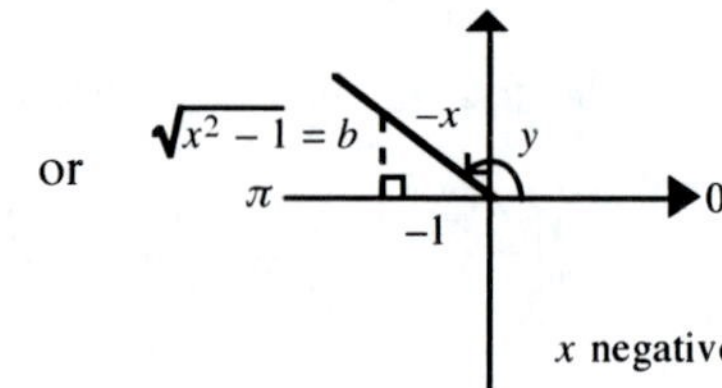

In either case, $b = \sqrt{x^2 - 1}$. Thus, if $x > 0$, $\csc(\sec^{-1} x) = \csc y = \frac{r}{b} = \frac{x}{\sqrt{x^2 - 1}}$.

If $x < 0$, $\csc(\sec^{-1} x) = \csc y = \frac{r}{b} = \frac{-x}{\sqrt{x^2 - 1}}$. A convenient notation for the quantity x if $x > 0$, $-x$ if $x < 0$, is $|x|$. Hence, we can write $\csc(\sec^{-1} x) = \frac{|x|}{\sqrt{x^2 - 1}}$.

79. Let $y = \cot^{-1} x \quad 0 < y < \pi$ $\quad$ or, equivalently,

$x = \cot y \quad 0 < y < \pi$

Then, $\sin(2 \cot^{-1} x) = \sin(2y) = 2 \sin y \cos y$. From Problem 75, we have

$$\sin y = \sin(\cot^{-1} x) = \frac{1}{\sqrt{x^2 + 1}}.$$

From the figure in Problem 65, we also have $\cos y = \cos(\cot^{-1} x) = \frac{a}{r} = \frac{x}{\sqrt{x^2 + 1}}$. Thus,

$\sin(2 \cot^{-1} x) = 2 \cdot \frac{1}{\sqrt{x^2 + 1}} \cdot \frac{x}{\sqrt{x^2 + 1}} = \frac{2x}{x^2 + 1}$.

81. Let $y = \sec^{-1} x \qquad x \le -1$ or $x \ge 1$

Then, $\sec y = x$	$0 \le y \le \pi, y \ne \frac{\pi}{2}$	Definition of $\sec^{-1}$
$\frac{1}{\cos y} = x$	$0 \le y \le \pi, y \ne \frac{\pi}{2}$	Reciprocal identity
$\cos y = \frac{1}{x}$	$0 \le y \le \pi, y \ne \frac{\pi}{2}$	Algebra
$y = \cos^{-1} \frac{1}{x}$	$0 \le y \le \pi, y \ne \frac{\pi}{2}$	Definition of $\cos^{-1}$

Thus, $\sec^{-1} x = \cos^{-1} \frac{1}{x}$ for $x \le -1$ and $x \ge 1$

83.

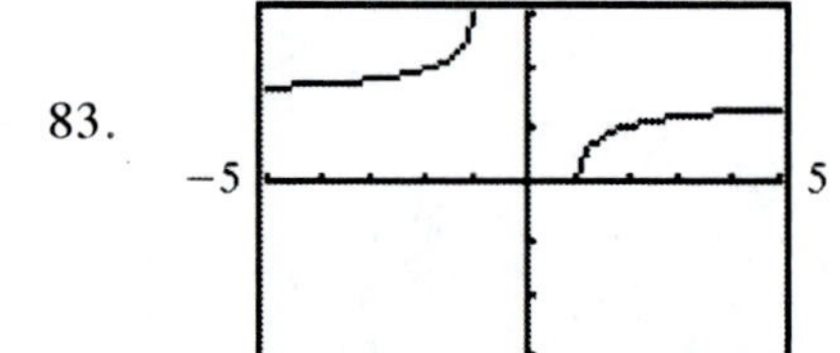

Use the identity $\sec^{-1} x = \cos^{-1} \frac{1}{x}$.

85.

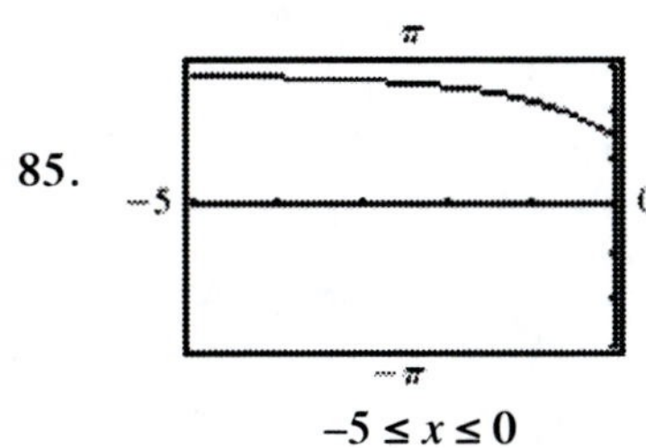

$-5 \le x \le 0$

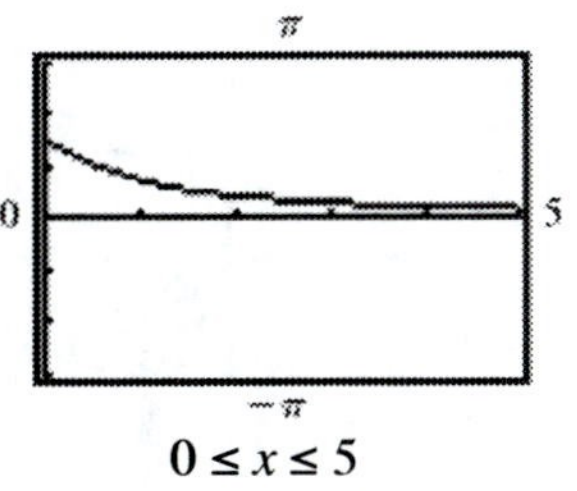

$0 \le x \le 5$

For the case $-5 \le x < 0$, use the identity $\cot^{-1} x = \pi + \tan^{-1}\left(\frac{1}{x}\right)$.

For the case $0 < x \le 5$, use the identity $\cot^{-1} x = \tan^{-1}\left(\frac{1}{x}\right)$.

87. This is an identity. First consider the case $x \ge 1$. Let

$u = \sec^{-1} x \quad 0 \le u < \frac{\pi}{2} \qquad v = \csc^{-1} x \quad 0 < v \le \frac{\pi}{2}$

$x = \sec u \quad 0 \le u < \frac{\pi}{2} \qquad x = \csc v \quad 0 < v \le \frac{\pi}{2}$

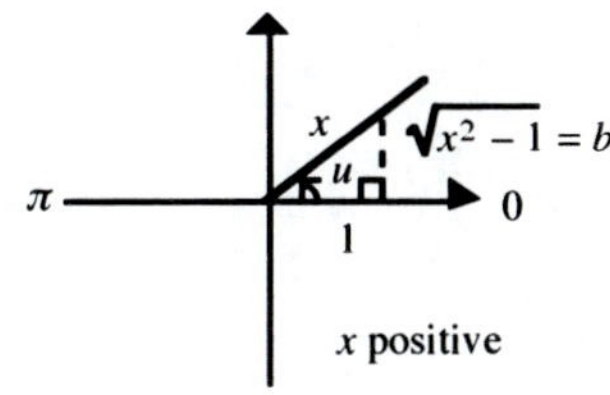

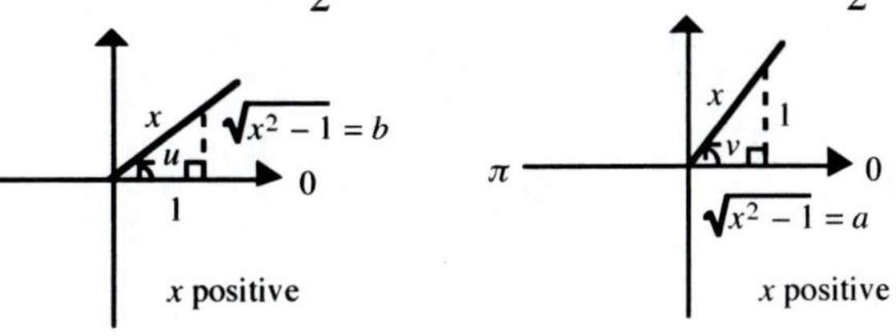

$0 \le u + v = \sec^{-1} x + \csc^{-1} x \le \frac{\pi}{2} + \frac{\pi}{2} = \pi$

From the figures, $\cos u = \cos(\sec^{-1} x) = \frac{1}{x}$

$\sin u = \sin(\sec^{-1} x) = \frac{\sqrt{x^2-1}}{x}$

$\cos v = \cos(\csc^{-1} x) = \frac{\sqrt{x^2-1}}{x}$

$\sin v = \sin(\csc^{-1} x) = \frac{1}{x}$

Then $\cos(\sec^{-1} x + \csc^{-1} x) = \cos(u+v)$

$= \cos u \cos v - \sin u \sin v$

$= \frac{1}{x} \cdot \frac{\sqrt{x^2-1}}{x} - \frac{\sqrt{x^2-1}}{x} \cdot \frac{1}{x} = 0$

Since $\cos(\sec^{-1} x + \csc^{-1} x) = 0$ and $0 \le \sec^{-1} x + \csc^{-1} x \le \pi$, it is necessary that $\sec^{-1} x + \csc^{-1} x = \frac{\pi}{2}$.

Now consider the case $x \le -1$. Let

$u = \sec^{-1} x \quad \frac{\pi}{2} < u \le \pi \qquad v = \csc^{-1} x \quad -\frac{\pi}{2} \le v < 0$

$x = \sec u \quad \frac{\pi}{2} < u \le \pi \qquad x = \csc v \quad -\frac{\pi}{2} \le v < 0$

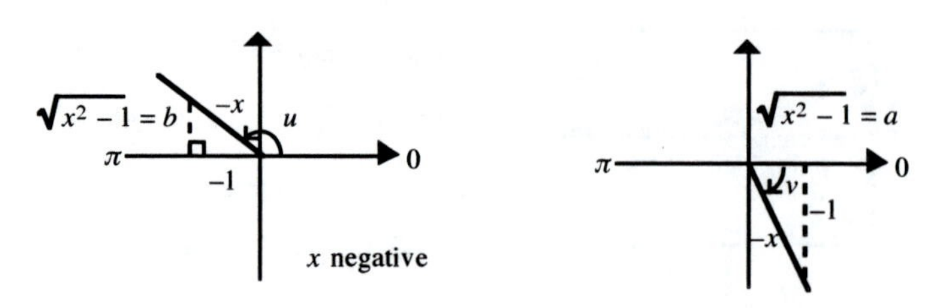

$$0 = \frac{\pi}{2} + \left(-\frac{\pi}{2}\right) \le u + v = \sec^{-1} x + \csc^{-1} x \le \pi + 0 = \pi$$

From the figures, $\cos u = \cos(\sec^{-1} x) = \dfrac{1}{x}$

$$\sin u = \sin(\sec^{-1} x) = \frac{\sqrt{x^2 - 1}}{-x}$$

$$\cos v = \cos(\csc^{-1} x) = \frac{\sqrt{x^2 - 1}}{-x}$$

$$\sin v = \sin(\csc^{-1} x) = \frac{1}{x}$$

Then
$$\begin{aligned}\cos(\sec^{-1} x + \csc^{-1} x) &= \cos(u + v)\\ &= \cos u \cos v - \sin u \sin v\\ &= \frac{1}{x} \cdot \frac{\sqrt{x^2 - 1}}{-x} - \frac{\sqrt{x^2 - 1}}{-x} \cdot \frac{1}{x} = 0\end{aligned}$$

Since $\cos(\sec^{-1} x + \csc^{-1} x) = 0$ and $0 \le \sec^{-1} x + \csc^{-1} x \le \pi$, it is necessary that $\sec^{-1} x + \csc^{-1} x = \dfrac{\pi}{2}$.

89. This is not an identity. For example, if $x = -1$, $\csc(\cot^{-1} x) = \csc(\cot^{-1}(-1)) = \csc\left(\frac{3}{4}\pi\right) = \sqrt{2}$, but $\sqrt{x^2 - 1} = \sqrt{(-1)^2 - 1} = 0$.

91. This is an identity. However, note that the left side is only defined if x, and thus $\csc^{-1} x$, are positive, since the restricted cotangent function $\cot x$ has domain $0 < x < \pi$.

Let $u = \csc^{-1} x \qquad 0 < u \le \dfrac{\pi}{2}$

$x = \csc u \qquad 0 < u \le \dfrac{\pi}{2}$

$a = \sqrt{x^2 - 1}$. Thus, $\cot(\csc^{-1} x) = \dfrac{a}{b} = \dfrac{\sqrt{x^2 - 1}}{1} = \sqrt{x^2 - 1}$.

93. This is not an identity. For example, if $x = -2$, $\sec^{-1} x = \sec^{-1}(-2) = \dfrac{2\pi}{3}$,

but $\csc^{-1} \dfrac{x}{\sqrt{x^2 - 1}} = \csc^{-1} \dfrac{-2}{\sqrt{(-2)^2 - 1}} = \csc^{-1} \dfrac{-2}{\sqrt{3}} = -\dfrac{\pi}{3}$.

EXERCISE 5.3 Trigonometric Equations: An Algebraic Approach

1. Here is a graph of $y = \sin x$, $y = \frac{2}{3}$, and $y = \frac{3}{2}$.

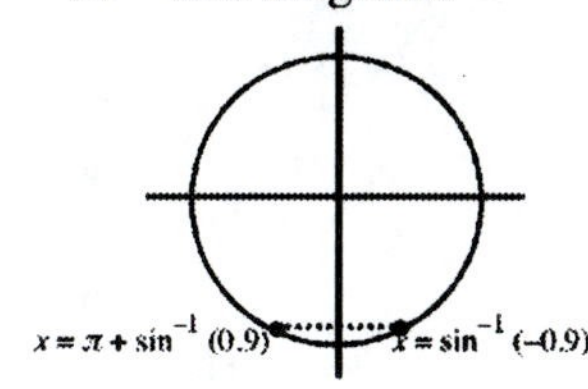

It shows clearly that the line $y = \frac{2}{3}$ intersects the graph of $y = \sin x$ at 4 places on the interval shown, which implies that $\sin x = \frac{2}{3}$ has solutions; by the periodicity of $y = \sin x$ there must be infinite solutions. The line $y = \frac{3}{2}$ does not intersect the graph of $y = \sin x$, thus the equation $\sin x = \frac{3}{2}$ has no solutions.

3. See diagram.

$x = \pi + \sin^{-1}(0.9)$ $x = \sin^{-1}(-0.9)$

The dashed line at $y = -0.9$ intersects the unit circle, and, because sine is periodic with period 2π, these solutions lead to an infinite number of solutions for $\sin x = -0.9$. The line at $y = -1.1$ does not intersect the unit circle, and there are no points on the unit circle corresponding to $y = -1.1$; thus, there are no solutions to $\sin x = -1.1$.

5. $\cos x = \frac{\sqrt{3}}{2}$ all real x

One solution is $x = \cos^{-1}\frac{\sqrt{3}}{2} = \frac{\pi}{6}$. A second point on the unit circle with x coordinate $\frac{\sqrt{3}}{2}$ is $x = \frac{11\pi}{6}$. Therefore, all solutions are given by $x = \frac{\pi}{6} + 2k\pi$, $x = \frac{11\pi}{6} + 2k\pi$, k any integer.

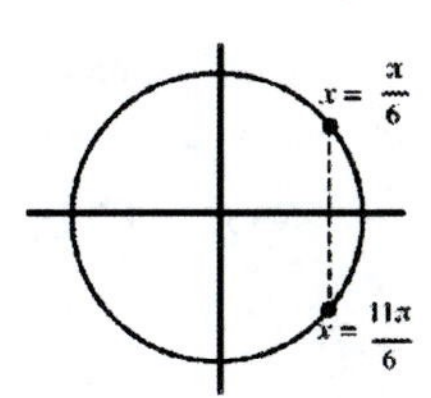

7. $\tan x = 1$ all real x

One solution is $x = \tan^{-1} 1 = \frac{\pi}{4}$. Therefore, all solutions are given by $x = \frac{\pi}{4} + k\pi$, k any integer.

9. $\sin x = -1$ all real x

The only point on the unit circle between 0 and 2π for which the y coordinate is -1 is $x = \frac{3\pi}{2}$.

Therefore, all solutions are given by $x = \frac{3\pi}{2} + 2k\pi$, k any integer.

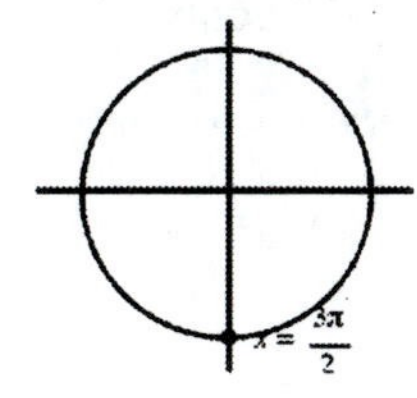

11. $2 \cos x + 1 = 0 \qquad 0 \le x < 2\pi$

Solve for cos x: $2 \cos x = -1$

$$\cos x = -\frac{1}{2}$$

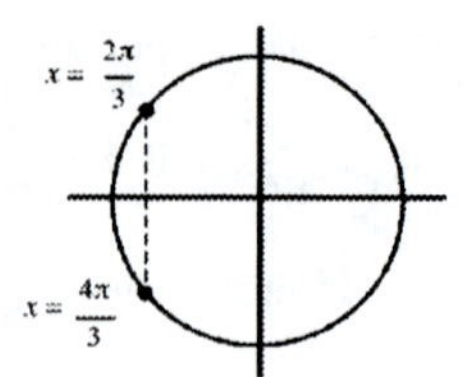

Solve over $0 \le x < 2\pi$: One solution is

$x = \cos^{-1}\left(-\frac{1}{2}\right) = \frac{2\pi}{3}$.

A second point on the unit circle with x coordinate $-\frac{1}{2}$ is $x = \frac{4\pi}{3}$. Therefore, the solutions in this interval are $x = \frac{2\pi}{3}$ and $x = \frac{4\pi}{3}$.

13. We have found all solutions of $2 \cos x + 1 = 0$ over one period in Problem 11. Because the cosine function is periodic with period 2π, all solutions of $2 \cos x + 1 = 0$ are given by

$$\frac{2\pi}{3} + 2k\pi, \ \frac{4\pi}{3} + 2k\pi, \quad k \text{ any integer}$$

15. $\sqrt{2} \sin \theta - 1 = 0 \qquad 0° \le \theta < 360°$

Solve for sin θ: $\sqrt{2} \sin \theta = 1$

$$\sin \theta = \frac{1}{\sqrt{2}}$$

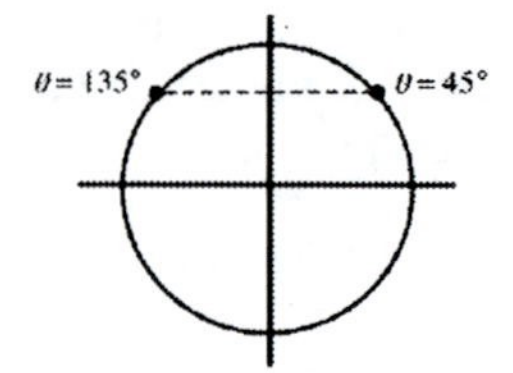

Solve over $0° \le \theta < 360°$: The location on the unit circle where $\sin \theta = \frac{1}{\sqrt{2}}$ corresponds to degree measures of 45° and 135°.

17. We have found all solutions of $\sqrt{2} \sin \theta - 1 = 0$ over one period in Problem 15. Because the sine function is periodic with period 360°, all solutions of $\sqrt{2} \sin \theta - 1 = 0$ are given by

$$45° + k(360°), 135° + k(360°), k \text{ any integer}$$

19. $3 \cos x - 6 = 0$ all real x

Solve for cos x: $3 \cos x = 6$

$$\cos x = 2$$

Impossible. No solution.

21. $4 \tan \theta + 15 = 0 \qquad 0° \le \theta < 180°$

Solve for tan θ: $\tan \theta = -\frac{15}{4}$

Solve over $0° \le \theta < 180°$: The location on the unit circle where $\theta = \tan^{-1}\left(-\frac{15}{4}\right)$ is at an angle of –75.0686°.

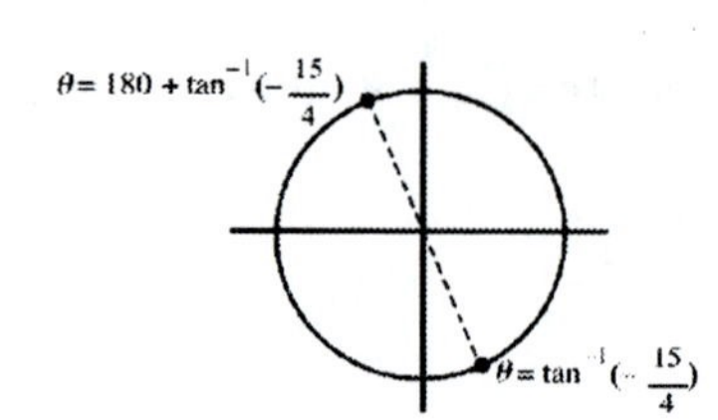

Therefore, the only solution in this interval is given by

$\theta = 180 + \tan^{-1}\left(-\frac{15}{4}\right) = 180° + (-75.0686°) = 104.9314°$

23. $5\cos x - 2 = 0 \qquad 0 \le x < 2\pi$

Solve for cos x: $\cos x = \dfrac{2}{5}$

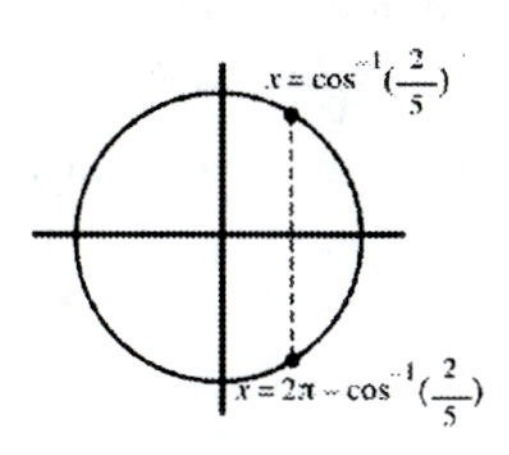

Solve over $0 \le x < 2\pi$: From the unit circle, we see that the solutions are in the first and fourth quadrants.

$$x = \cos^{-1}\frac{2}{5} = 1.1593$$
$$x = 2\pi - 1.1593 = 5.1239$$

25. $5.0118\sin x - 3.1105 = 0 \qquad 0 \le x < 2\pi$

Solve for sin x:
$$\sin x = \frac{3.1105}{5.0118}$$
$$\sin x = 0.620635$$

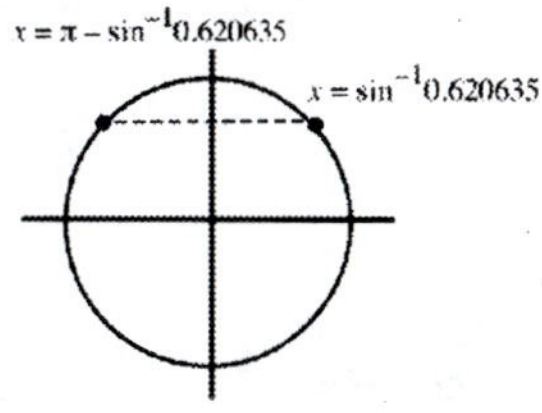

Solve over $0 \le x < 2\pi$: From the unit circle, we see that the solutions are in the first and second quadrants.

$$x = \sin^{-1} 0.620635 = 0.6696$$
$$x = \pi - 0.6696 = 2.4720$$

Write an expression for all solutions: Because the sine function is periodic with period 2π, all solutions are given by

$$x = 0.6696 + 2k\pi, \quad x = 2.4720 + 2k\pi, \quad k \text{ any integer}$$

27. $\cos x = \cot x \qquad 0 \le x < 2\pi$

Solve for sin x and/or cos x:

$$\cos x = \frac{\cos x}{\sin x} \qquad \text{Use quotient identity}$$
$$\cos x \sin x = \cos x \qquad (\sin x \ne 0)$$
$$\cos x \sin x - \cos x = 0$$
$$\cos x(\sin x - 1) = 0$$

$$\cos x = 0 \quad \text{or} \quad \sin x = 1$$
$$x = \frac{\pi}{2}, \frac{3\pi}{2} \qquad x = \frac{\pi}{2}$$

$$x = \frac{\pi}{2}, \frac{3\pi}{2}$$

29. $\cos^2\theta = \dfrac{1}{2}\sin 2\theta$

Solve over $0° \le \theta < 360°$:

$$\cos^2\theta = \frac{1}{2}\cdot 2\sin\theta\cos\theta \qquad \text{Use double-angle identity}$$
$$\cos^2\theta = \sin\theta\cos\theta$$
$$\cos^2\theta - \sin\theta\cos\theta = 0$$
$$\cos\theta(\cos\theta - \sin\theta) = 0$$
$$\cos\theta = 0 \quad \text{or} \quad \cos\theta - \sin\theta = 0$$
$$\theta = 90°, 270° \qquad \cos\theta = \sin\theta$$
$$1 = \frac{\sin\theta}{\cos\theta}$$
$$1 = \tan\theta$$
$$\theta = 45°, 225°$$

$$\theta = 45°, 90°, 225°, 270°$$

Write an expression for all solutions: Because the cosine function is periodic with period 360° and the tangent function is periodic with period 180°, all solutions are given by

$$45° + k180°, 90° + k360°, 225° + k180°, 270° + k360°, k \text{ any integer}$$

This can be written more compactly as $45° + k180°, 90° + k180°$, k any integer.

31. $\tan \frac{x}{2} - 1 = 0 \qquad 0 \le x < 2\pi$

This is equivalent to

$\tan \frac{x}{2} - 1 = 0 \qquad 0 \le \frac{x}{2} < \pi$

Solve over $0 \le \frac{x}{2} < \pi$: $\tan \frac{x}{2} = 1$

$$\frac{x}{2} = \frac{\pi}{4}$$

$$x = \frac{\pi}{2}$$

33. $\sin^2 \theta + 2\cos\theta = -2 \qquad 0° \le \theta < 360°$

Solve for $\sin\theta$ *and/or* $\cos\theta$: $1 - \cos^2\theta + 2\cos\theta = -2$ Use Pythagorean identity

$$-\cos^2\theta + 2\cos\theta + 3 = 0$$

$$\cos^2\theta - 2\cos\theta - 3 = 0$$

$$(\cos\theta - 3)(\cos\theta + 1) = 0$$

$\cos\theta - 3 = 0$ or $\cos\theta + 1 = 0$

$\cos\theta = 3 \qquad \cos\theta = -1$

No solution $\qquad \theta = 180°$

35. $\cos 2\theta + \sin^2\theta = 0 \quad 0° \le \theta < 360°$

Solve for $\sin\theta$ *and/or* $\cos\theta$: $1 - 2\sin^2\theta + \sin^2\theta = 0$ Use double-angle identity

$$1 - \sin^2\theta = 0$$

$$\cos^2\theta = 0$$

$$\cos\theta = 0$$

$$\theta = 90°, 270°$$

37. $\cos 2x = \sin x$

Solve for $\sin x$: $1 - 2\sin^2 x = \sin x$ Use double-angle identity

$$2\sin^2 x + \sin x - 1 = 0$$

$$(2\sin x - 1)(\sin x + 1) = 0$$

$2\sin x - 1 = 0$ or $\sin x + 1 = 0$

$\sin x = \frac{1}{2} \qquad \sin x = -1$

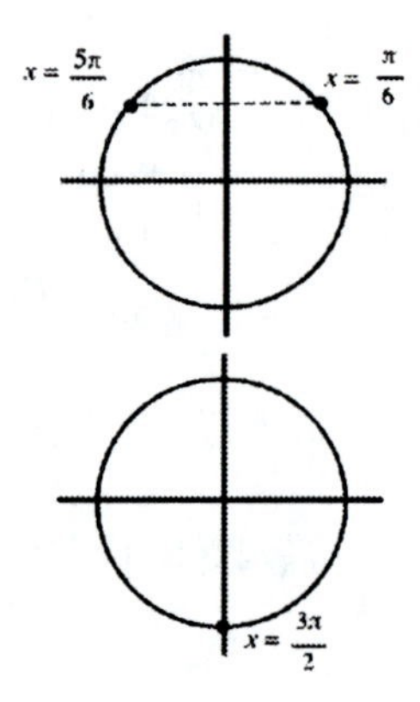

Solve over $0 \le x < 2\pi$: From unit circle graphs, we see that the solutions are in the first and second quadrants, as well as at $x = \frac{3\pi}{2}$.

$\sin x = \frac{1}{2} \qquad \sin x = -1$

$x = \frac{\pi}{6}$ or $x = \frac{5\pi}{6} \qquad x = \frac{3\pi}{2}$

Solve for all real solutions: Because the sine function is periodic with period 2π, all solutions are given by

$\frac{\pi}{6} + 2k\pi$, k any integer

$\frac{5\pi}{6} + 2k\pi$, k any integer

$\frac{3\pi}{2} + 2k\pi$, k any integer

39. $\cos 2x = \cos x - 1$

Solve for cos x: $2\cos^2 x - 1 = \cos x - 1$ Use double-angle identity

$$2\cos^2 x - \cos x = 0$$
$$\cos x(2\cos x - 1) = 0$$
$$\cos x = 0 \quad \text{or} \quad 2\cos x - 1 = 0$$
$$\cos x = \frac{1}{2}$$

Solve over $0 \le x < 2\pi$: From the unit circle graphs, we see that the solutions are in the first and fourth quadrants, as well as at $x = \frac{\pi}{2}$ and $x = \frac{3\pi}{2}$.

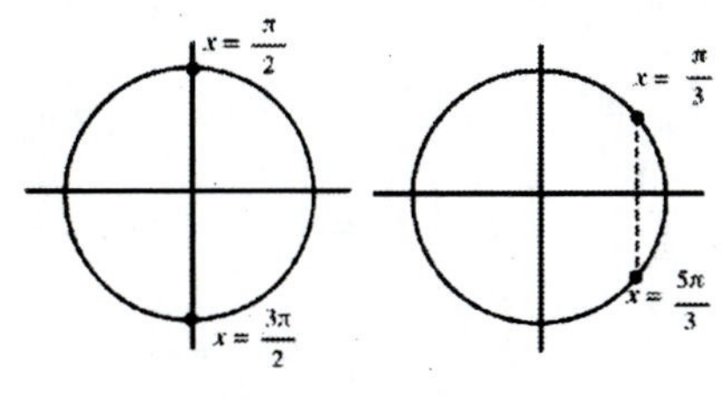

$\cos x = 0$ $\quad$ $\cos x = \frac{1}{2}$

$x = \frac{\pi}{2}$ or $\frac{3\pi}{2}$ $\quad$ $x = \frac{\pi}{3}$ or $\frac{5\pi}{3}$

Solve for all real solutions: Because the cosine function is periodic with period 2π, all solutions are given by

$\frac{\pi}{3} + 2k\pi$, k any integer

$\frac{\pi}{2} + 2k\pi$, k any integer

$\frac{3\pi}{2} + 2k\pi$, k any integer

$\frac{5\pi}{3} + 2k\pi$, k any integer

41. $2\cos\frac{\theta}{2} = \cos\theta + 1$

Solve for $\cos\frac{\theta}{2}$: $2\cos\frac{\theta}{2} = 2\cos^2\frac{\theta}{2} - 1 + 1$ Use double-angle identity

$$0 = 2\cos^2\frac{\theta}{2} - 2\cos\frac{\theta}{2}$$
$$0 = 2\cos\frac{\theta}{2}\left(\cos\frac{\theta}{2} - 1\right)$$
$$2\cos\frac{\theta}{2} = 0 \quad \text{or} \quad \cos\frac{\theta}{2} - 1 = 0$$
$$\cos\frac{\theta}{2} = 0 \qquad \cos\frac{\theta}{2} = 1$$

Solve over $0° \le \frac{\theta}{2} < 360°$: From unit circle graphs, we see that the solutions are given by

$$\frac{\theta}{2} = 0°, \ \frac{\theta}{2} = 90°, \ \frac{\theta}{2} = 270°$$

Thus, $\theta = 0°$, $\theta = 180°$, $\theta = 540°$.

Solve over all real θ: Since $\cos \frac{\theta}{2}$ is periodic with period 720°, all solutions are given by

$$\theta = 0° + k720°, k \text{ any integer}$$
$$\theta = 180° + k720°, k \text{ any integer}$$
$$\theta = 540° + k720°, k \text{ any integer}$$

The last two can be written more compactly as $\theta = 180° + k360°$, k any integer.

43. $\sin \frac{\theta}{2} = \cos \theta \quad 0° \le \theta < 720° \text{ or } 0° \le \frac{\theta}{2} < 360°$

Solve for $\sin \frac{\theta}{2}$: $\sin \frac{\theta}{2} = 1 - 2\sin^2 \frac{\theta}{2}$ — Use double-angle identity

$$2\sin^2 \frac{\theta}{2} + \sin \frac{\theta}{2} - 1 = 0$$

$$\left(2\sin\frac{\theta}{2} - 1\right)\left(\sin\frac{\theta}{2} + 1\right) = 0$$

$$2\sin \frac{\theta}{2} - 1 = 0 \quad \text{or} \quad \sin \frac{\theta}{2} + 1 = 0$$

$$\sin \frac{\theta}{2} = \frac{1}{2} \qquad \sin \frac{\theta}{2} = -1$$

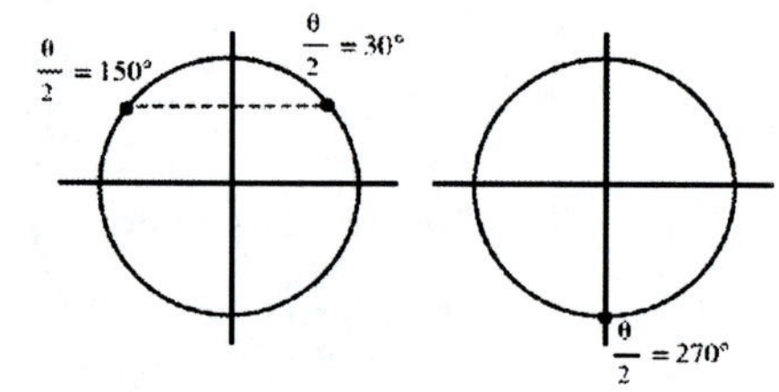

Solve over $0° \le \frac{\theta}{2} < 360°$: From the unit circle graphs, we see that the solutions are given by $\frac{\theta}{2} = 30°$, $\frac{\theta}{2} = 150°$, and $\frac{\theta}{2} = 270°$.

Thus, $\theta = 60°$, $\theta = 300°$, $\theta = 540°$.

45. $2\sqrt{2} \cos \frac{x}{2} = \cos x + 2$

Solve for $\cos \frac{x}{2}$: $2\sqrt{2}\cos \frac{x}{2} = 2\cos^2 \frac{x}{2} - 1 + 2$ — Use double-angle identity

$$2\cos^2 \frac{x}{2} - 2\sqrt{2}\cos \frac{x}{2} + 1 = 0$$

$$\left(\sqrt{2}\cos\frac{x}{2} - 1\right)^2 = 0$$

$$\sqrt{2}\cos \frac{x}{2} - 1 = 0$$

$$\cos \frac{x}{2} = \frac{1}{\sqrt{2}}$$

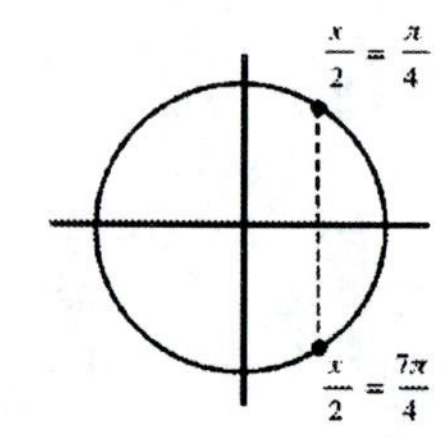

Solve over $0 \le x < 2\pi$: From the unit circle graphs, we see that the solutions are in the first and fourth quadrants.

$$\frac{x}{2} = \frac{\pi}{4} \quad \text{or} \quad \frac{x}{2} = \frac{7\pi}{4}$$

$$x = \frac{\pi}{2} \qquad\qquad x = \frac{7\pi}{2}$$

Solve for all real x: Because $\cos \frac{x}{2}$ is periodic with period 4π, all solutions are given by

$$x = \frac{\pi}{2} + 4k\pi,\ k \text{ any integer}$$

$$x = \frac{7\pi}{2} + 4k\pi,\ k \text{ any integer}$$

47. $4\cos^2\theta = 7\cos\theta + 2 \qquad 0° \le \theta < 180°$

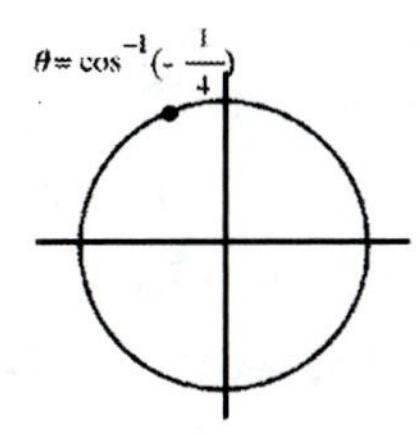

Solve for $\cos\theta$:

$$4\cos^2\theta - 7\cos\theta - 2 = 0$$

$$(4\cos\theta + 1)(\cos\theta - 2) = 0$$

$$4\cos\theta + 1 = 0 \quad \text{or} \quad \cos\theta - 2 = 0$$

$$\cos\theta = -\frac{1}{4} \qquad\qquad \cos\theta = 2$$

Solve over $0° \le \theta \le 180°$:

$\cos\theta = 2$ No solution (2 is not in the range of the cosine function.)

$\cos\theta = -\frac{1}{4}$ From the unit circle graph, we see that the solution is in the second quadrant.

$$\theta = \cos^{-1}\left(-\frac{1}{4}\right) = 104.5°$$

49. $\cos 2x + 10\cos x = 5 \qquad 0 \le x < 2\pi$

Solve for $\cos x$:

$2\cos^2 x - 1 + 10\cos x = 5$ Use double-angle identity

$2\cos^2 x + 10\cos x - 6 = 0$

$\cos^2 x + 5\cos x - 3 = 0$ Use quadratic formula

$$\cos x = \frac{-5 \pm \sqrt{5^2 - 4(1)(-3)}}{2(1)}$$

$$\cos x = \frac{-5 \pm \sqrt{37}}{2}$$

$$\cos x = -5.54138 \quad \text{or} \quad \cos x = 0.54138$$

Solve over $0 \le x < 2\pi$:

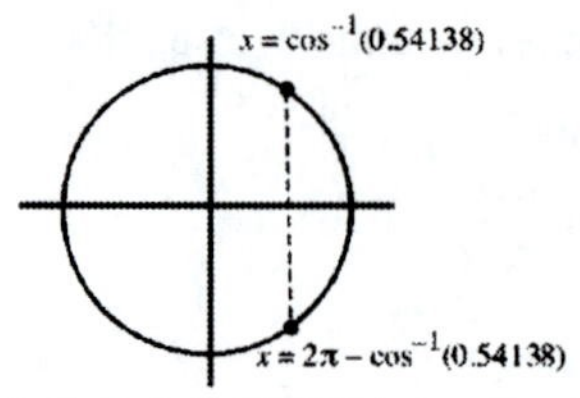

$\cos x = -5.54138$ No solution (-5.54138 is not in the range of the cosine function.)

$\cos x = 0.54138$ From the unit circle graph, we see that the solutions are in the first and fourth quadrants.

$x = \cos^{-1}(0.54138) = 0.9987$

$x = 2\pi - 0.9987 = 5.284$

51. $\cos^2 x = 3 - 5\cos x$

$\cos^2 x + 5\cos x - 3 = 0$

We have found all solutions of this equation over one period in Problem 49. Because the cosine function is periodic with period 2π, all solutions are given by

$0.9987 + 2k\pi,\quad 5.284 + 2k\pi,\quad k$ any integer

If 5.284 is replaced by $5.284 - 2\pi = -0.9987$, this answer is seen to be equivalent to the text answer,

$0.9987 + 2k\pi,\quad -0.9987 + 2k\pi,\quad k$ any integer

53. $\sin^2 x = -2 + 3\cos x$

Solve for cos x: $1 - \cos^2 x = -2 + 3\cos x$ Use Pythagorean identity

$0 = \cos^2 x + 3\cos x - 3$ Use quadratic formula

$$\cos x = \frac{-3 \pm \sqrt{3^2 - 4(1)(-3)}}{2(1)}$$

$$\cos x = \frac{-3 \pm \sqrt{21}}{2}$$

$\cos x = 0.79129$ or $\cos x = -3.79129$

Solve over $0 \le x < 2\pi$:

$\cos x = -3.79129$ No solution (-3.79129 is not in the range of the cosine function.)

$\cos x = 0.79129$ From the unit circle graph, we see that the solutions are in the first and fourth quadrants.

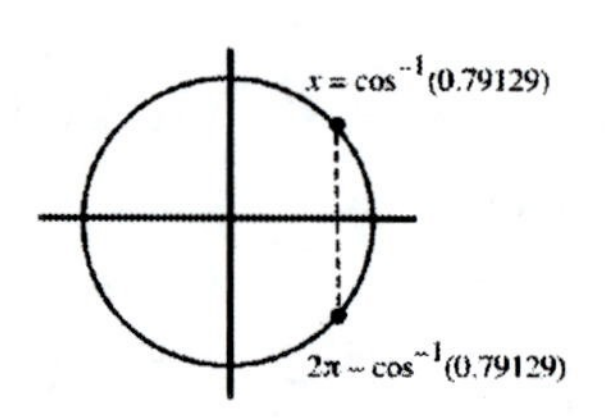

$x = \cos^{-1}(0.79129) = 0.6579$

$x = 2\pi - 0.6579 = 5.625$

Solve for all real solutions: Because the cosine function is periodic with period 2π, all solutions are given by

$0.6579 + 2k\pi$, k any integer

$5.625 + 2k\pi$, k any integer

55. $\cos 2x = 2\cos x$

Solve for cos x: $2\cos^2 x - 1 = 2\cos x$ Use double-angle identity

$2\cos^2 x - 2\cos x - 1 = 0$ Use quadratic formula

$$\cos x = \frac{-(-2) \pm \sqrt{(-2)^2 - 4(2)(-1)}}{2(2)}$$

$$\cos x = \frac{2 \pm \sqrt{12}}{4}$$

$\cos x = 1.3660$ or $\cos x = -0.3660$

Solve over $0 \le x < 2\pi$:

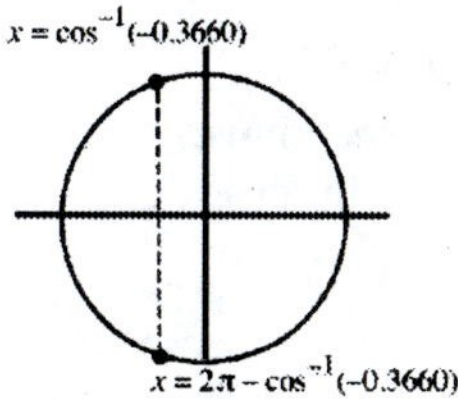

$\cos x = 1.3660$ No solution (1.3660 is not in the range of the cosine function.)

$\cos x = -0.3660$ From the unit circle graph, we see that the solutions are in the second and third quadrants.

$x = \cos^{-1}(-0.3660) = 1.946$

$x = 2\pi - 1.946 = 4.338$

Solve for all real solutions: Because the cosine function is periodic with period 2π, all solutions are given by

$1.946 + 2k\pi$, k any integer

$4.338 + 2k\pi$, k any integer

57. $\cos^{-1}(-0.7334)$ has exactly one value, 2.3941; the equation $\cos x = -0.7334$ has infinitely many solutions, which are found by adding $2\pi k$, k any integer, to each solution in one period of $\cos x$. Thus, $2.3941 + 2\pi k$ and $2\pi - 2.3941 + 2\pi k = 3.8891 + 2\pi k$ are all the solutions to $\cos x = -0.7334$.

59. $\sin x + \cos x = 1 \quad 0 \le x < 2\pi$

$$\pm\sqrt{1-\cos^2 x} + \cos x = 1$$

$$\pm\sqrt{1-\cos^2 x} = 1 - \cos x$$

$$1 - \cos^2 x = (1-\cos x)^2 \quad \text{Squaring both sides}$$

$$1 - \cos^2 x = 1 - 2\cos x + \cos^2 x$$

$$0 = -2\cos x + 2\cos^2 x$$

$$0 = 2\cos x(-1 + \cos x)$$

$2\cos x = 0$ $\qquad -1 + \cos x = 0$

$\cos x = 0$ $\qquad \cos x = 1$

$x = \frac{\pi}{2}, \frac{3\pi}{2}$ $\qquad x = 0$

In squaring both sides, we may have introduced extraneous solutions; hence, it is necessary to check solutions of these equations in the original equation.

Check:

$x = \frac{\pi}{2}$	$x = \frac{3\pi}{2}$	$x = 0$
$\sin x + \cos x = 1$	$\sin x + \cos x = 1$	$\sin x + \cos x = 1$
$\sin\frac{\pi}{2} + \cos\frac{\pi}{2} = 1$	$\sin\frac{3\pi}{2} + \cos\frac{3\pi}{2} = 1$	$\sin 0 + \cos 0 = 1$
$1 + 0 = 1$	$-1 + 0 \ne 1$	$0 + 1 = 1$
A solution	Not a solution	A solution

Solutions: $x = 0, \frac{\pi}{2}$

61. $\sec x + \tan x = 1 \quad 0 \le x < 2\pi$

$$\begin{aligned}
\pm\sqrt{1+\tan^2 x} + \tan x &= 1 \\
\pm\sqrt{1+\tan^2 x} &= 1 - \tan x \\
1 + \tan^2 x &= (1 - \tan x)^2 \qquad \text{Squaring both sides} \\
1 + \tan^2 x &= 1 - 2\tan x + \tan^2 x \\
0 &= -2\tan x \\
\tan x &= 0 \\
x &= 0, \pi
\end{aligned}$$

In squaring both sides, we may have introduced extraneous solutions; hence, it is necessary to check solutions of these equations in the original equation.

$$\begin{aligned}
x &= 0 & x &= \pi \\
\sec x + \tan x &= 1 & \sec x + \tan x &= 1 \\
\sec 0 + \tan 0 &= 1 & \sec \pi + \tan \pi &= 1 \\
1 + 0 &= 1 & -1 + 0 &\ne 1
\end{aligned}$$

A solution Not a solution

Solution: $x = 0$

63. The highest point is represented by $y = 10$, so we are to solve

$$\begin{aligned}
10 &= -10\cos 3t \\
\cos 3t &= -1 \\
3t &= \pi, 3\pi, 5\pi, 7\pi \\
t &= \frac{\pi}{3}, \pi, \frac{5\pi}{3}, \frac{7\pi}{3} \\
t &\approx 1.05, 3.14, 5.24, 7.33 \text{ seconds}
\end{aligned}$$

65. $I = 30\sin 120\pi t \qquad I = 25$

$$\begin{aligned}
25 &= 30\sin 120\pi t \\
\sin 120\pi t &= \frac{25}{30} \\
120\pi t &= \sin^{-1}\frac{25}{30} \text{ will yield the least positive solution of the equation} \\
t &= \frac{1}{120\pi}\sin^{-1}\frac{25}{30} = 0.002613 \text{ sec}
\end{aligned}$$

67. $y = 63 - 60\cos(10\pi t) \quad y = 80$

$80 = 63 - 60\cos(10\pi t)$

$17 = -60\cos(10\pi t)$

$\cos(10\pi t) = -\frac{17}{60}$

A first solution is given by $10\pi t = \cos^{-1}\left(-\frac{17}{60}\right)$

$$t = \frac{1}{10\pi}\cos^{-1}\left(-\frac{17}{60}\right) = 0.059 \text{ min}\left(\times 60\frac{\text{sec}}{\text{min}}\right) = 3.5 \text{ seconds.}$$

A second solution is given by $10\pi t = 2\pi - \cos^{-1}\left(-\frac{17}{60}\right)$

$$t = \frac{2\pi - \cos^{-1}\left(-\frac{17}{60}\right)}{10\pi} = 0.141 \text{ min } \left(\times 60\frac{\text{sec}}{\text{min}}\right) = 8.5 \text{ seconds.}$$

The remaining solutions less than one-half minute (30 seconds) are found by adding $\frac{2\pi}{10\pi} = 0.2$ min or 12 seconds the period of $\cos(10\pi t)$, to obtain $3.5 + 12 = 15.5, 8.5 + 12 = 20.5, 15.5 + 12 = 27.5$ seconds.

69. Following the hint, we solve:

$$I\cos^2\theta = 0.70I \qquad 0° \le \theta \le 180°$$
$$\cos^2\theta = 0.70$$
$$\cos\theta = \pm\sqrt{0.70}$$

$\theta = \cos^{-1}\sqrt{0.70}$ will yield the least positive solution of the equation

$$\theta = 33.21°$$

71. We are to solve $3.78 \times 10^7 = \frac{3.44 \times 10^7}{1 - 0.206\cos\theta}$. For convenience, we can divide both sides of this equation by 10^7 :

$$3.78 = \frac{3.44}{1 - 0.206\cos\theta}$$
$$3.78(1 - 0.206\cos\theta) = 3.44$$
$$3.78 - (3.78)(0.206)\cos\theta = 3.44$$
$$-(3.78)(0.206)\cos\theta = 3.44 - 3.78$$
$$\cos\theta = \frac{3.44 - 3.78}{-(3.78)(0.206)}$$
$$\theta = \cos^{-1}\frac{3.44 - 3.78}{-(3.78)(0.206)} = 64.1°$$

73. $r = 2\sin\theta \qquad 0° \le \theta \le 360°$
$r = 2(1 - \sin\theta)$

We solve this system of equations by equating the right sides:

$$2\sin\theta = 2(1 - \sin\theta) = 2 - 2\sin\theta$$
$$4\sin\theta = 2$$
$$\sin\theta = \frac{1}{2},\ \theta = 30° \text{ or } 150°$$

If we substitute these values of θ in either of the original equations, we obtain

$$r = 2\sin 30° = 1 \qquad r = 2\sin 150° = 1$$

Thus, the solutions of the system of equations are

$$(r, \theta) = (1, 30°) \text{ and } (r, \theta) = (1, 150°)$$

75. $xy = -2$

$$(u\cos\theta - v\sin\theta)(u\sin\theta + v\cos\theta) = -2 \quad \text{Substitution}$$
$$u\cos\theta\, u\sin\theta + u\cos\theta\, v\cos\theta - v\sin\theta\, u\sin\theta - v\sin\theta\, v\cos\theta = -2 \quad \text{Multiplication}$$
$$u^2\cos\theta\sin\theta + uv\cos^2\theta - uv\sin^2\theta - v^2\sin\theta\cos\theta = -2$$
$$u^2\sin\theta\cos\theta + uv(\cos^2\theta - \sin^2\theta) - v^2\sin\theta\cos\theta = -2$$

We are to find the least positive θ so that the coefficient of the uv term will be zero.

$$\cos^2\theta - \sin^2\theta = 0$$
$$\cos 2\theta = 0$$
$$2\theta = \cos^{-1} 0 \quad \text{yields the least positive } \theta$$
$$2\theta = 90°$$
$$\theta = 45°$$

77. The maximum value shown in the table is $y_{\max} = 15$.

The minimum value is $y_{\min} = 9 + \frac{20}{60} \approx 9.333$. Therefore, set

$$|A| = \frac{y_{\max} - y_{\min}}{2} = \frac{15 - 9.333}{2} = 2.833$$
$$B = \frac{2\pi}{\text{Period}} = \frac{2\pi}{12} = \frac{\pi}{6}$$
$$k = |A| + y_{\min} = 2.833 + 9.333 = 12.17$$

Thus, $y = 12.17 + 2.83\cos\frac{\pi}{6}x$ or $y = 12.17 - 2.83\cos\frac{\pi}{6}x$. The second one is the better model since $y = 10 + \frac{42}{60} = 10.7$ when $x = 2$, $y = 12.17 - 2.83\cos\frac{\pi}{6}x = 12.17 - 2.83\cos\left(\frac{\pi}{6} - 2\right) = 10.75$.

To find when Columbus has exactly 12 hours of daylight, solve $y = 12$. Set

$$12 = 12.17 - 2.83\cos\frac{\pi}{6}x$$
$$-0.17 = -2.83\cos\frac{\pi}{6}x$$
$$\cos\frac{\pi}{6}x = \frac{0.17}{2.83}$$
$$\frac{\pi}{6}x = \cos^{-1}\frac{0.17}{2.83} \quad \text{or} \quad \frac{\pi}{6}x = 2\pi - \cos^{-1}\frac{0.17}{2.83}$$
$$x = \frac{6}{\pi}\cos^{-1}\frac{0.17}{2.83} \quad \text{or} \quad x = 12 - \frac{6}{\pi}\cos^{-1}\frac{0.17}{2.83}$$
$$x = 2.9 \text{ months} \quad \text{or} \quad x = 9.1 \text{ months}$$

79. Applying the SinReg feature to the data in the table (using $x = 1$ for January, and so on) yields

$y = 2.818\sin(0.5108x - 1.605) + 12.14$

To find when Columbus has exactly 12 hours of daylights, solve $y = 12$. From the automatic intersection routine, the solutions are found to be $x = 3.0$ months and $x = 9.4$ months.

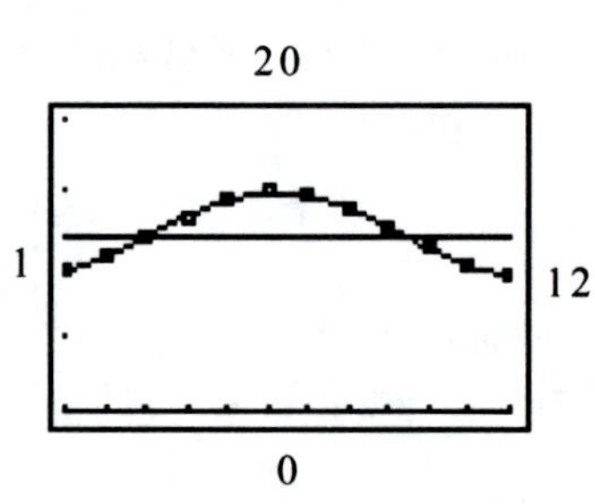

EXERCISE 5.4 Trigonometric Equations and Inequalities: A Graphing Calculator Approach

1. Graph $y_1 = f(x)$ and $y_2 = g(x)$ in an appropriate viewing window (for example, $-4\pi \le x \le 4\pi$) and find the point(s) of intersection of the two graphs using the INTERSECT command. Depending on $f(x)$ and $g(x)$, other windows and/or periodicity considerations may be necessary to find other solutions.

3. There are many examples. $\sin x = 2x - 1$ is one of many.

5. Graph $y1 = 2x$ and $y2 = \cos x$ in the same viewing window in a graphing calculator and find the point(s) of intersection using an automatic intersection routine. The single intersection point is found to be 0.4502.

   Check: $2(0.4502) = 0.9004$
   $\cos(0.4502) = 0.9004$

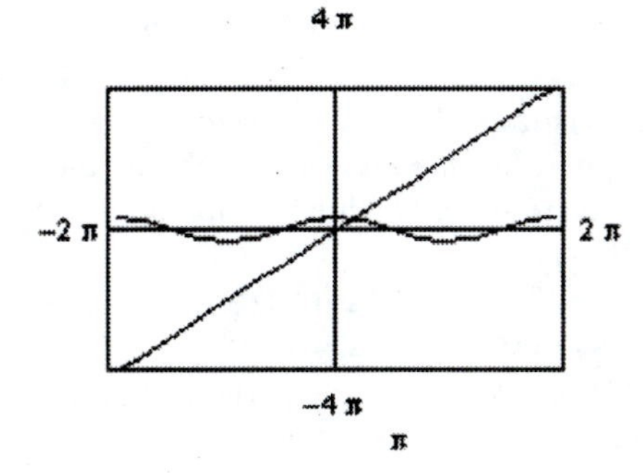

7. Graph $y1 = 3x + 1$ and $y2 = \tan 2x$ in the same viewing window in a graphing calculator and find the point(s) of intersection using an automatic intersection routine. The single intersection point is found to be 0.6167.

   Check: $3(0.6167) + 1 = 2.8501$
   $\tan 2(0.6167) = 2.8505$

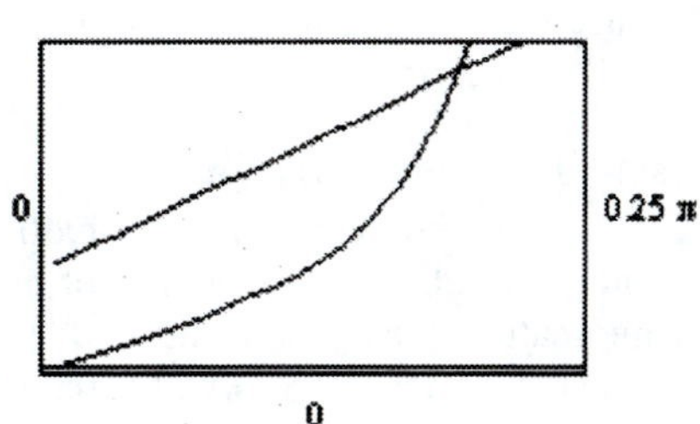

9. Graph $y1 = \cos x$ and $y2 = \sin x$ in the same viewing window in a graphing calculator. Finding the two points of intersection using an automatic intersection routine, we see that the graph of y1 is below the graph of y2 for the interval (0.7854, 3.9270).

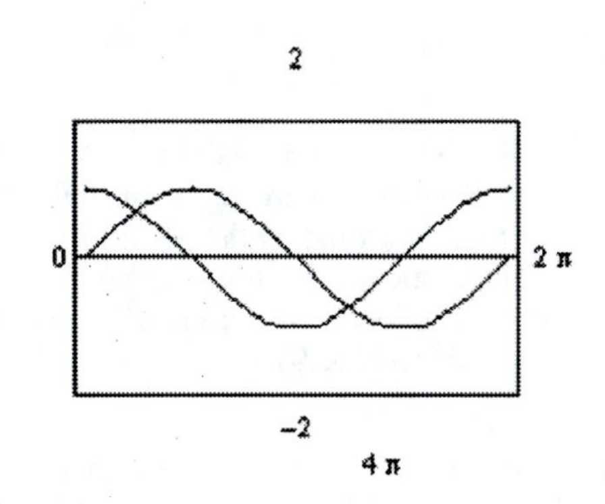

11. Graph $y1 = x$ and $y2 = \cos x$ in the same viewing window in a graphing calculator. Finding the point of intersection using an automatic intersection routine, we see that the graph of y1 is not below the graph of y2 for the interval $[0.7391, \infty)$.

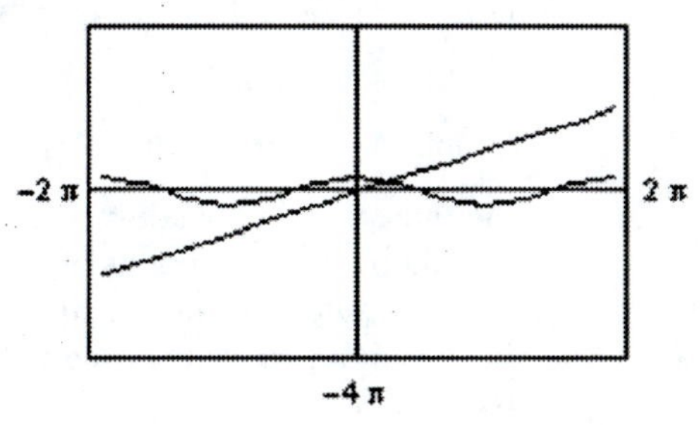

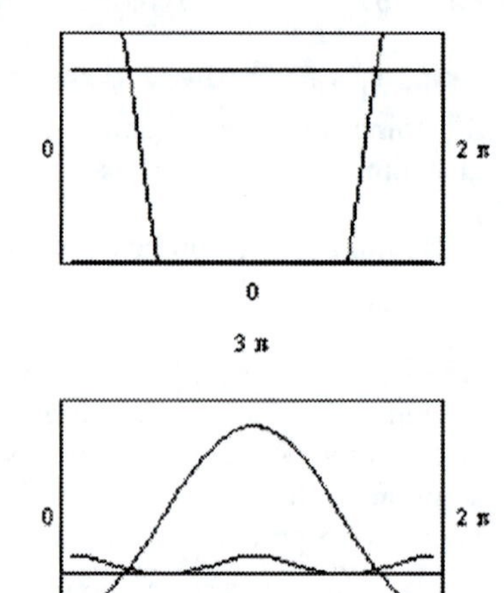

13. Graph y1 = cos $2x$ + 10 cos x and y2 = 5 in the same viewing window in a graphing calculator and find the points of intersection using an automatic intersection routine. The intersection points are found to be 0.9987 and 5.2845.

    Check:
    cos 2(0.9987) + 10 cos(0.9987) = 5.0002
    cos 2(5.2845) + 10 cos(5.2845) = 5.0003

15. Graph y1 = $\cos^2 x$ and y2 = 3 − 5 cos x in the same viewing window, $0 \le x < 2\pi$, in a graphing calculator and find the points of intersection using an automatic intersection routine. The intersection points are found to be 0.9987 and 5.2845.
    Check: $\cos^2 (0.9987) = 0.2931$
    $3 - 5 \cos (0.9987) = 0.2930$
    $\cos^2 (5.2845) = 0.2931$
    $3 - 5 \cos(5.2845) = 0.2930$

    Because the cosine function is periodic with period 2π, all solutions are given by $x = 0.9987 + 2k\pi$, $x = 5.2845 + 2k\pi$, k any integer.

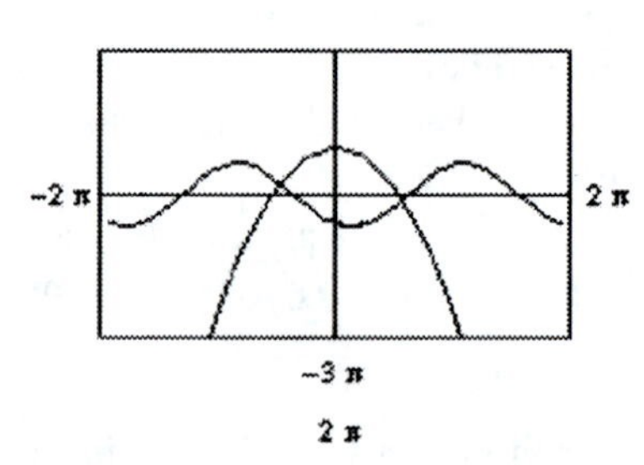

17. Graph y1 = 2 sin(x − 2) and y2 = 3 − x^2 in the same viewing window in a graphing calculator. Finding the two points of intersection using an automatic intersection routine, we see that the graph of y1 is below the graph of y2 for the interval (−1.5099, 1.8281).

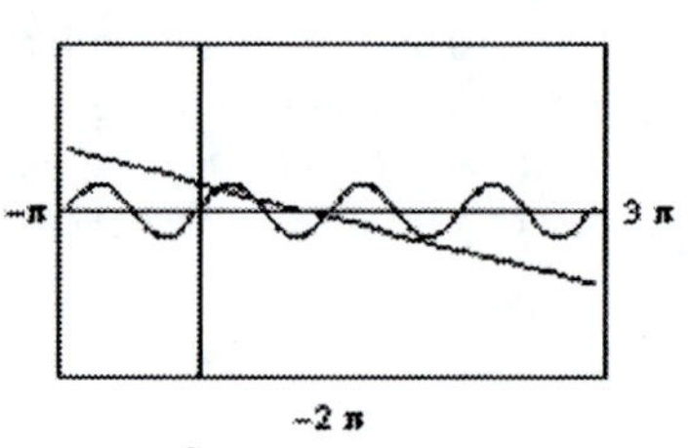

19. Graph y1 = sin(3 − $2x$) and y2 = 1 − $0.4x$ in the same viewing window in a graphing calculator. Finding the points of intersection using an automatic intersection routine, we see that the graph of y1 is not below the graph of y2 for the intervals [0.4204, 1.2346] and [2.9752, ∞).

21. $\sin^2 x - 2 \sin x + 1$ is greater than or equal to 0 for all real x because $\sin^2 x - 2 \sin x + 1 = (\sin x - 1)^2$, and the latter is greater than or equal to 0 for all real x.

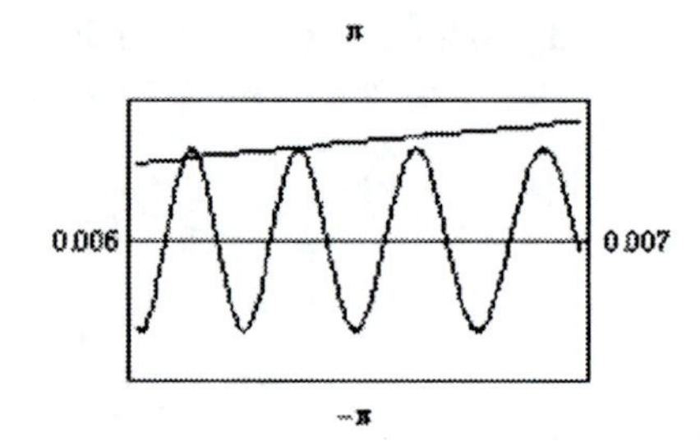

23. Graph y1 = $2 \cos \dfrac{1}{x}$ and y2 = $950x - 4$ in the same viewing window in a graphing calculator. Find the points of intersection using an automatic intersection routine. The intersection points are found to be 0.006104 and 0.006137.

25. Graph y1 = cos(sin x) and y2 = sin(cos x) in the same viewing window in a graphing calculator. Since y1 is above y2 for the entire width of the window, we conclude that on the interval $[-2\pi, 2\pi]$, the inequality cos(sin x) > sin(cos x) holds everywhere.

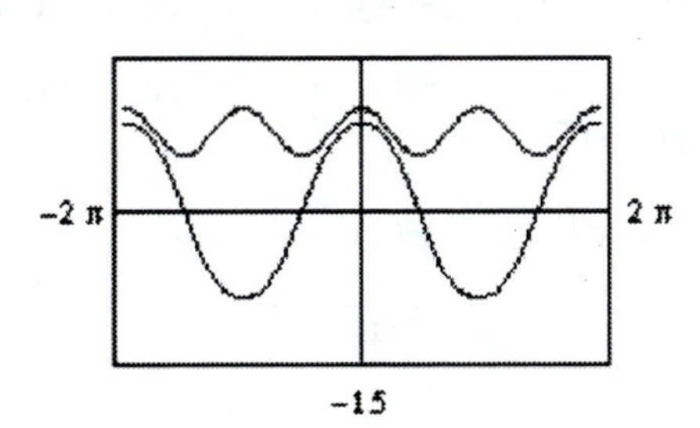

27. Graph y1 = $\tan^{-1} x$ and y2 = $0.5x$ in the same viewing window in a graphing calculator and find the points of intersection using an automatic intersection routine. The intersection points are found to be –2.331, 0, and 2.331.

Check: $\tan^{-1}(-2.331) = -1.1655$
$0.5(-2.331) = -1.1655$
$\tan^{-1}(0) = 0 = 0.5(0)$
$\tan^{-1}(2.331) = 1.1655$
$0.5(2.331) = 1.1655$

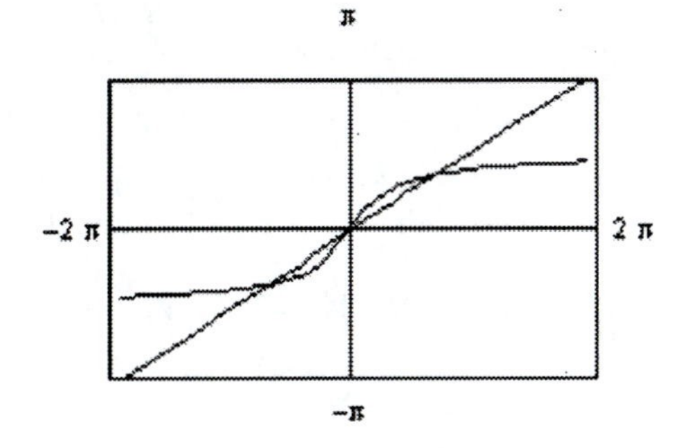

29. Graph y1 = $\sin^2 x$ and y2 = $\sin(x^2)$ in the same viewing window in a graphing calculator. Finding the points of intersection using an automatic intersection routine, we see that the graph of y1 is above the graph of y2 for the intervals (1.364, 2.566) and $(3.069, \pi]$.

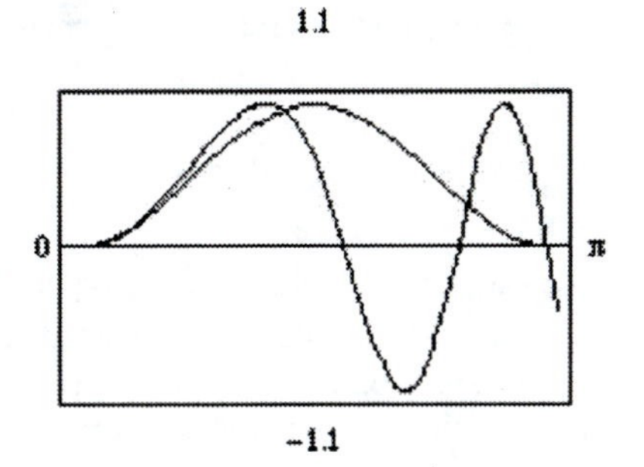

31. $\sqrt{3\sin(2x) - 2\cos(x/2) + x}$ is a real number as long as $3 \sin 2x - 2 \cos \frac{x}{2} + x \geq 0$.

Graph y1 = $3 \sin 2x - 2 \cos \frac{x}{2} + x$ and find the zeros of y1 using an automatic intersection routine. The zeros are found to be 0.2974, 1.6073, and 2.7097. The value 3.1416 is the right-hand endpoint of the stated interval and the desired inequality holds in the intervals [0.2974, 1.6073] and [2.7097, 3.1416].

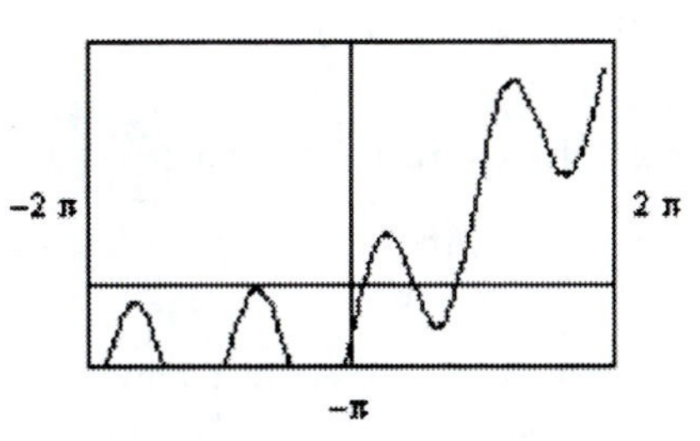

Check: $\sqrt{3\sin[2(0.2974)] - 2\cos(0.2974/2) + 0.2974} = 0.0022$

The remaining checking is left to the student.

33. Graph y1 = $\sin^{-1} x$ and y2 = $\dfrac{1}{\sin x}$ in the same viewing window in a graphing calculator and find the points of intersection using an automatic intersection routine. The intersection points are found to be ±0.9440.

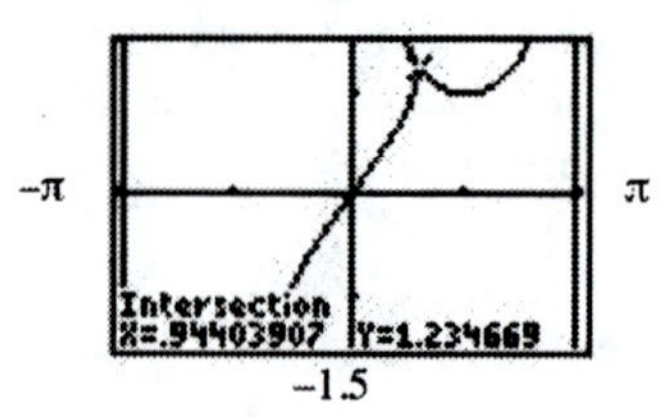

Check: $\sin^{-1}(0.9440) = 1.2345$

$$\frac{1}{\sin(0.9440)} = 1.2347$$

35. Graph y1 = $\tan^{-1} x$ and y2 = $\dfrac{1}{\tan x}$ in the same viewing window in a graphing calculator and find the points of intersection using an automatic intersection routine. The intersection points are found to be ±0.9284. There are actually an infinite number of solutions, but these are the two required solutions.

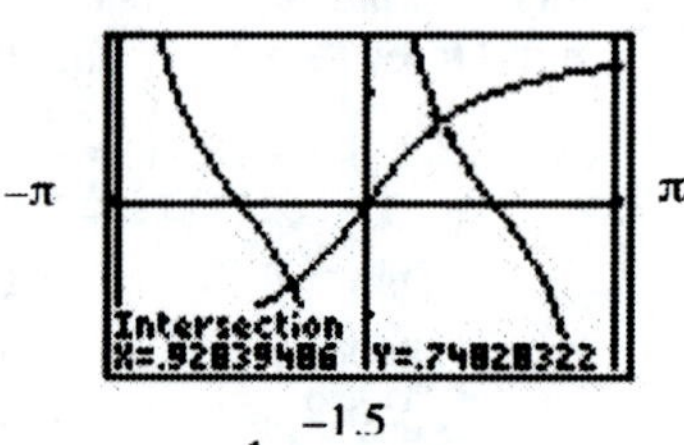

Check: $\tan^{-1}(0.9284) = 0.7483$

$$\frac{1}{\tan(0.9284)} = 0.7483$$

37. We use the given formula $A = \frac{1}{2}r^2(\theta - \sin\theta)$ with $r = 10$ and $A = 40$.

$$40 = \frac{1}{2}(10)^2(\theta - \sin\theta) = 50(\theta - \sin\theta)$$

$$\theta - \sin\theta = 0.8$$

We graph y1 = $x - \sin x$ and y2 = 0.8 on the interval from 0 to π. From the figure, we see that y1 and y2 intersect once on the given interval. Using the automatic intersection routine, the solution is found to be 1.78 radians.

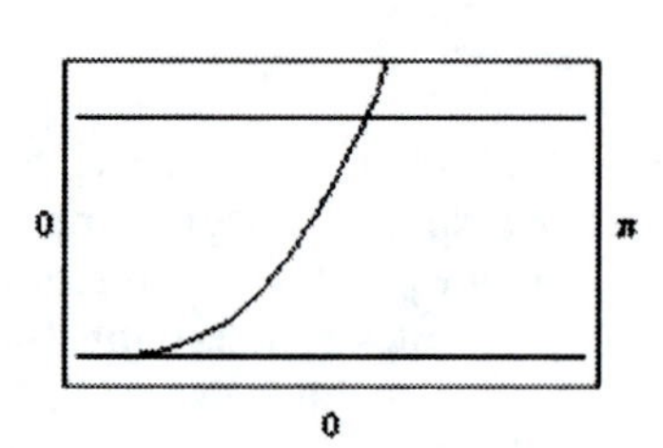

39. For weekly sales of 27,000 gallons, $y = 27$. To solve

$$27 = 12 + \frac{x}{8} - 9\cos\left(\frac{\pi x}{26}\right)$$

we graph y1 = $12 + \dfrac{x}{8} - 9\cos\left(\dfrac{\pi x}{26}\right)$ and y2 = 27 on the interval from 0 to 100. Using the automatic intersection routine, the solution is found to be 72 weeks to the nearest week.

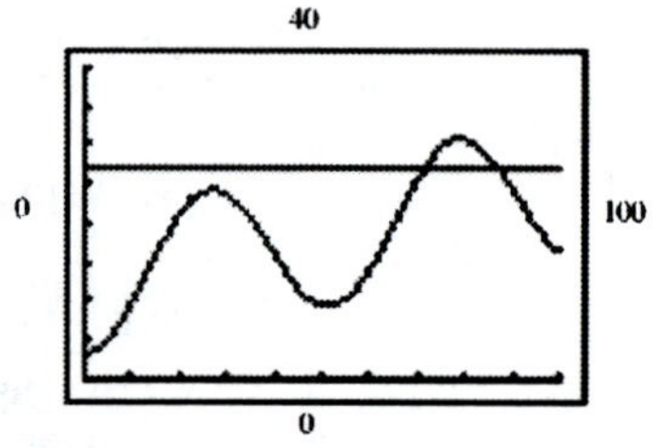

41. The doorway consists of a rectangle of area 4 × 8 = 32 square feet plus a segment *ABCDA* as shown. To find the area of the segment, we need to find θ and r. In triangle *OCD*, we have $\sin\dfrac{\theta}{2} = \dfrac{2}{r}$.

Since $s = r\theta$, we can also write $r = \dfrac{s}{\theta} = \dfrac{5}{\theta}$.

Therefore, θ must satisfy the relation

$$\sin\frac{\theta}{2} = 2 \div r = 2 \div \frac{5}{\theta} = \frac{2\theta}{5};\ \sin\frac{\theta}{2} = \frac{2\theta}{5}$$

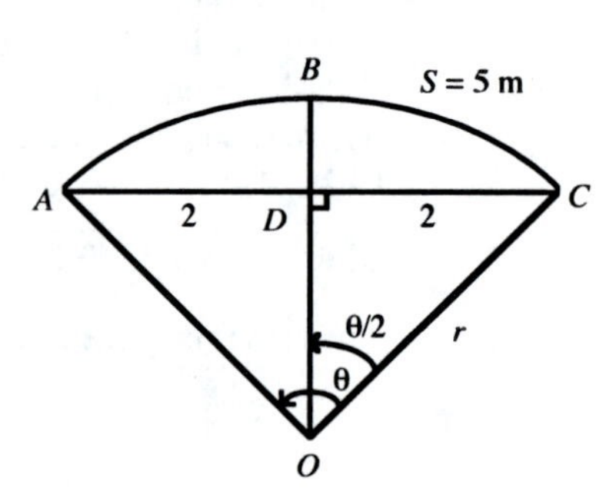

We graph y1 = sin $\frac{x}{2}$ and y2 = $\frac{2x}{5}$. From the figure, we see that y1 and y2 intersect once for positive x on the interval 0 to π. Using the automatic intersection routine, the solution is found to be $\theta = 2.262205$ radians. Then,

$$r = \frac{5}{\theta} = 2.21023 \text{ ft. Finally,}$$

$$A = \frac{1}{2} r^2 (\theta - \sin \theta)$$

$$= \frac{1}{2}(2.21023)^2 (2.262205 - \sin 2.262205)$$

$$= 3.64 \text{ sq. ft.}$$

Hence, the area of the doorway $= 32 + 3.64 = 35.64$ square feet.

43. (A) Analyzing triangle OAB, we see the following:

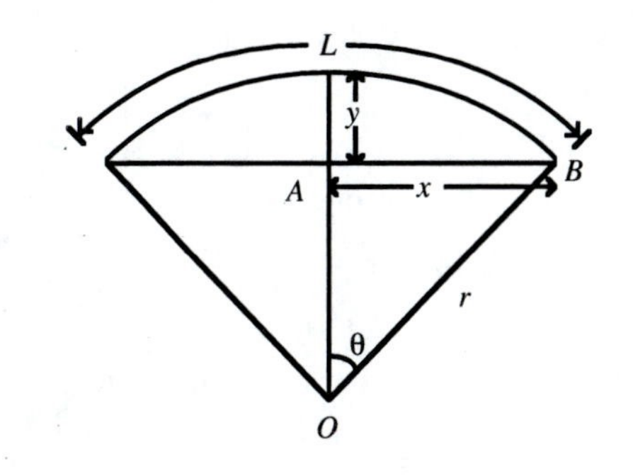

$$OA = r - y$$

$$OA^2 + AB^2 = r^2$$

$$(r - y)^2 + x^2 = r^2$$

$$r^2 - 2ry + y^2 + x^2 = r^2$$

$$y^2 + x^2 = 2ry$$

$$r = \frac{y^2 + x^2}{2y} = \frac{2.5^2 + 5.5^2}{2(2.5)} = 7.3$$

$$\sin \theta = \frac{x}{r}$$

$$\theta = \sin^{-1} \frac{x}{r} = \sin^{-1} \frac{5.5}{7.3} = 0.85325 \text{ rad.}$$

$$L = 2\theta \cdot r = 2(0.85325)(7.3) = 12.4575 \text{ mm}$$

(B) From part (A), we can write $r = \frac{y^2 + x^2}{2y}$, $\theta = \sin^{-1} \frac{x}{r} = \sin^{-1} \frac{2xy}{x^2 + y^2}$,

$$L = 2r\theta = \frac{y^2 + x^2}{y} \sin^{-1} \frac{2xy}{x^2 + y^2}$$

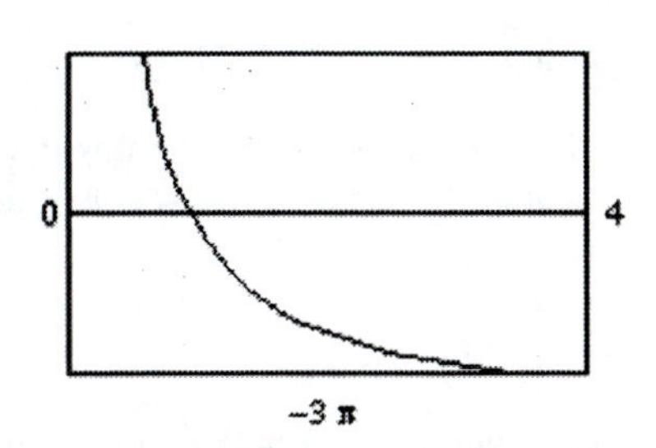

We want to find y if $L = 12.4575$ and $x = 5.4$. Substituting, we find that y must satisfy

$$12.4575 = \frac{y^2 + 29.16}{y} \sin^{-1} \frac{10.8y}{29.16 + y^2}$$

We graph

$$\text{y1} = \frac{y^2 + 29.16}{y} \sin^{-1} \frac{10.8y}{29.16 + y^2} - 12.4575.$$

From the figure, we see that y1 has one zero on the interval from 0 to 4. Using the automatic intersection routine, the solution is found to be $y = 2.6495$ mm.

45. (A) 160 yards = 480 feet.
We solve:

$$480 = \frac{130^2}{16}\sin\alpha\cos\alpha$$

We graph $y1 = \frac{130^2}{16}\sin\alpha\cos\alpha$ and $y2 = 480$ in the same window in a graphing calculator, on the interval from 0 to 90°. Using the automatic intersection routine, the solutions are found at 32.7° and 57.3°.

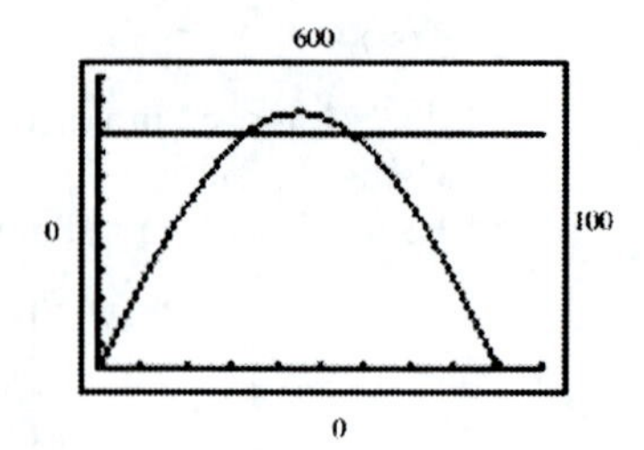

(B) 200 yards = 600 feet.

The equation $600 = \frac{130^2}{16}\sin\alpha\cos\alpha$ has no solution. No, she would not be able to hit a green 200 yards away.

47. The maximum value shown in the table is $y_{max} = 82.6$. The minimum value is $y_{min} = 33.6$. Therefore, set

$$|A| = \frac{y_{max} - y_{min}}{2} = \frac{82.6 - 33.6}{2} = 24.5$$

$$B = \frac{2\pi}{\text{Period}} = \frac{2\pi}{12} = \frac{\pi}{6}$$

$$k = |A| + y_{min} = 24.5 + 33.6 = 58.1$$

Use trial and error to find 4.2 to be best phase shift for the graph, or, if a negative phase shift is desired, take –12. Substitute $B = \frac{\pi}{6}$ and x into the phase-shift equation $x = -\frac{C}{B}, \frac{-C}{\pi/6}, C$.

Thus, the equation required is

$$y = 58.1 + 24.5\sin\left(\frac{\pi}{6}x\right) \quad \text{or} \quad y = 58.1 - 24.5\sin\left(\frac{\pi}{6}x\right).$$

Since $y = 33.6$ when $x = 1$, take $y = 58.1 + 24.5\sin\left(\frac{\pi}{6}x\right)$.

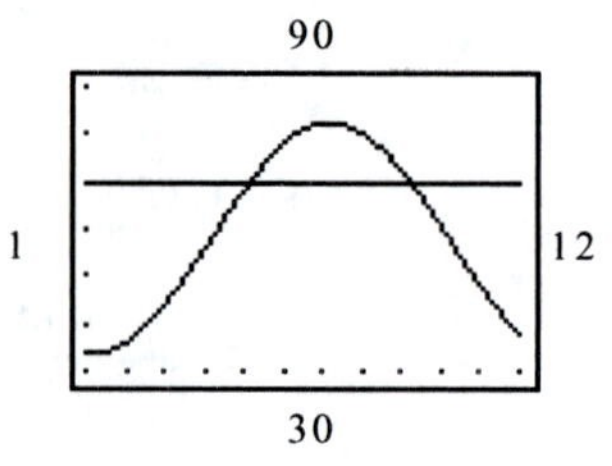

Graph $y1 = 58.1 + 24.5\sin\left(\frac{\pi}{6}x\right)$ and $y2 = 70$ in the same viewing window in a graphing calculator. Finding the points of intersection using an automatic intersection routine, we see that the graph of y1 is above the graph of y2 for $5.1 < x < 9.2$ months.

49. Entering the data and applying the SinReg feature yields the model $y = 57.5 + 25.5\sin(0.478x - 1.83)$. Graph $y1 = y$ and $y2 = 50$ in the same viewing window in a graphing calculator. Finding the points of intersection using an automatic intersection routine, we see that the graph of y1 is below the graph of y2 for $x < 3.2$ or $x > 11.0$ months.

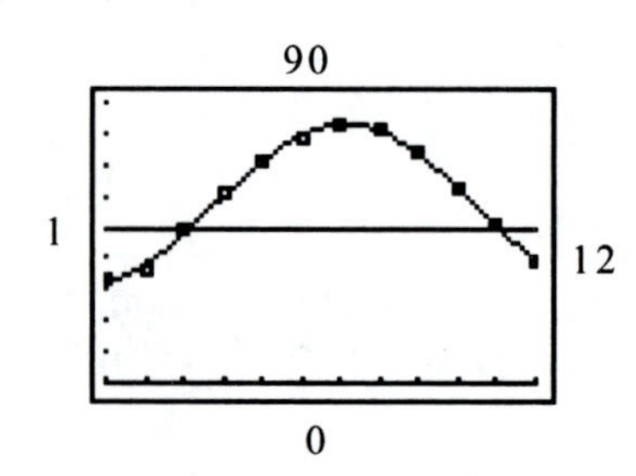

CHAPTER 5 REVIEW EXERCISE

1. $y = \tan^{-1}(-1)$ is equivalent to $\tan y = -1$. What y between $-\frac{\pi}{2}$ and $\frac{\pi}{2}$ has tangent equal to -1? y must be associated with a reference triangle in the fourth quadrant. Reference triangle is a special 45° triangle.

$$y = -\frac{\pi}{4}, \quad \tan^{-1}(-1) = -\frac{\pi}{4}$$

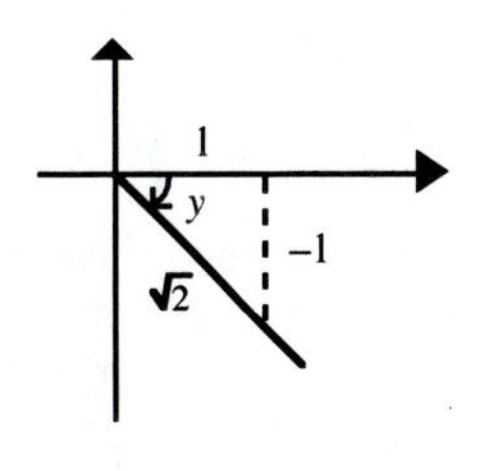

2. $y = \sin^{-1}\left(\frac{\sqrt{3}}{2}\right)$ is equivalent to $\sin y = \frac{\sqrt{3}}{2}$. What y between $-\frac{\pi}{2}$ and $\frac{\pi}{2}$ has sine equal to $\frac{\sqrt{3}}{2}$? y must be associated with a reference triangle in the first quadrant. Reference triangle is a special 30°–60° triangle.

$$y = \frac{\pi}{3}, \quad \sin^{-1}\left(\frac{\sqrt{3}}{2}\right) = \frac{\pi}{3}$$

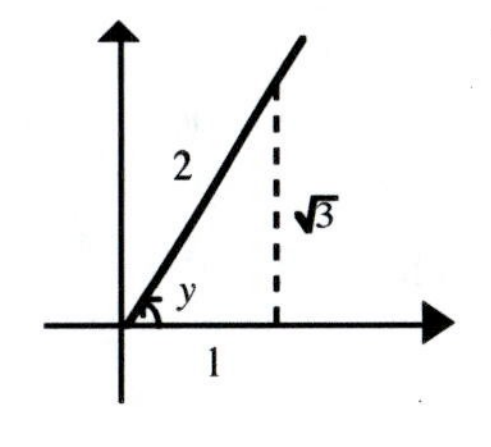

3. $y = \arccos 1$ is equivalent to $\cos y = 1$. No reference triangle can be drawn, but the only y between 0 and π that has cosine equal to 1 is $y = 0$. Thus, $\arccos 1 = 0$.

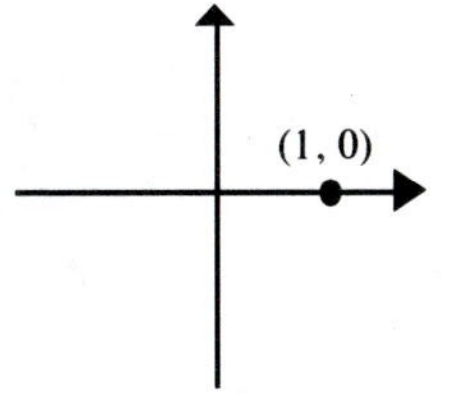

4. $y = \arctan\left(-\frac{1}{\sqrt{3}}\right)$ is equivalent to $\tan y = -\frac{1}{\sqrt{3}}$. What y between $-\frac{\pi}{2}$ and $\frac{\pi}{2}$ has tangent equal to $-\frac{1}{\sqrt{3}}$? y must be associated with a reference triangle in the fourth quadrant. Reference triangle is a special 30°–60° triangle.

$$y = -\frac{\pi}{6}, \quad \arctan\left(-\frac{1}{\sqrt{3}}\right) = -\frac{\pi}{6}$$

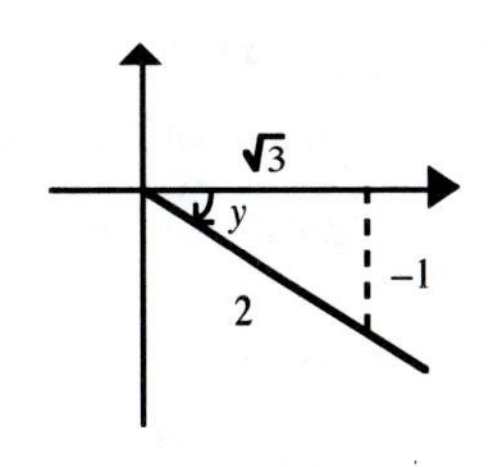

5. $\sqrt{2}$ is not in the domain of the inverse sine function. $-1 \le \sqrt{2} \le 1$ is false. $\arcsin \sqrt{2}$ is not defined.

6. $y = \cos^{-1}\left(-\frac{1}{\sqrt{2}}\right)$ is equivalent to $\cos y = -\frac{1}{\sqrt{2}}$. What y between 0 and π has cosine equal to $-\frac{1}{\sqrt{2}}$? y must be associated with a reference triangle in the second quadrant. Reference triangle is a special 45° triangle.

$$y = \frac{3\pi}{4}, \quad \cos^{-1}\left(-\frac{1}{\sqrt{2}}\right) = \frac{3\pi}{4}$$

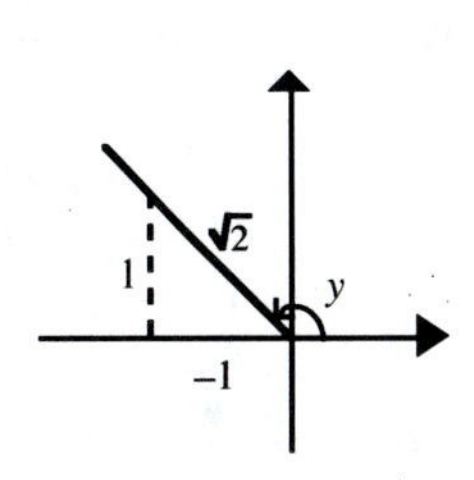

7. $y = \csc^{-1}\left(-\frac{2}{\sqrt{3}}\right)$ is equivalent to $\csc y = -\frac{2}{\sqrt{3}}$. What y between $-\frac{\pi}{2}$ and $\frac{\pi}{2}$ has cosecant equal to $-\frac{2}{\sqrt{3}}$? y must be associated with a reference triangle in the fourth quadrant. Reference triangle is a special 30°–60° triangle.

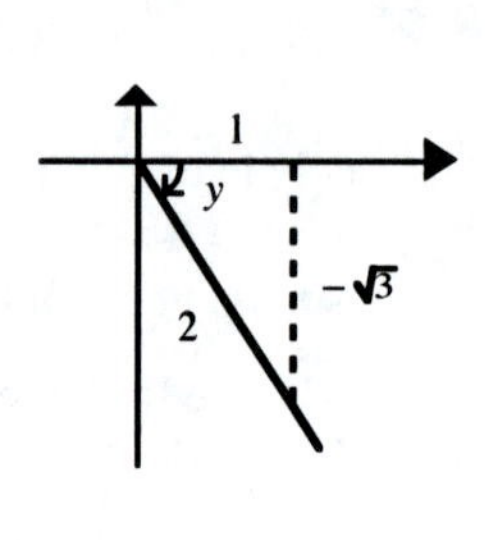

$$y = -\frac{\pi}{3}, \quad \csc^{-1}\left(-\frac{2}{\sqrt{3}}\right) = -\frac{\pi}{3}$$

8. $y = \cot^{-1} 0$ is equivalent to $\cot y = 0$. No reference triangle can be drawn, but the only y between 0 and π that has cotangent equal to 0 is $\frac{\pi}{2}$. Thus, $\cot^{-1} 0 = \frac{\pi}{2}$.

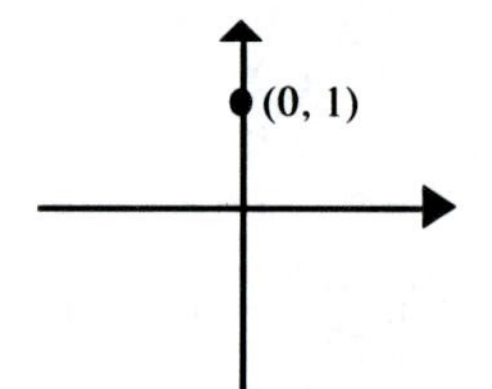

9. $y = \sec^{-1}(-2)$ is equivalent to $\sec y = -2$. What y between 0 and π has secant equal to -2? y must be associated with a reference triangle in the second quadrant. Reference triangle is a special 30°–60° triangle.

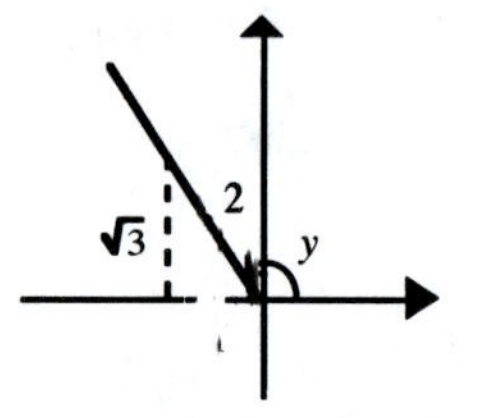

$$y = \frac{2\pi}{3}, \quad \sec^{-1}(-2) = \frac{2\pi}{3}$$

10. $\frac{1}{2}$ is not in the domain of the inverse cosecant function. Neither $\frac{1}{2} \le -1$ nor $\frac{1}{2} \ge 1$ is true. $\csc^{-1}\frac{1}{2}$ is not defined.

11. Calculator in radian mode: $\sin^{-1}(0.6298) = 0.6813$

12. Calculator in radian mode: $\arccos(-0.9704) = 2.898$

13. Calculator in radian mode: $\tan^{-1} 23.55 = 1.528$

14. Calculator in radian mode: $\cot^{-1}(-1.414) = \tan^{-1}\frac{1}{-1.414} + \pi = 2.526$

15. -1.025 is not in the domain of the inverse cosine function. $-1 \le -1.025 \le 1$ is false. $\cos^{-1}(-1.025)$ is not defined.

16. Calculator in degree mode: $\theta = \arctan 8.333 = 83.16°$

17. Calculator in degree mode: $\theta = \sin^{-1}(-0.1010) = -5.80°$

18. $\sin x = -\frac{\sqrt{3}}{2}$

One solution is $x = \frac{4\pi}{3}$. A second point on the unit circle with y coordinate $-\frac{\sqrt{3}}{2}$ is $x = \frac{5\pi}{3}$.

Therefore, all solutions are given by $x = \frac{4\pi}{3} + 2k\pi$, $x = \frac{5\pi}{3} + 2k\pi$, k any integer.

19. $\tan x = -\sqrt{3}$

One solution is $x = \tan^{-1}(\sqrt{3}) = -\frac{\pi}{3}$. Therefore all solutions are given by $x = -\frac{\pi}{3} + k\pi$, k any integer. This is equivalent to the text answer of $x = \frac{5\pi}{3} + k\pi$, k any integer. (Why?)

20. $\cos x = \frac{1}{\sqrt{2}}$

One solution is $x = \cos^{-1}\left(\frac{1}{\sqrt{2}}\right) = \frac{\pi}{4}$. A second point on the unit circle with x coordinate $\frac{1}{\sqrt{2}}$ is $x = \frac{7\pi}{4}$. Therefore, all solutions are given by $x = \frac{\pi}{4} + 2k\pi$, $x = \frac{7\pi}{4} + 2k\pi$, k any integer.

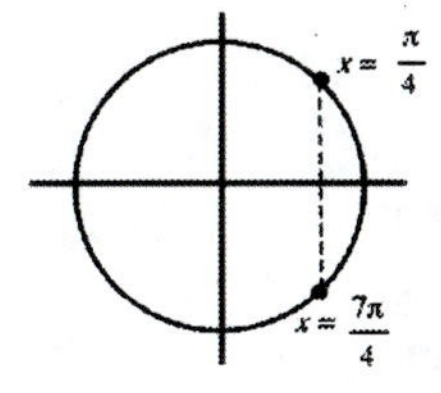

21. $-3\cos x + 10 = 0$

$$\cos x = \frac{10}{3}$$

$\frac{10}{3}$ is not in the range of the cosine function. No solution.

22. $2\cos x - \sqrt{3} = 0, \quad 0 \le x < 2\pi$

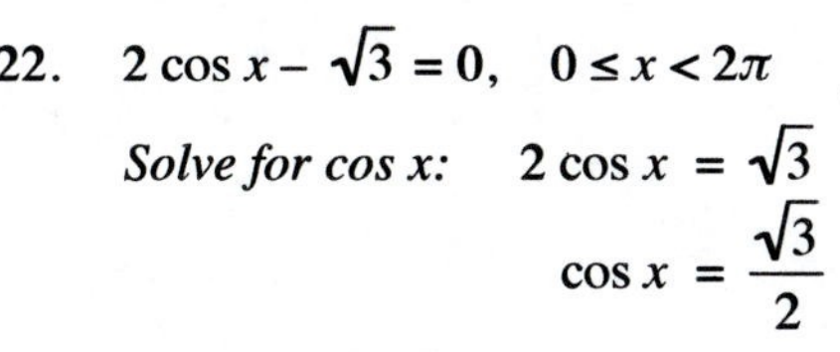

Solve for cos x: $2\cos x = \sqrt{3}$

$$\cos x = \frac{\sqrt{3}}{2}$$

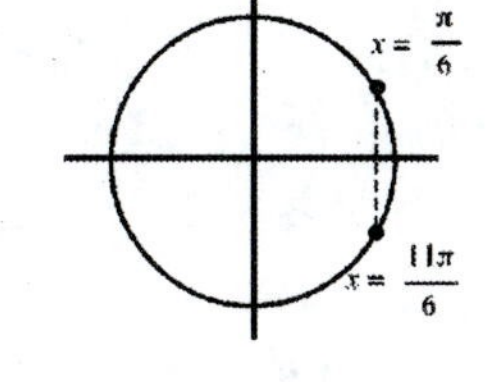

Solve over $0 \le x < 2\pi$: One solution is $x = \cos^{-1}\left(\frac{\sqrt{3}}{2}\right) = \frac{\pi}{6}$.

A second point on the unit circle with x coordinate $\frac{\sqrt{3}}{2}$ is $\frac{11\pi}{6}$. Therefore, the solutions in this interval are $x = \frac{\pi}{6}, \frac{11\pi}{6}$.

23. $2\sin^2\theta = \sin\theta \qquad 0° \le \theta < 360°$

Solve for sin θ:

$$2\sin^2\theta - \sin\theta = 0$$
$$\sin\theta(2\sin\theta - 1) = 0$$

$\sin\theta = 0$ $2\sin\theta - 1 = 0$

$\theta = 0°, 180°$ $\sin\theta = \frac{1}{2}$

$\theta = 30°, 150°$

Solutions: $\theta = 0°, 30°, 150°, 180°$

24. $4\cos^2 x - 3 = 0,\ \ 0 \le x < 2\pi$

Solve for cos x: $\cos^2 x = \frac{3}{4}$

$$\cos x = \pm\sqrt{\frac{3}{4}}$$

$$\cos x = \pm\frac{\sqrt{3}}{2}$$

$$\cos x = \pm\frac{\sqrt{3}}{2} \qquad\qquad \cos x = -\frac{\sqrt{3}}{2}$$

$$x = \frac{\pi}{6}, \frac{11\pi}{6} \qquad\qquad x = \frac{5\pi}{6}, \frac{7\pi}{6}$$

Solutions: $x = \frac{\pi}{6}, \frac{5\pi}{6}, \frac{7\pi}{6}, \frac{11\pi}{6}$

25. $2\cos^2\theta + 3\cos\theta + 1 = 0,\ \ 0° \le \theta < 360°$

Solve for cos θ: $(2\cos\theta + 1)(\cos\theta + 1) = 0$

$$2\cos\theta + 1 = 0 \quad \text{or} \quad \cos\theta + 1 = 0$$

$$\cos\theta = -\frac{1}{2} \qquad\qquad \cos\theta = -1$$

$$\theta = 120°, 240° \qquad\qquad \theta = 180°$$

Solutions: $\theta = 120°, 180°, 240°$

26. $\sqrt{2}\sin 4x - 1 = 0,\ \ 0 \le x < \frac{\pi}{2}$ is equivalent to

$\sqrt{2}\sin 4x - 1 = 0,\ \ 0 \le 4x < 2\pi$

Solve for sin 4x: $\sqrt{2}\sin 4x = 1$

$$\sin 4x = \frac{1}{\sqrt{2}}$$

Solve over $0 \le 4x < 2\pi$: One solution is given by $4x = \sin^{-1}\left(\frac{1}{\sqrt{2}}\right) = \frac{\pi}{4}$.

From the unit circle graph, a second point with y coordinate $\frac{1}{\sqrt{2}}$ is given by $4x = \frac{3\pi}{4}$.

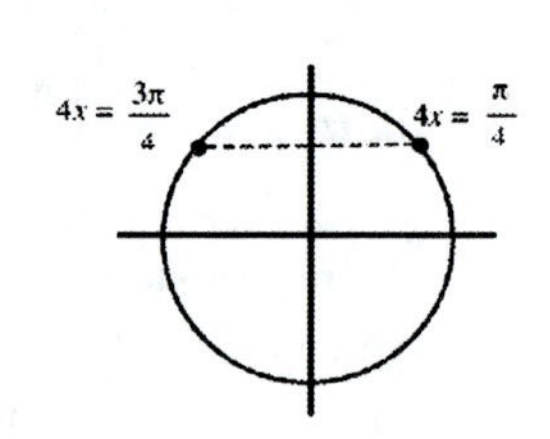

Thus:

$$4x = \frac{\pi}{4}, \frac{3\pi}{4}$$

Solutions: $x = \frac{\pi}{16}, \frac{3\pi}{16}$

27. $\tan\frac{\theta}{2} + \sqrt{3} = 0, \ -180° < \theta < 180°$ is equivalent to

$\tan\frac{\theta}{2} + \sqrt{3} = 0, \ -90° < \frac{\theta}{2} < 90°$

Solve for tan $\frac{\theta}{2}$: $\tan\frac{\theta}{2} = -\sqrt{3}$

Solve over $-90° < \frac{\theta}{2} < 90°$: The location on the unit circle where $\theta = \tan^{-1}(-\sqrt{3})$ is at an angle of $-60°$.

Therefore: $\frac{\theta}{2} = -60°$

$\theta = -120°$

28. $\cos^{-1} x = 25$ is equivalent to $\cos 25 = x$. Since the calculator is in degree mode, calculate $\cos 25° = 0.906308$.

29. $\cos(\cos^{-1} 0.315) = 0.315$ from the cosine-inverse cosine identity.

30. $\tan^{-1}[\tan(-1.5)] = -1.5$ from the tangent-inverse tangent identity.

31. Let $y = \tan^{-1}\left(-\frac{3}{4}\right)$, then $\tan y = -\frac{3}{4}, -\frac{\pi}{2} < y < \frac{\pi}{2}$.

Draw the reference triangle associated with y, then $\sin y = \sin\left[\tan^{-1}\left(-\frac{3}{4}\right)\right]$ can be determined directly from the triangle.

$a^2 + b^2 = c^2 \qquad c = \sqrt{4^2 + (-3)^2} = 5$

$\sin\left[\tan^{-1}\left(-\frac{3}{4}\right)\right] = \sin y = \frac{-3}{5} = -\frac{3}{5}$

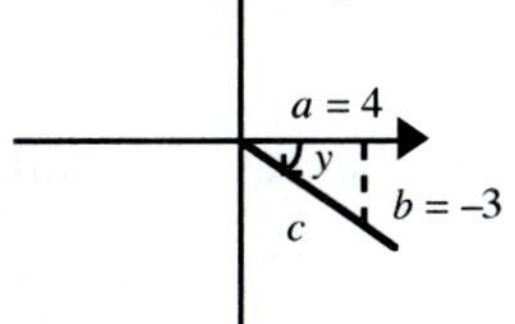

32. Let $y = \arccos\left(-\frac{2}{3}\right)$, then $\cos y = -\frac{2}{3}, 0 \le y \le \pi$.

Sketch the reference triangle associated with y, then

$\cot y = \cot\left[\cos^{-1}\left(-\frac{2}{3}\right)\right]$

can be determined directly from the triangle.

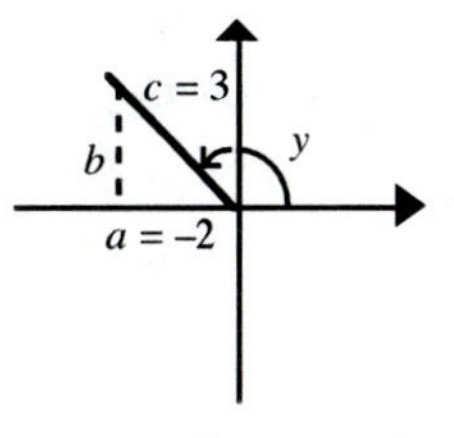

$a^2 + b^2 = c^2 \qquad b = \sqrt{3^2 - (-2)^2} = \sqrt{5}$

$\cot\left[\arccos\left(-\frac{2}{3}\right)\right] = \cot y = \frac{-2}{\sqrt{5}} = -\frac{2}{\sqrt{5}}$

33. Let $y = \cot^{-1}\left(-\frac{1}{3}\right)$, then $\cot y = -\frac{1}{3}, 0 < y < \pi$.

Draw the reference triangle associated with y, then

$\csc y = \csc\left[\cot^{-1}\left(-\frac{1}{3}\right)\right]$ can be determined directly from the triangle.

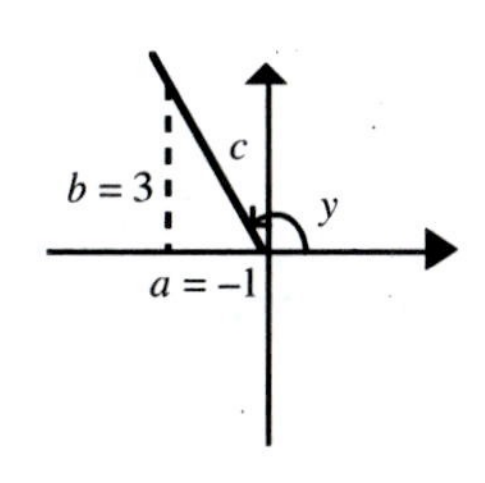

$a^2 + b^2 = c^2 \qquad c = \sqrt{(-1)^2 + 3^2} = \sqrt{10}$

$\csc\left[\cot^{-1}\left(-\frac{1}{3}\right)\right] = \csc y = \frac{\sqrt{10}}{3}$

34. Let $y = \text{arccsc } 5$, then $\csc y = 5, -\frac{\pi}{2} \le y \le \frac{\pi}{2}$.
Draw the reference triangle associated with y, then
$\cos y = \cos(\text{arccsc } 5)$ can be determined directly from the triangle.

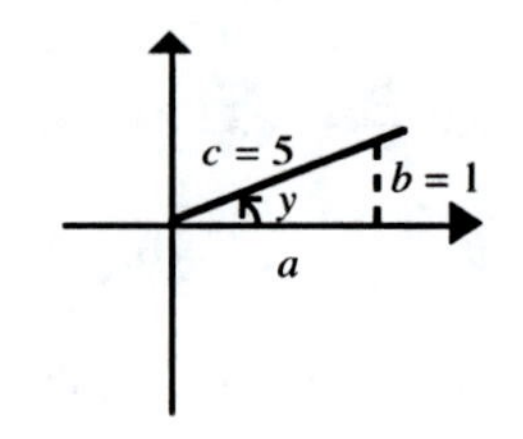

$$a^2 + b^2 = c^2 \qquad a = \sqrt{5^2 - 1^2} = \sqrt{24} = 2\sqrt{6}$$

$$\cos(\text{arccsc } 5) = \cos y = \frac{2\sqrt{6}}{5}$$

35. Calculator in radian mode: $\sin^{-1}(\cos 22.37) = -1.192$

36. Calculator in radian mode: $\sin^{-1}(\tan 1.345) = \text{error}$
$\sin^{-1}(\tan .345) = \sin^{-1}(4.353)$ is not defined.
4.353 is not in the domain of the inverse sine function.

37. Calculator in radian mode: $\sin[\tan^{-1}(-14.00)] = -0.9975$

38. Calculator in radian mode: $\csc[\cos^{-1}(-0.4081)] = \dfrac{1}{\sin[\cos^{-1}(-0.4081)]} = 1.095$

39. Calculator in radian mode: $\cos(\cot^{-1} 6.823) = \cos\left(\tan^{-1}\dfrac{1}{6.823}\right) = 0.9894$

40. Calculator in radian mode: $\sec[\text{arccsc}(-25.52)] = \dfrac{1}{\cos[\text{arccsc}(-25.52)]}$

$$= \frac{1}{\cos\left[\arcsin\left(\dfrac{1}{-25.52}\right)\right]} = 1.001$$

41. (a) is in degree mode and illustrates a sine-inverse sine identity, since 2 (meaning 2°) is in the domain for this identity, $-90° \le \theta \le 90°$. (b) is in radian mode and does not illustrate this identity, since 2 (meaning 2 radians) is not in the domain for the identity, $-\frac{\pi}{2} \le x \le \frac{\pi}{2}$.

42. $\sin^2 \theta = -\cos 2\theta, \quad 0° \le \theta \le 360°$

Solve for $\sin \theta$ and/or $\cos \theta$:
$$\begin{aligned} \sin^2 \theta &= -(1 - 2\sin^2 \theta) = -1 + 2\sin^2 \theta \quad \text{Use double-angle identity} \\ 0 &= -1 + \sin^2 \theta \\ \sin^2 \theta &= 1 \\ \sin \theta &= \pm 1 \end{aligned}$$

$$\sin \theta = 1 \qquad \sin \theta = -1$$
$$\theta = 90° \qquad \theta = 270°$$

Solutions: $\theta = 90°, 270°$

43. $\sin 2x = \frac{1}{2},\ 0 \le x < \pi$ is equivalent to

$\sin 2x = \frac{1}{2},\ 0 \le 2x < 2\pi$

Solve over $0 \le 2x < 2\pi$: From the unit circle graph, we see that the solutions are in the first and second quadrants.

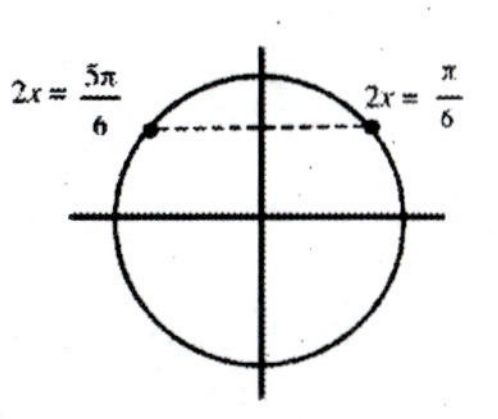

$$2x = \frac{\pi}{6}, \frac{5\pi}{6}$$

Solutions: $x = \frac{\pi}{12}, \frac{5\pi}{12}$

44. $2\cos x + 2 = -\sin^2 x,\quad -\pi \le x < \pi$

Solve for sin x and/or cos x: $2\cos x + 2 = -(1 - \cos^2 x) = -1 + \cos^2 x$ Use Pythagorean identity

$$0 = \cos^2 x - 2\cos x - 3$$
$$0 = (\cos x - 3)(\cos x + 1)$$

$\cos x - 3 = 0 \qquad \cos x + 1 = 0$

$\cos x = 3 \qquad \cos x = -1$

No solution $\qquad x = -\pi$

Solution: $x = -\pi$

45. $2\sin^2\theta - \sin\theta = 0$ All θ

Solve over $0° \le \theta < 360°$: $\sin\theta(2\sin\theta - 1) = 0$

$\sin\theta = 0 \qquad 2\sin\theta - 1 = 0$

$\theta = 0°, 180° \qquad \sin\theta = \frac{1}{2}$

$\theta = 30°, 150°$

Thus, the solutions over one period are 0°, 180°, 30°, 150°.

Write an expression for all solutions: Because the sine function is periodic with period 360°, all solutions are given by

$$\theta = 0° + k360°,\ 180° + k360°,\ 30° + k360°,\ 150° + k\,360°,\ k \text{ any integer.}$$

46. $\sin 2x = \sqrt{3}\sin x$ All real x

Solve over $0 \le x < 2\pi$: $2\sin x\cos x = \sqrt{3}\sin x$ Use double-angle identity

$$2\sin x\cos x - \sqrt{3}\sin x = 0$$
$$\sin x(2\cos x - \sqrt{3}) = 0$$

$\sin x = 0 \qquad 2\cos x - \sqrt{3} = 0$

$x = 0, \pi \qquad \cos x = \frac{\sqrt{3}}{2}$

$x = \frac{\pi}{6}, \frac{11\pi}{6}$

Thus, the solutions over one period are $0, \pi, \frac{\pi}{6}, \frac{11\pi}{6}$.

Write an expression for all solutions: Because the sine and cosine functions are periodic with period 2π, all solutions are given by

$$x = 0 + 2k\pi,\ \pi + 2k\pi,\ \frac{\pi}{6} + 2k\pi,\ \frac{11\pi}{6} + 2k\pi,\ k \text{ any integer.}$$

(The last set of solutions is completely equivalent to the text answer $-\frac{\pi}{6} + 2k\pi$. Why?)

47. $2\sin^2\theta + 5\cos\theta + 1 = 0 \qquad 0° \le \theta < 360°$

Solve for sin θ and/or cos θ:

$$2(1 - \cos^2\theta) + 5\cos\theta + 1 = 0 \qquad \text{Use Pythagorean identity}$$
$$2 - 2\cos^2\theta + 5\cos\theta + 1 = 0$$
$$-2\cos^2\theta + 5\cos\theta + 3 = 0$$
$$2\cos^2\theta - 5\cos\theta - 3 = 0$$
$$(2\cos\theta + 1)(\cos\theta - 3) = 0$$

$$2\cos\theta + 1 = 0 \qquad\qquad \cos\theta - 3 = 0$$
$$\cos\theta = -\frac{1}{2} \qquad\qquad \cos\theta = 3$$
$$\theta = 120°, 240° \qquad\qquad \text{No solution}$$

Solutions: $\theta = 120°, 240°$

48. $3\sin 2x = -2\cos^2 2x \qquad 0 \le x \le \pi$ is equivalent to

$3\sin 2x = -2\cos^2 2x \qquad 0 \le 2x \le 2\pi$

Solve for sin 2x and/or cos 2x:

$$3\sin 2x = -2(1 - \sin^2 2x) = -2 + 2\sin^2 2x \qquad \text{Use Pythagorean identity}$$
$$0 = 2\sin^2 2x - 3\sin 2x - 2$$
$$0 = (2\sin 2x + 1)(\sin 2x - 2)$$

$$2\sin 2x + 1 = 0 \qquad\qquad \sin 2x - 2 = 0$$
$$\sin 2x = -\frac{1}{2} \qquad 0 \le 2x \le 2\pi \qquad \sin 2x = 2$$
$$2x = \frac{7\pi}{6}, \frac{11\pi}{6} \qquad\qquad \text{No solution}$$
$$x = \frac{7\pi}{12}, \frac{11\pi}{12}$$

Solutions: $x = \frac{7\pi}{12}, \frac{11\pi}{12}$

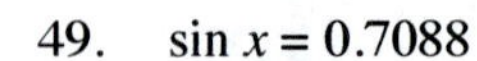

49. $\sin x = 0.7088$

Solve over $0 \le x < 2\pi$: From the unit circle graph, we see that the solutions are in the first and second quadrants.

$x = \pi - 0.7878$ $x = 0.7878$

$x = \sin^{-1} 0.7088 = 0.7878$
$x = \pi - 0.7878 = 2.354$

Write an expression for all solutions: Because the sine function is periodic with period 2π, all solutions are given by

$x = 0.7878 + 2k\pi,\; x = 2.354 + 2k\pi$, k any integer.

50. $\tan x = -4.318$

Solve over $-\frac{\pi}{2} < x < \frac{\pi}{2}$*:* The only solution is in the fourth quadrant.

$x = \tan^{-1}(-4.318) = -1.343$

Write an expression for all solutions: Because the tangent function is periodic with period π, all solutions are given by

$x = -1.343 + k\pi$, k any integer.

51. $\sin^2 x + 2 = 4\sin x$

Solve over $0 \le x < 2\pi$*:*

$\sin^2 x - 4\sin x + 2 = 0$, Quadratic in $\sin x$

$$\sin x = \frac{-(-4) \pm \sqrt{(-4)^2 - 4(1)(2)}}{2(1)}$$

$$= \frac{4 \pm \sqrt{8}}{2}$$

$\sin x = 3.4142$ or $\sin x = 0.5858$

$\sin x = 3.4142$ No solution (3.4142 is not in the range of the sine function.)

$\sin x = 0.5858$ The solutions are in the first and second quadrants.

$x = \sin^{-1}(0.5858) = 0.6259$

$x = \pi - 0.6259 = 2.516$

Write an expression for all solutions: Because the sine function is periodic with period 2π, all solutions are given by

$x = 0.6259 + 2k\pi$, $x = 2.516 + 2k\pi$, k any integer.

52. $\tan^2 x = 2\tan x + 1$

Solve over $-\frac{\pi}{2} < x < \frac{\pi}{2}$*:*

$\tan^2 x - 2\tan x - 1 = 0$ Quadratic in $\tan x$

$$\tan x = \frac{-(-2) \pm \sqrt{(-2)^2 - 4(1)(-1)}}{2(1)}$$

$$= \frac{2 \pm \sqrt{8}}{2}$$

$\tan x = 2.4142$ or $\tan x = -0.4142$

$\tan x = 2.4142$ $x = \tan^{-1} 2.4142 = 1.178$

$\tan x = -0.4142$ $x = \tan^{-1}(-0.4142) = -0.3927$

Because the tangent function is periodic with period π, all solutions are given by

$x = 1.178 + k\pi$, $x = -0.3927 + k\pi$, k any integer.

53. Graph y1 = sin x and y2 = 0.25 in the same viewing window in a graphing calculator and find the point(s) of intersection using an automatic intersection routine. The intersection points are found to be 0.253 and 2.889.

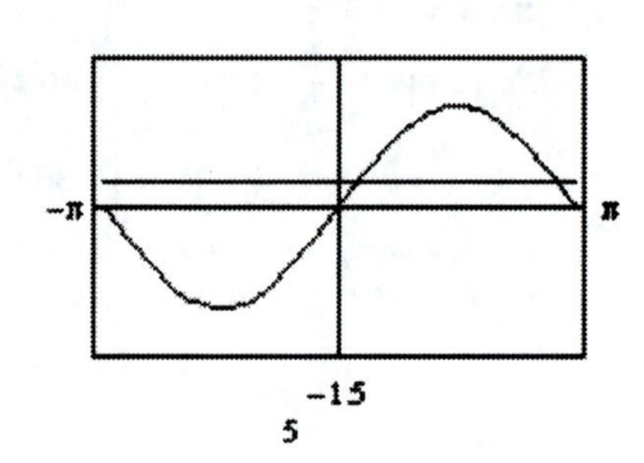

Check: sin(0.253) = 0.250
sin(2.889) = 0.250

54. Graph y1 = cot $x = \frac{1}{\tan x}$ and y2 = –4 in the same viewing window in a graphing calculator and find the point(s) of intersection using an automatic intersection routine. The intersection points are found to be –0.245 and 2.897.

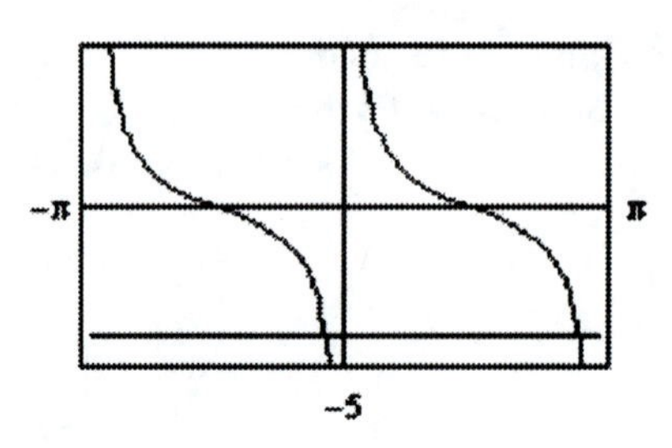

Check: $\cot(-0.245) = \frac{1}{\tan(-0.245)} = -4.000$

$\cot(2.897) = \frac{1}{\tan 2.897} = -4.007$

55. Graph y1 = sec $x = \frac{1}{\cos x}$ and y2 = 2 in the same viewing window in a graphing calculator and find the point(s) of intersection using an automatic intersection routine. The intersection points are found to be ±1.047.

Check: $\sec(\pm 1.047) = \frac{1}{\cos(\pm 1.047)} = 1.999$

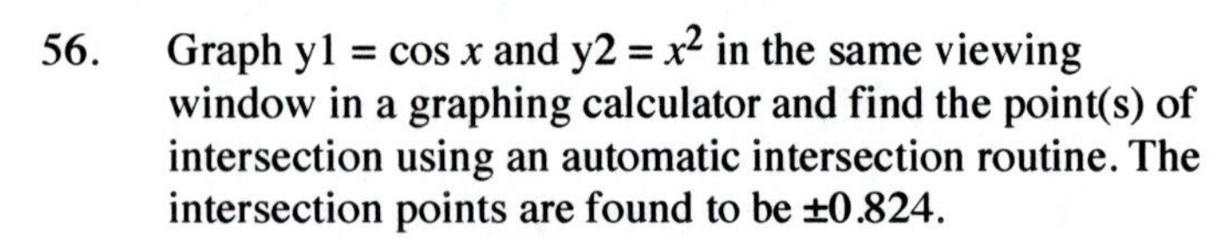

56. Graph y1 = cos x and y2 = x^2 in the same viewing window in a graphing calculator and find the point(s) of intersection using an automatic intersection routine. The intersection points are found to be ±0.824.

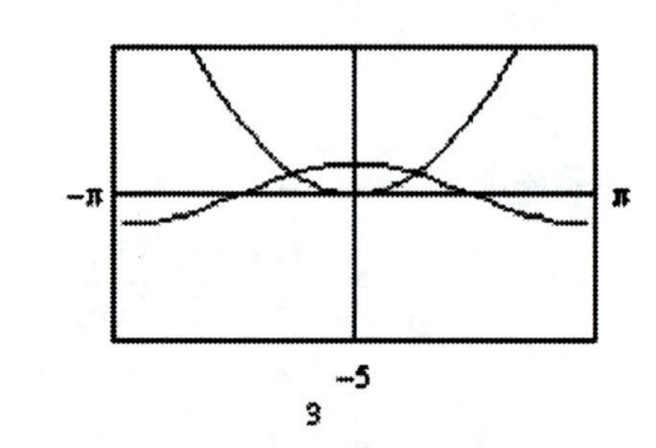

Check: cos(±0.824) = 0.679
$(\pm 0.824)^2 = 0.679$

Since $|\cos x| \le 1$, while $x^2 > 1$ for real x not shown, there can be no other solutions.

57. Graph y1 = sin x and y2 = $\sqrt{x}$ in the same viewing window in a graphing calculator and find the points of intersection using an automatic intersection routine. The single point of intersection appears to be 0.

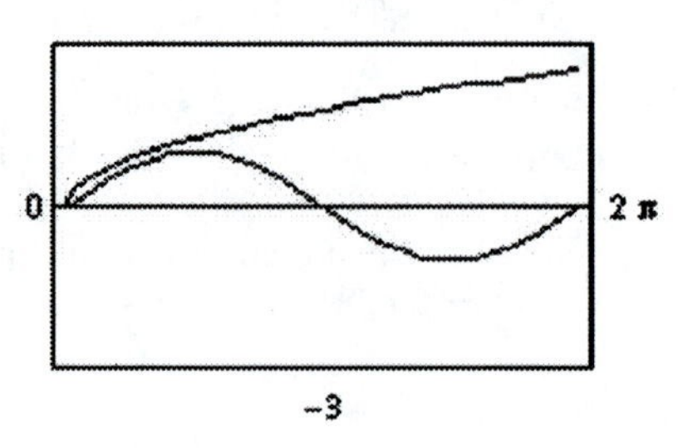

Check: $\sin 0 = 0 = \sqrt{0}$.

Since $|\sin x| \le 1$, while $\sqrt{x} > 1$ for real x not shown, there can be no other solutions.

58. Graph $y1 = 2 \sin x \cos 2x$ and $y2 = 1$ in the same viewing window in a graphing calculator and find the points of intersection using an automatic intersection routine. The intersection points are found to be 4.227 and 5.197.

Check: $2 \sin 4.227 \cos 2(4.227) = 0.999$
$2 \sin 5.197 \cos 2(5.197) = 1.002$

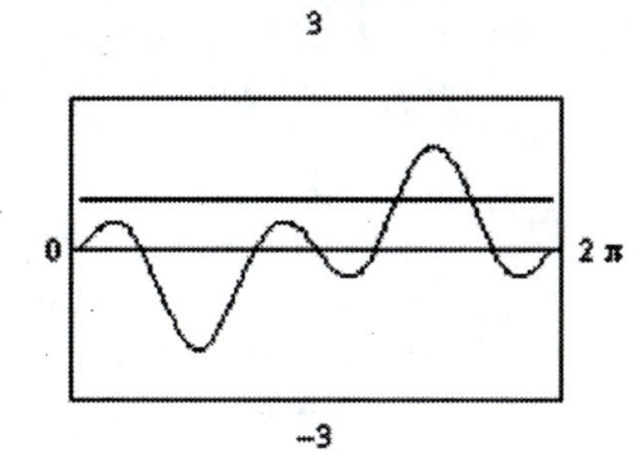

59. Graph $y1 = \sin \frac{x}{2} + 3 \sin x$ and $y2 = 2$ in the same viewing window in a graphing calculator and find the points of intersection using an automatic intersection routine. The intersection points are found to be 0.604, 2.797, 7.246, and 8.203.

Check: $\sin \frac{0.604}{2} + 3 \sin 0.604 = 2.001$

$\sin \frac{2.797}{2} + 3 \sin 2.797 = 1.999$

$\sin \frac{7.246}{2} + 3 \sin 7.246 = 1.999$

$\sin \frac{8.203}{2} + 3 \sin 8.203 = 2.000$

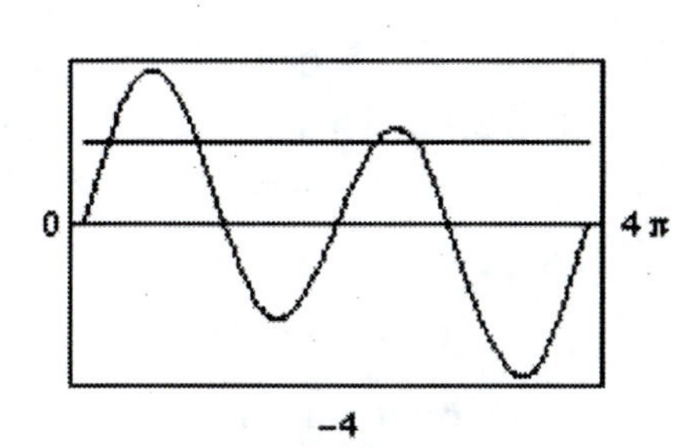

60. Graph $y1 = \sin x + 2 \sin 2x + 3 \sin 3x$ and $y2 = 3$ in the same viewing window in a graphing calculator and find the points of intersection using an automatic intersection routine. The intersection points are found to be 0.228 and 1.008.

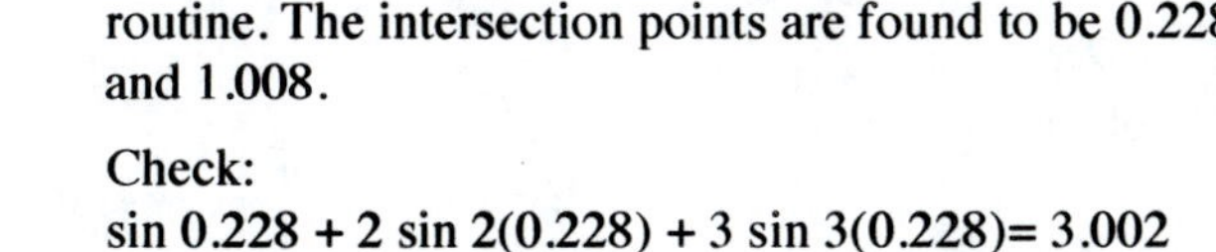

Check:
$\sin 0.228 + 2 \sin 2(0.228) + 3 \sin 3(0.228) = 3.002$
$\sin 1.008 + 2 \sin 2(1.008) + 3 \sin 3(1.008) = 3.003$

61. The domain for y1 is the domain for $\sin^{-1} x, -1 \le x \le 1$.

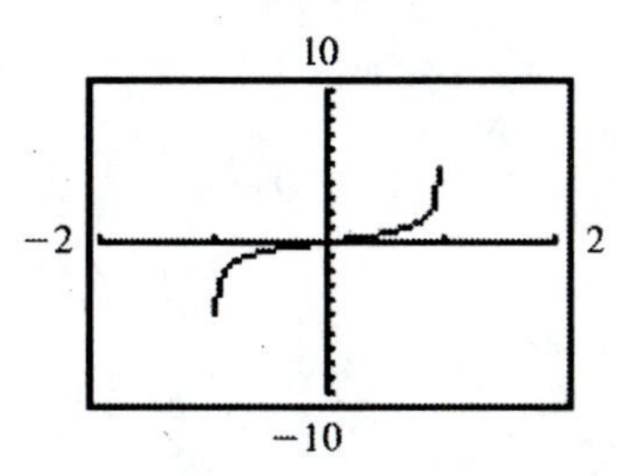

62. (A) $\sin^{-1} x$ has domain $-1 \le x \le 1$, therefore,

$\sin^{-1}\left(\frac{x-2}{2}\right)$ has domain given by

$$-1 \le \frac{x-2}{2} \le 1$$
$$-2 \le x-2 \le 2$$
$$0 \le x \le 4$$

(B) The graph appears only for the domain values $0 \le x \le 4$.

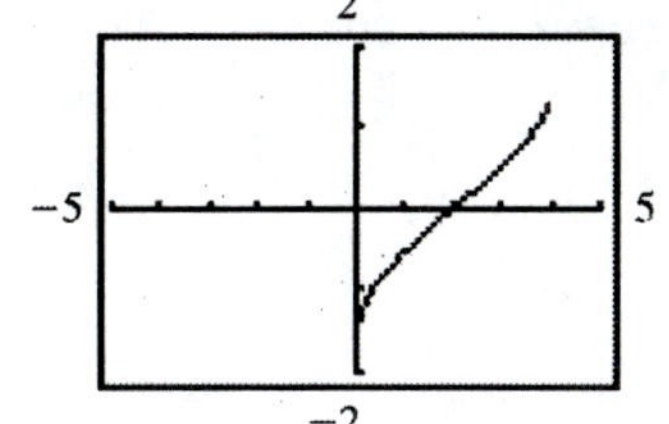

63. No, $\tan^{-1} 23.255$ represents only one number and one solution. The equation has infinitely many solutions, which are given by $x = \tan^{-1} 23.255 + k\pi$, k any integer.

64. $\cos x = 1 - \sin x \qquad 0 \le x < 2\pi$

$$
\begin{aligned}
\cos^2 x &= (1 - \sin x)^2 && \text{Squaring both sides} \\
\cos^2 x &= 1 - 2\sin x + \sin^2 x \\
1 - \sin^2 x &= 1 - 2\sin x + \sin^2 x && \text{Pythagorean identity} \\
0 &= 2\sin^2 x - 2\sin x \\
0 &= 2\sin x(\sin x - 1)
\end{aligned}
$$

$$
\begin{aligned}
2\sin x &= 0 & \sin x - 1 &= 0 \\
\sin x &= 0 & \sin x &= 1 \\
x &= 0, \pi & x &= \frac{\pi}{2}
\end{aligned}
$$

In squaring both sides, we may have introduced extraneous solutions; hence, it is necessary to check solutions of these equations in the original equation.

$x = 0$	$x = \pi$	$x = \frac{\pi}{2}$
$\cos x = 1 - \sin x$	$\cos x = 1 - \sin x$	$\cos x = 1 - \sin x$
$\cos 0 = 1 - \sin 0$	$\cos \pi = 1 - \sin \pi$	$\cos \frac{\pi}{2} = 1 - \sin \frac{\pi}{2}$
$1 = 1 - 0$	$-1 \ne 1 - 0$	$0 = 1 - 1$
A solution	Not a solution	A solution

Solutions: $x = 0, \frac{\pi}{2}$

65. $\cos^2 2x = \cos 2x + \sin^2 2x \qquad 0 \le x < \pi$ is equivalent to

$\cos^2 2x = \cos 2x + \sin^2 2x \qquad 0 \le 2x < 2\pi$

$\cos^2 2x = \cos 2x + 1 - \cos^2 2x \qquad$ Pythagorean identity

$$
\begin{aligned}
2\cos^2 2x - \cos 2x - 1 &= 0 \\
(2\cos 2x + 1)(\cos 2x - 1) &= 0
\end{aligned}
$$

$$
\begin{aligned}
2\cos 2x + 1 &= 0 \quad 0 \le 2x < 2\pi & \cos 2x - 1 &= 0 \quad 0 \le 2x < 2\pi \\
\cos 2x &= -\frac{1}{2} & \cos 2x &= 1 \\
2x &= \frac{2\pi}{3}, \frac{4\pi}{3} & 2x &= 0 \\
x &= \frac{\pi}{3}, \frac{2\pi}{3} & x &= 0
\end{aligned}
$$

Solutions: $x = 0, \frac{\pi}{3}, \frac{2\pi}{3}$

66. $2 + 2\sin x = 1 + 2\cos^2 x \qquad 0 \le x \le 2\pi$

Solve for sin x and/or cos x:

$$
\begin{aligned}
2 + 2\sin x &= 1 + 2(1 - \sin^2 x) && \text{Pythagorean identity} \\
2 + 2\sin x &= 1 + 2 - 2\sin^2 x \\
2\sin^2 x + 2\sin x - 1 &= 0 && \text{Quadratic in } \sin x
\end{aligned}
$$

$$
\sin x = \frac{-2 \pm \sqrt{(2)^2 - 4(2)(-1)}}{2(2)} = \frac{-2 \pm \sqrt{12}}{4}
$$

$\sin x = -1.3660 \quad \sin x = 0.3660$

Solve over $0 \le x \le 2\pi$*:*

$\sin x = -1.3660$ No solution (-1.3660 is not in the range of the sine function.)

$\sin x = 0.3660$ The solutions are in the first and second quadrants.

$x = \sin^{-1}(0.3660) = 0.375$

$x = \pi - 0.375 = 2.77$

67. Let $y = \tan^{-1}\left(-\frac{3}{4}\right)$. Then, $\tan y = -\frac{3}{4}$. We are asked to evaluate $\sin(2y)$, which is $2 \sin y \cos y$ from the double-angle identity. Sketch a reference triangle associated with y; then,

$\cos y = \cos\left[\tan^{-1}\left(-\frac{3}{4}\right)\right]$ and $\sin y = \sin\left[\tan^{-1}\left(-\frac{3}{4}\right)\right]$ can be determined directly from the triangle. Note: $-\frac{\pi}{2} < y < \frac{\pi}{2}$.

$c = \sqrt{4^2 + (-3)^2} = 5$ $\quad \cos y = \frac{4}{5}$ $\quad \sin y = \frac{-3}{5} = -\frac{3}{5}$

Thus, $\sin\left[2\tan^{-1}\left(-\frac{3}{4}\right)\right] = 2\sin\left[\tan^{-1}\left(-\frac{3}{4}\right)\right]\cos\left[\tan^{-1}\left(-\frac{3}{4}\right)\right] = 2\left(-\frac{3}{5}\right)\left(\frac{4}{5}\right) = -\frac{24}{25}$

68. Let $u = \sin^{-1}\left(\frac{3}{5}\right)$ and $v = \cos^{-1}\left(\frac{4}{5}\right)$. Then we are asked to evaluate $\sin(u + v)$, which is $\sin u \cos v + \cos u \sin v$ from the sum identity. We know $\sin u = \sin\left[\sin^{-1}\left(\frac{3}{5}\right)\right] = \frac{3}{5}$ and $\cos v = \cos\left[\cos^{-1}\left(\frac{4}{5}\right)\right] = \frac{4}{5}$ from the function-inverse function identities. It remains to find $\cos u$ and $\sin v$. Note: $-\frac{\pi}{2} \le u \le \frac{\pi}{2}$ and $0 \le v \le \pi$.

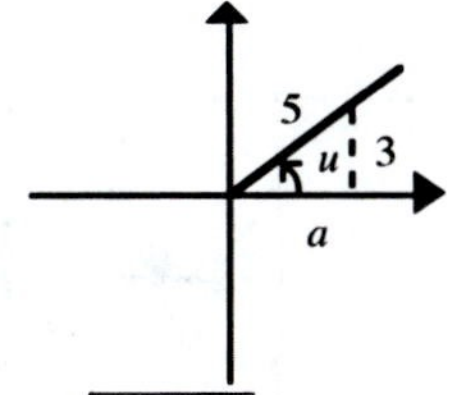

$a = \sqrt{5^2 - 3^2} = 4$ $\quad \cos u = \frac{4}{5}$ $\qquad b = \sqrt{5^2 - 4^2} = 3$ $\quad \sin v = \frac{3}{5}$ $(u = v)$

Then, $\sin\left[\sin^{-1}\left(\frac{3}{5}\right) + \cos^{-1}\left(\frac{4}{5}\right)\right] = \sin(u + v) = \sin u \cos v + \cos u \sin v$

$$= \left(\frac{3}{5}\right)\left(\frac{4}{5}\right) + \left(\frac{4}{5}\right)\left(\frac{3}{5}\right) = \frac{12}{25} + \frac{12}{25} = \frac{24}{25}$$

69. Let $y = \sin^{-1} x$ $\quad -\frac{\pi}{2} \le y \le \frac{\pi}{2}$ $\quad$ or, equivalently, $\quad \sin y = x$ $\quad -\frac{\pi}{2} \le y \le \frac{\pi}{2}$

Geometrically,

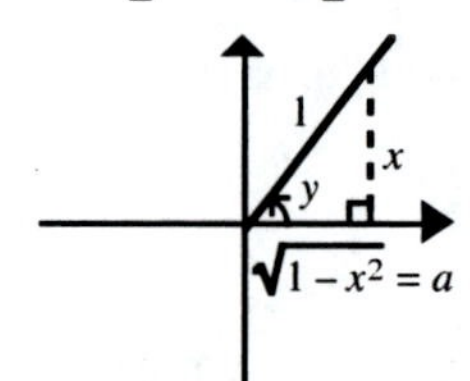

or

√1 − x² = a
y
1
x

In either case, $a = \sqrt{1 - x^2}$. Thus, $\tan(\sin^{-1} x) = \tan y = \frac{b}{a} = \frac{x}{\sqrt{1-x^2}}$.

70. Let $y = \tan^{-1} x$ $\quad -\frac{\pi}{2} < y < \frac{\pi}{2}$ $\quad$ or, equivalently, $\quad x = \tan y$ $\quad -\frac{\pi}{2} < y < \frac{\pi}{2}$

Geometrically,

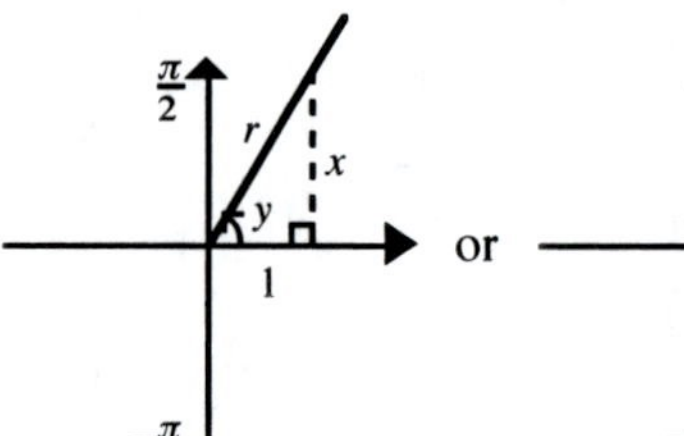

In either case, $r = \sqrt{x^2 + 1}$. Thus, $\cos(\tan^{-1} x) = \cos y = \frac{a}{r} = \frac{1}{\sqrt{x^2+1}}$.

71. (A)

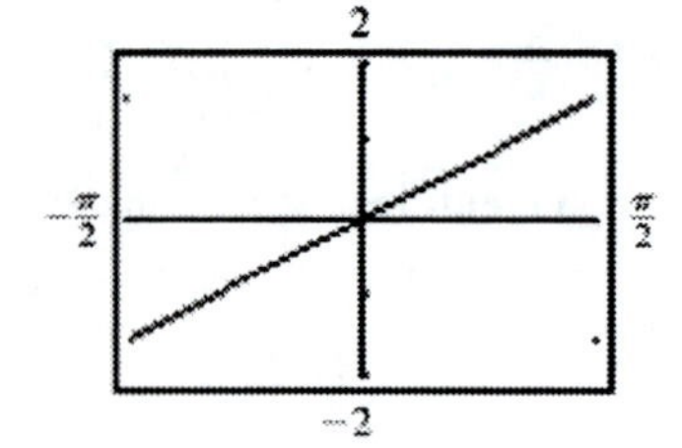

(B) The domain for $\tan x$ is the set of all real numbers, except $x = \frac{\pi}{2} + k\pi$, k an integer. The range is R, which is the domain for $\tan^{-1} x$. Thus, $y = \tan^{-1}(\tan x)$ has a graph for all real x, except $x = \frac{\pi}{2} + k\pi$, k an integer. But $\tan^{-1}(\tan x) = x$ only on the restricted domain of $\tan x$, $-\frac{\pi}{2} < x < \frac{\pi}{2}$.

72. We are to solve $0.05 = 0.08 \sin 880\pi t$.

$$\sin 880\pi t = \frac{0.05}{0.08}$$

$$880\pi t = \sin^{-1} \frac{0.05}{0.08} \quad \text{will yield the smallest positive } t$$

$$t = \frac{1}{880\pi} \sin^{-1} \frac{0.05}{0.08} \quad \text{(calculator in radian mode)}$$

$$= 0.00024 \text{ sec}$$

73. We are to solve $20 = 30 \sin 120\pi t$.

$$\sin 120\pi t = \frac{20}{30}$$

$$120\pi t = \sin^{-1}\frac{20}{30} \text{ will yield the smallest positive } t$$

$$t = \frac{1}{120\pi}\sin^{-1}\frac{20}{30} \qquad \text{(calculator in radian mode)}$$

$$= 0.001936 \text{ sec}$$

74. (A) From the figure in the text we can see that $\tan\frac{\theta}{2} = \frac{500}{x}$.

Thus, $\frac{\theta}{2} = \arctan\frac{500}{x}$ $\quad \theta = 2\arctan\frac{500}{x}$

(B) We use the above formula with $x = 1200$.

$$\theta = 2\arctan\frac{500}{1200} = 45.2°$$

75.

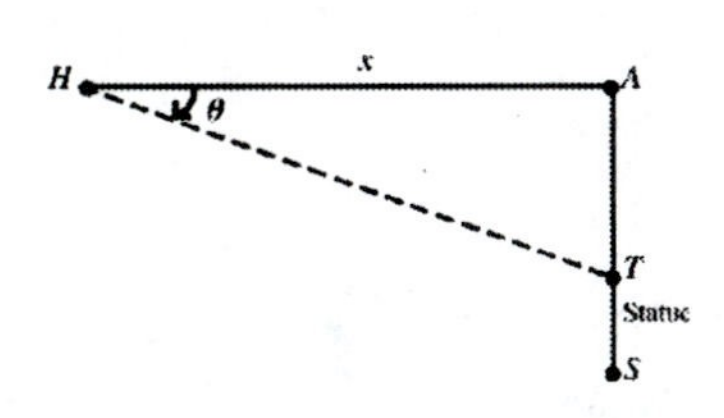

(A) In the figure, triangle THA, where H represents the position of the helicopter, is a right triangle. SA, the height of the helicopter is given at 3,500 feet; ST, the height of the statue as 305 feet. Therefore, AT = difference in height between the helicopter and the torch = 3,500 – 305 = 3,195 feet.

$$\tan\theta = \frac{AT}{x} = \frac{3{,}195}{x}$$

$$\theta = \tan^{-1}\left(\frac{3{,}195}{x}\right)$$

(B) Substitute $x = 5{,}280$ feet to obtain

$$\theta = \tan^{-1}\left(\frac{3{,}195}{5{,}280}\right) = 31.2°$$

76. (A) θ will increase rapidly from 0° to some maximum value, then decrease slowly after that.

(B) We redraw the text figure in a side view. We note:

$$\tan\angle BPC = \frac{BC}{x} = \frac{20-5}{x} = \frac{15}{x}$$

$$\tan\angle BPD = \frac{BD}{x} = \frac{15+1.5}{x} = \frac{16.5}{x}$$

$$\theta = \angle BPD - \angle BPC$$

Hence, $\tan\theta = \tan(\angle BPD - \angle BPC)$

$$= \frac{\tan\angle BPD \tan\angle BPC}{1+\tan\angle BPD\tan\angle BPC} \text{ by the subtraction identity for tangent.}$$

$$\tan\theta = \frac{\frac{16.5}{x} - \frac{15}{x}}{1+\frac{16.5}{x}\cdot\frac{15}{x}} = \frac{16.5x - 15x}{x^2 + (16.5)(15)} = \frac{1.5x}{x^2 + 247.5}$$

$$\theta = \tan^{-1}\frac{1.5x}{x^2+247.5}$$

(C)

x ft	0	5	10	15	20	25	30
$\theta°$	0.00	1.58	2.47	2.73	2.65	2.46	2.25

The maximum value of θ from the table is 2.73° when x is 15 feet.

(D)

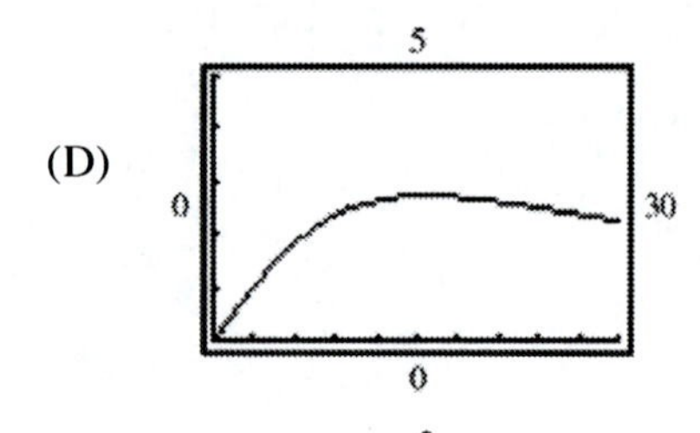

As x increases from 0 ft to 30 ft, θ increases rapidly at first to a maximum of about 2.73° at 15 feet then decreases more slowly.

(E)

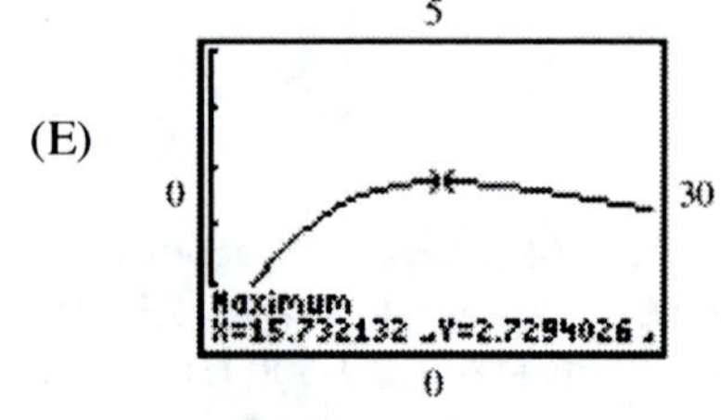

From the automatic maximum routine, the maximum value of θ is 2.73° when $x = 15.73$ feet.

(F) We are to solve $2.5 = \tan^{-1}\dfrac{1.5x}{x^2 + 247.5}$. We graph y1 = 2.5 and y2 = $\tan^{-1}\dfrac{1.5x}{x^2 + 247.5}$ in the same viewing window. From the automatic intersection routine, $\theta = 2.5°$ when $x = 10.28$ feet or when $x = 24.08$ feet.

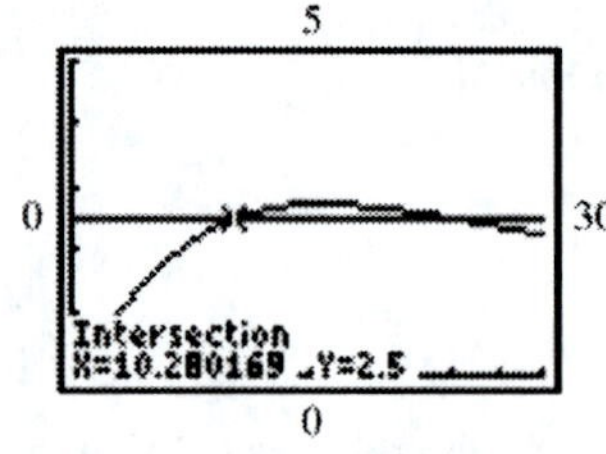

5
0 30
Intersection
X=24.075479 Y=2.5
0

77. Label the figure in the text as shown:

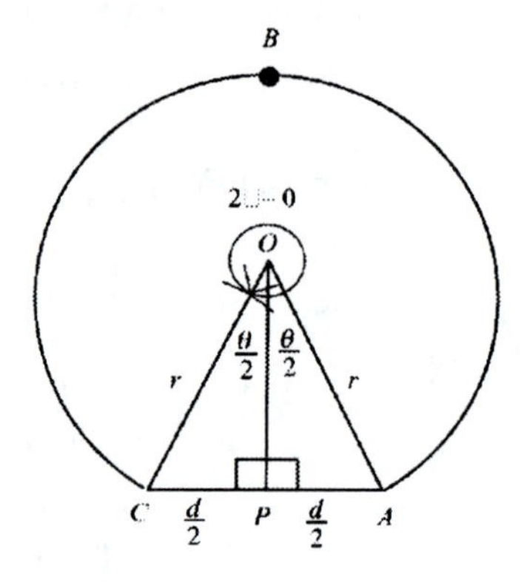

(A) The area of segment $PABCP$ = Area of sector $OABC$ + Area of triangle OAC. Area of sector $OABC$ =

$\frac{1}{2} r^2 (2\pi - \theta)$

Since $\sin\frac{\theta}{2} = \frac{AP}{OA} = \frac{d}{2} \div r = \frac{d}{2r}$, $\frac{\theta}{2} = \sin^{-1}\frac{d}{2r}$,

$\theta = 2\sin^{-1}\frac{d}{2r}$. Hence,

area of sector $OABC = \frac{1}{2}r^2\left(2\pi - 2\sin^{-1}\frac{d}{2r}\right)$

$= \pi r^2 - r^2\sin^{-1}\frac{d}{2r}$

Area of triangle $OAC = \frac{1}{2}$(base)(altitude) $= \frac{1}{2}(AC)(OP)$. $AC = d$. From the Pythagorean theorem applied to triangle OAP, $OP^2 + AP^2 = OA^2$, hence

$OP^2 = OA^2 - AP^2 = r^2 - \left(\frac{d}{2}\right)^2$. $OP = \sqrt{r^2 - \frac{d^2}{4}} = \sqrt{\frac{4r^2 - d^2}{4}} = \frac{1}{2}\sqrt{4r^2 - d^2}$

Hence, area of triangle $AOC = \frac{1}{2} \cdot d \cdot \frac{1}{2}\sqrt{4r^2 - d^2} = \frac{d}{4}\sqrt{4r^2 - d^2}$.

Thus, area of segment $PABCP$ = area of sector $OABC$ + Area of triangle OAC

$= \pi r^2 - r^2 \sin^{-1} \frac{d}{2r} + \frac{d}{4}\sqrt{4r^2 - d^2}$.

(B)

d ft	8	10	12	14	16	18	20
A ft^2	704	701	697	690	682	670	654

(C) We are to solve

$$675 = \pi 15^2 - 15^2 \sin^{-1} \frac{d}{2 \cdot 15} + \frac{d}{4}\sqrt{4 \cdot 15^2 - d^2}$$

or

$$675 = 225\pi - 225 \sin^{-1} \frac{d}{30} + \frac{d}{4}\sqrt{900 - d^2}$$

We graph y1 = 675 and y2 = $225\pi - 225 \sin^{-1} \frac{d}{30} + \frac{d}{4}\sqrt{900 - d^2}$ in the same viewing window.

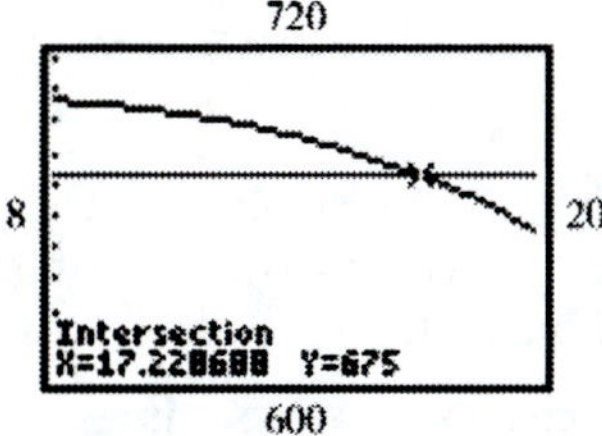

From the automatic intersection routine, $d = 17.2$ ft for $A = 675$ ft^2.

78. Analyzing triangle OAB, we see the following:

$$OA = r - h$$
$$OA^2 + AB^2 = r^2$$
$$(r - h)^2 + 5^2 = r^2$$
$$r^2 - 2hr + h^2 + 25 = r^2$$
$$h^2 + 25 = 2hr$$
$$r = \frac{h^2 + 25}{2h}$$

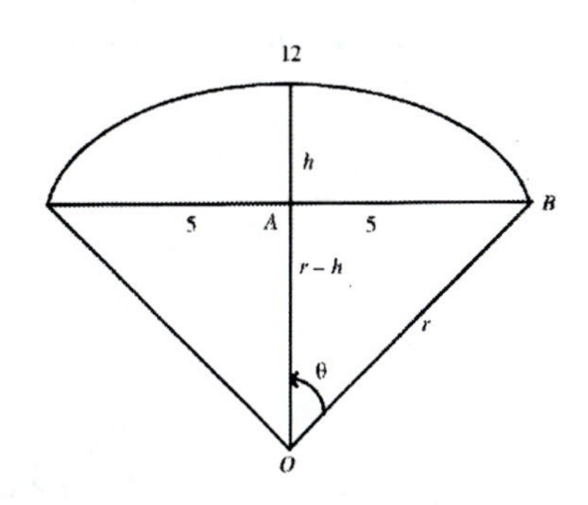

$\sin\theta = \frac{5}{r}$ $\quad \theta = \sin^{-1}\frac{5}{r}$. Using the formula $s = r\theta$, we can write $12 = 2\theta \cdot r$; $\theta = \frac{6}{r}$.

Hence, $\frac{6}{r} = \sin^{-1}\frac{5}{r}$.

$6 \div \frac{h^2 + 25}{2h} = \sin^{-1}\left(5 \div \frac{h^2 + 25}{2h}\right)$. Thus, $\frac{12h}{h^2 + 25} = \sin^{-1} \frac{10h}{h^2 + 25}$ is the equation that must be satisfied by h.

We graph y1 = $\frac{12h}{h^2+25}$ and y2 = $\sin^{-1}\frac{10h}{h^2+25}$. From the figure, we see that y1 and y2 intersect once for nonzero h on the interval from 0 to 4. Using zoom and trace procedures or the automatic intersection routine, the intersection is found to be $h = 2.82$ ft.

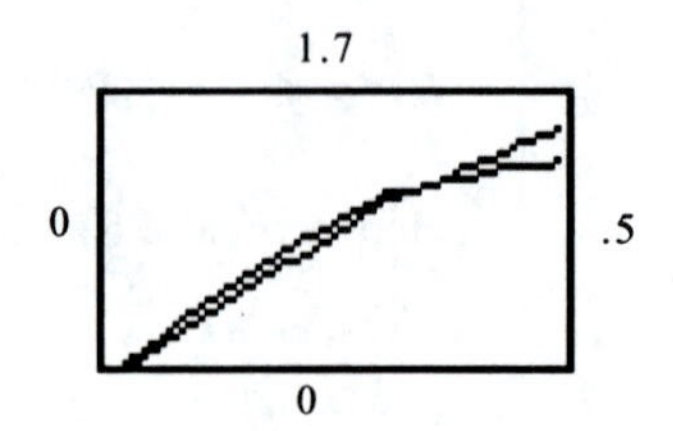

79. We are to solve

$$50 - \frac{40}{x+2}\cos(0.815x) \geq 52.$$

Graph y1 = $50 - \frac{40}{x+2}\cos(0.815x)$ and y2 = 52 on the interval from 0 to 30 seconds. We see that there are definitely four points of intersection. Possibly there are two more, but the trace procedure or a graph on a smaller interval of interest, from 17 to 21 seconds, shows that there are not. Using the automatic intersection routine, the intersections are found at 2.186, 5.322, 10.462, and 12.496 seconds. Thus, the graph of y1 is above the graph of y2 for $(5.322 - 2.186) + (12.496 - 10.462) = 3.136 + 2.034 = 5.2$ seconds.

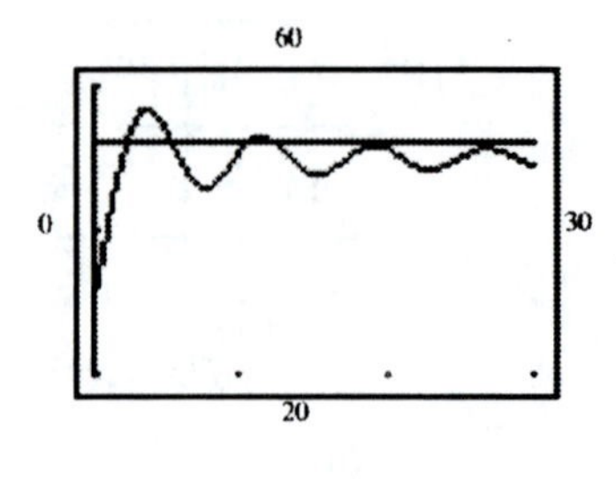

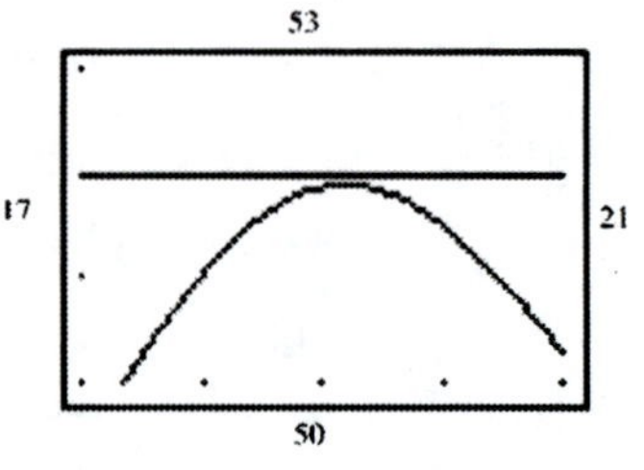

80. (A)

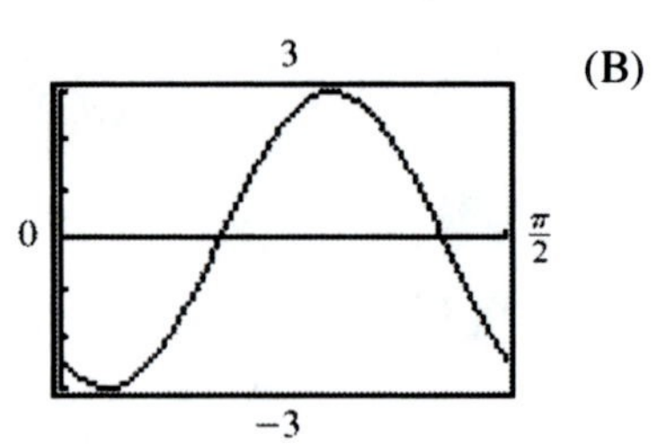

(B) We graph y1 = $-1.8 \sin 4t - 2.4 \cos 4t$ and y2 = 2. From the figure, we see that y1 and y2 intersect twice on the indicated interval.

Using zoom and trace procedures or the automatic intersection routine, the intersections are found to be $t = 0.74$ sec and $t = 1.16$ sec. Thus, the weight is 2 in. above the equilibrium position if $0.74 < t < 1.16$ sec.

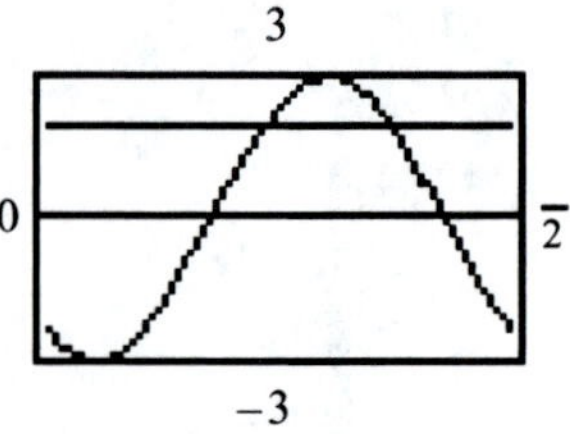

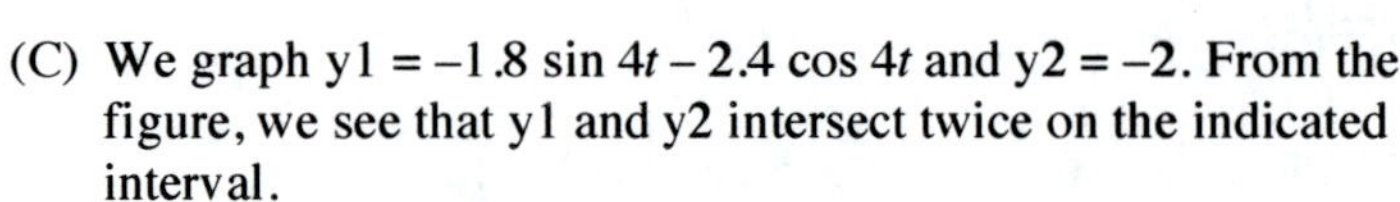

(C) We graph y1 = $-1.8 \sin 4t - 2.4 \cos 4t$ and y2 = −2. From the figure, we see that y1 and y2 intersect twice on the indicated interval.

Using zoom and trace procedures or the automatic intersection routine, the intersections are found to be $t = 0.37$ sec and $t = 1.52$ sec.

Thus, the weight is 2 in. below the equilibrium position if $0 \leq t < 0.37$ sec or $1.52 < t \leq \frac{\pi}{2}$ sec.

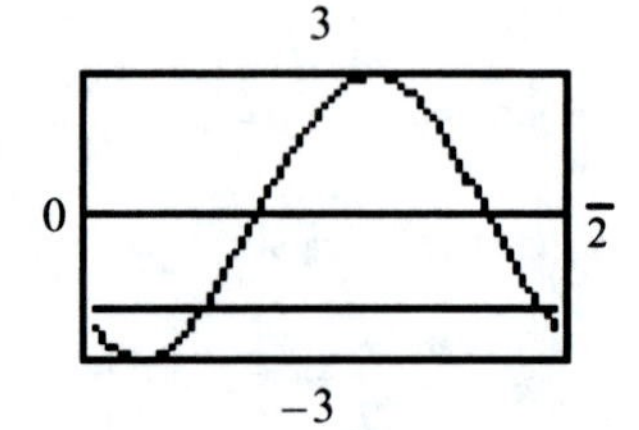

CUMULATIVE REVIEW EXERCISE CHAPTERS 1—5

1. $\theta_d = \dfrac{180°}{\pi \text{ rad}}\theta_r$. Thus, if $\theta_r = 4.21$, $\theta_d = \dfrac{180}{\pi}(4.21) = 241.22°$

2. $\theta_r = \dfrac{\pi \text{ rad}}{180°}\theta_d$. Thus, if $\theta_d = 505°42'$, or $\left(505+\dfrac{42}{60}\right)°$,

$$\theta_r = \frac{\pi}{180}\left(505+\frac{42}{60}\right) = 8.83 \text{ rad}$$

3. An angle of radian measure 0.5 is the central angle of a circle subtended by an arc with measure half that of the radius of the circle.

4. Let $c = 7.6$ m, $b = 4.5$ m

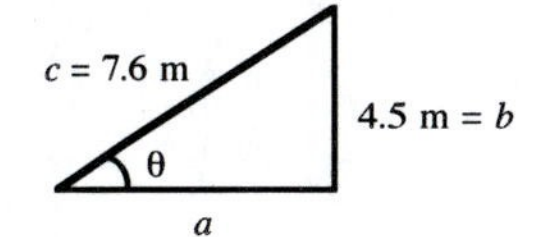

Solve for θ: We will use the sine: $\sin\theta = \dfrac{b}{c} = \dfrac{4.5\text{m}}{7.6\text{m}} = 0.5921$

$\theta = \sin^{-1}(0.5921) = 36°$

Solve for the complementary angle: $90° - \theta = 90° - 36° = 54°$

Solve for a: We choose the cosine to find a. $\cos\theta = \dfrac{a}{c}$

$a = c\cos\theta = (7.6 \text{ m})(\cos 36°) = 6.1$ m

5. The distance from P to the origin is

$$r = \sqrt{(-5)^2+(-12)^2} = \sqrt{25+144} = \sqrt{169} = 13$$

Apply the third remark following Definition 1, Section 2.3, with $a = -5$, $b = -12$ and $r = 13$.

$$\cos\theta = \frac{a}{r} = -\frac{5}{13} \qquad \cot\theta = \frac{a}{b} = \frac{5}{12}$$

6. (A) Degree mode: $\cos 67°45' = \cos 67.75° = 0.3786$ Convert to decimal degrees, if necessary

(B) Degree mode: $\csc 176.2° = \dfrac{1}{\sin 176.2°} = 15.09$

(C) Radian mode: $\cot 2.05 = \dfrac{1}{\tan 2.05} = -0.5196$

7. 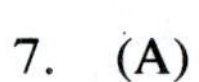
(A)
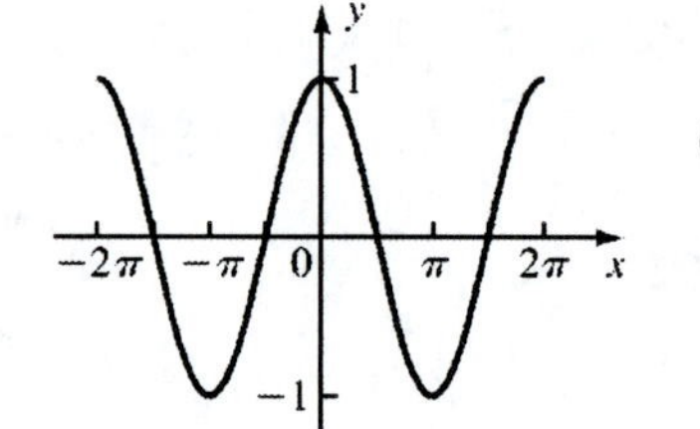

(B)
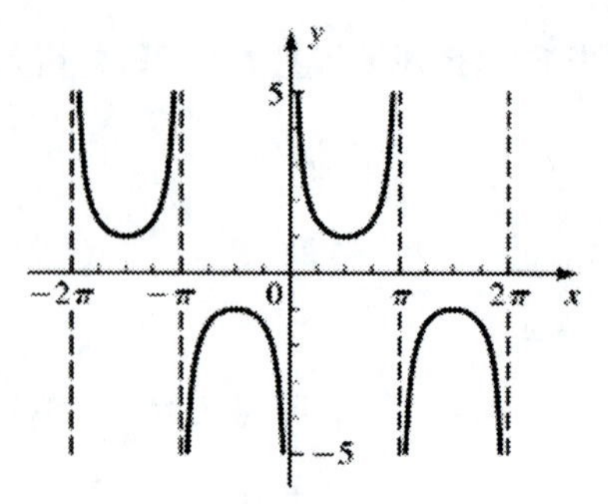

(C)
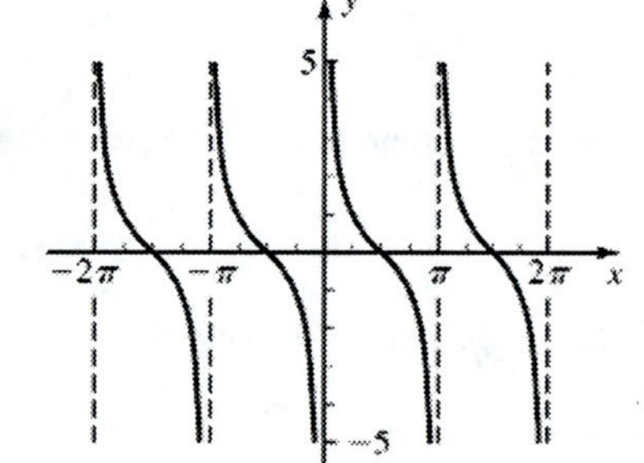

8. No, because $\sin\theta = \dfrac{1}{\csc\theta}$ and either both are positive or both are negative.

9.

$\tan x \csc x = \dfrac{\sin x}{\cos x}\,\dfrac{1}{\sin x}$	Quotient and reciprocal identities
$= \dfrac{1}{\cos x}$	Algebra
$= \sec x$	Reciprocal identity

10.

$\csc\theta - \sin\theta = \dfrac{1}{\sin\theta} - \sin\theta$	Reciprocal identity
$= \dfrac{1}{\sin\theta} - \dfrac{\sin^2\theta}{\sin\theta}$	Algebra
$= \dfrac{1-\sin^2\theta}{\sin\theta}$	Algebra
$= \dfrac{\cos^2\theta}{\sin\theta}$	Pythagorean identity
$= \cos\theta \cdot \dfrac{\cos\theta}{\sin\theta}$	Algebra
$= \cos\theta\cot\theta$	Quotient identity

11.

$(\sin^2 u)(\tan^2 u + 1) = \sin^2 u \sec^2 u$	Pythagorean identity
$= \sin^2 u\,\dfrac{1}{\cos^2 u}$	Reciprocal identity
$= \dfrac{\sin^2 u}{\cos^2 u}$	Algebra
$= \tan^2 u$	Quotient identity
$= \sec^2 u - 1$	Pythagorean identity

12. $$\frac{\sin^2\alpha - \cos^2\alpha}{\sin\alpha\cos\alpha} = \frac{\dfrac{\sin^2\alpha}{\sin\alpha\cos\alpha} - \dfrac{\cos^2\alpha}{\sin\alpha\cos\alpha}}{\dfrac{\sin\alpha\cos\alpha}{\sin\alpha\cos\alpha}}$$ Algebra

$$= \frac{\dfrac{\sin\alpha}{\cos\alpha} - \dfrac{\cos\alpha}{\sin\alpha}}{\dfrac{\sin\alpha}{\cos\alpha}\,\dfrac{\cos\alpha}{\sin\alpha}}$$ Algebra

$$= \frac{\tan\alpha - \cot\alpha}{\tan\alpha\cot\alpha}$$ Quotient identities

13. $\cos x = \sin x$ is not an identity, because $\sin x$ and $\cos x$ are not equal for other values of x for which both sides are defined, for example, they are not equal for $x = 0$ or $x = \frac{\pi}{2}$.

14. Locate the 45° reference triangle, determine (a, b) and r, then evaluate.

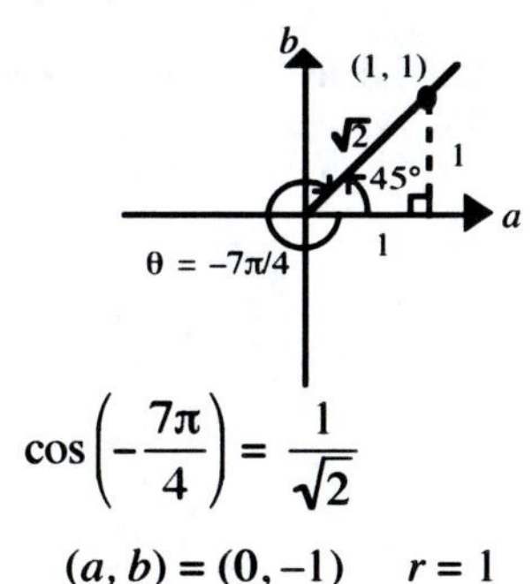

$$\cos\left(-\frac{7\pi}{4}\right) = \frac{1}{\sqrt{2}}$$

15. Locate the 30°–60° reference triangle, determine (a, b) and r, then evaluate.

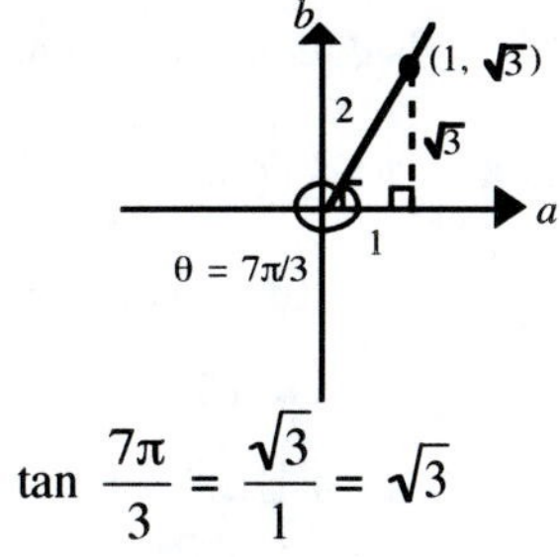

$$\tan\frac{7\pi}{3} = \frac{\sqrt{3}}{1} = \sqrt{3}$$

16. $(a, b) = (0, -1)$ $r = 1$

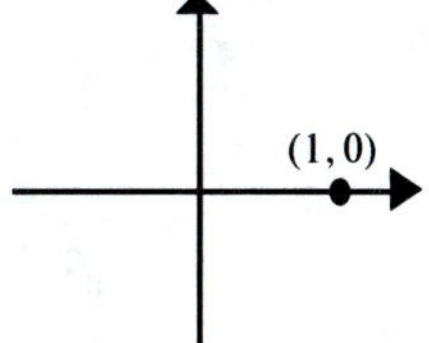

$$\sec\frac{3\pi}{2} = \frac{1}{0}$$
Not defined

17. $y = \arctan 0$ is equivalent to $\tan y = 0$. No reference triangle can be drawn, but the only y between $-\frac{\pi}{2}$ and $\frac{\pi}{2}$ which has tangent equal to 0 is $y = 0$. Thus, $\arctan 0 = 0$.

18. $y = \cos^{-1}\left(-\frac{\sqrt{3}}{2}\right)$ is equivalent to $\cos y = -\frac{\sqrt{3}}{2}$. What y between 0 and π has cosine equal to $-\frac{\sqrt{3}}{2}$? y must be associated with a reference triangle in the second quadrant.
Reference triangle is a special 30°–60° triangle.

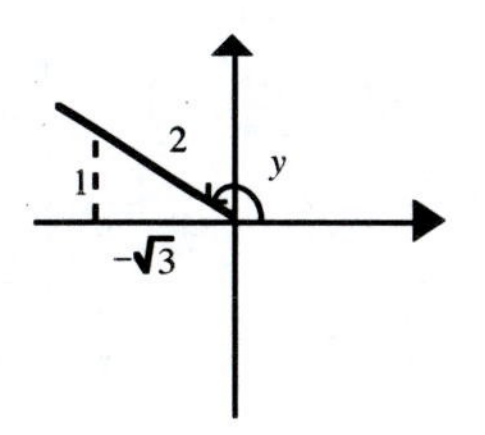

$$y = \frac{5\pi}{6} \qquad \cos^{-1}\left(-\frac{\sqrt{3}}{2}\right) = \frac{5\pi}{6}$$

19. 3 is not in the domain of the inverse sine function. $-1 \le 3 \le 1$ is false. arcsin 3 is not defined.

20. $y = \text{arccot}(-\sqrt{3})$ is equivalent to $\cot y = -\sqrt{3}$ and $0 < y < \pi$.
What number between 0 and π has cotangent equal to $-\sqrt{3}$?
y must be positive and in the second quadrant.

$$\cot y = -\sqrt{3} = \frac{-\sqrt{3}}{1} \qquad \alpha = \frac{\pi}{6} \quad y = \frac{5\pi}{6}$$

Thus, $\cot^{-1}(-\sqrt{3}) = \dfrac{5\pi}{6}$

21. Calculator in radian mode: $\sin^{-1}(0.0505) = 0.0505$

22. Calculator in radian mode: $\cos^{-1}(-0.7228) = 2.379$

23. Calculator in radian mode: $\arctan(-9) = -1.460$

24. Calculator in radian mode: $\text{arccot}\ 3 = \arctan \dfrac{1}{3} = 0.3218$

25. Calculator in radian mode: $\sec^{-1} 2.6 = \cos^{-1} \dfrac{1}{2.6} = 1.176$

26. $\tan^{-1} x = -1.000000$ is equivalent to $\tan(-1.000000) = x$. Since the calculator is in radian mode, calculate
$x = \tan(-1.000000) = -1.557408$.

27. $2 \sin \theta - 1 = 0 \qquad 0° \le \theta < 360°$

Solve for sin θ: $\quad 2 \sin \theta = 1$

$$\sin \theta = \frac{1}{2}$$

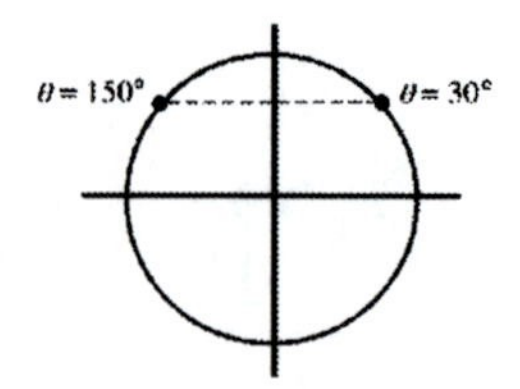

Solve over $0° \le \theta < 360°$: From the unit circle graph, we see that the solutions are in the first and second quadrants.
$\theta = 30°, 150°$

28. $3 \tan x + \sqrt{3} = 0 \qquad -\dfrac{\pi}{2} < x < \dfrac{\pi}{2}$

Solve for tan x: $\quad 3 \tan x = -\sqrt{3}$

$$\tan x = -\frac{\sqrt{3}}{3}$$

Solve over $-\dfrac{\pi}{2} < x < \dfrac{\pi}{2}$: The single solution is in the fourth quadrant.

$$x = -\frac{\pi}{6}$$

29. $2 \cos x + 2 = 0 \qquad -\pi \le x < \pi$

Solve for cos x: $\quad 2 \cos x = -2$

$$\cos x = -1$$

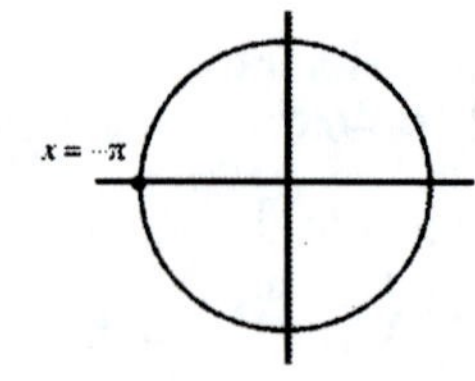

Solve over $-\pi \le x < \pi$: From the unit circle graph, we see that the single solution is $x = -\pi$.

30. $\sin x \cos y = \frac{1}{2}[\sin(x+y) + \sin(x-y)]$. Let $x = 7u$ and $y = 3u$

$\sin 7u \cos 3u = \frac{1}{2}[\sin(7u+3u) + \sin(7u-3u)] = \frac{1}{2}(\sin 10u + \sin 4u)$

31. $\cos x - \cos y = -2 \sin \frac{x+y}{2} \sin \frac{x-y}{2}$. Let $x = 5w$ and $y = w$

$\cos 5w - \cos w = -2 \sin \frac{5w+w}{2} \sin \frac{5w-w}{2} = -2 \sin 3w \sin 2w$

32. Yes, because, using the formula $\theta_d = \frac{360°}{2\pi}\theta_r$, if θ_r is halved, then θ_d will also be halved.

33. $92.462° = 92°(0.462 \times 60)' = 92°27.72' = 92°27'(0.72 \times 60)'' = 92°27'43''$

34. Since $\frac{s}{C} = \frac{\theta°}{360°}$ then $\frac{12\text{ in}}{30\text{ in}} = \frac{\theta}{360°}$ $\theta = \frac{12}{30} \cdot 360° = 144°$

35. Since the two right triangles are similar, we can write:

$$\frac{4}{x} = \frac{x}{x+3}$$
$$x(x+3)\frac{4}{x} = x(x+3)\frac{x}{x+3}$$
$$(x+3)4 = x^2$$
$$4x + 12 = x^2$$
$$0 = x^2 - 4x - 12$$
$$0 = (x-6)(x+2)$$

$x - 6 = 0$ or $x + 2 = 0$

$x = 6$ $\quad x = -2$ We discard the negative answer.

Since $\tan\theta = \frac{4}{x}$, we can write $\tan\theta = \frac{4}{6}$ $\quad \theta = \tan^{-1}\frac{4}{6} = 33.7°$

36. $\theta_r = \frac{\pi \text{ rad}}{180°}\theta_d$ Thus, if $\theta_d = 72°$, $\theta_r = \frac{\pi}{180} \cdot 72 = \frac{2\pi}{5}$

37. Because $\sin\theta$ is negative and $\tan\theta$ is positive, the terminal side of θ must lie in the third quadrant. Because $\tan\theta = \frac{b}{a} = \frac{1}{2}$, we let $a = -2$, $b = -1$. Then the distance of $P(a, b)$ from the origin is given by

$$r = \sqrt{(-2)^2 + (-1)^2} = \sqrt{4+1} = \sqrt{5}$$

Apply the third remark following Definition 1, Section 2.3, with $a = -2$, $b = -1$, and $r = \sqrt{5}$.

$\sin\theta = \frac{b}{r} = -\frac{1}{\sqrt{5}}$ $\quad \csc\theta = \frac{r}{b} = \frac{\sqrt{5}}{-1} = -\sqrt{5}$

$\cos\theta = \frac{a}{r} = -\frac{2}{\sqrt{5}}$ $\quad \sec\theta = \frac{r}{a} = -\frac{\sqrt{5}}{2}$

$\left(\tan\theta = \frac{b}{a} = \frac{1}{2}\right)$ $\quad \cot\theta = \frac{a}{b} = \frac{2}{1} = 2$

38. $y = 2 - 2\sin\frac{x}{2}$. Amplitude $= |-2| = 2$. Period $= 2\pi \div \frac{1}{2} = 4\pi$. This graph is the graph of $y = -2\sin\frac{x}{2}$ moved up 2 units. We start by drawing a horizontal broken line 2 units above the x axis, then graph $y = -2\sin\frac{x}{2}$ (an upside down sine curve with amplitude 2 and period 4π) relative to the broken line and the original y axis.

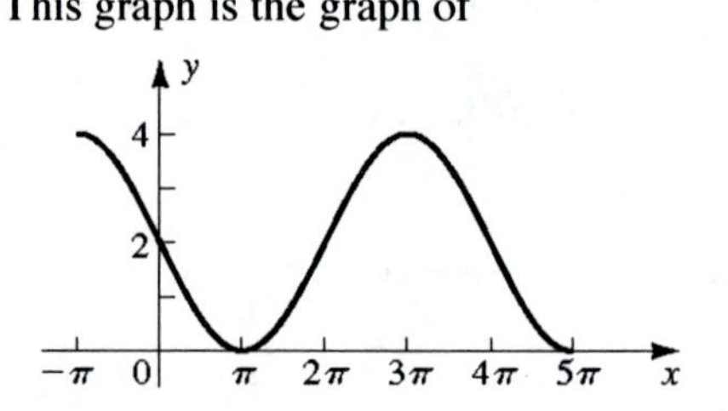

39. Amplitude $= |A| = |3| = 3$. Phase Shift and Period: Solve

$$\begin{aligned} Bx + C &= 0 \\ 2x - \pi &= 0 \\ 2x &= \pi \\ x &= \frac{\pi}{2} \end{aligned} \quad \text{and} \quad \begin{aligned} Bx + C &= 2\pi \\ 2x - \pi &= 2\pi \\ 2x &= \pi + 2\pi \\ x &= \frac{\pi}{2} + \pi \end{aligned}$$

$$\text{Phase Shift} = \frac{\pi}{2} \qquad \text{Period} = \pi$$

Graph one cycle over the interval from $\frac{\pi}{2}$ to $\frac{\pi}{2} + \pi = \frac{3\pi}{2}$. Then extend the graph from $-\pi$ to 2π.

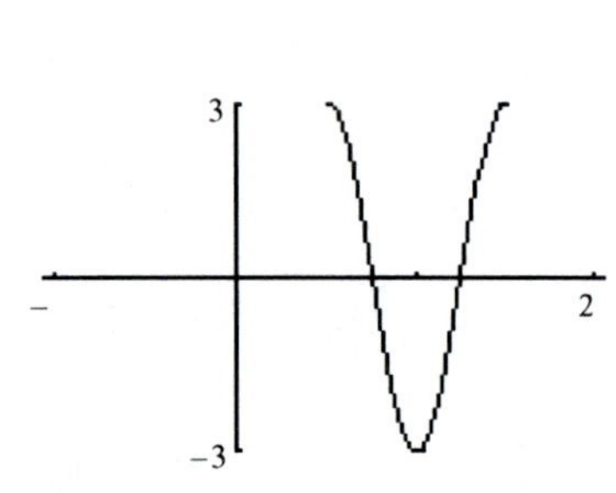

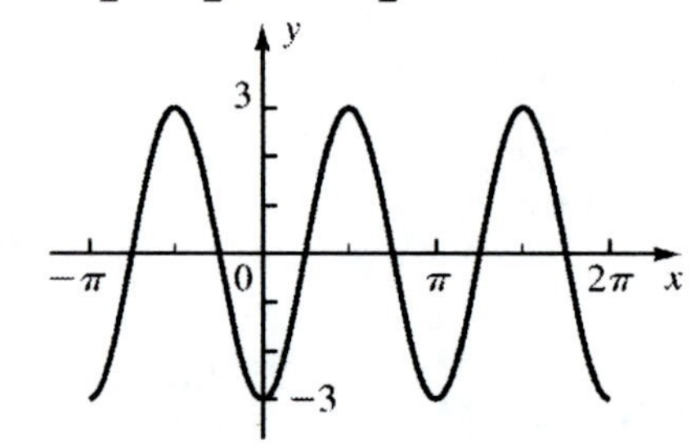

40. To find the asymptotes, set $\pi x - \frac{\pi}{4}$ equal to $-\frac{\pi}{2}$ and $\frac{\pi}{2}$ and solve for x.

$$\begin{aligned} \pi x - \frac{\pi}{4} &= -\frac{\pi}{2} \\ \pi x &= \frac{\pi}{4} - \frac{\pi}{2} \\ x &= \frac{1}{4} - \frac{1}{2} \\ x &= -\frac{1}{4} \end{aligned} \qquad \begin{aligned} \pi x - \frac{\pi}{4} &= \frac{\pi}{2} \\ \pi x &= \frac{\pi}{4} + \frac{\pi}{2} \\ x &= \frac{1}{4} + \frac{1}{2} \\ x &= \frac{3}{4} \end{aligned}$$

The two asymptotes, $x = -\frac{1}{4}$ and $x = \frac{3}{4}$, are 1 unit apart, so the period is 1. The phase shift is $-\frac{C}{B} = \frac{1}{4}$. Thus, the asymptotes in the required region $0 \le x \le 3$ are $x = \frac{3}{4}$, $x = \frac{3}{4} + 1 = \frac{7}{4}$, and $x = \frac{7}{4} + 1 = \frac{11}{4}$. Sketch these asymptotes and fill in the curve, noting that the coefficient of 2 makes the tangent curve steeper.

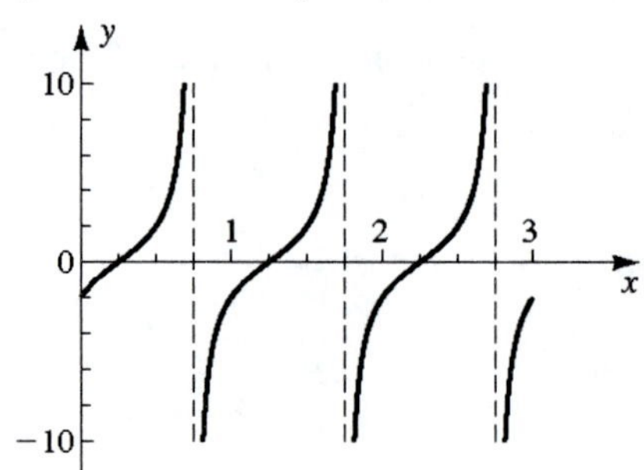

41. Amplitude $= \frac{1}{2}(y_{max} - y_{min}) = \frac{1}{2}(4-(-2)) = 3$. $k = \frac{1}{2}(y_{max} + y_{min}) = \frac{1}{2}(4+(-2)) = 1$.
Period $= \frac{2\pi}{B} = 2$. Thus, $B = \frac{2\pi}{2} = \pi$. The form of the graph is that of the basic sine curve.
Thus, $y = A \sin Bx + k = 3 \cos \pi x + 1$.

42. Amplitude $= \frac{1}{2}(y_{max} - y_{min}) = \frac{1}{2}(1-(-3)) = 2$. $k = \frac{1}{2}(y_{max} + y_{min}) = \frac{1}{2}(1+(-3)) = -1$.
Period $= \frac{2\pi}{B} = 4$. Thus, $B = \frac{2\pi}{4} = \frac{\pi}{2}$. The form of the graph is that of the basic sine curve.
Thus, $y = A \sin Bx + k = 2 \sin \frac{\pi}{2}x - 1$.

43. Let $a = 19.4$ cm and $b = 41.7$ cm.

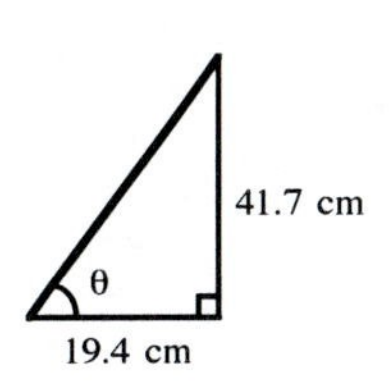

Solve for c: We will use the Pythagorean theorem

$$c^2 = a^2 + b^2$$
$$c = \sqrt{a^2+b^2} = \sqrt{(19.4)^2+(41.7)^2} = 46.0 \text{ cm}$$

Solve for θ: We will use the tangent $\tan\theta = \frac{b}{a} = \frac{41.7\,\text{cm}}{19.4\,\text{cm}} = 2.1495$

$$\theta = \tan^{-1}(2.1495) = 65°0'$$

Solve for the complementary angle: $90° - \theta = 90° - 65°0' = 25°0'$

44.

$\frac{\cos x}{1+\sin x} + \tan x = \frac{\cos x}{1+\sin x} + \frac{\sin x}{\cos x}$	Quotient identity
$= \frac{\cos x \cos x}{\cos x(1+\sin x)} + \frac{\sin x(1+\sin x)}{\cos x(1+\sin x)}$	Algebra
$= \frac{\cos^2 x + \sin x + \sin^2 x}{\cos x(1+\sin x)}$	Algebra
$= \frac{\cos^2 x + \sin^2 x + \sin x}{\cos x(1+\sin x)}$	Algebra
$= \frac{1+\sin x}{\cos x(1+\sin x)}$	Pythagorean identity
$= \frac{1}{\cos x}$	Algebra
$= \sec x$	Reciprocal identity

45. In this problem, it is more straightforward to start with the right-hand side of the identity to be verified. The student can confirm that the steps would be valid if reversed.

$$(\sec\theta - \tan\theta)(\sec\theta + 1) = \left(\frac{1}{\cos\theta} - \frac{\sin\theta}{\cos\theta}\right)\left(\frac{1}{\cos\theta} + 1\right)$$ Quotient and Reciprocal identities

$$= \left(\frac{1-\sin\theta}{\cos\theta}\right)\left(\frac{1}{\cos\theta} + \frac{\cos\theta}{\cos\theta}\right)$$ Algebra

$$= \frac{1-\sin\theta}{\cos\theta} \cdot \frac{1+\cos\theta}{\cos\theta}$$ Algebra

$$= \frac{(1-\sin\theta)(1+\cos\theta)}{\cos^2\theta}$$ Algebra

$$= \frac{(1-\sin\theta)(1+\cos\theta)}{1-\sin^2\theta}$$ Pythagorean identity

$$= \frac{(1-\sin\theta)(1+\cos\theta)}{(1-\sin\theta)(1+\sin\theta)}$$ Algebra

$$= \frac{1+\cos\theta}{1+\sin\theta}$$ Algebra

46. $\cot\frac{u}{2} = 1 \div \tan\frac{u}{2}$ Reciprocal identity

$$= 1 \div \frac{\sin u}{1+\cos u}$$ Half-angle identity for tangent

$$= 1 \cdot \frac{1+\cos u}{\sin u}$$ Algebra

$$= \frac{1+\cos u}{\sin u}$$ Algebra

$$= \frac{1}{\sin u} + \frac{\cos u}{\sin u}$$ Algebra

$$= \csc u + \cot u$$ Quotient and Reciprocal identities

47. $$\frac{2}{1+\sec 2\theta} = \frac{2}{1+\dfrac{1}{\cos 2\theta}}$$ Reciprocal identity

$$= \frac{2\cos 2\theta}{2\cos 2\theta + 1}$$ Algebra

$$= \frac{2(2\cos^2\theta - 1)}{2\cos^2\theta - 1 + 1}$$ Double-angle identity

$$= \frac{4\cos^2\theta - 2}{2\cos^2\theta}$$ Algebra

$$= \frac{4\cos^2\theta}{2\cos^2\theta} - \frac{2}{2\cos^2\theta}$$ Algebra

$= 2 - \dfrac{1}{\cos^2 \theta}$ — Algebra

$= 2 - \sec^2 \theta$ — Reciprocal identity

$= 2 - (\tan^2 \theta + 1)$ — Pythagorean identity

$= 1 - \tan^2 \theta$ — Algebra

48. $\dfrac{\sin x - \sin y}{\cos x + \cos y}$

$= \dfrac{(\sin x - \sin y)\sin(x - y)}{(\cos x + \cos y)\sin(x - y)}$ — Algebra

$= \dfrac{(\sin x - \sin y)(\sin x \cos y - \cos x \sin y)}{(\cos x + \cos y)\sin(x - y)}$ — Difference identity for sine

$= \dfrac{\sin^2 x \cos y - \sin x \cos x \sin y - \sin x \sin y \cos y + \cos x \sin^2 y}{(\cos x + \cos y)\sin(x - y)}$ — Algebra

$= \dfrac{(1 - \cos^2 x)\cos y - \sin x \cos x \sin y - \sin x \sin y \cos y + \cos x(1 - \cos^2 y)}{(\cos x + \cos y)\sin(x - y)}$ — Pythagorean identity

$= \dfrac{\cos y - \cos^2 x \cos y - \sin x \cos x \sin y - \sin x \sin y \cos y + \cos x - \cos x \cos^2 y}{(\cos x + \cos y)\sin(x - y)}$

Algebra

$= \dfrac{(\cos x - \cos^2 x \cos y - \sin x \cos x \sin y) + (\cos y - \sin x \sin y \cos y - \cos x \cos^2 y)}{(\cos x + \cos y)\sin(x - y)}$

Algebra

$= \dfrac{\cos x(1 - \cos x \cos y - \sin x \sin y) + \cos y(1 - \sin x \sin y - \cos x \cos y)}{(\cos x + \cos y)\sin(x - y)}$ — Algebra

$= \dfrac{(\cos x + \cos y)(1 - \cos x \cos y - \sin x \sin y)}{(\cos x + \cos y)\sin(x - y)}$ — Algebra

$= \dfrac{1 - (\cos x \cos y + \sin x \sin y)}{\sin(x - y)}$ — Algebra

$= \dfrac{1 - \cos(x - y)}{\sin(x - y)}$ — Difference identity for cosine

$= \tan \dfrac{x - y}{2}$ — Half-angle identity

49. (A) From the identities for negatives,

$\sin(-x) = -\sin x$

Hence

$\sin(-x) = -0.4969$

(B) From the Pythagorean identity, $\cos^2 x + \sin^2 x = 1$, thus

$\cos^2 x = 1 - \sin^2 x$

Hence

$(\cos x)^2 = 1 - 0.2469 = 0.7531$

50. $\sin 3x \cos x - \cos 3x \sin x = \sin(3x - x)$ — Difference identity for sine
$= \sin 2x$ — Algebra

Graph $y1 = \sin 3x \cos x - \cos 3x \sin x$ and $y2 = \sin 2x$.

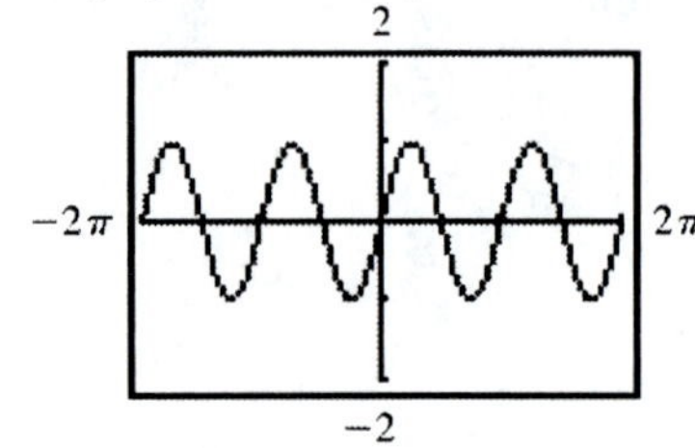

51. First draw a reference triangle in the third quadrant and find $\cos x$.

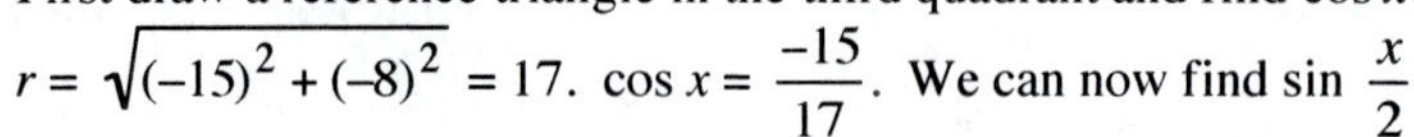

$r = \sqrt{(-15)^2 + (-8)^2} = 17$. $\cos x = \dfrac{-15}{17}$. We can now find $\sin \dfrac{x}{2}$ from the half-angle identity, after determining its sign, as follows:

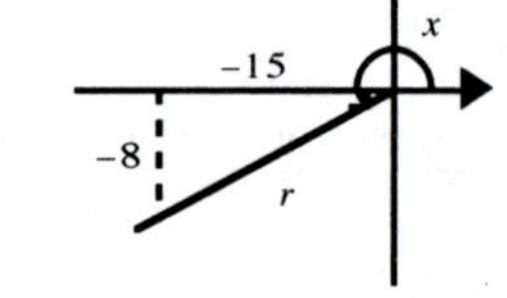

If $\pi < x < \dfrac{3\pi}{2}$, then $\dfrac{\pi}{2} < \dfrac{x}{2} < \dfrac{3\pi}{4}$.

Thus, $\dfrac{x}{2}$ is in the second quadrant, where sine is positive.

Using half-angle identities, we obtain

$$\sin \frac{x}{2} = \sqrt{\frac{1-\cos x}{2}} = \sqrt{\frac{1-(-15/17)}{2}} = \sqrt{\frac{32/17}{2}} = \sqrt{\frac{16}{17}} \text{ or } \frac{4}{\sqrt{17}}$$

To find $\cos 2x$, we use a double-angle identity.

$$\cos 2x = 2\cos^2 x - 1 = 2\left(-\frac{15}{17}\right)^2 - 1 = \frac{2}{1} \cdot \frac{225}{289} - 1 = \frac{450-289}{289} = \frac{161}{289}$$

52. $[-2\pi, -\pi] \cup [\pi, 2\pi]$

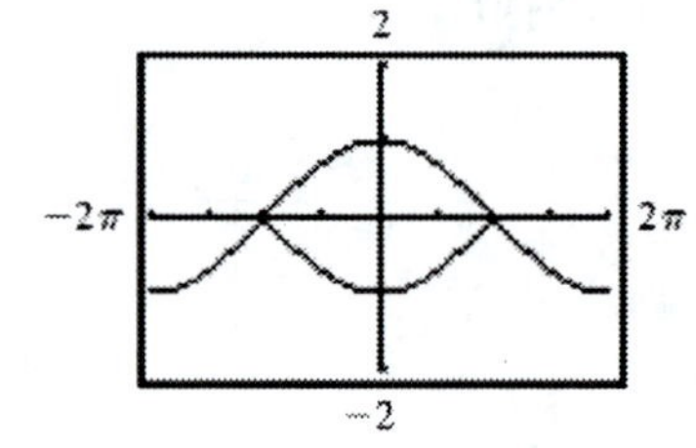

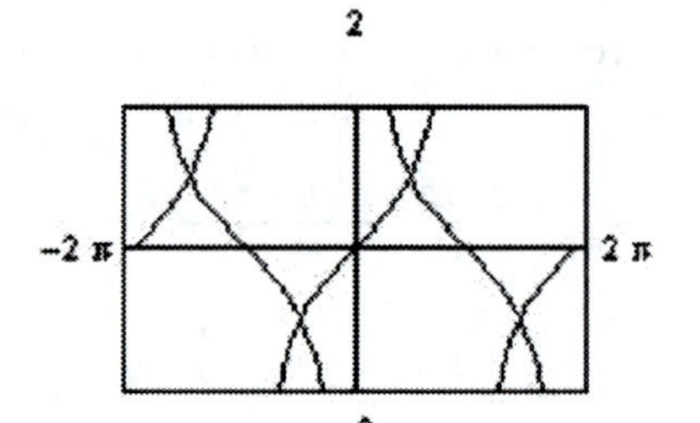

53. (A) Graph both sides of the equation in the same viewing window.

$\dfrac{\sin x}{1+\cos x} = \dfrac{1+\cos x}{\sin x}$ is not an identity, since the graphs do not match.

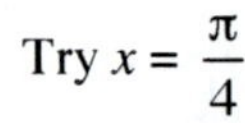

Try $x = \dfrac{\pi}{4}$

Left side: $\dfrac{\sin\left(\frac{\pi}{4}\right)}{1+\cos\left(\frac{\pi}{4}\right)} = \dfrac{\frac{1}{\sqrt{2}}}{1+\frac{1}{\sqrt{2}}} = \dfrac{1}{\sqrt{2}+1} = \sqrt{2} - 1$

Right side: $\dfrac{1+\cos\left(\frac{\pi}{4}\right)}{\sin\frac{\pi}{4}} = \dfrac{1+\frac{1}{\sqrt{2}}}{\frac{1}{\sqrt{2}}} = \sqrt{2} + 1$

(B) Graph both sides of the equation in the same viewing window.

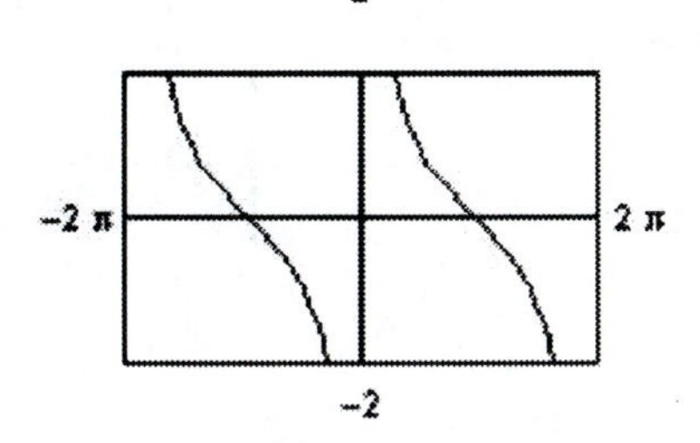

$$\frac{\sin x}{1-\cos x} = \frac{1+\cos x}{\sin x}$$ appears to be an identity, which we verify.

$$\frac{\sin x}{1-\cos x} = \frac{\sin x(1+\cos x)}{(1-\cos x)(1+\cos x)}$$ Algebra

$$= \frac{\sin x(1+\cos x)}{1-\cos^2 x}$$ Algebra

$$= \frac{\sin x(1+\cos x)}{\sin^2 x}$$ Pythagorean identity

$$= \frac{1+\cos x}{\sin x}$$ Algebra

54. (A) Let $y = \cos^{-1} 0.4$, then $\cos y = 0.4 = \frac{2}{5}, 0 \le y \le \pi$.

Draw the reference triangle associated with y. Then, $\sin y = \sin(\cos^{-1} 0.4)$ can be determined directly from the triangle.

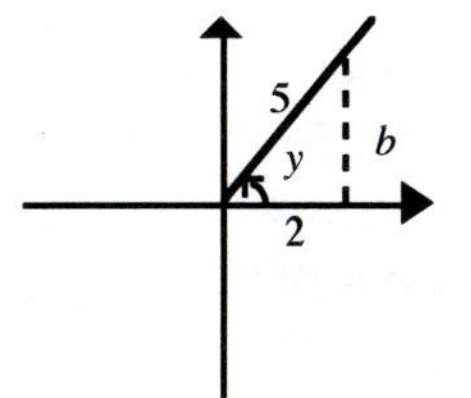

$$b = \sqrt{5^2 - 2^2} = \sqrt{21}$$

$$\sin(\cos^{-1} 0.4) = \sin y = \frac{b}{r} = \frac{\sqrt{21}}{5}$$

(B) Let $y = \arctan(-\sqrt{5})$, then $\tan y = -\sqrt{5}, -\frac{\pi}{2} < y < \frac{\pi}{2}$.

Draw the reference triangle associated with y. Then, $\sec y = \sec[\arctan(-\sqrt{5})]$ can be determined directly from the triangle.

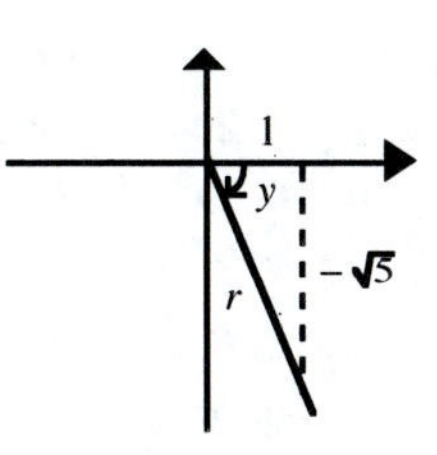

$$r = \sqrt{1^2 + (-\sqrt{5})^2} = \sqrt{6}$$

$$\sec[\arctan(-\sqrt{5})] = \sec y = \frac{r}{a} = \frac{\sqrt{6}}{1} = \sqrt{6}$$

(C) Let $y = \sin^{-1} \frac{1}{3}$, then $\sin y = \frac{1}{3}, -\frac{\pi}{2} \le y \le \frac{\pi}{2}$.

Then, $\csc y = \csc\left(\sin^{-1}\frac{1}{3}\right) = \frac{1}{\sin y} = \frac{1}{\frac{1}{3}} = 3$

(D) Let $y = \sec^{-1} 4$, then $\sec y = 4, 0 \le y \le \pi$. Draw the reference triangle associated with y. Then, $\tan y = \tan(\sec^{-1} 4)$ can be determined directly from the triangle.

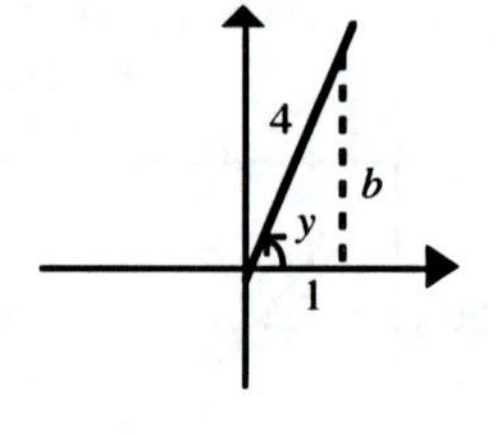

$$b = \sqrt{4^2 - 1^2} = \sqrt{15}$$

$$\tan(\sec^{-1} 4) = \tan y = \frac{b}{a} = \frac{\sqrt{15}}{1} = \sqrt{15}$$

55. $\theta = \tan^{-1} \sqrt{3}$ is equivalent to $\tan \theta = \sqrt{3}$ $\quad -90° < \theta < 90°$. Thus, $\theta = 60°$.

56. Calculator in degree mode: $\theta = \sin^{-1} 0.8989 = 64.01°$

57. $2 + 3 \sin x = \cos 2x \; 0 \le x \le 2\pi$

Solve for sin x and/or cos x: $\quad 2 + 3 \sin x = 1 - 2 \sin^2 x \quad$ Use double-angle identity

$$2 \sin^2 x + 3 \sin x + 1 = 0$$
$$(2 \sin x + 1)(\sin x + 1) = 0$$

$$2 \sin x + 1 = 0 \qquad \sin x + 1 = 0$$
$$\sin x = -\frac{1}{2} \qquad \sin x = -1$$
$$x = \frac{7\pi}{6}, \frac{11\pi}{6} \qquad x = \frac{3\pi}{2}$$

Solutions: $x = \frac{7\pi}{6}, \frac{3\pi}{2}, \frac{11\pi}{6}$

58. First solve for θ over one period, $0 \le \theta < 360°$. Then add integer multiples of 360° to find all solutions.

$\sin 2\theta = 2 \cos \theta \quad 0 \le \theta \le 360°$

$$2 \sin \theta \cos \theta = 2 \cos \theta \qquad \text{Use double-angle identity}$$
$$2 \sin \theta \cos \theta - 2 \cos \theta = 0$$
$$2 \cos \theta (\sin \theta - 1) = 0$$

$$2 \cos \theta = 0 \qquad \sin \theta - 1 = 0$$
$$\cos \theta = 0 \qquad \sin \theta = 1$$
$$\theta = 90°, 270° \qquad \theta = 90°$$

Thus, all solutions over one period, $0° \le \theta < 360°$, are $x = 90°, 270°$. Thus, the solutions, if θ is allowed to range over all degree values, can be written as

$\theta = 90° + k360°, 270° + k360°$, k any integer

More concisely, since the solutions are 90°, 270°, 450°, and so on, we can write

$\theta = 90° + k180° \qquad k$ any integer

59. First solve for x over one period of $\tan x$, $0 \le x < \pi$. Then add integer multiples of π to find all solutions.

$$\begin{aligned} 4\tan^2 x - 3\sec^2 x &= 0 \\ 4\tan^2 x - 3(\tan^2 x + 1) &= 0 \quad \text{Use Pythagorean identity} \\ 4\tan^2 x - 3\tan^2 x - 3 &= 0 \\ \tan^2 x - 3 &= 0 \\ \tan^2 x &= 3 \\ \tan x &= \pm\sqrt{3} \\ x &= \frac{\pi}{3}, \frac{2\pi}{3} \end{aligned}$$

Thus, all solutions over one period, $0 \le x < \pi$, are $x = \frac{\pi}{3}, \frac{2\pi}{3}$.

Because the tangent function is periodic with period π, all solutions are given by

$x = \frac{\pi}{3} + k\pi$, $x = \frac{2\pi}{3} + k\pi$, k any integer

60. $\sin x = -0.5678$

Solve over $0 \le x < 2\pi$: The solutions are in the third and fourth quadrants.

$x = 2\pi + \sin^{-1}(-0.5678) = 5.679$

$x = \pi + \sin^{-1}(0.5678) = 3.745$

Write an expression for all solutions: Because the sine function is periodic with period 2π, all solutions are given by

$x = 3.745 + 2k\pi$, $x = 5.679 + 2k\pi$,

k any integer.

61. $\sec x = 2.345$

Solve for $\cos x$:

$$\begin{aligned} \frac{1}{\cos x} &= 2.345 \\ \cos x &= \frac{1}{2.345} \\ \cos x &= 0.4264 \end{aligned}$$

Solve over $0 \le x < 2\pi$: The solutions are in the first and fourth quadrants.

$x = \cos^{-1} 0.4264 = 1.130$

$x = 2\pi - 1.130 = 5.153$

Write an expression for all solutions: Because the cosine function is periodic with period 2π, all solutions are given by:

$x = 1.130 + 2k\pi$, $x = 5.153 + 2k\pi$, k any integer.

62. First solve for x over one period, $0 \le x < 2\pi$. Then add integer multiples of 2π to find all solutions.

$2 \cos 2x = 7 \cos x$

$$\begin{aligned} 2(2\cos^2 x - 1) &= 7\cos x && \text{Use double-angle identity} \\ 4\cos^2 x - 2 &= 7\cos x \\ 4\cos^2 x - 7\cos x - 2 &= 0 \\ (4\cos x + 1)(\cos x - 2) &= 0 \end{aligned}$$

$$4\cos x + 1 = 0 \quad 0 \le x < 2\pi \qquad \cos x - 2 = 0$$

$$4\cos x = -1 \qquad \cos x = 2$$

$$\cos x = -\frac{1}{4} \qquad \text{No solution}$$

$$x = \cos^{-1}\left(-\frac{1}{4}\right) \text{ or } \pi + \cos^{-1}\left(\frac{1}{4}\right)$$

$$x = 1.823 \text{ or } 4.460$$

Thus, all solutions over one period, $0 \le x < 2\pi$, are $x = 1.823, 4.460$.

Because the cosine function is periodic with period 2π, all solutions are given by $x = 1.823 + 2k\pi$, $x = 4.460 + 2k\pi$, k any integer.

63. (a) does not illustrate a cosine-inverse cosine identity, because -3 is not in $0 \le x \le \pi$, the restricted domain of the cosine function.

(b) does illustrate a cosine-inverse cosine identity, because 2.51 is in $0 \le x \le \pi$, the restricted domain of the cosine function.

64. Graph y1 = tan x and y2 = 3 in the same viewing window in a graphing calculator and find the points of intersection using an automatic intersection routine. The intersection points are found to be -1.893 and 1.249.

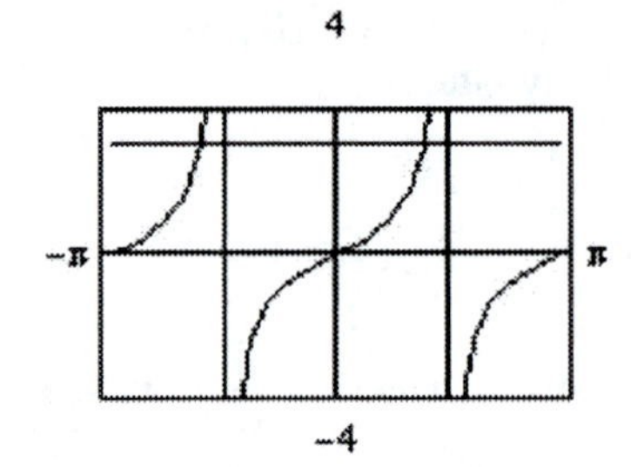

Check: $\tan(-1.893) = 2.995$
$\tan(1.249) = 3.000$

65. Graph y1 = cos x and y2 = $\sqrt{x}$ in the same viewing window in a graphing calculator and find the points of intersection using an automatic intersection routine. The intersection point is found to be 0.642.

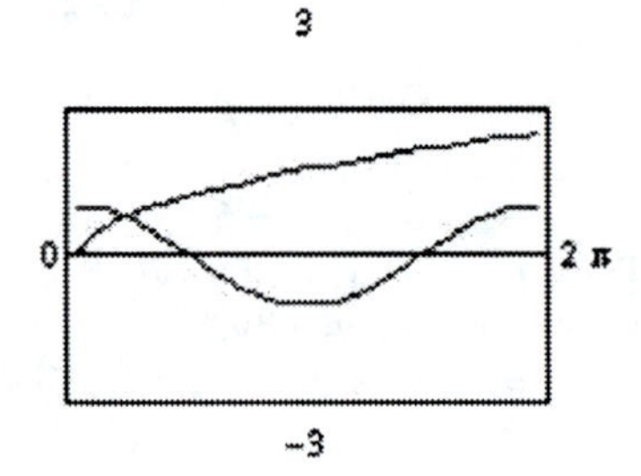

Check: $\cos(0.642) = 0.801$
$\sqrt{0.642} = 0.801$

Since $|\cos x| \le 1$, while $\sqrt{x} > 1$ for real x not shown, there can be no other solutions.

66. Graph y1 = $\cos\frac{x}{2} - 2\sin x$ and y2 = 1 in the same viewing window in a graphing calculator and find the points of intersection using an automatic intersection routine. The intersection points are found to be 0, 3.895, 5.121, 9.834, 12.566.

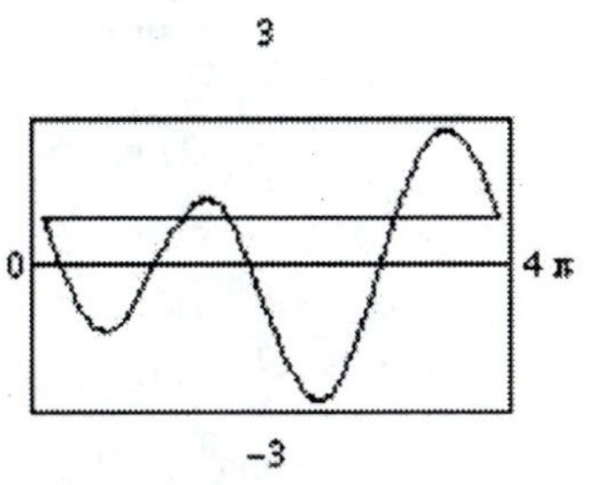

Check: $\cos\frac{0}{2} - 2\sin 0 = 1 - 0 = 1$

$\cos\frac{3.895}{2} - 2\sin 3.895 = 1.000$

(The remaining checking is left to the student.)

67. The coordinates are (cos 28.703, sin 28.703) = (–0.9095, –0.4157). The point P is in the third quadrant, since both coordinates are negative.

68. Since (x, y) is on a unit circle with $(x, y) = (0.5313, 0.8472) = (\cos s, \sin s)$, we can solve $\cos s = 0.5313$ or $\sin s = 0.8472$. Then $s = \cos^{-1}(0.5313) = 1.0107$ or $s = \sin^{-1}(0.8472) = 1.0107$

69. Draw a reference triangle and label what we know. Since $\cos\theta = a = \frac{a}{1}$, we can write $r = 1$. Use the Pythagorean theorem to find b.

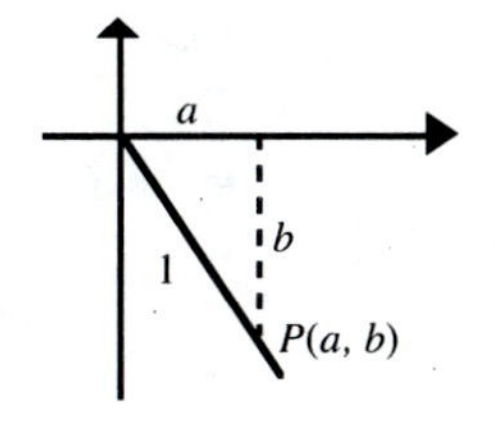

$$a^2 + b^2 = 1$$
$$b^2 = 1 - a^2$$
$$b = -\sqrt{1-a^2}$$

b is negative since $P(a, b)$ is in quadrant IV. We can now find the other five functions.

$$\sin\theta = \frac{b}{r} = \frac{-\sqrt{1-a^2}}{1} = -\sqrt{1-a^2}$$

$$\cot\theta = \frac{a}{b} = \frac{a}{-\sqrt{1-a^2}} = -\frac{a}{\sqrt{1-a^2}}$$

$$\tan\theta = \frac{b}{a} = \frac{-\sqrt{1-a^2}}{a}$$

$$\sec\theta = \frac{r}{a} = \frac{1}{a}$$

$$\csc\theta = \frac{r}{b} = \frac{1}{-\sqrt{1-a^2}} = -\frac{1}{\sqrt{1-a^2}}$$

70.

$\tan 3x = \tan(x + 2x)$	Algebra
$= \dfrac{\tan x + \tan 2x}{1 - \tan x \tan 2x}$	Sum identity for tangent
$= \dfrac{\tan x + \dfrac{2\tan x}{1-\tan^2 x}}{1 - \tan x \cdot \dfrac{2\tan x}{1-\tan^2 x}}$	Double-angle identity
$= \dfrac{(1-\tan^2 x)\tan x + 2\tan x}{(1-\tan^2 x) - \tan x \cdot 2\tan x}$	Algebra
$= \dfrac{\tan x - \tan^3 x + 2\tan x}{1 - \tan^2 x - 2\tan^2 x}$	Algebra
$= \dfrac{3\tan x - \tan^3 x}{1 - 3\tan^2 x}$	Algebra
$= \tan x\, \dfrac{3 - \tan^2 x}{1 - 3\tan^2 x}$	Algebra
$= \tan x\, \dfrac{3 - \dfrac{\sin^2 x}{\cos^2 x}}{1 - 3\dfrac{\sin^2 x}{\cos^2 x}}$	Quotient identity
$= \tan x\, \dfrac{3\cos^2 x - \sin^2 x}{\cos^2 x - 3\sin^2 x}$	Algebra
$= \tan x\, \dfrac{3\left(\dfrac{2\cos^2 x - 1 + 1}{2}\right) - \left(\dfrac{1 - (1 - 2\sin^2 x)}{2}\right)}{\dfrac{2\cos^2 x - 1 + 1}{2} - 3\left(\dfrac{1 - (1 - 2\sin^2 x}{2}\right)}$	Algebra
$= \tan x\, \dfrac{3\left(\dfrac{\cos 2x + 1}{2}\right) - \dfrac{1 - \cos 2x}{2}}{\dfrac{\cos 2x + 1}{2} - 3\left(\dfrac{1 - \cos 2x}{2}\right)}$	Double-angle identities
$= \tan x\, \dfrac{\frac{3}{2}\cos 2x + \frac{3}{2} - \frac{1}{2} + \frac{1}{2}\cos 2x}{\frac{1}{2}\cos 2x + \frac{1}{2} - \frac{3}{2} + \frac{3}{2}\cos 2x}$	Algebra
$= \tan x\, \dfrac{2\cos 2x + 1}{2\cos 2x - 1}$	Algebra

71. Let $y = \sin^{-1}\frac{1}{3}$. Then we are asked to evaluate $\cos(2y)$ which is $1 - 2\sin^2 y$ from the double-angle identity. Since $\sin y = \sin\left(\sin^{-1}\frac{1}{3}\right) = \frac{1}{3}$ from the sine-inverse sine identity, we can write

$$\cos\left(2\sin^{-1}\frac{1}{3}\right) = \cos 2y = 1 - 2\sin^2 y = 1 - 2\left(\frac{1}{3}\right)^2 = 1 - 2\cdot\frac{1}{9} = 1 - \frac{2}{9} = \frac{7}{9}$$

72. Let $u = \cos^{-1} x$ and $v = \tan^{-1} x$ $\quad -1 \le x \le 1$ $\quad$ or, equivalently

$x = \cos u$ $\quad 0 \le u \le \pi$ $\quad$ and $\quad x = \tan v$ $\quad -\frac{\pi}{2} < v < \frac{\pi}{2}$

Then, $\sin(\cos^{-1} x - \tan^{-1} x) = \sin(u - v) = \sin u \cos v - \cos u \sin v$

For u, geometrically, we have

or

In either case, $b = \sqrt{1 - x^2}$ $\quad \sin u = \frac{b}{r} = \sqrt{1 - x^2}$ $\quad \cos u = x$

For v, geometrically, we have

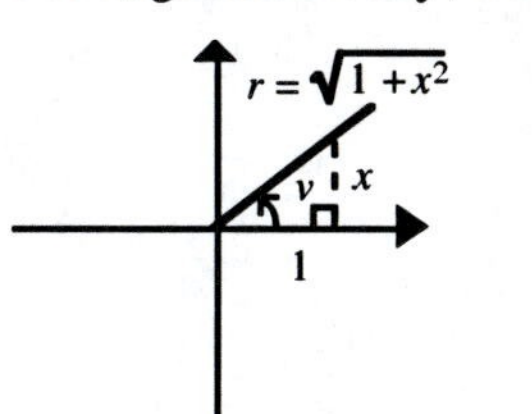

or

In either case, $r = \sqrt{1 + x^2}$; $\quad \sin v = \frac{x}{\sqrt{1 + x^2}}$; $\quad \cos v = \frac{1}{\sqrt{1 + x^2}}$

Thus,
$$\sin(\cos^{-1} x - \tan^{-1} x) = \sin(u - v) = \sin u \cos v - \cos u \sin v$$
$$= \sqrt{1 - x^2}\cdot\frac{1}{\sqrt{1 + x^2}} - x\cdot\frac{x}{\sqrt{1 + x^2}} = \frac{\sqrt{1 - x^2} - x^2}{\sqrt{1 + x^2}}$$

73. First solve for x over one period, $0 \le x < 2\pi$. Then add integer multiples of 2π to find all solutions.

$$\begin{aligned} \sin x &= 1 + \cos x \\ \pm\sqrt{1-\cos^2 x} &= 1 + \cos x \\ 1 - \cos^2 x &= (1 + \cos x)^2 \qquad \text{Squaring both sides} \\ 1 - \cos^2 x &= 1 + 2\cos x + \cos^2 x \\ 0 &= 2\cos x + 2\cos^2 x \\ 0 &= 2\cos x(1 + \cos x) \end{aligned}$$

$$\begin{aligned} 2\cos x &= 0 & 1 + \cos x &= 0 \\ \cos x &= 0 & \cos x &= -1 \\ x &= \frac{\pi}{2}, \frac{3\pi}{2} & x &= \pi \end{aligned}$$

In squaring both sides, we may have introduced extraneous solutions; hence, it is necessary to check solutions of these equations in the original equation.

Check:

$x = \frac{\pi}{2}$	$x = \frac{3\pi}{2}$	$x = \pi$
$\sin x = 1 + \cos x$	$\sin x = 1 + \cos x$	$\sin \pi = 1 + \cos x$
$\sin \frac{\pi}{2} = 1 + \cos \frac{\pi}{2}$	$\sin \frac{3\pi}{2} = 1 + \cos \frac{3\pi}{2}$	$\sin \pi = 1 + \cos \pi$
$1 = 1 + 0$	$-1 \ne 1 + 0$	$0 = 1 + (-1)$
A solution	Not a solution	A solution

Thus, all solutions over one period, $0 \le x < 2\pi$, are $x = \frac{\pi}{2}, \pi$.

Thus, the solutions, if x is allowed to range over all real numbers are

$$x = \begin{cases} \frac{\pi}{2} + 2k\pi \\ \pi + 2k\pi \end{cases} \quad k \text{ any integer}$$

74. $\sin x = \cos^2 x \qquad 0 \le x \le 2\pi$

Solve for sin x and/or cos x: $\sin x = 1 - \sin^2 x$ Use Pythagorean identity

$\sin^2 x + \sin x - 1 = 0$ Quadratic in $\sin x$

$$\sin x = \frac{-1 \pm \sqrt{1^2 - 4(1)(-1)}}{2\cdot 1} = \frac{-1 \pm \sqrt{5}}{2}$$

$\sin x = -1.6180 \qquad \sin x = 0.6180$

$\sin x = -1.6180$ No solution (-1.618 is not in the range of the sine function)

$\sin x = 0.6180$ The solutions are in the first and second quadrants.

$x = \sin^{-1}(0.6180) = 0.6662$

$x = \pi - 0.6662 = 2.475$

75. The graph of $f(x)$ is shown in the figure.

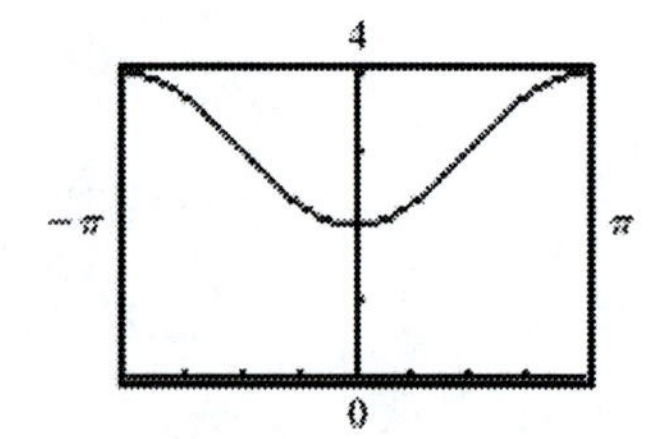

The graph appears to be an upside down cosine curve with period 2π, amplitude = $\frac{1}{2}(y_{max} - y_{min})$ $= \frac{1}{2}(4-2) = 1$, displaced upward by $k = 3$ units. It appears that $g(x) = 3 - \cos x$ would be an appropriate choice.

We verify $f(x) = g(x)$ as follows:

$f(x) = \dfrac{\sin^2 x}{1-\cos x} + \dfrac{2\tan^2 x \cos^2 x}{1+\cos x}$	
$= \dfrac{1-\cos^2 x}{1-\cos x} + \dfrac{2\dfrac{\sin^2 x}{\cos^2 x}\cos^2 x}{1+\cos x}$	Pythagorean and quotient identities
$= \dfrac{1-\cos^2 x}{1-\cos x} + \dfrac{2\sin^2 x}{1+\cos x}$	Algebra
$= \dfrac{1-\cos^2 x}{1-\cos x} + \dfrac{2(1-\cos^2 x)}{1+\cos x}$	Pythagorean identity
$= \dfrac{(1-\cos x)(1+\cos x)}{1-\cos x} + \dfrac{2(1-\cos x)(1+\cos x)}{1+\cos x}$	Algebra
$= 1 + \cos x + 2(1 - \cos x)$	Algebra
$= 1 + \cos x + 2 - 2\cos x$	Algebra
$= 3 - \cos x = g(x)$	Algebra

76. The graph of $f(x)$ is shown in the figure.

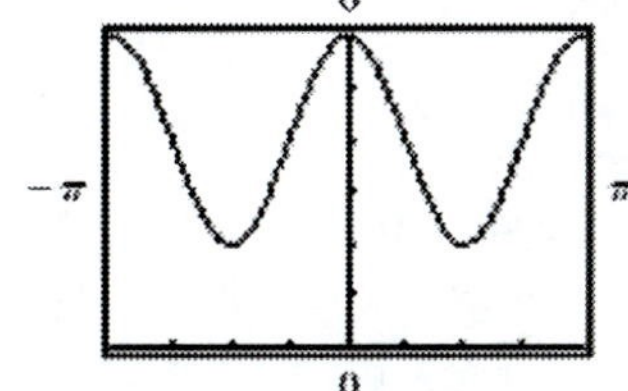

The graph appears to be a basic cosine curve with period π, amplitude = $\frac{1}{2}(y_{max} - y_{min})$ $= \frac{1}{2}(6-2) = 2$, displaced upward by $k = 4$ units. It appears that $g(x) = 4 + 2\cos 2x$ would be an appropriate choice.

We verify $f(x) = g(x)$ as follows:

$f(x) = 2\sin^2 x + 6\cos^2 x$	
$= 2(1 - \cos^2 x) + 6\cos^2 x$	Pythagorean identity
$= 2 - 2\cos^2 x + 6\cos^2 x$	Algebra
$= 2 + 4\cos^2 x$	Algebra
$= 2 + 2(2\cos^2 x - 1) + 2$	Algebra
$= 4 + 2(2\cos^2 x - 1)$	Algebra
$= 4 + 2\cos 2x = g(x)$	Double-angle identity

77. The graph of $f(x)$ is shown in the figure.

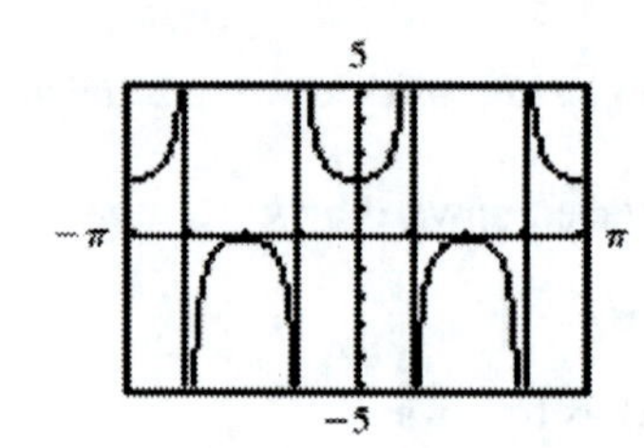

The graph appears to have vertical asymptotes $x = -\frac{3\pi}{4}, -\frac{\pi}{4}, \frac{\pi}{4}$, and $\frac{3\pi}{4}$ and period π. It appears to have high and low points with y coordinates 0 and 2, respectively.

It appears that $g(x) = 1 + \sec 2x$ would be an appropriate choice.

We verify $f(x) = g(x)$ as follows:

$$f(x) = \frac{2-2\sin^2 x}{2\cos^2 x - 1}$$

$$= \frac{1+1-2\sin^2 x}{2\cos^2 x - 1} \qquad \text{Algebra}$$

$$= \frac{1+\cos 2x}{\cos 2x} \qquad \text{Double-angle identity}$$

$$= \frac{1}{\cos 2x} + \frac{\cos 2x}{\cos 2x} \qquad \text{Algebra}$$

$$= \sec 2x + 1 = g(x) \qquad \text{Quotient identity, algebra}$$

78. The graph of $f(x)$ is shown in the figure.

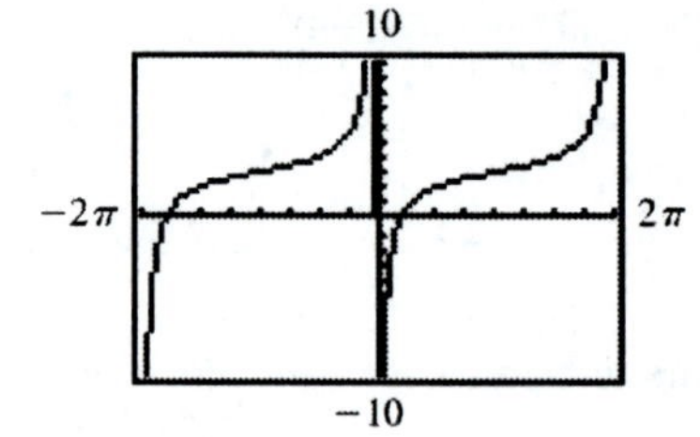

The graph appears to have vertical asymptotes $x = -2\pi$, $x = 0$, and $x = 2\pi$, and period 2π. Its x intercepts are difficult to determine, but since there appears to be symmetry with respect to the points where the curve crosses the line $y = 3$, it appears to be an upside down cotangent curve displaced upward by $k = 3$ units.

It appears that $g(x) = 3 - \cot \frac{x}{2}$ would be an appropriate choice.

We verify $f(x) = g(x)$ as follows:

$$f(x) = \frac{3\cos x + \sin x - 3}{\cos x - 1}$$

$$= \frac{3(\cos x - 1) + \sin x}{\cos x - 1} \qquad \text{Algebra}$$

$$= \frac{3(\cos x - 1)}{\cos x - 1} - \frac{\sin x}{1 - \cos x} \qquad \text{Algebra}$$

$$= 3 - \frac{\sin x}{1 - \cos x} \qquad \text{Algebra}$$

$= 3 - 1 \div \dfrac{1 - \cos x}{\sin x}$ Algebra

$= 3 - 1 \div \tan \dfrac{x}{2}$ Half-angle identity

$= 3 - \cot \dfrac{x}{2} = g(x)$ Reciprocal identity

79. Graph $y1 = 2 \sin \dfrac{\pi x}{2} - 3 \cos \dfrac{\pi x}{2}$ and $y2 = 2x - 1$ in the same viewing window in a graphing calculator. Finding the points of intersection using an automatic intersection routine, we see that the graph of y1 is not below the graph of y2 for the interval [0.694, 2.000].

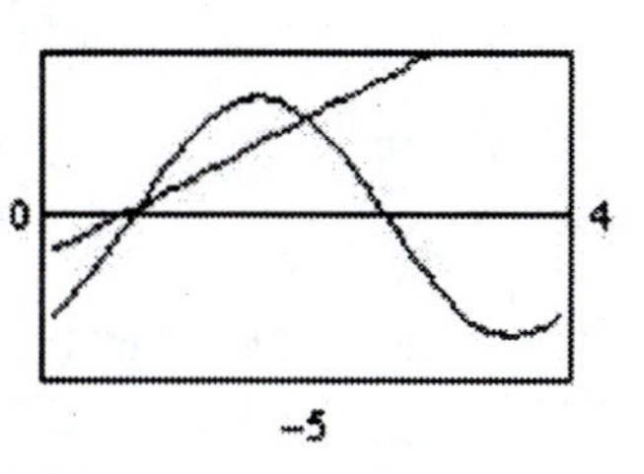

80. Labeling the diagram, we note: In triangle ACD, $\cot 32° = \dfrac{1000 + x}{h}$

In triangle BCD, $\cot 48° = \dfrac{x}{h}$

We solve the system of equations

$$\cot 32° = \frac{1000 + x}{h} \qquad \cot 48° = \frac{x}{h}$$

by clearing of fractions, then eliminating x.

$$h \cot 32° = 1000 + x \qquad h \cot 48° = x$$

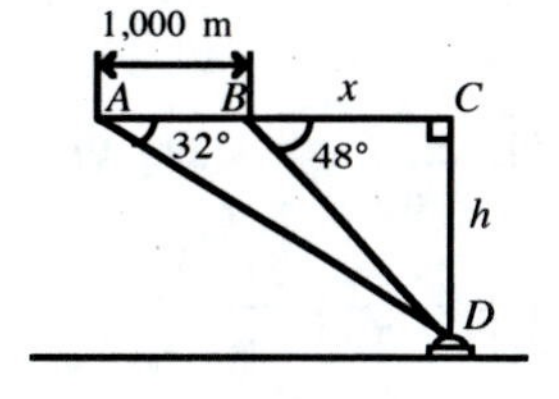

$$\begin{aligned} h \cot 32° &= 1000 + h \cot 48° \\ h \cot 32° - h \cot 48° &= 1000 \\ h(\cot 32° - \cot 48°) &= 1000 \\ h &= \frac{1000}{\cot 32° - \cot 48°} = 1429 \text{ meters} \end{aligned}$$

81. $\sqrt{u^2 - a^2} = \sqrt{(a \csc x)^2 - a^2}$ Using the given substitution

$= \sqrt{a^2 \csc^2 x - a^2}$ Algebra

$= \sqrt{a^2(\csc^2 x - 1)}$ Algebra

$= \sqrt{a^2 \cot^2 x}$ Pythagorean identity

$= |a|\ |\cot x|$ Algebra

$= a \cot x$ Since $a > 0$ and x is in quadrant I $\left(\text{given } 0 < x < \dfrac{\pi}{2}\right)$, thus, $\cot x > 0$.

82. We note that $\tan\theta = \frac{2}{x}$ and $\tan 2\theta = \frac{2+4}{x} = \frac{6}{x}$ (see figure).

Then, $$\tan 2\theta = \frac{2\tan\theta}{1-\tan^2\theta}$$

$$\frac{6}{x} = \frac{2\left(\frac{x}{2}\right)}{1-\left(\frac{2}{x}\right)^2} = \frac{\frac{4}{x}}{1-\left(\frac{2}{x}\right)^2} = \frac{x^2\cdot\frac{4}{x}}{x^2\cdot 1 - x^2\cdot\frac{4}{x^2}} = \frac{4x}{x^2-4}$$

$$x(x^2-4)\cdot\frac{6}{x} = x(x^2-4)\cdot\frac{4x}{x^2-4}$$

$$6(x^2-4) = x\cdot 4x$$

$$6x^2 - 24 = 4x^2$$

$$2x^2 - 24 = 0$$

$$x^2 = 12$$

$$x = \sqrt{12} \text{ (we discard the negative solution)}$$

$$= 2\sqrt{3}$$

Since $\tan\theta = \frac{2}{x} = \frac{2}{2\sqrt{3}} = \frac{1}{\sqrt{3}}$, $\theta = 30°$

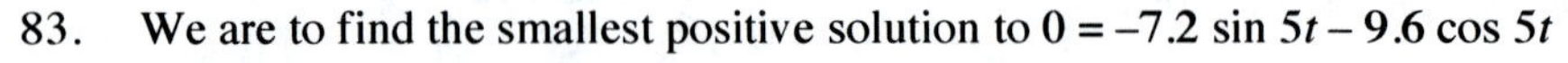

83. We are to find the smallest positive solution to $0 = -7.2\sin 5t - 9.6\cos 5t$

$$7.2\sin 5t = -9.6\cos 5t$$

$$\frac{7.2\sin 5t}{7.2\cos 5t} = \frac{9.6\cos 5t}{7.2\cos 5t}$$

$$\tan 5t = -\frac{9.6}{7.2}$$

$$5t = \tan^{-1}\left(-\frac{9.6}{7.2}\right) + \pi$$

$$t = \frac{1}{5}\left[\tan^{-1}\left(-\frac{9.6}{7.2} + \pi\right)\right] = 0.443 \text{ sec}$$

84. We graph $y1 = -7.2\sin 5t - 9.6\cos 5t$ and $y2 = 5$. From the figure, we see that y1 and y2 intersect twice on the indicated interval. From the automatic intersection routine, the intersections are found to be $t = 0.529$ sec and $t = 0.985$ sec.

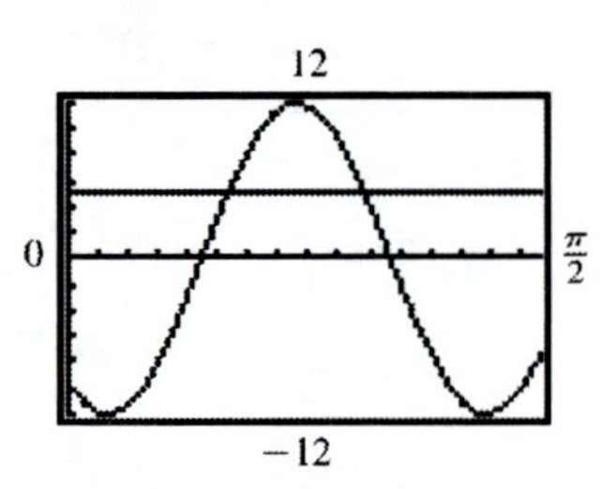

85. Following the hint, label the text diagram as shown at the right.

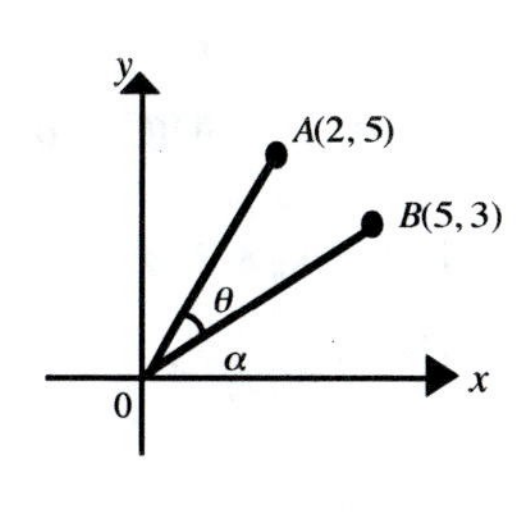

Then $\tan\alpha = \frac{y_B}{x_B} = \frac{3}{5}$ $\qquad \tan(\alpha+\theta) = \frac{y_A}{x_A} = \frac{5}{2}$

$$\tan\theta = \tan(\alpha+\theta-\alpha) = \tan[(\alpha+\theta)-\alpha]$$

$$= \frac{\tan(\alpha+\theta)-\tan\alpha}{1+\tan(\alpha+\theta)\tan\alpha}$$

$$= \frac{\frac{5}{2}-\frac{3}{5}}{1+\frac{5}{2}\cdot\frac{3}{5}} = \frac{10\cdot\frac{5}{2}-10\cdot\frac{3}{5}}{10\cdot 1+10\cdot\frac{5}{2}\cdot\frac{3}{5}} = \frac{25-6}{10+15} = \frac{19}{25}$$

Therefore, $\theta = \tan^{-1}\frac{19}{25} = 0.650$

86. $I = 60\sin\left(90\pi t - \frac{\pi}{2}\right)$

(A) We compute amplitude, period, frequency, and phase shift as follows:
Amplitude = $|A| = |60| = 60$ amperes
Phase Shift and Period: Solve $Bx + C = 0$ and $Bx + C = 2\pi$

$$90\pi t - \frac{\pi}{2} = 0 \qquad 90\pi t - \frac{\pi}{2} = 2\pi$$

$$90\pi t = \frac{\pi}{2} \qquad 90\pi t = \frac{\pi}{2} + 2\pi$$

$$t = \frac{1}{180} \qquad t = \frac{1}{180} + \frac{1}{45}$$

↑ Phase Shift ↑ Period in seconds

$$\text{Frequency} = \frac{1}{\text{Period}} = \frac{1}{1/45} = 45 \text{ Hz.}$$

(B)

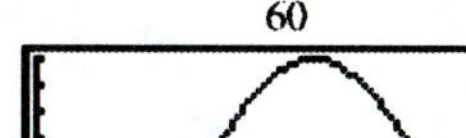

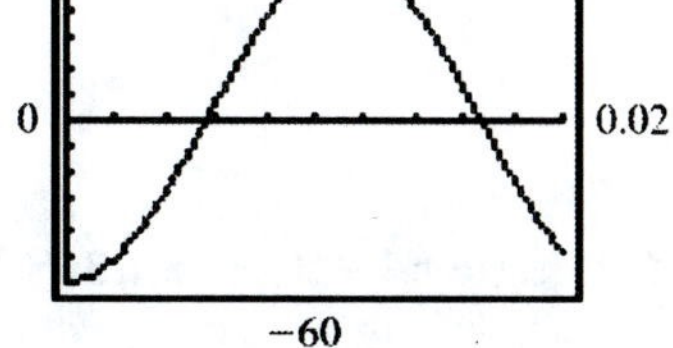

(C) $t = 0.0078$ sec

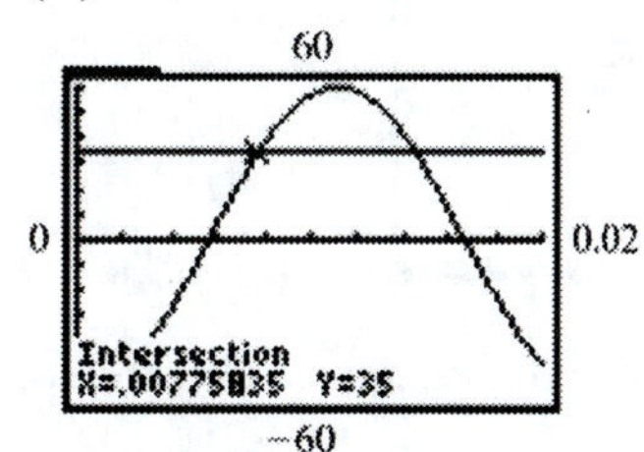

87. (A) From the figure, it should be clear that $\cot\theta = \frac{d}{200}$.
Thus, $d = 200\cot\theta$.

(B) Period = π
One period of the graph would therefore extend from 0 to π, with vertical asymptotes at $t = 0$ and $t = \pi$. We sketch half of one period, since the required interval is from 0 to $\frac{\pi}{2}$ only.
Ordinates can be determined from a calculator, thus:

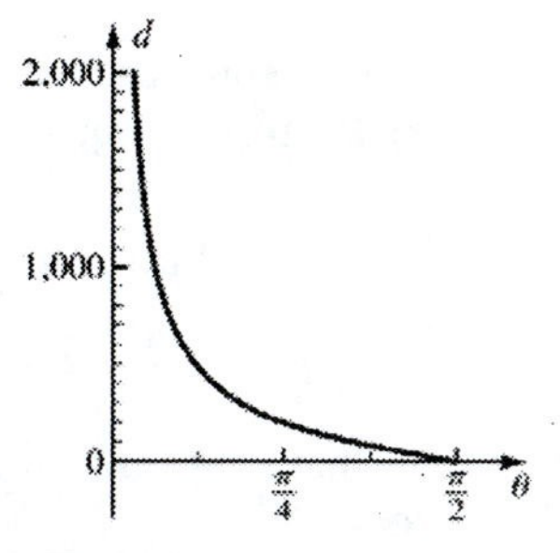

θ	$\frac{\pi}{20}$	$\frac{\pi}{10}$	$\frac{3\pi}{20}$	$\frac{\pi}{5}$	$\frac{\pi}{4}$	$\frac{3\pi}{10}$	$\frac{7\pi}{20}$	$\frac{2\pi}{5}$	$\frac{9\pi}{20}$	$\frac{\pi}{2}$
$200\cot\theta$	1263	616	393	275	200	145	101	65	32	0

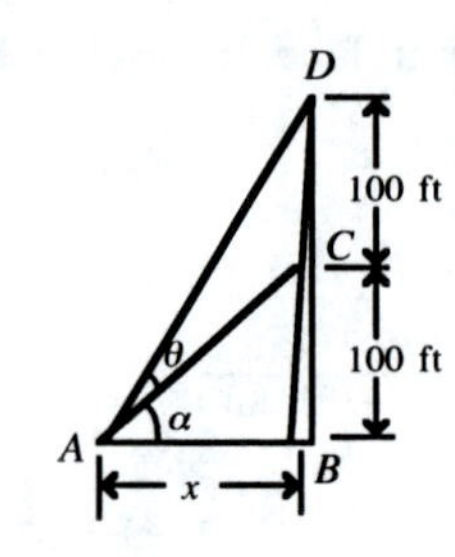

88. (A) Labeling the diagram as shown, we note

In triangle ABC, $\tan\alpha = \dfrac{100}{x}$

In triangle ABD, $\tan(\theta+\alpha) = \dfrac{200}{x}$

Hence,

$$\tan\theta = \tan(\theta+\alpha-\alpha) = \frac{\tan(\theta+\alpha)-\tan\alpha}{1+\tan(\theta+\alpha)\tan\alpha} = \frac{\frac{200}{x}-\frac{100}{x}}{1+\frac{200}{x}\frac{100}{x}}$$

$$= \frac{x^2\cdot\frac{200}{x}-x^2\cdot\frac{100}{x}}{x^2+x^2\cdot\frac{200}{x}\cdot\frac{100}{x}} = \frac{200x-100x}{x^2+20{,}000} = \frac{100x}{x^2+20{,}000}$$

$$\theta = \arctan\frac{100x}{x^2+20{,}000}$$

(B) We are given $x = 50$ feet. Thus, $\theta = \arctan\dfrac{100\cdot 50}{50^2+20{,}000} = 12.5°$

89. We graph $y1 = \tan^{-1}\dfrac{100x}{x^2+20{,}000}$ and $y2 = 15$ on the interval from 0 to 400. We see that the curves intersect twice on the interval.

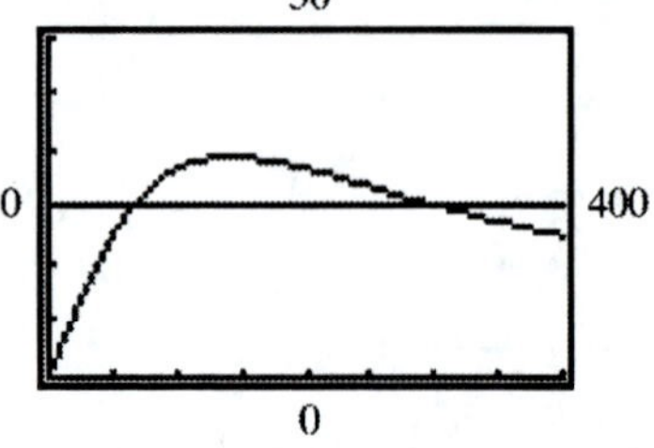

From the automatic intersection routine, the solutions are found to be $x = 64.9$ ft and $x = 308.3$ ft.

90. Since the sprockets are connected by the bicycle chain, the distance (arc length) that the rear sprocket turns is equal to the distance that the pedal sprocket turns. Let R_1 = radius of rear sprocket and

R_2 = radius of pedal sprocket.

$$s = R_1\theta_1 \qquad s = R_2\theta_2 \qquad R_1\theta_1 = R_2\theta_2 \qquad \theta_1 = \frac{R_2}{R_1}\theta_2$$

Note that the angle through which the rear wheel turns is equal to the angle through which the rear sprocket turns.

Thus, $\theta_1 = \dfrac{11.0}{4.0}18\pi = 49.5\,\pi$ rad

91. We use the formula $\theta_1 = \frac{R_2}{R_1}\theta_2$ from the previous problem, and note:

$\omega_1 = \frac{\theta_1}{t}$ angular velocity of rear wheel and sprocket

$\omega_2 = \frac{\theta_2}{t}$ angular velocity of pedal sprocket.

Hence, $\frac{\theta_1}{t} = \frac{R_2}{R_1}\frac{\theta_2}{t}$, $\omega_1 = \frac{R_2}{R_1}\omega_2 = \frac{11.0}{4.0}60.0$ rpm $= 165$ rpm or $165 \cdot 2\pi$ rad/min

Then, the linear velocity of the rear wheel (i.e., that of the bicycle), v, is given by

$$v = R_{\text{wheel}}\omega_1 = \frac{70.0\text{ cm}}{2} \cdot 165 \cdot 2\pi \text{ rad/min} = 11{,}550\pi \text{ cm/min} \approx 36{,}300 \text{ cm/min}$$

92. (A) For small θ (near 0°), L is extremely large. As θ increases from 0°, L decreases to some minimum value, then increases again beyond all bounds as θ approaches 90°.

(B) Label the text figure as shown. Note that ABC and CDE are right triangles, and that

$AE = AC + CE = a + b$.

Then, $\csc\theta = \frac{a}{50}$ from triangle ABC, so

$a = 50\csc\theta$, $\sec\theta = \frac{b}{25}$ from triangle CDE,

so $b = 25\sec\theta$. Thus, $AE = 50\csc\theta + 25\sec\theta$.

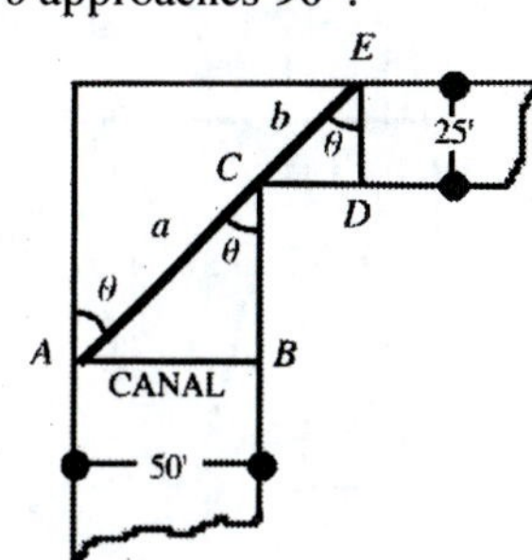

(C)

θ°	35	40	45	50	55	60	65
L ft	117.7	110.4	106.1	104.2	104.6	107.7	114.3

(D) According to the table, min $L = 104.2$ ft for $\theta = 50°$. The length of the longest log that will go around the corner is the minimum length L.

(E) According to the graph, min $L = 104.0$ ft for $\theta = 51.6°$.

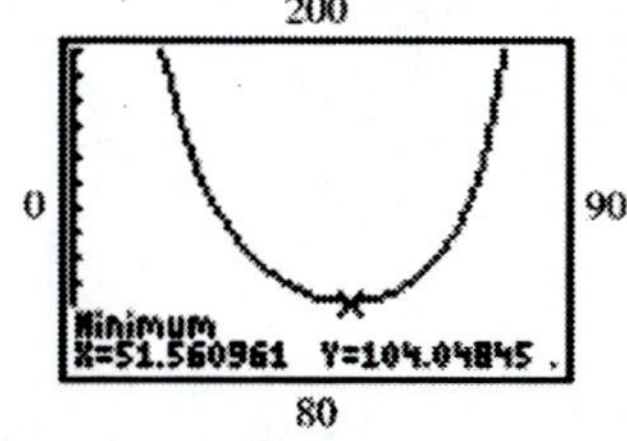

93. (A)

x(months)	1, 13	2, 14	3, 15	4, 16	5, 17	6, 18	7, 19	8, 20	9, 21
y (daylight duration)	6.52	9.17	11.83	14.60	17.55	19.27	18.45	15.85	12.95

x(months)	10, 22	11, 23	12, 24
y (daylight duration)	10.15	7.32	5.60

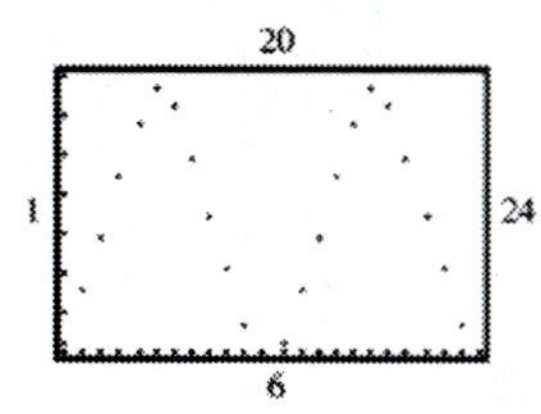

(B) From the table, max $x = 19.27$ and min $y = 5.60$. Then,

$$A = \frac{\max y - \min y}{2} = \frac{19.27 - 5.60}{2} = 6.835$$

$$B = \frac{2\pi}{\text{Period}} = \frac{2\pi}{12} = \frac{\pi}{6}$$

$$k = \min y + A = 5.60 + 6.835 = 12.435$$

From the plot in (A) or the table, we estimate the smallest value of x for which $y = k = 12.435$ to be approximately 3.1. Then, this is the phase-shift for the graph. Substitute $B = \frac{\pi}{6}$ and $x = 3.1$ into the phase-shift equation $x = -\frac{C}{B}$, $3.1 = \frac{-C}{\pi/6}$, $C = \frac{-3.1\pi}{6} \approx -1.6$. Thus, the equation required is $y = 12.435 + 6.835 \sin\left(\frac{\pi x}{6} - 1.6\right)$.

(C)

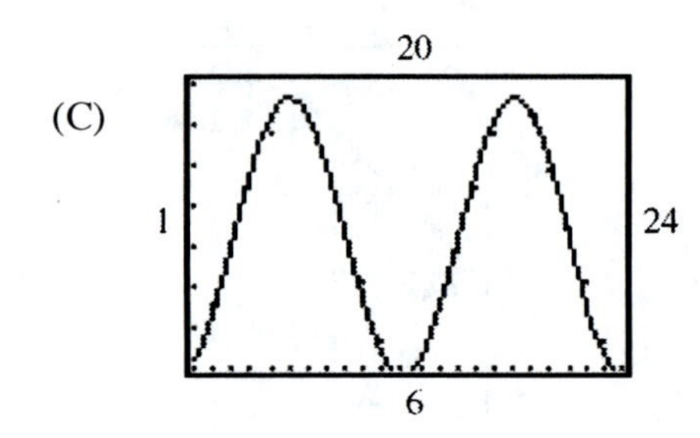

Chapter 6 Additional Topics: Triangles and Vectors

EXERCISE 6.1 Law of Sines

1. Two angles and the included side are given.

3. In order to set up an equation with only one unknown, you need to know at least one side opposite a known angle. This information is lacking in the *SAS* case.

5. Since $\frac{\sin\alpha}{a}$ is given and b is also given, the law of sines can be used to find β, then γ, then c.

7. None of the ratios in the law of sines can be found and the law of sines cannot be used at this point.

9. None of the ratios in the law of sines can be found and the law of sines cannot be used at this point.

11. *Solve for α:* $\alpha + \beta + \gamma = 180°$

$$\alpha = 180° - (36° + 43°) = 101°$$

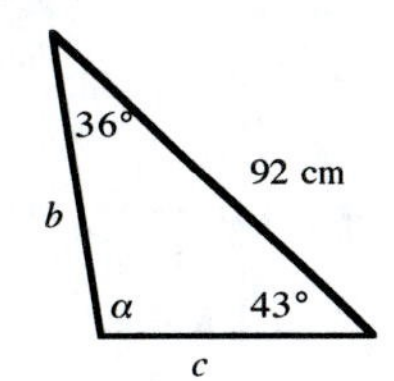

Solve for b: $\frac{\sin\alpha}{a} = \frac{\sin\beta}{b}$

$$b = \sin\beta\ \frac{a}{\sin\alpha} = (\sin 43°)$$

$$= 64 \text{ cm}$$

Solve for c: $\frac{\sin\alpha}{a} = \frac{\sin\gamma}{c}$

$$c = \sin\gamma\ \frac{a}{\sin\alpha} = (\sin 36°)\ \frac{92\text{ cm}}{\sin 101°}$$

$$= 55 \text{ cm}$$

13. *Solve for α:* $\alpha + \beta + \gamma = 180°$

$$\alpha = 180° - (27.5° + 54.5°) = 98.0°$$

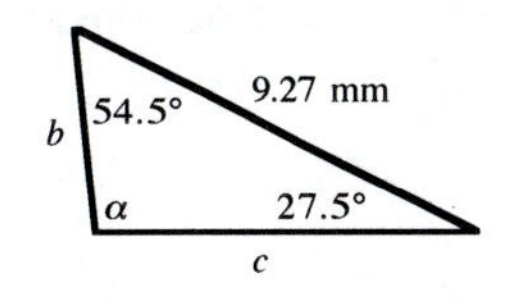

Solve for b: $\frac{\sin\alpha}{a} = \frac{\sin\beta}{b}$

$$b = \sin\beta\ \frac{a}{\sin\alpha} = (\sin 27.5°)\ \frac{9.27 \text{ mm}}{\sin 98.0°}$$

$$= 4.32 \text{ mm}$$

Solve for c: $\frac{\sin\alpha}{a} = \frac{\sin\gamma}{c}$

$$c = \sin\gamma\ \frac{a}{\sin\alpha} = (\sin 54.5°)\ \frac{9.27 \text{ mm}}{\sin 98.0°}$$

$$= 7.62 \text{ mm}$$

15. *Solve for γ:* $\alpha + \beta + \gamma = 180°$

$\gamma = 180° - (122.7° + 34.4°) = 22.9°$

Solve for a: $\dfrac{\sin\alpha}{a} = \dfrac{\sin\beta}{b}$

$a = \sin\alpha \dfrac{b}{\sin\beta} = (\sin 122.7°)\dfrac{18.3\text{cm}}{\sin 34.4°}$

$= 27.3$ cm

Solve for c: $\dfrac{\sin\beta}{b} = \dfrac{\sin\gamma}{c}$

$c = \sin\gamma \dfrac{b}{\sin\beta} = (\sin 22.9°)\dfrac{18.3\text{cm}}{\sin 34.4°}$

$= 12.6$ cm

17. *Solve for α:* $\alpha + \beta + \gamma = 180°$

$\alpha = 180° - (100°0' + 12°40') = 67°20'$

Solve for a: $\dfrac{\sin\alpha}{a} = \dfrac{\sin\beta}{b}$

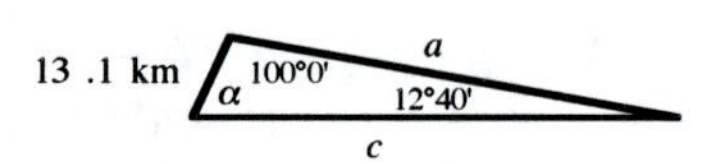

$a = \sin\alpha \dfrac{b}{\sin\beta} = (\sin 67°20')\dfrac{13.1\text{km}}{\sin 12°40'}$

$= 55.1$ km

Solve for c: $\dfrac{\sin\beta}{b} = \dfrac{\sin\gamma}{c}$

$c = \sin\gamma \dfrac{b}{\sin\beta} = (\sin 100°0')\dfrac{13.1\text{km}}{\sin 12°40'} = 58.8$ km

19.

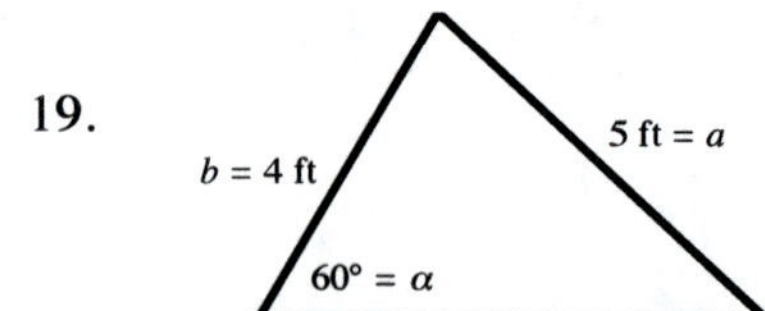

This is the case (d) where α is acute and $a \geq b$. 1 triangle can be constructed.

21.

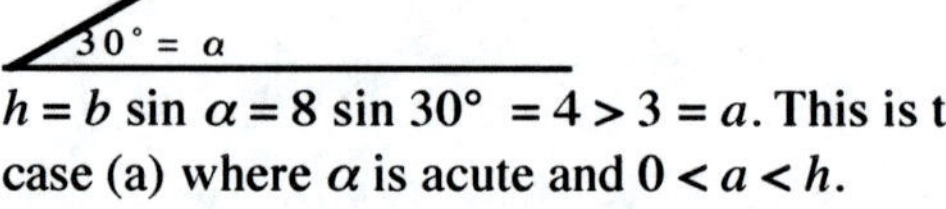

$h = b \sin\alpha = 8 \sin 30° = 4 > 3 = a$. This is the case (a) where α is acute and $0 < a < h$.
0 triangles can be constructed.

23.

$h = b \sin\alpha = 6 \sin 45° = 3\sqrt{2} < 5 = a$.
This is the case (c) where α is acute and $h < a < b$. 2 triangles can be constructed.

25.

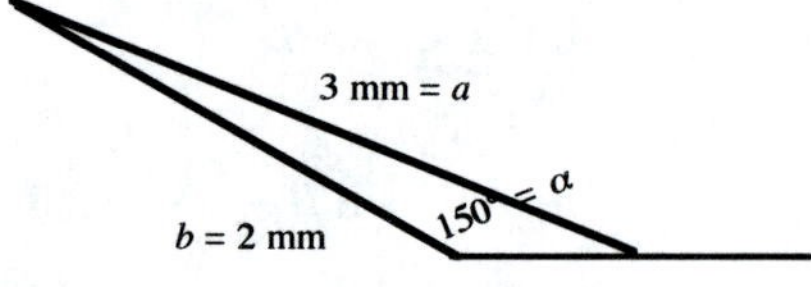

This is the case (f) where α is obtuse and $a > b$.
1 triangle can be constructed.

27. 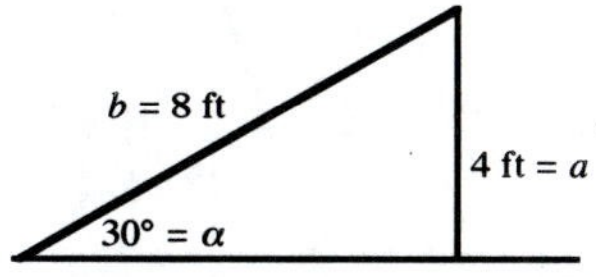

$h = b \sin \alpha = 8 \sin 30° = 4 = a$. This is the case (b) where α is acute and $a = h$.
1 triangle can be constructed.

29. 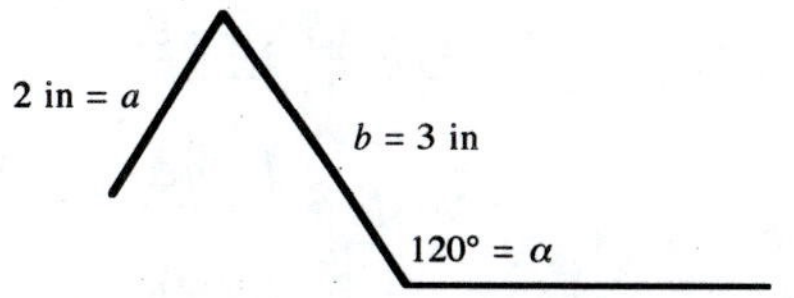

This is the case (e) where α is obtuse and $0 < a < b$.
0 triangles can be constructed.

31. First, draw a horizontal line of reasonable length which will eventually represent c. Draw an angle of 30° with this c as one side. The other side will represent b, and must be 2 units long. Then, noting that

$$\frac{\sin\beta}{b} = \frac{\sin\alpha}{a}$$

$$\sin\beta = \frac{b\sin\alpha}{a} = \frac{2\sin 30°}{\sqrt{2}} = \frac{\sqrt{2}}{2}$$

Angle β therefore can be either acute or obtuse, either $\beta = \sin^{-1}\frac{\sqrt{2}}{2} = 45°$ or

$\beta' = 180° - \sin^{-1}\frac{\sqrt{2}}{2} = 135°$. These two possibilities must be drawn as the two possible triangles.

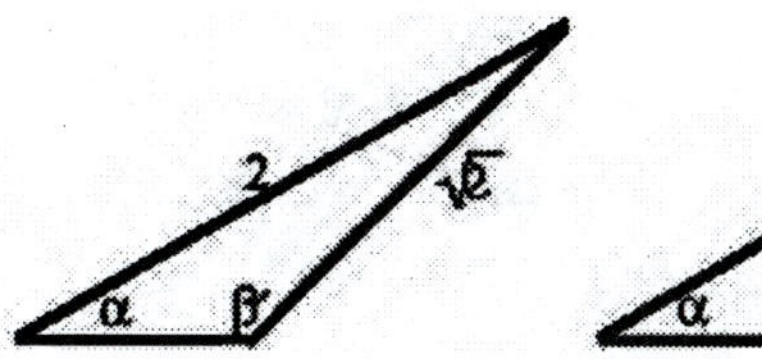

33. *Solve for β:*

$$\frac{\sin\beta}{b} = \frac{\sin\alpha}{a}$$

$$\sin\beta = \frac{b\sin\alpha}{a} = \frac{(24\text{ cm})\sin 134°}{(38\text{ cm})} = 0.4543$$

$$\beta = \sin^{-1} 0.4543 = 27°$$

(There is another solution of $\sin\beta = 0.4543$ that deserves brief consideration:

$\beta' = 180° - \sin^{-1} 0.4543 = 153°$. However, there is not enough room in a triangle for two obtuse angles.)

Solve for γ:

$$\alpha + \beta + \gamma = 180°$$

$$\gamma = 180° - (134° + 27°) = 19°$$

Solve for c:

$$\frac{\sin\alpha}{a} = \frac{\sin\gamma}{c}$$

$$c = \frac{a\sin\gamma}{\sin\alpha} = \frac{(38\text{ cm})\sin 19°}{\sin 134°} = 17\text{ cm}$$

35. *Solve for β:*
$$\frac{\sin\beta}{b} = \frac{\sin\alpha}{a}$$
$$\sin\beta = \frac{b\sin\alpha}{a} = \frac{(91\text{ ft})(\sin 69°)}{86\text{ ft}}$$
$$= 0.9879$$

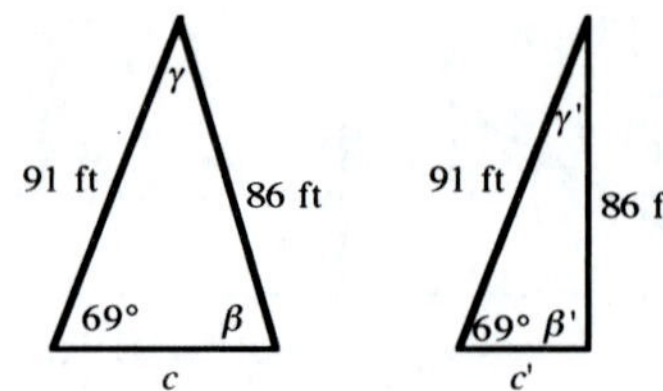

Angle β can be either acute or obtuse.

$\beta = \sin^{-1}(0.9879) = 81°$ $\qquad\qquad$ $\beta' = 180° - \sin^{-1}(0.9879) = 99°$

Solve for γ and γ':
$$\gamma = 180° - (\alpha + \beta) = 180° - (69° + 81°) = 30°$$
$$\gamma' = 180° - (\alpha + \beta') = 180° - (69° + 99°) = 12°$$

Solve for c and c':
$$\frac{\sin\alpha}{a} = \frac{\sin\gamma}{c}$$
$$c = \frac{a\sin\gamma}{\sin\alpha} = \frac{(86\text{ ft})(\sin 30°)}{\sin 69°} = 46\text{ ft}$$
$$\frac{\sin\alpha}{a} = \frac{\sin\gamma'}{c'}$$
$$c' = \frac{a\sin\gamma'}{\sin\alpha} = \frac{(86\text{ ft})(\sin 12°)}{\sin 69°} = 19\text{ ft}$$

37. *Solve for β:*
$$\frac{\sin\beta}{b} = \frac{\sin\alpha}{a}$$
$$\sin\beta = \frac{b\sin\alpha}{a} = \frac{(6.2\text{ in})(\sin 21°)}{4.7\text{ in}}$$
$$= 0.4727$$

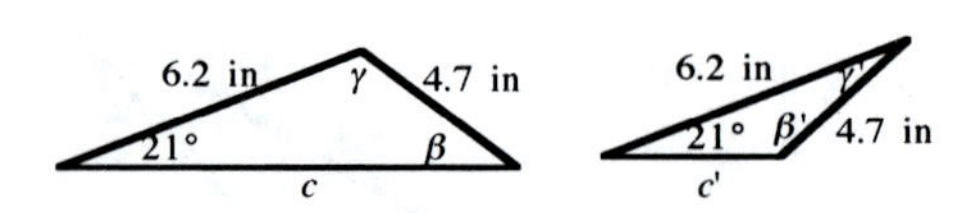

Angle β can be either acute or obtuse.

$\beta = \sin^{-1}(0.4727) = 28°$ $\qquad\qquad$ $\beta' = 180° - (\sin^{-1} 0.4727) = 152°$

Solve for γ and γ':
$$\gamma = 180° - (\alpha + \beta) = 180° - (21° + 28°) = 131°$$
$$\gamma' = 180° - (\alpha + \beta') = 180° - (21° + 152°) = 7°$$

Solve for c and c':
$$\frac{\sin\alpha}{a} = \frac{\sin\gamma}{c}$$
$$c = \frac{a\sin\gamma}{\sin\alpha} = \frac{(4.7\text{ in})(\sin 131°)}{\sin 21°} = 9.9\text{ in}$$
$$\frac{\sin\alpha}{a} = \frac{\sin\gamma'}{c'}$$
$$c' = \frac{a\sin\gamma'}{\sin\alpha} = \frac{(4.7\text{ in})(\sin 7°)}{\sin 21°} = 1.6\text{ in}$$

39.

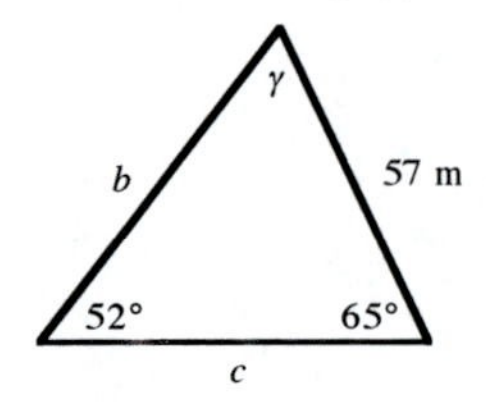

Solve for γ: $\alpha + \beta + \gamma = 180°$
$$\gamma = 180° - (\alpha + \beta) = 180° - (52° + 65°) = 63°$$

Solve for b: $\dfrac{\sin\alpha}{a} = \dfrac{\sin\beta}{b}$
$$b = \sin\beta\,\frac{a}{\sin\alpha} = (\sin 65°)\,\frac{57\text{ m}}{\sin 52°} = 66\text{ m}$$

Solve for c: $\dfrac{\sin\alpha}{a} = \dfrac{\sin\gamma}{c}$
$$c = \sin\gamma\,\frac{a}{\sin\alpha} = (\sin 63°)\,\frac{57\text{ m}}{\sin 52°} = 64\text{ m}$$

41. Solve for α: $\dfrac{\sin\alpha}{a} = \dfrac{\sin\beta}{b}$

$$\sin\alpha = \frac{a\sin\beta}{b} = \frac{(54\text{ cm})\sin 30°}{27\text{ cm}} = 1$$

$$\alpha = 90°$$

Solve for γ: $\gamma = 180° - (\alpha + \beta)$

$$= 180° - (90° + 30°) = 60°$$

Solve for c: $\dfrac{c}{a} = \cos 30°$ (using right-triangle relationships)

$$c = (54\text{ cm})(\cos 30°) = 47\text{ cm}$$

43. β is acute $b > a$ Only one triangle is possible.

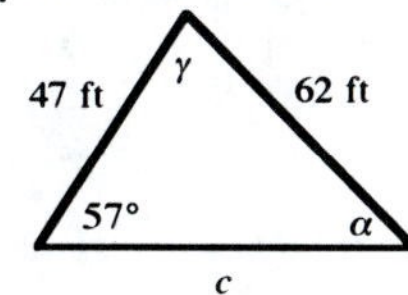

Solve for α: $\dfrac{\sin\alpha}{a} = \dfrac{\sin\beta}{b}$

$$\sin\alpha = \frac{a\sin\beta}{b} = \frac{(47\text{ ft})(\sin 57°)}{62\text{ ft}} = 0.6358$$

$$\alpha = \sin^{-1} 0.6358 = 39°$$

Solve for γ: $\alpha + \beta + \gamma = 180°$

$$\gamma = 180° - (57° + 39°) = 84°$$

Solve for c: $\dfrac{\sin\beta}{b} = \dfrac{\sin\gamma}{c}$

$$c = \frac{b\sin\gamma}{\sin\beta} = \frac{(62\text{ ft})(\sin 84°)}{\sin 57°} = 74\text{ ft}$$

45.

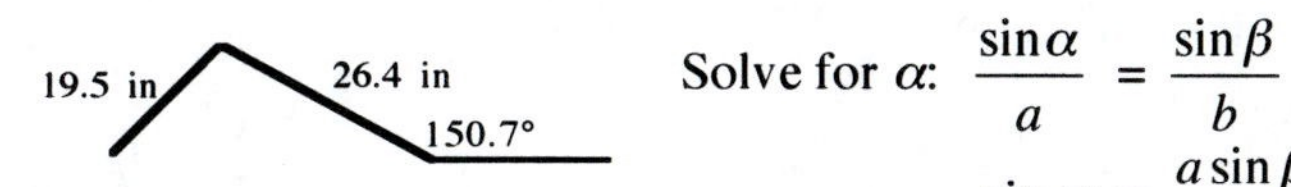

Solve for α: $\dfrac{\sin\alpha}{a} = \dfrac{\sin\beta}{b}$

$$\sin\alpha = \frac{a\sin\beta}{b} = \frac{(26.4\text{ in})(\sin 150.7°)}{19.5\text{ in}} = 0.6625$$

$$\alpha = \sin^{-1}(0.6625) = 41.5°$$

Since there is not enough room in a triangle for an angle of 150.7° and an angle of 41.5° (their sum is more than 180°) no triangle exists with the given measurements. No solution.

47.

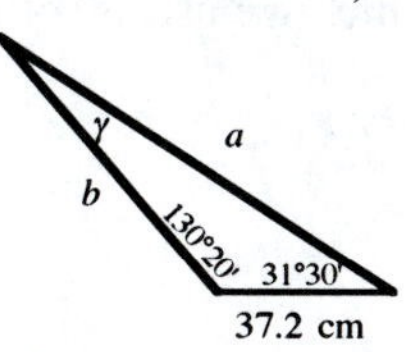

Solve for γ: $\alpha + \beta + \gamma = 180°$

$$\gamma = 180° - (130°20' + 31°30') = 18°10'$$

Solve for b: $\dfrac{\sin\beta}{b} = \dfrac{\sin\gamma}{c}$

$$b = \sin\beta\,\frac{c}{\sin\gamma} = (\sin 31°30')\,\frac{37.2\text{ cm}}{\sin 18°10'} = 62.3\text{ cm}$$

Solve for a: $\dfrac{\sin\alpha}{a} = \dfrac{\sin\gamma}{c}$

$$a = \sin\alpha\,\frac{c}{\sin\gamma} = (\sin 130°20')\,\frac{37.2\text{ cm}}{\sin 18°10'} = 91.0\text{ cm}$$

49.

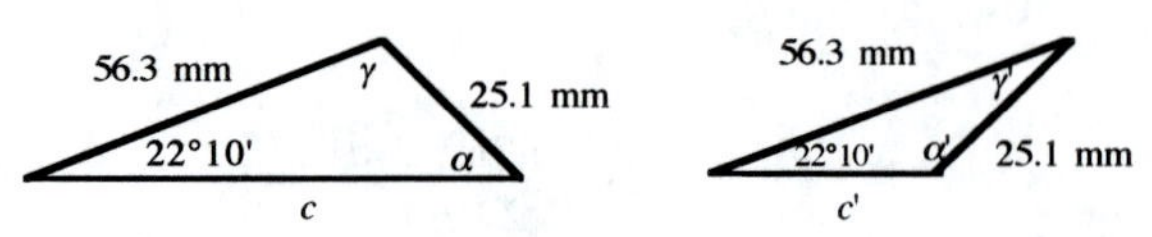

Solve for α: $\dfrac{\sin\alpha}{a} = \dfrac{\sin\beta}{b}$

$$\sin\alpha = \frac{a\sin\beta}{b} = \frac{(56.3\,\text{mm})(\sin 22°10')}{25.1\,\text{mm}} = 0.8463$$

Angle α can be either acute or obtuse.

$\alpha = \sin^{-1}(0.8463) = 57°50'$ $\quad$ $\alpha' = 180° - \sin^{-1}(0.8463) = 122°10'$

Solve for γ and γ':

$$\gamma = 180° - (\alpha + \beta) = 180° - (57°50' + 22°10') = 100°$$

$$\gamma' = 180° - (\alpha' + \beta) = 180° - (122°10' + 22°10') = 35°40'$$

Solve for c and c':

$$\frac{\sin\gamma}{c} = \frac{\sin\beta}{b} \qquad c = \frac{b\sin\gamma}{\sin\beta} = \frac{(25.1\,\text{mm})(\sin 100°)}{\sin 22°10'} = 65.5\text{ mm}$$

$$\frac{\sin\gamma'}{c'} = \frac{\sin\beta}{b} \qquad c' = \frac{b\sin\gamma'}{\sin\beta} = \frac{(25.1\,\text{mm})(\sin 35°40')}{\sin 22°10'} = 38.8\text{ mm}$$

51. Using the given and the calculated data, we have

$$(a - b)\cos\frac{\gamma}{2} = c\sin\frac{\alpha - \beta}{2}$$

$$(92 - 64)\cos\frac{36°}{2} = 55\sin\frac{101° - 43°}{2}$$

$$26.6 = 26.7$$

53. Using the law of sines in the form $\dfrac{a}{\sin\alpha} = \dfrac{b}{\sin\beta}$, we have $b = \dfrac{a\sin\beta}{\sin\alpha}$, which we use as follows:

$$\frac{a-b}{a+b} = \frac{a - \dfrac{a\sin\beta}{\sin\alpha}}{a + \dfrac{a\sin\beta}{\sin\alpha}} \qquad \text{Law of sines as above}$$

$$= \frac{a\sin\alpha - a\sin\beta}{a\sin\alpha + a\sin\beta} \qquad \text{Algebra}$$

$$= \frac{a(\sin\alpha - \sin\beta)}{a(\sin\alpha + \sin\beta)} \qquad \text{Algebra}$$

$$= \frac{\sin\alpha - \sin\beta}{\sin\alpha + \sin\beta} \qquad \text{Algebra}$$

$$= \frac{2\cos\frac{\alpha+\beta}{2}\sin\frac{\alpha-\beta}{2}}{2\sin\frac{\alpha+\beta}{2}\cos\frac{\alpha-\beta}{2}}$$ Sum-product identities

$$= \frac{\frac{\sin\frac{\alpha-\beta}{2}}{\cos\frac{\alpha-\beta}{2}}}{\frac{\sin\frac{\alpha+\beta}{2}}{\cos\frac{\alpha+\beta}{2}}}$$ Algebra

$$= \frac{\tan\frac{\alpha-\beta}{2}}{\tan\frac{\alpha+\beta}{2}}$$ Quotient identities

55. From the diagram, we can see that k = altitude of any possible triangle.
Thus, $\sin\beta = \frac{k}{a}$, $k = a\sin\beta$.
$k = (66.8 \text{ yd}) \sin 46.8° = 48.7$ yd
If $0 < b < k$, there is no solution: (1) in the diagram.
If $b = k$, there is one solution.
If $k < b < a$, there are two solutions: (2) in the diagram.

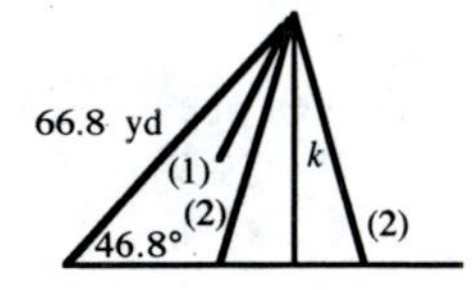

57. $\angle BAC + \angle ABC + \angle ACB = 180°$
$\angle ACB = 180° - (\angle BAC + \angle ABC) = 180° - (118.1° + 58.1°) = 3.8°$

Now apply the law of sines to find AC

$$\frac{\sin\angle ABC}{AC} = \frac{\sin\angle ACB}{AB}$$
$$AC = \frac{AB\sin\angle ABC}{\sin\angle ACB} = \frac{1.00\sin 58.1°}{\sin 3.8°} = 12.8 \text{ mi}$$

59. First draw a figure and label known information.
$\angle ABF + \angle BAF + \angle BFA = 180°$
$\angle BFA = 180° - (\angle ABF + \angle BAF)$
$= 180° - (52.6° + 25.3°) = 102.1°$

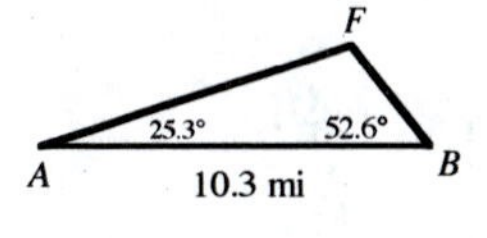

Now apply the law of sines to find AF and BF:

$$\frac{\sin\angle ABF}{AF} = \frac{\sin\angle BFA}{AB}$$
$$AF = \frac{AB\sin\angle ABF}{\sin\angle BFA} = \frac{(10.3\text{ mi})(\sin 52.6°)}{\sin 102.1°} = 8.37 \text{ mi}$$

$$\frac{\sin\angle BAF}{BF} = \frac{\sin\angle BFA}{AB}$$
$$BF = \frac{AB\sin\angle BAF}{\sin\angle BFA} = \frac{(10.3\text{ mi})(\sin 25.3°)}{\sin 102.1°} = 4.50 \text{ mi}$$

The fire is 8.37 mi from A, 4.50 mi from B.

61. Label known information in the figure.

$$\angle ABC + \angle CAB + \angle ACB = 180°$$
$$\angle ACB = 180° - (\angle ABC + \angle CAB)$$
$$= 180° - (19.2° + 118.4°) = 42.4°$$

C
112 m
118.4° 19.2°
A B

Now apply the law of sines to find AB.

$$\frac{\sin \angle ACB}{AB} = \frac{\sin \angle ABC}{AC}$$
$$AB = \frac{AC \sin \angle ACB}{\sin \angle ABC} = \frac{(112\,\text{m}) \sin 42.4°}{\sin 19.2°} = 230 \text{ m}$$

63. First draw a figure and label known information. Based on the given information, find angle L first.

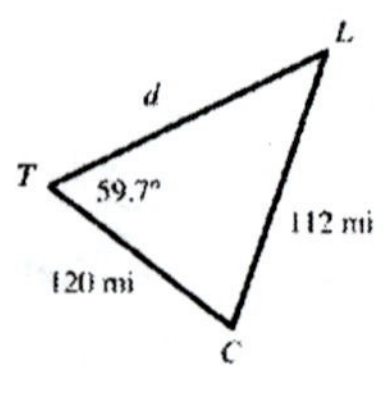

$$\frac{\sin 59.7°}{112} = \frac{\sin L}{120}$$
$$\sin L = \frac{120 \sin 59.7°}{112}$$
$$L = \sin^{-1}\left(\frac{120 \sin 59.7°}{112}\right)$$
$$= 67.7°$$

Now find angle C, then use the law of sines again to find d.

$$C = 180° - (67.7° - 59.7°) = 52.6°$$
$$\frac{\sin 59.7°}{112} = \frac{\sin 52.6°}{d}$$
$$d \sin 59.7° = 112 \sin 52.6°$$
$$d = \frac{112 \sin 52.6°}{\sin 59.7°}$$
$$= 103 \text{ mi}$$

65. In the figure, note: Triangle TBC_2 is a right triangle, hence, $\frac{h}{x} = \sin 43.5°$. Triangle TC_1C_2 is not a right triangle; but, from the law of sines,

$$\frac{\sin \alpha}{C_1C_2} = \frac{\sin \beta}{x}.$$

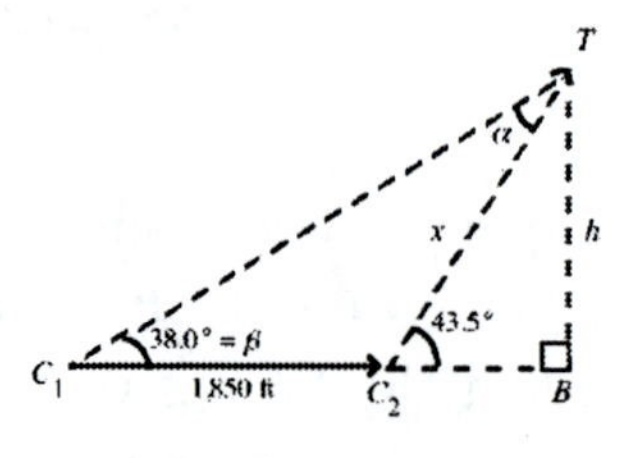

Hence, $x = \dfrac{C_1C_2 \sin \beta}{\sin \alpha}$

$$h = x \sin 43.5° = \frac{C_1C_2 \sin \beta \sin 43.5°}{\sin \alpha}$$

Given $C_1C_2 = 1{,}850$ ft, $\beta = 38.0°$, we can find α since the exterior angle of a triangle has measure equal to the sum of the two nonadjacent interior angles.

Hence, $\alpha + \beta = 43.5°$, $\alpha = 43.5° - \beta = 43.5° - 38.0° = 5.5°$.

Hence, $h = \dfrac{(1{,}850\,\text{ft})(\sin 38.0°)(\sin 43.5°)}{\sin 5.5°} = 8{,}180$ ft

67. In the figure, note: Triangle ADC is a right triangle, hence $\angle ACB = 90° - 62.0° = 28.0°$. Triangle ABC is not a right triangle; but, from the law of sines,

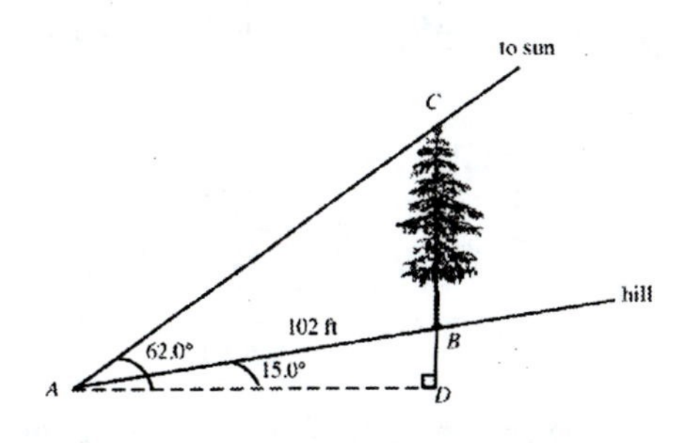

$$\frac{\sin\angle CAB}{BC} = \frac{\sin\angle ACB}{AB}$$

$$\angle CAB = \angle CAD - \angle BAD = 62.0° - 15.0° = 47.0°$$

$$BC = \frac{AB\sin\angle CAB}{\sin\angle ACB} = \frac{(102\text{ ft})\sin 47.0°}{\sin 28.0°} = 159\text{ ft}$$

69. Labeling the figure as shown, we have, from the law of sines,

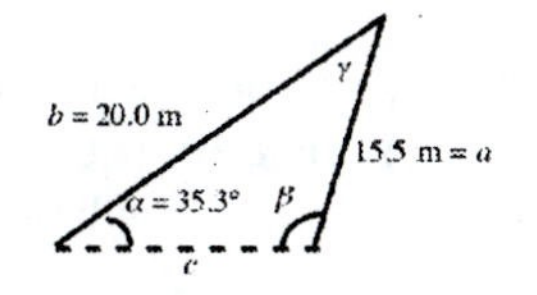

$$\frac{\sin\beta}{b} = \frac{\sin\alpha}{a}$$

$$\sin\beta = \frac{b\sin\alpha}{a} = \frac{(20.0\text{ m})(\sin 35.3°)}{15.5\text{ m}} = 0.7456$$

$$\beta = 180° - \sin^{-1}(0.7456) \quad (\text{since } \beta \text{ is obtuse})$$
$$= 131.8°$$

Since $\gamma = 180° - (\alpha + \beta) = 180° - (35.3° + 131.8°) = 12.9°$, we have, from the law of sines,

$$\frac{\sin\gamma}{c} = \frac{\sin\alpha}{a}$$

$$c = \frac{a\sin\gamma}{\sin\alpha} = \frac{(15.5\text{ m})(\sin 12.9°)}{\sin 35.3°} = 5.99\text{ m}$$

71. Labeling the diagram as shown, we note: triangle OAB is isosceles, hence $\alpha = \beta$. Thus,

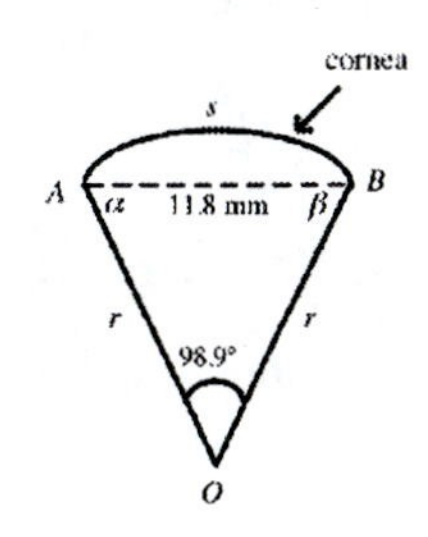

$$\alpha + \alpha + 98.9° = 180°$$
$$2\alpha = 81.1°$$
$$\alpha = 40.55°$$

Now apply the law of sines to find r.

$$\frac{\sin\alpha}{r} = \frac{\sin\angle AOB}{AB}$$

$$r = \frac{AB\sin\alpha}{\sin\angle AOB} = \frac{(11.8\text{ mm})(\sin 40.55°)}{\sin 98.9°} = 7.76\text{ mm}$$

To find s, we use the formula $s = \frac{\pi}{180°}\theta r$ from Chapter 2, thus

$$s = \frac{\pi}{180°}(98.9°)(7.76\text{ mm}) = 13.4\text{ mm}$$

73. Following the hint, we find all angles for triangle ACS first.

$$\angle SAC = \theta + 90° = 26.2° + 90° = 116.2°$$

For $\angle ACS$, we use the formula $s = \frac{\pi}{180°}\theta R$ from Chapter 2, with $s = 632$ miles, $\theta = \angle ACS$, $R = 3{,}964$ miles. Then,

$$\angle ACS = \frac{180° s}{\pi R} = \frac{180°(632\text{ mi})}{\pi(3{,}964\text{ mi})} = 9.13°$$

Since $\angle SAC + \angle ACS + \angle ASC = 180°$, we have $\angle ASC = 180° - (\angle ACS + \angle SAC)$
$= 180° - (116.2° + 9.13°) = 54.67°$

Now apply the law of sines to find side CS:

$$\frac{\sin \angle SAC}{CS} = \frac{\sin \angle ASC}{AC}$$

$$CS = \frac{AC \sin \angle SAC}{\sin \angle ASC} = \frac{(3{,}964 \text{ mi}) \sin 116.2°}{\sin 54.67°} = 4{,}360 \text{ miles}$$

Hence, the height of the satellite above $B = BS = CS - BC$.
$BS = 4{,}360 \text{ mi} - 3{,}964 \text{ mi} = 396 \text{ miles}$

75. We know from the hint that α is maximum when the line EV (see figure) is tangent to the orbit of Venus. We know from geometry that triangle EVS will then be a right triangle. Hence

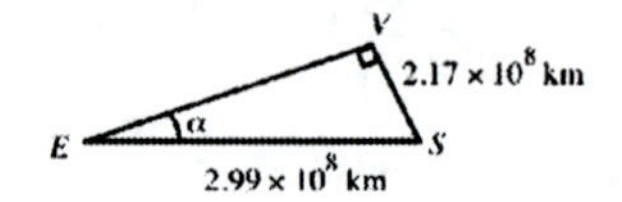

$$\sin \alpha = \frac{VS}{ES} = \frac{2.17 \times 10^8}{2.99 \times 10^8}$$

$$\alpha = \sin^{-1}\left(\frac{2.17}{2.99}\right) = 46°30'$$

77. In the diagram, we are to calculate c based on the given information. There are two possible triangles, but the requirement that the distance c be as short as possible leads to the choice of α obtuse.

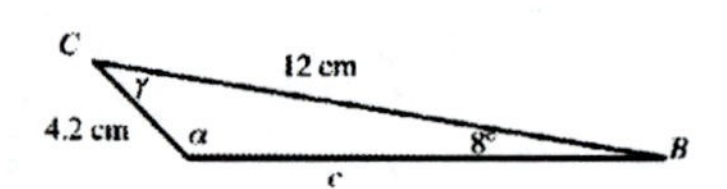

Solve for α: $\dfrac{\sin \alpha}{a} = \dfrac{\sin \beta}{b}$

$$\sin \alpha = \frac{a \sin \beta}{b} = \frac{(12 \text{ cm})(\sin 8°)}{4.2 \text{ cm}} = 0.3976$$

$$\alpha = 180° - \sin^{-1} 0.397 = 156.6°$$

Solve for γ: $\alpha + \beta + \gamma = 180°$ $\quad\gamma = 180° - (8° + 156.6°) = 15.4°$

Solve for c: $\dfrac{\sin \beta}{b} = \dfrac{\sin \gamma}{c}$ $\quad c = \dfrac{b \sin \gamma}{\sin \beta} = \dfrac{(4.2 \text{ cm})(\sin 15.4°)}{\sin 8°} = 8.0 \text{ cm}$

EXERCISE 6.2 Law of Cosines

1. Two sides and the included angle are given.

3. If all three sides are known, the law of cosines can be used to find any angle. If two sides are known, then the third side can be found by using the law of cosines; however, the law then requires knowing the angle between the two sides; any other angle does not fit into the equations provided by the law of cosines.

5. Three sides are given, the *SSS* case. Use the law of cosines.

7. Two sides and a non-included angle are given, the *SSA* case. Use the law of sines.

9. Two sides and the included angle are given, the *SAS* case. Use the law of cosines.

11. Two angles and a non-included side are given, the *AAS* case. Use the law of sines.

13. A triangle can have at most one obtuse angle. Since β is acute, then, if the triangle has an obtuse angle it must be the angle opposite the longer of the two sides, a and c. Thus, γ, the angle opposite the shorter of the two sides, c, must be acute.

15. *Solve for a:* We use the law of cosines.

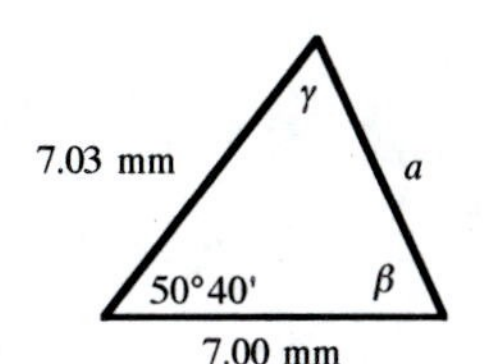

$$\begin{aligned} a^2 &= b^2 + c^2 - 2bc \cos \alpha \\ &= (7.03)^2 + (7.00)^2 - 2(7.03)(7.00) \cos 50°40' \\ &= 36.039253\ldots \\ a &= 6.00 \text{ mm} \end{aligned}$$

Since c is the shorter of the remaining sides, γ, the angle opposite c, must be acute.

Solve for γ: We use the law of sines.

$$\frac{\sin\alpha}{a} = \frac{\sin\gamma}{c}$$

$$\sin \gamma = \frac{c\sin\alpha}{a}$$

$$\gamma = \sin^{-1} \frac{c\sin\alpha}{a} = \sin^{-1} \frac{(7.00\,\text{mm})(\sin 50°40')}{6.00\,\text{m}} = 64°20'$$

Solve for β: $\beta = 180° - (\alpha + \gamma) = 180° - (50°40' + 64°20') = 65°0'$

17. *Solve for c:* We use the law of cosines.

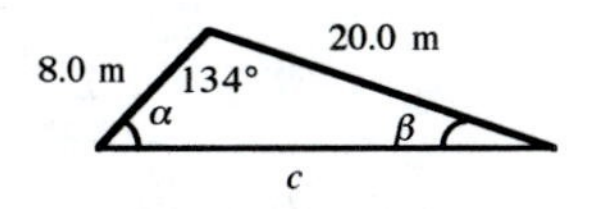

$$\begin{aligned} c^2 &= a^2 + b^2 - 2ab \cos \gamma \\ &= (20.0)^2 + (8.00)^2 - 2(20.0)(8.00) \cos 134° \\ &= 686.29068\ldots \\ c &= 26.2 \text{ m} \end{aligned}$$

Since b is the shorter of the remaining sides, β, the angle opposite b, must be acute.

Solve for β: We use the law of sines. $\dfrac{\sin\beta}{b} = \dfrac{\sin\gamma}{c}$

$$\sin \beta = \frac{b\sin\gamma}{c}$$

$$\beta = \sin^{-1} \frac{b\sin\gamma}{c} = \sin^{-1} \frac{(8.00\,\text{m})(\sin 134.0°)}{26.2\,\text{m}} = 12.7°$$

Solve for α: $\alpha = 180° - (\beta + \gamma) = 180° - (134.0° + 12.7°) = 33.3°$

19. If the triangle has an obtuse angle, then it must be the angle opposite the longest side; in this case, α.

21. Find the measure of the angle opposite the longest side first, using the law of cosines. In this problem this is angle γ.

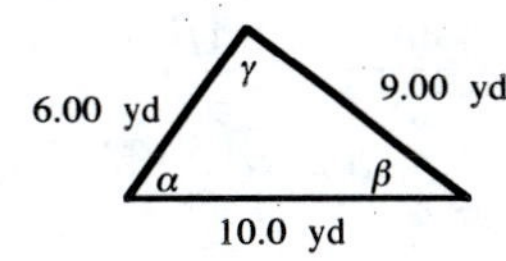

$$c^2 = a^2 + b^2 - 2ab \cos \gamma$$

$$\cos \gamma = \frac{a^2 + b^2 - c^2}{2ab}$$

$$\begin{aligned} \gamma &= \cos^{-1} \frac{a^2 + b^2 - c^2}{2ab} \\ &= \cos^{-1} \frac{(9.00)^2 + (6.00)^2 - (10.0)^2}{2(9.00)(6.00)} \\ &= 80.9° \end{aligned}$$

Solve for β: Both α and β must be acute, since they are smaller than γ. We choose to solve for β using the law of sines.

$$\frac{\sin\beta}{b} = \frac{\sin\gamma}{c} \qquad \sin\beta = \frac{b\sin\gamma}{c}$$

$$\beta = \sin^{-1}\frac{b\sin\gamma}{c}$$
$$= \sin^{-1}\frac{6.00\sin 80.9°}{10.0}$$
$$= 36.3°$$

Solve for α: $\alpha = 180° - (\beta + \gamma) = 180° - (36.3° + 80.9°) = 62.8°$

23. Find the measure of the angle opposite the longest side first, using the law of cosines. In this problem this is angle γ.

$$c^2 = a^2 + b^2 - 2ab\cos\gamma$$

$$\cos\gamma = \frac{a^2 + b^2 - c^2}{2ab}$$

$$\gamma = \cos^{-1}\frac{a^2 + b^2 - c^2}{2ab}$$
$$= \cos^{-1}\frac{(420.0)^2 + (770.0)^2 - (860.0)^2}{2(420.0)(770.0)} = 87°22'$$

Solve for β: Both α and β must be acute, since they are smaller than γ. We choose to solve for β using the law of sines.

$$\frac{\sin\beta}{b} = \frac{\sin\gamma}{c} \qquad \sin\beta = \frac{b\sin\gamma}{c}$$

$$\beta = \sin^{-1}\frac{b\sin\gamma}{c}$$
$$= \sin^{-1}\frac{770.0\sin 87°22'}{860.0} = 63°26'$$

Solve for α: $\alpha = 180° - (\beta + \gamma) = 180° - (63°26' + 87°22') = 29°12'$

25. We are given two angles and a non-included side (*AAS*).

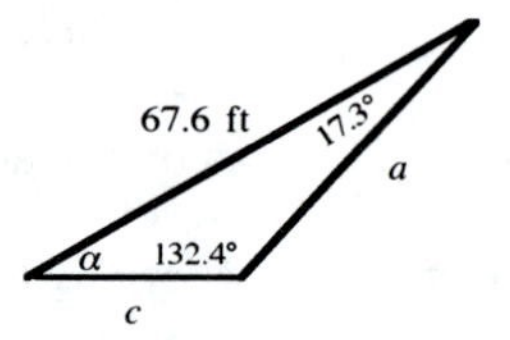

Solve for α: $\alpha + \beta + \gamma = 180°$

$$\alpha = 180° - (17.3° + 132.4°) = 30.3°$$

Now use the law of sines to find the remaining sides.

Solve for a: $\dfrac{\sin\alpha}{a} = \dfrac{\sin\beta}{b}$

$$a = \frac{b\sin\alpha}{\sin\beta} = \frac{(67.6\text{ ft})(\sin 30.3°)}{\sin 132.4°} = 46.2\text{ ft}$$

Solve for c: $\dfrac{\sin\beta}{b} = \dfrac{\sin\gamma}{c} \qquad c = \dfrac{b\sin\gamma}{\sin\beta} = \dfrac{(67.6\text{ ft})(\sin 17.3°)}{\sin 132.4°} = 27.2\text{ ft}$

27. We are given two sides and the included angle (*SAS*). We use the law of cosines to find the third side, then the law of sines to find a second angle.

Solve for b: $b^2 = a^2 + c^2 - 2ac\cos\beta$

$= (13.7)^2 + (20.1)^2 - 2(13.7)(20.1)\ \cos 66.5°$

$= 372.09294\ldots$

$b = 19.3$ m

Since a is the shorter of the remaining sides, α, the angle opposite a, must be acute.

Solve for α: $\dfrac{\sin\alpha}{a} = \dfrac{\sin\beta}{b}$

$\sin\alpha = \dfrac{a\sin\beta}{b} = \dfrac{(13.7\,\text{m})(\sin 66.5°)}{19.3\,\text{m}} = 0.6513$

$\alpha = \sin^{-1} 0.6513 = 40.6°$

Solve for γ: $\alpha + \beta + \gamma = 180°$

$\gamma = 180° - (66.5° + 40.6°) = 72.9°$

29. It is impossible to draw or form a triangle with this data, since angles $\beta + \gamma$ together add up to more than 180°. No solution.

31. We are given three sides (*SSS*). We solve for the largest angle, α (largest because it is opposite the largest side, a) using the law of cosines. We then solve for a second angle using the law of sines, because it involves simpler calculations.

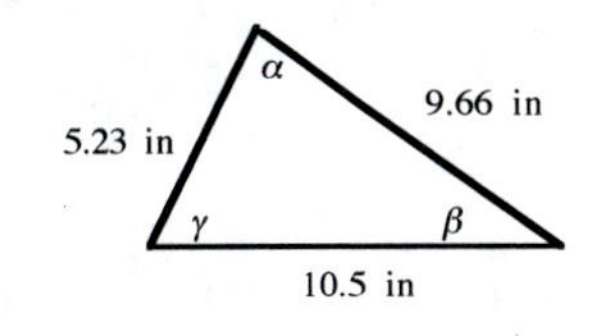

Solve for α: $a^2 = b^2 + c^2 - 2bc\cos\alpha$

$\cos\alpha = \dfrac{b^2 + c^2 - a^2}{2bc}$

$\alpha = \cos^{-1} \dfrac{(5.23)^2 + (9.66)^2 - (10.5)^2}{2(5.23)(9.66)}$

$= \cos^{-1} 0.1031 = 84.1°$

Both β and γ must be acute, since they are smaller than α.

Solve for β: $\dfrac{\sin\alpha}{a} = \dfrac{\sin\beta}{b}$

$\sin\beta = \dfrac{b\sin\alpha}{a} = \dfrac{(5.23\,\text{in})(\sin 84.1°)}{10.5\,\text{in}} = 0.4954$

$\beta = \sin^{-1} 0.4954 = 29.7°$

Solve for γ: $\gamma = 180° + (\alpha + \beta) = 180° - (84.1° + 29.7°) = 66.2°$

33. We are given two sides and a non-included angle (*SSA*).

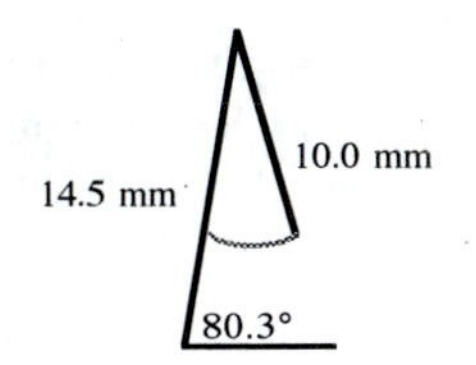

Solve for α: $\dfrac{\sin\alpha}{a} = \dfrac{\sin\gamma}{c}$

$\sin\alpha = \dfrac{a\sin\gamma}{c} = \dfrac{(14.5\,\text{mm})(\sin 80.3°)}{10.0\,\text{mm}} = 1.429$

Since $\sin\alpha = 1.429$ has no solution, no triangle exists with the given measurements. No solution.

35. We are given two angles and the included side (*ASA*). We use the law of sines.

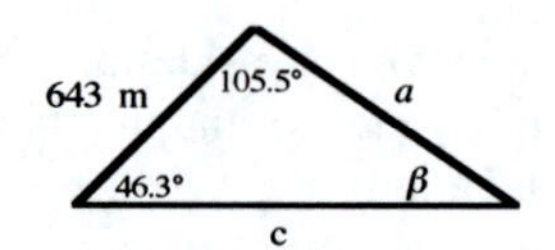

Solve for β: $\alpha + \beta + \gamma = 180°$

$\beta = 180° - (46.3° + 105.5) = 28.2°$

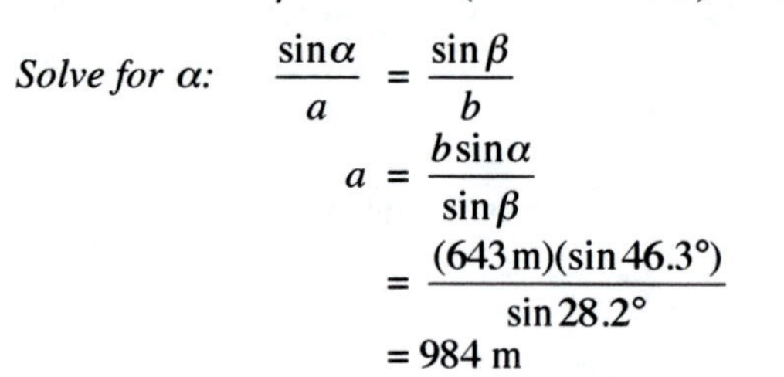

Solve for α:

$$\frac{\sin\alpha}{a} = \frac{\sin\beta}{b}$$
$$a = \frac{b\sin\alpha}{\sin\beta} = \frac{(643\text{ m})(\sin 46.3°)}{\sin 28.2°} = 984\text{ m}$$

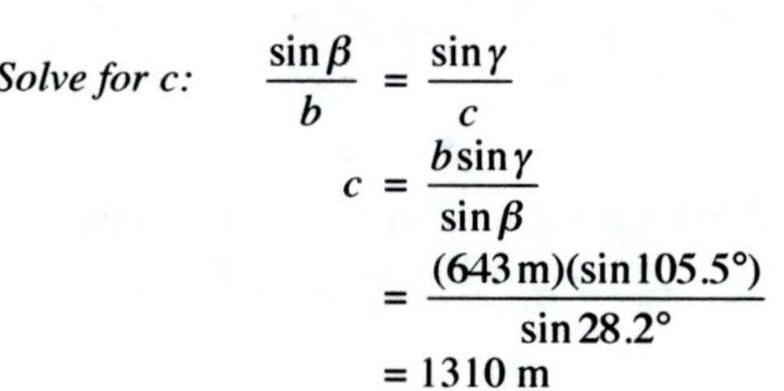

Solve for c:

$$\frac{\sin\beta}{b} = \frac{\sin\gamma}{c}$$
$$c = \frac{b\sin\gamma}{\sin\beta} = \frac{(643\text{ m})(\sin 105.5°)}{\sin 28.2°} = 1310\text{ m}$$

37. It is impossible to form a triangle with this data, since the triangle inequality ($a + b > c$) is not satisfied. No solution.

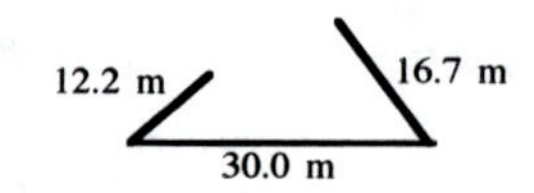

39. We are given two sides and a non-included angle (*SSA*). Two triangles are possible. There are two possible values for β. We use the law of sines.

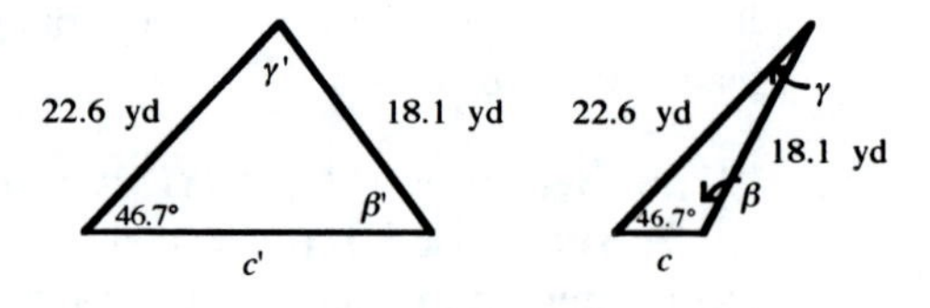

Solve for β:

$$\frac{\sin\alpha}{a} = \frac{\sin\beta}{b}$$
$$\sin\beta = \frac{b\sin\alpha}{a} = \frac{(22.6\text{ yd})(\sin 46.7°)}{18.1\text{ yd}} = 0.9087$$

Angle β can be either obtuse or acute.

$\beta = 180° - \sin^{-1} 0.9087 = 114.7°$ $\qquad$ $\beta' = \sin^{-1} 0.9087 = 65.3°$

Solve for γ and γ':

$\gamma = 180° - (\alpha + \beta) = 180° - (46.7° + 114.7°) = 18.6°$

$\gamma' = 180° - (\alpha' + \beta') = 180° - (46.7° + 65.3°) = 68.0°$

Solve for c and c':

$$\frac{\sin\alpha}{a} = \frac{\sin\gamma}{c}$$
$$c = \frac{a\sin\gamma}{\sin\alpha} = \frac{(18.1\text{ yd})(\sin 18.6°)}{\sin 46.7°} = 7.93\text{ yd}$$

$$\frac{\sin\alpha'}{a'} = \frac{\sin\gamma'}{c'}$$
$$c' = \frac{a'\sin\gamma'}{\sin\alpha'} = \frac{(18.1\text{ yd})(\sin 68.0°)}{\sin 46.7°} = 23.1\text{ yd}$$

41. We are given three angles (*AAA*). An infinite number of triangles, all similar, can be drawn from the given values, but no one triangle is determined. No solution.

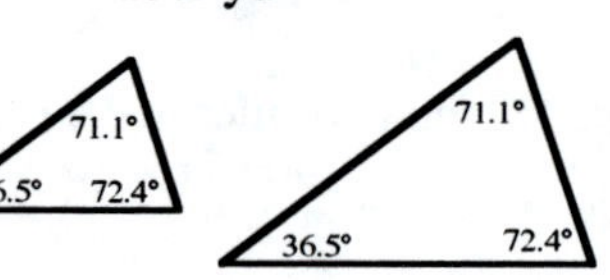

43. The law of cosines states that $b^2 = c^2 + a^2 - 2ac\cos\beta$ for any triangle.
If $\beta = 90°$, $\cos\beta = 0$; hence, $b^2 = c^2 + a^2 - 2ac\cos(90°) = c^2 + a^2 - 0 = c^2 + a^2$

45. Using the given and the calculated data, we have:

$$(a-b)\cos\frac{\gamma}{2} = c\sin\frac{(\alpha-\beta)}{2}$$

$$(6.00-7.03)\cos\frac{64°20'}{2} = 7.00\sin\frac{(50°40'-65°0')}{2}$$

$$-0.872 \approx -0.873$$

47. We can write the law of cosines two different ways as follows:

(1) $a^2 = b^2 + c^2 - 2bc\cos\alpha$

(2) $b^2 = a^2 + c^2 - 2ac\cos\beta$

Adding (1) and (2), we have

$$a^2 + b^2 = a^2 + b^2 + 2c^2 - 2bc\cos\alpha - 2ac\cos\beta$$

$$0 = 2c^2 - 2bc\cos\alpha - 2a\cos\beta$$

$$-2c^2 = -2bc\cos\alpha - 2ac\cos\beta$$

Dividing both sides by $-2c$ (which is never 0), we obtain $c = b\cos\alpha + a\cos\beta$.

49. From the given data, using the law of cosines

$$a^2 = b^2 + c^2 - 2bc\cos\alpha$$

$$x^2 = x^2 + (1.5x)^2 - 2x(1.5x)\cos\alpha$$

$$x^2 = 3.25x^2 - 3x^2\cos\alpha$$

$$\cos\alpha = \frac{x^2 - 3.25x^2}{-3x^2} = \frac{2.25}{3} = 0.75$$

$$\alpha = \cos^{-1}(0.75) = 41.4°$$

51. We are given two sides and an included angle. We use the law of cosines to find side BC.

$$BC^2 = AB^2 + AC^2 - 2(AB)(AC)\cos\angle CAB = 425^2 + 384^2 - 2(425)(384)\cos 98.3°$$

$$= 375{,}198.864\ldots$$

$$BC = 613 \text{ m}$$

53. In triangle OAB, we are given $OA = OB = 8.26$ cm and $AB = 13.8$ cm. From the law of cosines, we can determine the central angle AOB.

$$\cos\angle AOB = \frac{OA^2 + OB^2 - AB^2}{2(OA)(OB)}$$

$$\angle AOB = \cos^{-1}\frac{OA^2 + OB^2 - AB^2}{2(OA)(OB)}$$

$$= \cos^{-1}\frac{(8.26)^2 + (8.26)^2 - (13.8)^2}{2(8.26)(8.26)} = \cos^{-1}(-0.3956) = 113.3°$$

55. First, complete and label the figure. From the given information, we know:

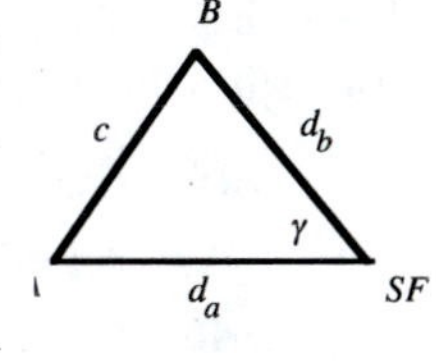

γ = angle between west and northwest = 45°

d_a = (rate of plane A)(time of plane A) = (250 km/hr)(1 hr) = 250 km

d_b = (rate of plane B)(time of plane B) = (210 km/hr)(1 hr) = 210 km

Hence, from the law of cosines,

$$c^2 = d_a^2 + d_b^2 - 2d_ad_b\cos\gamma = (250)^2 + (210)^2 - 2(250)(210)\cos 45° = 32353.78798\ldots$$

$$c = 180 \text{ km}$$

57. First, find how far the ship has traveled after 6 hours at 26 miles per hour, that is $26\frac{\text{mi}}{\text{hr}} \cdot 6\text{ hr} = 156\text{ mi}$. Now draw a diagram.

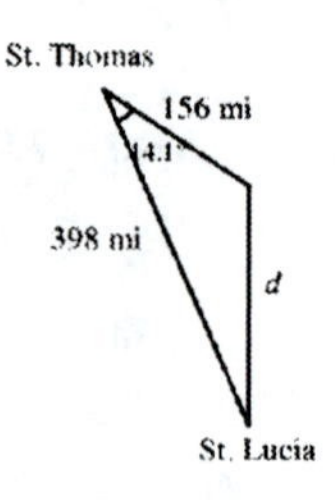

Then find d from the law of cosines.

$$d^2 = (156)^2 + (398)^2 - 2(156)(398)\cos 14.1°$$
$$d = \sqrt{(156)^2 + (398)^2 - 2(156)(398)\cos 14.1°}$$
$$= 249.6\text{ mi}$$

At 26 miles per hour, the trip will take $249.6\text{ mi} \div 26\frac{\text{mi}}{\text{hr}} = 9.6\text{ hr}$.

59.

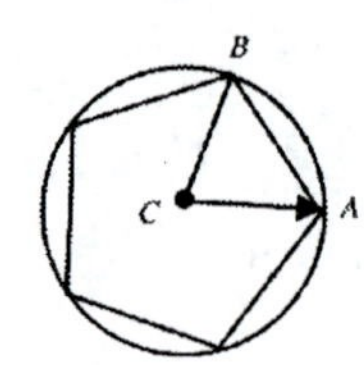

Labeling the figure in the text, we have: $AC = 5$ cm, $BC = 5$ cm, $\angle BCA = \frac{1}{5}(360°) = 72°$. Therefore, from the law of cosines

$$AB^2 = 5^2 + 5^2 - 2 \cdot 5 \cdot 5\cos 72°$$
$$AB = \sqrt{5^2 + 5^2 - 2 \cdot 5 \cdot 5\cos 72°}$$
$$= 5.878\text{ cm}$$

The perimeter of the pentagon = $5(AB) = 5(5.878\text{ cm}) = 29$ cm.

61. Label the text figure as follows:

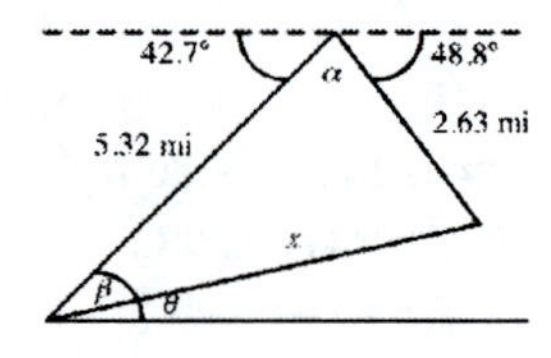

(A) Then $42.7° + \alpha + 48.8° = 180°$

$$\alpha = 180° - (42.7° + 48.8°) = 88.5°$$

Therefore, from the law of cosines

$$x^2 = 5.32^2 + 2.63^2 - 2(5.32)(2.63)\cos 88.5°$$
$$x = \sqrt{5.32^2 + 2.63^2 - 2(5.32)(2.63)\cos 88.5°}$$

$x = 5.87$ mi is the length of the tunnel

(B) Apply the law of sines to find β.

$$\frac{\sin\beta}{2.63} = \frac{\sin 88.5°}{5.87}$$
$$\beta = \sin^{-1}\left(2.63\frac{\sin 88.5°}{5.87}\right)$$
$$\beta = 26.6°$$

Since $\beta + \theta = 42.7°$, $\theta = 42.7° - \beta = 42.7° - 26.6° = 16.1°$ is the angle the tunnel makes with the horizontal.

63. In triangle ACD, angle $ADC = 11° + 90° = 101°$ (Why?) $DC = 12.0$ ft and $AD = 18.0$ ft. We know two sides and the included angle; hence we can apply the law of cosines to find side AC.

$$AC^2 = AD^2 + DC^2 - 2(AD)(DC)\cos\angle ADC$$
$$= (18.0)^2 + (12.0)^2 - 2(18.0)(12.0)\cos 101° = 550.429\ldots$$
$$AC = 23.5\text{ ft}$$

To find side AB, we need to find angle ADB, then apply the law of cosines again, in triangle ADB. Since angle ADB + angle $BDC = 101°$; angle $ADB = 101°$ – angle BDC. By symmetry, angle BDC = angle ACD of triangle ACD, which we can find from the law of sines. Thus,

$$\frac{\sin\angle ACD}{AD} = \frac{\sin\angle ADC}{AC} \qquad \sin\angle ACD = \frac{(AD)(\sin\angle ADC)}{AC}$$

$$\text{angle } ACD = \sin^{-1}\frac{(AD)(\sin\angle ADC)}{AC} = \sin^{-1}\frac{18.0\sin 101°}{23.5} = 48.86°$$

Hence, angle $ADB = 101° - 48.87° = 52.14°$. Then, applying the law of cosines to triangle ADB, we have ($AC = BD$ by symmetry.)

$$AB^2 = AD^2 + BD^2 - 2(AD)(BD)\cos\angle ADB$$
$$= (18.0)^2 + (23.5)^2 - 2(18.0)(23.5)\cos 52.13° = 356.04\ldots$$
$$AB = 18.9 \text{ ft}$$

65. Labeling the figure in the text, we have:

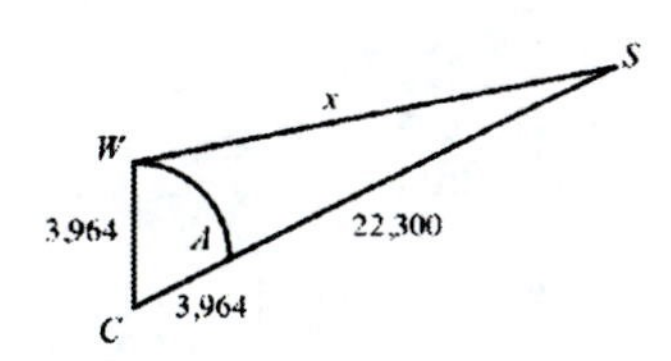

In triangle CWS, $CW = 3{,}964$ mi, $CS = CA + AS = 3{,}964 + 22{,}300 = 26{,}264$ mi, $\angle WCS = 65°$. Applying the law of cosines, we have:

$$x^2 = (3{,}694)^2 + (26{,}264)^2 - 2(3{,}964)(26{,}264)\cos 65°$$
$$x = \sqrt{(3{,}964)^2 + (26{,}264)^2 - 2(3{,}964)(26{,}264)\cos 65°}$$
$x = 24{,}800$ mi from the satellite to White Sands

67. The three sides of triangle ABC are each in turn the hypotenuse of a right triangle formed with two edges of the solid. Hence, by the Pythagorean theorem.

$$AB^2 = 6.0^2 + 3.0^2 = 45.00 \qquad AB = 6.7 \text{ cm}$$
$$AC^2 = 6.0^2 + 4.0^2 = 52.00 \qquad AC = 7.2 \text{ cm}$$
$$BC^2 = 3.0^2 + 4.0^2 = 25.00 \qquad BC = 5.0 \text{ cm}$$

To find $\angle ACB$, we apply the law of cosines.

$$AB^2 = AC^2 + BC^2 - 2(AC)(BC)\cos\angle ACB \qquad \cos\angle ACB = \frac{AC^2 + BC^2 - AB^2}{2(AC)(BC)}$$

$$\angle ABC = \cos^{-1}\frac{AC^2 + BC^2 - AB^2}{2(AC)(BC)} = \cos^{-1}\frac{7.2^2 + 5.0^2 - 6.7^2}{2(7.2)(5.0)} = 64°$$

69. We will first find BD by applying the law of sines to triangle ABD, in which we know all angles and a side. We then will find CD by applying the law of sines to triangle ACD, in which we also know all angles and a side. We can then apply the law of cosines to triangle BCD to find BC. In triangle ABD:

$$\frac{\sin\angle DAB}{BD} = \frac{\sin\angle DBA}{AD} \qquad BD = \frac{AD\sin\angle DAB}{\sin\angle DBA} = \frac{320\sin(87° + 15°)}{\sin 37°} = 520 \text{ m}$$

In triangle ACD:

$$\frac{\sin\angle DAC}{CD} = \frac{\sin\angle DCA}{AD} \qquad CD = \frac{AD\sin\angle DAC}{\sin\angle DCA} = \frac{320\sin 87°}{\sin 36°} = 543.7 \text{ m}$$

In triangle BCD:

$$BC^2 = BD^2 + CD^2 - 2(BD)(CD)\cos\angle BDC$$
$$= (520)^2 + (543.7)^2 - 2(520)(543.7)\cos 16°$$
$$BC = \sqrt{(520)^2 + (543.7)^2 - 2(520)(543.7)\cos 16°}$$
$$= 150 \text{ m}$$

EXERCISE 6.3 Areas of Triangles

1. The formula $A = \frac{1}{2}ab\sin\theta$ gives the area of the triangle directly upon substituting the given information.

3. The area cannot be found directly from this information. The law of sines can be used to find a second angle (there may be two possible solutions; this is the ambiguous case) and a third angle. This third angle is the included angle for the given two sides, so now the formula $A = \frac{1}{2}ab\sin\theta$ can be applied.

5. The base and the height of the triangle are given; hence, we use the formula
$$A = \frac{1}{2}bh = \frac{1}{2}(17.0\text{ m})(12.0\text{ m}) = 102\text{ m}^2$$

7. The base and the height of the triangle are given; hence, we use the formula
$$A = \frac{1}{2}bh = \frac{1}{2}(3\sqrt{5})\left(\frac{\sqrt{5}}{2}\right) = \frac{15}{4}\text{m}^2$$

9. The given information consists of two sides and the included angle; hence, we use the formula
$$A = \frac{ab}{2}\sin\theta \text{ in the form } A = \frac{1}{2}bc\sin\alpha = \frac{1}{2}(6.0\text{ cm})(8.0\text{ cm})\sin 30° = 12\text{ cm}^2$$

11. The given information consists of three sides; hence, we use Heron's formula. First, find the semiperimeter s:
$$s = \frac{a+b+c}{2} = \frac{4.00+6.00+8.00}{2} = 9.00\text{ in.}$$

Then,
$$s - a = 9.00 - 4.00 = 5.00$$
$$s - b = 9.00 - 6.00 = 3.00$$
$$s - c = 9.00 - 8.00 = 1.00$$

Thus,
$$A = \sqrt{s(s-a)(s-b)(s-c)} = \sqrt{(9.00)(5.00)(3.00)(1.00)} = \sqrt{135} = 11.6\text{ in}^2$$

13. The given information consists of two sides and the included angle; hence, we use the formula
$$A = \frac{ab}{2}\sin\theta \text{ in the form } A = \frac{1}{2}bc\sin\alpha = \frac{1}{2}(403)(512)\sin 23°20' = 40{,}900\text{ ft}^2$$

15. The given information consists of two sides and the included angle; hence, we use the formula
$A = \frac{ab}{2}\sin\theta$ in the form $A = \frac{1}{2}bc\sin\alpha$. α is obtuse; this does not alter the use of the formula
$$= \frac{1}{2}(12.1)(10.2)\sin 132.67° = 45.4\text{ cm}^2$$

17. The given information consists of three sides; hence, we use Heron's formula. First, find the semiperimeter s:

$$s = \frac{a+b+c}{2} = \frac{12.7+20.3+24.4}{2} = 28.7 \text{ m}$$

Then,
$$s-a = 28.7 - 12.7 = 16.0$$
$$s-b = 28.7 - 20.3 = 8.4$$
$$s-c = 28.7 - 24.4 = 4.3$$

Thus,
$$A = \sqrt{s(s-a)(s-b)(s-c)} = \sqrt{(28.7)(16.0)(8.4)(4.3)} = 129 \text{ m}^2$$

19. First note that since $\alpha + \beta + \gamma = 180°$, $\gamma = 180° - (\alpha + \beta) = 180° - (15° + 35°) = 130°$. Then from the law of sines side a can be determined

$$\frac{\sin\alpha}{a} = \frac{\sin\gamma}{c}$$

$$a = \sin\alpha \frac{c}{\sin\gamma}$$

$$= (\sin 15°)\frac{4.5\text{ft}}{\sin 130°}$$

$$= 1.5 \text{ ft}$$

β is the included angle for sides a and c, so

$$A = \frac{ac}{2}\sin\beta = \frac{1}{2}(1.5)(4.5)\sin 35°$$

$$= 2.0 \text{ ft}^2$$

21. By the law of sines

$$\frac{\sin\beta}{b} = \frac{\sin\alpha}{a}$$

$$\sin\beta = \frac{b\sin\alpha}{a}$$

$$= \frac{29\sin 72°}{38} = 0.7258$$

Therefore, $\beta = 47°$ and the third angle α is given by

$$\gamma = 180° - (\alpha + \beta) = 180° - (72° + 47°) = 61°$$

γ is the included angle for sides a and b, so

$$A = \frac{ab}{2}\sin\gamma = \frac{1}{2}(38)(29)\sin 61°$$

$$= 480 \text{ in}^2$$

23. By the law of sines

$$\frac{\sin\alpha}{a} = \frac{\sin\beta}{b}$$

$$\sin\alpha = \frac{a\sin\beta}{b}$$

$$= \frac{3.2\sin 175°}{4.5} = 0.0620$$

Therefore $\alpha = 3.6°$ and the third angle γ is given by

$$\gamma = 180° - (\alpha + \beta) = 180° - (3.6° + 175°) = 1.4°$$

γ is the included angle for sides a and b, so

$$A = \frac{ab}{2}\sin\gamma = \frac{1}{2}(3.2)(4.5)\sin 1.4° = 0.18 \text{ cm}^2$$

25. This statement is true because by the SSS axiom of geometry, triangles with the same side lengths are congruent.

27. This statement is false. Consider for example a 3-4-5 right triangle. It has semiperimeter $\frac{1}{2}(3 + 4 + 5) = 6$ units and area $\frac{1}{2} \cdot 3 \cdot 4 = 6$ square units. Now consider an isosceles triangle with base 2 units and equal sides 5 and 5 units. It also has semiperimeter 6 units, but since its altitude forming a perpendicular bisector of the base has length $\sqrt{5^2 - 1^2} = \sqrt{24}$, it has area $\frac{1}{2} \cdot \sqrt{24} \cdot 2 = 2\sqrt{6}$ square units. Alternatively, use Heron's formula to obtain area = $\sqrt{6(6-2)(6-5)(6-5)} = \sqrt{24} = 2\sqrt{6}$ square units.

29. All four triangles have two sides a and b given. For triangles with areas A_1 and A_3, the included angle is $180° - \theta$; hence, $A_1 = A_3 = \frac{1}{2}ab\sin(180° - \theta)$.

For triangles with areas A_2 and A_4, the included angle is θ; hence, $A_2 = A_4 = \frac{1}{2}ab\sin\theta$

But, $\sin(180° - \theta) = \sin 180° \cos\theta - \cos 180° \sin\theta = 0\cos\theta - (-1)\sin\theta = \sin\theta$

Hence, $A_1 = A_3 = \frac{1}{2}ab\sin(180° - \theta) = \frac{1}{2}ab\sin\theta = A_2 = A_4$.

31. Sketch a figure as follows:

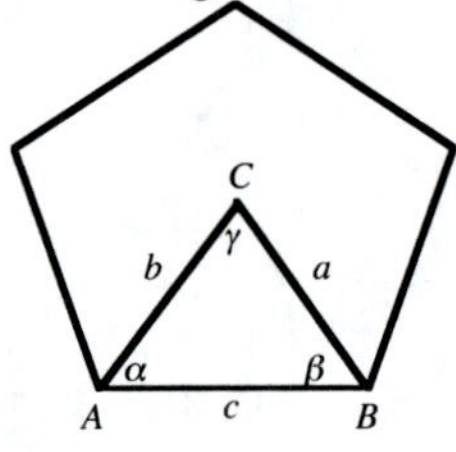

Side c is one-fifth the perimeter of the pentagon, 7 feet.
Angle γ is one-fifth of a complete circle, 72°.

Triangle ABC has one-fifth the area of the pentagon; it is isosceles, hence, $\alpha = \beta = \frac{1}{2}(180° - 72°) = 54°$.

By the law of sines:

$$\frac{\sin\alpha}{a} = \frac{\sin\gamma}{c}$$

$$a = \sin\alpha \frac{c}{\sin\gamma}$$

$$= \sin 54° \frac{7}{\sin 72°} = 5.95 \text{ ft.}$$

β is the included angle for sides a and c, so the area of ABC

$$A = \frac{ac}{2}\sin\beta = \frac{1}{2}(5.95)(7)\sin 54° = 16.9 \text{ ft}^2$$

Then the area of the pentagon = $5A$ = 84 ft^2.

33. Sketch a figure as follows:

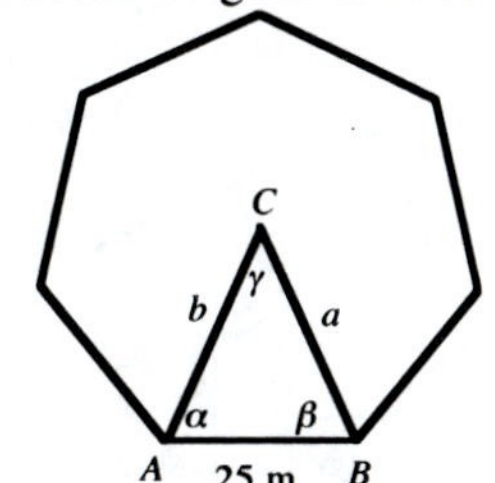

Angle γ is one-seventh of a complete circle, $\frac{360°}{7}$.

Triangle ABC has one-seventh the area of the polygon; it is isosceles, hence, $\alpha = \beta = \frac{1}{2}\left(180° - \frac{360°}{7}\right) = \frac{450°}{7}$.

By the law of sines:

$$\frac{\sin\alpha}{a} = \frac{\sin\gamma}{c}$$

$$a = \sin\alpha \frac{c}{\sin\gamma}$$

$$= \sin\frac{450°}{7}\frac{25}{\sin\left(\frac{360°}{7}\right)} = 28.8 \text{ m}$$

β is the included angle for sides a and c, so the area of ABC

$$A = \frac{ac}{2}\sin\beta = \frac{1}{2}(28.8)(25)\sin\frac{450°}{7} = 324 \text{ m}^2$$

Then the area of the polygon = $7A$ = 2300 m^2.

35. Label the sides: $a = 42, b = 51\frac{1}{3}, c = 37\frac{3}{4}$. The given information consists of three sides; hence, we use Heron's formula. First, find the semiperimeter s:

$$s = \frac{a+b+c}{2} = \frac{42 + 51\frac{1}{3} + 37\frac{3}{4}}{2} = 65.54 \text{ ft}$$

Then,

$$s - a = 65.54 - 42 = 23.54 \text{ ft}$$

$$s - b = 65.54 - 51\tfrac{1}{3} = 14.21 \text{ ft}$$

$$s - c = 65.54 - 37\tfrac{3}{4} = 27.79 \text{ ft}$$

Thus, $A = \sqrt{s(s-a)(s-b)(s-c)} = \sqrt{(65.54)(23.54)(14.21)(27.79)} = 780.56 \text{ ft}^2$

Then the cost of the patio = $780.56 \text{ ft}^2 \times 15.50\frac{\text{dollars}}{\text{ft}^2}$ = \$12,100.

37. Following the hint, divide the plot into two triangles as shown:

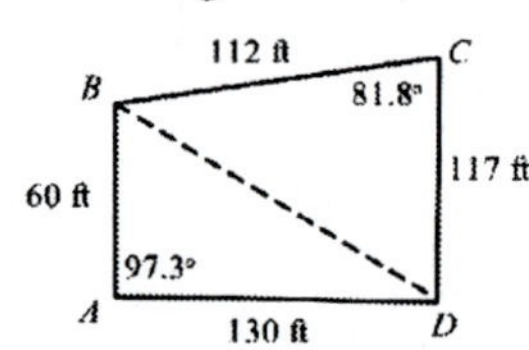

Then in each triangle the given information consists of two sides and the included angle; hence, we use the formula $A = \frac{ab}{2}\sin\theta$.

Triange ADB: $A_1 = \frac{1}{2}(60)(130)\sin 97.3° = 3{,}868.4 \text{ ft}^2$

Triangle BCD: $A_2 = \frac{1}{2}(112)(117)\sin 81.8° = 6{,}485 \text{ ft}^2$

The area of the plot $= A = A_1 + A_2 = 3{,}868.4 + 6{,}485 = 10{,}353.4 \text{ ft}^2$.

Thus the price of the plot $= 10{,}353.4 \text{ ft}^2 \times 4.50 \frac{\text{dollars}}{\text{ft}^2} = \$46{,}600$.

EXERCISE 6.4 Vectors

1. A vector is a quantity that has both magnitude and direction; a scalar is a quantity that has magnitude only.

3. The magnitude of $\langle a, b\rangle$ is given by $\sqrt{a^2+b^2}$.

5. The coordinates of $P(x, y)$ are given by

$$x = x_b - x_a = 5 - 2 = 3$$
$$y = y_b - y_a = 1 - (-3) = 4$$

Thus, $P(x, y) = P(3, 4)$

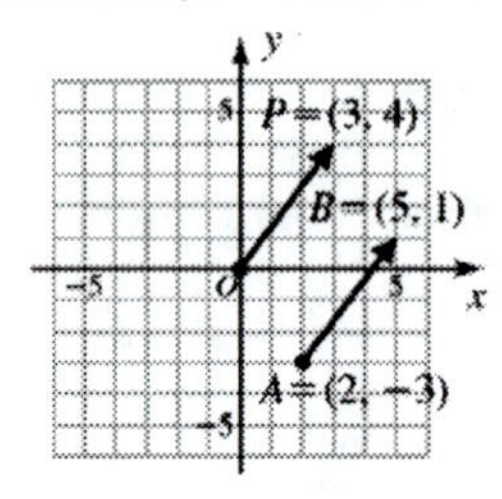

7. The coordinates of $P(x, y)$ are given by

$$x = x_b - x_a = (-3) - (-1) = -2$$
$$y = y_b - y_a = (-1) - 3 = -4$$

Thus, $P(x, y) = P(-2, -4)$

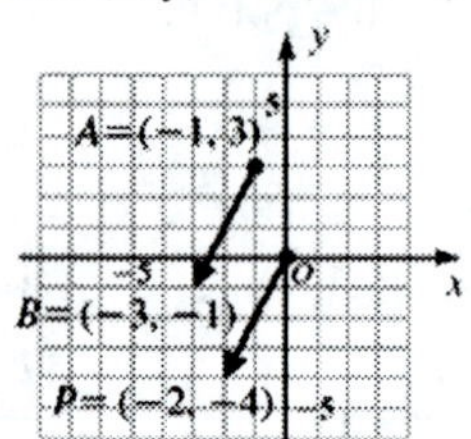

9. The algebraic vector $\langle a, b\rangle$ has coordinates given by

$$a = x_b - x_a = 6 - 3 = 3 \qquad b = y_b - y_a = 12 - 4 = 8$$

Hence, $\langle a, b\rangle = \langle 3, 8\rangle$

11. The algebraic vector $\langle a, b\rangle$ has coordinates given by

$$a = x_b - x_a = (-10) - (-5) = -5 \qquad b = y_b - y_a = (-2) - 6 = -8$$

Hence, $\langle a, b\rangle = \langle -5, -8\rangle$

13. $|\langle a, b\rangle| = \sqrt{a^2+b^2} = \sqrt{1^2+2^2} = \sqrt{5}$

15. $|\langle a, b\rangle| = \sqrt{a^2+b^2} = \sqrt{12^2+5^2} = 13$

17. $|\langle a, b\rangle| = \sqrt{a^2+b^2} = \sqrt{0^2+(-8)^2} = 8$

19. (A) $\mathbf{u} + \mathbf{v} = \langle 1, 4\rangle + \langle -3, 2\rangle = \langle -2, 6\rangle$
    (B) $\mathbf{u} - \mathbf{v} = \langle 1, 4\rangle - \langle -3, 2\rangle = \langle 4, 2\rangle$
    (C) $2\mathbf{u} - 3\mathbf{v} = 2\langle 1, 4\rangle - 3\langle -3, 2\rangle = \langle 2, 8\rangle + \langle 9, -6\rangle = \langle 11, 2\rangle$
    (D) $3\mathbf{u} - \mathbf{v} + 2\mathbf{w} = 3\langle 1, 4\rangle - \langle -3, 2\rangle + 2\langle 0, 4\rangle = \langle 3, 12\rangle + \langle 3, -2\rangle + \langle 0, 8\rangle = \langle 6, 18\rangle$

21. (A) $\mathbf{u} + \mathbf{v} = \langle 2, -3\rangle + \langle -1, -3\rangle = \langle 1, -6\rangle$
    (B) $\mathbf{u} - \mathbf{v} = \langle 2, -3\rangle - \langle -1, -3\rangle = \langle 3, 0\rangle$
    (C) $2\mathbf{u} - 3\mathbf{v} = 2\langle 2, -3\rangle - 3\langle -1, -3\rangle = \langle 4, -6\rangle + \langle 3, 9\rangle = \langle 7, 3\rangle$
    (D) $3\mathbf{u} - \mathbf{v} - 2\mathbf{w} = 3\langle 2, -3\rangle - \langle -1, -3\rangle + 2\langle -2, 0\rangle = \langle 6, -9\rangle + \langle 1, 3\rangle + \langle -4, 0\rangle = \langle 3, -6\rangle$

23. $|\mathbf{v}| = \sqrt{0^2 + (-4)^2} = 4 \qquad \mathbf{u} = \frac{1}{|\mathbf{v}|}\mathbf{v} = \frac{1}{4}\langle 0, -4\rangle = \langle 0, -1\rangle$

25. $|\mathbf{v}| = \sqrt{3^2 + 1^2} = \sqrt{10} \qquad \mathbf{u} = \frac{1}{|\mathbf{v}|}\mathbf{v} = \frac{1}{\sqrt{10}}\langle 3, 1\rangle = \left\langle \frac{3}{\sqrt{10}}, \frac{1}{\sqrt{10}}\right\rangle$

27. $|\mathbf{v}| = \sqrt{(-24)^2 + 7^2} = 25 \qquad \mathbf{u} = \frac{1}{|\mathbf{v}|}\mathbf{v} = \frac{1}{25}\langle -24, 7\rangle = \left\langle -\frac{24}{25}, \frac{7}{25}\right\rangle$

29. $|\mathbf{v}| = \sqrt{4^2 + 0^2} = 4 \qquad \mathbf{u} = -\frac{1}{|\mathbf{v}|}\mathbf{v} = -\frac{1}{4}\langle 4, 0\rangle = \langle -1, 0\rangle$

31. $|\mathbf{v}| = \sqrt{(-3)^2 + 4^2} = 5 \qquad \mathbf{u} = -\frac{1}{|\mathbf{v}|}\mathbf{v} = -\frac{1}{5}\langle -3, 4\rangle = \left\langle \frac{3}{5}, -\frac{4}{5}\right\rangle$

33. $\mathbf{v} = \langle 2, -7\rangle = \langle 2, 0\rangle + \langle 0, -7\rangle = 2\langle 1, 0\rangle - 7\langle 0, 1\rangle = 2\mathbf{i} - 7\mathbf{j}$

35. $\mathbf{v} = \langle 4, 5\rangle = \langle 4, 0\rangle + \langle 0, 5\rangle = 4\langle 1, 0\rangle + 5\langle 0, 1\rangle = 4\mathbf{i} + 5\mathbf{j}$

37. $\mathbf{v} = 3\langle 0, -18\rangle = \langle 0, -54\rangle = -54\langle 0, 1\rangle = -54\mathbf{j}$

39. $\mathbf{v} = \overrightarrow{AB} = \langle (-2) - 1, 13 - (-6)\rangle = \langle -3, 19\rangle = \langle -3, 0\rangle + \langle 0, 19\rangle$
    $= -3\langle 1, 0\rangle + 19\langle 0, 1\rangle = -3\mathbf{i} + 19\mathbf{j}$

41. $\mathbf{v} = \overrightarrow{AB} = \langle 0 - (-9), 0 - 1\rangle = \langle 9, -1\rangle = \langle 9, 0\rangle + \langle 0, -1\rangle$
    $= 9\langle 1, 0\rangle - \langle 0, 1\rangle = 9\mathbf{i} - \mathbf{j}$

43. First find the horizontal and vertical components **H** and **V** for vector **u**.

$$\cos 41° = \frac{|\mathbf{H}|}{12}$$
$$|\mathbf{H}| = 12\cos 41° = 9.1$$
$$\sin 41° = \frac{|\mathbf{v}|}{12}$$
$$|\mathbf{v}| = 12\sin 41° = 7.9$$

We can write **H** as $9.1\mathbf{i}$ and **v** as $7.9\mathbf{j}$, so $\mathbf{u} = 9.1\mathbf{i} + 7.9\mathbf{j}$.

45. First find the horizontal and vertical components **H** and **V** for vector **w**.

$$\cos 78° = \frac{|\mathbf{H}|}{11}$$
$$|\mathbf{H}| = 11 \cos 78° = 2.3$$
$$\sin 78° = \frac{|\mathbf{v}|}{11}$$
$$|\mathbf{v}| = 11 \sin 78° = 10.8$$

We can write **H** as 2.3**i** and **v** as –10.8**j**, so **w** = 2.3**i** – 10.8**j**.

47. $5\mathbf{u} + \mathbf{w} = 5(2\mathbf{i} - 3\mathbf{j}) + 5\mathbf{j} = 10\mathbf{i} - 15\mathbf{j} + 5\mathbf{j} = 10\mathbf{i} - 10\mathbf{j}$

49. $\mathbf{u} - \mathbf{v} + \mathbf{w} = 2\mathbf{i} - 3\mathbf{j} - (3\mathbf{i} + 4\mathbf{j}) + 5\mathbf{j} = 2\mathbf{i} - 3\mathbf{j} - 3\mathbf{i} - 4\mathbf{j} + 5\mathbf{j} = -\mathbf{i} - 2\mathbf{j}$

51. $2\mathbf{u} + 4\mathbf{v} - 6\mathbf{w} = 2(2\mathbf{i} - 3\mathbf{j}) + 4(3\mathbf{i} + 4\mathbf{j}) - 6(5\mathbf{j}) = 4\mathbf{i} - 6\mathbf{j} + 12\mathbf{i} + 16\mathbf{j} - 30\mathbf{j} = 16\mathbf{i} - 20\mathbf{j}$

53. The sum of two unit vectors is never a unit vector. For example, **i** and **j** are unit vectors, but $|\mathbf{i} + \mathbf{j}| = \sqrt{1^2 + 1^2} = \sqrt{2}$.

55. Any one of the force vectors must have the same magnitude as the resultant of the other two force vectors and be oppositely directed to the resultant of the other two.

57. Let $\mathbf{v} = \langle 3, 2\rangle$. $|\mathbf{v}| = \sqrt{3^2 + 2^2} = \sqrt{13}$. A unit vector in **v**'s direction would be $\mathbf{u} = \frac{1}{|\mathbf{v}|}\mathbf{v} = \frac{1}{\sqrt{13}}\langle 3, 2\rangle$. Then the required vector is $5\mathbf{u} = \frac{5}{\sqrt{13}}\langle 3, 2\rangle = \left\langle \frac{15}{\sqrt{13}}, \frac{10}{\sqrt{13}} \right\rangle$.

59. Let $\mathbf{v} = \langle -8, 6\rangle$. $|\mathbf{v}| = \sqrt{(-8)^2 + 6^2} = 10$. A unit vector in **v**'s direction would be $\mathbf{u} = \frac{1}{|\mathbf{v}|}\mathbf{v} = \frac{1}{10}\langle -8, 6\rangle$. A unit vector in the opposite direction would be $-\mathbf{u} = -\frac{1}{10}\langle -8, 6\rangle$.

Then the required vector is $-2\mathbf{u} = -\frac{2}{10}\langle -8, 6\rangle = \left\langle \frac{8}{5}, -\frac{6}{5} \right\rangle$.

61. To show that $\langle -4, -5\rangle$ is a linear combination of $\langle 1, 0\rangle$ and $\langle 3, 1\rangle$, set

$$\langle -4, -5\rangle = c_1\langle 1, 0\rangle + c_2\langle 3, 1\rangle$$

Then

$$\langle -4, -5\rangle = \langle c_1 + 3c_2, c_2\rangle$$
$$-4 = c_1 + 3c_2$$
$$-5 = c_2$$

This system of equations has a solution $c_1 = 11, c_2 = -5$, hence $\langle -4, -5\rangle = 11\langle 1, 0\rangle + (-5)\langle 3, 1\rangle$ and $\langle -4, -5\rangle$ is a linear combination of $\langle 1, 0\rangle$ and $\langle 3, 1\rangle$.

63. To show that $\langle 1, 2\rangle$ is not a linear combination of $\langle 8, 4\rangle$ and $\langle 10, 5\rangle$, set

$$\langle 1, 2\rangle = c_1\langle 8, 4\rangle + c_2\langle 10, 5\rangle$$

Then

$$\langle 1, 2\rangle = \langle 8c_1 + 10c_2, 4c_1 + 5c_2\rangle$$
$$1 = 8c_1 + 10c_2$$
$$2 = 4c_1 + 5c_2$$

This system of equations is equivalent, in turn, to

$$\begin{aligned} 1 &= 8c_1 + 10c_2 \\ 4 &= 8c_1 + 10c_2 \end{aligned}$$

and

$$\begin{aligned} 1 &= 8c_1 + 10c_2 \\ 3 &= 0 \end{aligned}$$

which has no solution. Hence $\langle 1, 2\rangle$ is not a linear combination of $\langle 8, 4\rangle$ and $\langle 10, 5\rangle$.

65. $\mathbf{u} + \mathbf{v} = \langle a, b\rangle + \langle c, d\rangle$
$= \langle a + c, b + d\rangle$ Definition of vector addition
$= \langle c + a, d + b\rangle$ Commutative property for addition of real numbers*
$= \langle c, d\rangle + \langle a, b\rangle$ Definition of vector addition
$= \mathbf{v} + \mathbf{u}$

67. $\mathbf{v} + (-\mathbf{v}) = \langle c, d\rangle + (-\langle c, d\rangle)$
$= \langle c, d\rangle + \langle -c, -d\rangle$ Definition of scalar multiplication
$= \langle c + (-c), d + (-d)\rangle$ Definition of vector addition
$= \langle 0, 0\rangle$ Additive inverse property for real numbers*
$= \mathbf{0}$

69. $m(\mathbf{u} + \mathbf{v}) = m(\langle a, b\rangle + \langle c, d\rangle)$
$= m\langle a + c, b + d\rangle$ Definition of vector addition
$= \langle m(a + c), m(b + d)\rangle$ Definition of scalar multiplication
$= \langle ma + mc, mb + md\rangle$ Distributive property for real numbers*
$= \langle ma, mb\rangle + \langle mc, md\rangle$ Definition of vector addition
$= m\langle a, b\rangle + m\langle c, d\rangle$ Definition of scalar multiplication
$= m\mathbf{u} + m\mathbf{v}$

71. $1\mathbf{v} = 1\langle a, b\rangle$
$= \langle 1a, 1b\rangle$ Definition of scalar multiplication
$= \langle a, b\rangle$ Multiplicative identity property for real numbers*
$= \mathbf{v}$

* The basic properties of the set of real numbers are listed in the text, Appendix A.1.

73. The actual velocity **v** of the boat is the vector sum of the apparent velocity **B** of the boat and the velocity **R** of the river.

$|\mathbf{B}| = 4.0$ km/hr $\qquad |\mathbf{R}| = 3.0$ km/hr

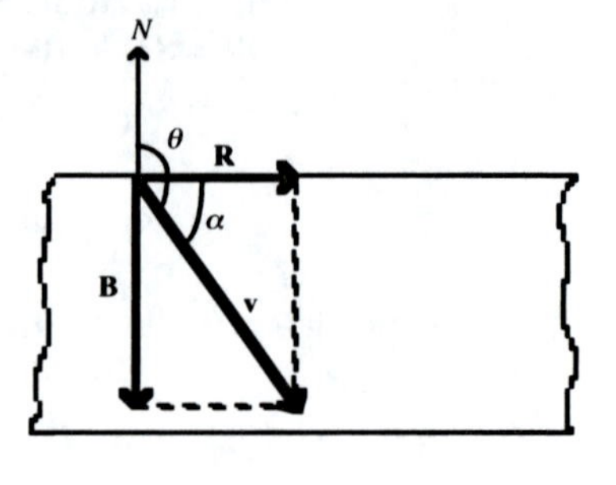

Using the Pythagorean theorem, we find the magnitude of the resultant vector to be

$$|\mathbf{v}| = \sqrt{4.0^2 + 3.0^2} = 5.0 \text{ km/hr}$$

To find θ, we see that

$$\tan\alpha = \frac{|\mathbf{B}|}{|\mathbf{R}|} = \frac{4.0}{3.0} \qquad \alpha = \tan^{-1}\frac{4.0}{3.0} = 53°$$

θ = actual heading = $90° + \alpha = 90° + 53° = 143°$.

75. We require θ such that the actual velocity **R** will be the resultant of the apparent velocity **v** and the wind velocity **w**. The heading α will then be $360° - \theta$. From the diagram it should be clear that

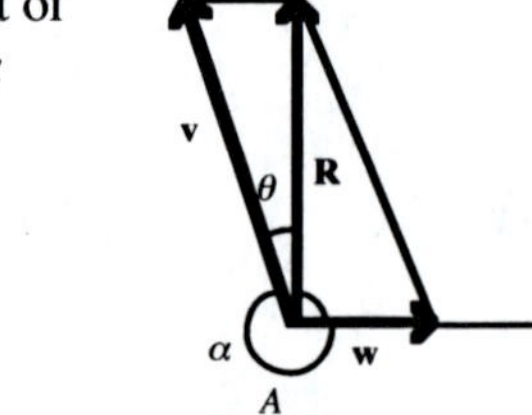

$$\sin\theta = \frac{|\mathbf{w}|}{|\mathbf{v}|} = \frac{46}{255}$$

$$\theta = \sin^{-1}\frac{46}{255} = 10° \qquad \alpha = 360° - 10° = 350°$$

The ground speed for this course will be the magnitude $|\mathbf{R}|$ of the actual velocity. In the right triangle, ABC, we have

$$\cos\theta = \frac{|\mathbf{R}|}{|\mathbf{v}|} \qquad |\mathbf{R}| = |\mathbf{v}|\cos\theta = 255\cos\left(\sin^{-1}\frac{46}{255}\right)$$

$$= 255\sqrt{1-\left(\frac{46}{255}\right)^2} = 250 \text{ mi/hr}$$

77. In triangle ABC, $\beta = 180° - 32° = 148°$. We can find the magnitude of M of the resulting force using the law of cosines.

$$M^2 = |\mathbf{F}_1|^2 + |\mathbf{F}_2|^2 - 2|\mathbf{F}_1||\mathbf{F}_2|\cos\beta = 1{,}500^2 + 1{,}100^2 - 2(1{,}500)(1{,}100)\cos 148°$$

$$= 6{,}258{,}558.7\ldots$$

$$M = \sqrt{6{,}258{,}558.7\ldots} = 2{,}500 \text{ lb}$$

To find α, we use the law of sines.

$$\frac{\sin\alpha}{|\mathbf{F}_2|} = \frac{\sin\beta}{M}$$

$$\frac{\sin\alpha}{1{,}100} = \frac{\sin 148°}{2{,}500}$$

$$\sin\alpha = \frac{1{,}100}{2{,}500}\sin 148° \qquad \alpha = \sin^{-1}\left(\frac{1{,}100}{2{,}500}\sin 148°\right) = 13° \text{ (relative to } \mathbf{F}_1\text{)}$$

79. The force parallel to the hill is the component of **W** parallel to the hill, that is, the magnitude of **CD**.

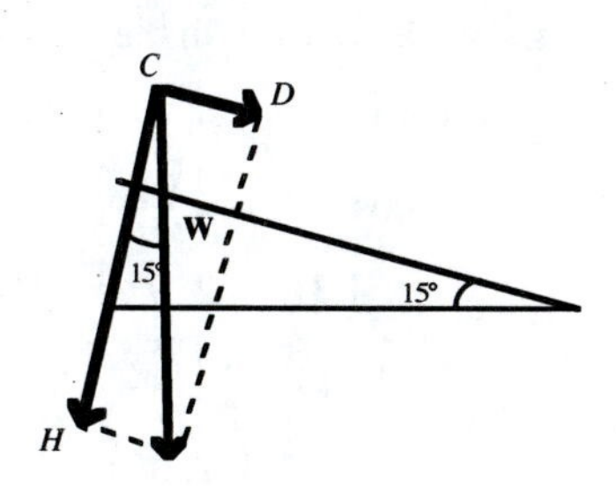

$$\frac{|\mathbf{CD}|}{|\mathbf{W}|} = \sin 15°$$

$$|\mathbf{CD}| = |\mathbf{W}| \sin 15° = 2500 \sin 15° = 650 \text{ lb}$$

The force perpendicular to the hill is the component of **W** perpendicular to the hill, that is, the magnitude of **CH**.

$$\frac{|\mathbf{CH}|}{|\mathbf{W}|} = \cos 15° \qquad |\mathbf{CH}| = |\mathbf{W}| \cos 15° = 2500 \cos 15° = 2400 \text{ lb}$$

81. From the figure it should be clear that in triangle OAB,

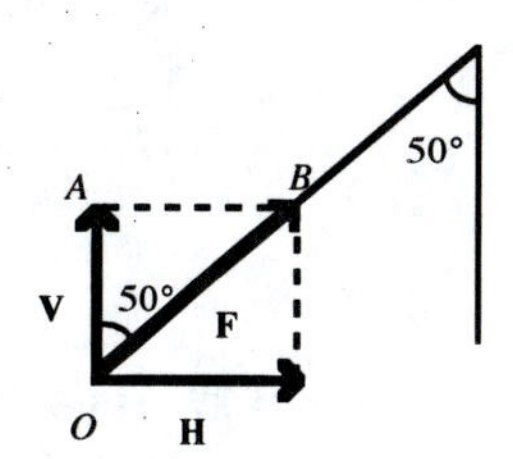

$$\sin 50° = \frac{|\mathbf{H}|}{|\mathbf{F}|} \qquad \cos 50° = \frac{|\mathbf{V}|}{|\mathbf{F}|}$$

Thus

$$|\mathbf{H}| = |\mathbf{F}| \sin 50° = 52 \sin 50° = 40 \text{ lb}$$

$$|\mathbf{V}| = |\mathbf{F}| \cos 50° = 52 \cos 50° = 33 \text{ lb}$$

83. First, form a force diagram with all force vectors in standard position at the origin.

Let $\mathbf{F}_1$ = the tension in one rope

$\mathbf{F}_2$ = the tension in the other rope

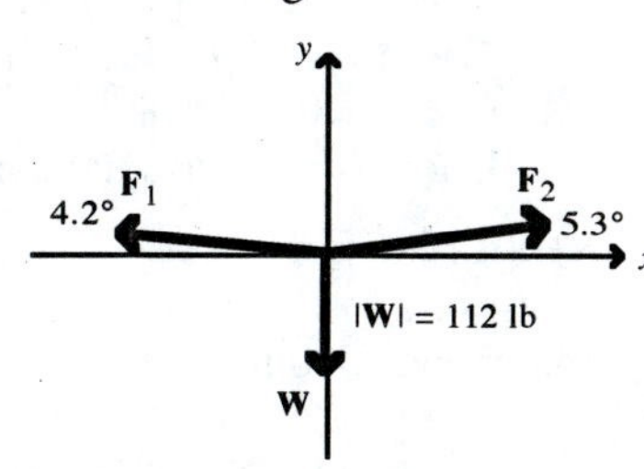

Write each force vector in terms of **i** and **j** unit vectors.

$$\mathbf{F}_1 = |\mathbf{F}_1|(-\cos 4.2°)\mathbf{i} + |\mathbf{F}_1|(\sin 4.2°)\mathbf{j}$$

$$\mathbf{F}_2 = |\mathbf{F}_2|(\cos 5.3°)\mathbf{i} + |\mathbf{F}_2|(\sin 5.3°)\mathbf{j}$$

$$\mathbf{W} = -112\mathbf{j}$$

For the system to be in static equilibrium, we must have $\mathbf{F}_1 + \mathbf{F}_2 + \mathbf{W} = \mathbf{0}$ which becomes, on addition,

$$[-|\mathbf{F}_1|(\cos 4.2°) + |\mathbf{F}_2|(\cos 5.3°)]\,\mathbf{i} + [\,|\mathbf{F}_1|(\sin 4.2°) + |\mathbf{F}_2|(\sin 5.3°) - 112]\,\mathbf{j} = 0\mathbf{i} + 0\mathbf{j}$$

Since two vectors are equal if and only if their corresponding components are equal, we are led to the following system of equations in $|\mathbf{F}_1|$ and $|\mathbf{F}_2|$:

$$-|\mathbf{F}_1| \cos 4.2° + |\mathbf{F}_2| \cos 5.3° = 0$$

$$|\mathbf{F}_1| \sin 4.2° + |\mathbf{F}_2| \sin 5.3° - 112 = 0$$

Solving, $|\mathbf{F}_2| = |\mathbf{F}_1| \dfrac{\cos 4.2°}{\cos 5.3°}$

$$|\mathbf{F}_1| \sin 4.2° + |\mathbf{F}_1| \frac{\cos 4.2°}{\cos 5.3°} \sin 5.3° = 112$$

$$|\mathbf{F}_1| \; [\sin 4.2° + \cos 4.2° \tan 5.3°] = 112$$

$$|\mathbf{F}_1| = \frac{112}{\sin 4.2° + \cos 4.2° \tan 5.3°} = 676 \text{ lb}$$

$$|\mathbf{F}_2| = 676 \frac{\cos 4.2°}{\cos 5.3°} = 677 \text{ lb}$$

85. Label the tension in the left side of the strap **L** and the tension in the right side of the strap **R**.

Find the horizontal and vertical components $\mathbf{H}_1$ and $\mathbf{V}_1$ for vector **L**.

$$\cos 49° = \frac{|\mathbf{H}_1|}{230}$$
$$|\mathbf{H}_1| = 230 \cos 49° = 151$$
$$\sin 49° = \frac{|\mathbf{V}_1|}{230}$$
$$|\mathbf{V}_1| = 230 \sin 49° = 173.6$$

We can write $\mathbf{H}_1$ as $-151\mathbf{i}$ and $\mathbf{V}_1$ as $173.6\mathbf{j}$ so $\mathbf{L} = -151\mathbf{i} + 173.6\mathbf{j}$.

Now find the horizontal and vertical components $\mathbf{H}_2$ and $\mathbf{V}_2$ for vector **R**.

$$\cos 40° = \frac{|\mathbf{H}_2|}{197}$$
$$|\mathbf{H}_2| = 197 \cos 40° = 151$$
$$\sin 40° = \frac{|\mathbf{V}_2|}{197}$$
$$|\mathbf{V}_2| = 197 \sin 40° = 126.6$$

We can write $\mathbf{H}_2$ as $151\mathbf{i}$ and $\mathbf{V}_2$ as $126.6\mathbf{j}$ so $\mathbf{R} = 151\mathbf{i}$ and $126.6\mathbf{j}$.

(A) The resultant force on the piano is $\mathbf{L} + \mathbf{R}$.

$$\mathbf{L} + \mathbf{R} = -151\mathbf{i} + 173.6\mathbf{j} + 151\mathbf{i} + 126.6\mathbf{j}$$
$$= 0\mathbf{i} + 300\mathbf{j}$$

There is no horizontal component of force on the piano, therefore it is not moving horizontally.

(B) The outside force on the piano, $\mathbf{L} + \mathbf{R} = 300\mathbf{j}$. Since the piano is not moving vertically, there must be a force equal and opposite to $300\mathbf{j}$, that is, $-300\mathbf{j}$. This is the weight of the piano, and has magnitude 300 lbs.

87. First, form a force diagram with all force vectors in standard position at the origin.

Let $\mathbf{F}_1$ = the force on the horizontal member BC
$\mathbf{F}_2$ = the force on the supporting member AB
$\mathbf{W}$ = the downward force (5,000 lb)

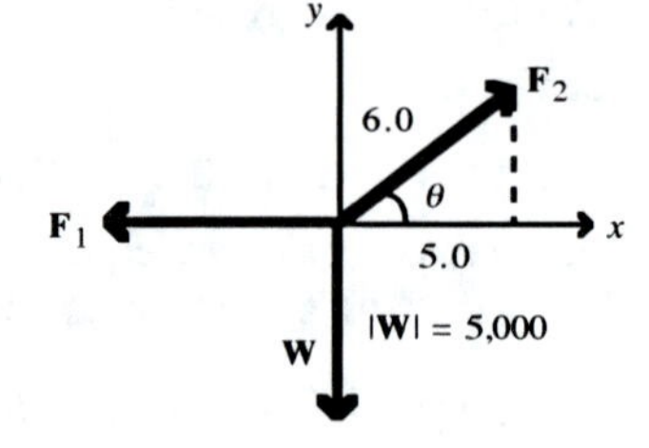

We note: $\cos\theta = \dfrac{5.0}{6.0}$ $\qquad \theta = \cos^{-1}\dfrac{5.0}{6.0} = 33.6°$

Then write each force vector in terms of **i** and **j** unit vectors.

$$\mathbf{F}_1 = -|\mathbf{F}_1|\,\mathbf{i}$$
$$\mathbf{F}_2 = |\mathbf{F}_2|(\cos 33.6°)\,\mathbf{i} + |\mathbf{F}_2|(\sin 33.6°)\,\mathbf{j}$$
$$\mathbf{W} = -5{,}000\mathbf{j}$$

For the system to be in static equilibrium, we must have $\mathbf{F}_1 + \mathbf{F}_2 + \mathbf{W} = \mathbf{0}$ which becomes, on addition,

$$[-|\mathbf{F}_1| + |\mathbf{F}_2|(\cos 33.6°)]\,\mathbf{i} + [\,|\mathbf{F}_2|(\sin 33.6°) - 5{,}000]\,\mathbf{j} = 0\mathbf{i} + 0\mathbf{j}$$

Since two vectors are equal if and only if their corresponding components are equal, we are led to the following system of equations in $|\mathbf{F}_1|$ and $|\mathbf{F}_2|$:

$$-|\mathbf{F}_1| + |\mathbf{F}_2|(\cos 33.6°) = 0$$
$$|\mathbf{F}_1|(\sin 33.6°) - 5{,}000 = 0$$

Solving, $|\mathbf{F}_2| = \dfrac{5{,}000}{\sin 33.6°} = 9{,}000$ lb

$|\mathbf{F}_1| = |\mathbf{F}_2| \cos 33.6° = 7{,}500$ lb

The force in the member *AB* is directed oppositely to the diagram—a compression of 9,000 lb. The force in the member *BC* is also directed oppositely to the diagram—a tension of 7,500 lb.

EXERCISE 6.5 The Dot Product

1. The length of a vector is the square root of its dot product with itself.

3. Tow vectors **u** and **v** are orthogonal if and only if $\mathbf{u} \cdot \mathbf{v} = 0$.

5. $\langle 5, 3\rangle \cdot \langle -2, 6\rangle = 5(-2) + 3 \cdot 6 = 8$

7. $\langle 8, 1\rangle \cdot \langle -8, 0\rangle = 8(-8) + 1 \cdot 0 = -64$

9. $\langle 0, 6\rangle \cdot \langle 0, 6\rangle = 0 \cdot 0 + 6 \cdot 6 = 36$

11. $8\mathbf{i} \cdot 3\mathbf{j} = (8\mathbf{i} + 0\mathbf{j}) \cdot (0\mathbf{i} + 3\mathbf{j}) = 8 \cdot 0 + 0 \cdot 3 = 0$

13. $(-2\mathbf{i} - 5\mathbf{j}) \cdot 6\mathbf{i} = (-2\mathbf{i} - 5\mathbf{j}) \cdot (6\mathbf{i} + 0\mathbf{j}) = (-2)6 + (-5)0 = -12$

15. $(7\mathbf{i} - 6\mathbf{j}) \cdot (2\mathbf{i} + 3\mathbf{j}) = 7 \cdot 2 + (-6)3 = -4$

17. $|\mathbf{u}| = \sqrt{0^2 + 1^2} = 1$ $\quad |\mathbf{v}| = \sqrt{5^2 + 5^2} = \sqrt{50}$

$\cos\theta = \dfrac{\mathbf{u} \cdot \mathbf{v}}{|\mathbf{u}||\mathbf{v}|} = \dfrac{\langle 0,1\rangle \cdot \langle 5,5\rangle}{(1)\sqrt{50}} = \dfrac{0+5}{\sqrt{50}} = \dfrac{5}{\sqrt{50}}$ $\quad \theta = \cos^{-1}\dfrac{5}{\sqrt{50}} = 45.0°$

19. $|\mathbf{u}| = \sqrt{2^2 + 9^2} = \sqrt{85}$ $\quad |\mathbf{v}| = \sqrt{2^2 + 10^2} = \sqrt{104}$

$\cos\theta = \dfrac{\mathbf{u} \cdot \mathbf{v}}{|\mathbf{u}||\mathbf{v}|} = \dfrac{\langle 2,9\rangle \cdot \langle 2,10\rangle}{\sqrt{85}\sqrt{104}} = \dfrac{4+90}{\sqrt{85}\sqrt{104}} = \dfrac{94}{\sqrt{85}\sqrt{104}}$ $\quad \theta = \cos^{-1}\dfrac{94}{\sqrt{85}\sqrt{104}} = 1.2°$

21. $|\mathbf{u}| = \sqrt{8^2 + 1^2} = \sqrt{65}$ $\quad |\mathbf{v}| = \sqrt{(-8)^2 + 1^2} = \sqrt{65}$

$\cos\theta = \dfrac{\mathbf{u} \cdot \mathbf{v}}{|\mathbf{u}||\mathbf{v}|} = \dfrac{(8\mathbf{i} + \mathbf{j}) \cdot (-8\mathbf{i} + \mathbf{j})}{\sqrt{65}\sqrt{65}} = \dfrac{8(-8) + 1 \cdot 1}{65} = \dfrac{-63}{65}$ $\quad \theta = \cos^{-1}\left(-\dfrac{63}{65}\right) = 165.7°$

23. Since $|\mathbf{u}|$ and $|\mathbf{v}|$ are never negative, the sign of the dot product depends only on $\cos\theta$. Since θ is an obtuse angle, $\cos\theta$ and the dot product are negative.

25. $\mathbf{u} \cdot \mathbf{v} = (2\mathbf{i} + \mathbf{j}) \cdot (\mathbf{i} - 2\mathbf{j}) = 2 - 2 = 0$ Thus, **u** and **v** are orthogonal.

27. $\mathbf{u} \cdot \mathbf{v} = \langle 1, 3\rangle \cdot \langle -3, -1\rangle = -3 - 3 = -6 \neq 0$ Thus, **u** and **v** are not orthogonal.

29. $|\mathbf{u}| = \sqrt{(-2)^2 + 7^2} = \sqrt{53}$ $\quad |\mathbf{v}| = \sqrt{2^2 + 7^2} = \sqrt{53}$

$\cos\theta = \dfrac{\mathbf{u} \cdot \mathbf{v}}{|\mathbf{u}||\mathbf{v}|} = \dfrac{\langle -2,7\rangle \cdot \langle 2,7\rangle}{\sqrt{53}\sqrt{53}} = \dfrac{-4+49}{53} = \dfrac{45}{53}$

For **u** and **v** to be parallel, θ must be 0 or π, $\cos\theta$ must be 1 or -1. Since this is not the case here, **u** and **v** are not parallel.

31. $|\mathbf{u}| = \sqrt{4^2 + (-3)^2} = 5$ $\qquad |\mathbf{v}| = \sqrt{(-20)^2 + 15^2} = 25$

$\cos\theta = \dfrac{\mathbf{u}\cdot\mathbf{v}}{|\mathbf{u}||\mathbf{v}|} = \dfrac{(4\mathbf{i} - 3\mathbf{j})\cdot(-20\mathbf{i} + 15\mathbf{j})}{5(25)} = \dfrac{4(-20) - 3(15)}{5(25)} = -1$

Since $\cos\theta = -1$, $\theta = \cos^{-1}(-1) = \pi$. These vectors are parallel.

33. $\text{comp}_{\mathbf{v}}\ \mathbf{u} = \dfrac{\mathbf{u}\cdot\mathbf{v}}{|\mathbf{v}|} = \dfrac{\langle 5,12\rangle\cdot\langle 0,1\rangle}{|\langle 0,1\rangle|} = \dfrac{0+12}{\sqrt{0^2+1^2}} = \dfrac{12}{1} = 12.0$

35. $\text{comp}_{\mathbf{v}}\ \mathbf{u} = \dfrac{\mathbf{u}\cdot\mathbf{v}}{|\mathbf{v}|} = \dfrac{\langle -6,4\rangle\cdot\langle 2,3\rangle}{|\langle 2,3\rangle|} = \dfrac{-12+12}{\sqrt{2^2+3^2}} = \dfrac{0}{\sqrt{13}} = 0$

37. $\text{comp}_{\mathbf{v}}\ \mathbf{u} = \dfrac{\mathbf{u}\cdot\mathbf{v}}{|\mathbf{v}|} = \dfrac{(-3\mathbf{i}+\mathbf{j})\cdot(5\mathbf{i}+14\mathbf{j})}{|5\mathbf{i}+14\mathbf{j}|} = \dfrac{-15+14}{\sqrt{5^2+14^2}} = \dfrac{-1}{\sqrt{221}} = -0.0673$

39. $\mathbf{u}\cdot\mathbf{u} = \langle a, b\rangle \cdot \langle a, b\rangle = a^2 + b^2 = (\sqrt{a^2+b^2})^2 = |\mathbf{u}|^2$

41.
$\mathbf{u}\cdot(\mathbf{v}+\mathbf{w}) = \langle a, b\rangle\cdot(\langle c, d\rangle + \langle e, f\rangle)$
$= \langle a, b\rangle\cdot\langle c+e, d+f\rangle$ — Definition of vector addition
$= a(c+e) + b(d+f)$ — Definition of dot product
$= ac + ae + bd + bf$ — Distributive property of real numbers
$= (ac+bd) + (ae+bf)$ — Commutative and associative properties of real numbers
$= \langle a, b\rangle\cdot\langle c, d\rangle + \langle a, b\rangle\cdot\langle e, f\rangle$ — Definition of dot product
$= \mathbf{u}\cdot\mathbf{v} + \mathbf{u}\cdot\mathbf{w}$

43.
$k(\mathbf{u}\cdot\mathbf{v}) = k(\langle a, b\rangle\cdot\langle c, d\rangle)$
$= k(ac+bd)$ — Definition of dot product
$= k(ac) + k(bd)$ — Distributive property of real numbers
$= (ka)c + (kb)d$ — Associative property for multiplication of real numbers
$= \langle ka, kb\rangle\cdot\langle c, d\rangle$ — Definition of dot product
$= (k\langle a, b\rangle)\cdot\langle c, d\rangle$ — Definition of scalar multiplication
$= (k\mathbf{u})\cdot\mathbf{v}$

Also, $k(\mathbf{u}\cdot\mathbf{v}) = k(\langle a, b\rangle\cdot\langle c, d\rangle)$
$= k(ac+bd)$ — Definition of dot product
$= k(ac) + k(bd)$ — Distributive property of real numbers
$= a(kc) + b(kd)$ — Commutative and associative properties of real numbers
$= \langle a, b\rangle\cdot\langle kc, kd\rangle$ — Definition of dot product
$= \langle a, b\rangle\cdot(k\langle c, d\rangle)$ — Definition of scalar multiplication
$= \mathbf{u}\cdot(k\mathbf{v})$

45. $\text{Proj}_{\mathbf{v}}\ \mathbf{u} = \dfrac{\mathbf{u}\cdot\mathbf{v}}{\mathbf{v}\cdot\mathbf{v}}\mathbf{v} = \dfrac{\langle 3,4\rangle\cdot\langle 4,0\rangle}{\langle 4,0\rangle\cdot\langle 4,0\rangle}\langle 4, 0\rangle = \dfrac{12+0}{16+0}\langle 4, 0\rangle = \dfrac{3}{4}\langle 4, 0\rangle = \langle 3, 0\rangle$

47. $\text{Proj}_{\mathbf{v}}\,\mathbf{u} = \dfrac{\mathbf{u}\cdot\mathbf{v}}{\mathbf{v}\cdot\mathbf{v}}\,\mathbf{v} = \dfrac{(-6\mathbf{i}+3\mathbf{j})\cdot(-3\mathbf{i}-2\mathbf{j})}{(-3\mathbf{i}-2\mathbf{j})\cdot(-3\mathbf{i}-2\mathbf{j})}(-3\mathbf{i}-2\mathbf{j}) = \dfrac{18-6}{9+4}(-3\mathbf{i}-2\mathbf{j}) = \dfrac{12}{13}(-3\mathbf{i}-2\mathbf{j})$

$= -\dfrac{36}{13}\,\mathbf{i} - \dfrac{24}{13}\,\mathbf{j}$

49. If $\mathbf{u}\cdot\mathbf{v} = |\mathbf{u}||\mathbf{v}|$
then the angle θ between **u** and **v** is given by

$$\cos\theta = \frac{\mathbf{u}\cdot\mathbf{v}}{|\mathbf{u}||\mathbf{v}|}$$

$$\cos\theta = \frac{|\mathbf{u}||\mathbf{v}|}{|\mathbf{u}||\mathbf{v}|}$$

$$\cos\theta = 1$$

The only relevant solution of this equation is $\theta = 0°$; therefore the two vectors have the same direction.

51. If $0 < \mathbf{u}\cdot\mathbf{v} < |\mathbf{u}||\mathbf{v}|$

then $\dfrac{0}{|\mathbf{u}||\mathbf{v}|} < \dfrac{\mathbf{u}\cdot\mathbf{v}}{|\mathbf{u}||\mathbf{v}|} < \dfrac{|\mathbf{u}||\mathbf{v}|}{|\mathbf{u}||\mathbf{v}|}$

$$0 < \cos\theta < 1,$$

where θ is the angle between the two vectors. Therefore, θ must be acute.

53. The statement is false. Consider for example $\mathbf{u} = \langle 1, 0\rangle$ and $\mathbf{v} = \langle 0, 1\rangle$. Neither is a zero vector; however, $\mathbf{u}\cdot\mathbf{v} = 1\cdot 0 + 0\cdot 1 = 0$.

55. The statement is obviously true if $\overline{u} = \overline{0}$ or $\overline{v} = \overline{0}$, so we may assume that $\overline{u} \neq \overline{0}$ and $\overline{v} \neq \overline{0}$. Since

$$\cos\theta = \frac{\mathbf{u}\cdot\mathbf{v}}{|\mathbf{u}||\mathbf{v}|}$$

it follows that

$$\left|\frac{\mathbf{u}\cdot\mathbf{v}}{|\mathbf{u}||\mathbf{v}|}\right| = |\cos\theta|$$

$$\frac{|\mathbf{u}\cdot\mathbf{v}|}{|\mathbf{u}||\mathbf{v}|} = |\cos\theta|$$

Since $0 \le |\cos\theta| \le 1$ for all θ, it follows that

$$0 \le \frac{\mathbf{u}\cdot\mathbf{v}}{|\mathbf{u}||\mathbf{v}|} \le 1$$

Hence, $0 \le |\mathbf{u}\cdot\mathbf{v}| \le |\mathbf{u}||\mathbf{v}|$ for all vectors **u** and **v**.

57. $W = \begin{pmatrix}\text{component of force in}\\ \text{the direction of motion}\end{pmatrix}(\text{displacement}) = [(15\text{ lb})\cos 42°](440\text{ ft}) = 4{,}900\text{ ft-lb}$

59. Now, $W = \begin{pmatrix}\text{component of force in}\\ \text{the direction of motion}\end{pmatrix}(\text{displacement}) = [(15\text{ lb})\cos 30°](440\text{ ft}) = 5{,}700\text{ ft-lb}$

$5{,}700 - 4{,}900 = 800$ ft-lb more work is done.

61. $\mathbf{d} = \langle 8, 1\rangle$ $\quad W = \mathbf{F}\cdot\mathbf{d} = \langle 10, 5\rangle\cdot\langle 8, 1\rangle = 85$ ft-lb

63. $\mathbf{d} = \langle -3, 1\rangle = -3\mathbf{i}+\mathbf{j}$ $\quad W = \mathbf{F}\cdot\mathbf{d} = (-2\mathbf{i}+3\mathbf{j})\cdot(-3\mathbf{i}+\mathbf{j}) = 9$ ft-lb

65. $\mathbf{d} = \langle 1, 1\rangle = \mathbf{i}+\mathbf{j}$ $\quad W = \mathbf{F}\cdot\mathbf{d} = (10\mathbf{i}+10\mathbf{j})\cdot(\mathbf{i}+\mathbf{j}) = 20$ ft-lb

67. To prove that $\angle ACB$, an angle inscribed in a semicircle, is a right angle, we need only show that $\mathbf{c} - \mathbf{a}$ and $\mathbf{c} + \mathbf{a}$ are orthogonal.

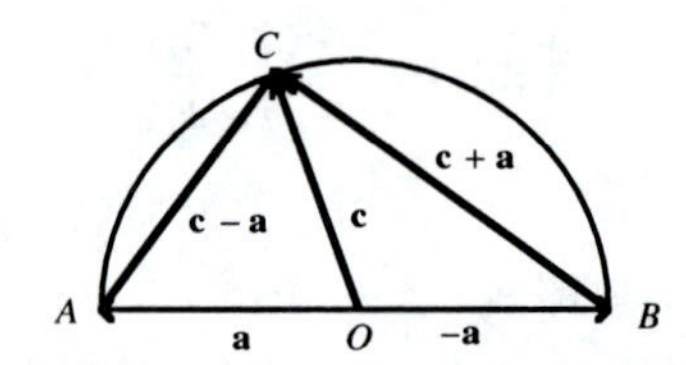

But $(\mathbf{c} - \mathbf{a}) \cdot (\mathbf{c} + \mathbf{a}) = (\mathbf{c} - \mathbf{a}) \cdot \mathbf{c} + (\mathbf{c} - \mathbf{a}) \cdot \mathbf{a}$ Distributive property of dot product

$= \mathbf{c} \cdot \mathbf{c} - \mathbf{a} \cdot \mathbf{c} + \mathbf{c} \cdot \mathbf{a} - \mathbf{a} \cdot \mathbf{a}$ Distributive property of dot product

$= \mathbf{c} \cdot \mathbf{c} - \mathbf{a} \cdot \mathbf{a} - \mathbf{a} \cdot \mathbf{c} + \mathbf{c} \cdot \mathbf{a}$ Commutative and associative properties of real numbers

$= \mathbf{c} \cdot \mathbf{c} - \mathbf{a} \cdot \mathbf{a} - \mathbf{a} \cdot \mathbf{c} + \mathbf{a} \cdot \mathbf{c}$ Commutative property of dot product

$= \mathbf{c} \cdot \mathbf{c} - \mathbf{a} \cdot \mathbf{a}$ Additive inverse and additive identity properties of real numbers

$= |\mathbf{c}|^2 - |\mathbf{a}|^2$ Magnitude property of dot product (proved in problem 33)

$= (\text{Radius})^2 - (\text{Radius})^2$

$= 0$

Therefore, $\mathbf{c} - \mathbf{a}$ and $\mathbf{c} + \mathbf{a}$ are orthogonal, and an arbitrary angle ACB, inscribed in a semicircle, is a right angle.

CHAPTER 6 REVIEW EXERCISE

1. The law of sines needs to have an angle and a side opposite the angle given, which is not the case here.
2. The two forces are oppositely directed, that is, the angle between the two forces is 180°.
3. The forces are acting in the same direction, that is, the angle between the two forces is 0°.
4. (A)

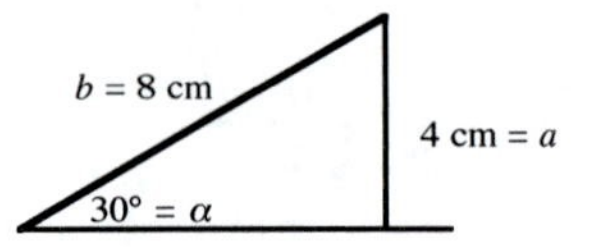

$h = b \sin \alpha = 8 \sin 30°$
$= 4 = a$

This is the case of SSA where α is acute and $a = h$.
1 triangle can be constructed.

(B)

$h = b \sin \alpha = 7 \sin 30°$
$= 3.5 < 5 = a$

This is the case of SSA where α is acute and $h < a < b$.
2 triangles can be constructed.

(C)

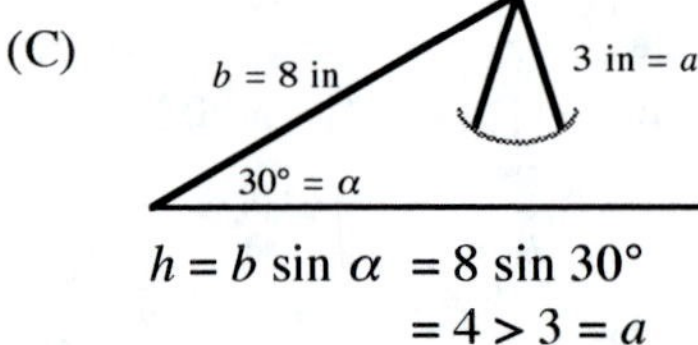

$h = b \sin \alpha = 8 \sin 30°$
$= 4 > 3 = a$

This is the case where α is acute and $0 < a < h$.
0 triangles can be constructed.

5. We are given two angles and the included side (*ASA*). We use the law of sines.

Solve for β: $\alpha + \beta + \gamma = 180°$

$\beta = 180° - (105° + 53°) = 22°$

Solve for β: $\dfrac{\sin\alpha}{a} = \dfrac{\sin\beta}{b}$

$a = \dfrac{b\sin\alpha}{\sin\beta} = \dfrac{(42\text{ cm})(\sin 53°)}{\sin 22°} = 90\text{ cm}$

Solve for c: $\dfrac{\sin\beta}{b} = \dfrac{\sin\gamma}{c}$

$c = \dfrac{b\sin\gamma}{\sin\beta} = \dfrac{(42\text{ cm})(\sin 105°)}{\sin 22°} = 110\text{ cm}$

6. 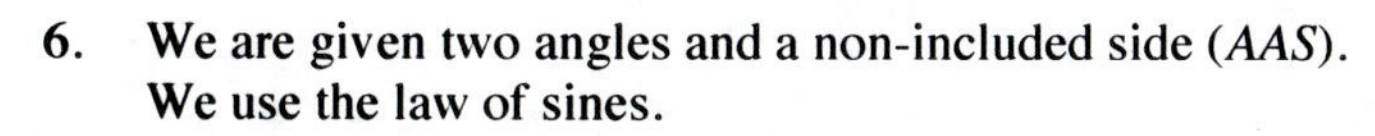
We are given two angles and a non-included side (*AAS*). We use the law of sines.

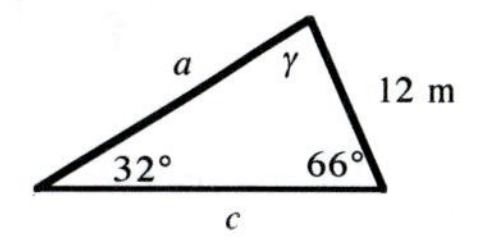

Solve for γ: $\alpha + \beta + \gamma = 180°$

$\gamma = 180° - (66° + 32°) = 82°$

Solve for a: $\dfrac{\sin\alpha}{a} = \dfrac{\sin\beta}{b}$

$a = \dfrac{b\sin\alpha}{\sin\beta} = \dfrac{(12\text{ m})(\sin 66°)}{\sin 32°} = 21\text{ m}$

Solve for c: $\dfrac{\sin\gamma}{c} = \dfrac{\sin\beta}{b}$

$c = \dfrac{b\sin\gamma}{\sin\beta} = \dfrac{(12\text{ m})(\sin 82°)}{\sin 32°} = 22\text{ m}$

7. We are given two sides and the included angle (*SAS*). We use the law of cosines to find the third side, then the law of sines to find a second angle.

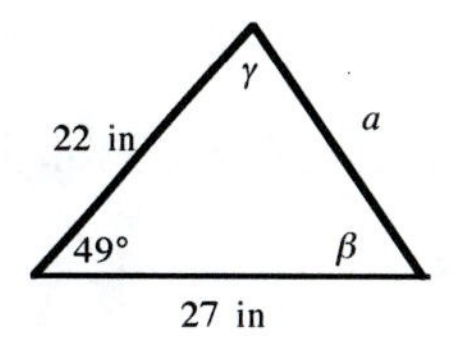

Solve for a: $a^2 = b^2 + c^2 - 2bc\cos\alpha$

$= 22^2 + 27^2 - 2(22)(27)\cos 49° = 433.60187\ldots$

$a = \sqrt{433.60187\ldots} = 21\text{ in}$

Since b is shorter than side c, β, the angle opposite b, must be acute.

Solve for β: We use the law of sines. $\dfrac{\sin\beta}{b} = \dfrac{\sin\alpha}{a}$

$\sin\beta = \dfrac{b\sin\alpha}{a} = \dfrac{(22\text{ in})(\sin 49°)}{21\text{ in}} = 0.7974$

$\beta = \sin^{-1} 0.7974 = 53°$

Solve for γ: $\gamma = 180° - (\alpha + \beta) = 180° - (49° + 53°) = 78°$

8. We are given two sides and a non-included angle (*SSA*). α is acute. $a > b$. One triangle is possible. We use the law of sines.

Solve for β: $\dfrac{\sin\alpha}{a} = \dfrac{\sin\beta}{b}$

$$\sin\beta = \frac{b\sin\alpha}{\sin a} = \frac{(12\text{ cm})(\sin 62°)}{14\text{ cm}} = 0.7568$$

$$\beta = \sin^{-1} 0.7568 = 49°$$

(There is another solution of $\sin\beta = 0.7568$ that deserves brief consideration: $\beta' = 180° - \sin^{-1} 0.7568 = 131°$. However, there is not enough room in a triangle for an angle of 62° and an angle of 131°, since their sum is greater than 180°.)

Solve for γ: $\alpha + \beta + \gamma = 180°$ $\qquad \gamma = 180° - (62° + 49°) = 69°$

Solve for c: $\dfrac{\sin\beta}{b} = \dfrac{\sin\gamma}{c}$

$$c = \frac{b\sin\gamma}{\sin\beta} = \frac{(12\text{ cm})(\sin 69°)}{\sin 49°} = 15\text{ cm}$$

9. The given information consists of two sides and the included angle; hence, we use the formula

$$A = \frac{ab}{2}\sin\theta \text{ in the form } A = \frac{1}{2}bc\sin\alpha = \frac{1}{2}(22\text{ in})(27\text{ in})\sin 49° = 224\text{ in}^2$$

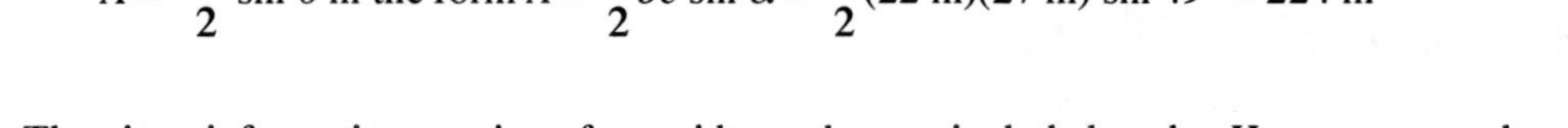

10. The given information consists of two sides and a non-included angle. Hence, we use the information computed in Problem 8 in the formula

$$A = \frac{ab}{2}\sin\theta \text{ in the form } A = \frac{1}{2}ab\sin\gamma = \frac{1}{2}(14\text{ cm})(12\text{ cm})\sin 69° = 79\text{ cm}^2$$

11. To find $|\mathbf{u} + \mathbf{v}|$: Apply the Pythagorean theorem to triangle *OCB*.

$$|\mathbf{u} + \mathbf{v}|^2 = OB^2 = OC^2 + BC^2 = 8.0^2 + 5.0^2 = 89.00$$

$$|\mathbf{u} + \mathbf{v}| = \sqrt{89.00} = 9.4$$

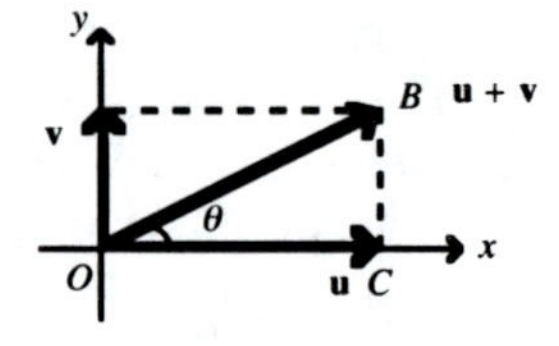

Solve triangle *OCB* for θ: $\tan\theta = \dfrac{BC}{OC} = \dfrac{|\mathbf{v}|}{|\mathbf{u}|}$ $\qquad \theta = \tan^{-1}\dfrac{|\mathbf{v}|}{|\mathbf{u}|}$ $\qquad \theta$ is acute

$$\theta = \tan^{-1}\frac{5.0}{8.0} = 32°$$

12. $|\mathbf{v}| = 12$ $\qquad \theta = 35°$

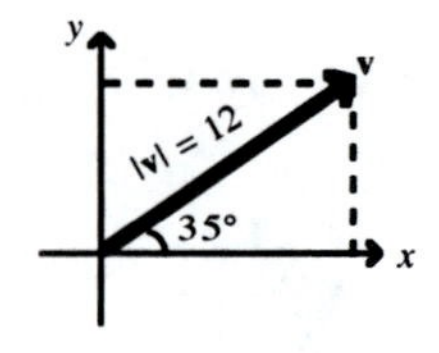

Horizontal component H: $\cos 35° = \dfrac{H}{12}$ $\qquad H = 12\cos 35° = 9.8$

Vertical component V: $\sin 35° = \dfrac{V}{12}$ $\qquad V = 12\sin 35° = 6.9$

Then $\mathbf{v} = H\mathbf{i} + V\mathbf{j} = 9.8\mathbf{i} + 6.9\mathbf{j}$.

13. The algebraic vector $\langle a, b\rangle$ has coordinates given by

$$a = x_b - x_a = (-1) - (-3) = 2 \qquad b = y_b - y_a = (-3) - 2 = -5$$

Hence, $\langle a, b\rangle = \langle 2, -5\rangle$

14. Magnitude of $\langle -5, 12\rangle = |\langle -5, 12\rangle| = \sqrt{a^2+b^2} = \sqrt{(-5)^2+12^2} = 13$

15. $\langle 2,-1\rangle \cdot \langle -3,2\rangle = 2\cdot(-3) + (-1)\cdot 2 = -8$

16. $(2\mathbf{i}+\mathbf{j})\cdot(3\mathbf{i}-2\mathbf{j}) = 2\cdot 3 + 1\cdot(-2) = 4$

17. $|\mathbf{u}| = \sqrt{4^2+3^2} = 5$ $\quad |\mathbf{v}| = \sqrt{3^2+0^2} = 3$

$\cos\theta = \dfrac{\mathbf{u}\cdot\mathbf{v}}{|\mathbf{u}||\mathbf{v}|} = \dfrac{\langle 4,3\rangle\cdot\langle 3,0\rangle}{(5)(3)} = \dfrac{12+0}{15} = \dfrac{12}{15}$ $\quad \theta = \cos^{-1}\dfrac{12}{15} = 36.9°$

18. $|\mathbf{u}| = \sqrt{5^2+1^2} = \sqrt{26}$ $\quad |\mathbf{v}| = \sqrt{(-2)^2+2^2} = \sqrt{8}$

$\cos\theta = \dfrac{\mathbf{u}\cdot\mathbf{v}}{|\mathbf{u}||\mathbf{v}|} = \dfrac{(5\mathbf{i}+\mathbf{j})\cdot(2\mathbf{i}+2\mathbf{j})}{\sqrt{26}\sqrt{8}} = \dfrac{5(-2)+1\cdot 2}{\sqrt{26}\sqrt{8}} = \dfrac{-8}{\sqrt{26}\sqrt{8}}$ $\quad \theta = \cos^{-1}\dfrac{-8}{\sqrt{26}\sqrt{8}} = 123.7°$

19. (A) Let $\mathbf{u} = \langle 4,-3\rangle$ and $\mathbf{v} = \langle 8,6\rangle$

$|\mathbf{u}| = \sqrt{4+(-3)^2} = 5$ $\quad |\mathbf{v}| = \sqrt{8^2+6^2} = 10$

$\cos\theta = \dfrac{\mathbf{u}\cdot\mathbf{v}}{|\mathbf{u}||\mathbf{v}|} = \dfrac{\langle 4,-3\rangle\cdot\langle 8,6\rangle}{5(10)} = \dfrac{4(8)-3(6)}{5(10)} = \dfrac{14}{50}$

For $\mathbf{u}$ and $\mathbf{v}$ to be parallel, θ must be 0 or π, that is, $\cos\theta = \pm 1$.

For $\mathbf{u}$ and $\mathbf{v}$ to be orthogonal, θ must be $\dfrac{\pi}{2}$, that is, $\cos\theta = 0$.

Since neither of these holds here, $\mathbf{u}$ and $\mathbf{v}$ are neither parallel nor orthogonal.

(B) Let $\mathbf{u} = \left\langle \dfrac{5}{2}, -\dfrac{1}{2}\right\rangle$ and $\mathbf{v} = \langle -10,2\rangle$

$|\mathbf{u}| = \sqrt{\left(\dfrac{5}{2}\right)^2 + \left(-\dfrac{1}{2}\right)^2} = \dfrac{\sqrt{26}}{2}$ $\quad |\mathbf{v}| = \sqrt{(-10)^2+2^2} = \sqrt{104}$

$\cos\theta = \dfrac{\mathbf{u}\cdot\mathbf{v}}{|\mathbf{u}||\mathbf{v}|} = \dfrac{\left\langle \frac{5}{2}, -\frac{1}{2}\right\rangle\cdot\langle -10,2\rangle}{\frac{\sqrt{26}}{2}\cdot\sqrt{104}} = \dfrac{\frac{5}{2}(-10) - \frac{1}{2}(2)}{\frac{\sqrt{26}}{2}\sqrt{104}} = -1$

$\theta = \cos^{-1}(-1) = \pi$, hence these vectors are parallel.

(C) Let $\mathbf{u} = \langle 10,-6\rangle$ and $\mathbf{v} = \langle 3,5\rangle$

$|\mathbf{u}| = \sqrt{10^2+(-6)^2} = \sqrt{136}$ $\quad |\mathbf{v}| = \sqrt{3^2+5^2} = \sqrt{34}$

$\cos\theta = \dfrac{\mathbf{u}\cdot\mathbf{v}}{|\mathbf{u}||\mathbf{v}|} = \dfrac{\langle 10,-6\rangle\cdot\langle 3,5\rangle}{\sqrt{136}\sqrt{34}} = \dfrac{10(3)-6(5)}{\sqrt{136}\sqrt{34}} = 0$

$\theta = \cos^{-1}(0) = \dfrac{\pi}{2}$, hence these vectors are orthogonal.

20. No triangle is possible, since $\sin\beta$ cannot exceed one.

21. We are given two sides and the included angle (*SAS*). We use the law of cosines to find the third side, then the law of sines to find a second angle.

β
a
72.4 m
65.0°
γ
103 m

Solve for a:

$$a^2 = b^2 + c^2 - 2bc \cos \alpha$$
$$= (103)^2 + (72.4)^2 - 2(103)(72.4) \cos 65.0° = 9{,}547.6622\ldots$$
$$a = \sqrt{9{,}547.6622\ldots} = 97.7 \text{ m}$$

Since c is the shorter of the remaining sides, γ, the angle opposite c, must be acute.

Solve for γ: $\dfrac{\sin \gamma}{c} = \dfrac{\sin \alpha}{a}$

$$\sin \gamma = \frac{c \sin \alpha}{a} = \frac{(72.4 \text{ m})(\sin 65.0°)}{97.7 \text{ m}} = 0.6715$$
$$\gamma = \sin^{-1} 0.6715 = 42.2°$$

Solve for β: $\alpha + \beta + \gamma = 180°$ $\qquad \beta = 180° - (\alpha + \gamma) = 180° - (65.0° + 42.2°) = 72.8°$

22. We are given two sides and a non-included angle (*SSA*). α is acute. $h = b \sin \alpha = 15.7 \sin 35°20' = 9.08$, $h < a < b$. Two triangles are possible, but β is specified acute. We use the law of sines.

Solve for β: $\dfrac{\sin \alpha}{a} = \dfrac{\sin \beta}{b}$

$$\sin \beta = \frac{b \sin \alpha}{a} = \frac{(15.7 \text{ in})(\sin 35°20')}{13.2 \text{ in}} = 0.6879$$

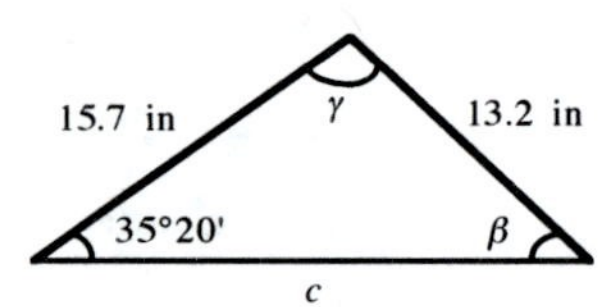

Since β is specified acute, we choose the acute angle solution to this equation,

$$\beta = \sin^{-1} 0.6879 = 43°30'$$

Solve for γ: $\alpha + \beta + \gamma = 180°$ $\qquad \gamma = 180° - (\alpha + \beta) = 180° - (35°20' + 43°30') = 101°10'$

Solve for c: $\dfrac{\sin \gamma}{c} = \dfrac{\sin \alpha}{a}$

$$c = \frac{a \sin \gamma}{\sin \alpha} = \frac{(13.2 \text{ in})(\sin 101°10')}{\sin 35°20'} = 22.4 \text{ in}$$

23. We are given the same information as in Problem 22, except that β is specified obtuse.

Solve for β: $\dfrac{\sin \alpha}{a} = \dfrac{\sin \beta}{b}$

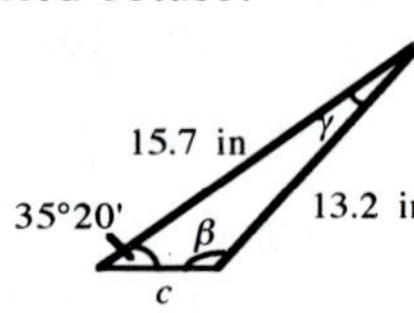

$$\sin \beta = \frac{b \sin \alpha}{a} = \frac{(15.7 \text{ in})(\sin 35°20')}{13.2 \text{ in}} = 0.6879$$

Since β is specified obtuse, we choose the obtuse angle solution to this equation,

$$\beta = 180° - \sin^{-1} 0.6879 = 136°30'$$

Solve for γ: $\alpha + \beta + \gamma = 180°$ $\qquad \gamma = 180° - (\alpha + \beta) = 180° - (35°20' + 136°30') = 8°10'$

Solve for c: $\frac{\sin \gamma}{c} = \frac{\sin \alpha}{a}$

$$c = \frac{a \sin \gamma}{\sin \alpha} = \frac{(13.2 \text{ in})(\sin 8°10')}{\sin 35°20'} = 3.24 \text{ in}$$

24. We are given three sides (*SSS*). We solve for the largest angle, γ (largest because it is opposite the largest side, c) using the law of cosines. We then solve for a second angle using the law of sines, because it involves simpler calculations.

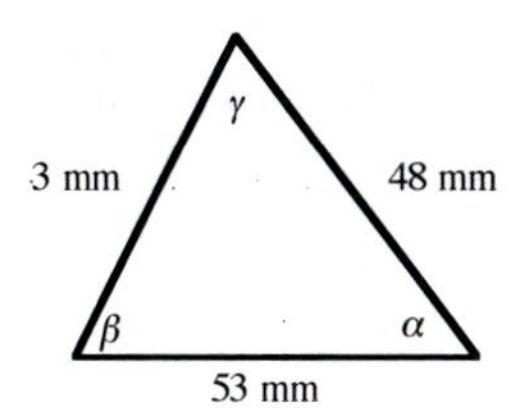

Solve for γ: $c^2 = a^2 + b^2 - 2ab \cos \gamma$

$$\cos \gamma = \frac{a^2 + b^2 - c^2}{2ab}$$

$$\gamma = \cos^{-1} \frac{a^2 + b^2 - c^2}{2ab} = \cos^{-1} \frac{43^2 + 48^2 - 53^2}{2 \cdot 43 \cdot 48} = 71°$$

Both β and α must be acute, since they are smaller than γ.

Solve for β: $\frac{\sin \beta}{b} = \frac{\sin \gamma}{c}$

$$\sin \beta = \frac{b \sin \gamma}{c}$$

$$\beta = \sin^{-1} \frac{b \sin \gamma}{c}$$

$$= \sin^{-1} \frac{(48 \text{ mm}) \sin 71°}{53 \text{ mm}} = 59°$$

Solve for α: $\alpha + \beta + \gamma = 180°$ $\qquad \alpha = 180° - (\beta + \gamma) = 180° - (59° + 71°) = 50°$

25. The given information consists of two sides and the included angle; hence, we use the formula $A = \frac{ab}{2} \sin \theta$ in the form $A = \frac{1}{2} bc \sin \alpha = \frac{1}{2}(103 \text{ m})(72.4 \text{ m}) \sin 65.0° = 3{,}380 \text{ m}^2$

26. The given information consists of three sides; hence, we use Heron's formula. First, find the semiperimeter s:

$$s = \frac{a + b + c}{2} = \frac{43 + 48 + 53}{2} = 72 \text{ mm}$$

Then, $s - a = 72 - 43 = 29$

$s - b = 72 - 48 = 24$

$s - c = 72 - 53 = 19$

Thus, $A = \sqrt{s(s-a)(s-b)(s-c)} = \sqrt{72(29)(24)(19)} = 980 \text{ mm}^2$

27. **u** and **v** have the same magnitudes, since $|\mathbf{u}| = |\mathbf{v}|$. However, if they have different directions, then **u** and **v** are not equal, since two geometric vectors are equal if and only if their magnitudes are equal and their directions are equal.

28. $\langle a, b \rangle = \langle c, d \rangle$ if and only if $a = c$ and $b = d$.

29. $\angle DOC = 45.0°$. Hence, $\angle OCB = 180° - 45.0° = 135.0°$.
We can find the magnitude of $\mathbf{u} + \mathbf{v}$ using the law of cosines:

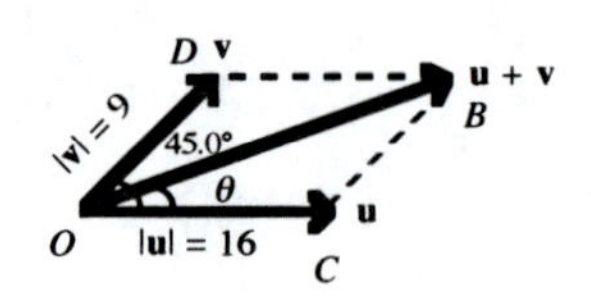

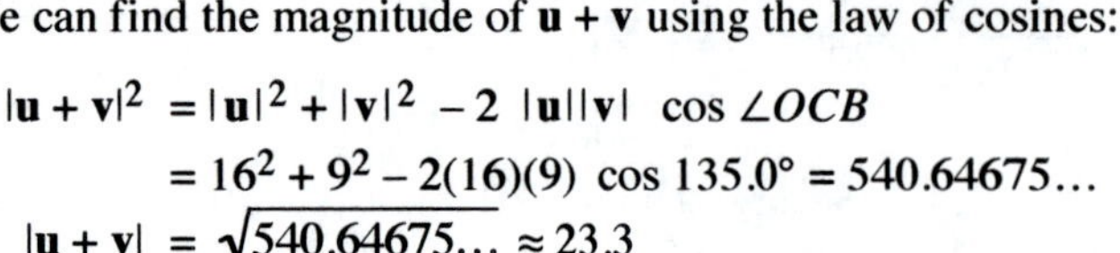

$$|\mathbf{u} + \mathbf{v}|^2 = |\mathbf{u}|^2 + |\mathbf{v}|^2 - 2\,|\mathbf{u}||\mathbf{v}|\,\cos\angle OCB$$
$$= 16^2 + 9^2 - 2(16)(9)\cos 135.0° = 540.64675\ldots$$
$$|\mathbf{u} + \mathbf{v}| = \sqrt{540.64675\ldots} \approx 23.3$$

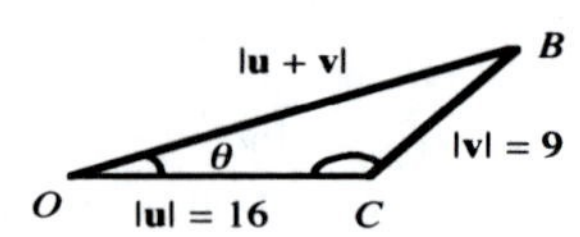

To find θ, we use the law of sines:
$$\frac{\sin\theta}{|\mathbf{v}|} = \frac{\sin\angle OCB}{|\mathbf{u}+\mathbf{v}|}$$
$$\frac{\sin\theta}{9} = \frac{\sin 135.0°}{23.3}$$
$$\sin\theta = \frac{9\sin 135.0°}{23.3}$$
$$\theta = \sin^{-1}\left(\frac{9\sin 135.0°}{23.3}\right) \approx 15.9°$$

30. (A) $\mathbf{u} + \mathbf{v} = \langle 4, 0\rangle + \langle -2, -3\rangle = \langle 2, -3\rangle$
(B) $\mathbf{u} - \mathbf{v} = \langle 4, 0\rangle - \langle -2, -3\rangle = \langle 6, 3\rangle$
(C) $3\mathbf{u} - 2\mathbf{v} = 3\langle 4, 0\rangle - 2\langle -2, -3\rangle = \langle 12, 0\rangle + \langle 4, 6\rangle = \langle 16, 6\rangle$
(D) $2\mathbf{u} - 3\mathbf{v} + \mathbf{w} = 2\langle 4, 0\rangle - 3\langle -2, -3\rangle + \langle 1, -1\rangle = \langle 8, 0\rangle + \langle 6, 9\rangle + \langle 1, -1\rangle = \langle 15, 8\rangle$

31. (A) $\mathbf{u} + \mathbf{v} = (3\mathbf{i} - \mathbf{j}) + (2\mathbf{i} - 3\mathbf{j}) = 3\mathbf{i} + 2\mathbf{i} - \mathbf{j} - 3\mathbf{j} = 5\mathbf{i} - 4\mathbf{j}$
(B) $\mathbf{u} - \mathbf{v} = (3\mathbf{i} - \mathbf{j}) - (2\mathbf{i} - 3\mathbf{j}) = 3\mathbf{i} - \mathbf{j} - 2\mathbf{i} + 3\mathbf{j} = \mathbf{i} + 2\mathbf{j}$
(C) $3\mathbf{u} - 2\mathbf{v} = 3(3\mathbf{i} - \mathbf{j}) - 2(2\mathbf{i} - 3\mathbf{j}) = 9\mathbf{i} - 3\mathbf{j} - 4\mathbf{i} + 6\mathbf{j} = 5\mathbf{i} + 3\mathbf{j}$
(D) $2\mathbf{u} - 3\mathbf{v} + \mathbf{w} = 2(3\mathbf{i} - \mathbf{j}) - 3(2\mathbf{i} - 3\mathbf{j}) + (-2\mathbf{j}) = 6\mathbf{i} - 2\mathbf{j} - 6\mathbf{i} + 9\mathbf{j} - 2\mathbf{j} = 0\mathbf{i} + 5\mathbf{j}$ or $5\mathbf{j}$

32. $|\mathbf{v}| = \sqrt{(-8)^2 + 15^2} = 17 \qquad \mathbf{u} = \frac{1}{|\mathbf{v}|}\mathbf{v} = \frac{1}{17}\langle -8, 15\rangle = \left\langle -\frac{8}{17}, \frac{15}{17}\right\rangle$

33. (A) $\mathbf{v} = \langle -5, 7\rangle = \langle -5, 0\rangle + \langle 0, 7\rangle = -5\langle 1, 0\rangle + 7\langle 0, 1\rangle = -5\mathbf{i} + 7\mathbf{j}$
(B) $\mathbf{v} = \langle 0, -3\rangle = -3\langle 0, 1\rangle = -3\mathbf{j}$
(C) $\mathbf{v} = \overrightarrow{AB} = \langle 0 - 4, (-3) - (-2)\rangle = \langle -4, -1\rangle = \langle -4, 0\rangle + \langle 0, -1\rangle = -4\langle 1, 0\rangle - \langle 0, 1\rangle = -4\mathbf{i} - \mathbf{j}$

34. (A) $\mathbf{u} \cdot \mathbf{v} = \langle -12, 3\rangle \cdot \langle 2, 8\rangle = -24 + 24 = 0$ Thus, **u** and **v** are orthogonal.
(B) $\mathbf{u} \cdot \mathbf{v} = (-4\mathbf{i} + \mathbf{j}) \cdot (-\mathbf{i} + 4\mathbf{j}) = 4 + 4 = 8 \neq 0$ Thus, **u** and **v** are not orthogonal.

35. (A) $\text{comp}_{\mathbf{v}}\,|\mathbf{u}| = \frac{\mathbf{u}\cdot\mathbf{v}}{|\mathbf{v}|} = \frac{\langle 4,5\rangle\cdot\langle 3,1\rangle}{|\langle 3,1\rangle|} = \frac{12+5}{\sqrt{3^2+1^2}} = \frac{17}{\sqrt{10}} \approx 5.38$

(B) $\text{comp}_{\mathbf{v}}\,|\mathbf{u}| = \frac{\mathbf{u}\cdot\mathbf{v}}{|\mathbf{v}|} = \frac{(-\mathbf{i}+4\mathbf{j})\cdot(3\mathbf{i}-\mathbf{j})}{|3\mathbf{i}-\mathbf{j}|} = \frac{-3-4}{\sqrt{3^2+(-1)^2}} = -\frac{7}{\sqrt{10}} \approx -2.21$

36. Any one of the force vectors must have the same magnitude as the resultant of the other two force vectors and be oppositely directed to the resultant of the other two.

37.
$$(a-b)\cos\frac{\gamma}{2} = c\sin\frac{\alpha-\beta}{2}$$
$$(8.42-11.5)\cos\frac{59.1°}{2} = 10.2\sin\frac{45.1°-75.8°}{2}$$
$$-2.68 \approx -2.70$$
The results agree to two significant digits; the results check.

38. From the diagram, we can see that k = altitude of any possible triangle.
Thus, $\sin\alpha = \frac{k}{b}$, $k = b\sin\alpha$.

$k = (12.7\text{ cm})\ \sin 52.3° = 10.0\text{ cm}$
If $0 < a < k$, there is no solution: (1) in the diagram.

If $a = k$, there is one solution.

If $k < a < b$, there are two solutions: (2) in the diagram.

39.
$\mathbf{u}+\mathbf{v} = \langle a, b\rangle + \langle c, d\rangle$
$= \langle a+c, b+d\rangle$ — Definition of vector addition
$= \langle c+a, d+b\rangle$ — Commutative property for addition of real numbers
$= \langle c, d\rangle + \langle a, b\rangle$ — Definition of vector addition
$= \mathbf{v}+\mathbf{u}$

40.
$\mathbf{u}\cdot\mathbf{v} = \langle a, b\rangle \cdot \langle c, d\rangle$
$= ac + bd$ — Definition of dot product
$= ca + db$ — Commutative property for multiplication of real numbers
$= \langle c, d\rangle \cdot \langle a, b\rangle$ — Definition of dot product
$= \mathbf{v}\cdot\mathbf{u}$

41.
$(mn)\mathbf{v} = (mn)\langle c, d\rangle$
$= \langle (mn)c, (mn)d\rangle$ — Definition of scalar multiplication
$= \langle m(nc), m(nd)\rangle$ — Associative property for multiplication of real numbers
$= m\langle nc, nd\rangle$ — Definition of scalar multiplication
$= m(n\langle c, d\rangle)$ — Definition of scalar multiplication
$= m(n\mathbf{v})$

42.
$m(\mathbf{u}\cdot\mathbf{v}) = m(\langle a, b\rangle \cdot \langle c, d\rangle)$
$= m(ac + bd)$ — Definition of dot product
$= m(ac) + m(bd)$ — Distributive property of real numbers
$= (ma)c + (mb)d$ — Associative property for multiplication of real numbers
$= \langle ma, mb\rangle \cdot \langle c, d\rangle$ — Definition of dot product
$= (m\langle a, b\rangle)\cdot\langle c, d\rangle$ — Definition of scalar multiplication
$= (m\mathbf{u})\cdot\mathbf{v}$

Also, $m(\mathbf{u} \cdot \mathbf{v}) = m(\langle a, b\rangle \cdot \langle c, d\rangle)$

$= m(ac + bd)$	Definition of dot product
$= m(ac) + m(bd)$	Distributive property of real numbers
$= a(mc) + b(md)$	Commutative and associative properties of real numbers
$= \langle a, b\rangle \cdot \langle mc, md\rangle$	Definition of dot product
$= \langle a, b\rangle \cdot (m\langle c, d\rangle)$	Definition of scalar multiplication
$= \mathbf{u} \cdot (m\mathbf{v})$	

43. $\mathbf{u} \cdot (\mathbf{v} + \mathbf{w}) = \langle a, b\rangle \cdot (\langle c, d\rangle + \langle e, f\rangle)$

$= \langle a, b\rangle \cdot \langle c + e,\ d + f\rangle$	Definition of vector addition
$= a(c + e) + b(d + f)$	Definition of dot product
$= ac + ae + bd + bf$	Distributive property of real numbers
$= (ac + bd) + (ae + bf)$	Commutative and associative properties of real numbers
$= \langle a, b\rangle \cdot \langle c, d\rangle + \langle a, b\rangle \cdot \langle e, f\rangle$	Definition of dot product
$= \mathbf{u} \cdot \mathbf{v} + \mathbf{u} \cdot \mathbf{w}$	

44. $\mathbf{u} \cdot \mathbf{u} = \langle a, b\rangle \cdot \langle a, b\rangle$

$= a^2 + b^2$	Definition of dot product
$= (\sqrt{a^2 + b^2})^2$	Definition of the square root of a real number
$= \lvert\mathbf{u}\rvert^2$	Definition of magnitude

45. Since the diagonals of a parallelogram bisect each other, we see that

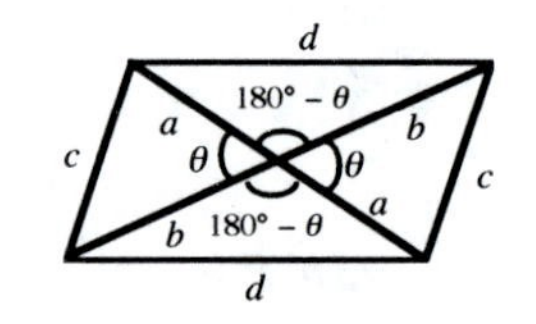

$$a = \frac{1}{2}(16.0 \text{ cm}) = 8.0 \text{ cm} \qquad b = \frac{1}{2}(20.0 \text{ cm}) = 10.0 \text{ cm}$$

$$\theta = 36.4° \qquad 180° - \theta = 143.6°$$

To find c and d, we use the law of cosines.

$$c^2 = a^2 + b^2 - 2ab \cos \theta$$
$$= (8.0)^2 + (10.0)^2 - 2(8.0)(10.0) \cos 36.4°$$
$$= 35.216992\ldots$$
$$c = \sqrt{35.216992\ldots} = 5.9 \text{ cm}$$
$$d^2 = a^2 + b^2 - 2ab \cos(180° - \theta) = (8.0)^2 + (10.0)^2 - 2(8.0)(10.0) \cos(143.6°) = 292.78301\ldots$$
$$d = \sqrt{292.78301\ldots} = 17 \text{ cm}$$

46. In triangle ACD, we note $\frac{AD}{h} = \cot 31°20'$.

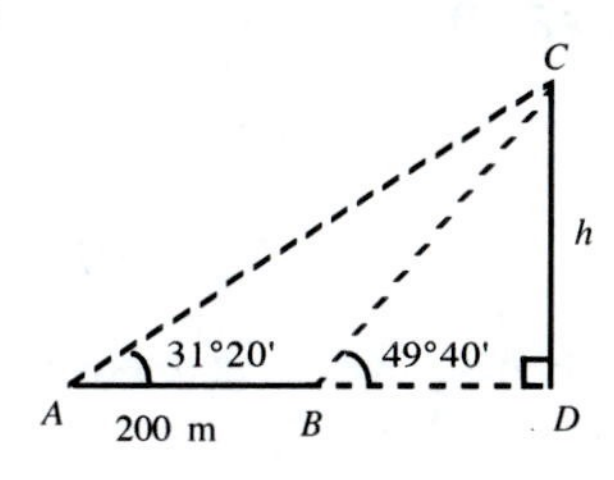

In triangle BCD, we note $\frac{BD}{h} = \cot 49°40'$.

Hence, $AD = h \cot 31°20'$, $BD = h \cot 49°40'$.

Since $AD - BD = AB = 200$ m, we have

200 m $= h \cot 31°20' - h \cot 49°40'$ or

$$h = \frac{200}{\cot 31°20' - \cot 49°40'} = 252 \text{ m}$$

Alternatively, we can note $\angle ABC = 180° - 49°40' = 130°20'$, and use the law of sines to determine BC.

$$\frac{\sin \angle CAB}{BC} = \frac{\sin \angle BCA}{AB}$$

$$\angle BCA = 180° - (\angle CAB + \angle ABC) = 180° - (31°20' + 130°20') = 18°20'$$

$$BC = \frac{AB \sin \angle CAB}{\sin \angle BCA} = \frac{(200 \text{ m})(\sin 31°20')}{\sin 18°20'} = 331 \text{ m}$$

Then, in right triangle BCD, we have

$$\frac{h}{BC} = \sin \angle CBD$$

$$h = BC \sin \angle CBD = (331 \text{ m})(\sin 49°40') = 252 \text{ m}$$

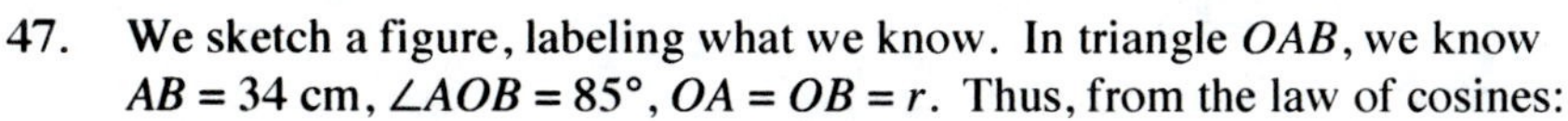

47. We sketch a figure, labeling what we know. In triangle OAB, we know $AB = 34$ cm, $\angle AOB = 85°$, $OA = OB = r$. Thus, from the law of cosines:

$$AB^2 = OA^2 + OB^2 - 2(OA)(OB) \cos AOB$$

$$34^2 = r^2 + r^2 - 2r^2 \cos 85° = 2r^2 - 2r^2 \cos 85° = r^2 (2 - 2 \cos 85°)$$

$$r^2 = \frac{34^2}{2 - 2\cos 85°} \qquad r = \sqrt{\frac{34^2}{2 - 2\cos 85°}} \quad \text{We discard the negative solution}$$

$$= 25 \text{ cm}$$

48. We will first find AD by applying the law of sines to triangle ABD, in which we know all angles and a side. We will then find AC by applying the law of sines to triangle ABC, in which we also know all angles and a side. We can then apply the law of cosines to triangle ADC to find DC.

In triangle ABD: $\frac{\sin \angle ABD}{AD} = \frac{\sin \angle ADB}{AB}$

$$AD = \frac{AB \sin \angle ABD}{\sin \angle ADB} = \frac{(450 \text{ ft}) \sin 72°}{\sin 50°} = 558.7 \text{ ft}$$

In triangle ABC: $\frac{\sin \angle ABC}{AC} = \frac{\sin \angle ACB}{AB}$

$$AC = \frac{AB \sin \angle ABC}{\sin \angle ACB} = \frac{(450 \text{ ft}) \sin (72° + 65°)}{\sin 20°} = 897.3 \text{ ft}$$

In triangle ACD:

$$CD^2 = AC^2 + AD^2 - 2(AC)(AD) \cos \angle CAD$$

$$= (558.7)^2 + (897.3)^2 - 2(558.7)(897.3) \cos 35°$$

$$= 295993.95\ldots$$

$$CD = 540 \text{ ft}$$

49. Area of $ABCD$ = Area of triangle ADC + Area of triangle ABC

$$= \frac{1}{2}(AD)(AC) \sin \angle DAC + \frac{1}{2}(AC)(AB) \sin \angle CAB$$

$$= \frac{1}{2}(558.7)(897.3) \sin 35° + \frac{1}{2}(897.3)(450) \sin 23° = 220{,}000 \text{ ft}^2$$

50. In triangle ABC, we are given two sides and the included angle; hence, we can apply the law of cosines to find side BC.

B
5.72 km
38.2°
A
Tunnel
6.37 km
C

$$BC^2 = AB^2 + AC^2 - 2(AB)(AC) \cos \angle BAC$$
$$= (5.72)^2 + (6.37)^2 - 2(5.72)(6.37) \cos 38.2°$$
$$= 16.027708\ldots$$
$$BC = 4.00 \text{ km}$$

51. In triangle CST, we are given $TS = 1{,}147$ miles and $TC = 3{,}964$ miles.

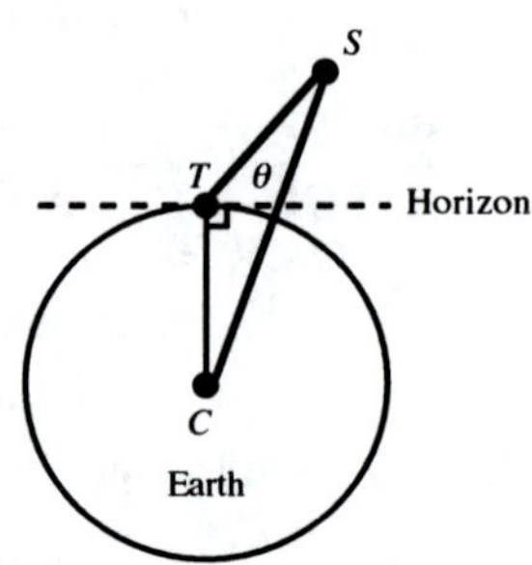

$$\angle STC = \theta + 90° = 28.6° + 90° = 118.6°.$$

We are given two sides and the included angle; hence, we can apply the law of cosines to find side SC.

$$SC^2 = TS^2 + TC - 2(TS)(TC) \cos STC$$
$$= (1{,}147)^2 + (3{,}964)^2 - 2(1{,}147)(3{,}964) \cos 118.6°$$
$$= 21{,}381{,}849\ldots$$
$$SC = 4{,}624 \text{ mi}$$

Hence, the height of the satellite = SC – radius of the earth = 4,624 mi – 3,964 mi = 660 mi

52. Following the hint, we find all angles for triangle ACS first.

$$\angle SAC = \theta + 90° = 21.7° + 90° = 111.7°$$

For $\angle ACS$, we use the formula $s = \dfrac{\pi}{180}\theta R$ from Chapter 2, with

$s = 632$ mi $\qquad \theta = \angle ACS \qquad R = 3{,}964$ mi

Then, $\angle ACS = \dfrac{180° s}{\pi R} = \dfrac{180°(632 \text{ mi})}{\pi(3{,}964 \text{ mi})} = 9.13°$. Since $\angle SAC + \angle ACS + \angle ASC = 180°$, we have

$$\angle ASC = 180° - (\angle ACS + \angle SAC) = 180° - (111.7° + 9.13°) = 59.2°$$

Now, apply the law of sines to find side CS:

$$\frac{\sin \angle SAC}{CS} = \frac{\sin \angle ASC}{AC}$$

$$CS = \frac{AC \angle \sin SAC}{\sin \angle ASC} = \frac{(3{,}964 \text{ mi}) \sin 111.7°}{\sin 59.2°} = 4{,}289 \text{ mi}$$

Hence, the height of the satellite above $B = BS = CS - BC$.

$$BS = 4{,}289 \text{ mi} - 3{,}964 \text{ mi} = 325 \text{ mi}$$

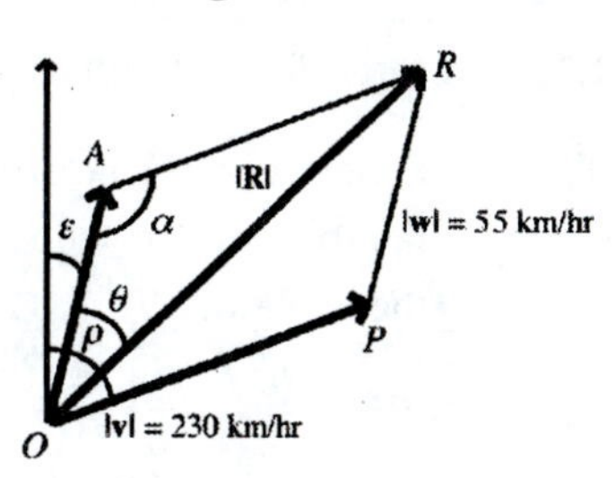

53. We are given $\varepsilon = 5°$ (wind heading), $\rho = 68°$ (plane heading). We want to find $|\mathbf{R}|$, where $\mathbf{R}$ is the resultant sum of $\mathbf{v}$, the plane's velocity, and $\mathbf{w}$, the wind velocity. We also want θ, then $\theta + \varepsilon = \theta + 5°$ will be the plane's actual direction relative to north.

Solve for $|\mathbf{R}|$:
In triangle ORA, since $\angle AOR = \rho - \varepsilon = 68° - 5° = 63°$, α must equal

$180° - \angle AOR = 180° - 63° = 117°$ $\quad AO = |\mathbf{w}| = 55$ km/hr $\quad AR = |\mathbf{v}| = 230$ km/hr

Now, apply the law of cosines.

$$|\mathbf{R}|^2 = |\mathbf{w}|^2 + |\mathbf{v}|^2 - 2|\mathbf{w}||\mathbf{v}| \cos\alpha = 55^2 + 230^2 - 2(55)(230)\cos 117° = 67{,}410.96\ldots$$
$$|\mathbf{R}| = \sqrt{67{,}410.96\ldots} \approx 260 \text{ km/hr}$$

Solve for θ: $\dfrac{\sin\theta}{AR} = \dfrac{\sin\alpha}{|\mathbf{R}|}$ $\qquad \dfrac{\sin\theta}{230} = \dfrac{\sin 117°}{260}$

$$\sin\theta = \frac{230}{260}\sin 117° \qquad \theta = \sin^{-1}\left(\frac{230}{260}\sin 117°\right) \approx 52°$$

The actual direction $= \theta + \varepsilon = 52° + 5° = 57°$.

54. Note that the plane has traveled 125 mi in the 1 hour. Draw a diagram.

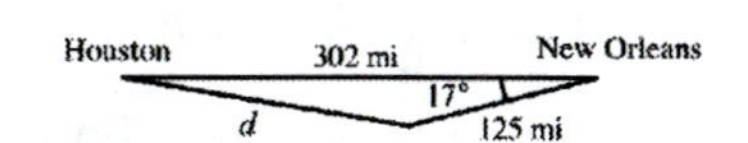

Then find d from the law of cosines

$$d^2 = (125)^2 + (302)^2 - 2(125)(302)\cos 17°$$
$$d = \sqrt{(125)^2 + (302)^2 - 2(125)(302)\cos 17°}$$
$$= 186 \text{ mi}$$

At 125 miles per hour, the trip will take 186 mi $\div$ 125 $\dfrac{\text{mi}}{\text{hr}}$ = 1.5 hr.

55. $\angle OCB = 180° - 42.3° = 137.7°$ $\qquad |\mathbf{F}_1| = 352$ lb $\qquad |\mathbf{F}_2| = 168$ lb

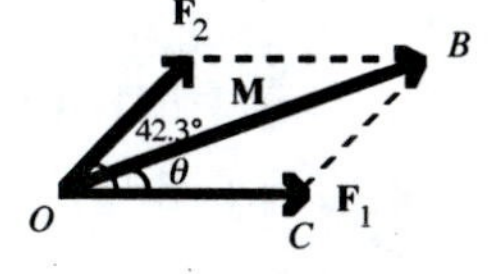

We can find the magnitude M of $\mathbf{F}_1 + \mathbf{F}_2$ using the law of cosines:

$$M^2 = |\mathbf{F}_1|^2 + |\mathbf{F}_2|^2 - 2\,|\mathbf{F}_1||\mathbf{F}_2| \cos OCB$$
$$= 352^2 + 168^2 - 2(352)(168)\cos 137.7° = 239{,}605.65\ldots$$
$$M = \sqrt{239{,}605.65\ldots} = 489 \text{ lb}$$

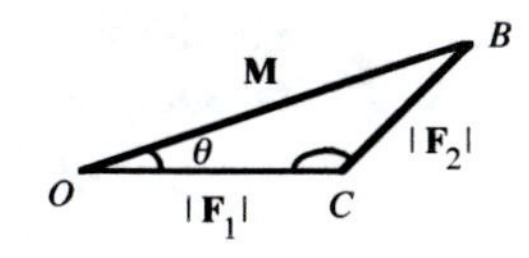

To find θ, we use the law of sines: $\dfrac{\sin\theta}{|\mathbf{F}_2|} = \dfrac{\sin\angle OCB}{M}$ $\qquad \dfrac{\sin\theta}{168} = \dfrac{\sin 137.7°}{489}$

$$\sin\theta = \frac{168}{489}\sin 137.7° \qquad \theta = \sin^{-1}\left(\frac{168}{489}\sin 137.7°\right) \approx 13.4° \text{ (relative to } \mathbf{F}_1\text{)}$$

56. $W = Fd = (489 \text{ lb})(22 \text{ ft}) = 10{,}800$ ft-lb

57. First, form a force diagram with all force vectors in standard position at the origin.

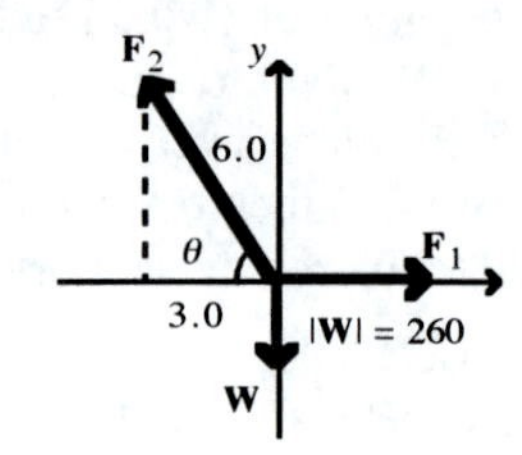

Let $\mathbf{F}_1$ = the force on the horizontal member *AB*

$\mathbf{F}_2$ = the force on the supporting member *BC*

$\mathbf{W}$ = the downward force (260 lb)

We note: $\cos\theta = \frac{3.0}{6.0}$ $\quad\theta = \cos^{-1}\frac{3.0}{6.0} = 60°$

Then write each force vector in terms of **i** and **j** unit vectors:

$\mathbf{F}_1 = |\mathbf{F}_1|\,\mathbf{i}$

$\mathbf{F}_2 = -|\mathbf{F}_2|(\cos 60°)\,\mathbf{i} + |\mathbf{F}_2|(\sin 60°)\,\mathbf{j}$

$\mathbf{W} = -260\mathbf{j}$

For the system to be in static equilibrium, we must have $\mathbf{F}_1 + \mathbf{F}_2 + \mathbf{W} = \mathbf{0}$ which becomes, on addition,

$$[\,|\mathbf{F}_1| - |\mathbf{F}_2|(\cos 60°)]\,\mathbf{i} + [\,|\mathbf{F}_2|(\sin 60°) - 260]\mathbf{j} = 0\mathbf{i} + 0\mathbf{j}.$$

Since two vectors are equal if and only if their corresponding components are equal, we are led to the following system of equations in $|\mathbf{F}_1|$ and $|\mathbf{F}_2|$:

$|\mathbf{F}_1| - |\mathbf{F}_2|(\cos 60°) = 0$

$|\mathbf{F}_2|(\sin 60°) - 260 = 0$

Solving, $|\mathbf{F}_2| = \frac{260}{\sin 60°} = 300$ lb

$|\mathbf{F}_1| = |\mathbf{F}_2|\cos 60° = 150$ lb

The force in the member *AB* is directed oppositely to the diagram, a compression of 150 lb.

The force in the member *BC* is also directed oppositely to the diagram, a tension of 300 lb.

58. (A) We write each force vector in terms of **i** and **j** unit vectors.

Note: $\alpha = 90° - \theta$.

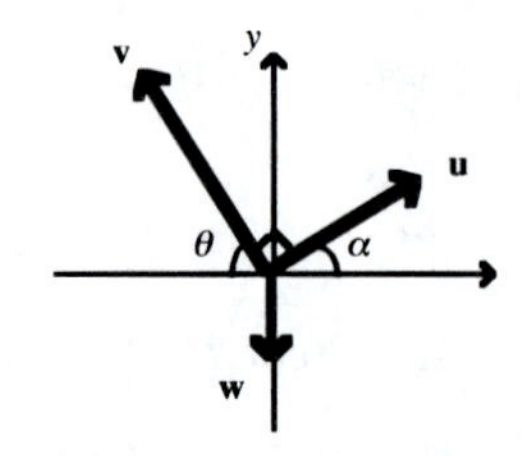

$\mathbf{u} = |\mathbf{u}|\cos\alpha\mathbf{i} + |\mathbf{u}|\sin\alpha\mathbf{j}$

$= |\mathbf{u}|\sin\theta\mathbf{i} + |\mathbf{u}|\cos\theta\mathbf{j}$

$\mathbf{v} = -|\mathbf{v}|\cos\theta\mathbf{i} + |\mathbf{v}|\sin\theta\mathbf{j}$

$\mathbf{w} = -|\mathbf{w}|\,\mathbf{j}$

For the system to be in static equilibrium, we must have $\mathbf{u} + \mathbf{v} + \mathbf{w} = \mathbf{0}$ which becomes, on addition,

$$[|\mathbf{u}|\sin\theta - |\mathbf{v}|\cos\theta]\mathbf{i} + [|\mathbf{u}|\cos\theta + |\mathbf{v}|\sin\theta - |\mathbf{w}|]\mathbf{j} = 0\mathbf{i} + 0\mathbf{j}$$

Since two vectors are equal if and only if their corresponding components are equal, we are let to the following system of equations in $|\mathbf{u}|$ and $|\mathbf{v}|$:

(1) $|\mathbf{u}|\sin\theta - |\mathbf{v}|\cos\theta = 0$

(2) $|\mathbf{u}|\cos\theta + |\mathbf{v}|\sin\theta - |\mathbf{w}| = 0$

Solving, we have:

(3) $\lvert\mathbf{u}\rvert \sin^2\theta - \lvert\mathbf{v}\rvert \sin\theta\cos\theta = 0$		Multiplying (1) by $\sin\theta$
(4) $\lvert\mathbf{u}\rvert \cos^2\theta + \lvert\mathbf{v}\rvert \sin\theta\cos\theta - \lvert\mathbf{w}\rvert\cos\theta = 0$		Multiplying (2) by $\cos\theta$
$\lvert\mathbf{u}\rvert(\sin^2\theta + \cos^2\theta) - \lvert\mathbf{w}\rvert\cos\theta = 0$		Adding (3) and (4)

Since $\sin^2\theta + \cos^2\theta = 1$ by the Pythagorean identity, we have $\lvert\mathbf{u}\rvert = \lvert\mathbf{w}\rvert\cos\theta$.

Substituting this result into (1), we have
$\lvert\mathbf{w}\rvert\cos\theta\sin\theta - \lvert\mathbf{v}\rvert\cos\theta = 0$

Hence, $\lvert\mathbf{v}\rvert = \dfrac{\lvert\mathbf{w}\rvert\cos\theta\sin\theta}{\cos\theta}$ $\qquad \lvert\mathbf{v}\rvert = \lvert\mathbf{w}\rvert\sin\theta$

(B) We are given $\lvert\mathbf{w}\rvert = 130$ lb and $\theta = 72°$. Hence,
$\lvert\mathbf{u}\rvert = \lvert\mathbf{w}\rvert\cos\theta = (130\text{ lb})\cos 72° = 40$ lb
and
$\lvert\mathbf{v}\rvert = \lvert\mathbf{w}\rvert\sin\theta = (130\text{ lb})\sin 72° = 124$ lb

(C) We are given $\lvert\mathbf{u}\rvert = \frac{1}{6}\lvert\mathbf{w}\rvert$. Hence

$\frac{1}{6}\lvert\mathbf{w}\rvert = \lvert\mathbf{w}\rvert\cos\theta \qquad \cos\theta = \frac{1}{6} \qquad \theta = \cos^{-1}\frac{1}{6} = 80°$

59. $W = \mathbf{F}\cdot\mathbf{d} = \langle -5, 8\rangle \cdot \langle -8, 2\rangle = 56$ ft-lb

60. First, form a force diagram with all force vectors in standard position at the origin.

Let $\mathbf{F}_1$ = the force exerted by the weight (20 lb)
$\mathbf{F}_2$ = the tension on the line
$\mathbf{W}$ = the weight of the leg (12 lb)

Then write each force vector in terms of **i** and **j** unit vectors.

$\mathbf{F}_1 = 20\cos 20°\mathbf{i} + 20\sin 20°\mathbf{j}$

$\mathbf{F}_2 = -\lvert\mathbf{F}_2\rvert\cos\theta\,\mathbf{i} + \lvert\mathbf{F}_2\rvert\sin\theta\,\mathbf{j}$

$\mathbf{W} = -12\mathbf{j}$

For the system to be in static equilibrium, we must have $\mathbf{F}_1 + \mathbf{F}_2 + \mathbf{W} = \mathbf{0}$ which becomes, on addition,

$[20\cos 20° - \lvert\mathbf{F}_2\rvert\cos\theta]\,\mathbf{i} + [20\sin 20° + \lvert\mathbf{F}_2\rvert\sin\theta - 12]\,\mathbf{j} = \mathbf{0}$

Since two vectors are equal if and only if their corresponding components are equal, we are led to the following system of equations in $\lvert\mathbf{F}_2\rvert$ and θ:

$20\cos 20° - \lvert\mathbf{F}_2\rvert\cos\theta = 0$
$20\sin 20° + \lvert\mathbf{F}_2\rvert\sin\theta - 12 = 0$

Solving,

$$|\mathbf{F}_2| \cos\theta = 20\cos 20°$$

$$|\mathbf{F}_2| = 20\frac{\cos 20°}{\cos\theta}$$

$$20\sin 20° + 20\frac{\cos 20°}{\cos\theta}\sin\theta - 12 = 0$$

$$20\cos 20° \tan\theta = 12 - 20\sin 20°$$

$$\tan\theta = \frac{12 - 20\sin 20°}{20\cos 20°}$$

$$\theta = \tan^{-1}\frac{12 - 20\sin 20°}{20\cos 20°} = 15.4°$$

$$|\mathbf{F}_2| = 20\frac{\cos 20°}{\cos\theta} = 19.5 \text{ lb}$$

Chapter 7 Polar Coordinates; Complex Numbers

EXERCISE 7.1 Polar and Rectangular Coordinates

1. r gives the distance of P from the pole (origin).

3. Q is the reflection of P across the polar axis (x axis).

5.

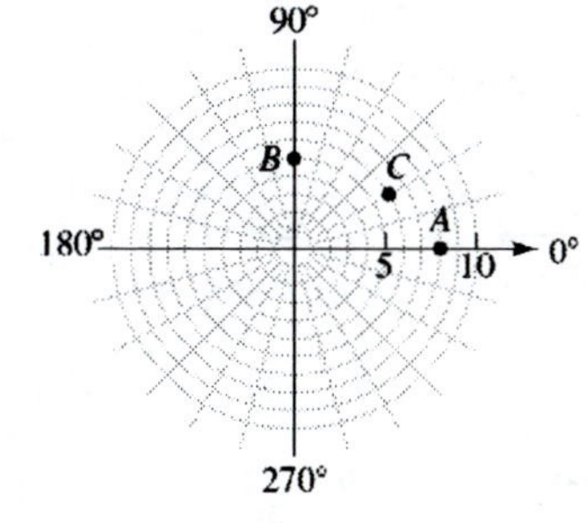

7.

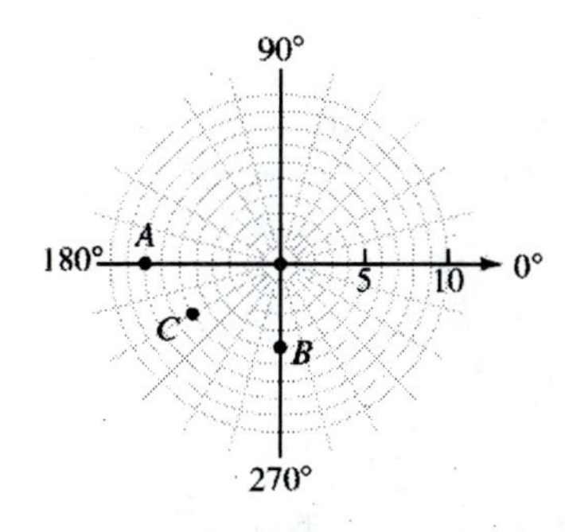

9.

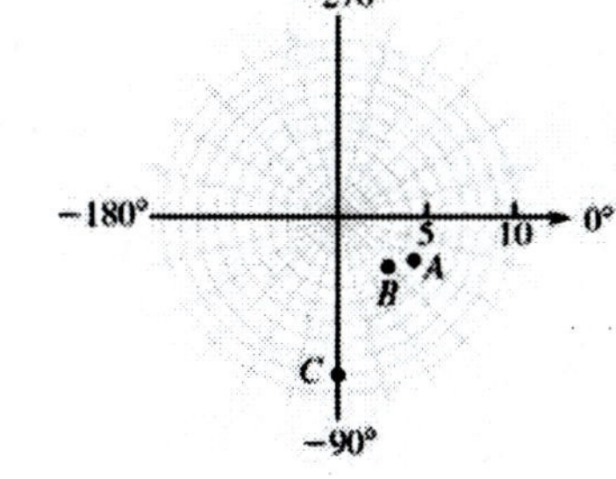

11.

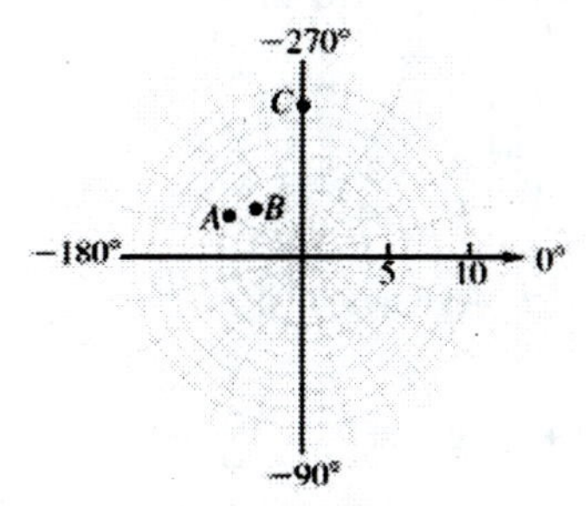

13.

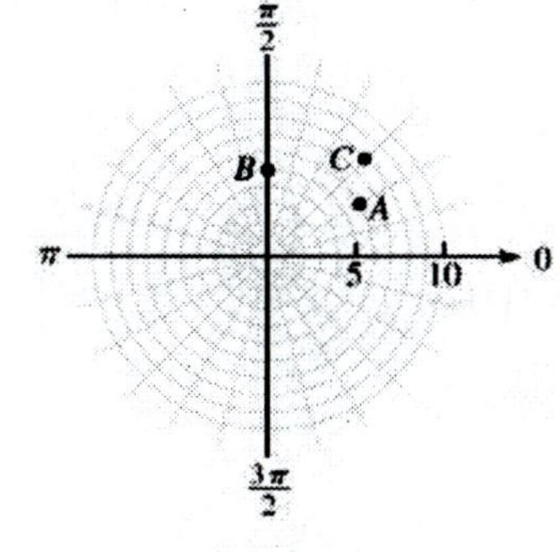

15.

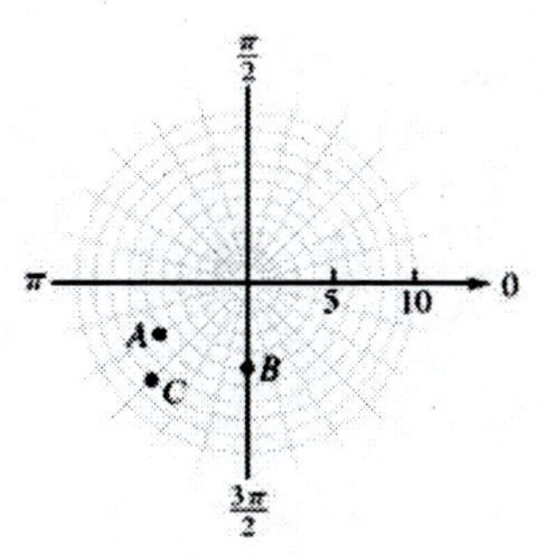

17.

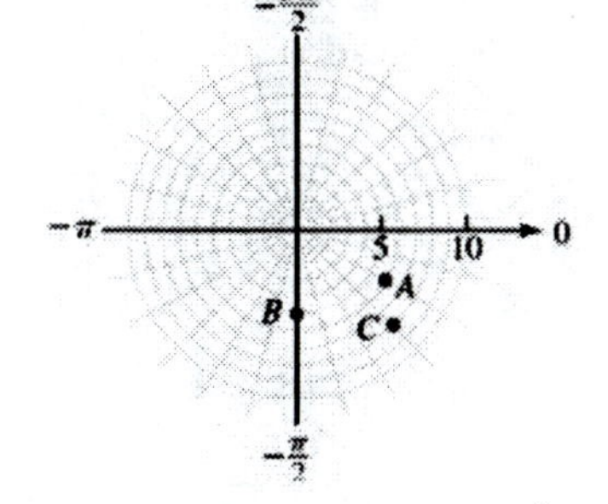

19. Yes. Any point can be represented by (r, θ) and by $(r, \theta + 2\pi)$, among other possibilities.

21. Given $P(x, y)$, then one pair of polar coordinates can be found by taking $r = \sqrt{x^2 + y^2}$ and if $x \neq 0$ and $y \neq 0$, θ as a solution of $\tan\theta = \frac{y}{x}$. The appropriate solution is determined by the quadrant in which P lies. If $x = 0$, then if P is on the positive y axis, $\theta = \frac{\pi}{2}$, and if P is on the negative y axis, $\theta = \frac{3\pi}{2}$. If $y = 0$, then if P is on the positive x axis, $\theta = 0$ and if P is on the negative x axis, $\theta = \pi$.

23. Use $x = r\cos\theta$, $y = r\sin\theta$

$x = 5\cos\pi = -5$ $\quad y = 5\sin\pi = 0$

Rectangular coordinates: $(-5, 0)$

25. Use $x = r\cos\theta$, $y = r\sin\theta$

$x = 3\sqrt{2}\cos\left(-\frac{\pi}{4}\right) = 3\sqrt{2}\left(\frac{1}{\sqrt{2}}\right) = 3$ $\quad y = 3\sqrt{2}\sin\left(-\frac{\pi}{4}\right) = 3\sqrt{2}\left(-\frac{1}{\sqrt{2}}\right) = -3$

Rectangular coordinates: $(3, -3)$

27. Use $x = r\cos\theta, y = r\sin\theta$

$x = 6\cos\frac{2\pi}{3} = 6\left(-\frac{1}{2}\right) = -3$ $\quad y = 6\sin\frac{2\pi}{3} = 6\left(\frac{\sqrt{3}}{2}\right) = 3\sqrt{3}$

Rectangular coordinates: $(-3, 3\sqrt{3})$

29. Use $x = r\cos\theta, y = r\sin\theta$

$x = -3\cos 90° = 0$ $\quad y = -3\sin 90° = -3$

Rectangular coordinates: $(0, -3)$

31. Use $r^2 = x^2 + y^2$ and $\tan\theta = \frac{y}{x}$

$r^2 = 25^2 + 0^2 = 625$ $\quad r = 25$ $\quad \tan\theta = \frac{0}{25} = 0$ $\quad \theta = 0$

Polar coordinates: $(25, 0)$

33. Use $r^2 = x^2 + y^2$ and $\tan\theta = \frac{y}{x}$

$r^2 = 0^2 + 9^2 = 81$ $\quad r = 9$ $\quad \tan\theta = \frac{9}{0} = 0$ is undefined $\quad \theta = \frac{\pi}{2}$ since the point is on the positive y axis

Polar coordinates: $\left(9, \frac{\pi}{2}\right)$

35. Use $r^2 = x^2 + y^2$ and $\tan\theta = \frac{y}{x}$

$r^2 = (-\sqrt{2})^2 + (-\sqrt{2})^2 = 4$ $\quad r = 2$ $\quad \tan\theta = \frac{-\sqrt{2}}{-\sqrt{2}} = 1$ $\quad \theta = -\frac{3\pi}{4}$ since the point is in the third quadrant

Polar coordinates: $\left(2, -\frac{3\pi}{4}\right)$

37. Use $r^2 = x^2 + y^2$ and $\tan\theta = \frac{y}{x}$

$r^2 = (-4)^2 + (-4\sqrt{3})^2 = 64$ $\quad r = 8$ $\quad \tan\theta = \frac{-4\sqrt{3}}{-4} = \sqrt{3}$ $\quad \theta = -\frac{2\pi}{3}$ since the point is in the third quadrant

Polar coordinates: $\left(8, -\frac{2\pi}{3}\right)$

39. See figure at the right. The point with coordinates $(6, -30°)$ can equally well be described as $(-6, -210°)$ or $(-6, 150°)$ or $(6, 330°)$. Thus, $(-6, -210°)$: The polar axis is rotated 210° clockwise (negative direction) and the point is located 6 units from the pole along the negative polar axis. $(-6, 150°)$: The polar axis is rotated 150° counterclockwise (positive direction) and the point is located 6 units from the pole along the negative polar axis. $(6, 330°)$: The polar axis is rotated 330° counterclockwise (positive direction) and the point is located 6 units along the positive polar axis.

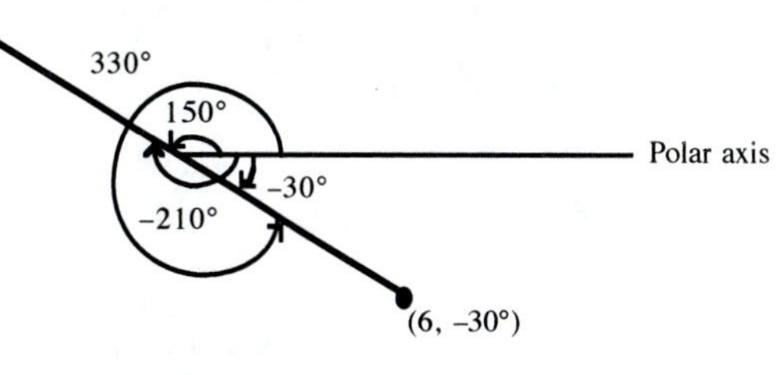

41. Use $r^2 = x^2 + y^2$ and $\tan\theta = \dfrac{y}{x}$

$r^2 = (1.625)^2 + (3.545)^2 \qquad r = \sqrt{(1.625)^2 + (3.545)^2} = 3.900$

$\tan\theta = \dfrac{3.545}{1.625} \qquad \theta = \tan^{-1}\dfrac{3.545}{1.625} = 65.374°$ since the point is in the first quadrant

Polar coordinates: $(3.900, 65.374°)$

43. Use $r^2 = x^2 + y^2$ and $\tan\theta = \dfrac{y}{x}$

$r^2 = (9.984)^2 + (-1.102)^2 \qquad r = \sqrt{(9.984)^2 + (-1.102)^2} = 10.045$

$\tan\theta = \dfrac{-1.102}{9.984} \qquad \theta = \tan^{-1}\dfrac{-1.102}{9.984} = -6.299°$ since the point is in the fourth quadrant

Polar coordinates: $(10.045, -6.299°)$

45. Use $r^2 = x^2 + y^2$ and $\tan\theta = \dfrac{y}{x}$

$r^2 = (-3.217)^2 + (-8.397)^2 \qquad r = \sqrt{(-3.217)^2 + (-8.397)^2} = 8.992$

$\tan\theta = \dfrac{-8.397}{-3.217} \qquad \theta = \tan^{-1}\left(\dfrac{-8.397}{-3.217}\right) - 180° = -110.962°$ since the point is in the third quadrant

Polar coordinates: $(8.992, -110.962°)$

47. Use $x = r\cos\theta,\ y = r\sin\theta$

$x = 2.718\cos(-31.635°) \qquad y = 2.718\sin(-31.635°)$

$ = 2.314 \qquad\qquad\qquad\quad = -1.426$

Rectangular coordinates: $(2.314, -1.426)$

49. Use $x = r\cos\theta,\ y = r\sin\theta$

$x = -4.256\cos 3.085 \qquad y = -4.256\sin 3.085$

$ = 4.249 \qquad\qquad\qquad\quad = -0.241$

Rectangular coordinates: $(4.249, -0.241)$

51. Use $x = r\cos\theta,\ y = r\sin\theta$

$x = 0.903\cos 1.514 \qquad y = 0.903\sin 1.514$

$ = 0.051 \qquad\qquad\qquad\quad = 0.902$

Rectangular coordinates: $(0.051, 0.902)$

53. $6x - x^2 = y^2$

$6x = x^2 + y^2$

Use $x = r\cos\theta$ and $x^2 + y^2 = r^2$

$6r\cos\theta = r^2$

$0 = r^2 - 6r\cos\theta = r(r - 6\cos\theta)$

$r = 0$ or $r - 6\cos\theta = 0$

The graph of $r = 0$ is the pole, and since the pole is included as a solution of $r - 6\cos\theta = 0$ $\left(\text{let } \theta = \frac{\pi}{2}\right)$, we can discard $r = 0$ and keep only

$r - 6\cos\theta = 0$ or $r = 6\cos\theta$

55. $2x + 3y = 5$

Use $x = r\cos\theta$ and $y = r\sin\theta$

$2r\cos\theta + 3r\sin\theta = 5$

$r(2\cos\theta + 3\sin\theta) = 5$

57. $x^2 + y^2 = 9$

Use $x^2 + y^2 = r^2$

$r^2 = 9$ or $r = \pm 3$

59. $2xy = 1$

Use $x = r\cos\theta$ and $y = r\sin\theta$

$2r\cos\theta\, r\sin\theta = 1$

$r^2(2\sin\theta\cos\theta) = 1$

$r^2 = \frac{1}{2\sin\theta\cos\theta}$

$= \frac{1}{\sin 2\theta}$

61. $4x^2 - y^2 = 4$

Use $x = r\cos\theta$ and $y = r\sin\theta$

$4(r\cos\theta)^2 - (r\sin\theta)^2 = 4$

$4r^2\cos^2\theta - r^2\sin^2\theta = 4$

$r^2(4\cos^2\theta - \sin^2\theta) = 4$

$r^2[4(1 - \sin^2\theta) - \sin^2\theta] = 4$

$r^2(4 - 4\sin^2\theta - \sin^2\theta) = 4$

$r^2(4 - 5\sin^2\theta) = 4$

$r^2 = \frac{4}{4 - 5\sin^2\theta}$

63. $x = 3$

Use $x = r\cos\theta$

$r\cos\theta = 3$

$r = \frac{3}{\cos\theta}$

65. $y = -7$

Use $y = r\sin\theta$

$r\sin\theta = -7$

$r = \frac{-7}{\sin\theta}$

67. $r(2\cos\theta + \sin\theta) = 4$

$2r\cos\theta + r\sin\theta = 4$

Use $x = r\cos\theta$ and $y = r\sin\theta$ $\quad 2x + y = 4$

69. $r = 8\cos\theta$

We multiply both sides by r, which adds the pole to the graph. But the pole is already part of the graph $\left(\text{let } \theta = \frac{\pi}{2}\right)$, so we have changed nothing. $r^2 = 8r\cos\theta$

But $r^2 = x^2 + y^2$, $r\cos\theta = x$. Hence, $x^2 + y^2 = 8x$.

71. $r = 2\cos\theta + 3\sin\theta$

We multiply both sides by r, which adds the pole to the graph. But the pole is already part of the graph $\left(\text{let } \tan\theta = -\frac{2}{3}\right)$, so we have changed nothing. $r^2 = 2r\cos\theta + 3r\sin\theta$

But $r^2 = x^2 + y^2$, $r\cos\theta = x$, $r\sin\theta = y$. Hence, $x^2 + y^2 = 2x + 3y$.

73. $r^2 \cos 2\theta = 4$

$$r^2(\cos^2\theta - \sin^2\theta) = 4$$
$$r^2\cos^2\theta - r^2\sin^2\theta = 4$$
$$(r\cos\theta)^2 - (r\sin\theta)^2 = 4$$

Use $r\cos\theta = x$ and $r\sin\theta = y$

$$x^2 - y^2 = 4$$

75. $r^2 = 3\cos 2\theta$

We multiply both sides by r^2, which adds the pole to the graph. But the pole is already part of the graph $\left(\text{let } \theta = \frac{\pi}{4}\right)$, so we have changed nothing. $r^2r^2 = 3r^2\cos 2\theta$

$$(r^2)^2 = 3r^2(\cos^2\theta - \sin^2\theta)$$
$$(r^2)^2 = 3r^2\cos^2\theta - 3r^2\sin^2\theta$$
$$(r^2)^2 = 3(r\cos\theta)^2 - 3(r\sin\theta)^2$$

Use $r^2 = x^2 + y^2$, $r\cos\theta = x$, and $r\sin\theta = y$

$$(x^2 + y^2)^2 = 3x^2 - 3y^2$$

77. $r = 4$, so $r^2 = 16$

Use $r^2 = x^2 + y^2$

$x^2 + y^2 = 16$

79. $\theta = 30°$ $\quad \tan\theta = \tan 30°$

$\tan\theta = \dfrac{1}{\sqrt{3}}$ $\quad$ Use $\tan\theta = \dfrac{y}{x}$

$\dfrac{y}{x} = \dfrac{1}{\sqrt{3}}$, $\quad y = \dfrac{1}{\sqrt{3}}x$

81. $r = \dfrac{3}{\sin\theta - 2}$

Multiply both sides by $\sin\theta - 2$, which is never 0 since $\sin\theta$ is never 2.

$r(\sin\theta - 2) = 3 \qquad r\sin\theta - 2r = 3$

Use $r\sin\theta = y$ and $r = -\sqrt{x^2 + y^2} \qquad y + 2\sqrt{x^2 + y^2} = 3$

$y - 3 = -2\sqrt{x^2 + y^2}$ or $(y - 3)^2 = 4(x^2 + y^2)$

Note: The unusual choice of $r = -\sqrt{x^2 + y^2}$ is not a misprint. It is necessary to correspond with the fact that in the original polar equation, $\sin\theta < 2$; hence, $\sin\theta - 2$ is negative; hence, r is negative.

83. If P is on the positive x axis, then $(a, 0)$ can represent both its rectangular coordinates and its polar coordinates. If P is on the negative x axis with rectangular coordinate $(a, 0)$, then $(a, 0)$ can also represent both its rectangular coordinates and its polar coordinates, although $(-a, \pi)$ is often a more natural choice. The origin, also, can be represented by $(0, 0)$ in both rectangular and polar coordinates.

85. $d = \sqrt{r_1^2 + r_2^2 - 2r_1r_2\cos(\theta_2 - \theta_1)} \qquad (r_1, \theta_1) = (2, 30°) \qquad (r_2, \theta_2) = (3, 60°)$

$= \sqrt{2^2 + 3^2 - 2(2)(3)\cos(60° - 30°)}$

$= \sqrt{2.6076952\ldots} \approx 1.615$ units

EXERCISE 7.2 Sketching Polar Equations

1. The graph is a circle with center the pole and radius 4.

3. $r + 10 = 0$ is equivalent to $r = -10$. The graph is a circle with center the pole and radius 10.

5.

	θ	0	$\frac{\pi}{6}$	$\frac{\pi}{4}$	$\frac{\pi}{3}$	$\frac{\pi}{2}$
Exact value	r	10	$5\sqrt{3}$	$5\sqrt{2}$	5	0
Calculator value	r	10	8.7	7.1	5	0

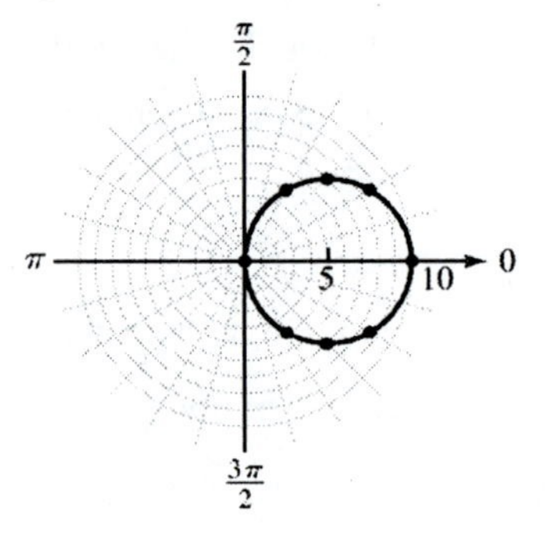

	θ	$\frac{2\pi}{3}$	$\frac{3\pi}{4}$	$\frac{5\pi}{6}$	π
Exact value	r	-5	$-5\sqrt{2}$	$-5\sqrt{3}$	-10
Calculator value	r	-5	-7.1	-8.7	-10

7.

	θ	0°	30°	60°	90°	120°	150°	180°	210°
Exact value	r	6	$\left(3+\frac{3\sqrt{3}}{2}\right)$	$\frac{9}{2}$	3	$\frac{3}{2}$	$\left(3-\frac{3\sqrt{3}}{2}\right)$	0	$\left(3-\frac{3\sqrt{3}}{2}\right)$
Calculator value	r	6	5.6	4.5	3	1.5	0.4	0	0.4

	θ	240°	270°	300°	330°	360°
Exact value	r	$\frac{3}{2}$	3	$\frac{9}{2}$	$\left(3+\frac{3\sqrt{3}}{2}\right)$	6
Calculator value	r	1.5	3	4.5	5.6	6

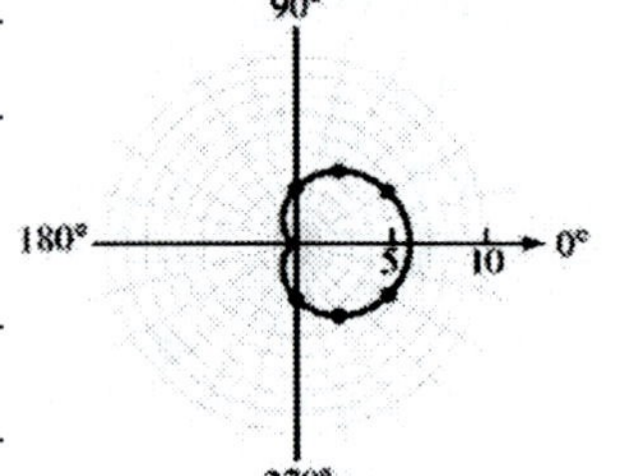

9.

θ	0	$\frac{\pi}{6}$	$\frac{\pi}{3}$	$\frac{\pi}{2}$	$\frac{2\pi}{3}$	$\frac{5\pi}{6}$	π	$\frac{7\pi}{6}$	$\frac{4\pi}{3}$	$\frac{3\pi}{2}$	$\frac{5\pi}{3}$	$\frac{11\pi}{6}$	2π
r	0	0.5	1.0	1.6	2.1	2.6	3.1	3.7	4.2	4.7	5.2	5.8	6.3

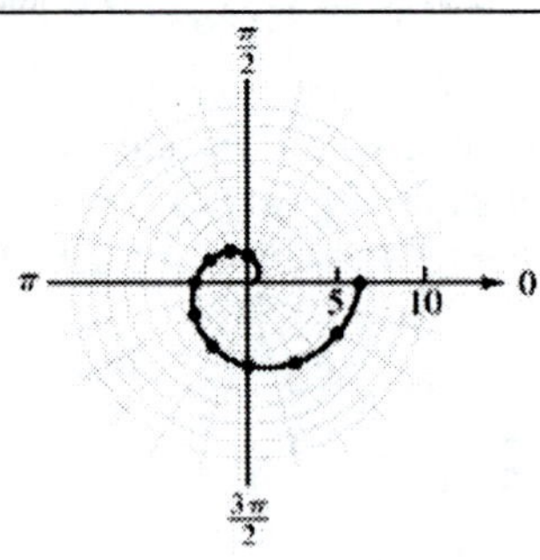

11. The graph consists of all points whose distance from the pole is 5, a circle with center at the pole, and radius 5.

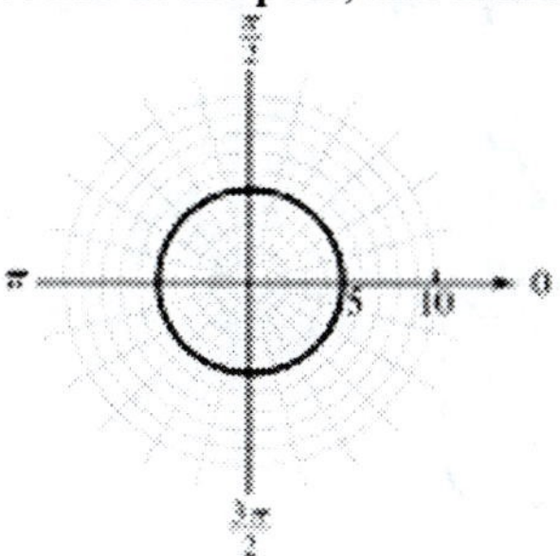

13. The graph consists of all points on a line that forms an angle of $\frac{\pi}{4}$ with the polar axis and passes through the pole.

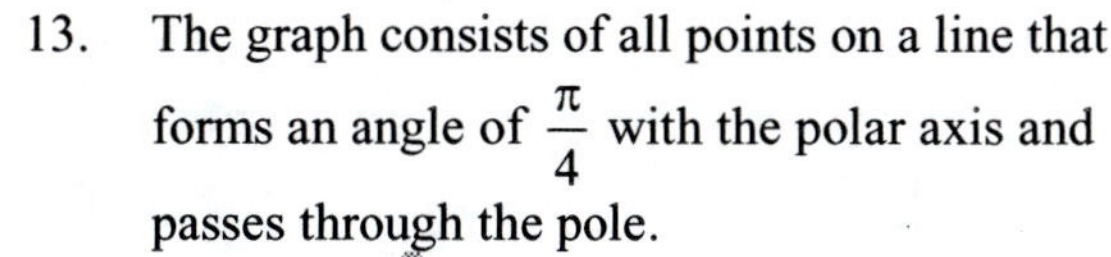

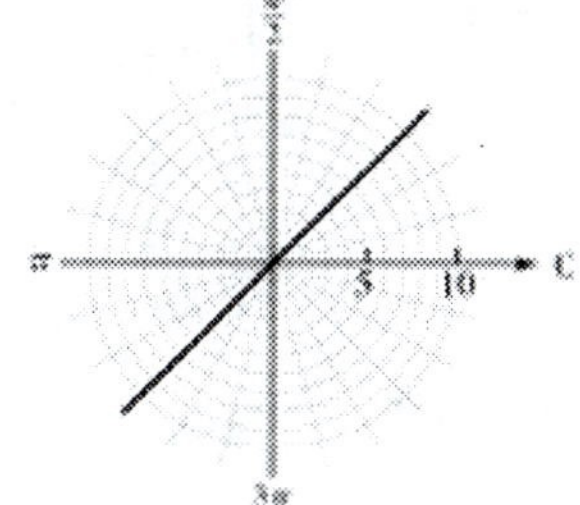

15. Since $\theta^2 = \frac{\pi^2}{9}$, $\theta = \frac{\pi}{3}$ or $\theta = -\frac{\pi}{3}$. The graph consists of all points on a line that forms an angle of $\frac{\pi}{3}$ with the polar axis as well as all points on a line that forms an angle of $-\frac{\pi}{3}$ $\left(\text{or } \frac{2\pi}{3}\right)$ with the polar axis, both passing through the pole.

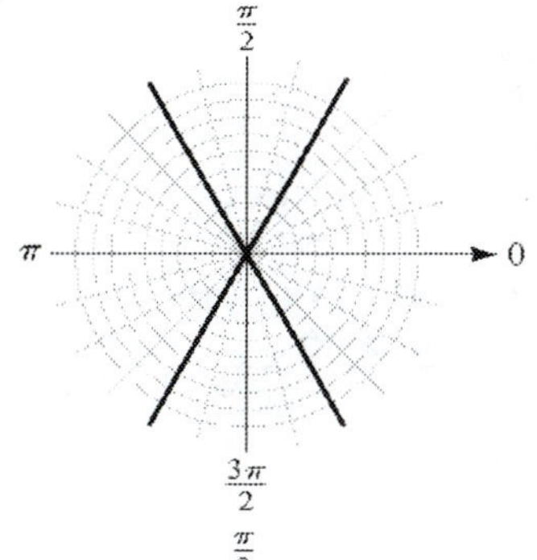

17. Since $r^2 = 16$, $r = 4$ or $r = -4$. Both equations describe a graph which consists of all points whose distance from the axis is 4, a circle with center at the pole, and radius 4.

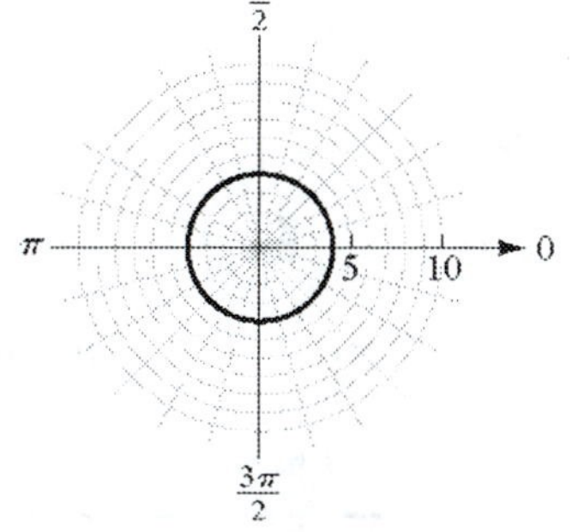

19. Set up a table that shows how r varies as θ varies through each set of quadrant values, then sketch the polar curve from the information in the table.

θ	$4\cos\theta$
0 to $\frac{\pi}{2}$	4 to 0
$\frac{\pi}{2}$ to π	0 to –4
π to $\frac{3\pi}{2}$	–4 to 0
$\frac{3\pi}{2}$ to 2π	0 to 4

(for π to 2π:) Curve is traced out a second time in this region, although coordinate pairs are different

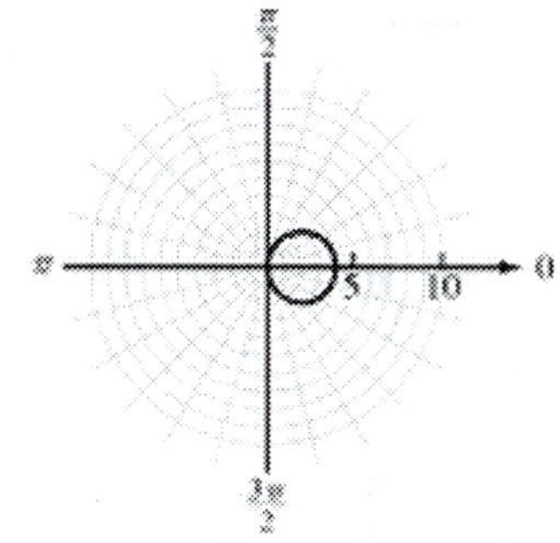

Exercise 7.2 Sketching Polar Equations

21. Set up a table that shows how r varies as 2θ varies through each set of quadrant values, then sketch the polar curve from the information in the table.

θ	2θ	$\cos 2\theta$	$8\cos 2\theta$
0 to $\frac{\pi}{4}$	0 to $\frac{\pi}{2}$	1 to 0	8 to 0
$\frac{\pi}{4}$ to $\frac{\pi}{2}$	$\frac{\pi}{2}$ to π	0 to –1	0 to –8
$\frac{\pi}{2}$ to $\frac{3\pi}{4}$	π to $\frac{3\pi}{2}$	–1 to 0	–8 to 0
$\frac{3\pi}{4}$ to π	$\frac{3\pi}{2}$ to 2π	0 to 1	0 to 8
π to $\frac{5\pi}{4}$	2π to $\frac{5\pi}{2}$	1 to 0	8 to 0
$\frac{5\pi}{4}$ to $\frac{3\pi}{2}$	$\frac{5\pi}{2}$ to 3π	0 to –1	0 to –8
$\frac{3\pi}{2}$ to $\frac{7\pi}{4}$	3π to $\frac{7\pi}{2}$	–1 to 0	–8 to 0
$\frac{7\pi}{4}$ to 2π	$\frac{7\pi}{2}$ to 4π	0 to 1	0 to 8

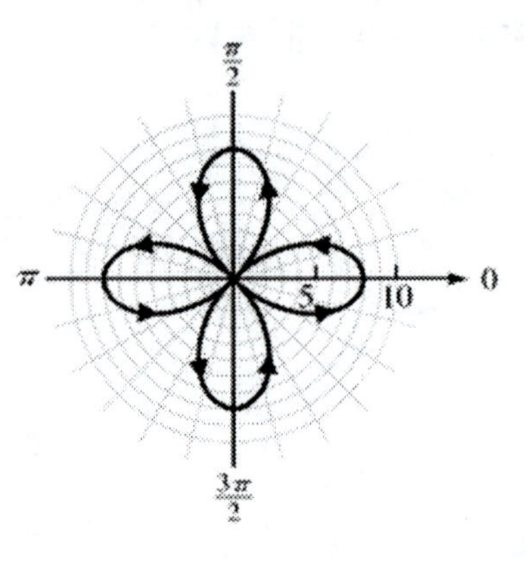

23. Set up a table that shows how r varies as 3θ varies through each set of quadrant values, then sketch the polar curve from the information in the table.

θ	3θ	$\sin 3\theta$	$6\sin 3\theta$
0 to $\frac{\pi}{6}$	0 to $\frac{\pi}{2}$	0 to 1	0 to 6
$\frac{\pi}{6}$ to $\frac{\pi}{3}$	$\frac{\pi}{2}$ to π	1 to 0	6 to 0
$\frac{\pi}{3}$ to $\frac{\pi}{2}$	π to $\frac{3\pi}{2}$	0 to –1	0 to –6
$\frac{\pi}{2}$ to $\frac{2\pi}{3}$	$\frac{3\pi}{2}$ to 2π	–1 to 0	–6 to 0
$\frac{2\pi}{3}$ to $\frac{5\pi}{6}$	2π to $\frac{5\pi}{3}$	0 to 1	0 to 6
$\frac{5\pi}{6}$ to π	$\frac{5\pi}{3}$ to 3π	1 to 0	6 to 0

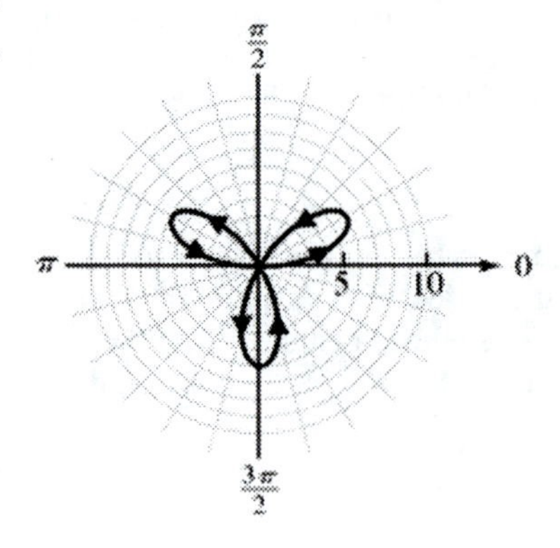

25. Set up a table that shows how r varies as θ varies through each set of quadrant values, then sketch the polar curve from the information in the table.

θ	$\cos\theta$	$3\cos\theta$	$3+3\cos\theta$
0 to $\frac{\pi}{2}$	1 to 0	3 to 0	6 to 3
$\frac{\pi}{2}$ to π	0 to –1	0 to –3	3 to 0
π to $\frac{3\pi}{2}$	–1 to 0	–3 to 0	0 to 3
$\frac{3\pi}{2}$ to π	0 to 1	0 to 3	3 to 6

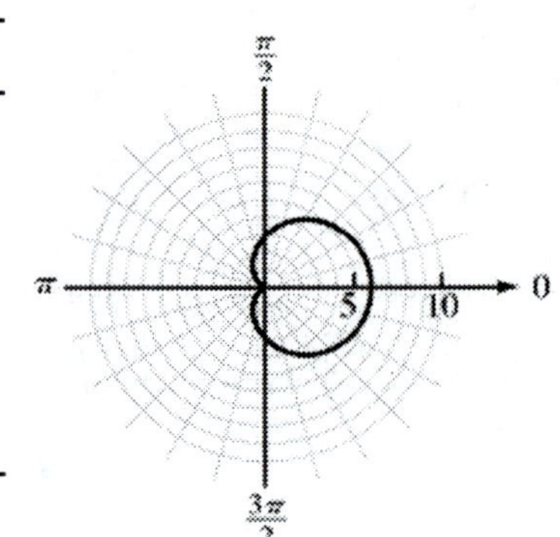

27. Note that $r = 2 + 4\cos\theta$ will equal 0 when $\cos\theta = -\frac{1}{2}$, that is, when $\theta = \frac{2\pi}{3}, \frac{4\pi}{3}$ to list values between 0 and 2π only). Set up a table that shows how θ varies over particular intervals, then sketch the polar curve from the information in the table.

θ	$\cos\theta$	$4\cos\theta$	$2 + 4\cos\theta$
0 to $\frac{\pi}{2}$	1 to 0	4 to 0	6 to 2
$\frac{\pi}{2}$ to $\frac{2\pi}{3}$	0 to $-\frac{1}{2}$	0 to –2	2 to 0
$\frac{2\pi}{3}$ to π	$-\frac{1}{2}$ to –1	–2 to –4	0 to –2
π to $\frac{4\pi}{3}$	–1 to $-\frac{1}{2}$	–4 to –2	–2 to 0
$\frac{4\pi}{3}$ to $\frac{3\pi}{2}$	$-\frac{1}{2}$ to 0	–2 to 0	0 to 2
$\frac{3\pi}{2}$ to 2π	0 to 1	0 to 4	2 to 6

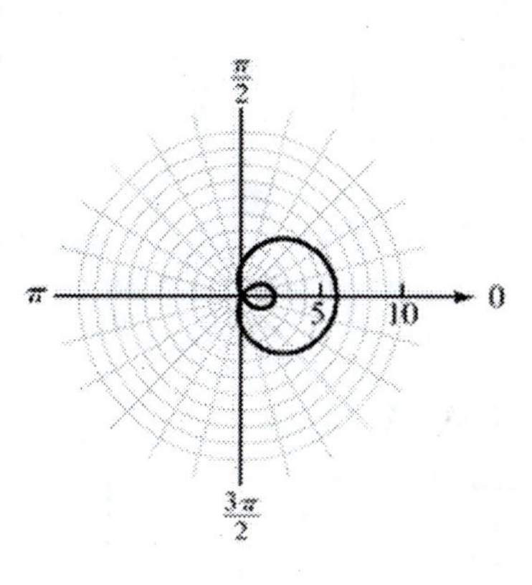

29. The simplest way to graph $r = \frac{7}{\cos\theta}$ is to notice that it is the same as $r\cos\theta = 7$ or $x = 7$, a vertical line through (7, 0).

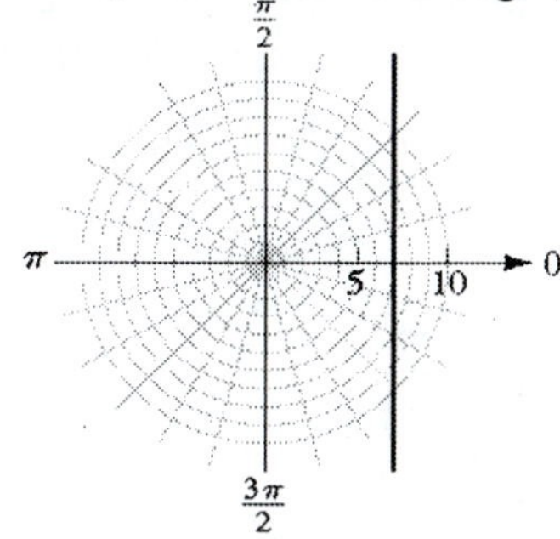

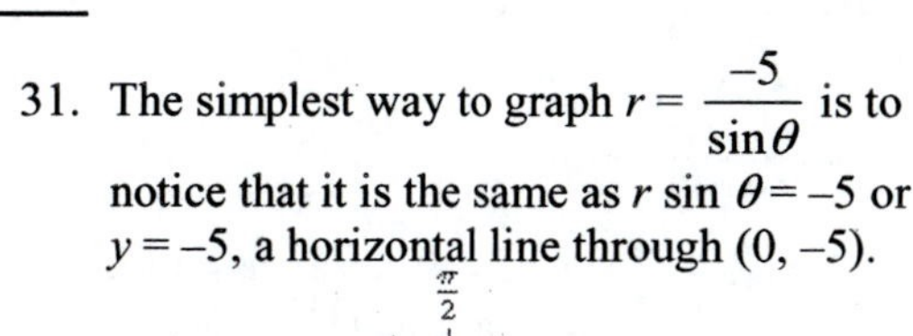

31. The simplest way to graph $r = \frac{-5}{\sin\theta}$ is to notice that it is the same as $r\sin\theta = -5$ or $y = -5$, a horizontal line through (0, –5).

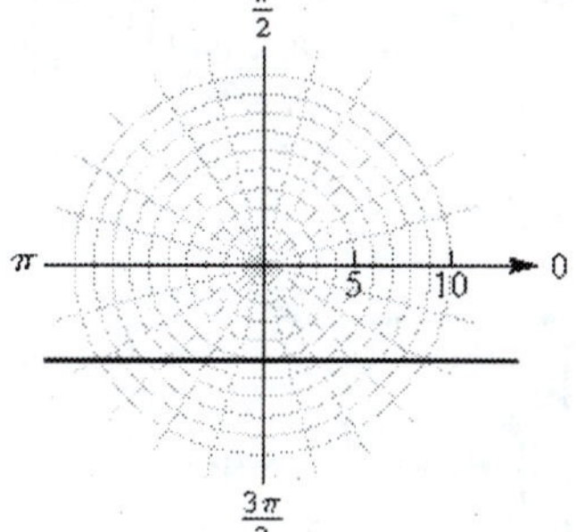

33. (A) $r = 5 + 5\cos\theta$

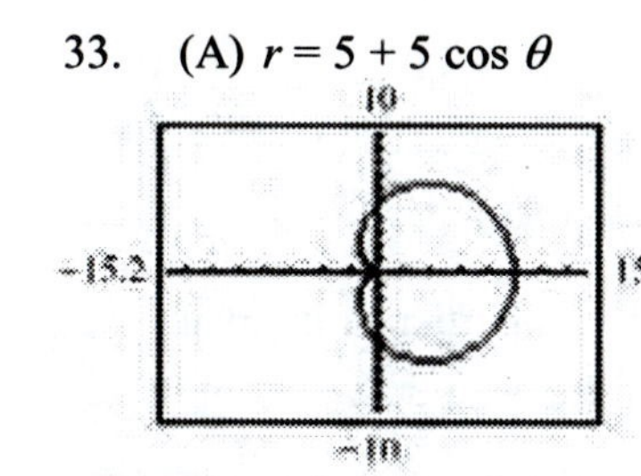

(B) $r = 5 + 4\cos\theta$

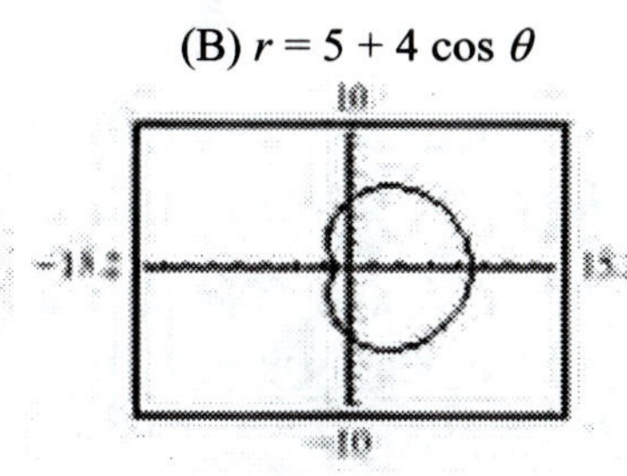

(C) $r = 4 + 5\cos\theta$

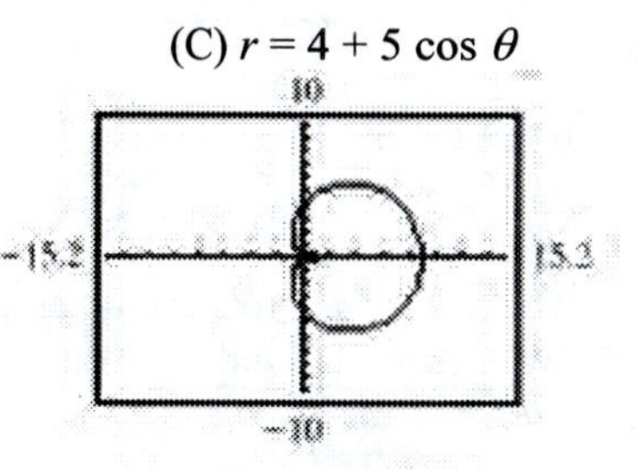

(D) If $a = b$, the graph will touch but not pass through the origin. If $a > b$, the graph will not touch nor pass through the origin. If $a < b$, the graph will go through the origin and part of the graph will be inside the other part.

35. (A) $r = 9\cos\theta$ $\qquad r = 9\cos 3\theta$ $\qquad r = 9\cos 5\theta$

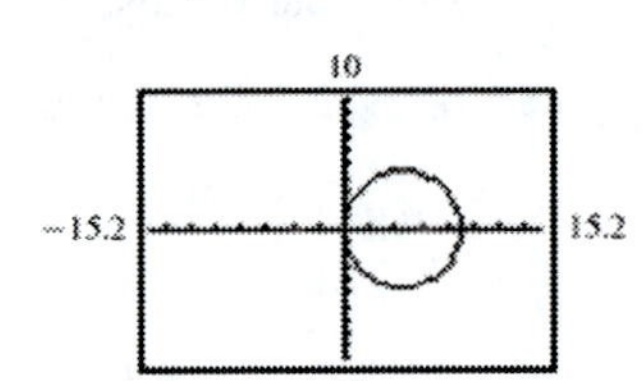

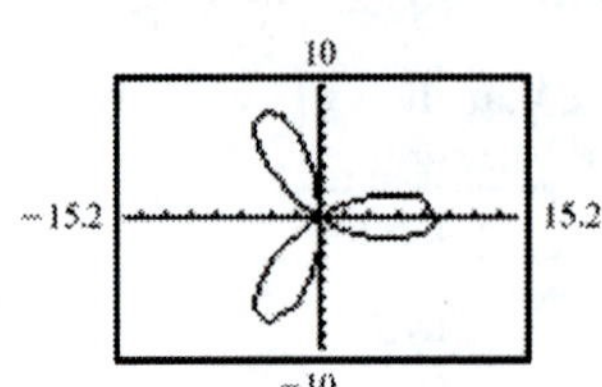

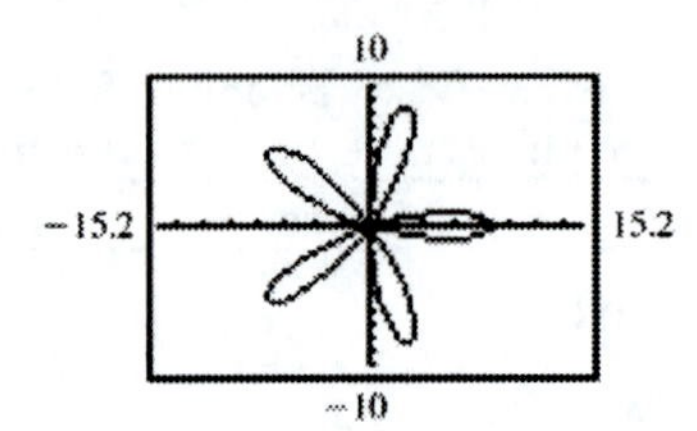

(B) and (C) Since $r = 9\cos\theta$ has one leaf, $r = 9\cos 3\theta$ has 3 leaves, and $r = 9\cos 5\theta$ has 5 leaves, a reasonable guess for $r = 9\cos 7\theta$ would be 7 leaves. A reasonable guess for $r = 9\cos n\theta$ would be n leaves (n odd).

37. (A) $r = 9\cos 2\theta$ $\qquad r = 9\cos 4\theta$ $\qquad r = 9\cos 6\theta$

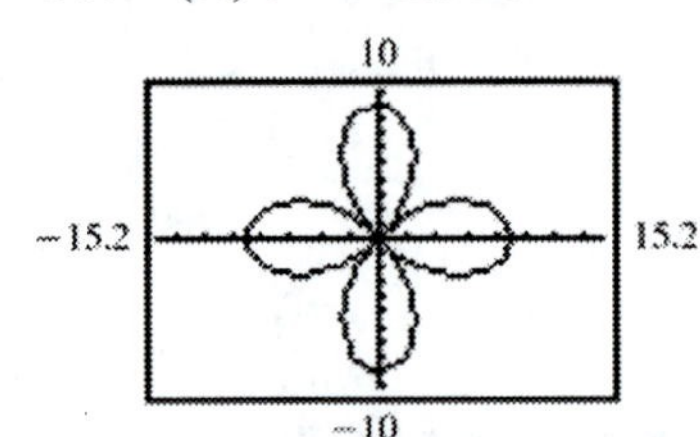

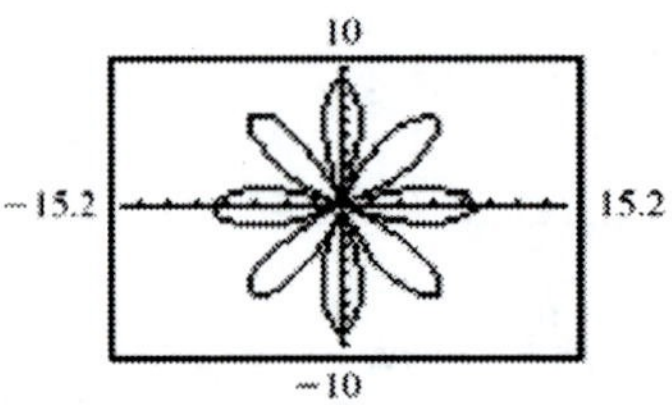

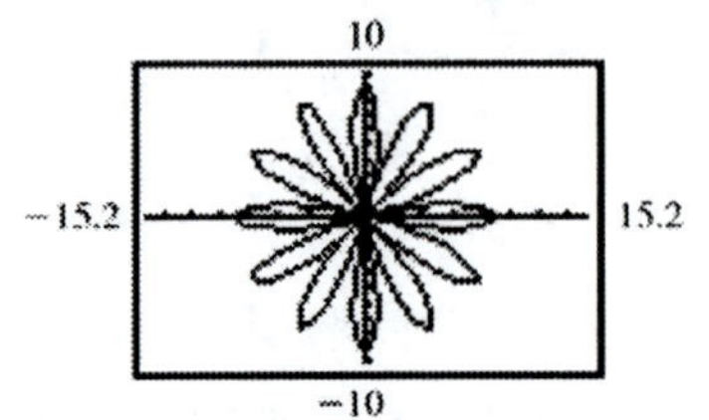

(B) and (C) Since $r = 9\cos 2\theta$ has 4 leaves, $r = 9\cos 4\theta$ has 8 leaves, and $r = 9\cos 6\theta$ has 12 leaves, a reasonable guess for $r = 9\cos 8\theta$ would be 16 leaves. A reasonable guess for $r = 9\cos n\theta$ would be $2n$ leaves (n even).

39. (A) $r = 9\cos\dfrac{\theta}{2}$ $\qquad$ (B) $r = 9\cos\dfrac{\theta}{4}$

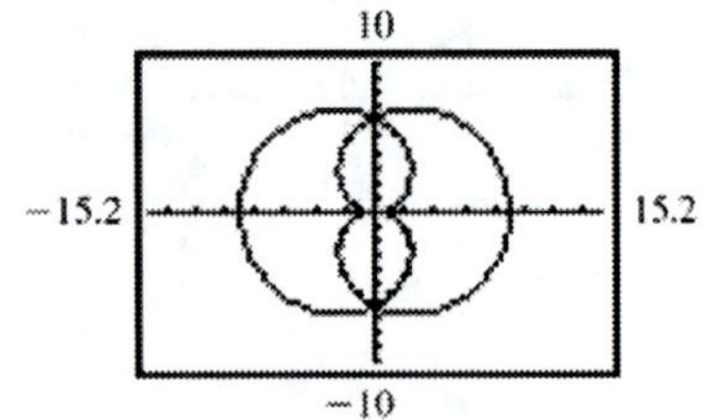

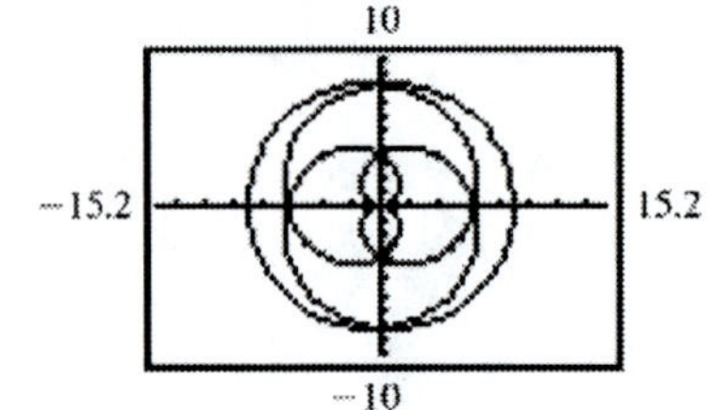

(C) n times.

41. The following table can be used to investigate how r varies as θ varies over particular intervals. We then sketch the polar curve from the information in the table.

θ	2θ	$\cos 2\theta$	$64\cos 2\theta$	$r = \pm\sqrt{64\cos 2\theta}$
0 to $\dfrac{\pi}{4}$	0 to $\dfrac{\pi}{2}$	1 to 0	64 to 0	$\begin{cases} 8 \text{ to } 0 \\ -8 \text{ to } 0 \end{cases}$ The two branches of the curve are reflections of each other in the origin
$\dfrac{\pi}{4}$ to $\dfrac{\pi}{2}$	$\dfrac{\pi}{2}$ to π	0 to −1	0 to −64	No curve; no real square root of a negative no.
$\dfrac{\pi}{2}$ to $\dfrac{3\pi}{4}$	π to $\dfrac{3\pi}{2}$	−1 to 0	−64 to 0	
$\dfrac{3\pi}{4}$ to π	$\dfrac{3\pi}{2}$ to 2π	0 to 1	0 to 64	$\begin{cases} 0 \text{ to } 8 \\ 0 \text{ to } -8 \end{cases}$ The two branches of the curve are reflections of each other in the origin

θ	2θ	$\cos 2\theta$	$64\cos 2\theta$	$r = \pm\sqrt{64\cos 2\theta}$	
π to $\dfrac{5\pi}{4}$	2π to $\dfrac{5\pi}{2}$	1 to 0	64 to 0	$\begin{cases} 8 \text{ to } 0 \\ -8 \text{ to } 0 \end{cases}$	This repeats the curve already traced out
$\dfrac{5\pi}{4}$ to $\dfrac{3\pi}{2}$	$\dfrac{5\pi}{2}$ to 3π	0 to −1	0 to −64	No curve	
$\dfrac{3\pi}{2}$ to $\dfrac{7\pi}{4}$	3π to $\dfrac{7\pi}{2}$	−1 to 0	−64 to 0		
$\dfrac{7\pi}{4}$ to 2π	$\dfrac{7\pi}{2}$ to 4π	0 to 1	0 to 64	$\begin{cases} 0 \text{ to } 8 \\ -8 \text{ to } 0 \end{cases}$	

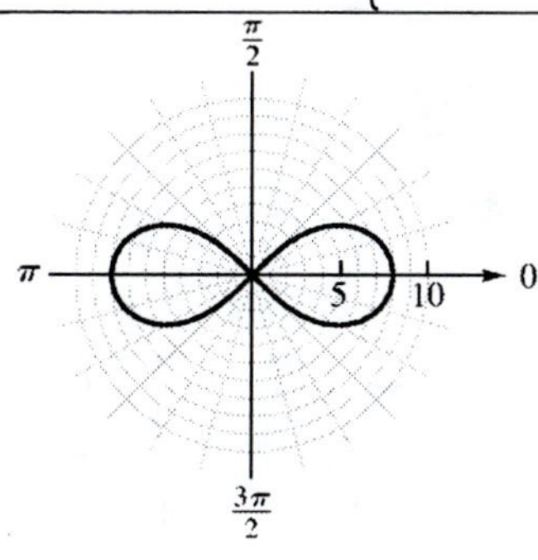

43. Here are graphs of $r = 1 + 2\cos(n\theta)$ for $n = 1, 2, 3$, and 4.

$n = 1$

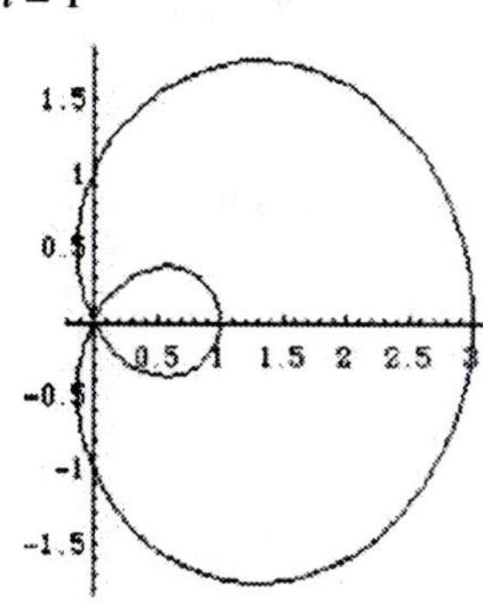

$n = 2$

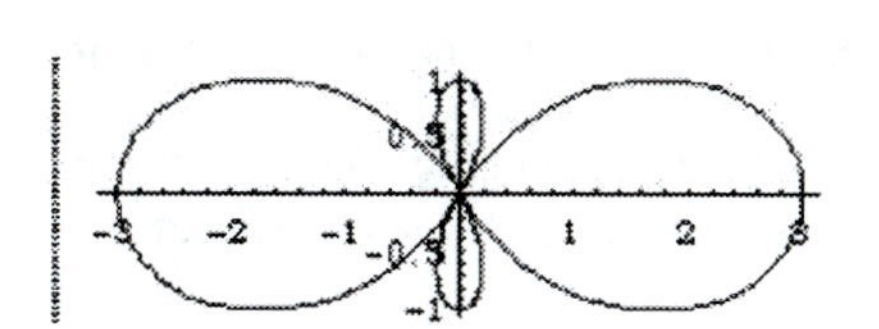

$n = 3$

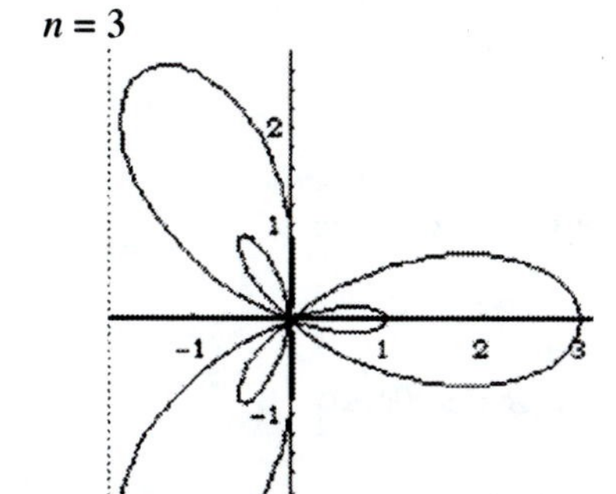

$n = 4$

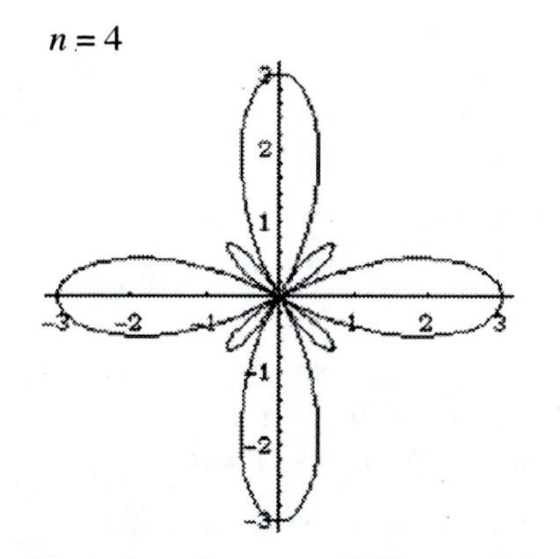

Generalizing from these graphs, we conclude that for each n, there are n large petals and n small petals. For n odd, the small petals are within the large petals, for n even the small petals are between the large petals.

45. We sketch the two graphs using rapid sketching techniques. Tables of how r varies for each curve as θ varies over particular intervals can be readily constructed.

θ	$2\cos\theta$	$2\sin\theta$
0 to $\frac{\pi}{2}$	2 to 0	0 to 2
$\frac{\pi}{2}$ to π	0 to –2	2 to 0
π to $\frac{3\pi}{2}$	–2 to 0	0 to –2
$\frac{3\pi}{2}$ to 2π	0 to 2	–2 to 0

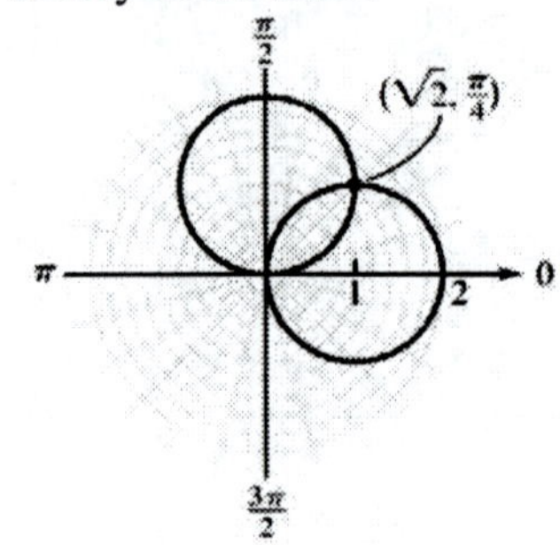

We solve the system, $r = 2\cos\theta$, $r = 2\sin\theta$, by equating the right sides:

$$2\cos\theta = 2\sin\theta$$
$$\cos\theta = \sin\theta$$
$$1 = \tan\theta$$

The only solution of this equation, $0 \le \theta \le \pi$, is $\theta = \frac{\pi}{4}$. If we substitute this into either of the original equations, we get $r = 2\cos\frac{\pi}{4} = 2\sin\frac{\pi}{4} = 2\left(\frac{\sqrt{2}}{2}\right) = \sqrt{2}$. Solution: $\left(\sqrt{2}, \frac{\pi}{4}\right)$

The sketch shows that the pole is on both graphs, however, the pole has no ordered pairs of coordinates that simultaneously satisfy both equations. As $\left(0, \frac{\pi}{2}\right)$, it satisfies the first; as $(0, 0)$, it satisfies the second; it is not a solution of the system.

47. We graph the system using rapid sketching techniques. See Problem 21 for a table for $r = 8\cos 2\theta$; a table for $r = 8\sin\theta$ can be readily constructed.

θ	$8\sin\theta$
0 to $\frac{\pi}{2}$	0 to 8
$\frac{\pi}{2}$ to π	8 to 0
π to $\frac{3\pi}{2}$	0 to –8
$\frac{3\pi}{2}$ to 2π	–8 to 0

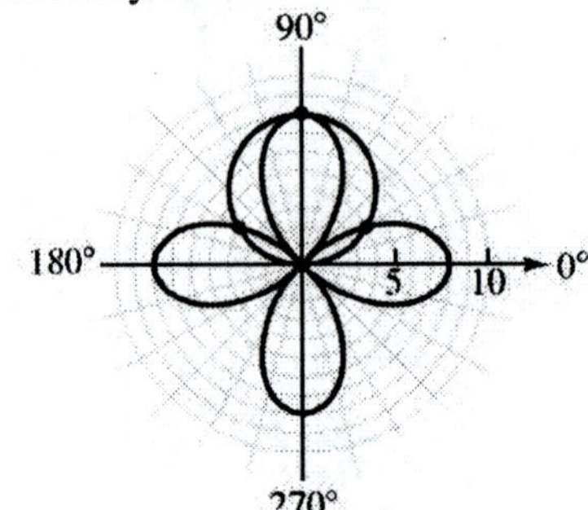

We solve the system: $r = 8\sin\theta$, $r = 8\cos 2\theta$, by equating the right-hand sides:

$$8\sin\theta = 8\cos 2\theta$$
$$\sin\theta = \cos 2\theta$$
$$\sin\theta = 1 - 2\sin^2\theta \qquad \text{Double-angle identity}$$
$$2\sin^2\theta + \sin\theta - 1 = 0$$
$$(2\sin\theta - 1)(\sin\theta + 1) = 0 \qquad \text{Factoring}$$

Therefore, $2\sin\theta - 1 = 0$ or $\sin\theta + 1 = 0$

$$2\sin\theta - 1 = 0 \qquad\qquad \sin\theta + 1 = 0$$
$$\sin\theta = \frac{1}{2} \qquad\qquad \sin\theta = -1$$
$$\theta = 30°, 150° \qquad\qquad \theta = 270°$$

If we substitute these values into either of the original equations, we get (choosing $r = 8 \sin \theta$ for ease of calculation):

$\theta = 30°$ $\quad r = 8 \sin 30° = 8\left(\frac{1}{2}\right) = 4$ $\quad$ Solution: $(4, 30°)$

$\theta = 150°$ $\quad r = 8 \sin 150° = 8\left(\frac{1}{2}\right) = 4$ $\quad$ Solution: $(4, 150°)$

$\theta = 270°$ $\quad r = 8 \sin 270° = 8(-1) = -8$ $\quad$ Solution: $(-8, 270°)$

The sketch shows that all three solutions are on both graphs; it also shows that the pole is on both graphs. However, the pole has no ordered pairs of coordinates that simultaneously satisfy both equations. As $(0, 0°)$, it satisfies the first; as $(0, 45°)$, it satisfies the second; it is not a solution of the system.

49. (a), (d), (e), (g), (i), (j), (k) are symmetric with respect to the x axis.

51. (A) (a) Since $\frac{a}{\cos(-\theta)} = \frac{a}{\cos\theta}$, the equation $r = \frac{a}{\cos\theta}$ is unchanged after substitution and simplification.

(d) Since there is no occurrence of θ in this equation, it remains unchanged by substitution for θ.

(e) Since $a \cos(-\theta) = a \cos \theta$, the equation $r = a \cos \theta$ is unchanged after substitution and simplification.

(g) $r = a + a \cos \theta$

Substitute to obtain

$r = a + a \cos(-\theta)$

Since $\cos(-\theta) = \cos \theta$, this simplifies to

$r = a + a \cos \theta$

The equation is unchanged after substitution and simplification.

(i) $r = a \cos 3\theta$

Substitute to obtain

$r = a \cos 3(-\theta)$

$r = a \cos(-3\theta)$

Since $\cos(-3\theta) = \cos 3\theta$, this simplifies to

$r = a \cos 3\theta$

The equation is unchanged after substitution and simplification.

(j) $r = a \cos 2\theta$

Substitute to obtain

$r = a \cos 2(-\theta)$

$r = a \cos(-2\theta)$

Since $\cos(-2\theta) = \cos 2\theta$, this simplifies to

$r = a \cos 2\theta$

The equation is unchanged after substitution and simplification.

(k) $r^2 = a^2 \cos 2\theta$

Substitute to obtain

$r^2 = a^2 \cos 2(-\theta)$

$r^2 = a^2 \cos(-2\theta)$

Since $\cos(-2\theta) = \cos 2\theta$, this simplifies to

$r^2 = a^2 \cos 2\theta$

The equation is unchanged after substitution and simplification.

(B) The points (r, θ) and $(r, -\theta)$ are reflections of one another in the polar axis (x axis). A curve is symmetric with respect to the x axis if whenever a point is on the curve, its reflection in the x axis is also on the curve. The algebraic representation of this is that whenever the coordinates (in this case (r, θ)) of a point satisfy the equation of the curve, the coordinates (in this case $(r, -\theta)$) also satisfy the equation. The substitution and simplification tests whether this is true.

53. (A) The America's Cup curve appears to pass through the point with polar coordinates (9.5, 30°). Speed = 9.5 knots.
(B) The America's Cup curve appears to pass through the point with polar coordinates (12.0, 60°). Speed = 12.0 knots.
(C) The America's Cup curve appears to pass through the point with polar coordinates (13.5, 90°). Speed = 13.5 knots.
(D) The America's Cup curve appears to pass through the point with polar coordinates (12.0, 120°). Speed = 12.0 knots.

55. (A) $r = \dfrac{8}{1 - 0.5\cos\theta}$ (B) $r = \dfrac{8}{1 - \cos\theta}$ (C) $r = \dfrac{8}{1 - 2\cos\theta}$

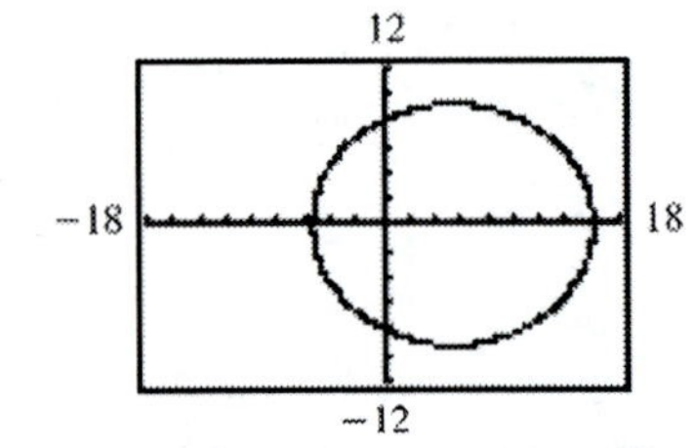

The graph is an ellipse.

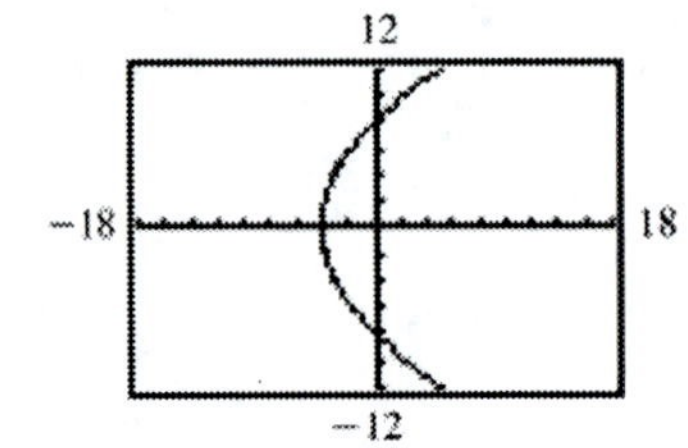

The graph is a parabola.

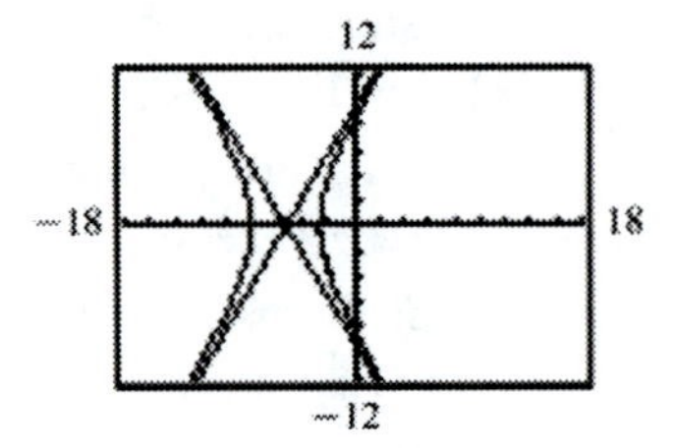

The graph is a hyperbola.

57. (A) At aphelion, $\theta = 0°$, hence

$$r = \frac{3.44 \times 10^7 \text{ mi}}{1 - 0.206 \cos 0°} = 4.33 \times 10^7 \text{ mi}$$

At perihelion, $\theta = 180°$, hence

$$r = \frac{3.44 \times 10^7 \text{ mi}}{1 - 0.206 \cos 180°} = 2.85 \times 10^7 \text{ mi}$$

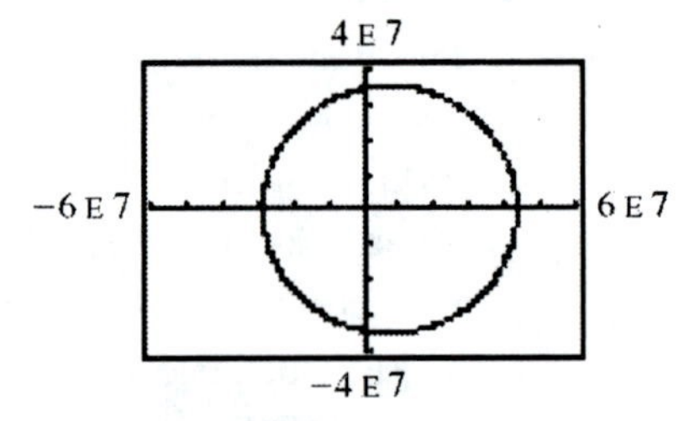

(B) The diagram in the text shows the areas swept out when the planet is near aphelion or perihelion as, approximately, triangles, whose areas would be proportional to their base (distance traveled by planet) and altitude (distance from planet to sum). Since at aphelion the distance from planet to sun is greater, the distance traveled by planet must be smaller in order to have equal area. Thus, the planet travels more slowly at aphelion, faster at perihelion.

EXERCISE 7.3 Complex Numbers in Rectangular and Polar Forms

1. Yes, every real number x can be written as $x + 0i$ and therefore is a complex number.

3. The modulus of a complex number z is the distance from z to the origin.

5. $x + yi$.

7.

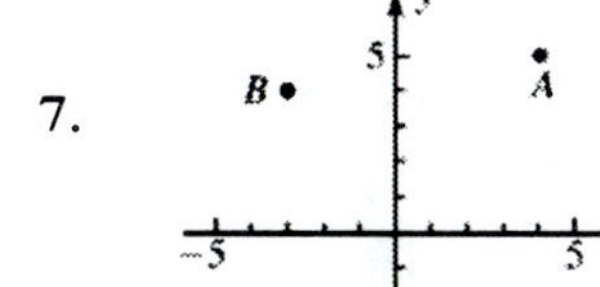

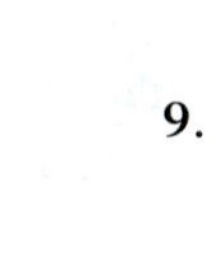

9.

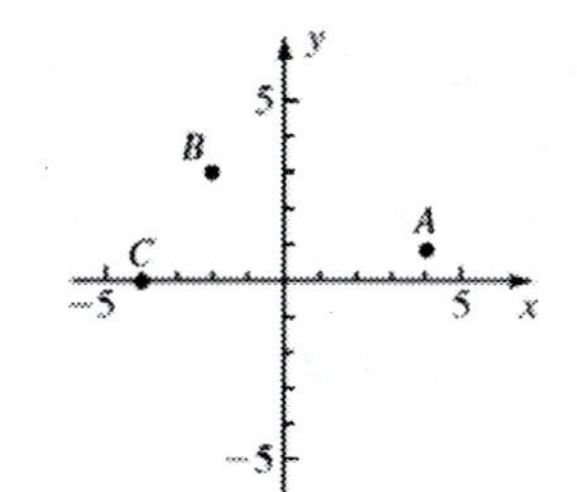

11.

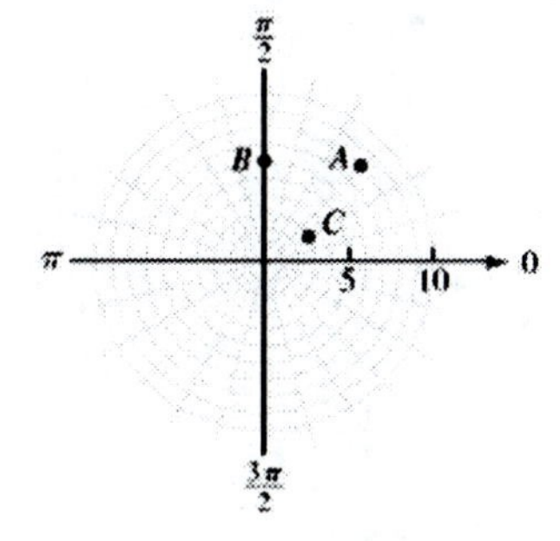

13.

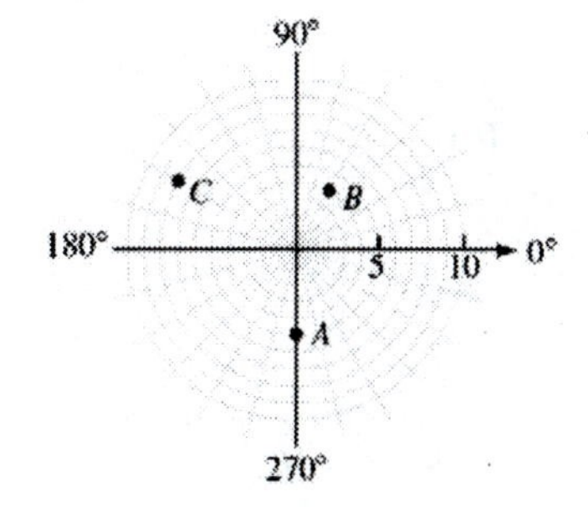

15. A sketch shows that $7i$ is located on the positive y axis.
So $\text{mod}(7i) = r = 7$, $\arg(7i) = \theta = \dfrac{\pi}{2}$, and the polar form is $7e^{(\pi/2)i}$.

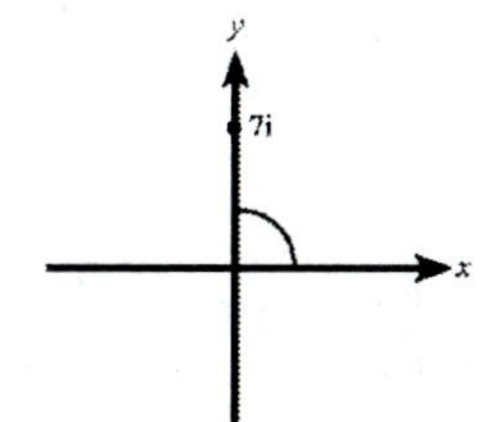

17. A sketch shows that $5 + 5i\sqrt{3}$ is associated with a special 30°–60°–90° reference triangle in the first quadrant.
So $\text{mod}(5 + 5i\sqrt{3}) = \sqrt{5^2 + (5\sqrt{3})^2} = 10$,
$\arg(5 + 5i\sqrt{3}) = \dfrac{\pi}{3}$, and the polar form is $10e^{(\pi/3)i}$.

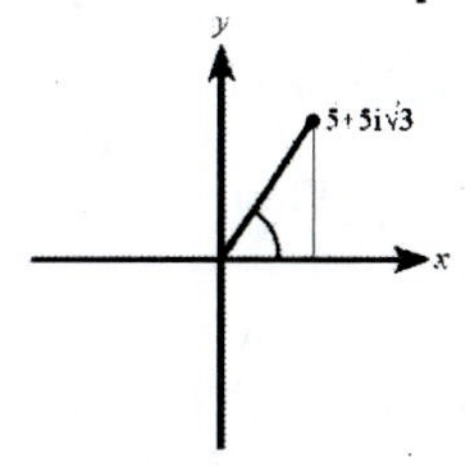

19.
$$
\begin{aligned}
x + iy &= 32e^{\pi i} \\
&= 32(\cos \pi + i \sin \pi) \\
&= 32(-1 + i \cdot 0) \\
&= -32
\end{aligned}
$$

21.

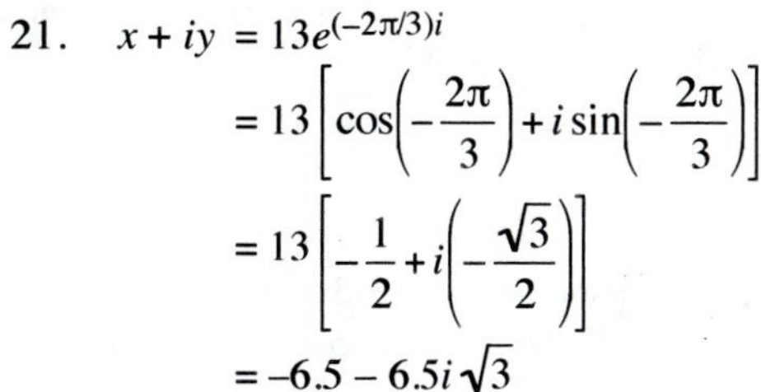

$$
\begin{aligned}
x + iy &= 13e^{(-2\pi/3)i} \\
&= 13\left[\cos\left(-\frac{2\pi}{3}\right) + i \sin\left(-\frac{2\pi}{3}\right)\right] \\
&= 13\left[-\frac{1}{2} + i\left(-\frac{\sqrt{3}}{2}\right)\right] \\
&= -6.5 - 6.5i\sqrt{3}
\end{aligned}
$$

23. A sketch shows that $18\sqrt{2} + 18i\sqrt{2}$ is associated with a special 45°–45°–90° reference triangle in the first quadrant. So $\text{mod}(18\sqrt{2} + 18i\sqrt{2}) = \sqrt{(18\sqrt{2})^2 + (18\sqrt{2})^2} = 36$,
$\arg(18\sqrt{2} + 18i\sqrt{2}) = 45°$,
and the polar form is $36e^{45° i}$.

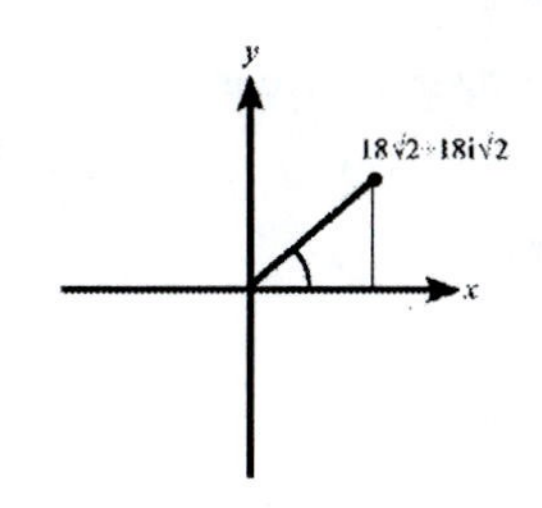

25. A sketch shows that $-6 + 2i\sqrt{3}$ is associated with a special 30°–60°–90° reference triangle in the second quadrant. So $\text{mod}(-6 + 2i\sqrt{3}) = \sqrt{(-6)^2 + (2\sqrt{3})^2}$ $= 4\sqrt{3}$, $\arg(-6 + 2i\sqrt{3}) = 150°$, and the polar form is $4\sqrt{3}\,e^{(150°)i}$.

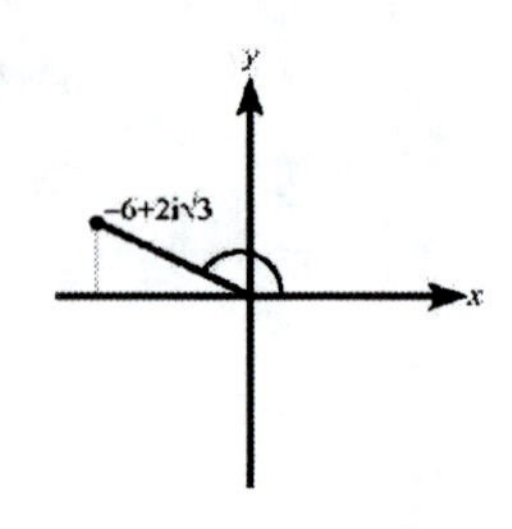

27.
$$\begin{aligned} x + iy &= 6e^{(-90°)i} \\ &= 6[\cos(-90°) + i\sin(-90°)] \\ &= 6[0 + i(-1)] \\ &= -6i \end{aligned}$$

29.

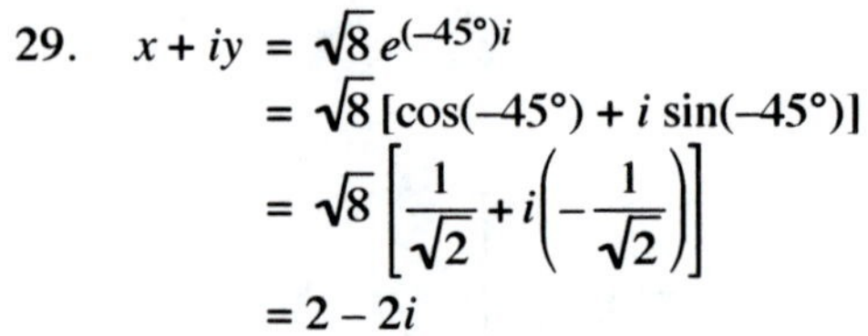

$$\begin{aligned} x + iy &= \sqrt{8}\,e^{(-45°)i} \\ &= \sqrt{8}\,[\cos(-45°) + i\sin(-45°)] \\ &= \sqrt{8}\left[\frac{1}{\sqrt{2}} + i\left(-\frac{1}{\sqrt{2}}\right)\right] \\ &= 2 - 2i \end{aligned}$$

31. A sketch shows that -50 is located on the negative x axis. So $r = |-50| = 50$, $\theta = \pi$, and the trigonometric form is $50(\cos \pi + i \sin \pi)$.

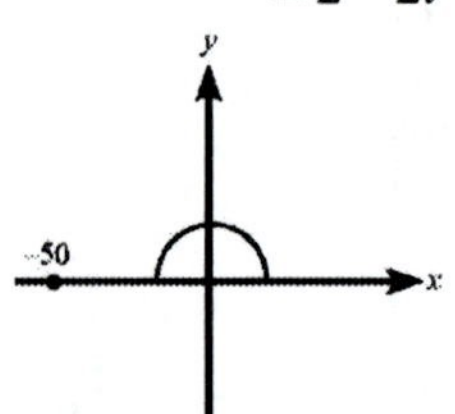

33. A sketch shows that $28 - 28i$ is associated with a special 45°–45°–60° reference triangle in the fourth quadrant. So $r = \sqrt{(28)^2 + (-28)^2} = 28\sqrt{2}$, $\theta = -\frac{\pi}{4}$, and the trigonometric form is

$$28\sqrt{2}\left[\cos\left(-\frac{\pi}{4}\right) + i\sin\left(-\frac{\pi}{4}\right)\right].$$

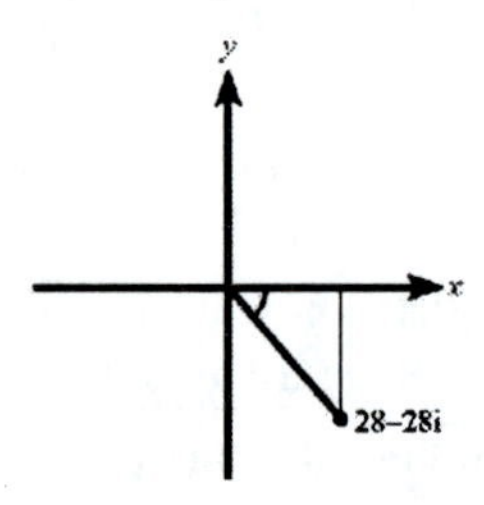

35. A sketch shows that 13 is located on the positive x axis. So $r = 13$, $\theta = 0°$, and the trigonometric form is $13(\cos 0° + i \sin 0°)$.

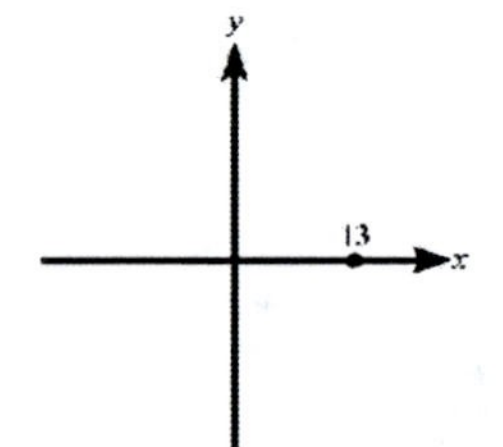

37. A sketch shows that $-\sqrt{3} + 3i$ is associated with a special 30°–60°–90° reference triangle in the second quadrant. So $r = \sqrt{(-\sqrt{3})^2 + 3^2} = 2\sqrt{3}$, $\theta = 120°$, and the trigonometric form is $2\sqrt{3}(\cos 120° + i \sin 120°)$.

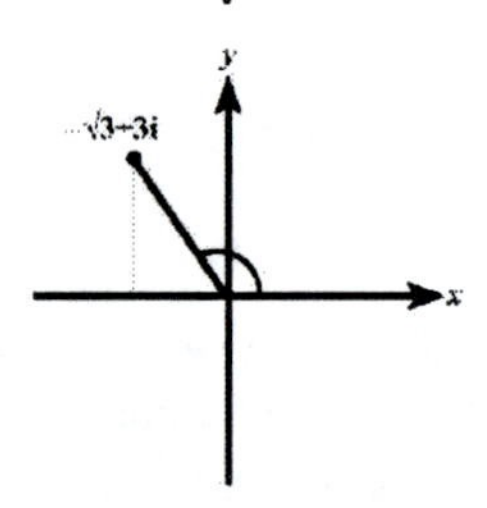

39. A sketch shows that $2 + 5i$ is not associated with a special reference triangle.

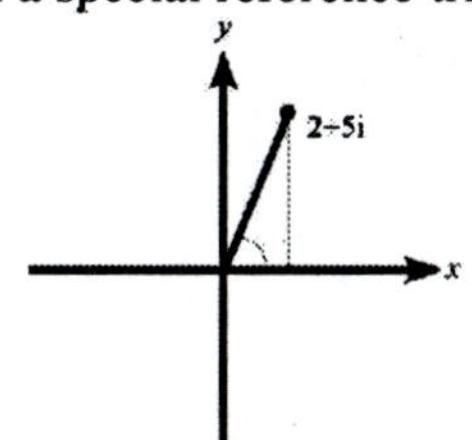

$\text{mod}(2 + 5i) = r = \sqrt{2^2 + 5^2} = 5.39$

$\arg(2 + 5i) = \theta = \tan^{-1}\frac{5}{2} = 1.19$

Therefore, the polar form is $5.39e^{1.19i}$.

41. A sketch shows that $8 - 7i$ is not associated with a special reference triangle.

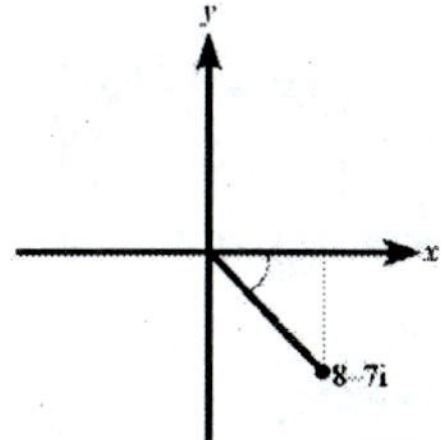

$\text{mod}(8 - 7i) = r = \sqrt{8^2 + (-7)^2} = 10.63$

$\arg(8 - 7i) = \theta = \tan^{-1}\frac{-7}{8} = -0.72$

Therefore, the polar form is $10.63e^{(-0.72)i}$.

43. A sketch shows that $-12 + 15i$ is not associated with a special reference triangle.

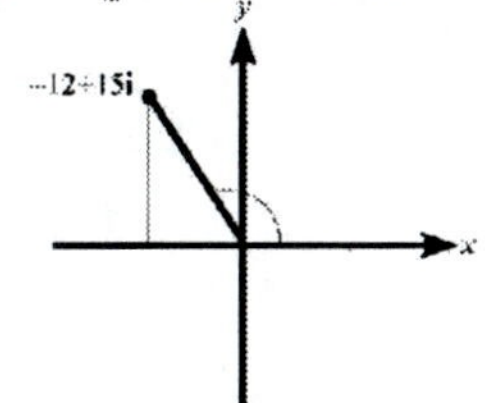

$\text{mod}(-12 + 15i) = \sqrt{(-12)^2 + 15^2} = 19.21$

$\arg(-12 + 15i) = 180° + \tan^{-1}\frac{15}{-12} = 128.66°$

because $-12 + 15i$ is in Quadrant II. Therefore, the polar form is $19.21e^{(128.66°)i}$.

45. A sketch shows that $-8 - i$ is not associated with a special reference triangle.

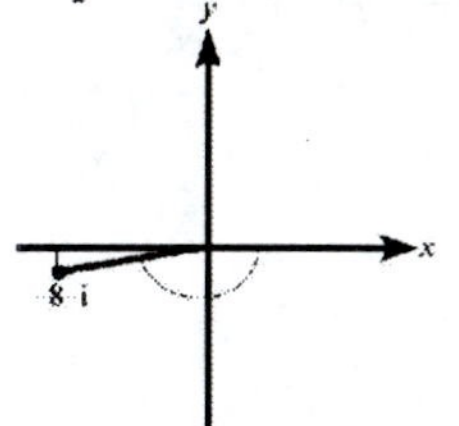

$\text{mod}(-8 - i) = \sqrt{(-8)^2 + (-1)^2} = 8.06$

$\arg(-8 - i) = \tan^{-1}\frac{-1}{-8} - 180° = -172.87°$

because $-8 - i$ is in Quadrant III. Therefore, the polar form is $8.06e^{(-172.87°)i}$.

47.
$$\begin{aligned} x + yi &= 5e^{1.7i} \\ &= 5(\cos 1.7 + i \sin 1.7) \\ &= 5 \cos 1.7 + 5i \sin 1.7 \\ &= -0.64 + 4.96i \end{aligned}$$

49.
$$\begin{aligned} x + yi &= 10e^{-0.78i} \\ &= 10[\cos(-0.78) + i \sin(-0.78)] \\ &= 10 \cos(-0.78) + 10i \sin(-0.78) \\ &= 7.11 - 7.03i \end{aligned}$$

51.
$$\begin{aligned} x + yi &= 25e^{(-125°)i} \\ &= 25[\cos(-125°) + i \sin(-125°)] \\ &= 25 \cos(-125°) + 25i \sin(-125°) \\ &= -14.34 - 20.48i \end{aligned}$$

53.
$$\begin{aligned} x + yi &= 13e^{(29°)i} \\ &= 13(\cos 29° + i \sin 29°) \\ &= 13 \cos 29° + 13i \sin 29° \\ &= 11.37 + 6.30i \end{aligned}$$

55.
$$\begin{aligned} z_1 z_2 &= 4e^{(25°)i} \cdot 8e^{(12°)i} \\ &= 4 \cdot 8e^{i(25°+12°)} = 32e^{(37°)i} \end{aligned}$$

$$\begin{aligned} \frac{z_1}{z_2} &= \frac{4e^{(25°)i}}{8e^{(12°)i}} \\ &= \frac{4}{8}e^{i(25°-12°)} = 0.5e^{(13°)i} \end{aligned}$$

57.
$$\begin{aligned} z_1 z_2 &= 6e^{(108°)i} \cdot 2e^{(-120°)i} \\ &= 6 \cdot 2e^{i(108°-120°)} = 12e^{(-12°)i} \end{aligned}$$

$$\begin{aligned} \frac{z_1}{z_2} &= \frac{6e^{(108°)i}}{2e^{(-120°)i}} \\ &= \frac{6}{2}e^{[108°-(-120°)]i} = 3e^{(228°)i} \\ &= 3e^{(228°-360°)i} = 3e^{(-132°)i} \end{aligned}$$

59. $z_1z_2 = 4.36e^{1.27i} \cdot 1.69e^{0.91i}$

$= (4.36)(1.69)e^{i(1.27+0.91)} = 7.37e^{2.18i}$

$\dfrac{z_1}{z_2} = \dfrac{4.36e^{1.27i}}{1.69e^{0.91i}}$

$= \dfrac{4.36}{1.69}e^{i(1.27-0.91)} = 2.58e^{0.36i}$

61. $z_1z_2 = 4e^{(\pi/3)i} \cdot 5e^{(-\pi/4)i}$

$= 4 \cdot 5e^{(\pi/3 - \pi/4)i}$

$= 20e^{(\pi/12)i}$

$\dfrac{z_1}{z_2} = \dfrac{4e^{(\pi/3)i}}{5e^{(-\pi/4)i}}$

$= \dfrac{4}{5}e^{[\pi/3 - (-\pi/4)]i}$

$= 0.8e^{(7\pi/12)i}$

63. Directly, $(1 + i\sqrt{3})(\sqrt{3} + i) = \sqrt{3} + i + 3i + i^2\sqrt{3} = \sqrt{3} + 4i - \sqrt{3} = 4i$

Using polar forms,

$1 + i\sqrt{3} = 2e^{60°i}$

$\sqrt{3} + i = 2e^{30°i}$

Hence, $(1 + i\sqrt{3})(\sqrt{3} + i) = 2e^{60°i} \cdot 2e^{30°i} = 2 \cdot 2e^{i(60° + 30°)} = 4e^{90°i}$

65. Directly, $(1 + i)^2 = 1 + 2i + i^2 = 1 + 2i + (-1) = 2i$

Using polar forms,

$1 + i = \sqrt{2}\, e^{45°i}$

Hence, $(1 + i)^2 = (1 + i)(1 + i) = \sqrt{2}\, e^{45°i} \cdot \sqrt{2}\, e^{45°i} = \sqrt{2} \cdot \sqrt{2}\, e^{i(45° + 45°)} = 2e^{90°i}$

67. Directly, $(1 + i)^3 = (1 + i)(1 + i)^2 = (1 + i)2i$ (from the previous problem)

$= 2i + 2i^2 = 2i + 2(-1) = -2 + 2i$

Using polar forms,

$1 + i = \sqrt{2}\, e^{45°i}$

$(1 + i)^2 = 2e^{90°i}$ (from the previous problem)

Hence, $(1 + i)^3 = (1 + i)(1 + i)^2 = \sqrt{2}\, e^{45°i} \cdot 2e^{90°i} = \sqrt{2} \cdot 2e^{i(45 + 90°)} = 2\sqrt{2}\, e^{135°i}$

69. $z^2 = zz = re^{i\theta} \cdot re^{i\theta} = rre^{i(\theta + \theta)} = r^2e^{2i\theta}$

71. $(r^{1/2}e^{i\theta/2})^2 = r^{1/2}e^{i\theta/2}r^{1/2}e^{i\theta/2} = r^{1/2}r^{1/2}e^{i(\theta/2 + \theta/2)} = re^{i\theta}$

Since $(r^{1/2}e^{i\theta/2})^2 = re^{i\theta}$, $r^{1/2}e^{i\theta/2}$ is a square root of $re^{i\theta}$.

73. Generalizing from the results of Problems 69 and 70, z^n appears to be $r^ne^{n\theta i}$.

75. Since $z + \bar{z} = (x + yi) + (x - yi) = 2x$, this point must lie on the x axis.

77. For z, mod $|z| = r_1 = \sqrt{x^2 + y^2}$. For $\bar{z}$, mod $|\bar{z}| = r_2 = \sqrt{x^2 + (-y)^2} = \sqrt{x^2 + y^2}$.
Clearly $r_1 = r_2$; the numbers have the same modulus.

79. If $z = x + yi = re^{i\theta}$ where $x \neq 0$, then $\tan\theta = \dfrac{y}{x}$. Then $\dfrac{-y}{x} = -\dfrac{y}{x} = -\tan\theta = \tan(-\theta)$, hence $\bar{z} = x - yi$ $= re^{i(-\theta)} = re^{-i\theta}$, since the two numbers have the same modulus r. If $x = 0$, then $z = yi$, so it is obvious that $\bar{z} = re^{-i\theta}$.

81\. (A) $8(\cos 0° + i \sin 0°) = 8(1 + 0i) = 8 + 0i$

$6(\cos 30° + i \sin 30°) = 6\left(\frac{\sqrt{3}}{2} + i\frac{1}{2}\right) = 3\sqrt{3} + 3i$

$(8 + 0i) + (3\sqrt{3} + 3i) = (8 + 3\sqrt{3}) + 3i$

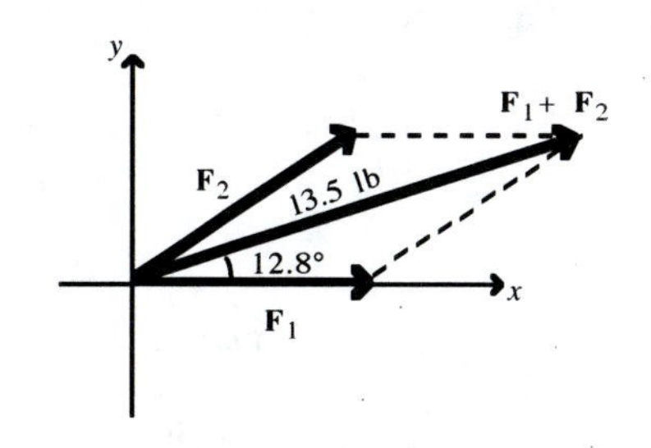

(B) $r = \sqrt{(8+3\sqrt{3})^2 + 3^2} \approx 13.5$

$\tan\theta = \frac{3}{8+3\sqrt{3}}$ $\qquad \theta = 12.8°$

$(8 + 3\sqrt{3}) + 3i = 13.5(\cos 12.8° + i \sin 12.8°)$

(C) We can interpret $13.5(\cos 12.8° + i \sin 12.8°)$ as a force of 13.5 lb at an angle of 12.8° with respect to the force $\mathbf{F}_1$.

EXERCISE 7.4 DeMoivre's Theorem and the *n*th Root Theorem

1\. No. Positive real numbers have two real square roots; negative real numbers have two imaginary square roots.

3\. The two square roots of z are real numbers. They lie on the real axis, one on the positive real axis at $(\sqrt{z}, 0)$, the other on the negative real axis at $(-\sqrt{z}, 0)$.

5\. $(3e^{15°i})^3 = 3^3e^{3\cdot 15°i} = 27e^{45°i}$

7\. $(\sqrt{2}\,e^{45°i})^{10} = (\sqrt{2})^{10}e^{(10\cdot 45°)i} = (2^{1/2})^{10}\,e^{450°i} = 2^5e^{450°i} = 32e^{450°i}$ or $32e^{(450° - 360°)i} = 32e^{90°i}$

9\. $(2e^{(\pi/5)i})^4 = 2^4e^{(4\cdot\pi/5)i} = 16e^{(4\pi/5)i}$

11\. $(5e^{(11\pi/6)i})^2 = 5^2e^{(2\cdot 11\pi/6)i} = 25e^{(11\pi/3)i}$

13\. $(-1 + i)^4 = (\sqrt{2}\,e^{135°i})^4 = (\sqrt{2})^4\,e^{(4\cdot 135°)i} = (2^{1/2})^4\,e^{540°i} = 2^2e^{540°i}$

$= 4e^{540°i} = 4(\cos 540° + i \sin 540°) = 4(-1 + 0i) = -4$

15\. $(-\sqrt{3} + i)^5 = (2e^{150°i})^5 = 2^5\,e^{(5\cdot 150°)i} = 32e^{750°i}$

$= 32(\cos 750° + i \sin 750°) = 32\left(\frac{\sqrt{3}}{2} + i\cdot\frac{1}{2}\right) = 16\sqrt{3} + 16i$

17\. $\left(-\frac{1}{2} - \frac{\sqrt{3}}{2}i\right)^3 = (1e^{-120°i})^3 = 1^3\,e^{3(-120°)i} = e^{-360°i}$

$= \cos(-360°) + i\sin(-360°) = 1 + 0i = 1$

19\. First, write 100 in polar form.

$100 = 100e^{0°i}$

From the *n*th root theorem, both roots are given by

$100^{1/2}e^{(0°/2 + k360°/2)i} \qquad k = 0, 1$

Thus,

$w_1 = 100^{1/2}e^{(0° + 0\cdot 180°)i} = 10e^{(0°)i} = 10$

$w_2 = 100^{1/2}e^{(0° + 1\cdot 180°)i} = 10e^{(180°)i} = -10$

21. First, write –361 in polar form.

$$-361 = 361e^{180°i}$$

From the *n*th root theorem, both roots are given by

$$361^{1/2}e^{(180°/2 + k360°/2)i} \quad k = 0, 1$$

Thus,

$$w_1 = 361^{1/2}e^{(90° + 0 \cdot 180°)i} = 19e^{(90°)i} = 19i$$

$$w_2 = 361^{1/2}e^{(90° + 1 \cdot 180°)i} = 19e^{(270°)i} = -19i$$

23. First, write $81i$ in polar form.

$$81i = 81e^{(\pi/2)i}$$

From the *n*th root theorem, both roots are given by

$$81^{1/2}e^{[(\pi/2)\div 2 + k \cdot 2\pi/2]i} \quad k = 0, 1$$

Thus,

$$w_1 = 81^{1/2}e^{(\pi/4 + 0\pi)i} = 9e^{(\pi/4)i} = \frac{9}{\sqrt{2}} + i\frac{9}{\sqrt{2}}$$

$$w_2 = 81^{1/2}e^{(\pi/4 + 1\pi)i} = 9e^{(5\pi/4)i} = -\frac{9}{\sqrt{2}} - i\frac{9}{\sqrt{2}}$$

25. First, write $-169i$ in polar form.

$$-169i = 169e^{(3\pi/2)i}$$

From the *n*th root theorem, both roots are given by

$$169^{1/2}e^{[(3\pi/2)\div 2 + k \cdot 2\pi/2]i} \quad k = 0, 1$$

Thus,

$$w_1 = 169^{1/2}e^{(3\pi/4 + 0\pi)i} = 13e^{(3\pi/4)i} = -\frac{13}{\sqrt{2}} + i\frac{13}{\sqrt{2}}$$

$$w_2 = 169^{1/2}e^{(3\pi/4 + 1\pi)i} = 13e^{(7\pi/4)i} = \frac{13}{\sqrt{2}} - i\frac{13}{\sqrt{2}}$$

27. First, write 125 in polar form.

$$125 = 125e^{0°i}$$

From the *n*th root theorem, all three roots are given by

$$125^{1/3}e^{(0°/3 + k360°/3)i} \quad k = 0, 1, 2$$

Thus,

$$w_1 = 125^{1/3}e^{(0° + 0 \cdot 120°)i} = 5e^{(0°)i} = 5$$

$$w_2 = 125^{1/3}e^{(0° + 1 \cdot 120°)i} = 5e^{(120°)i}$$

$$w_3 = 125^{1/3}e^{(0° + 2 \cdot 120°)i} = 5e^{(240°)i}$$

29. First, write –216 in polar form.

$$-216 = 216e^{(180°)i}$$

From the *n*th root theorem, all three roots are given by

$$216^{1/3}e^{(180°/3 + k360°/3)i} \quad k = 0, 1, 2$$

Thus,

$$w_1 = 216^{1/3}e^{(60° + 0 \cdot 120°)i} = 6e^{(60°)i}$$

$$w_2 = 216^{1/3}e^{(60° + 1 \cdot 120°)i} = 6e^{(180°)i} = -6$$

$$w_3 = 216^{1/3}e^{(60° + 2 \cdot 120°)i} = 6e^{(300°)i}$$

31. First, write $343i$ in polar form.

$$343i = 343e^{(\pi/2)i}$$

From the nth root theorem, all three roots are given by

$$343^{1/3}e^{[(\pi/2)\div 3 + k \cdot 2\pi/3]i} \qquad k = 0, 1, 2$$

Thus,

$$w_1 = 343^{1/3}e^{(\pi/6 + 0 \cdot 2\pi/3]i} = 7e^{(\pi/6)i}$$
$$w_2 = 343^{1/3}e^{(\pi/6 + 1 \cdot 2\pi/3]i} = 7e^{(5\pi/6)i}$$
$$w_3 = 343^{1/3}e^{(\pi/6 + 2 \cdot 2\pi/3]i} = 7e^{(3\pi/2)i}$$

33. First, write $-\dfrac{i}{8}$ in polar form.

$$-\frac{i}{8} = \frac{1}{8}e^{(3\pi/2)i}$$

From the nth root theorem, all three roots are given by

$$\left(\frac{1}{8}\right)^{1/3}e^{[(3\pi/2)\div 3 + k \cdot 2\pi/3]i} \qquad k = 0, 1, 2$$

Thus,

$$w_1 = \left(\frac{1}{8}\right)^{1/3}e^{(\pi/2 + 0 \cdot 2\pi/3]i} = \frac{1}{2}e^{(\pi/2)i}$$
$$w_2 = \left(\frac{1}{8}\right)^{1/3}e^{(\pi/2 + 1 \cdot 2\pi/3]i} = \frac{1}{2}e^{(7\pi/6)i}$$
$$w_3 = \left(\frac{1}{8}\right)^{1/3}e^{(\pi/2 + 2 \cdot 2\pi/3]i} = \frac{1}{2}e^{(11\pi/6)i}$$

35. (A) Since mod $z = 1024$, mod $z^{1/10} = 1024^{1/10} = 2$ = the radius.

(B) The angle between consecutive 10th roots is $360°/10 = 36°$.

37. (A) Since mod $z = 49$, mod $z^{1/4} = 49^{1/4} = \sqrt{7}$ = the radius.

(B) The angle between consecutive fourth roots is $360°/4 = 90°$.

39. (A) Since mod $z = 1000$, mod $z^{1/3} = 1000^{1/3} = 10$ = the radius.

(B) The angle between consecutive cube roots is $2\pi/3$.

41. (A) Since mod $z = 3$, mod $z^{1/7} = 3^{1/7}$.

(B) The angle between consecutive 7th roots is $2\pi/7$.

43. From the nth root theorem, both roots are given by

$$4^{1/2}\, e^{(30°/2 + k360°/2)i} \qquad k = 0, 1$$

Thus,

$$w_1 = 4^{1/2}\, e^{(15° + 0 \cdot 180°)i} = 2e^{15°i}$$
$$w_2 = 4^{1/2}\, e^{(15° + 1 \cdot 180°)i} = 2e^{195°i}$$

45. From the nth root theorem, all three roots are given by

$$8^{1/3}\, e^{(90°/3 + k360°/3)i} \qquad k = 0, 1, 2$$

Thus,

$$w_1 = 8^{1/3}\, e^{(30° + 0 \cdot 120°)i} = 2e^{30°i}$$
$$w_2 = 8^{1/3}\, e^{(30° + 1 \cdot 120°)i} = 2e^{150°i}$$
$$w_3 = 8^{1/3}\, e^{(30° + 2 \cdot 120°)i} = 2e^{270°i}$$

47. First, write $-1 + i$ in polar form.

$$-1 + i = 2^{1/2}\, e^{135°i}$$

From the *n*th root theorem, all five roots are given by

$$(2^{1/2})^{1/5}\, e^{(135°/5 + k360°/5)i} \qquad k = 0, 1, 2, 3, 4$$

Thus,

$$w_1 = 2^{1/10}\, e^{(27° + 0 \cdot 72°)i} = 2^{1/10}\, e^{27°i}$$
$$w_2 = 2^{1/10}\, e^{(27° + 1 \cdot 72°)i} = 2^{1/10}\, e^{99°i}$$
$$w_3 = 2^{1/10}\, e^{(27° + 2 \cdot 72°)i} = 2^{1/10}\, e^{171°i}$$
$$w_4 = 2^{1/10}\, e^{(27° + 3 \cdot 72°)i} = 2^{1/10}\, e^{243°i}$$
$$w_5 = 2^{1/10}\, e^{(27° + 4 \cdot 72°)i} = 2^{1/10}\, e^{315°i}$$

49. First, write -8 in polar form.

$$-8 = 8e^{180°i}$$

From the *n*th root theorem, all three roots are given by

$$8^{1/3}\, e^{(180°/3 + k360°/3)i} \qquad k = 0, 1, 2$$

Thus,

$$w_1 = 8^{1/3}\, e^{(60° + 0 \cdot 120°)i} = 2e^{60°i}$$
$$w_2 = 8^{1/3}\, e^{(60° + 1 \cdot 120°)i} = 2e^{180°i}$$
$$w_3 = 8^{1/3}\, e^{(60° + 2 \cdot 120°)i} = 2e^{300°i}$$

The three roots are equally spaced around a circle with radius 2 at an angular increment of 120° from one root to the next.

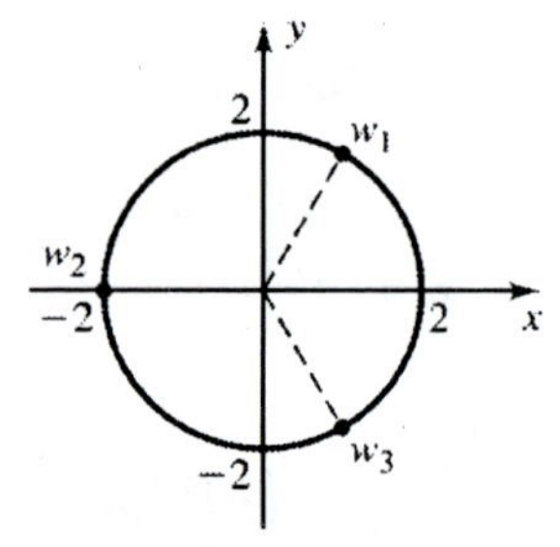

51. First, write 1 in polar form.

$$1 = 1e^{0°i}$$

From the *n*th root theorem, all four roots are given by

$$1^{1/4}\, e^{(0°/4 + k360°/4)i} \qquad k = 0, 1, 2, 3$$

Thus,

$$w_1 = 1^{1/4}\, e^{(0° + 0 \cdot 90°)i} = 1e^{0°i}$$
$$w_2 = 1^{1/4}\, e^{(0° + 1 \cdot 90°)i} = 1e^{90°i}$$
$$w_3 = 1^{1/4}\, e^{(0° + 2 \cdot 90°)i} = 1e^{180°i}$$
$$w_4 = 1^{1/4}\, e^{(0° + 3 \cdot 90°)i} = 1e^{270°i}$$

The four roots are equally spaced around a circle with radius 1 at an angular increment of 90° from one root to the next.

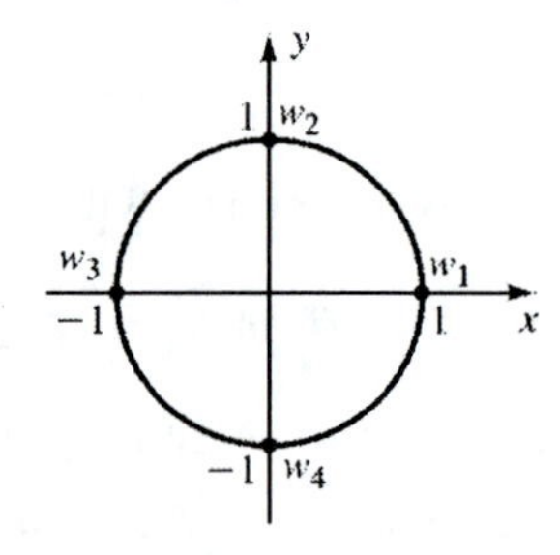

53. First, write $-i$ in polar form.

$$-i = 1e^{(-90°)i}$$

From the *n*th root theorem, all five roots are given by

$$1^{1/5}\, e^{(-90°/5 + k360°/5)i} \qquad k = 0, 1, 2, 3, 4$$

Thus,

$$w_1 = 1^{1/5}\, e^{(-18° + 0 \cdot 72°)i} = 1e^{-18°i}$$
$$w_2 = 1^{1/5}\, e^{(-18° + 1 \cdot 72°)i} = 1e^{54°i}$$
$$w_3 = 1^{1/5}\, e^{(-18° + 2 \cdot 72°)i} = 1e^{126°i}$$
$$w_4 = 1^{1/5}\, e^{(-18° + 3 \cdot 72°)i} = 1e^{198°i}$$
$$w_5 = 1^{1/5}\, e^{(-18° + 4 \cdot 72°)i} = 1e^{270°i}$$

The five roots are equally spaced around a circle with radius 1 at an angular increment of 72° from one root to the next.

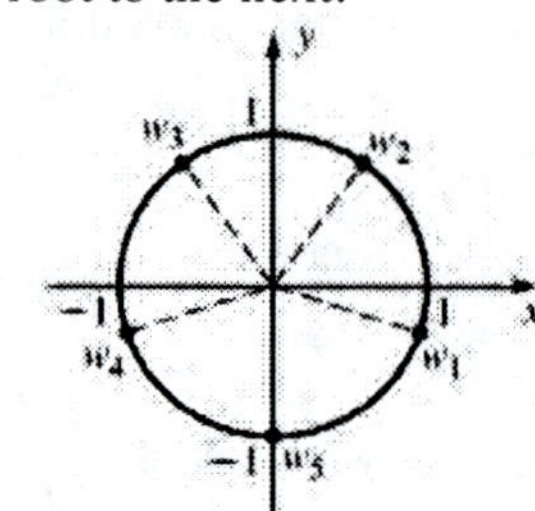

55. (A) $(-\sqrt{2}+i\sqrt{2})^4+16 = ((-\sqrt{2}+i\sqrt{2})^2)^2+16$

$$= (2-4i+2i^2)^2+16$$
$$= (-4i)^2+16$$
$$= 16i^2+16$$
$$= -16+16$$
$$= 0$$

Thus, $-\sqrt{2}+i\sqrt{2}$ is a root of $x^4+16=0$.

Since the polynomial has degree 4, there are three other roots.

(B) The four roots are equally spaced around the circle. Since there are 4 roots, the angle between successive roots on the circle is $\frac{360°}{4} = 90°$.

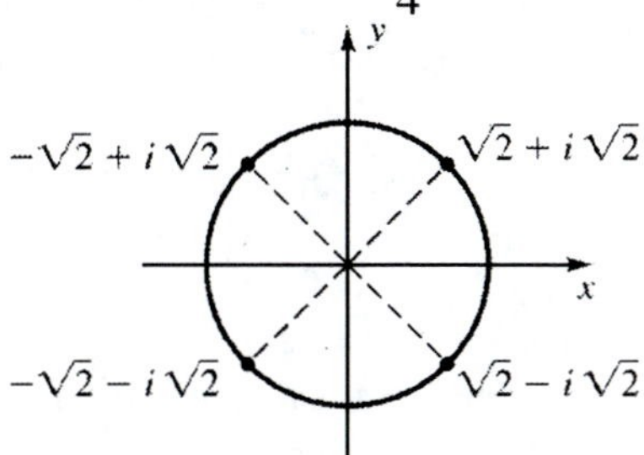

(C) $(-\sqrt{2}-i\sqrt{2})^4+16 = ((-\sqrt{2}-i\sqrt{2})^2)^2+16$

$$= (2+4i+2i^2)^2+16$$
$$= (4i)^2+16 = 16i^2+16 = -16+16 = 0$$

$(\sqrt{2}-i\sqrt{2})^4+16 = ((\sqrt{2}-i\sqrt{2})^2)^2+16$

$$= (2-4i+2i^2)^2+16$$
$$= (-4i)^2+16 = 16i^2+16 = -16+16 = 0$$

$(\sqrt{2}+i\sqrt{2})^4+16 = ((\sqrt{2}+i\sqrt{2})^2)^2+16$

$$= (2+4i+2i^2)^2+16$$
$$= (4i)^2+16 = 16i^2+16 = -16+16 = 0$$

57. $x^2 - i = 0$

$x^2 = i$

Therefore x is a square root of i, and there are two of them. First, write i in polar form and use the nth root theorem.

$$i = 1e^{90°i}$$

The two square roots of i are given by

$$1^{1/2}\,e^{(90°/2\,+\,k360°/2)i} \qquad k = 0, 1$$

Thus,

$$w_1 = 1^{1/2}e^{(45°+0\cdot 180°)i} = e^{45°i} = \cos 45° + i\sin 45° = \frac{\sqrt{2}}{2} + i\frac{\sqrt{2}}{2}$$

$$w_2 = 1^{1/2}e^{(45°+1\cdot 180°)i} = e^{225°i} = \cos 225° + i\sin 225° = -\frac{\sqrt{2}}{2} - i\frac{\sqrt{2}}{2}$$

59. $x^2 + 4i = 0$

$x^2 = -4i$

Therefore x is a square root of $-4i$, and there are two of them. First, write $-4i$ in polar form and use the nth root theorem.

$$-4i = 4e^{(-90°)i}$$

The two square roots of $-4i$ are given by

$$4^{1/2}\, e^{(-90°/2 + k360°/2)i} \qquad k = 0, 1$$

Thus,

$$w_1 = 4^{1/2}e^{(-45° + 0 \cdot 180°)i} = 2e^{-45°i} = 2[\cos(-45°) + i\sin(-45°)] = 2\left(\frac{\sqrt{2}}{2} - i\frac{\sqrt{2}}{2}\right) = \sqrt{2} - i\sqrt{2}$$

$$w_2 = 4^{1/2}e^{(-45° + 1 \cdot 180°)i} = 2e^{135°i} = 2[\cos 135° + i\sin 135°) = 2\left(-\frac{\sqrt{2}}{2} + i\frac{\sqrt{2}}{2}\right) = -\sqrt{2} + i\sqrt{2}$$

61. $x^3 + 27 = 0$

$x^3 = -27$

Therefore x is a cube root of -27, and there are three of them. First, write -27 in polar form and use the *n*th root theorem.

$$-27 = 27e^{180°i}$$

All three cube roots of -27 are given by

$$27^{1/3}\, e^{(180°/3 + k360°/3)i} \qquad k = 0, 1, 2$$

Thus,

$$w_1 = 27^{1/3}e^{(60° + 0 \cdot 120°)i} = 3e^{60°i} = 3(\cos 60° + i\sin 60°) = 3\left(\frac{1}{2} + i\frac{\sqrt{3}}{2}\right) = \frac{3}{2} + \frac{3\sqrt{3}}{2}i$$

$$w_2 = 27^{1/3}e^{(60° + 1 \cdot 120°)i} = 3e^{180°i} = 3(\cos 180° + i\sin 180°) = 3[(-1) + i0] = -3$$

$$w_3 = 27^{1/3}e^{(60° + 2 \cdot 120°)i} = 3e^{300°i} = 3(\cos 300° + i\sin 300°) = 3\left[\frac{1}{2} + i\left(\frac{\sqrt{3}}{2}\right)\right] = \frac{3}{2} - \frac{3\sqrt{3}}{2}i$$

63. $x^3 - 64 = 0$

$x^3 = 64$

Therefore x is a cube root of 64, and there are three of them. First, write 64 in polar form and use the *n*th root theorem.

$$64 = 64e^{0°i}$$

All three cube roots of 64 are given by

$$64^{1/3}\, e^{(0°/3 + k360°/3)i} \qquad k = 0, 1, 2$$

Thus,

$$w_1 = 64^{1/3}\, e^{(0° + 0 \cdot 120°)i} = 4e^{0°i} = 4(\cos 0° + i\sin 0°) = 4(1 + 0i) = 4$$

$$w_2 = 64^{1/3}\, e^{(0° + 1 \cdot 120°)i} = 4e^{120°i} = 4(\cos 120° + i\sin 120°) = 4\left[\left(-\frac{1}{2}\right) + i\frac{\sqrt{3}}{2}\right] = -2 + 2i\sqrt{3}$$

$$w_3 = 64^{1/3}\, e^{(0° + 2 \cdot 120°)i} = 4e^{240°i} = 4(\cos 240° + i\sin 240°) = 4\left[\left(-\frac{1}{2}\right) + i\left(-\frac{\sqrt{3}}{2}\right)\right] = -2 - 2i\sqrt{3}$$

65. For $k = 0$,

$$r^{1/n}\, e^{(\theta/n + k \cdot 360°/n)i} = r^{1/n}\, e^{\theta/n\, i} = r^{1/n}\left(\cos\frac{\theta}{n} + i\sin\frac{\theta}{n}\right)$$

For $k = n$,

$$\begin{aligned} r^{1/n}\, e^{(\theta/n + k \cdot 360°/n)i} &= r^{1/n}\, e^{(\theta/n + n \cdot 360°/n)i} \\ &= r^{1/n}\, e^{(\theta/n + 360°)i} \\ &= r^{1/n}\left[\cos\left(\frac{\theta}{n} + 360°\right) + i\sin\left(\frac{\theta}{n} + 360°\right)\right] \end{aligned}$$

$$= r^{1/n}\left(\cos\frac{\theta}{n} + i\sin\frac{\theta}{n}\right)$$

Thus, the two are the same number.

67. $x^5 - 1 = 0$

$x^5 = 1$

Therefore x is a fifth root of 1, and there are five of them. First, write 1 in polar form and use the nth root theorem.

$1 = 1e^{0°i}$

All five fifth roots of 1 are given by

$1^{1/5}\,e^{(0°/5 + k360°/5)i} \qquad k = 0, 1, 2, 3, 4$

Thus,

$w_1 = 1^{1/5}\,e^{(0° + 0\cdot 72°)i} = e^{0°i} = \cos 0° + i\sin 0° = 1 + i0 = 1$

$w_2 = 1^{1/5}\,e^{(0° + 1\cdot 72°)i} = e^{72°i} = \cos 72° + i\sin 72° = 0.309 + 0.951i$

$w_3 = 1^{1/5}\,e^{(0° + 2\cdot 72°)i} = e^{144°i} = \cos 144° + i\sin 144° = -0.809 + 0.588i$

$w_4 = 1^{1/5}\,e^{(0° + 3\cdot 72°)i} = e^{216°i} = \cos 216° + i\sin 216° = -0.809 - 0.588i$

$w_5 = 1^{1/5}\,e^{(0° + 4\cdot 72°)i} = e^{288°i} = \cos 288° + i\sin 288° = 0.309 - 0.951i$

69. $x^3 + 5 = 0$

$x^3 = -5$

Therefore x is a cube root of -5, and there are three of them. First, write -5 in polar form and use the nth root theorem.

$-5 = 5e^{180°i}$

All three cube roots of -5 are given by

$5^{1/3}\,e^{(180°/3 + k360°/3)i} \qquad k = 0, 1, 2$

Thus,

$w_1 = 5^{1/3}e^{(60° + 0\cdot 120°)i} = 5^{1/3}e^{60°i} = 5^{1/3}(\cos 60° + i\sin 60°) = 0.855 + 1.481i$

$w_2 = 5^{1/3}e^{(60° + 1\cdot 120°)i} = 5^{1/3}e^{180°i} = 5^{1/3}(\cos 180° + i\sin 180°) = -1.710$

$w_3 = 5^{1/3}e^{(60° + 2\cdot 120°)i} = 5^{1/3}e^{300°i} = 5^{1/3}(\cos 300° + i\sin 300°) = 0.855 - 1.481i$

71. (A) $z = e^{(60°)i}$

$z^k = [e^{(60°)i}]^k$

$= e^{(60°)ik}$

$= e^{(60°)ki}$

Now consider $(z^k)^6$. For any integer k, $1 \le k \le 6$,

$(z^k)^6 = (e^{(60°)ki})^6$

$= e^{(360°)ki}$

$= \cos(360°\,k) + i\sin(360°\,k)$

$= 1$

Thus z^k is a sixth root of 1.

(B) Since $(z^3)^2 = (e^{(180°)i})^2 = e^{(360°)i} = 1$, z^3 is a square root of 1.

Since $(z^6)^2 = 1^2 = 1$, z^6 is a square root of 1.

(C) Since $(z^2)^3 = (e^{(120°)i})^3 = e^{(360°)i} = 1$, z^2 is a cube root of 1.

Since $(z^4)^3 = (e^{(240°)i})^3 = e^{(720°)i} = 1$, z^4 is a cube root of 1.

Since $(z^6)^3 = 1^3 = 1$, z^6 is a cube root of 1.

73. (A)

$$\begin{aligned} z &= e^{(36°)i} \\ z^k &= [e^{(36°)i}]^k \\ &= e^{(36°)ik} \\ &= e^{(36°)ki} \end{aligned}$$

Now consider $(z^k)^{10}$. For any integer k, $1 \le k \le 10$,

$$\begin{aligned} (z^k)^{10} &= (e^{(36°)ki})^{10} \\ &= e^{(360°)ki} \\ &= \cos(360°\ k) + i \sin(360°\ k) \\ &= 1 \end{aligned}$$

Thus z^k is a tenth root of 1.

(B) Since $(z^5)^2 = (e^{(180°)i})^2 = e^{(360°)i} = 1$, z^5 is a square root of 1.

Since $(z^{10})^2 = 1^2 = 1$, z^{10} is a square root of 1.

(C) Since $(z^2)^5 = (e^{(72°)i})^5 = e^{(360°)i} = 1$, z^2 is a fifth root of 1.

Since $(z^4)^5 = (e^{(144°)i})^5 = e^{(720°)i} = 1$, z^4 is a fifth root of 1.

Since $(z^6)^5 = (e^{(216°)i})^5 = e^{(1080°)i} = 1$, z^6 is a fifth root of 1.

Since $(z^8)^5 = (e^{(288°)i})^5 = e^{(1440°)i} = 1$, z^8 is a fifth root of 1.

Since $(z^{10})^5 = 1^5 = 1$, z^{10} is a fifth root of 1.

CHAPTER 7 Review Exercise

1.

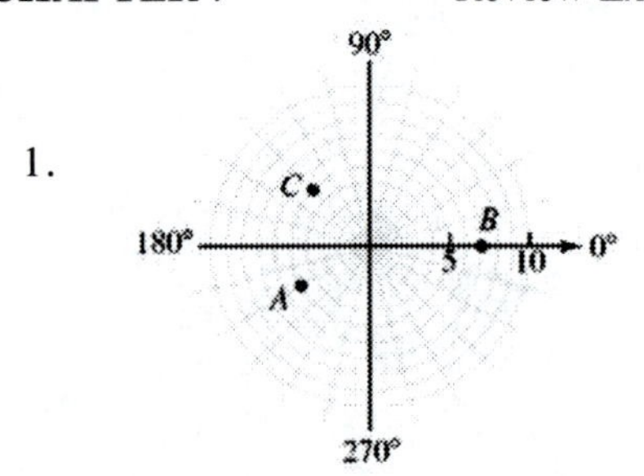

2. Set up a table that shows how r varies as θ varies through each set of quadrant values, then sketch the polar curve from the information in the table.

θ	$5\sin\theta$
0 to $\frac{\pi}{2}$	0 to 5
$\frac{\pi}{2}$ to π	5 to 0
π to $\frac{3\pi}{2}$	0 to –5
$\frac{3\pi}{2}$ to 2π	–5 to 0

Curve is traced out a second time in this region, although coordinate pairs are different

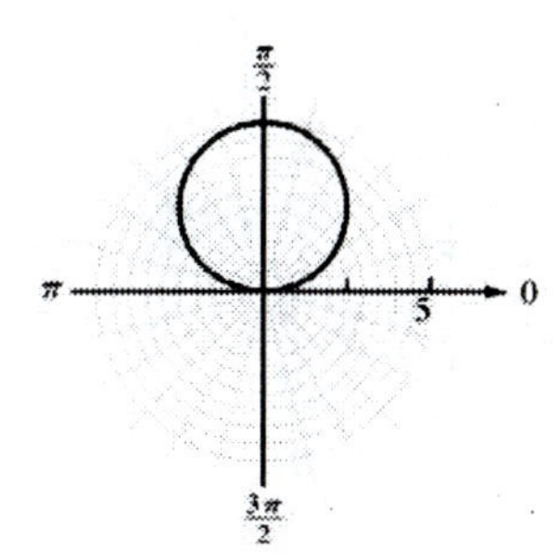

3. Set up a table that shows how r varies as θ varies through each set of quadrant values, then sketch the polar curve from the information in the table.

θ	$\cos\theta$	$4\cos\theta$	$4+4\cos\theta$
0 to $\frac{\pi}{2}$	1 to 0	4 to 0	8 to 4
$\frac{\pi}{2}$ to π	0 to –1	0 to –4	4 to 0
π to $\frac{3\pi}{2}$	–1 to 0	–4 to 0	0 to 4
$\frac{3\pi}{2}$ to 2π	0 to 1	0 to 4	4 to 8

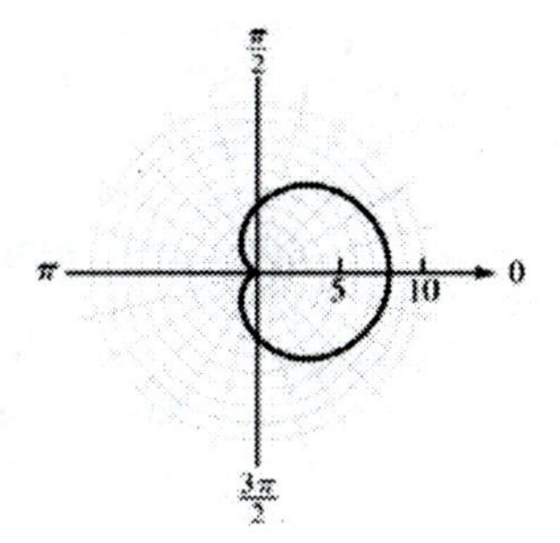

4. The graph consists of all points whose distance from the pole is 8, a circle with center at the pole, and radius 8.

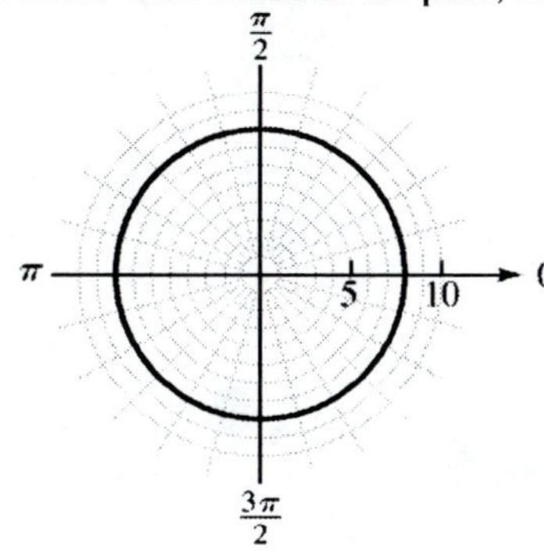

5.

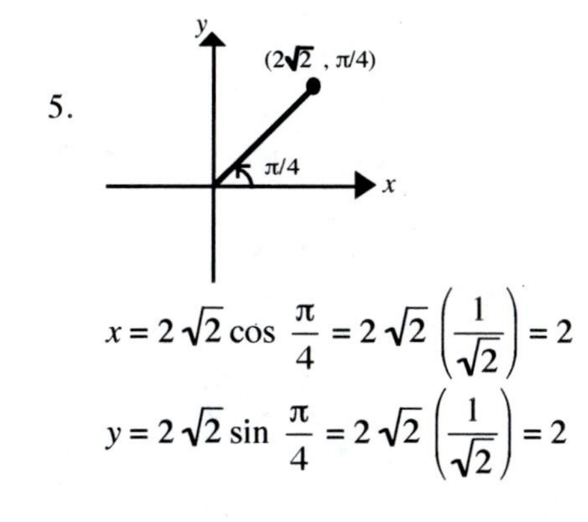

$$x = 2\sqrt{2}\cos\frac{\pi}{4} = 2\sqrt{2}\left(\frac{1}{\sqrt{2}}\right) = 2$$

$$y = 2\sqrt{2}\sin\frac{\pi}{4} = 2\sqrt{2}\left(\frac{1}{\sqrt{2}}\right) = 2$$

6. Use $r^2 = x^2 + y^2$ and $\tan\theta = \frac{y}{x}$

$r^2 = (-\sqrt{3})^2 + 1^2 = 4 \quad r = 2 \quad \tan\theta = \frac{1}{-\sqrt{3}} = -\frac{1}{\sqrt{3}} \quad \theta = \frac{5\pi}{6}$ since the point is in the second quadrant

Polar coordinates: $\left(2, \frac{5\pi}{6}\right)$

7. 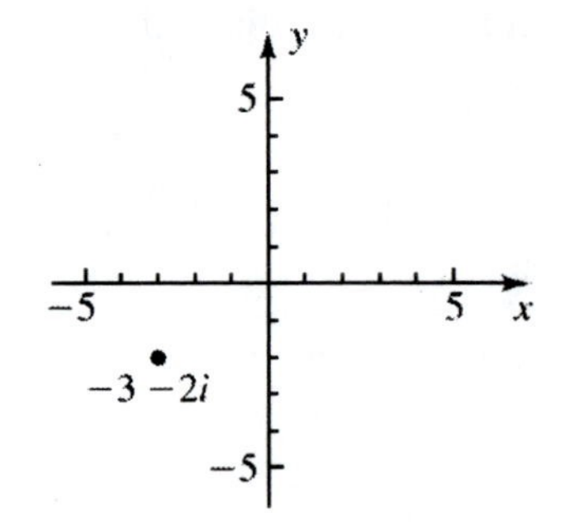

8. 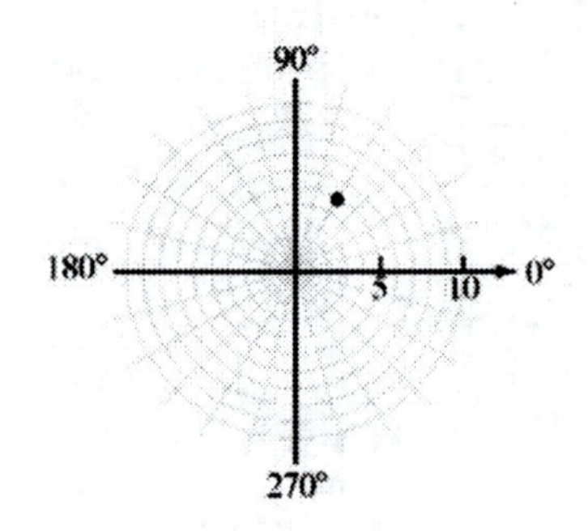

9. See figure at the right. The point with coordinates (–8, 30°) can equally well be described as (–8, –330°) or (8, –150°) or (8, 210°). Thus, (–8, –330°): The polar axis is rotated 330° clockwise (negative direction) and the point is located 8 units from the pole along the negative polar axis. (8, –150°): The polar axis is rotated 150° clockwise (negative direction) and the point is located 8 units from the pole along the positive polar axis. (8, 210°): The polar axis is rotated 210° counterclockwise (positive direction) and the point is located 8 units along the positive polar axis.

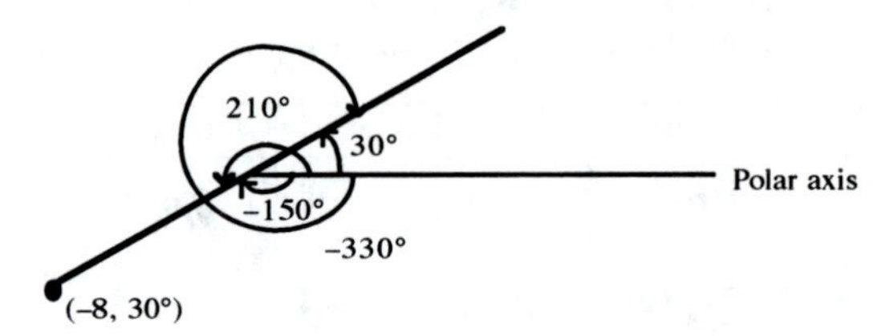

10. $z_1 z_2 = 9e^{42°i} \cdot 3e^{37°i}$

$= 9 \cdot 3e^{i(42° + 37°)} = 27e^{79°i}$

$\dfrac{z_1}{z_2} = \dfrac{9e^{42°i}}{3e^{37°i}} = \dfrac{9}{3}e^{i(42° - 37°)} = 3e^{5°i}$

11. $(2e^{10°i})^4 = 2^4 e^{4 \cdot 10°i} = 16e^{40°i}$

12. $y = 7$

Use $y = r \sin \theta$

$r \sin \theta = 7$

$r = \dfrac{7}{\sin\theta}$

13. 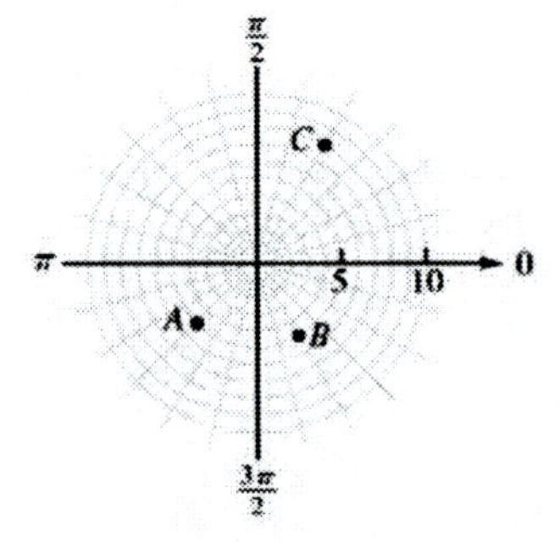

14. Set up a table that shows how r varies as 3θ varies through each set of quadrant values, then sketch the polar curve from the information in the table.

θ	3θ	$\sin 3\theta$	$8\sin 3\theta$
0 to $\frac{\pi}{6}$	0 to $\frac{\pi}{2}$	0 to 1	0 to 8
$\frac{\pi}{6}$ to $\frac{\pi}{3}$	$\frac{\pi}{2}$ to π	1 to 0	8 to 0
$\frac{\pi}{3}$ to $\frac{\pi}{2}$	π to $\frac{3\pi}{2}$	0 to –1	0 to –8
$\frac{\pi}{2}$ to $\frac{2\pi}{3}$	$\frac{3\pi}{2}$ to 2π	–1 to 0	–8 to 0
$\frac{2\pi}{3}$ to $\frac{5\pi}{6}$	2π to $\frac{5\pi}{2}$	0 to 1	0 to 8
$\frac{5\pi}{6}$ to π	$\frac{5\pi}{2}$ to 3π	1 to 0	8 to 0

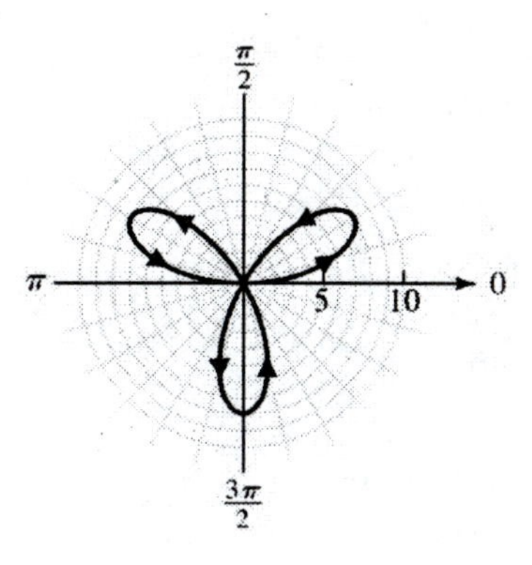

15. Set up a table that shows how r varies as 2θ varies through each set of quadrant values, then sketch the polar curve from the information in the table.

θ	2θ	$\sin 2\theta$	$4\sin 2\theta$
0 to $\frac{\pi}{4}$	0 to $\frac{\pi}{2}$	0 to 1	0 to 4
$\frac{\pi}{4}$ to $\frac{\pi}{2}$	$\frac{\pi}{2}$ to π	1 to 0	4 to 0
$\frac{\pi}{2}$ to $\frac{3\pi}{4}$	π to $\frac{3\pi}{2}$	0 to –1	0 to –4
$\frac{3\pi}{4}$ to π	$\frac{3\pi}{2}$ to 2π	–1 to 0	–4 to 0
π to $\frac{5\pi}{4}$	2π to $\frac{5\pi}{2}$	0 to 1	0 to 4
$\frac{5\pi}{4}$ to $\frac{3\pi}{2}$	$\frac{5\pi}{2}$ to 3π	1 to 0	4 to 0
$\frac{3\pi}{2}$ to $\frac{7\pi}{4}$	3π to $\frac{7\pi}{2}$	0 to –1	0 to –4
$\frac{7\pi}{4}$ to 2π	$\frac{7\pi}{2}$ to 4π	–1 to 0	–4 to 0

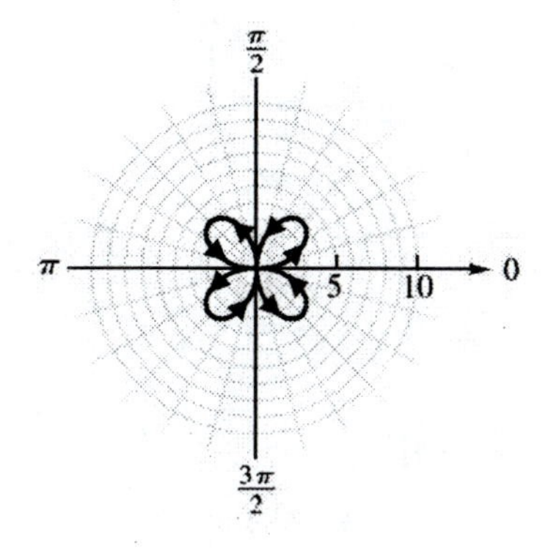

16. The graph consists of all points on a line that forms an angle of $\frac{\pi}{6}$ with the polar axis and passes through the pole.

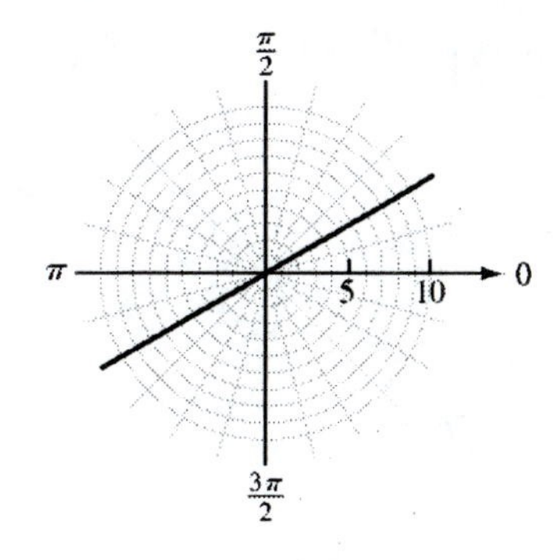

17. $8x - y^2 = x^2$
$8x = x^2 + y^2$

Use $x = r\cos\theta$ and $x^2 + y^2 = r^2$

$8r\cos\theta = r^2$

$0 = r^2 - 8r\cos\theta = r(r - 8\cos\theta)$

$r = 0$ or $r - 8\cos\theta = 0$

The graph of $r = 0$ is the pole, and since the pole is included as a solution of $r - 8\cos\theta = 0$ $\left(\text{let } \theta = \dfrac{\pi}{2}\right)$, we can discard $r = 0$ and keep only $r - 8\cos\theta = 0$ or $r = 8\cos\theta$

18. $r(3\cos\theta - 2\sin\theta) = -2$
$3r\cos\theta - 2r\sin\theta = -2$

Use $x = r\cos\theta$ and $y = r\sin\theta$ $\qquad 3x - 2y = -2$

19. $r = -3\cos\theta$

We multiply both sides by r, which adds the pole to the graph. But the pole is already part of the graph $\left(\text{let } \theta = \dfrac{\pi}{2}\right)$, so we have changed nothing. $r^2 = -3r\cos\theta$

But $r^2 = x^2 + y^2$, $r\cos\theta = x$. Hence, $x^2 + y^2 = -3x$

20.

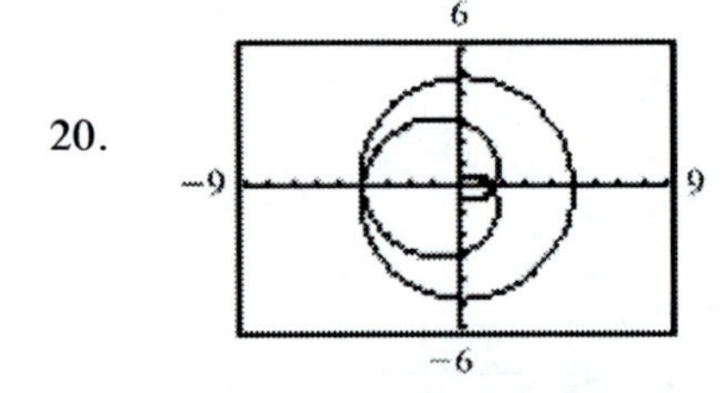

21.

22. $z_1 z_2 = 32e^{(3\pi/4)i} \cdot 8e^{(\pi/2)i}$

$= 32 \cdot 8e^{(3\pi/4 + \pi/2)I} = 256e^{(5\pi/4)i}$

$\dfrac{z_1}{z_2} = \dfrac{32e^{(3\pi/4)i}}{8e^{(\pi/2)i}}$

$= \dfrac{32}{8}e^{(3\pi/4 - \pi/2)i} = 4e^{(\pi/4)i}$

23. $(6e^{(5\pi/6)i})^3 = 6^3 e^{(3 \cdot 5\pi/6)i} = 216e^{(5\pi/2)i}$

24. A sketch shows that $-\sqrt{3} - i$ is associated with a special 30°–60° reference triangle in the third quadrant. Thus, $r = 2$ and $\theta = -150°$ and the polar form for $-\sqrt{3} - i$ is

$-\sqrt{3} - i = 2\,[\cos(-150°) + i\sin(-150°)]$

$= 2e^{(-150°)i}$

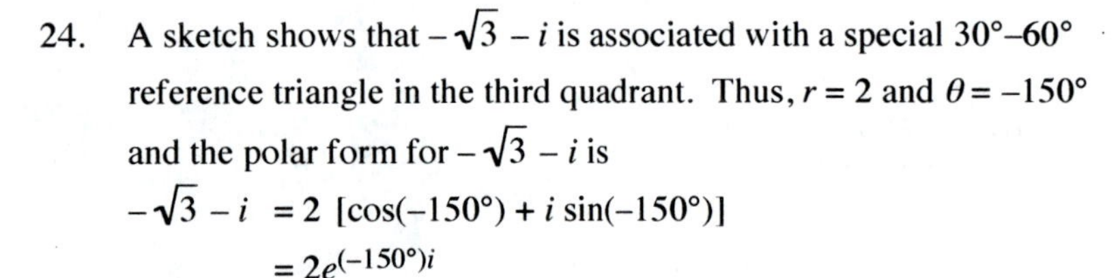

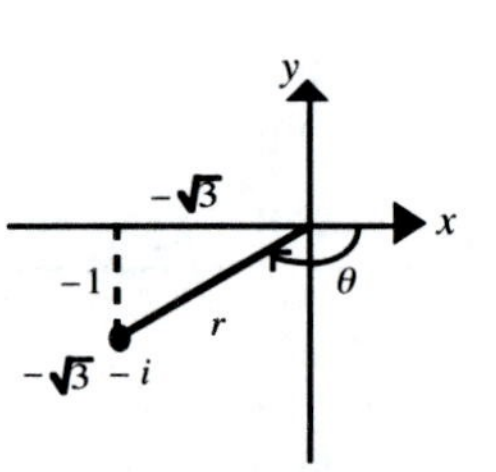

25. $x + iy = 3\sqrt{2}\,e^{(3\pi/4)i}$

$= 3\sqrt{2}\left(\cos\dfrac{3\pi}{4} + i\sin\dfrac{3\pi}{4}\right)$

$= 3\sqrt{2}\left(-\dfrac{1}{\sqrt{2}} + i \cdot \dfrac{1}{\sqrt{2}}\right)$

$= -3 + 3i$

26. A sketch shows that $2 + 2i\sqrt{3}$ is associated with a special 30°–60° reference triangle in the first quadrant and $-\sqrt{2} + i\sqrt{2}$ is associated with a special 45° triangle in the second quadrant.

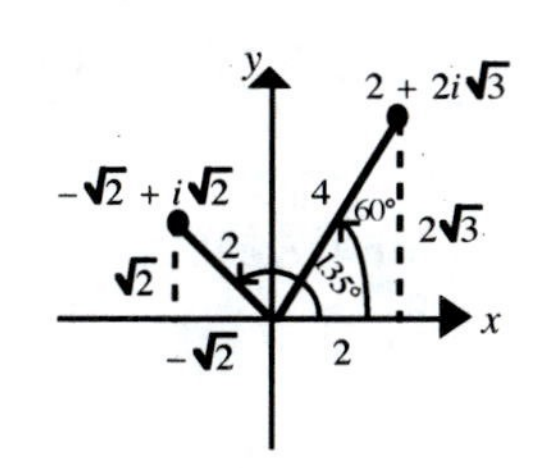

Thus,

$$2 + 2i\sqrt{3} = 4(\cos 60° + i \sin 60°) = 4e^{60°i}$$

$$-\sqrt{2} + i\sqrt{2} = 2(\cos 135° + i \sin 135°) = 2e^{135°i}$$

Therefore, $(2 + 2i\sqrt{3})(-\sqrt{2} + i\sqrt{2}) = 4e^{60°i} \cdot 2e^{135°i} = 4 \cdot 2e^{(60° + 135°)i} = 8e^{195°i}$

27. Using the results of the previous problem,

$$\frac{-\sqrt{2} + i\sqrt{2}}{2 + 2i\sqrt{3}} = \frac{2e^{135°i}}{4e^{60°i}} = \frac{2}{4}e^{(135° - 60°)i} = 0.5e^{75°i}$$

28. $$(-1 - i)^4 = (\sqrt{2}\,e^{225°i})^4 = (\sqrt{2}\,)^4 e^{(4 \cdot 225°)i} = (2^{1/2})^4\ e^{900°i} = 2^2 e^{900°i}$$
$$= 4e^{900°i} = 4(\cos 900° + i \sin 900°) = 4(-1 + 0i) = -4$$

29. $$x^2 + 9i = 0$$
$$x^2 = -9i$$

Therefore, x is a square root of $-9i$, and there are two of them. First write $-9i$ in polar form and use the nth root theorem.

$$-9i = 9e^{-90°i}$$

The two square roots of $-9i$ are given by

$$9^{1/2}e^{(-90°/2 + k360°/2)i} \qquad k = 0, 1$$

Thus,

$$w_1 = 9^{1/2}e^{(-45° + 0 \cdot 180°)i} = 3e^{-45°i} = 3[\cos(-45°) + i \sin(-45°)] = 3\left(\frac{\sqrt{2}}{2} - i\frac{\sqrt{2}}{2}\right) = \frac{3\sqrt{2}}{2} - \frac{3\sqrt{2}}{2}i$$

$$w_2 = 9^{1/2}\ e^{(-45° + 1 \cdot 180°)i} = 3e^{135°i} = 3(\cos 135° + i \sin 135°) = 3\left(-\frac{\sqrt{2}}{2} + i\frac{\sqrt{2}}{2}\right) = -\frac{3\sqrt{2}}{2} + \frac{3\sqrt{2}}{2}i$$

30. $$x^3 - 64 = 0$$
$$x^3 = 64$$

Therefore, x is a cube root of 64, and there are three of them. First write 64 in polar form and use the nth root theorem.

$$64 = 64e^{0°i}$$

All three cube roots of 64 are given by

$$64^{1/3}e^{(0°/3 + k360°/3)i} \qquad k = 0, 1, 2$$

Thus,

$$w_1 = 64^{1/3}e^{(0° + 0 \cdot 120°)i} = 4e^{0°i} = 4(\cos 0° + i \sin 0°) = 4(1 + 0i) = 4$$

$$w_2 = 64^{1/3}\, e^{(0° + 1 \cdot 120°)i} = 4e^{120°i} = 4(\cos 120° + i \sin 120°) = 4\left[\left(-\frac{1}{2}\right) + i\frac{\sqrt{3}}{2}\right] = -2 + 2i\sqrt{3}$$

$$w_3 = 64^{1/3}\, e^{(0° + 2 \cdot 120°)i} = 4e^{240°i} = 4(\cos 240° + i \sin 210°) = 4\left[\left(-\frac{1}{2}\right) + i\left(-\frac{\sqrt{3}}{2}\right)\right] = -2 - 2i\sqrt{3}$$

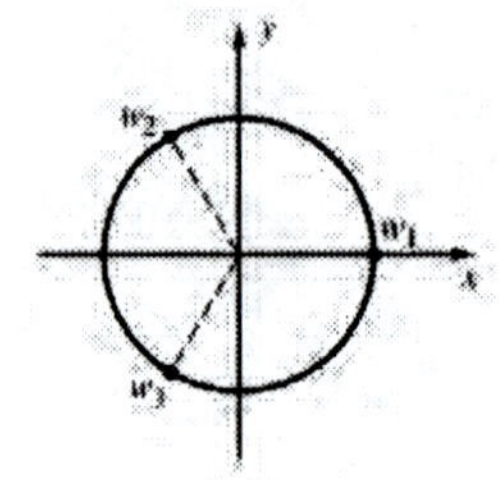

31. First, write $-4\sqrt{3} - 4i$ in polar form.

    $-4\sqrt{3} - 4i = 8e^{210°i}$

    From the nth root theorem, all three roots are given by

    $8^{1/3}\, e^{(210°/3 + k360°/3)i} \quad k = 0, 1, 2$

    Thus,

    $w_1 = 8^{1/3}\, e^{(70° + 0 \cdot 120°)i} = 2e^{70°i}$

    $w_2 = 8^{1/3}\, e^{(70° + 1 \cdot 120°)i} = 2e^{190°i}$

    $w_3 = 8^{1/3}\, e^{(70° + 2 \cdot 120°)i} = 2e^{310°i}$

32. First, write i in polar form.

    $i = 1\, e^{(\pi/2)i}$

    From the nth root theorem, all three cube roots are given by

    $1^{1/3} e^{[(\pi/2 + k \cdot 2\pi)/3]i} \quad k = 0, 1, 2$

    Thus,

    $w_1 = 1^{1/3} e^{[(\pi/2 + 0 \cdot 2\pi)/3]i} = 1e^{(\pi/6)i}$

    $w_2 = 1^{1/3} e^{[(\pi/2 + 1 \cdot 2\pi)/3]i} = 1e^{(5\pi/6)i}$

    $w_3 = 1^{1/3} e^{[(\pi/2 + 2 \cdot 2\pi)/3]i} = 1e^{(3\pi/2)i}$

33. To show that $2e^{30°i}$ is a square root of $2 + i\, 2\sqrt{3}$, we need only show that $(2e^{30°i})^2 = 2 + i\, 2\sqrt{3}$.

    But, by DeMoivre's theorem, $(2e^{30°i})^2 = 2^2\, e^{2 \cdot 30°i} = 4e^{60°i} = 4(\cos 60° + i \sin 60°) = 4\left(\frac{1}{2} + i \cdot \frac{\sqrt{3}}{2}\right)$

    $= 2 + i\, 2\sqrt{3}$

34. (A)

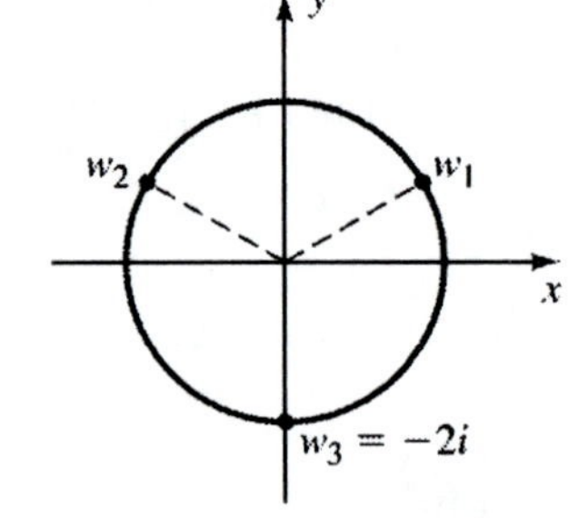

There are a total of three cube roots and they are spaced equally around a circle of radius 2.

(B) Since $w_3 = -2i = 2e^{-90°i}$, and each root is spaced 120° along the circle from the last,

$$w_1 = 2e^{(-90° + 120°)i} = 2e^{30°i} = 2(\cos 30° + i \sin 30°) = 2\left(\frac{\sqrt{3}}{2} + i \cdot \frac{1}{2}\right) = \sqrt{3} + i$$

$$w_2 = 2e^{(30° + 120°)i} = 2e^{150°i} = 2(\cos 150° + i \sin 150°) = 2\left[\left(-\frac{\sqrt{3}}{2}\right) + i \cdot \frac{1}{2}\right] = -\sqrt{3} + i$$

(C)

$$(-2i)^3 = (2e^{-90°i})^3 = 2^3 e^{3(-90°i)} = 8e^{-270°i} = 8[\cos(-270°) + i \sin(-270°)] = 8(0 + i) = 8i$$

$$(\sqrt{3} + i)^3 = (2e^{30°i})^3 = 2^3 e^{3 \cdot 30°i} = 8e^{90°i} = 8(\cos 90° + i \sin 90°) = 8(0 + i) = 8i$$

$$(-\sqrt{3} + i)^3 = (2e^{150°i})^3 = 2^3 e^{3 \cdot 150°i} = 8e^{450°i} = 8(\cos 450° + i \sin 450°) = 8(0 + i) = 8i$$

35. $n = 1$ $\quad$ $n = 2$ $\quad$ $n = 3$

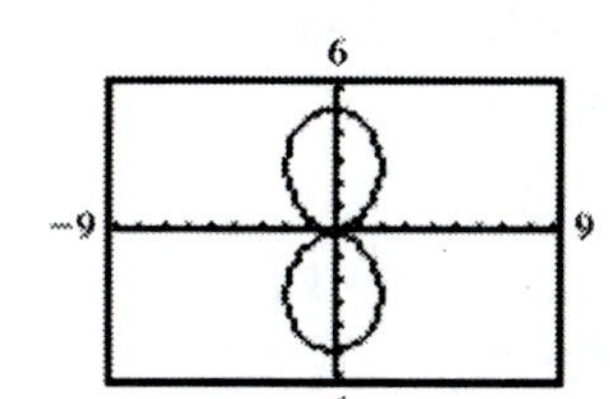

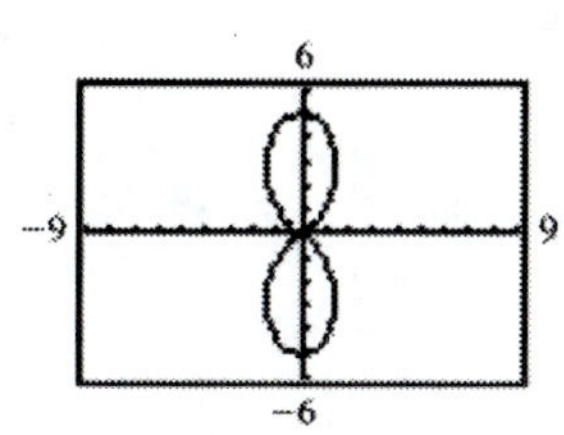

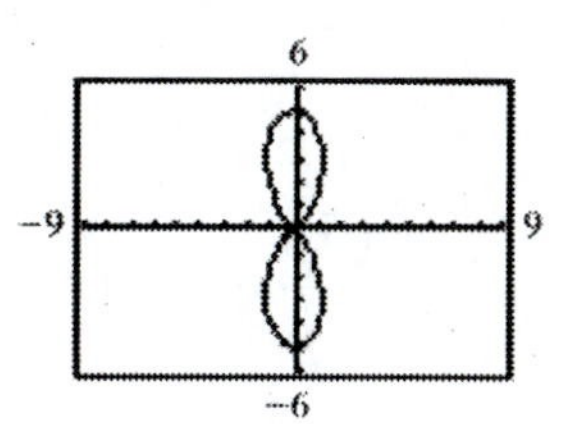

Since each graph has 2 leaves, we expect 2 leaves for arbitrary n.

36. (A) $r = \dfrac{2}{1 - 1.6\sin\theta}$ $\quad$ (B) $r = \dfrac{2}{1 - \sin\theta}$ $\quad$ (C) $r = \dfrac{2}{1 - 0.4\sin\theta}$

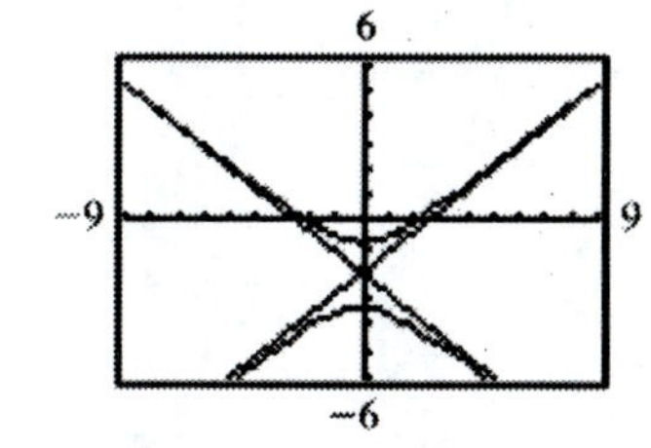

The graph is a hyperbola.

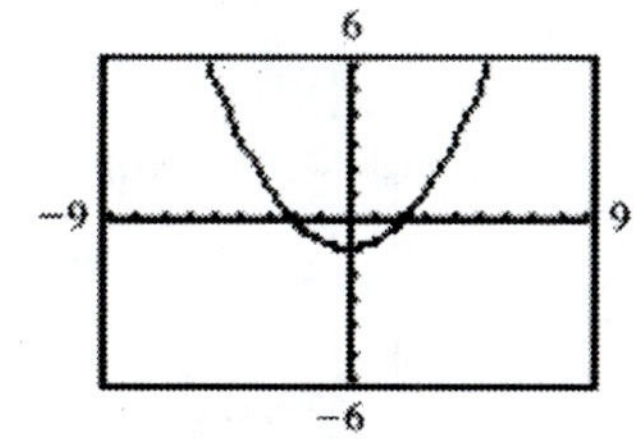

The graph is a parabola.

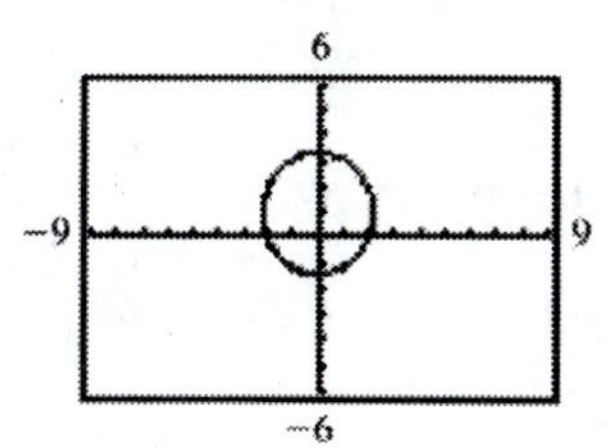

The graph is an ellipse.

37. $r(\sin\theta - 2) = 3$ $\quad$ $r\sin\theta - 2r = 3$

Use $r\sin\theta = y$ $\quad$ $r = -\sqrt{x^2 + y^2}$ $\quad$ $y + 2\sqrt{x^2 + y^2} = 3$

$y - 3 = -2\sqrt{x^2 + y^2}$ or $(y - 3)^2 = 4(x^2 + y^2)$

Note: See comment, Exercise 7.1, Problem 81.

38. $x^3 - 12 = 0$

$x^3 = 12$

Therefore, x is a cube root of 12, and there are three of them. First write 12 in polar form and use the nth root theorem.

$12 = 12e^{0°i}$

All three cube roots of 12 are given by

$12^{1/3}\ e^{(0°/3 + k360°/3)i} \qquad k = 0, 1, 2$

Thus,

$w_1 = 12^{1/3}\ e^{(0° + 0 \cdot 120°)i} = 12^{1/3}\ e^{0°i} = 12^{1/3}(\cos 0° + i \sin 0°) = 2.289$

$w_2 = 12^{1/3}\ e^{(0° + 1 \cdot 120°)i} = 12^{1/3}\ e^{120°i} = 12^{1/3}(\cos 120° + i \sin 120°) = -1.145 + 1.983i$

$w_3 = 12^{1/3}\ e^{(0° + 2 \cdot 120°)i} = 12^{1/3}\ e^{240°i} = 12^{1/3}(\cos 240° + i \sin 240°) = -1.145 - 1.983i$

39. $[r^{1/3}\ e^{(\theta/3 + k \cdot 120°)i}]^3 = (r^{1/3})^3\ e^{3(\theta/3 + k \cdot 120°)i}$ — DeMoivre's theorem

$= re^{(\theta + k \cdot 360°)i}$ — Algebra

$= r\,[\cos(\theta + k \cdot 360°) + i \sin(\theta + k \cdot 360°)]$ — Definition of $e^{i\theta}$

$= r(\cos \theta + i \sin \theta)$ — Periodic property of sine and cosine functions

$= re^{i\theta}$ — Definition of $e^{i\theta}$

40. (A) The coordinates of P represent a simultaneous solution.

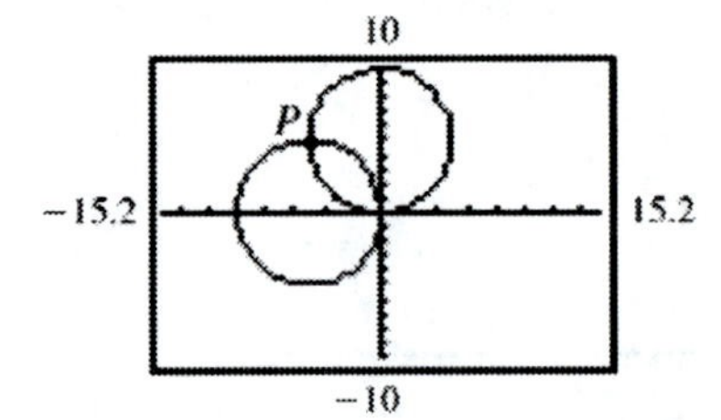

(B) We solve the system, $r = 10 \sin \theta$, $r = -10 \cos \theta$, by equating the right sides:

$10 \sin \theta = -10 \cos \theta$

$\sin \theta = -\cos \theta$

$\tan \theta = -1$

The only solution of this equation, $0 \le \theta \le \pi$, is $\theta = \frac{3\pi}{4}$. If we substitute this into either of the original equations, we get $r = 10 \sin \frac{3\pi}{4} = -10 \cos \frac{3\pi}{4} = 10\left(\frac{\sqrt{2}}{2}\right) = 5\sqrt{2}$

Solution: $\left(5\sqrt{2}, \frac{3\pi}{4}\right)$

(C) The pole has no ordered pairs of coordinates that simultaneously satisfy both equations. As (0, 0), it satisfies the first; as $\left(0, \frac{\pi}{2}\right)$, it satisfies the second; it is not a solution of the system.

41. The 360 complex solutions are equally spaced around a unit circle at an angular increment of 1° from one solution to the next. Two of them, 1 and −1, are real.

CUMULATIVE REVIEW EXERCISE CHAPTERS 1—7

1. We compare α and β by changing to decimal degrees.

   Since $\theta_d = \dfrac{180}{\pi}\theta_r$, $\alpha_d = \dfrac{180°}{\pi}\alpha_r = \dfrac{180°}{\pi} \cdot \dfrac{2\pi}{7} = 51.42857°\ldots$

   Since $25' = \dfrac{25°}{60}$ and $40'' = \dfrac{40°}{3600}$, then, $\beta = 51°25'40'' = 51.427777°\ldots.$ Thus, $\alpha > \beta$.

2. *Solve for the hypotenuse c:* $c^2 = a^2 + b^2$

   $$c = \sqrt{a^2 + b^2} = \sqrt{(1.27 \text{ cm})^2 + (4.65 \text{ cm})^2} = 4.82 \text{ cm}$$

   Solve for θ: We use the tangent. $\tan\theta = \dfrac{1.27}{4.65}$ $\theta = \tan^{-1}\dfrac{1.27}{4.65} = 15.3°$

   Solve for the complementary angle: $90° - \theta = 90° - 15.3° = 74.7°$

3. The distance from P to the origin is

   $$r = \sqrt{7^2 + (-24)^2} = \sqrt{49 + 576} = \sqrt{625} = 25$$

   Apply the third remark following Definition 1, Section 2.3, with $a = 7$, $b = -24$, and $r = 25$.

   $$\sec\theta = \frac{r}{a} = \frac{25}{7} \qquad \tan\theta = \frac{b}{a} = -\frac{24}{7}$$

4. (A)

$\cot x \sec x \sin x = \dfrac{\cos x}{\sin x} \cdot \dfrac{1}{\cos x} \cdot \sin x$	Quotient and reciprocal identities
$= 1$	Algebra

   (B)

$\tan\theta + \cot\theta = \dfrac{\sin\theta}{\cos\theta} + \dfrac{\cos\theta}{\sin\theta}$	Quotient identities
$= \dfrac{\sin^2\theta}{\sin\theta\cos\theta} + \dfrac{\cos^2\theta}{\sin\theta\cos\theta}$	Algebra
$= \dfrac{\sin^2\theta + \cos^2\theta}{\sin\theta\cos\theta}$	Algebra
$= \dfrac{1}{\sin\theta\cos\theta}$	Pythagorean identity
$= \dfrac{1}{\sin\theta} \cdot \dfrac{1}{\cos\theta}$	Algebra
$= \csc\theta\sec\theta$	Reciprocal identities
$= \sec\theta\csc\theta$	Algebra

5. Locate the 30°–60° reference triangle, determine (a, b) and r, then evaluate.

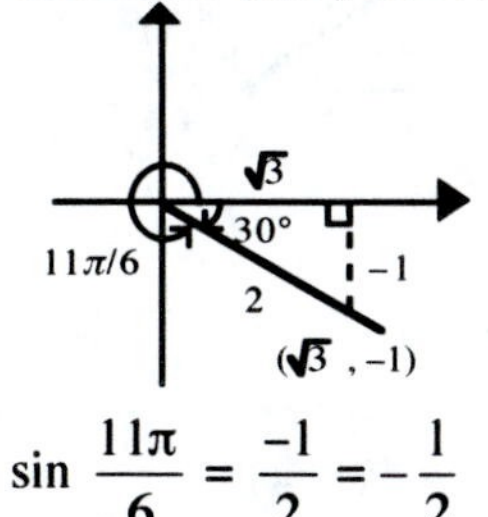

$$\sin\frac{11\pi}{6} = \frac{-1}{2} = -\frac{1}{2}$$

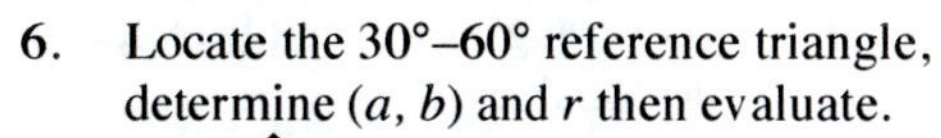

6. Locate the 30°–60° reference triangle, determine (a, b) and r then evaluate.

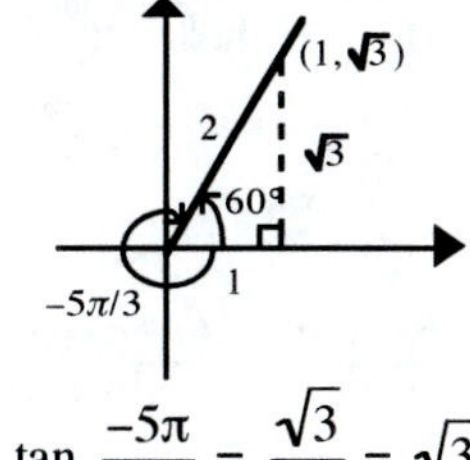

$$\tan\frac{-5\pi}{3} = \frac{\sqrt{3}}{1} = \sqrt{3}$$

7. $y = \cos^{-1}(-0.5)$ is equivalent to $\cos y = -0.5$. What y between 0 and π has cosine equal to -0.5? y must be associated with a reference triangle in the second quadrant. Reference triangle is a special 30°–60° triangle.

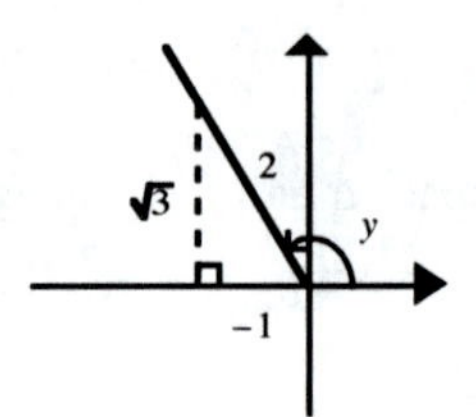

$$y = \frac{2\pi}{3} \qquad \cos^{-1}(-0.5) = \frac{2\pi}{3}$$

8. $y = \csc^{-1}(\sqrt{2})$ is equivalent to $\csc y = \sqrt{2}$ and $-\frac{\pi}{2} \le y \le \frac{\pi}{2}, y \ne 0$.

What number between $-\frac{\pi}{2}$ and $\frac{\pi}{2}$ has cosecant equal to $\sqrt{2}$?

y must be in the first quadrant.

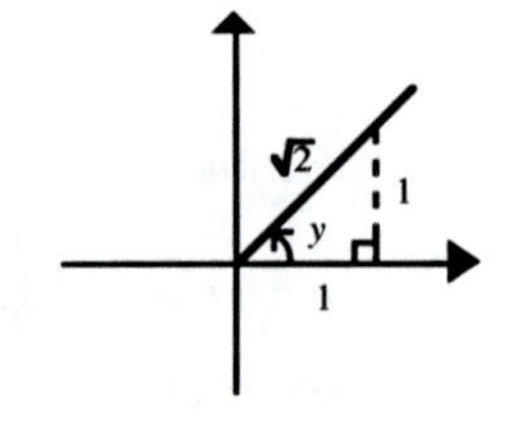

$$\csc y = \sqrt{2} = \frac{\sqrt{2}}{1} \qquad y = \frac{\pi}{4}$$

Thus, $\csc^{-1}(\sqrt{2}) = \frac{\pi}{4}$

9. Calculator in degree mode: $\sin 43°22' = \sin(43.366...°)$ Convert to decimal degrees, if necessary

$= 0.6867$

10. Use the reciprocal relationship $\cot\theta = \dfrac{1}{\tan\theta}$

Calculator in radian mode: $\cot\dfrac{2\pi}{5} = \dfrac{1}{\tan\dfrac{2\pi}{5}} = 0.3249$

11. Calculator in radian mode: $\sin^{-1}(0.8) = 0.9273$

12. Calculator in radian mode: $\sec^{-1}(4.5) = \cos^{-1}\dfrac{1}{4.5} = 1.347$

13. $\sin x + \sin y = 2\sin\dfrac{x+y}{2}\cos\dfrac{x-y}{2}$

$\sin 3t + \sin t = 2\sin\dfrac{3t+t}{2}\cos\dfrac{3t-t}{2} = 2\sin 2t\cos t$

14. An angle of radian measure 2.5 is the central angle of a circle subtended by an arc with measure 2.5 times that of the radius of the circle.

15. We are given two angles and a non-included side (*AAS*). We use the law of sines.

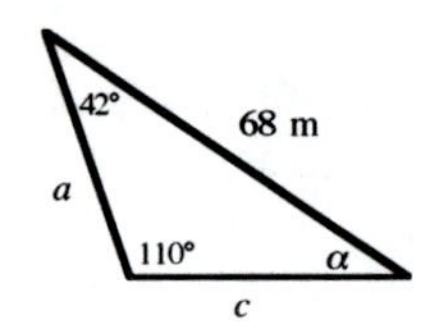

Solve for α: $\alpha + \beta + \gamma = 180°$

$\alpha = 180° - (42° + 110°) = 28°$

Solve for a: $\dfrac{\sin\alpha}{a} = \dfrac{\sin\beta}{b}$

$a = \dfrac{b\sin\alpha}{\sin\beta} = \dfrac{(68\text{ m})\sin 28°}{\sin 110°} = 34\text{ m}$

Solve for c: $\dfrac{\sin\gamma}{c} = \dfrac{\sin\beta}{b}, c = \dfrac{b\sin\gamma}{\sin\beta} = \dfrac{(68\text{ m})\sin 42°}{\sin 110°} = 48\text{ m}$

16. We are given two sides and the included angle (*SAS*). We use the law of cosines to find the third side, then the law of sines to find a second angle.

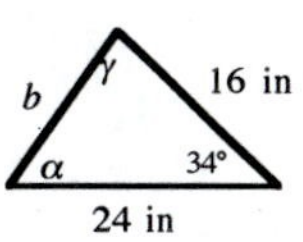

Solve for b:
$$\begin{aligned} b^2 &= a^2 + c^2 - 2ac \cos \beta \\ &= 16^2 + 24^2 - 2(16)(24) \cos 34° \\ &= 195.2991\ldots \\ b &= \sqrt{195.2991\ldots} = 14 \text{ in} \end{aligned}$$

Since a is the shorter of the remaining sides, α, the angle opposite a, must be acute.

Solve for α:
$$\frac{\sin \alpha}{a} = \frac{\sin \beta}{b}$$
$$\sin \alpha = \frac{a \sin \beta}{b} = \frac{(16 \text{ in})(\sin 34°)}{14 \text{ in}} = 0.6402$$
$$\alpha = \sin^{-1}(0.6402) = 40°$$

Solve for γ:
$$\alpha + \beta + \gamma = 180°$$
$$\gamma = 180° - (40° + 34°) = 106°$$

17. We are given three sides (*SSS*). We solve for the largest angle, γ (largest because it is opposite the largest side, c) using the law of cosines. We then solve for a second angle using the law of sines, because it involves simpler calculations.

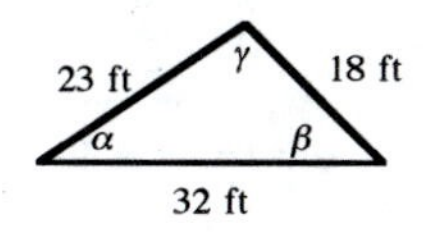

Solve for γ:
$$c^2 = a^2 + b^2 - 2ab \cos \gamma$$
$$\cos \gamma = \frac{a^2 + b^2 - c^2}{2ab}$$
$$\gamma = \cos^{-1} \frac{a^2 + b^2 - c^2}{2ab} = \cos^{-1} \frac{18^2 + 23^2 - 32^2}{2 \cdot 18 \cdot 23} = 102°$$

Both α and β must be acute, since α is obtuse.

Solve for β:
$$\frac{\sin\beta}{b} = \frac{\sin\gamma}{c} \qquad \sin \beta = \frac{b\sin\gamma}{c}$$
$$\beta = \sin^{-1} \frac{b\sin\gamma}{c} \quad \left\{ \begin{array}{l} \beta \text{ is acute, because there is room for} \\ \text{only one obtuse angle in a triangle.} \end{array} \right.$$
$$= \sin^{-1} \frac{(23\text{ ft})\sin 102°}{32\text{ ft}} = 45°$$

Solve for α:
$$\alpha + \beta + \gamma = 180°$$
$$\alpha = 180° - (\beta + \gamma) = 180° - (45° + 102°) = 33°$$

18. The given information consists of two sides and the included angle; hence, we use the formula
$$A = \frac{ab}{2} \sin \theta \text{ in the form } A = \frac{1}{2} ac \sin \beta = \frac{1}{2}(16 \text{ in})(24 \text{ in}) \sin 34° = 110 \text{ in}^2$$

19. The point with coordinates $(-7, 30°)$ can equally well be described as $(7, -150°)$. Thus: rotate the polar axis 150° clockwise (negative direction) and go 7 units along the positive polar axis.

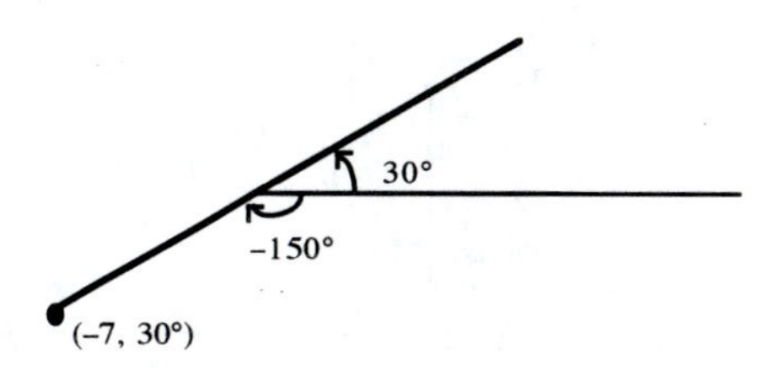

20. Horizontal component $|\mathbf{H}|$: $\cos 25° = \dfrac{|\mathbf{H}|}{13}$ $\qquad |\mathbf{H}| = 13 \cos 25° = 12$

Vertical component $|\mathbf{V}|$: $\sin 25° = \dfrac{|\mathbf{V}|}{13}$ $\qquad |\mathbf{V}| = 13 \sin 25° = 5.5$

21. To find $|\mathbf{u} + \mathbf{v}|$: Apply the Pythagorean theorem to triangle OCB.

$$|\mathbf{u} + \mathbf{v}|^2 = OB^2 = OC^2 + BC^2 = 6.4^2 + 3.9^2 = 56.17$$

$$|\mathbf{u} + \mathbf{v}| = \sqrt{56.17} = 7.5$$

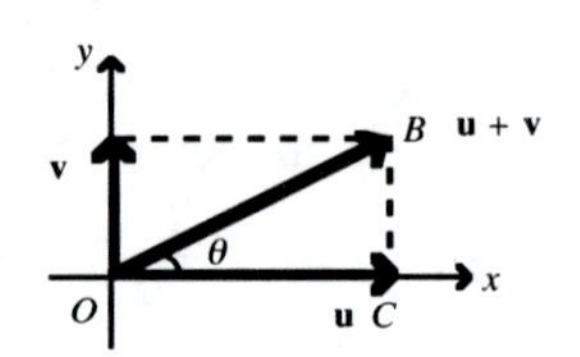

Solve triangle OCB for θ: $\tan \theta = \dfrac{BC}{OC} = \dfrac{|\mathbf{v}|}{|\mathbf{u}|}$ $\qquad \theta = \tan^{-1} \dfrac{|\mathbf{v}|}{|\mathbf{u}|}$ $\qquad \theta$ is acute

$$\theta = \tan^{-1} \frac{3.9}{6.4} = 31°$$

22. The algebraic vector $\langle a, b\rangle$ has coordinates given by

$a = x_b - x_a = (-3) - 4 = -7 \qquad b = y_b - y_a = 7 - (-2) = 9$

Hence, $\langle a, b\rangle = \langle -7, 9\rangle$

Magnitude of $\langle a, b\rangle = |\langle -7, 9\rangle| = \sqrt{a^2 + b^2} = \sqrt{(-7)^2 + 9^2} = \sqrt{130}$

23. $|\mathbf{u}| = \sqrt{2^2 + (-7)^2} = \sqrt{53} \qquad |\mathbf{v}| = \sqrt{3^2 + 8^2} = \sqrt{73}$

$$\cos \theta = \frac{\mathbf{u} \cdot \mathbf{v}}{|\mathbf{u}||\mathbf{v}|} = \frac{(2\mathbf{i} - 7\mathbf{j}) \cdot (3\mathbf{i} + 8\mathbf{j})}{\sqrt{53}\sqrt{73}} = \frac{2 \cdot 3 + (-7) \cdot 8}{\sqrt{53}\sqrt{73}} \qquad \theta = \cos^{-1} \frac{-50}{\sqrt{53}\sqrt{73}} = 143.5°$$

24.

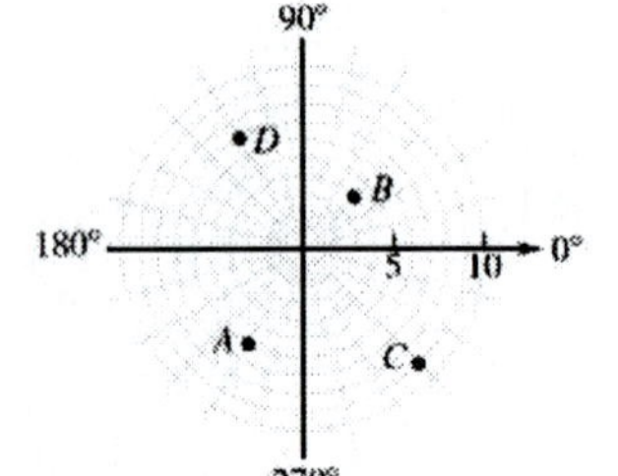

25. Set up a table that shows how r varies as θ varies through each set of quadrant values, then sketch the polar curve from the information in the table.

θ	$\sin \theta$	$5 \sin \theta$	$5 + 5 \sin \theta$
0 to $\frac{\pi}{2}$	0 to 1	0 to 5	5 to 10
$\frac{\pi}{2}$ to π	1 to 0	5 to 0	10 to 5
π to $\frac{3\pi}{2}$	0 to –1	0 to –5	5 to 0
$\frac{3\pi}{2}$ to 2π	–1 to 0	–5 to 0	0 to 5

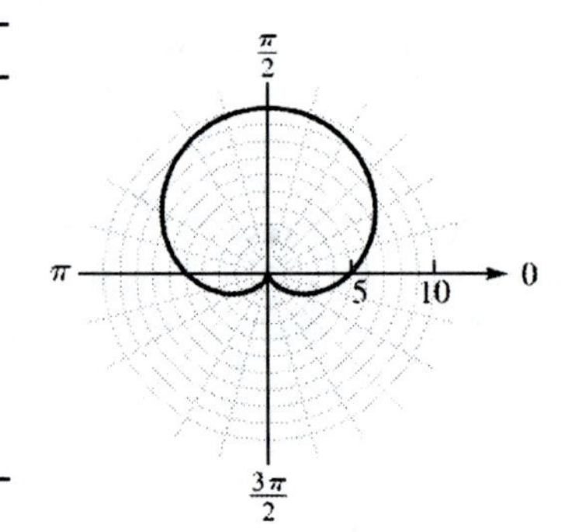

26.

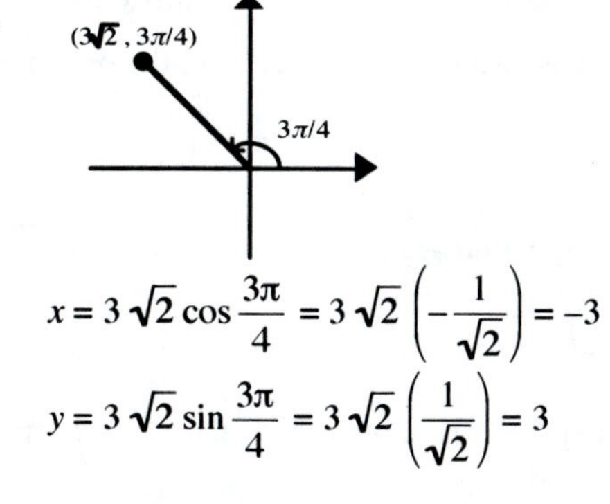

$$x = 3\sqrt{2}\cos\frac{3\pi}{4} = 3\sqrt{2}\left(-\frac{1}{\sqrt{2}}\right) = -3$$

$$y = 3\sqrt{2}\sin\frac{3\pi}{4} = 3\sqrt{2}\left(\frac{1}{\sqrt{2}}\right) = 3$$

27. Use $r^2 = x^2 + y^2$ and $\tan\theta = \dfrac{y}{x}$

$$r^2 = (-2\sqrt{3})^2 + 2^2 = 16$$
$$r = 4$$
$$\tan\theta = \frac{2}{-2\sqrt{3}} = -\frac{1}{\sqrt{3}}$$
$$\theta = \frac{5\pi}{6} \text{ since the point is in the second quadrant}$$

Polar coordinates: $\left(4, \dfrac{5\pi}{6}\right)$

28. A sketch shows that $2 - 2i$ is associated with a special 45° reference triangle in the fourth quadrant. Thus, $r = 2\sqrt{2}$ and $\theta = -\dfrac{\pi}{4}$ and the polar form for $2 - 2i$ is

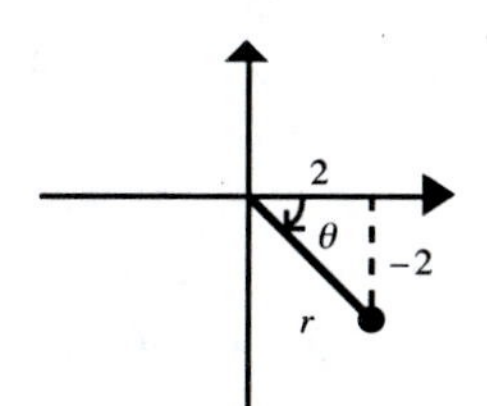

$$2 - 2i = 2\sqrt{2}\left[\cos\left(-\frac{\pi}{4}\right) + i\sin\left(-\frac{\pi}{4}\right)\right]$$
$$= 2\sqrt{2}\,e^{(-\pi/4)i}$$

29.
$$x + iy = 3e^{(3\pi/2)i}$$
$$= 3\left(\cos\frac{3\pi}{2} + i\sin\frac{3\pi}{2}\right)$$
$$= 3\ [0 + i(-1)\ ]$$
$$= -3i$$

30.
$$z_1z_2 = 3e^{50°i} \cdot 5e^{15°i}$$
$$= 3 \cdot 5e^{(50° + 15°)i} = 15e^{65°i}$$

$$\frac{z_1}{z_2} = \frac{3e^{50°i}}{5e^{15°i}}$$
$$= \frac{3}{5}e^{i(50° - 15°)i} = 0.6e^{35°i}$$

31. $(3e^{25°i})^4 = 3^4e^{4\cdot 25°i} = 81e^{100°i}$

32. Since the maximum value occurs at the end points of the interval, it would appear that A should be positive.

Since the maximum value of the function appears to be 3, and the minimum value appears to be −1,
$A = \dfrac{3-(-1)}{2} = 2$ and $k = \dfrac{3+(-1)}{2} = 1$

Since the maximum value is achieved at 0 and at 2, the period of the function is 2. Hence, $\dfrac{2\pi}{B} = 2$ and $B = \pi$.

Thus, the required function is $y = 1 + 2\cos\pi x$.

33. Because $\sin\theta$ is negative and $\cot\theta$ is positive, the terminal side of θ must lie in the third quadrant. Because $\cot\theta = \frac{a}{b} = 4 = \frac{4}{1}$, we let $a = -4$, $b = -1$. Then the distance of $P(a, b)$ from the origin is given by

$$r = \sqrt{(-4)^2 + (-1)^2} = \sqrt{16+1} = \sqrt{17}$$

Apply the third remark following Definition 1, Section 2.3, with $a = -4$, $b = -1$, and $r = \sqrt{17}$

$$\csc\theta = \frac{r}{b} = \frac{\sqrt{17}}{-1} = -\sqrt{17} \qquad \cos\theta = \frac{a}{r} = -\frac{4}{\sqrt{17}}$$

34. This graph is the graph of $y = 2\sin(2x + \pi)$ moved up one unit. The amplitude is $|2| = 2$. We first find the period and phase shift by solving

$$2x + \pi = 0 \quad \text{and} \quad 2x + \pi = 2\pi$$

$$x = -\frac{\pi}{2} \qquad\qquad x = -\frac{\pi}{2} + \pi$$

$$\text{Period} = \pi \qquad \text{Phase Shift} = -\frac{\pi}{2} \qquad \text{Frequency} = \frac{1}{\text{Period}} = \frac{1}{\pi}$$

We then sketch one period of the graph starting at $x = -\frac{\pi}{2}$ (the phase shift) and ending at $x = -\frac{\pi}{2} + \pi = \frac{\pi}{2}$ (the phase shift plus one period). The graph is a basic sine curve relative to the horizontal line $y = 1$ (shown as a broken line) and the y axis. We then extend the graph from $-\pi$ to 2π.

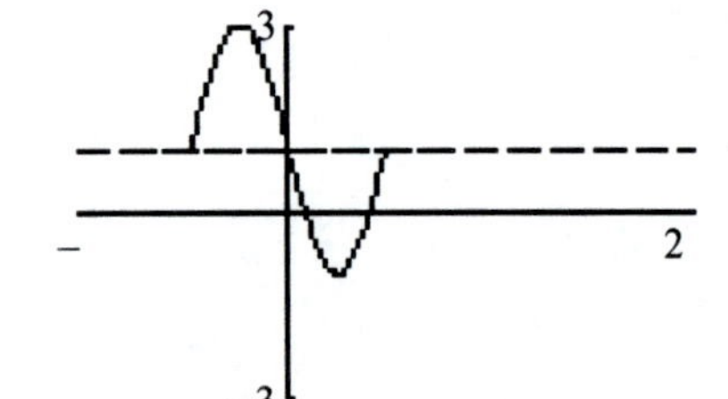

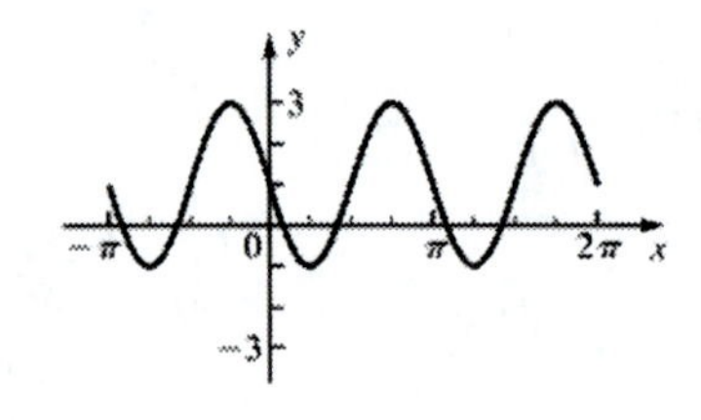

35.

$$\frac{\cos x}{1-\sin x} + \frac{\cos x}{1+\sin x} = \frac{\cos x(1+\sin x)}{(1-\sin x)(1+\sin x)} + \frac{\cos x(1-\sin x)}{(1-\sin x)(1+\sin x)} \quad \text{Algebra}$$

$$= \frac{\cos x(1+\sin x) + \cos x(1-\sin x)}{(1-\sin x)(1+\sin x)} \quad \text{Algebra}$$

$$= \frac{\cos x + \sin x\cos x + \cos x - \sin x\cos x}{1-\sin^2 x} \quad \text{Algebra}$$

$$= \frac{2\cos x}{1-\sin^2 x} \quad \text{Algebra}$$

$$= \frac{2\cos x}{\cos^2 x} \quad \text{Pythagorean identity}$$

$$= \frac{2}{\cos x} \quad \text{Algebra}$$

$$= 2\sec x \quad \text{Reciprocal identity}$$

36. $\tan\dfrac{\theta}{2} = \dfrac{\sin\theta}{1+\cos\theta}$ Half-angle identity

$= \dfrac{\dfrac{\sin\theta}{\sin\theta}}{\dfrac{1+\cos\theta}{\sin\theta}}$ Algebra

$= \dfrac{1}{\dfrac{1}{\sin\theta}+\dfrac{\cos\theta}{\sin\theta}}$ Algebra

$= \dfrac{1}{\csc\theta+\cot\theta}$ Reciprocal and quotient identities

37. $\dfrac{\cos x-\sin x}{\cos x+\sin x} = \dfrac{(\cos x-\sin x)(\cos x-\sin x)}{(\cos x+\sin x)(\cos x-\sin x)}$ Algebra

$= \dfrac{\cos^2 x - 2\cos x\sin x+\sin^2 x}{\cos^2 x-\sin^2 x}$ Algebra

$= \dfrac{1-2\cos x\sin x}{\cos^2 x-\sin^2 x}$ Pythagorean identity

$= \dfrac{1-\sin 2x}{\cos 2x}$ Double-angle identities

$= \dfrac{1}{\cos 2x} - \dfrac{\sin 2x}{\cos 2x}$ Algebra

$= \sec 2x - \tan 2x$ Reciprocal and quotient identities

38. First draw a reference triangle in the first quadrant and find $\sin x$.

$b = \sqrt{25^2-24^2} = 7 \qquad \sin x = \dfrac{b}{r} = \dfrac{7}{25}$

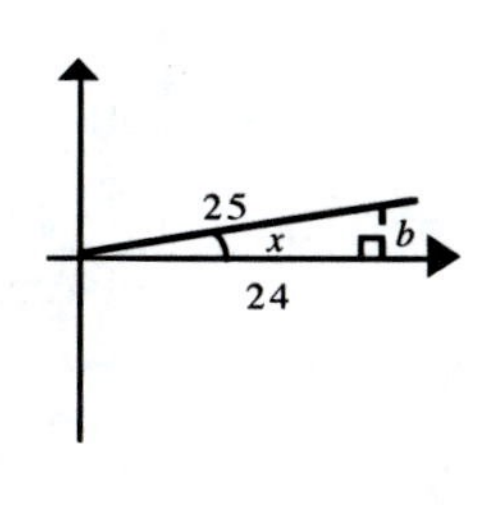

We can now find $\tan\dfrac{x}{2}$ from the half-angle identity.

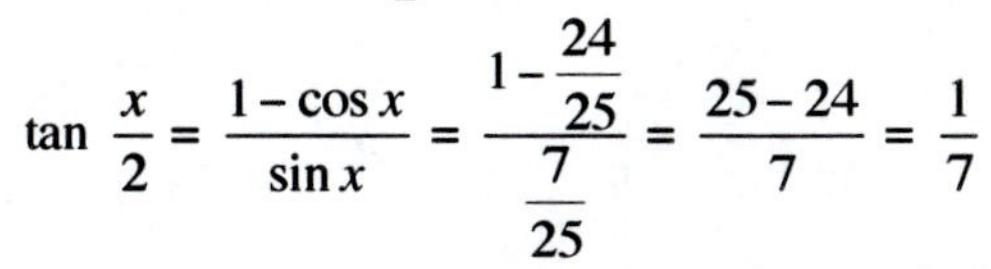

$$\tan\frac{x}{2} = \frac{1-\cos x}{\sin x} = \frac{1-\dfrac{24}{25}}{\dfrac{7}{25}} = \frac{25-24}{7} = \frac{1}{7}$$

To find $\sin 2x$, we use double-angle identity.

$$\sin 2x = 2\sin x\cos x = 2\cdot\frac{7}{25}\cdot\frac{24}{25} = \frac{336}{625}$$

39. (A) Graph both sides of the equation in the same viewing window.

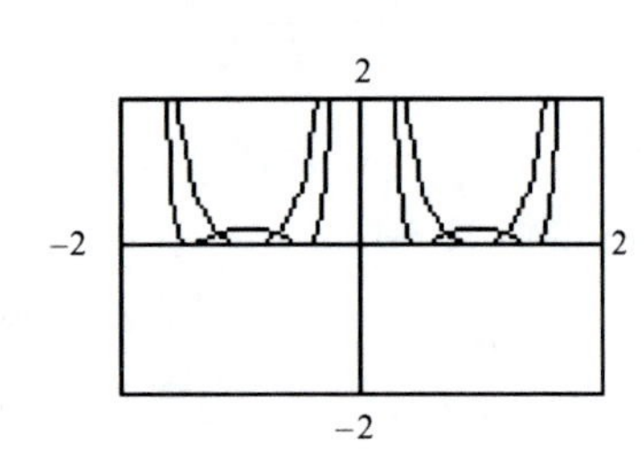

$\dfrac{\cos^2 x}{(\cos x-1)^2} = \dfrac{1+\cos x}{1-\cos x}$ is not an identity, since the graphs do not match.

Try $x = \dfrac{\pi}{2}$.

Left side: $\dfrac{\cos^2(\pi/2)}{[\cos(\pi/2)-1]^2} = \dfrac{0}{(0-1)^2} = 0$

Right side: $\dfrac{1+\cos(\pi/2)}{1-\cos(\pi/2)} = \dfrac{1+0}{1-0} = 1$

This verifies that the equation is not an identity.

(B) Graph both sides of the equation in the same viewing window.

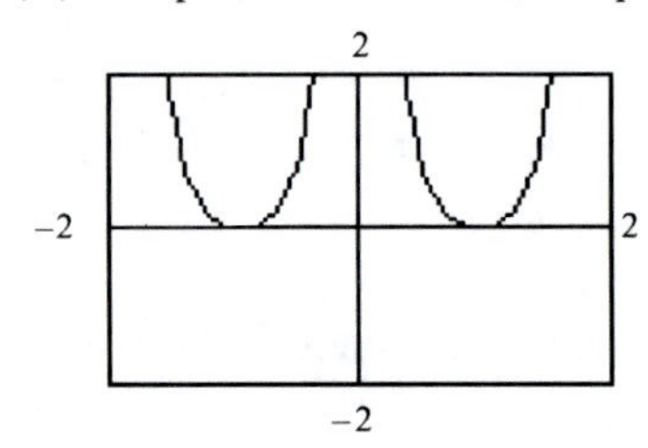

$\dfrac{\sin^2 x}{(\cos x-1)^2} = \dfrac{1+\cos x}{1-\cos x}$ appears to be an identity, which we verify.

$\dfrac{\sin^2 x}{(\cos x-1)^2} = \dfrac{1-\cos^2 x}{(\cos x-1)^2}$ Pythagorean identity

$= \dfrac{1-\cos^2 x}{(1-\cos x)^2}$ Algebra

$= \dfrac{(1-\cos x)(1+\cos x)}{(1-\cos x)(1-\cos x)}$ Algebra

$= \dfrac{1+\cos x}{1-\cos x}$ Algebra

40. Let $y = \sin^{-1}\dfrac{3}{4}$, then $\sin y = \dfrac{3}{4}, -\dfrac{\pi}{2} \le y \le \dfrac{\pi}{2}$. Sketch the reference triangle associated with y, then $\sec y = \sec\left(\sin^{-1}\dfrac{3}{4}\right)$, can be determined directly from the triangle.

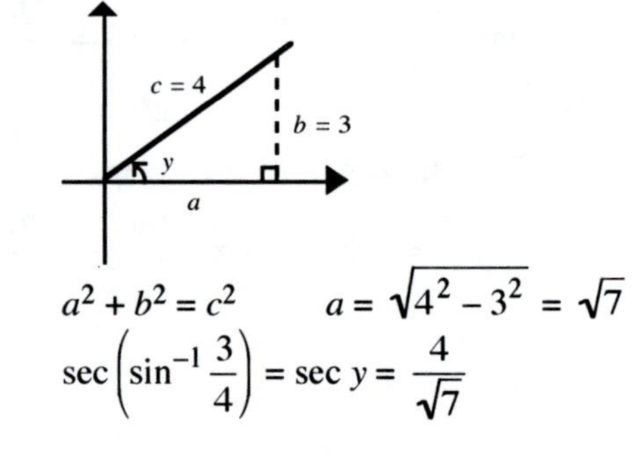

$a^2 + b^2 = c^2 \qquad a = \sqrt{4^2-3^2} = \sqrt{7}$

$\sec\left(\sin^{-1}\dfrac{3}{4}\right) = \sec y = \dfrac{4}{\sqrt{7}}$

41. (A)

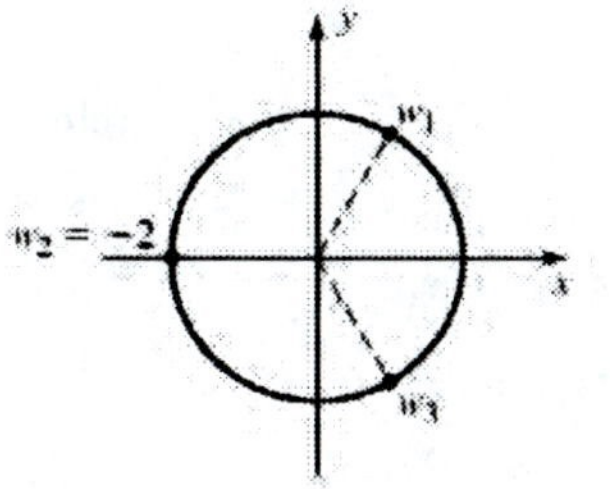

There are a total of three cube roots and they are spaced equally around a circle of radius 2.

(B) Since $w_2 = -2 = 2e^{180°i}$, and each root is spaced 120° along the circle from the last,

$$w_3 = 2e^{(180° + 120°)i} = 2e^{300°i} = 2(\cos 300° + i \sin 300°) = 2\left(\frac{1}{2} - i\frac{\sqrt{3}}{2}\right) = 1 - i\sqrt{3}$$

$$w_1 = 2e^{(180° - 120°)i} = 2e^{60°i} = 2(\cos 60° + i \sin 60°) = 2\left(\frac{1}{2} + i\frac{\sqrt{3}}{2}\right) = 1 + i\sqrt{3}$$

(C) $(1 - i\sqrt{3})^3 = (2e^{300°i})^3 = 2^3 e^{3 \cdot 300°i} = 8e^{900°i} = 8(\cos 900° + i \sin 900°) = 8(-1 + i0) = -8$

$(1 + i\sqrt{3})^3 = (2e^{60°i})^3 = 2^3 e^{3 \cdot 60°i} = 8e^{180°i} = 8(\cos 180° + i \sin 180°) = 8(-1 + i0) = -8$

42. First solve for x over one period, $0 \le x < 2\pi$. The add integer multiples of 2π to find all solutions.

$$\sin 2x + \sin x = 0 \qquad \text{Use double-angle identity}$$

$$2 \sin x \cos x + \sin x = 0$$

$$\sin x (2 \cos x + 1) = 0$$

$$\sin x = 0 \quad \text{or} \quad 2 \cos x + 1 = 0$$

$$x = 0, \pi \qquad \cos x = -\frac{1}{2}$$

$$x = \frac{2\pi}{3}, \frac{4\pi}{3}$$

Thus, the solutions over one period, $0 \le x < 2\pi$, are $0, \pi, \frac{2\pi}{3}, \frac{4\pi}{3}$. Thus, if x can range over all real numbers,

$$x = \begin{cases} \left.\begin{array}{l} 0 + 2k\pi \\ \pi + 2k\pi \end{array}\right\} \text{or } k\pi \\ \frac{2\pi}{3} + 2k\pi \\ \frac{4\pi}{3} + 2k\pi \end{cases} \qquad k \text{ any integer}$$

43. $2 \cos 2x = 5 \sin x - 4$

Solve for sin x and/or cos x:

$$\begin{aligned} 2(1 - 2\sin^2 x) &= 5 \sin x - 4 \quad \text{Use double-angle identity} \\ 2 - 4\sin^2 x &= 5 \sin x - 4 \\ 0 &= 4\sin^2 x + 5\sin x - 6 \\ 0 &= (4\sin x - 3)(\sin x + 2) \end{aligned}$$

$$4\sin x - 3 = 0 \qquad \sin x + 2 = 0$$
$$\sin x = \frac{3}{4} \qquad \sin x = -2$$

Solve over $0 \le x < 2\pi$:

$\sin x = -2$ No solution (–2 is not in the range of the sine function)

$\sin x = \frac{3}{4}$ From a unit circle graph, we see that the solutions are:

$x = \sin^{-1}\frac{3}{4} = 0.8481$

$x = \pi - 0.8481 = 2.294$

Because the sine function is periodic with period 2π, all solutions are given by:
$x = 0.8481 + 2k\pi,\ x = 2.294 + 2k\pi$, k any integer.

44. We are given two sides and a non-included angle (*SSA*).

(A) $a = 11.5$ cm

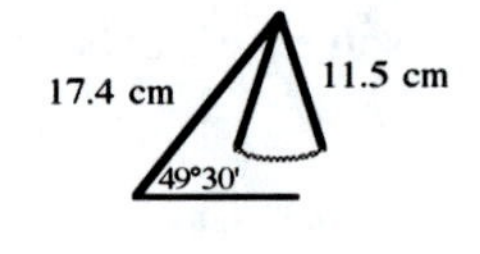

Solve for β: $\frac{\sin\beta}{b} = \frac{\sin\alpha}{a}$

$$\sin\beta = \frac{b\sin\alpha}{a} = \frac{17.4\sin 49°30'}{11.5} = 1.151$$

Since $\sin\beta = 1.151$ has no solution, no triangle exists with the given measurements. No solution.

(B) *Solve for β:* $\frac{\sin\beta}{b} = \frac{\sin\alpha}{a}$

$$\sin\beta = \frac{b\sin\alpha}{a} = \frac{(17.4 \text{ cm})\sin 49°30'}{14.7 \text{ cm}} = 0.9001$$

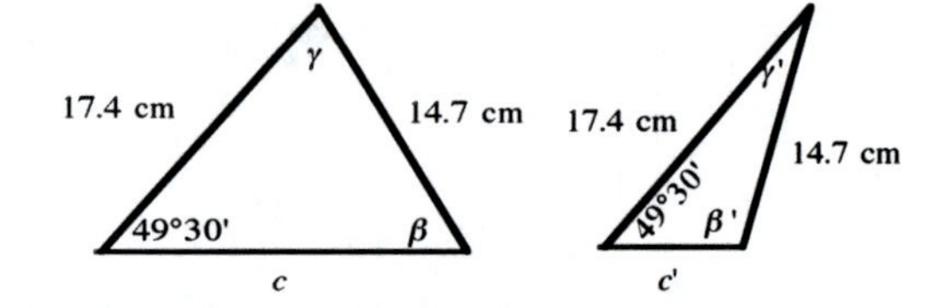

Two triangles are possible; angle β can be either acute or obtuse.

$\beta = \sin^{-1} 0.9001 = 64°10'$ $\qquad$ $\beta' = 180° - \sin^{-1} 0.9001 = 115°50°$

Solve for γ and γ':

$$\begin{aligned} \alpha + \beta + \gamma &= 180° \\ \gamma &= 180° - (49°30' + 64°10') \\ &= 66°20' \end{aligned}$$

$$\begin{aligned} \alpha' + \beta' + \gamma' &= 180° \\ \gamma' &= 180° - (49°30' + 115°50') \\ &= 14°40' \end{aligned}$$

Solve for c and c':

$$\frac{\sin\alpha}{a} = \frac{\sin\gamma}{c} \qquad\qquad \frac{\sin\alpha}{a} = \frac{\sin\gamma'}{c'}$$

$$c = \frac{a\sin\gamma}{\sin\alpha} = \frac{(14.7\text{ cm})\sin 66°20'}{\sin 49°30'} = 17.7\text{ cm} \qquad c = \frac{a\sin\gamma'}{\sin\alpha} = \frac{(14.7\text{ cm})\sin 14°40'}{\sin 49°30'} = 4.89\text{ cm}$$

(C) *Solve for β:*

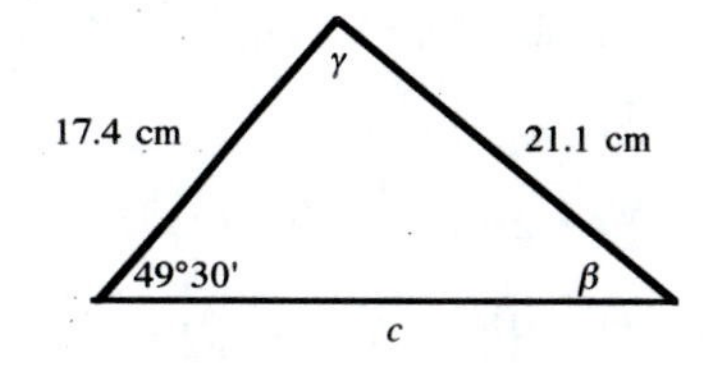

$$\frac{\sin\beta}{b} = \frac{\sin\alpha}{a}$$

$$\sin\beta = \frac{b\sin\alpha}{a} = \frac{(17.4\text{ cm})\sin 49°30'}{21.1} = 0.6271$$

$$\beta = \sin^{-1} 0.6271 = 38°50'$$

(There is another solution of $\sin\beta = 0.6271$ that deserves brief consideration:

$$\beta' = 180° - \sin^{-1} 0.6271 = 141°10'.$$

However, there is not enough room in a triangle for an angle of 141°10' and an angle of 49°30', since their sum is greater than 180°.)

Solve for γ: $\alpha + \beta + \gamma = 180°$

$$\gamma = 180° - (49°30' + 38°50') = 91°40'$$

Solve for c:

$$\frac{\sin\alpha}{a} = \frac{\sin\gamma}{c}$$

$$c = \frac{a\sin\gamma}{\sin\alpha} = \frac{(21.1\text{ cm})\sin 91°40'}{\sin 49°30'} = 27.7\text{ cm}$$

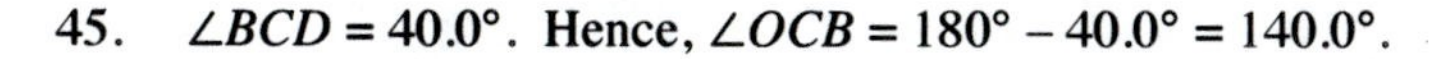

45. $\angle BCD = 40.0°$. Hence, $\angle OCB = 180° - 40.0° = 140.0°$.

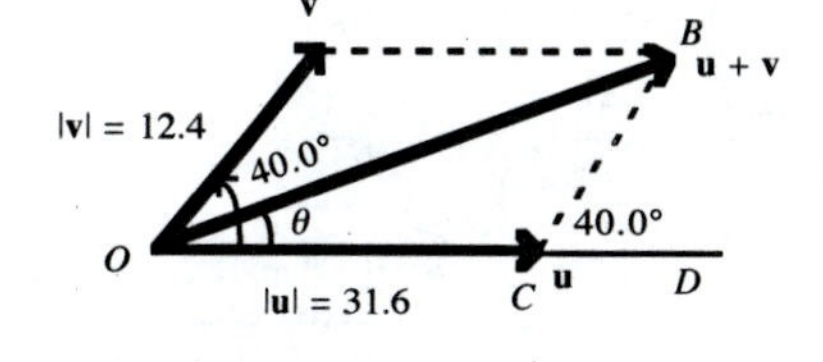

We can find $|\mathbf{u} + \mathbf{v}|$ using the law of cosines:

$$|\mathbf{u} + \mathbf{v}|^2 = |\mathbf{u}|^2 + |\mathbf{v}|^2 - 2\,|\mathbf{u}||\mathbf{v}|\cos(OCB)$$

$$= 31.6^2 + 12.4^2 - 2(31.6)(12.4)\cos 140.0°$$

$$= 1752.65370\ldots$$

$$|\mathbf{u} + \mathbf{v}| = \sqrt{1752.65370\ldots} = 41.9$$

To find θ, we use the law of sines:

$$\frac{\sin\theta}{|\mathbf{v}|} = \frac{\sin\angle OCB}{|\mathbf{u}+\mathbf{v}|}$$

$$\frac{\sin\theta}{12.4} = \frac{\sin 140.0°}{41.9}$$

$$\sin\theta = \frac{12.4}{41.9}\sin 140.0°$$

$$\theta = \sin^{-1}\left(\frac{12.4}{41.9}\sin 140.0°\right) = 11.0°$$

46. (A) $3\mathbf{u} - 4\mathbf{v} = 3\langle 1, -2\rangle - 4\langle 0, 3\rangle = \langle 3, -6\rangle + \langle 0, -12\rangle = \langle 3, -18\rangle$

(B) $3\mathbf{u} - 4\mathbf{v} = 3(2\mathbf{i} + 3\mathbf{j}) - 4(-\mathbf{i} + 5\mathbf{j}) = 6\mathbf{i} + 9\mathbf{j} + 4\mathbf{i} - 20\mathbf{j} = 10\mathbf{i} - 11\mathbf{j}$

47. $|\mathbf{v}| = \sqrt{7^2 + (24)^2} = 25$ $\qquad \mathbf{u} = \frac{1}{|\mathbf{v}|}\mathbf{v} = \frac{1}{25}\langle 7, -24\rangle = \left\langle \frac{7}{25}, -\frac{24}{25}\right\rangle$ or $\langle 0.28, -0.96\rangle$

48. The algebraic vector $\langle a, b\rangle$ has coordinates given by

$$a = x_b - x_a = (-1) - (-3) = 2 \qquad b = y_b - y_a = 5 - 2 = 3$$

Hence, $\langle a, b\rangle = \langle 2, 3\rangle = \langle 2, 0\rangle + \langle 0, 3\rangle = 2\langle 1, 0\rangle + 3\langle 0, 1\rangle = 2\mathbf{i} + 3\mathbf{j}$

49. (A) $\mathbf{u} \cdot \mathbf{v} = \langle 4, 0\rangle \cdot \langle 0, -5\rangle = 0 + 0 = 0$ Thus, **u** and **v** are orthogonal.

(B) $\mathbf{u} \cdot \mathbf{v} = \langle 3, 2\rangle \cdot \langle -3, 4\rangle = -9 + 8 = -1 \neq 0$ Thus, **u** and **v** are not orthogonal.

(C) $\mathbf{u} \cdot \mathbf{v} = (\mathbf{i} - 2\mathbf{j}) \cdot (6\mathbf{i} + 3\mathbf{j}) = 6 - 6 = 0$ Thus, **u** and **v** are orthogonal.

50. Set up a table that shows how r varies as 2θ varies through each set of quadrant values, then sketch the polar curve from the information in the table.

θ	2θ	$\cos\theta$	$8\cos 2\theta$
0 to $\frac{\pi}{4}$	0 to $\frac{\pi}{2}$	1 to 0	8 to 0
$\frac{\pi}{4}$ to $\frac{\pi}{2}$	$\frac{\pi}{2}$ to π	0 to –1	0 to –8
$\frac{\pi}{2}$ to $\frac{3\pi}{4}$	π to $\frac{3\pi}{2}$	–1 to 0	–8 to 0
$\frac{3\pi}{4}$ to π	$\frac{3\pi}{2}$ to 2π	0 to 1	0 to 8
π to $\frac{5\pi}{4}$	2π to $\frac{5\pi}{2}$	1 to 0	8 to 0
$\frac{5\pi}{4}$ to $\frac{3\pi}{2}$	$\frac{5\pi}{2}$ to 3π	0 to –1	0 to –8
$\frac{3\pi}{2}$ to $\frac{7\pi}{4}$	3π to $\frac{7\pi}{2}$	–1 to 0	–8 to 0
$\frac{7\pi}{4}$ to 2π	$\frac{7\pi}{2}$ to 4π	0 to 1	0 to 8

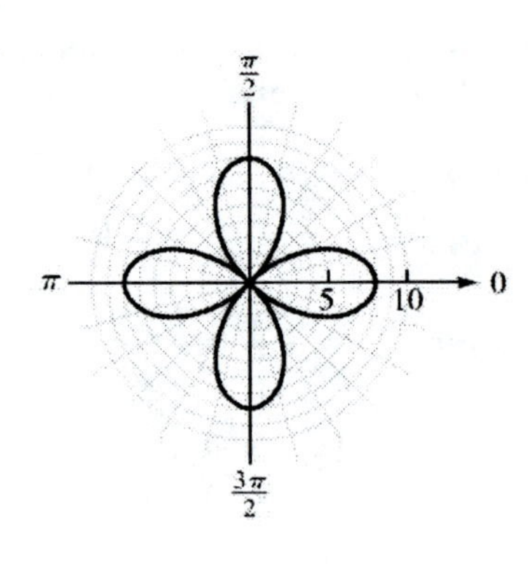

51. $x^2 = 6y$

Use $x = r\cos\theta$ and $y = r\sin\theta$

$$(r\cos\theta)^2 = 6(r\sin\theta)$$
$$r^2\cos^2\theta = 6r\sin\theta$$
$$r^2\cos^2\theta - 6r\sin\theta = 0$$
$$r(r\cos^2\theta - 6\sin\theta) = 0$$
$$r = 0 \text{ or } r\cos^2\theta - 6\sin\theta = 0$$
$$r\cos^2\theta = 6\sin\theta$$
$$r = \frac{6\sin\theta}{\cos^2\theta} = 6\,\frac{\sin\theta}{\cos\theta}\,\frac{1}{\cos\theta} = 6\tan\theta\sec\theta$$

The graph of $r = 0$ is the pole, and since the pole is included as a solution of $r = 6\tan\theta\sec\theta$ (let $\theta = 0$), we can discard $r = 0$ and keep only $r = 6\tan\theta\sec\theta$.

52. $r = 4 \sin \theta$

We multiply both sides by r, which adds the pole to the graph. But the pole is already part of the graph (let $\theta = 0$), so we have changed nothing. $r^2 = 4r \sin \theta$. But $r^2 = x^2 + y^2$, $r \sin \theta = y$. Hence, $x^2 + y^2 = 4y$.

53. A sketch shows that $3 + 3i$ is associated with a special 45° reference triangle in the first quadrant and $-1 + i\sqrt{3}$ is associated with a special 30°–60° reference triangle in the second quadrant. Thus,

$$\begin{aligned} 3 + 3i &= 3\sqrt{2}(\cos 45° + i \sin 45°) \\ &= 3\sqrt{2}\, e^{45°i} \\ -1 + i\sqrt{3} &= 2(\cos 120° + i \sin 120°) \\ &= 2e^{120°i} \end{aligned}$$

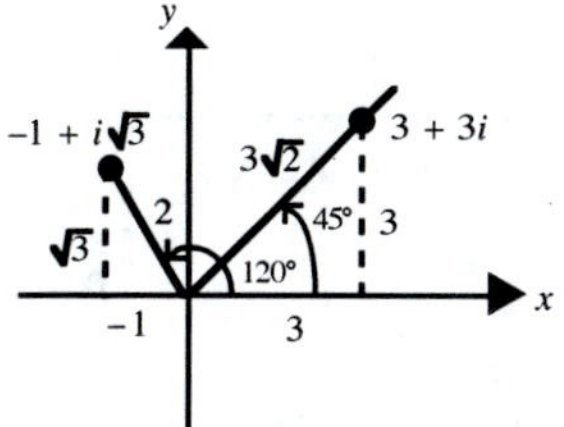

Therefore, $(3 + 3i)(-1 + i\sqrt{3}) = 3\sqrt{2}\, e^{45i} \cdot 2e^{120i} = 3\sqrt{2} \cdot 2e^{(45° + 120°)i} = 6\sqrt{2}\, e^{165°i}$

54. Using the results of the previous problem,

$$\frac{-1 + i\sqrt{3}}{3 + 3i} = \frac{2e^{120°i}}{3\sqrt{2}e^{45°i}} = \frac{2}{3\sqrt{2}} e^{(120° - 45°)i} = \frac{\sqrt{2}}{3} e^{75°i}$$

55.
$$\begin{aligned} (1 - i)^6 &= (\sqrt{2}\, e^{-45°i})^6 = (\sqrt{2})^6\, e^{6(-45°i)} = (2^{1/2})^6 e^{-270°i} = 2^3 e^{-270°i} \\ &= 8e^{-270°i} = 8[\cos(-270°) + i \sin(-270°)] = 8[0 + i \cdot 1] = 8i \end{aligned}$$

56. First, write $8i$ in polar form.

$$8i = 8e^{90i}$$

From the nth root theorem, all three roots are given by

$$8^{1/3} e^{(90°/3 + k360°/3)i} \qquad k = 0, 1, 2$$

Thus,

$$w_1 = 8^{1/3} e^{(30° + 0 \cdot 120°)i} = 2e^{30°i} = 2(\cos 30° + i \sin 30°) = 2\left(\frac{\sqrt{3}}{2} + i\frac{1}{2}\right) = \sqrt{3} + i$$

$$w_2 = 8^{1/3} e^{(30° + 1 \cdot 120°)i} = 2e^{150°i} = 2(\cos 150° + i \sin 150°) = 2\left(-\frac{\sqrt{3}}{2} + i\frac{1}{2}\right) = -\sqrt{3} + i$$

$$w_3 = 8^{1/3} e^{(30° + 2 \cdot 120°)i} = 2e^{270°i} = 2(\cos 270° + i \sin 270°) = 2(0 + i(-1)) = -2i$$

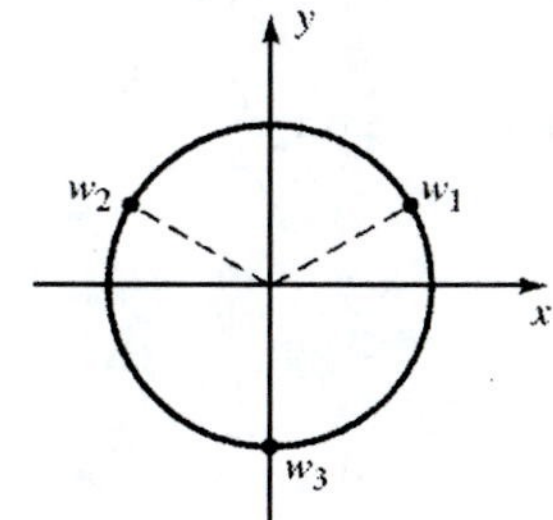

57. 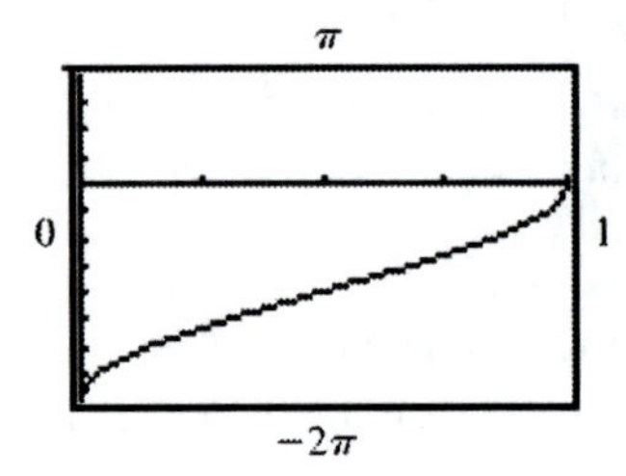

58. 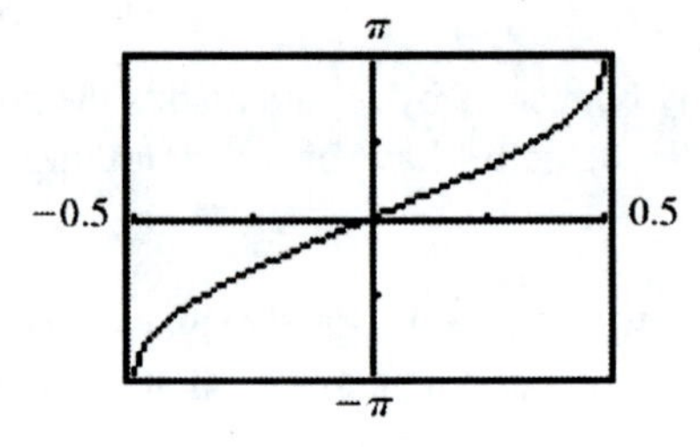

59. 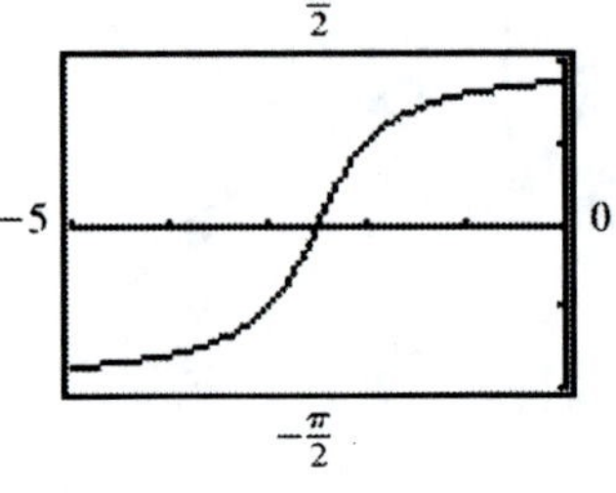

60. Graph y1 = tan x and y2 = 5 in the same viewing window in a graphing calculator and find the points of intersection using an automatic intersection routine. The intersection points are found to be –1.768 and 1.373.

Check: tan (–1.768) = 5.005
tan (1.373) = 4.990

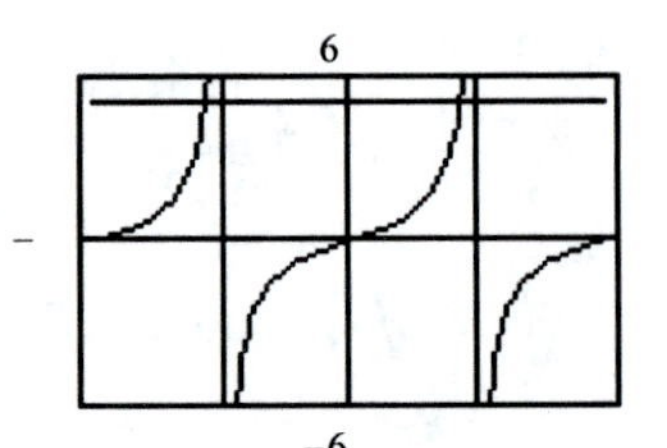

61. Graph y1 = cos x and y2 = $\sqrt[3]{x}$ in the same viewing window in a graphing calculator and find the points of intersection using an automatic intersection routine. The intersection point is found to be 0.582.

Check: cos (0.582) = 0.835
$\sqrt[3]{0.582}$ = 0.835

Since $|\cos x| \le 1$, while $|\sqrt[3]{x}| > 1$ for real x not shown, there can be no other solutions.

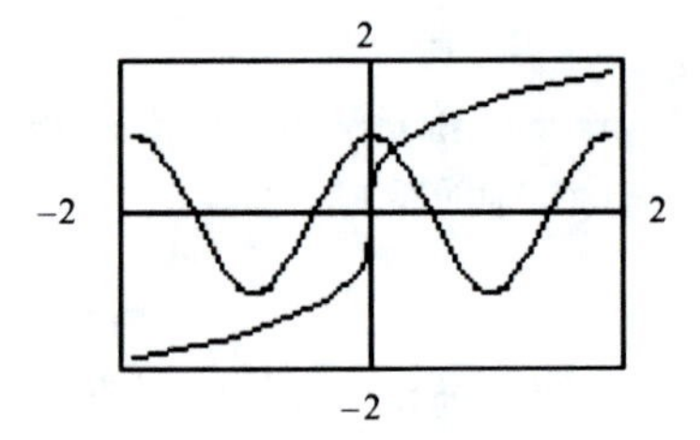

62. Graph y1 = 3 sin 2x cos 3x and y2 = 2 in the same viewing window in a graphing calculator and find the points of intersection using an automatic intersection routine. The intersection points are found to be 3.909, 4.313, 5.111, and 5.516.

Check: 3 sin [2(3.909)] cos [3(3.909)] = 2.002

(The remaining checking is left to the student.)

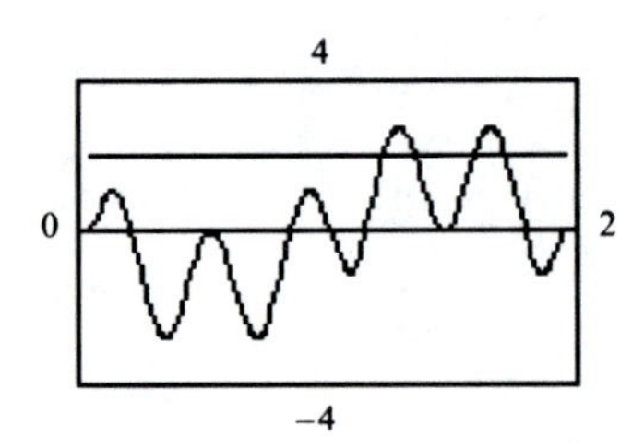

63. 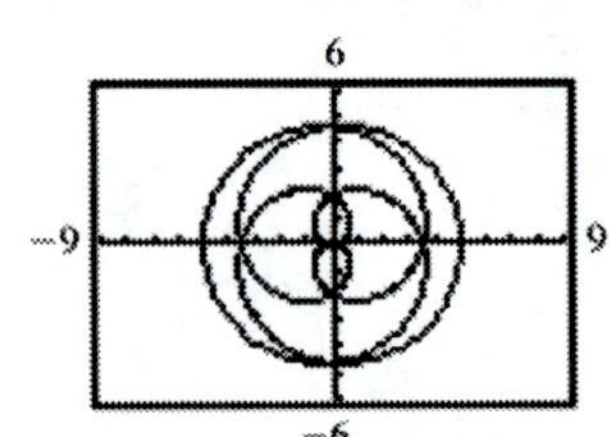

64. (A) $r = \dfrac{2}{1 - 0.7\sin(\theta + 0.6)}$

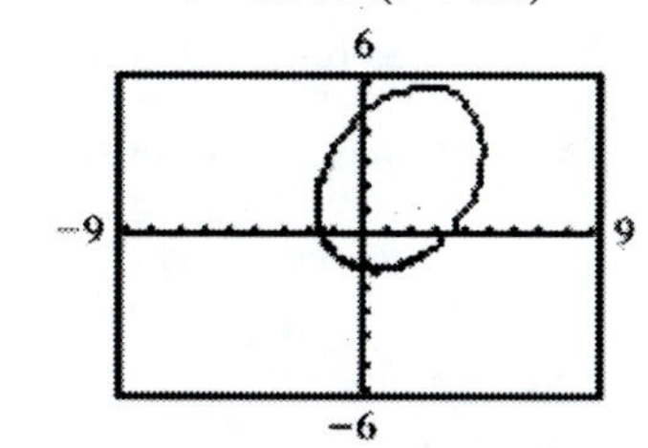

The graph is an ellipse.

(B) $r = \dfrac{2}{1 - \sin(\theta + 0.6)}$

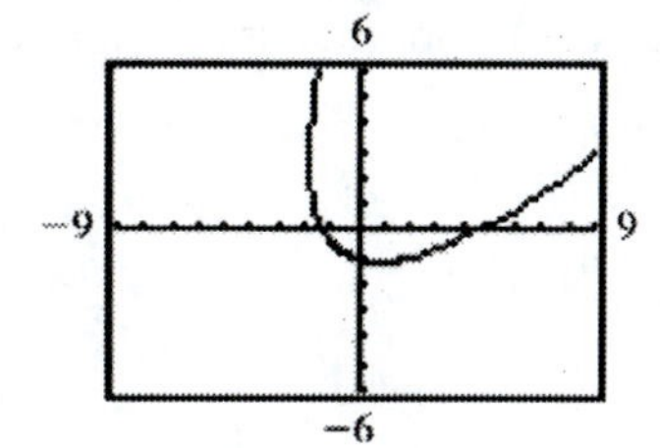

The graph is a parabola.

(C) $r = \dfrac{2}{1 - 1.5\sin(\theta + 0.6)}$

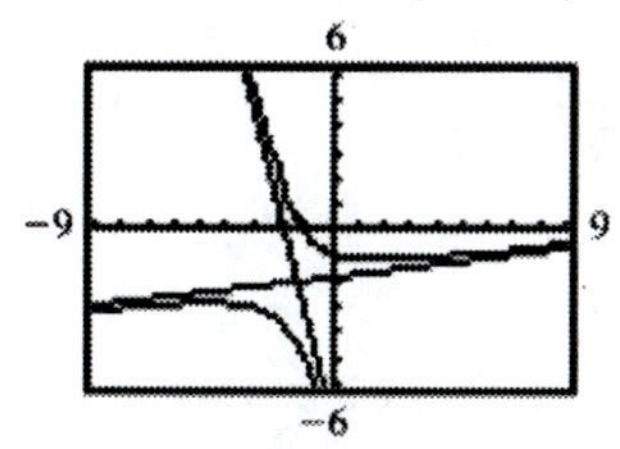

The graph is a hyperbola.

65. Since $\theta = \dfrac{s}{r}$, and $s = 8$, and $r = 2$, we have $\theta = \dfrac{8}{2} = 4$ rad.

Since $\cos\theta = \dfrac{a}{r}$ and $\sin\theta = \dfrac{b}{r}$, we have

$$a = r\cos\theta = 2\cos 4 \quad \text{and} \quad b = r\sin\theta = 2\sin 4$$

Thus, $(a, b) = (2\cos 4, 2\sin 4) = (-1.307, -1.514)$.

66. Since $\tan\theta = \dfrac{b}{a}$, we have $\tan\theta = \dfrac{-1.2}{-1.6} = 0.75$. Since (a, b) is in Quadrant III,

$$\theta = \tan^{-1}\theta + \pi = \tan^{-1} 0.75 + \pi = 3.785 \text{ rad}$$

Since $\theta = \dfrac{s}{r}$, we can write $3.785 = \dfrac{s}{2}$, $s = 2(3.785) = 7.570$ units

67. To find two asymptotes, set $\pi x + \dfrac{\pi}{4}$ equal to $-\dfrac{\pi}{2}$ and $\dfrac{\pi}{2}$ and solve for x.

$$\begin{aligned} \pi x + \frac{\pi}{4} &= -\frac{\pi}{2} \\ \pi x &= -\frac{\pi}{4} - \frac{\pi}{2} \\ x &= -\frac{1}{4} - \frac{1}{2} \\ x &= -\frac{3}{4} \end{aligned}$$

$$\begin{aligned} \pi x + \frac{\pi}{4} &= \frac{\pi}{2} \\ \pi x &= -\frac{\pi}{4} + \frac{\pi}{2} \\ x &= -\frac{1}{4} + \frac{1}{2} \\ x &= \frac{1}{4} \end{aligned}$$

There are asymptotes at $x = -\frac{3}{4}$ and $x = \frac{1}{4}$, and a portion of the graph that opens up is between them. The low point of this portion is at height 2. The remaining asymptotes in the required region are at $x = \frac{5}{4}$ and $x = \frac{9}{4}$. The period is 2 (twice the distance between successive asymptotes) and the phase shift is $-\frac{1}{4}\left(= -\frac{C}{B}\right)$.

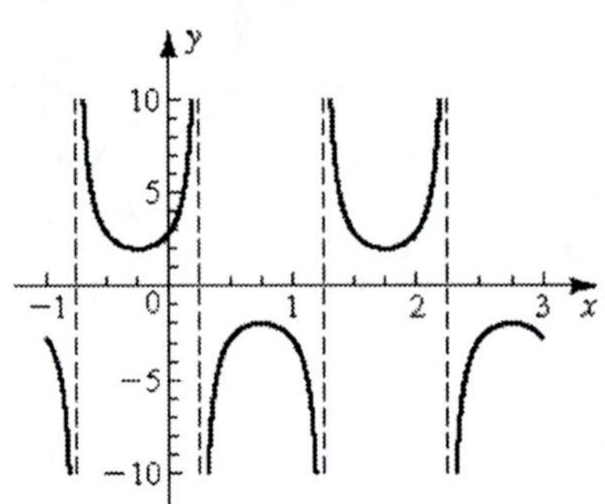

68. Let $y = \tan^{-1} x \quad -\frac{\pi}{2} < y < \frac{\pi}{2}$ or, equivalently, $x = \tan y \quad -\frac{\pi}{2} < y < \frac{\pi}{2}$

Geometrically,

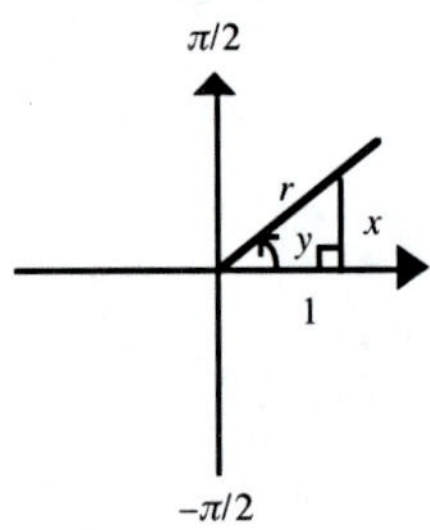

or

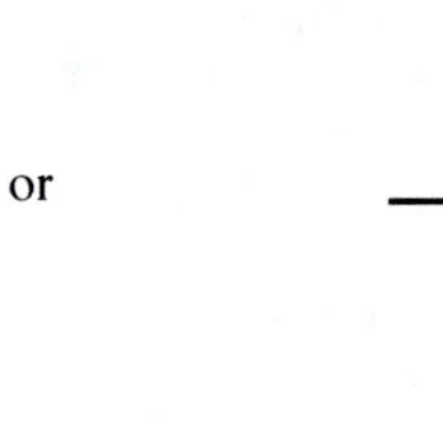

In either case, $r = \sqrt{x^2+1}$

$$\sec(2\tan^{-1} x) = \sec(2y) = \frac{1}{\cos 2y} = \frac{1}{\cos^2 y - \sin^2 y} = 1 \div (\cos^2 y - \sin^2 y)$$

$$= 1 \div \left[\left(\frac{1}{\sqrt{x^2+1}}\right)^2 - \left(\frac{x}{\sqrt{x^2+1}}\right)^2\right]$$

$$= 1 \div \left[\frac{1}{x^2+1} - \frac{x^2}{x^2+1}\right]$$

$$= 1 \div \frac{1-x^2}{x^2+1} = \frac{x^2+1}{1-x^2} \text{ or } \frac{1+x^2}{1-x^2}$$

69. $\tan 3x = \tan(x + 2x)$ — Algebra

$= \dfrac{\tan x + \tan 2x}{1 - \tan x \tan 2x}$ — Sum identity

$= \dfrac{\tan x + \dfrac{2\tan x}{1-\tan^2 x}}{1 - \tan x \cdot \dfrac{2\tan x}{1-\tan^2 x}}$ — Double-angle identity

$= \dfrac{(1-\tan^2 x)}{(1-\tan^2 x)} \cdot \dfrac{\tan x + \dfrac{2\tan x}{1-\tan^2 x}}{1 - \dfrac{2\tan^2 x}{1-\tan^2 x}}$ — Algebra

$= \dfrac{(1-\tan^2 x)\tan x + 2\tan x}{1 - \tan^2 x - 2\tan^2 x}$ — Algebra

$= \dfrac{\tan x - \tan^3 x + 2\tan x}{1 - 3\tan^2 x}$ — Algebra

$= \dfrac{3\tan x - \tan^3 x}{1 - 3\tan^2 x}$ — Algebra

$= \dfrac{\tan x(3 - \tan^2 x)}{1 - 3\tan^2 x}$ — Algebra

$= \dfrac{1}{\cot x}\dfrac{3 - \tan^2 x}{1 - 3\tan^2 x}$ — Reciprocal identity

$= \dfrac{3 - \tan^2 x}{\cot x - 3\tan^2 x \cot x}$ — Algebra

$= \dfrac{3 - \tan^2 x}{\cot x - 3\tan x \cdot (\tan x \cot x)}$ — Algebra

$= \dfrac{3 - \tan^2 x}{\cot x - 3\tan x \cdot 1}$ — Reciprocal identity

$= \dfrac{3 - \tan^2 x}{\cot x - 3\tan x}$ — Algebra

70. Use $r = \sqrt{x^2 + y^2}$ and $r\cos\theta = x$

$r(\cos\theta + 1) = 1$

$r\cos\theta + r = 1$

$x + \sqrt{x^2 + y^2} = 1$

$x = 1 - \sqrt{x^2 + y^2}$

$x - 1 = -\sqrt{x^2 + y^2}$ or, squaring both sides

$(x-1)^2 = (-\sqrt{x^2 + y^2})^2 = x^2 + y^2$

71. (A)

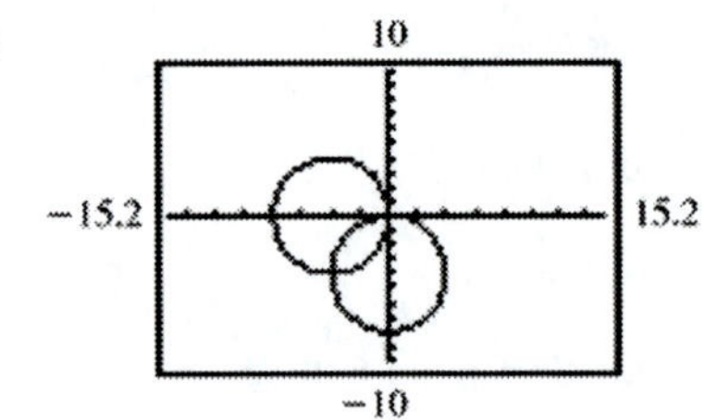

(B) The pole has no ordered pairs of solutions that simultaneously satisfy both equations. As (0, 0), it satisfies the first; as $\left(0, \frac{\pi}{2}\right)$ it satisfies the second; it is not a solution of the system.

72. $x^3 - 4 = 0$

$x^3 = 4$

Therefore x is a cube root of 4, and there are three of them. First, write 4 in polar form and use the nth root theorem.

$4 = 4e^{0°i}$

All three cube roots of 4 are given by

$4^{1/3}e^{(0°/3 + k360°/3)i} \qquad k = 0, 1, 2$

Thus,

$w_1 = 4^{1/3}e^{(0° + 0 \cdot 120°)i} = 4^{1/3}e^{0°i} = 4^{1/3}(\cos 0° + i \sin 0°) = 1.587$

$w_2 = 4^{1/3}e^{(0° + 1 \cdot 120°)i} = 4^{1/3}e^{120°i} = 4^{1/3}(\cos 120° + i \sin 120°) = -0.794 + 1.375i$

$w_3 = 4^{1/3}e^{(0° + 2 \cdot 120°)i} = 4^{1/3}e^{240°i} = 4^{1/3}(\cos 240° + i \sin 240°) = -0.794 - 1.375i$

73. (A) By DeMoivre's theorem, $(\cos\theta + i\sin\theta)^3 = (e^{\theta i})^3 = e^{3\theta i} = \cos 3\theta + i\sin 3\theta$

By the binomial theorem,

$$(\cos\theta + i\sin\theta)^3 = \cos^3\theta + 3\cos^2\theta(i\sin\theta) + 3\cos\theta(i\sin\theta)^2 + (i\sin\theta)^3$$
$$= \cos^3\theta + 3i\cos^2\theta\sin\theta - 3\cos\theta\sin^2\theta - i\sin^3\theta$$

Thus,

$$\cos 3\theta + i\sin 3\theta = \cos^3\theta - 3\cos\theta\sin^2\theta + i(3\cos^2\theta\sin\theta - \sin^3\theta)$$

Equating the real and imaginary parts of the left and right sides, we obtain

$$\cos 3\theta = \cos^3\theta - 3\cos\theta\sin^2\theta \text{ and } \sin 3\theta = 3\cos^2\theta\sin\theta - \sin^3\theta$$

(B)

$\cos 3\theta = \cos(\theta + 2\theta)$	Algebra
$= \cos\theta\cos 2\theta - \sin\theta\sin 2\theta$	Sum identity
$= \cos\theta(\cos^2\theta - \sin^2\theta) - \sin\theta(2\sin\theta\cos\theta)$	Double-angle identity
$= \cos^3\theta - \cos\theta\sin^2\theta - 2\cos\theta\sin^2\theta$	Algebra
$= \cos^3\theta - 3\cos\theta\sin^2\theta$	Algebra
$\sin 3\theta = \sin(\theta + 2\theta)$	Algebra
$= \sin\theta\cos 2\theta + \cos\theta\sin 2\theta$	Sum identity
$= \sin\theta(\cos^2\theta - \sin^2\theta) + \cos\theta(2\sin\theta\cos\theta)$	Double-angle identities
$= \sin\theta\cos^2\theta - \sin^3\theta + 2\sin\theta\cos^2\theta$	Algebra
$= 3\sin\theta\cos^2\theta - \sin^3\theta$	Algebra

74. The graph of $f(x)$ is shown in the figure.
The graph appears to be a basic cosine curve with period π, amplitude $= \frac{1}{2}(y_{max} - y_{min}) = \frac{1}{2}[2-(-4)] = 3$, displaced downward by $k = 1$ unit.
It appears that $g(x) = -1 + 3\cos 2x$ would be an appropriate choice.

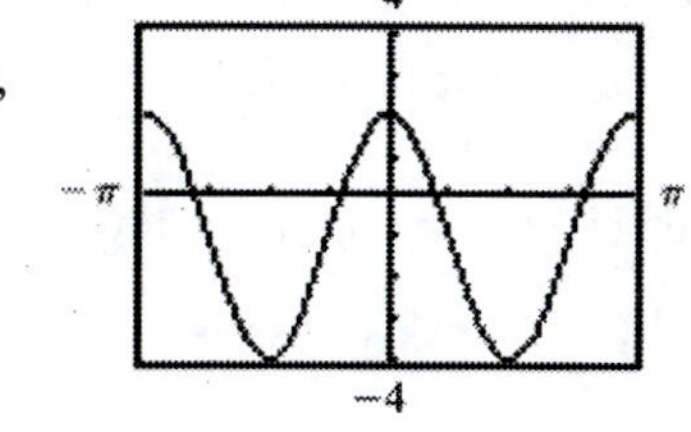

We verify $f(x) = g(x)$ as follows:

$f(x) = 2\cos^2 x - 4\sin^2 x$	
$= 2\cos^2 x - 4(1-\cos^2 x)$	Pythagorean identity
$= 2\cos^2 x - 4 + 4\cos^2 x$	Algebra
$= 6\cos^2 x - 4$	Algebra
$= 3(2\cos^2 x - 1) - 1$	Algebra
$= 3\cos 2x - 1$	Double-angle identity
$= -1 + 3\cos 2x = g(x)$	Algebra

75. The graph of $f(x)$ is shown in the figure.
The graph appears to be have vertical asymptotes $x = -\frac{3\pi}{4}, -\frac{\pi}{4}, \frac{\pi}{4}$, and $\frac{3\pi}{4}$ and period π. It appears to have high and low points with y coordinates -4 and -2, respectively. It appears that $g(x) = \sec 2x - 3$ would be an appropriate choice.

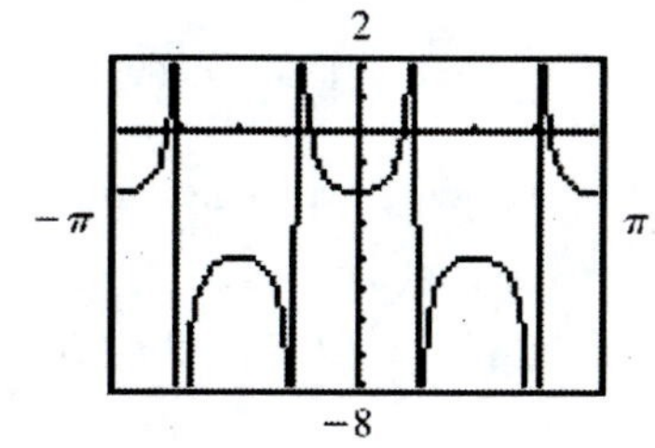

We verify $f(x) = g(x)$ as follows:

$f(x) = \dfrac{6\sin^2 x - 2}{2\cos^2 x - 1}$	
$= \dfrac{3(2\sin^2 x - 1) + 1}{2\cos^2 x - 1}$	Algebra
$= \dfrac{1 - 3(1 - 2\sin^2 x)}{2\cos^2 x - 1}$	Algebra
$= \dfrac{1 - 3\cos 2x}{\cos 2x}$	Double-angle identities
$= \dfrac{1}{\cos 2x} - 3$	Algebra
$= \sec 2x - 3 = g(x)$	Reciprocal identity

76. The graph of $f(x)$ is shown in the figure.

The graph appears to be have vertical asymptotes $x = -2\pi$, $x = 0$, and $x = -2\pi$, and period 2π.
It appears to have x intercepts $-\frac{3\pi}{2}$ and $\frac{\pi}{2}$, and symmetry with respect to points where the curve crosses the line $y = -1$.

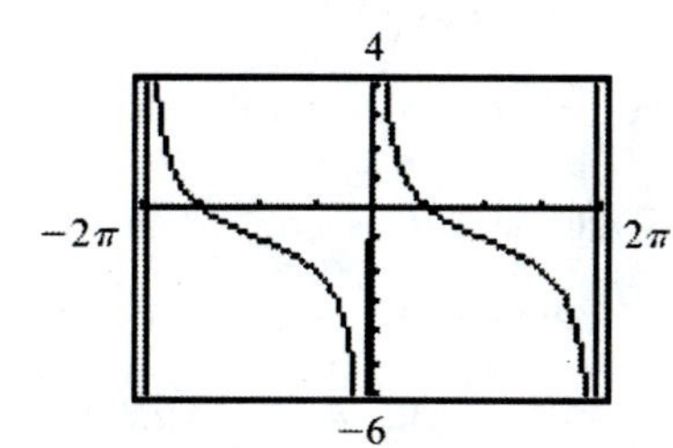

It appears to be a cotangent curve displaced downward by $|k| = 1$ unit.

It appears that $g(x) = -1 + \cot \dfrac{x}{2}$ would be an appropriate choice.

We verify $f(x) = g(x)$ as follows:

$$f(x) = \frac{\sin x + \cos x - 1}{1 - \cos x}$$

$$= \frac{\sin x - (1 - \cos x)}{1 - \cos x} \qquad \text{Algebra}$$

$$= \frac{\sin x}{1 - \cos x} - \frac{1 - \cos x}{1 - \cos x} \qquad \text{Algebra}$$

$$= 1 \div \frac{1 - \cos x}{\sin x} - 1 \qquad \text{Algebra}$$

$$= 1 \div \tan \frac{x}{2} - 1 \qquad \text{Half-angle identity}$$

$$= \cot \frac{x}{2} - 1 = g(x) \qquad \text{Reciprocal identity}$$

77. We are given two angles and the included side (*ASA*). We find the third angle, then apply the law of sines to find side *BC*.

$$\angle ABC + \angle BCA + \angle CAB = 180°$$

$$\angle ABC = 180° - (52° + 77°) = 51°$$

$$\frac{\sin \angle CAB}{BC} = \frac{\sin \angle ABC}{AC}$$

$$BC = \frac{AC \sin \angle CAB}{\sin \angle ABC} = \frac{(520 \text{ ft}) \sin 77°}{\sin 51°} = 650 \text{ ft}$$

78. Here we are given two sides and the included angle, hence we can use the law of cosines to find side *BC*.

$$BC^2 = AB^2 + AC^2 - 2(AB)(AC) \cos \angle CAB$$

$$= (580)^2 + (430)^2 - 2(580)(530) \cos 64° = 302{,}640.4\ldots$$

$$BC = 550 \text{ ft}$$

79. (A) Triangle *ABC* is a right triangle.

$$\tan \angle BAC = \frac{BC}{AC}$$

$$BC = AC \tan \angle BAC$$

$$= (35 \text{ ft}) \tan 54° = 48 \text{ ft.}$$

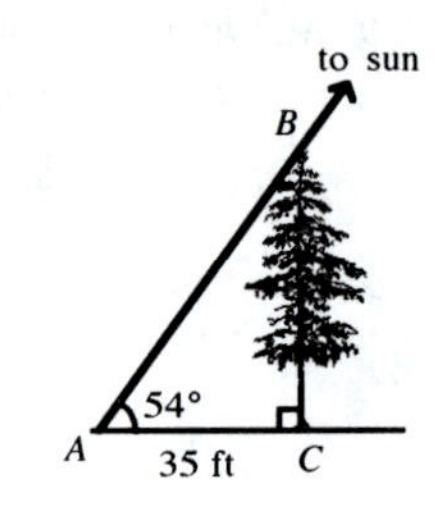

(B) Here triangle ABC is an oblique triangle.

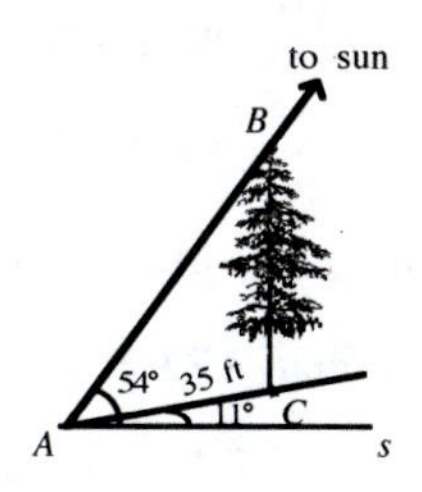

$$\angle BAC = 54° - 11° = 43°$$
$$\angle BCA = 90° + 11° = 101°.$$

We are given two angles and the included side.
We find the third angle, then apply the law of sines to find side BC.

$$\angle ABC + \angle ACB + \angle BAC = 180°$$
$$\angle ABC = 180° - (43° + 101°) = 36°$$
$$\frac{\sin \angle ABC}{AC} = \frac{\sin \angle BAC}{BC}$$
$$BC = \frac{AC \sin \angle BAC}{\sin \angle ABC} = \frac{(35\,\text{ft}) \sin 43°}{\sin 36°} = 41 \text{ ft.}$$

80. In previous exercises, we have solved similar problems using right triangle methods. (See Chapter 1, Review Exercise, Problem 40, for example.)

For comparison, we solve this problem using oblique triangle methods. We are given two angles, $\angle ABC = 180° - 67° = 113°$ and $\angle CAB = 42°$, and the included side, hence, we can find the third angle, then use the law of sines to find the other two sides.

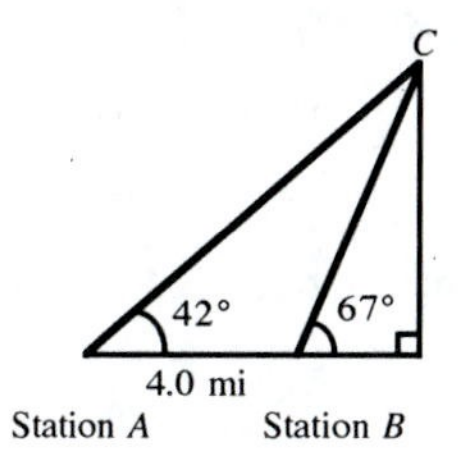

$$\angle ABC + \angle CAB + \angle BCA = 180°$$
$$\angle BCA = 180° - (113° + 42°) = 25°$$

Solve for BC:

$$\frac{\sin \angle CAB}{BC} = \frac{\sin \angle BCA}{AB}$$
$$BC = \frac{AB \sin \angle CAB}{\sin \angle BCA} = \frac{(4.0\,\text{mi}) \sin 42°}{\sin 25°}$$
$= 6.3$ mi from Station B

Solve for AC:

$$\frac{\sin \angle ABC}{AC} = \frac{\sin \angle BCA}{AB}$$
$$AC = \frac{AB \sin \angle ABC}{\sin \angle BCA} = \frac{(4.0\,\text{mi}) \sin 113°}{\sin 25°}$$
$= 8.7$ mi from Station A

81. (A) Period $= \dfrac{1}{\text{Frequency}} = \dfrac{1}{70\,\text{Hz}} = \dfrac{1}{70}$ sec. Since Period $= \dfrac{2\pi}{B}$, $B = \dfrac{2\pi}{\text{Period}} = \dfrac{2\pi}{1/70} = 140\pi$

(B) Frequency $= \dfrac{1}{\text{Period}} = \dfrac{1}{0.0125\,\text{sec}} = 80$ Hz. Since Period $= \dfrac{2\pi}{B}$, $B = \dfrac{2\pi}{\text{Period}} = \dfrac{2\pi}{0.0125} = 160\pi$

(C) Period $= \dfrac{2\pi}{B} = \dfrac{2\pi}{100\pi} = \dfrac{1}{50}$ sec. Frequency $= \dfrac{1}{\text{Period}} = \dfrac{1}{(1/50)\,\text{sec}} = 50$ Hz

82. The height of the wave from trough to crest is the difference in height between the crest (height A) and the trough (height $-A$). In this case, $A = 2$ ft.

$A - (-A) = 2A = 2(2\text{ ft}) = 4$ ft.

To find the wavelength λ, we note, $\lambda = 5.12T^2$ $\quad T = 4$ sec $\quad \lambda = 5.12(4)^2 \approx 82$ ft

To find the speed S, we use

$$S = \sqrt{\frac{g\lambda}{2\pi}} \qquad g = 32 \text{ ft/sec}^2 \qquad S = \sqrt{\frac{32(82)}{2\pi}} \approx 20 \text{ ft/sec}$$

83. Area $OCBA$ = Area of Sector OCB + Area of triangle OAB.

Area of Sector $OCB = \frac{1}{2}r^2\theta = \frac{1}{2} \cdot 1^2 \cdot \theta = \frac{1}{2}\theta$

Area of right triangle $OAB = \frac{1}{2}$ (base)(height) $= \frac{1}{2}xy$

Since OAB is a right triangle, we have

$\sin\theta = \frac{x}{1}$ $\quad x = \sin\theta$ $\quad \cos\theta = \frac{y}{1}$ $\quad y = \cos\theta$

Hence, area of triangle $OAB = \frac{1}{2}xy = \frac{1}{2}\sin\theta\cos\theta$.

Thus, Area of $OCBA = \frac{1}{2}\theta + \frac{1}{2}\sin\theta\cos\theta$

84. Since $x = \sin\theta$, $\theta = \sin^{-1} x$ (θ is acute)

Since OAB is a right triangle, applying the Pythagorean theorem, we have

$$x^2 + y^2 = 1^2$$
$$y^2 = 1 - x^2$$
$$y = \sqrt{1 - x^2}$$

Thus, Area of $OCBA$ $= \frac{1}{2}\theta + \frac{1}{2}xy$ (see previous problem)

$$= \frac{1}{2}\sin^{-1} x + \frac{1}{2}x\sqrt{1 - x^2}$$

85. We are to solve $\frac{1}{2}\theta + \frac{1}{2}\sin\theta\cos\theta = 0.5$. We graph $y1 = \frac{1}{2}y + \frac{1}{2}\sin x\cos x$ and $y2 = 0.5$ on the interval from 0 to $\frac{\pi}{2}$. From the figure, we see that y1 and y2 intersect once on the interval. Using an automatic intersection routine the solution is found to be $\theta = 0.553$.

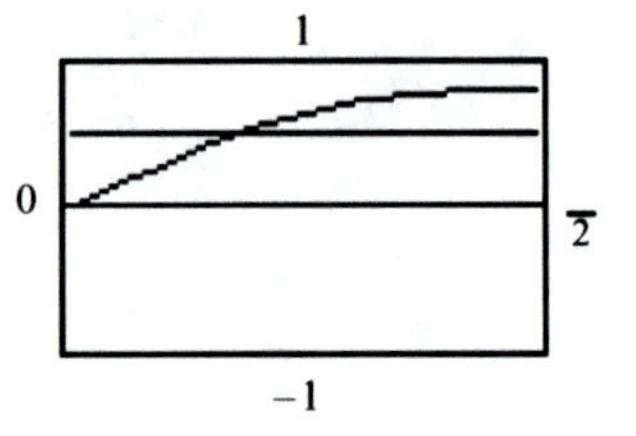

86. We are to solve $\frac{1}{2}\sin^{-1} x + \frac{1}{2}x\sqrt{1 - x^2} = 0.4$.

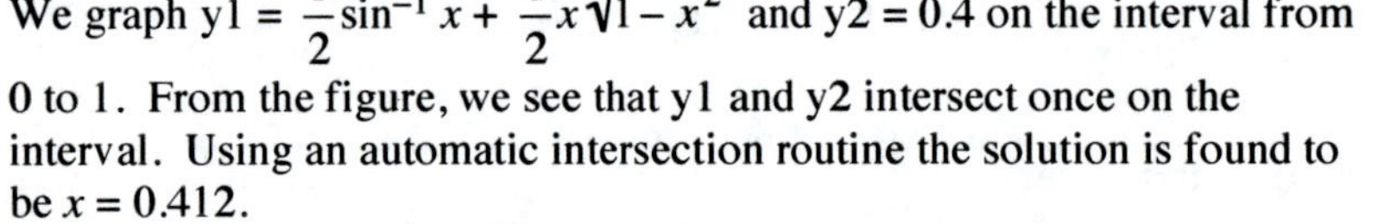

We graph $y1 = \frac{1}{2}\sin^{-1} x + \frac{1}{2}x\sqrt{1 - x^2}$ and $y2 = 0.4$ on the interval from 0 to 1. From the figure, we see that y1 and y2 intersect once on the interval. Using an automatic intersection routine the solution is found to be $x = 0.412$.

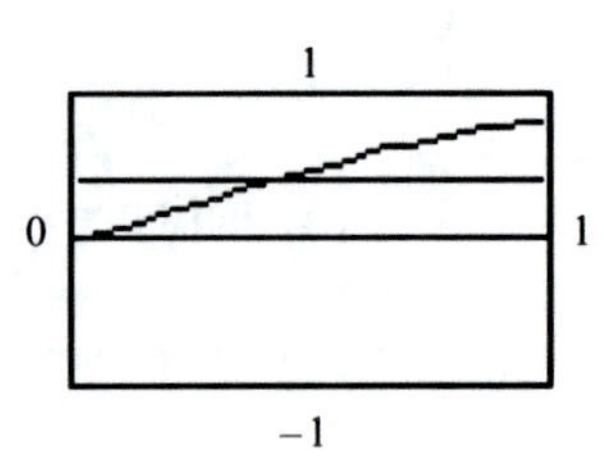

87. (A) Since the right triangles in the figure are similar, we can write $\frac{r}{h} = \frac{R}{H}$.

Since $\tan\alpha = \frac{r}{h} = \frac{R}{H}$, we can write $\frac{r}{h} = \tan\alpha$ $\quad r = h\tan\alpha$

$\frac{R}{H} = \tan\alpha$ $\quad R = H\tan\alpha$

Then, $R = H\tan\alpha = (H - h + h)\tan\alpha = (H - h)\tan\alpha + h\tan\alpha$

$R = (H - h)\tan\alpha + r$

(B) Solving the previous equation for tan α, we can write

$$R - r = (H - h)\tan\alpha \qquad \tan\alpha = \frac{R - r}{H - h}$$

Since α and β are complementary angles, we can write

$$\tan\beta = \cot\alpha = \frac{1}{\tan\alpha} = 1 \div \tan\alpha = 1 \div \frac{R - r}{H - h} = \frac{H - h}{R - r}$$

Thus, $\beta = \tan^{-1}\left(\frac{H - h}{R - r}\right)$.

88. We require θ such that the actual velocity **R** will be the resultant of the apparent velocity **v** and the wind velocity **w**. From the diagram it should be clear that

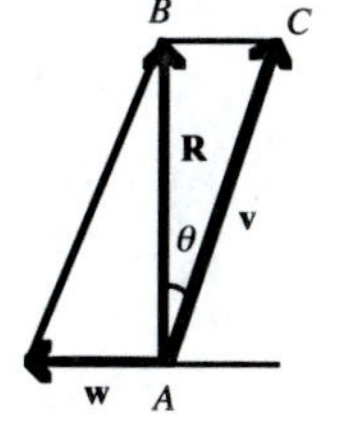

$$\sin\theta = \frac{|\mathbf{w}|}{|\mathbf{v}|} = \frac{81.5}{265} \qquad \theta = \sin^{-1}\frac{81.5}{265} = 18°$$

The ground speed for this course will be the magnitude $|\mathbf{R}|$ of the actual velocity. In the right triangle, ABC, we have

$$\cos\theta = \frac{|\mathbf{R}|}{|\mathbf{v}|} \qquad |\mathbf{R}| = |\mathbf{v}|\cos\theta = 265\cos\left(\sin^{-1}\frac{81.5}{265}\right)$$

$$= 265\sqrt{1 - \left(\frac{81.5}{265}\right)^2} = 252 \text{ mph}$$

89. (A) First, form a force diagram with all force vectors in standard position at the origin.

Let $\mathbf{F}_1$ = the tension in the left side
$\mathbf{F}_2$ = the tension in the right side

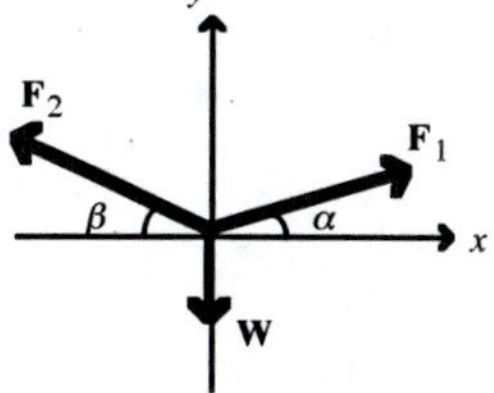

$$|\mathbf{F}_1| = T_L \qquad |\mathbf{F}_2| = T_R \qquad |\mathbf{W}| = w$$

Write each force vector in terms of **i** and **j** unit vectors.

$$\mathbf{F}_1 = T_R\cos\alpha\mathbf{i} + T_R\sin\alpha\mathbf{j} \qquad \mathbf{F}_2 = T_L(-\cos\beta)\mathbf{i} + T_L\sin\beta\mathbf{j} \qquad \mathbf{W} = -w\mathbf{j}$$

For the system to be in static equilibrium, we must have $\mathbf{F}_1 + \mathbf{F}_2 + \mathbf{W} = \mathbf{0}$ which becomes, on addition,

$$(T_R\cos\alpha - T_L\cos\beta)\mathbf{i} + (T_R\sin\alpha + T_L\sin\beta - w)\mathbf{j} = 0\mathbf{i} + 0\mathbf{j}$$

Since two vectors are equal if and only if their corresponding components are equal, we are led to the following system of equations in T_L and T_R :

$$T_R\cos\alpha - T_L\cos\beta = 0 \qquad T_R\sin\alpha + T_L\sin\beta - w = 0$$

Solving, $T_R = T_L\frac{\cos\beta}{\cos\alpha}$

$$T_L\sin\alpha\frac{\cos\beta}{\cos\alpha} + T_L\sin\beta = w$$

$$T_L\left(\frac{\sin\alpha\cos\beta}{\cos\alpha} + \sin\beta\right) = w$$

$$T_L\left(\frac{\sin\alpha\cos\beta + \cos\alpha\sin\beta}{\cos\alpha}\right) = w$$

$$T_L\frac{\sin(\alpha + \beta)}{\cos\alpha} = w$$

Thus, $T_L = \frac{w\cos\alpha}{\sin(\alpha + \beta)}$. Hence, $T_R = \frac{w\cos\alpha}{\sin(\alpha + \beta)}\frac{\cos\beta}{\cos\alpha} = \frac{w\cos\beta}{\sin(\alpha + \beta)}$

(B) If $\alpha = \beta$, then

$$T_L = T_R = \frac{w\cos\alpha}{\sin(\alpha+\alpha)} = \frac{w\cos\alpha}{\sin 2\alpha} = \frac{w\cos\alpha}{2\sin\alpha\cos\alpha} = \frac{w}{2\sin\alpha} = \frac{w}{2}\frac{1}{\sin\alpha}$$
$$= \frac{1}{2}w\csc\alpha$$

90. We can apply the law of cosines to the triangle shown in the figure. Then,

$$100^2 = r^2 + r^2 - 2r\cdot r\cdot\cos\theta$$
$$10{,}000 = 2r^2 - 2r^2\cos\theta = 2r^2(1-\cos\theta)$$
$$5000 = r^2(1-\cos\theta)$$

(A) Given $\theta = 10°$, then $5000 = r^2(1-\cos 10°)$

$$r^2 = \frac{5000}{1-\cos 10°}$$
$$r = \sqrt{\frac{5000}{1-\cos 10°}} = 574 \text{ ft}$$

(B) Given $r = 2000$, then $5000 = (2000)^2(1-\cos\theta)$

$$\frac{5000}{(2000)^2} = 1-\cos\theta$$
$$\cos\theta = 1 - \frac{5000}{(2000)^2}$$
$$\theta = \cos^{-1}\left[1-\frac{5000}{(2000)^2}\right] = 2.9°$$

91. (A) Since $s = r\theta$, we can write $r\theta = 50$. To determine θ, we note that triangle ABC is a right triangle, with side $AC = r - 1$. Then,

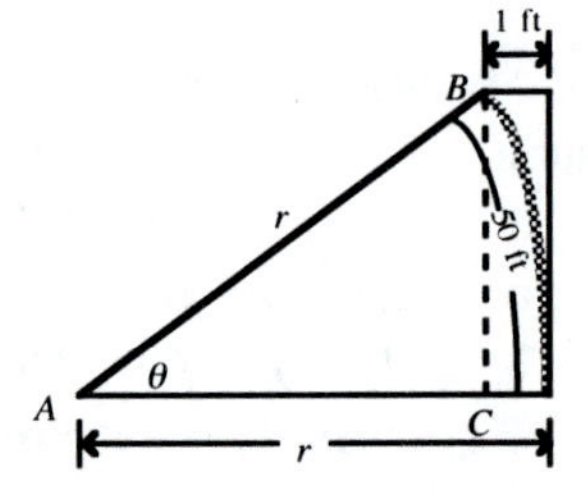

$$\cos\theta = \frac{AC}{AB} = \frac{r-1}{r}$$
$$\theta = \cos^{-1}\left(\frac{r-1}{r}\right)$$

Thus, $r\cos^{-1}\left(\frac{r-1}{r}\right) = 50$. To solve this, we graph $y1 = x\cos^{-1}\left(\frac{x-1}{x}\right)$ and $y2 = 50$ on the interval from 1000 to 2000. From the figure, we see that y1 and y2 intersect once on this interval.

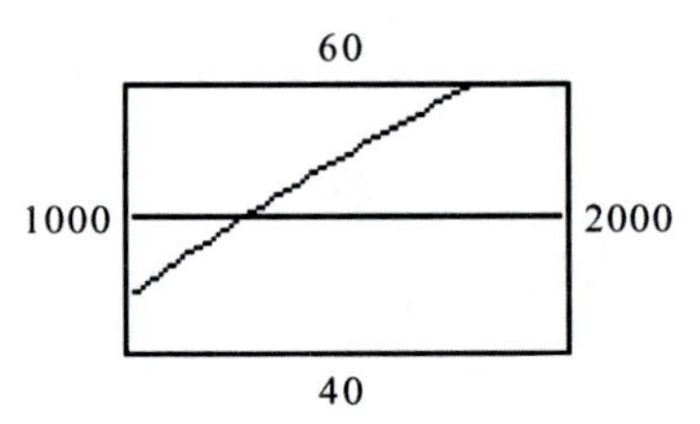

Using an automatic intersection routine, the solution is found to be $r = 1{,}250$ ft.

(B) From Problem 90, we have $5000 = r^2(1 - \cos\theta)$. If $r = 1250$, then

$$5000 = (1250)^2(1 - \cos\theta)$$

$$\frac{5000}{(1250)^2} = 1 - \cos\theta$$

$$\cos\theta = 1 - \frac{5000}{(1250)^2}$$

$$\theta = \cos^{-1}\left[1 - \frac{5000}{(1250)^2}\right] = 4.6°$$

92. The areas of the end pieces are given by the formula from Section 5.4, Problem 37 to be $\frac{1}{2}r^2(\theta - \sin\theta)$. Subtracting this twice from the area of the total cross-section, πr^2, we obtain

$$A = \pi r^2 - \frac{1}{2}r^2(\theta - \sin\theta) - \frac{1}{2}r^2(\theta - \sin\theta) = \pi r^2 - r^2(\theta - \sin\theta)$$

$$= \pi r^2 - r^2\theta + r^2\sin\theta = r^2(\pi - \theta + \sin\theta)$$

93. If all three pieces of the log have the same cross-sectional area, then each of these areas is one-third of the entire area, that is,

$$r^2(\pi - \theta + \sin\theta) = \frac{1}{3}\pi r^2$$

Thus, $\pi - \theta + \sin\theta = \frac{1}{3}\pi$. To solve this, we graph

$y1 = \pi - \theta + \sin\theta$ and $y2 = \frac{1}{3}\pi$ on the interval from 0 to π.

From the figure, we see that y1 and y2 intersect once on the interval.

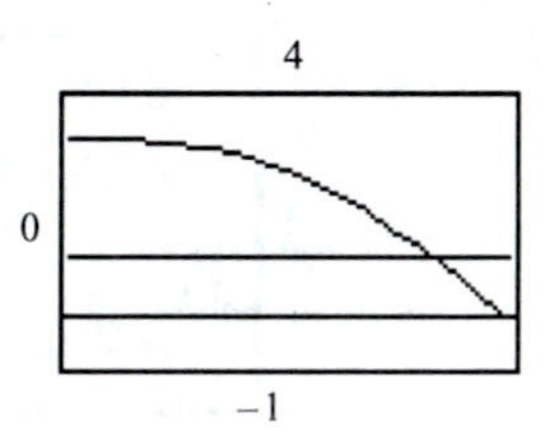

Using an automatic intersection routine, the solution is found to be $\theta = 2.6053$ rad.

94. (A)

x months	1, 13	2, 14	3, 15	4, 16	5, 17	6, 18	7, 19	8, 20
y (twilight duration)	1.62	1.82	2.35	2.98	3.55	4.12	4.05	3.50

x months	9, 21	10, 22	11, 23	12, 24
y (twilight duration)	2.80	2.22	1.80	1.57

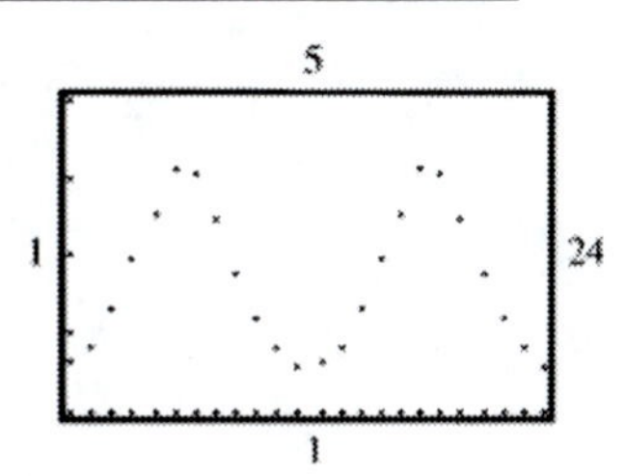

(B) From the table, Max $y = 4.12$ and Min $y = 1.57$. Then,

$$A = \frac{\text{Max } y - \text{Min } y}{2} = \frac{4.12 - 1.57}{2} = 1.275$$

$$B = \frac{2\pi}{\text{Period}} = \frac{2\pi}{12} = \frac{\pi}{6}$$

$$k = \text{Min } y + A = 1.57 + 1.275 = 2.845$$

From the plot in (A) or the table, we estimate the smallest value of x for which $y = k = 2.845$ to be approximately 3.4. Then, this is the phase shift for the graph. Substitute $B = \frac{\pi}{6}$ and $x = 3.4$ into the phase-shift equation.

$$x = -\frac{C}{B} \qquad 3.4 = -\frac{C}{\pi/6} \qquad C = \frac{-3.4\pi}{6} = -1.8$$

Thus, the equation required is $y = 2.845 + 1.275 \sin\left(\frac{\pi x}{6} - 1.8\right)$.

(C)

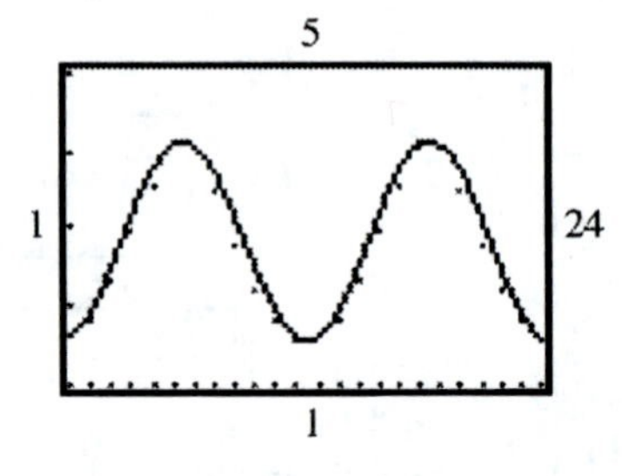

APPENDICES

APPENDIX A Comments on Numbers

EXERCISE A.1 Real Numbers

1. There are infinitely many negative integers. Examples include –3. The only integer that is neither positive nor negative is 0. There are infinitely many positive integers. Examples include 5.

3. There are infinitely many rational numbers that are not integers. Examples include $\frac{2}{3}$.

5. (A) True. (B) False (6 is a real number, but is not irrational.) (C) True.

7. (A) $0.3636\overline{36}$, rational (B) $0.77\overline{77}$, rational (C) 2.64575131..., irrational
   (D) $1.625\overline{00}$, rational

9. (A) Since $\frac{26}{9} = 2.8\overline{88}$, it lies between 2 and 3.
   (B) Since $-\frac{19}{5} = -3.8$, it lies between –4 and –3.
   (C) Since $-\sqrt{23} = -4.7958\ldots$, it lies between –5 and –4.

11. $y + 3$ 13. $(3 \cdot 2)x$ 15. $7x$ 17. $(5 + 7) + x$

19. $3m$ 21. $u + v$ 23. $(2 + 3)x$

25. (A) True. (This is the commutative property for addition of real numbers.)
   (B) False. For example, $4 - 2 \neq 2 - 4$.
   (C) True. (This is the commutative property for multiplication of real numbers.)
   (D) False. For example, $8 \div 2 \neq 2 \div 8$

EXERCISE A.2 Complex Numbers

1. $(3 - 2i) + (4 - 7i) = 3 - 2i + 4 + 7i = 3 + 4 - 2i + 7i = 7 + 5i$

3. $(3 - 2i) - (4 + 7i) = 3 - 2i - 4 - 7i = 3 - 4 - 2i + 7i = -1 - 9i$

5. $(6i)(3i) = 18i^2 = 18(-1) = -18$

7. $(2i)(3 - 4i) = 6i - 8i^2 = 6i - 8(-1) = 8 + 6i$

9. $(3 - 4i)(1 - 2i) = 3 - 10i + 18i^2 = 3 - 10i + 8(-1) = 3 - 10i - 8 = -5 - 10i$

11. $(3 + 5i)(3 - 5i) = 9 - 25i^2 = 9 - 25(-1) = 9 + 25 = 34$

13. $\frac{1}{2+i} = \frac{1}{2+i}\frac{2-i}{2-i} = \frac{2-i}{4-i^2} = \frac{2-i}{4+1} = \frac{2-i}{5} = \frac{2}{5} - \frac{1}{5}i$

15. $\frac{2-i}{3+2i} = \frac{2-i}{3+2i}\frac{3-2i}{3-2i} = \frac{6-7i+2i^2}{9-4i^2} = \frac{6-7i-2}{9+4} = \frac{4-7i}{13} = \frac{4}{13} - \frac{7}{13}i$

Exercise A.3

17. $\dfrac{-1+2i}{4+3i} = \dfrac{-1+2i}{4+3i}\dfrac{4-3i}{4-3i} = \dfrac{-4+11i-6i^2}{16-9i^2} = \dfrac{-4+11i+6}{16+9} = \dfrac{2+11i}{25} = \dfrac{2}{25} + \dfrac{11}{25}i$

19. $(3+\sqrt{-4}) + (2-\sqrt{-16}) = (3+i\sqrt{4}) + (2-i\sqrt{16}) = 3+2i+2-4i = 3+2+2i-4i = 5-2i$

21. $(5-\sqrt{-1}) - (2-\sqrt{-36}) = (5-i) - (2-i\sqrt{36}) = (5-i) - (2-6i) = 5-i-2+6i$
$= 5-2-i+6i = 3+5i$

23. $(-3-\sqrt{-1})(-2+\sqrt{-49}) = (-3-i)(-2+i\sqrt{49}) = (-3-i)(-2+7i) = 6-19i-7i^2$
$= 6-19i+7 = 13-19i$

25. $\dfrac{5-\sqrt{-1}}{2+\sqrt{-4}} = \dfrac{5-i}{2+i\sqrt{4}} = \dfrac{5-i}{2+2i} = \dfrac{5-i}{2+2i}\dfrac{2-2i}{2-2i} = \dfrac{10-12i+2i^2}{4-4i^2} = \dfrac{10-12i-2}{4+4} = \dfrac{8-12i}{8} = 1 - \dfrac{3}{2}i$

27. $(1-i)^2 = 2(1-i) + 2 = 1-2i+i^2-2+2i+2 = 1-2i-1-2+2i+2 = 0+0i = 0$

29. $$\left(\frac{-1}{2}+\frac{\sqrt{3}}{2}i\right)^3 = \left(-\frac{1}{2}+\frac{\sqrt{3}}{2}i\right)\left(-\frac{1}{2}+\frac{\sqrt{3}}{2}i\right)^2$$

$$= \left(-\frac{1}{2}+\frac{\sqrt{3}}{2}i\right)\left[\left(-\frac{1}{2}\right)^2+2\left(-\frac{1}{2}\right)\left(\frac{\sqrt{3}}{2}i\right)+\left(\frac{\sqrt{3}}{2}i\right)^2\right]$$

$$= \left(-\frac{1}{2}+\frac{\sqrt{3}}{2}i\right)\left[\frac{1}{4}-\frac{2\sqrt{3}}{4}i+\frac{3}{4}i^2\right] = \left(-\frac{1}{2}+\frac{\sqrt{3}}{2}i\right)\left(\frac{1}{4}-\frac{2\sqrt{3}}{4}i-\frac{3}{4}\right)$$

$$= \left(-\frac{1}{2}+\frac{\sqrt{3}}{2}i\right)\left(-\frac{1}{2}-\frac{2\sqrt{3}}{4}i\right)$$

$$= \left(-\frac{1}{2}\right)\left(-\frac{1}{2}\right)+\left(-\frac{1}{2}\right)\left(\frac{2\sqrt{3}}{4}i\right)+\left(\frac{\sqrt{3}}{2}i\right)\left(-\frac{1}{2}\right)+\left(\frac{\sqrt{3}}{2}i\right)\left(-\frac{2\sqrt{3}}{4}i\right)$$

$$= \frac{1}{4}+\frac{\sqrt{3}}{4}i-\frac{\sqrt{3}}{4}i-\frac{3}{4}i^2 = \frac{1}{4}+\frac{3}{4} = 1$$

EXERCISE A.3 Significant Digits

1. $640 = 6.40. \times 10^2 = 6.4 \times 10^2$
   2 places left
   |
   positive exponent

3. $5{,}460{,}000{,}000 = 5.460{,}000{,}000. \times 10^9 = 5.46 \times 10^9$
   9 places left
   |
   positive exponent

5. $0.73 = 0.7.3 \times 10^{-1} = 7.3 \times 10^{-1}$
   1 place right
   |
   negative exponent

7. $0.00000032 = 0.0000003.2 \times 10^{-7} = 3.2 \times 10^{-7}$
   7 places right
   |
   negative exponent

9. $0.0000491 = 0.00004.91 \times 10^{-5} = 4.91 \times 10^{-5}$

5 places right
|
negative exponent

11. $67{,}000{,}000{,}000 = 6.7{,}000{,}000{,}000. \times 10^{10}$

10 places left
|
positive exponent

13. $5.6 \times 10^4 = 5.6 \times 10{,}000 = 56{,}000$

15. $9.7 \times 10^{-3} = 9.7 \times 0.001 = 0.0097$

17. $4.61 \times 10^{12} = 4.61 \times 1{,}000{,}000{,}000{,}000 = 4{,}610{,}000{,}000{,}000$

19. $1.08 \times 10^{-1} = 1.08 \times 0.1 = 0.108$

21. 12.3 has a decimal point. From the first nonzero digit (1) to the last digit (3), there are 3 digits. 3 significant digits.

23. 12.300 has a decimal point. From the first nonzero digit (1) to the last digit (0), there are 5 digits. 5 significant digits.

25. 0.01230 has a decimal point. From the first nonzero digit (1) to the last digit (0), there are 4 digits. 4 significant digits.

27. 6.7×10^{-1} is in scientific notation. There are 2 digits in 6.7. 2 significant digits.

29. 6.700×10^{-1} is in scientific notation. There are 4 digits in 6.700. 4 significant digits.

31. 7.090×10^5 is in scientific notation. There are 4 digits in 7.090. 4 significant digits.

33. 635,000

35. 86.8 (convention of leaving the digit before the 5 alone, if it is even)

37. 0.00465

39. $734 = 7.34 \times 10^2 \approx 7.3 \times 10^2$

41. $0.040 = 4.0 \times 10^{-2}$

43. $0.000435 = 4.35 \times 10^{-4} \approx 4.4 \times 20^{-4}$ (convention of rounding the digit before the 5 up, if it is odd)

45. 3, since there are three significant digits in the number (32.8) with the least number of significant digits in the calculation.

47. 2, since there are two significant digits in the numbers (360 and 1,200) with the least number of significant digits in the calculation.

49. 1, since there is one significant digit in the number (6×10^4) with the least number of significant digits in the calculation.

51. $\dfrac{6.07}{0.5057}$ $= 12.0$

6.07 has the least number of significant digits (3).
Answer must have 3 significant digits.

53. $(6.14 \times 10^9)(3.154 \times 10^{-1})$ $= 1.94 \times 10^9$

6.14×10^9 has the least number of significant digits (3).
Answer must have 3 significant digits.

55. $\dfrac{6{,}730}{(2.30)(0.0551)}$ $= 53{,}100$

All numbers in the calculation have 3 significant digits.
Answer must have 3 significant digits.

Exercise A.3

57. $C = 2\pi(25.31 \text{ cm})$ There are 4 significant digits in 25.31 cm.

$= 159.0 \text{ cm}$ Answer must have 4 significant digits.

59. $A = \frac{1}{2}(22.4 \text{ ft})(8.6 \text{ ft})$ 8.6 has the least number of significant digits (2)

$= 96 \text{ ft}^2$ Answer must have 2 significant digits

61. $s = 4\pi(1.5 \text{ mm})^2$ There are 2 significant digits in 1.5 mm

$= 28 \text{ mm}^2$ Answer must have 2 significant digits

63.

$$V = \ell wh$$

$$h = \frac{V}{\ell w}$$

$$h = \frac{24.2cm^3}{(3.25cm)(4.50cm)}$$

All numbers in the calculation have 3 significant digits

$$= 1.65 \text{ cm}$$

Answer must have 3 significant digits

65.

$$V = \frac{1}{3}\pi r^2 h$$

$$3V = \pi r^2 h$$

$$r^2 = \frac{3V}{\pi h}$$

$$r = \sqrt{\frac{3V}{\pi h}} = \sqrt{\frac{3(1200 \text{ in}^3)}{\pi(6.55 \text{ in})}}$$

1200 in^3 has the least number of significant digits (2)

$$= 13 \text{ in}$$

Answer must have 2 significant digits

APPENDIX B Functions and Inverse Functions

EXERCISE B.1 Functions

1. $f(x) = 4x - 1$
$f(1) = 4(1) - 1$
$= 3$

3. $f(x) = 4x - 1$
$f(-1) = 4(-1) - 1$
$= -5$

5. $f(x) = 4x - 1$
$f(0) = 4(0) - 1$
$= -1$

7. $g(x) = x - x^2$
$g(1) = 1 - 1^2$
$= 0$

9. $g(x) = x - x^2$
$g(5) = 5 - 5^2$
$= -20$

11. $g(x) = x - x^2$
$g(-2) = -2 - (-2)^2$
$= -2 - 4 = -6$

13. $f(0) + g(0) = 1 - 2 \cdot 0 + 4 - 0^2 = 1 + 4 = 5$

15. $\dfrac{f(3)}{g(1)} = \dfrac{1-2\cdot 3}{4-1^2} = \dfrac{-5}{3} = -\dfrac{5}{3}$

17. $2f(-1) = 2[1 - 2(-1)] = 2(1 + 2) = 6$

19. $f(2 + h) = 1 - 2(2 + h) = 1 - 4 - 2h = -3 - 2h$

21. $\dfrac{f(2+h)-f(2)}{h} = \dfrac{[1-2(2+h)]-(1-2\cdot 2)}{h} = \dfrac{-3-2h-(-3)}{h} = \dfrac{-2h}{h} = -2$

23. $g[f(2)] = g(1 - 2 \cdot 2) = g(-3) = 4 - (-3)^2 = 4 - 9 = -5$

25. $x^2 + y^2 = 25$ does not specify a function, since both $(3, 4)$ and $(3, -4)$ are solutions, in which the same domain value corresponds to more than one range value.

27. $2x - 3y = 6$ specifies a function, since each domain value x corresponds to exactly one range value $\left(\dfrac{2x-6}{3}\right)$.

29. $y^2 = x$ does not specify a function, since both $(9, 3)$ and $(9, -3)$ are solutions, in which the same domain value corresponds to more than one range value.

31. $y = |x|$ specifies a function, since each domain value x corresponds to exactly one range value $(|x|)$.

33. $f(x) = x^2 - x + 1$ Domain $X = \{-2, -1, 0, 1, 2\}$
$f(-2) = (-2)^2 - (-2) + 1 = 7$
$f(-1) = (-1)^2 - (-1) + 1 = 3$
$f(0) = 0^2 - 0 + 1 = 1$
$f(1) = 1^2 - 1 + 1 = 1$
$f(2) = 2^2 - 2 + 1 = 3$ Range $Y = \{1, 3, 7\}$

35. G does not specify a function since the domain value -4 corresponds to more than one range value (3 and 0). F specifies a function.
Domain of F = Set of all first components = $\{-2, -1, 0\} = X$
Range of F = Set of all second components = $\{0, 1\} = Y$

Exercise B.2

37.
$$\begin{aligned} s(t) &= 4.88t^2 \\ s(0) &= 4.88(0)^2 = 0 \\ s(1) &= 4.88(1)^2 = 4.88 \text{ m} \\ s(2) &= 4.88(2)^2 = 19.52 \text{ m} \\ s(3) &= 4.88(3)^2 = 43.92 \text{ m} \end{aligned}$$

39.
$$\frac{s(2+h)-s(2)}{h} = \frac{4.88(2+h)^2 - 4.88(2)^2}{h} = \frac{4.88(4+4h+h^2) - 19.52}{h}$$
$$= \frac{19.52 + 19.52h + 4.88h^2 - 19.52}{h} = \frac{19.52h + 4.88h^2}{h} = 19.52 + 4.88h$$

As h gets closer to 0, this gets closer to 19.52; the average speed $\frac{s(2+h)-s(2)}{h}$ tends to a quantity called the speed at $t = 2$.

EXERCISE B.2 Graphs and Transformations

1. The graph of $y = f(x)$ is shifted downward 2 units.

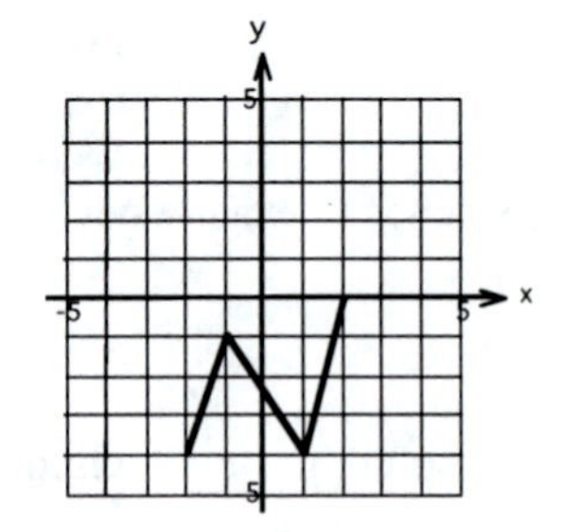

3. The graph of $y = f(x)$ is shifted 2 units to the right.

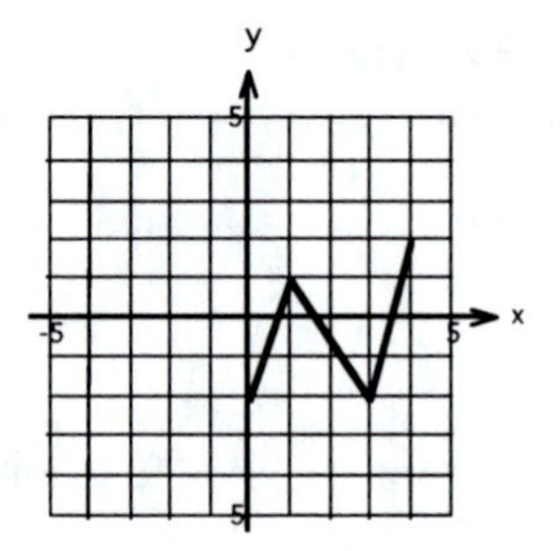

5. The graph of $y = g(x)$ is shifted 3 units to the left.

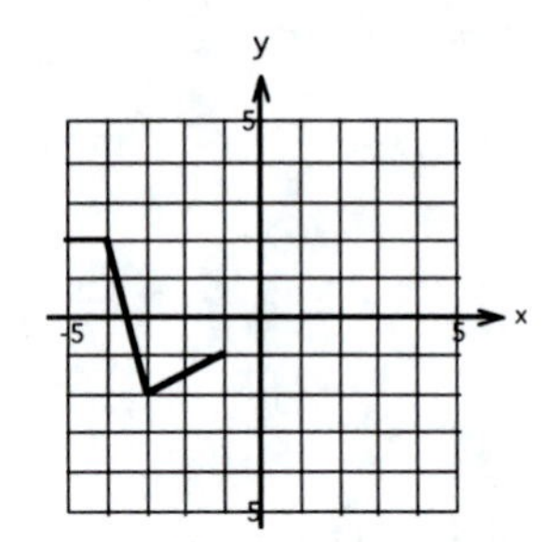

7. The graph of $y = g(x)$ is shifted upward 3 units.

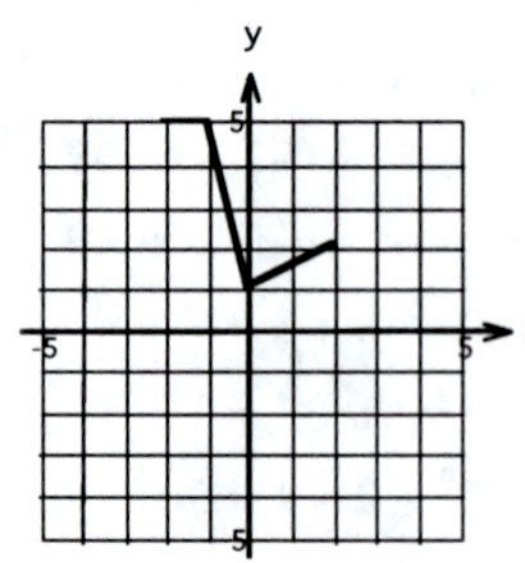

9. The graph of $y = f(x)$ is reflected in the x axis.

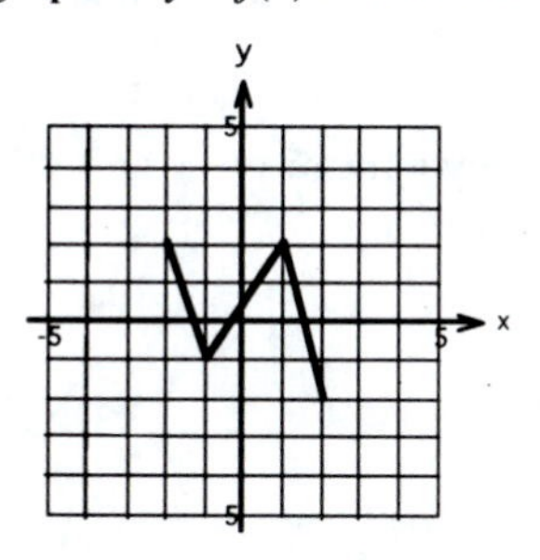

11. The graph of $y = f(x)$ is reflected in the y axis.

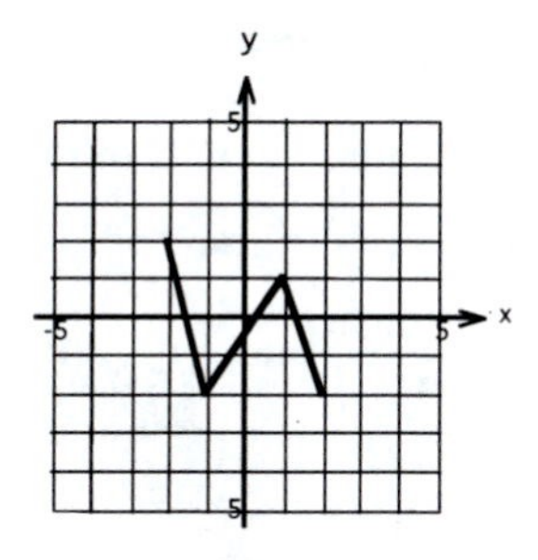

13. The graph of $y = g(x)$ is shrunk vertically by a factor of 0.5.

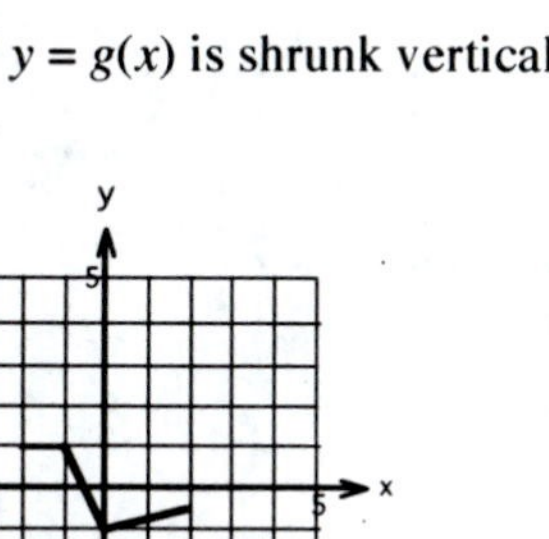

15. The graph of $y = g(x)$ is stretched horizontally by a factor of $1/0.5 = 2$.

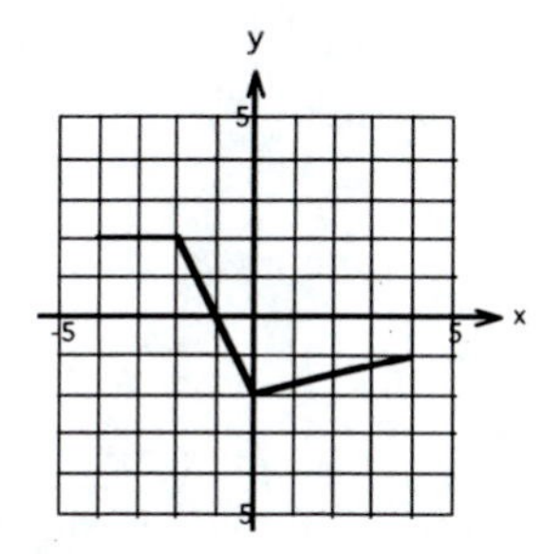

17. The graph of $g(x) = -|x - 3|$ is the graph of $y = |x|$ shifted right 3 units and reflected in the x axis.

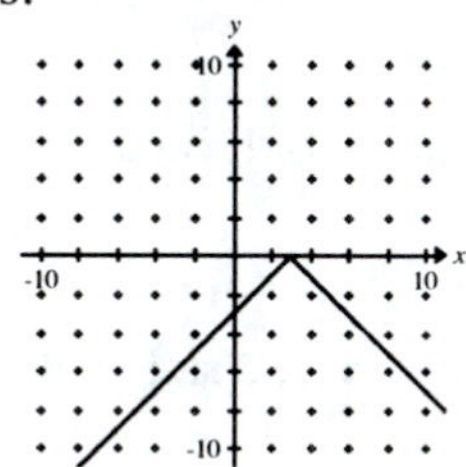

19. The graph of $f(x) = (x + 3)^2 + 3$ is the graph of $y = x^2$ shifted left 3 units and up 3 units.

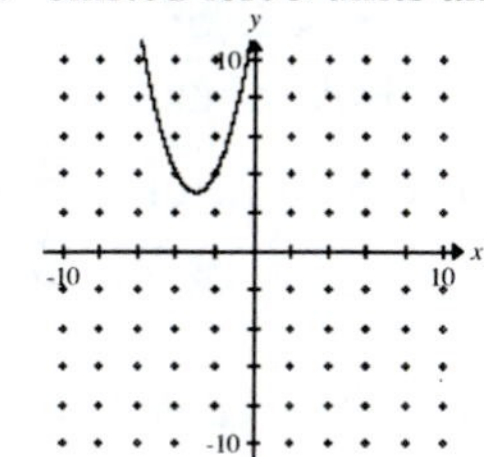

21. The graph of $f(x) = 5 - \sqrt{x}$ is the graph of $y = \sqrt{x}$ reflected in the x axis and shifted up 5 units.

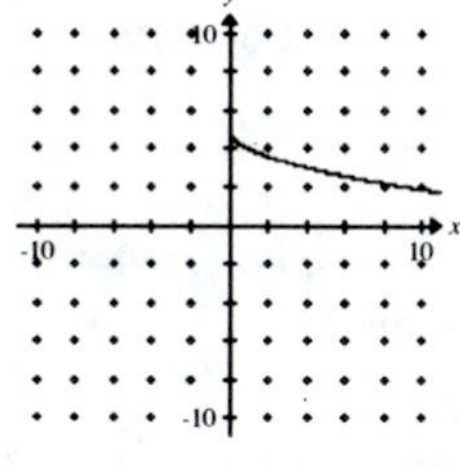

23. The graph of $h(x) = -\sqrt[3]{2x}$ is the graph of $y = \sqrt[3]{x}$ shrunk horizontally by a factor of ½ and reflected in the x axis.

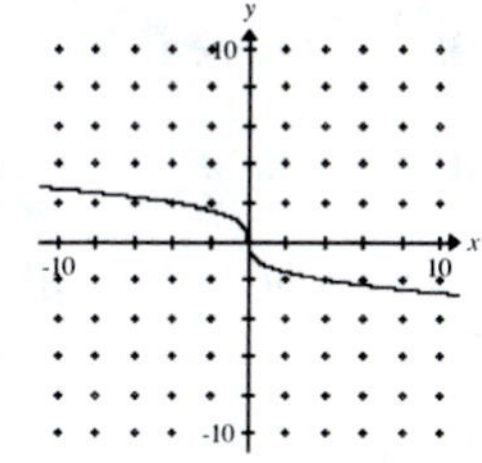

25. The basic function $y = x^2$ is shifted right 2 units and down 4 units to produce the graph of $y = (x - 2)^2 - 4$.

27. The basic function $y = x^2$ is shifted right 1 unit, reflected in the x axis, and shifted up 3 units to produce the graph of $y = 3 - (x - 1)^2$.

29. The basic function $y = \sqrt{x}$ is shrunk horizontally by a factor of ½ , reflected in the x axis, and shifted up 4 units to produce the graph of $y = 4 - \sqrt{2x}$.

31. The basic function $y = x^3$ is shifted right 1 unit, shrunk vertically by a factor of ½, and shifted 2 units down to produce the graph of $y = \frac{1}{2}(x-1)^3 - 2$.

33. $g(x) = 2\sqrt{x+3} - 4$,

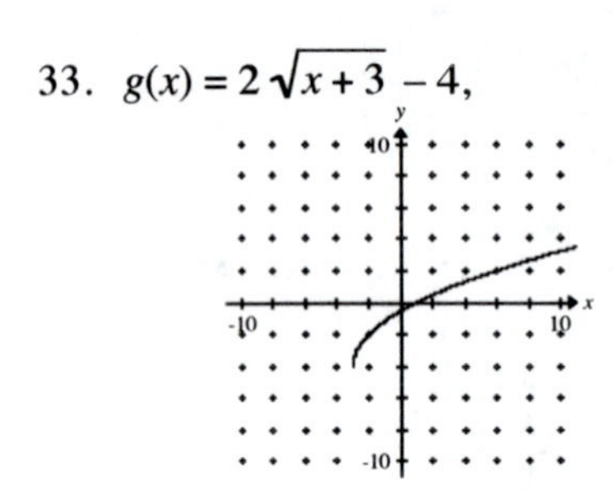

35. $g(x) = -\frac{1}{2}|x-1| + 4$,

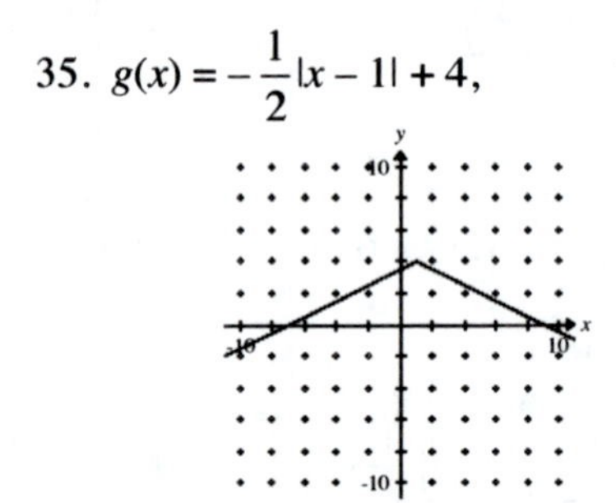

37. $g(x) = -(x+2)^3 - 1$,

39. The graph of $y = \sqrt{x}$ is reflected in the y axis, shifted right 3 units, stretched vertically by a factor of 2, and shifted down 2 units to produce the graph of $y = 2\sqrt{3-x} - 2$.

41. The graph of $y = x^2$ is shifted 1 unit to the right, shrunk vertically by a factor of ½, and shifted down 4 units to produce the graph of $y = \frac{1}{2}(x-1)^2 - 4$.

43. The graph of $y = x^3$ is shifted left 1 unit, shrunk vertically by a factor of ¼, and shifted down 2 units to produce the graph of $y = \frac{1}{4}(x+1)^3 - 2$.

EXERCISE B.3 Inverse Functions

1. This function is one-to-one since each range element corresponds to exactly one domain element.

3. This function is not one-to-one since the range element 9 corresponds to several domain elements.

5. This function passes the horizontal line test; it is one-to-one.

7. This function fails the horizontal line test; it is not one-to-one.

9. This function passes the horizontal line test; it is one-to-one.

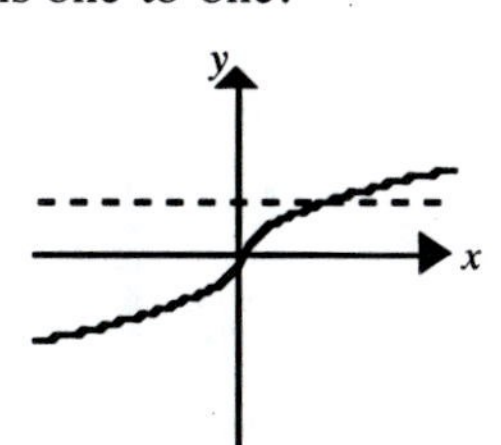

11. The function fails the horizontal line test; it is not one-to-one.

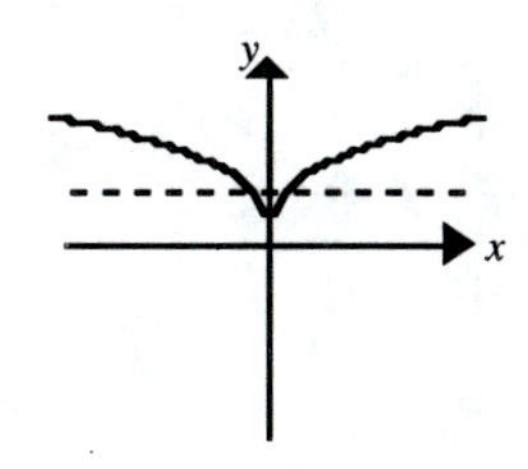

13. First, we note that f is not one-to-one, since the range element 0 corresponds to two domain elements, –1 and 2. The function g is one-to-one, since each range element corresponds to exactly one domain element. Reversing the ordered pairs in the function g produces the inverse function.

$$g^{-1} = \{(-8,-2), (1,1), (8,2)\}$$

Its domain is $\{-8, 1, 8\}$. Its range is $\{-2, 1, 2\}$.

15.

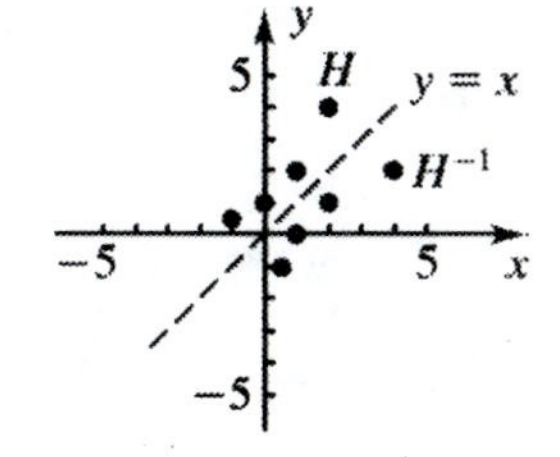

17. Replace $f(x)$ with y: $\quad f: \; y = 2x - 7$

Interchange the variables x and y to form f^{-1}: $\quad f^{-1}: x = 2y - 7$

Solve for y in terms of x:

$$x + 7 = 2y$$

$$y = \frac{x+7}{2}$$

Replace y with $f^{-1}(x)$: $f^{-1}(x) = \dfrac{x+7}{2}$

19. Replace $h(x)$ with y: $\quad h: \; y = \dfrac{x+3}{3}$

Interchange the variables x and y to form h^{-1}: $\quad h^{-1}: \; x = \dfrac{y+3}{3}$

Solve for y in terms of x:

$$3x = y + 3$$

$$y = 3x - 3$$

Replace y with $h^{-1}(x)$: $h^{-1}(x) = 3x - 3$

Exercise B.3

21.

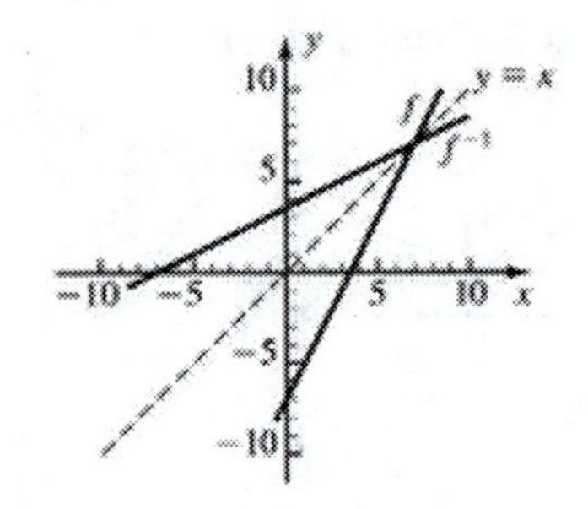

23. Replace $f(x)$ with y: $\quad f: \quad y = 2x - 7$

Interchange the variables x and y to form f^{-1}: $\quad f^{-1}: x = 2y - 7$

Solve for y in terms of x: $\quad x + 7 = 2y$

$$y = \frac{x+7}{2}$$

Replace y with $f^{-1}(x)$: $f^{-1}(x) = \dfrac{x+7}{2}$

Thus, $f^{-1}(3) = \dfrac{3+7}{2} \quad f^{-1}(3) = 5$

25. Replace $h(x)$ with y: $\quad h: \quad y = \dfrac{x}{3} + 1$

Interchange the variables x and y to form h^{-1}: $\quad h^{-1}: \; x = \dfrac{y}{3} + 1$

Solve for y in terms of x: $\quad x - 1 = \dfrac{y}{3}$

$$y = 3(x - 1) = 3x - 3$$

Replace y with $h^{-1}(x)$: $h^{-1}(x) = 3x - 3$

Thus, $h^{-1}(2) = 3 \cdot 2 - 3 \quad h^{-1}(2) = 3$

27. $f\,[f^{-1}(4)] = 2[f^{-1}(4)] - 7 = 2\left(\dfrac{4+7}{2}\right) - 7 = 4 + 7 - 7 = 4$

29. $h^{-1}\,[h(x)] = 3h(x) - 3 = 3\left(\dfrac{x}{3} + 1\right) - 3 = x + 3 - 3 = x$

31. $h[h^{-1}(x)] = \dfrac{h^{-1}(x)}{3} + 1 = \dfrac{3x-3}{3} + 1 = x - 1 + 1 = x$